Lecture Notes in Bioinformatics 9044

Subseries of Lecture Notes in Computer Science

T0214832

Francisco Ortuño   Ignacio Rojas (Eds.)

Bioinformatics and Biomedical Engineering

Third International Conference, IWBBIO 2015
Granada, Spain, April 15-17, 2015
Proceedings, Part II

 Springer

Volume Editors

Francisco Ortuño
Ignacio Rojas
Universidad de Granada
Dpto. de Arquitectura y Tecnología de Computadores (ATC)
E.T.S. de Ingenierías en Informática y Telecomunicación, CITIC-UGR
Granada, Spain
E-mail: {fortuno, irojas}@ugr.es

ISSN 0302-9743 e-ISSN 1611-3349
ISBN 978-3-319-16479-3 e-ISBN 978-3-319-16480-9
DOI 10.1007/978-3-319-16480-9
Springer Cham Heidelberg New York Dordrecht London

Library of Congress Control Number: 2015934926

LNCS Sublibrary: SL 8 – Bioinformatics

Typesetting: Camera-ready by author, data conversion by Scientific Publishing Services, Chennai, India

Printed on acid-free paper

Springer is part of Springer Science+Business Media (www.springer.com)

Preface

We are proud to present the set of final accepted full papers for the third edition of the IWBBIO conference "International Work-Conference on Bioinformatics and Biomedical Engineering" held in Granada (Spain) during April 15–17, 2015.

The IWBBIO 2015 (International Work-Conference on Bioinformatics and Biomedical Engineering) seeks to provide a discussion forum for scientists, engineers, educators, and students about the latest ideas and realizations in the foundations, theory, models, and applications for interdisciplinary and multidisciplinary research encompassing disciplines of computer science, mathematics, statistics, biology, bioinformatics, and biomedicine.

The aims of IWBBIO 2015 is to create a friendly environment that could lead to the establishment or strengthening of scientific collaborations and exchanges among attendees, and therefore, IWBBIO 2015 solicited high-quality original research papers (including significant work-in-progress) on any aspect of Bioinformatics, Biomedicine, and Biomedical Engineering.

New computational techniques and methods in machine learning; data mining; text analysis; pattern recognition; data integration; genomics and evolution; next generation sequencing data; protein and RNA structure; protein function and proteomics; medical informatics and translational bioinformatics; computational systems biology; modeling and simulation and their application in life science domain, biomedicine, and biomedical engineering were especially encouraged. The list of topics in the successive Call for Papers has also evolved, resulting in the following list for the present edition:

1. **Computational proteomics**. Analysis of protein–protein interactions. Protein structure modeling. Analysis of protein functionality. Quantitative proteomics and PTMs. Clinical proteomics. Protein annotation. Data mining in proteomics.
2. **Next generation sequencing and sequence analysis**. De novo sequencing, re-sequencing, and assembly. Expression estimation. Alternative splicing discovery. Pathway Analysis. Chip-seq and RNA-Seq analysis. Metagenomics. SNPs prediction.
3. **High performance in Bioinformatics**. Parallelization for biomedical analysis. Biomedical and biological databases. Data mining and biological text processing. Large-scale biomedical data integration. Biological and medical ontologies. Novel architecture and technologies (GPU, P2P, Grid,...) for Bioinformatics.
4. **Biomedicine**. Biomedical Computing. Personalized medicine. Nanomedicine. Medical education. Collaborative medicine. Biomedical signal analysis. Biomedicine in industry and society. Electrotherapy and radiotherapy.
5. **Biomedical Engineering**. Computer-assisted surgery. Therapeutic engineering. Interactive 3D modeling. Clinical engineering. Telemedicine.

Biosensors and data acquisition. Intelligent instrumentation. Patient Monitoring. Biomedical robotics. Bio-nanotechnology. Genetic engineering.

6. **Computational systems for modeling biological processes.** Inference of biological networks. Machine learning in Bioinformatics. Classification for biomedical data. Microarray Data Analysis. Simulation and visualization of biological systems. Molecular evolution and phylogenetic modeling.

7. **Healthcare and diseases.** Computational support for clinical decisions. Image visualization and signal analysis. Disease control and diagnosis. Genome-phenome analysis. Biomarker identification. Drug design. Computational immunology.

8. **E-Health.** E-Health technology and devices. E-Health information processing. Telemedicine/E-Health application and services. Medical Image Processing. Video techniques for medical images. Integration of classical medicine and E-Health.

After a careful peer review and evaluation process (268 submission were submitted and each submission was reviewed by at least 2, and on the average 2.7, Program Committee members or Additional Reviewer), 134 papers were accepted to be included in LNBI proceedings.

During IWBBIO 2015 several Special Sessions will be carried out. Special Sessions will be a very useful tool to complement the regular program with new and emerging topics of particular interest for the participating community. Special Sessions that emphasize on multidisciplinary and transversal aspects, as well as cutting-edge topics are especially encouraged and welcome, and in this edition of IWBBIO 2015 are the following:

1. **SS1: Expanding Concept of Chaperone Therapy for Inherited Brain Diseases**
   Chaperone therapy is a new concept of molecular therapeutic approach, first developed for lysosomal diseases, utilizing small molecular competitive inhibitors of lysosomal enzymes. This concept has been gradually targeted to many diseases of other categories, utilizing various compounds not necessarily competitive inhibitors but also non-competitive inhibitors or endogenous protein chaperones (heat-shock proteins).
   In this session, we discuss current trends of chaperone therapy targeting various types of neurological and non-neurological diseases caused by misfolded mutant proteins. This molecular approach will open a new therapeutic view for a wide variety of diseases, genetic and non-genetic, and neurological and non-neurological, in the near future.

   Organizer: Dr. Prof. Yaping Tian, Department of Clinical Biochemistry, Chinese PLA General Hospital, Beijing (China).
2. **SS2: Quantitative and Systems Pharmacology: Thinking in a wider "systems-level" context accelerates drug discovery and enlightens our understanding of drug action**
   "Quantitative and Systems Pharmacology (QSP) is an emerging discipline focused on identifying and validating drug targets, understanding existing

therapeutics and discovering new ones. The goal of QSP is to understand, in a precise, predictive manner, how drugs modulate cellular networks in space and time and how they impact human pathophysiology." (QSP White Paper - October, 2011)

Over the past three decades, the predominant paradigm in drug discovery was designing selective ligands for a specific target to avoid unwanted side effects. However, in the current postgenomic era, the aim is to design drugs that perturb biological networks rather than individual targets. The challenge is to be able to consider the complexity of physiological responses to treatments at very early stages of the drug development. In this way, current effort has been put into combining 0 chemogenomics with network biology to implement new network-pharmacology approaches to drug discovery; i.e., polypharmacology approaches combined with systems biology information, which advance further in both improving efficacy and predicting unwanted off-target effects. Furthermore, the use of network biology to understand drug action outputs treasured information, i.e., for pharmaceutical companies, such as alternative therapeutic indications for approved drugs, associations between proteins and drug side effects, drug–drug interactions, or pathways, and gene associations which provide leads for new drug targets that may drive drug development.

Following the line of QSP Workshops I and II (2008, 2010), the QSP White Paper (2011), or QSP Pittsburgh Workshop (2013), the goal of this symposium is to bring together interdisciplinary experts to help advance the understanding of how drugs act, with regard to their beneficial and toxic effects, by sharing new integrative, systems-based computational, or experimental approaches/tools/ideas which allow to increase the probability that the newly discovered drugs will prove therapeutically beneficial, together with a reduction in the risk of serious adverse events.

Organizer: Violeta I. Perez-Nueno, Ph.D., Senior Scientist, Harmonic Pharma, Nancy (France).

3. **SS3: Hidden Markov Model (HMM) for Biological Sequence Modeling** Sequence Modeling is one of the most important problems in bioinformatics. In the sequential data modeling, Hidden Markov Models(HMMs) have been widely used to find similarity between sequences. Some of the most important topics in this session are:

(a) Modeling of biological sequences in bioinformatics;
(b) The application of Hidden Markov Models(HMM);
(c) HMM in modeling of sequential data;
(d) The advantages of HMM in biological sequence modeling compared to other algorithms;
(e) The new algorithms of training HMM;
(f) Gene sequence modeling with HMM;

Organizer: Mohammad Soruri, Department of Electrical and Computer Engineering, University of Birjand, Birjand (Iran).

4. **SS4: Advances in Computational Intelligence for Bioinformatics and Biomedicine** Biomedicine and, particularly, Bioinformatics are increasingly and rapidly becoming data-based sciences, an evolution driven by technological advances in image and signal non-invasive data acquisition (exemplified by the 2014 Nobel Prize in Chemistry for the development of super-resolved fluorescence microscopy). In the Biomedical field, the large amount of data generated from a wide range of devices and patients is creating challenging scenarios for researchers, related to storing, processing, and even just transferring information in its electronic form, all these compounded by privacy and anonymity legal issues. This can equally be extended to Bioinformatics, with the burgeoning of the .omics sciences.

New data requirements require new approaches to data analysis, some of the most interesting ones are currently stemming from the fields of Computational Intelligence (CI) and Machine Learning (ML). This session is particularly interested in the proposal of novel CI and ML approaches to problems in the biomedical and bioinformatics domains.

Topics that are of interest in this session include (but are not necessarily limited to):

(a) Novel applications of existing CI and ML methods to biomedicine and bioinformatics.
(b) Novel CI and ML techniques for biomedicine and bioinformatics.
(c) CI and ML-based methods to improve model interpretability in biomedical problems, including data/model visualization techniques.
(d) Novel CI and ML techniques for dealing with nonstructured and heterogeneous data formats.

More information at
http://www.cs.upc.edu/ avellido/research/conferences/
IWBBIO15-CI-BioInfMed.html
Main Organizer: Alfredo Vellido, PhD, Department of Computer Science, Universitat Politécnica de Catalunya, BarcelonaTECH (UPC), Barcelona (Spain).

Co-organizers: Jesus Giraldo, PhD, Institut de Neurociències and Unitat de Bioestadística, Universitat Autònoma de Barcelona (UAB), Cerdanyola del Vallès, Barcelona (Spain).
René Alquézar, PhD, Department of Computer Science, Universitat Politécnica de Catalunya, BarcelonaTECH (UPC), Barcelona (Spain).

5. **SS5: Tools for Next Generation Sequencing data analysis** Next Generation Sequencing (NGS) is the main term used to describe a number of different modern sequencing technologies such as Illumina, Roche 454 Sequencing, Ion torrent, SOLiD sequencing, and Pacific Biosciences. These technologies allow us to sequence DNA and RNA more quickly and cheaply than Sanger sequencing and have opened new ways for the study of genomics, transcriptomics, and molecular biology, among others.

The continuous improvements on those technologies (longer read length, better read quality, greater throughput, etc.), and the broad application of NGS in several research fields, have produced (and still produce) a huge amount of software tools for the analysis of NGS genomic/transcriptomic data.

We invite authors to submit original research, pipelines, and review articles on topics related to software tools for NGS data analysis such as (but not limited to):

(a) Tools for data preprocessing (quality control and filtering).
(b) Tools for sequence alignment.
(c) Tools for de novo assembly.
(d) Tools for the analysis of genomic data: identification and annotation of genomic variants (variant calling, variant annotation).
(e) Tools for functional annotation to describe domains, orthologs, genomic variants, controlled vocabulary (GO, KEGG, InterPro...).
(f) Tools for the analysis of transcriptomic data: RNA-Seq data (quantification, normalization, filtering, differential expression) and transcripts and isoforms finding.
(g) Tools for Chip-Seq data.
(h) Tools for "big-data" analysis of reads and assembled reads.

Organizers: Javier Perez Florido, PhD, Genomics and Bioinformatics Platform of Andalusia (GBPA), Seville, (Spain).

Antonio Rueda Martin, Genomics and Bioinformatics Platform of Andalusia (GBPA), Seville, (Spain).

M. Gonzalo Claros Diaz, PhD, Department of Molecular Biology and Biochemistry, University of Malaga (Spain).

6. **SS6: Dynamics Networks in System Medicine**

Over the past two decades, It Is Increasingly Recognized that a biological function can only rarely be attributed to an individual molecule. Instead, most biological functions arise from signaling and regulatory pathways connecting many constituents, such as proteins and small molecules, allowing them to adapt to environmental changes. "Following on from this principle, a disease phenotype is rarely a consequence of an abnormality in a single effector gene product, but reflects various processes that interact in a complex network." Offering a unifying language to describe relations within such a complex system has made network science a central component of systems biology and recently system medicine. Despite the knowledge that biological networks can change with time and environment, much of the efforts have taken a static view. Time-varying networks support richer dynamics and may better reflect underlying biological changes in abnormal state versus normal state and this provides a powerful motivation and application domain for computational modeling. We introduce this session on the Dynamics Networks in System Medicine to encourage and support the development of computational methods that elucidate the Dynamics Networks and its application in medicine. We will discuss current trends and potential biological and clinical applications of network-based approaches to human disease. We aim to bring together experts in different fields in order to promote cross fertilization between different communities.

Organizer: Narsis Aftab Kiani, PhD, Computational Medicine Unit, Department of Medicine, Karolinska Institute (Sweden).

7. **SS7: Interdisciplinary puzzles of measurements in biological systems**

Natural sciences demand measurements of the subject of interest as a necessary part of the experimental process. Thus, for the proper understanding of the obtained datasets, it is the necessity to take into question all mathematical, biological, chemical, or technical conditions affecting the process of the measurement itself. While assumptions and recommendations within the field itself are usually concerned, some issues, especially discretization, quantization, experiment time, self-organization, and consequent anomalous statistics might cause puzzling behavior.

In this special section we describe particular examples across disciplines with joint systems theory-based approach, including noise and baseline filtration in mass spectrometry, image processing and analysis, and distributed knowledge database. The aim of this section is to present a general overview of the systemic approach.

Organizer: Jan Urban, PhD, Laboratory of Signal and Image Processing, Institute of Complex Systems, Faculty of Fisheries and protection of Waters, University of South Bohemia. (Czech republic).

8. **SS8: Biological Networks: Insight from interactions**

The complete sequencing of the human genome has shown us a new era of Systems Biology (SB) referred to as omics. From genomics to proteomics and furthermore, "Omics"-es existing nowadays integrate many areas of biology. This resulted in an essential ascent from Bioinformatics to Systems Biology leaving room for identifying the number of interactions in a cell. Tools have been developed to utilize evolutionary relationships toward understanding uncharacterized proteins, while there is a need to generate and understand functional interaction networks. A systematic understanding of genes and proteins in a regulatory network has resulted in the birth of Systems Biology (SB), there-by raising several unanswered questions. Through this conference, we will raise some questions on why and how interactions, especially protein–protein interactions (PPI), are useful while discussing methods to remove false positives by validating the data. The conference is aimed at the following two focal themes:

(a) Bioinformatics and systems biology for deciphering the known–unknown regions.

(b) Systems Biology of regulatory networks and machine learning.

Organizers: Prof. Alfredo Benso, PhD, Department of Control and Computer Engineering, Politecnico di Torino (Italy).

Dr. Prashanth Suravajhala, PhD, Founder of Bioclues.org and Director of Bioinformatics.org

9. **SS9: Tissue engineering for biomedical and pharmacological applications**

The concept of tissues appeared more than 200 years ago, since textures and attendant differences were described within the whole organism components. Instrumental developments in optics and biochemistry subsequently paved the way to transition from classical to molecular histology in order to decipher the molecular contexts associated with physiological or pathological development or function of a tissue. The aim of this special session is to provide an overview of the most cutting edge updates in tissue engineering technologies. This will cover the most recent developments for tissue proteomics, and the applications of the ultimate molecular histology method in pathology and pharmacology: MALDI Mass Spectrometry Imaging. This session will be of great relevance for people willing to have a relevant summary of possibilities in the field of tissue molecular engineering.

Organizer: Rémi Longuespée, PhD, Laboratoire de Spectrométrie de Masse, University of Liege (Belgium).

10. **SS10: Toward an effective telemedicine: an interdisciplinary approach**

In the last 20 years many resources have been spent in experimentation and marketing of telemedicine systems, but — as pointed by several researchers — no real product has been fully realized — neither in developed nor in underdeveloped countries. Many factors could be detected:

(a) lack of a decision support system in analyzing collected data;
(b) the difficulty of using the specific monitoring devices;
(c) the caution of patients and/or doctors toward E-health or telemedicine systems;
(d) the passive role imposed on the patient by the majority of experimented systems;
(e) the limits of profit-driven outcome measures;
(f) a lack of involvement of patients and their families as well as an absence of research on the consequences in the patient's life.

The constant improvement of ICT tools should be taken into account: at-home and mobile monitoring are both possible; virtual visits can be seen as a new way to perform an easier and more accepted style of patient-doctor communication (which is the basis of a new active role of patients in monitoring symptoms and evolution of the disease). The sharing of this new approach could be extended from patients to healthy people, obtaining tools for a real preventive medicine: a large amount of data could be gained, stored, and analyzed outside the sanitary structures, contributing to a low-cost approach to health.

The goal of this session is to bring together interdisciplinary experts to develop (discuss about) these topics:

(a) decision support systems for the analysis of collected data;
(b) customized monitoring based on the acuteness of the disease;
(c) integration of collected data with E-Health systems;
(d) attitudes of doctors and sanitary staff;

(e) patient-doctor communication;

(f) involvement of patients and of their relatives and care-givers;

(g) digital divide as an obstacle/hindrance;

(h) alternative measurements on the effectiveness of telemedicine (quality of life of patients and caregivers, etc.)

(i) mobile versus home monitoring (sensors, signal transmissions, etc.)

(j) technology simplification (auto-calibrating systems, patient interface, physician interface, bio-feedback for improving learning)

The session will also have the ambition of constituting a team of interdisciplinary research, spread over various countries, as a possible basis for effective participation in European calls.

Organizers: Maria Francesca Romano, Institute of Economics, Scuola Superiore Sant'Anna, Pisa (Italy).

Giorgio Buttazzo, Institute of Communication, Information and Perception Technologies (TeCIP), Scuola Superiore Sant'Anna, Pisa (Italy).

11. **SS11A: High Performance Computing in Bioinformatics, Computational Biology, and Computational Chemistry**

The goal of this special session is to explore the use of emerging parallel computing architectures as well as High-Performance Computing systems (Supercomputers, Clusters, Grids) for the simulation of relevant biological systems and for applications in Bioinformatics, Computational Biology, and Computational Chemistry. We welcome papers, not submitted elsewhere for review, with a focus on topics of interest ranging from but not limited to:

(a) Parallel stochastic simulation.

(b) Biological and Numerical parallel computing.

(c) Parallel and distributed architectures.

(d) Emerging processing architectures (e.g. GPUs, Intel Xeon Phi, FPGAs, mixed CPU-GPU, or CPU-FPGA, etc).

(e) Parallel Model checking techniques.

(f) Parallel algorithms for biological analysis.

(g) Cluster and Grid Deployment for system biology.

(h) Biologically inspired algorithms.

(i) Application of HPC developments in Bioinformatics, Computational Biology, and Computational Chemistry.

Organizers: Dr. Horacio Perez-Sanchez, Dr. Afshin Fassihi and Dr. Jose M. Cecilia, Universidad Católica San Antonio de Murcia (UCAM), (Spain).

12. **SS11B: High-Performance Computing for Bioinformatics Applications**

This Workshop has a focus on interdisciplinary nature and is designed to attract the participation of several groups including Computational Scientists, Bioscientists, and the fast growing group of Bioinformatics, researchers. It is primarily intended for computational scientists who are interested in Biomedical Research and the impact of high-performance computing in the analysis of biomedical data and in advancing Biomedical Informatics. Bioscientists with some background in computational concepts represent another group of intended participants. The interdisciplinary group of research groups with interests in Biomedical Informatics in general and Bioinformatics in particular will likely be the group attracted the most to the workshop. The Workshop topics include (but are not limited to) the following:

(a) HPC for the Analysis of Biological Data.
(b) Bioinformatics Tools for Health Care.
(c) Parallel Algorithms for Bioinformatics Applications.
(d) Ontologies in biology and medicine.
(e) Integration and analysis of molecular and clinical data.
(f) Parallel bioinformatics algorithms.
(g) Algorithms and Tools for Biomedical Imaging and Medical Signal Processing.
(h) Energy Aware Scheduling Techniques for Large-Scale Biomedical Applications.
(i) HPC for analyzing Biological Networks.
(j) HPC for Gene, Protein/RNA Analysis, and Structure Prediction.

For more information, you can see the Call for Paper for this special session.
Organizers: Prof. Hesham H. Ali, Department of Computer Science, College of Information Science and Technology, University of Nebraska at Omaha (EEUU).
Prof. Mario Cannataro, Informatics and Biomedical Engineering University "Magna Graecia" of Catanzaro (Italy).

13. **SS12: Advances in Drug Discovery**
We welcome papers, not submitted elsewhere for review, with a focus in topics of interest ranging from but not limited to:
(a) Target identification and validation.
(b) Chemoinformatics and Computational Chemistry: Methodological basis and applications to drug discovery of: QSAR, Docking, CoMFA-like methods, Quantum Chemistry and Molecular Mechanics (QM/MM), High-performance Computing (HPC), Cloud Computing, Biostatistics, Artificial Intelligence (AI), Machine Learning (ML), and Bio-inspired Algorithms like Artificial Neural Networks (ANN), Genetic Algorithms, or Swarm Intelligence.
(c) Bioinformatics and Biosystems: Methodological basis and applications to drug design, target or biomarkers discovery of: Alignment tools, Pathway analysis, Complex Networks, Nonlinear methods, Microarray analysis, Software, and Web servers.
(d) High Throughput Screening (HTS) of drugs; Fragment-Based Drug Discovery; Combinatorial chemistry, and synthesis.
Organizers: Dr. Horacio Perez-Sanchez and Dr. Afshin Fassihi, Universidad Católica San Antonio de Murcia (UCAM), (Spain).

14. **SS13: Deciphering the human genome**
Accomplishment of "1000 Genomes Project" revealed immense amount of information about variation, mutation dynamics, and evolution of the human DNA sequences. These genomes have been already used in a number

of bioinformatics studies, which added essential information about human populations, allele frequencies, local haplotype structures, distribution of common and rare genetic variants, and determination of human ancestry and familial relationships. Humans have modest intra-species genetic variations among mammals. Even so, the number of genetic variations between two persons from the same ethnic group is in the range of 3.4–5.2 million. This gigantic pool of nucleotide variations is constantly updating by 40–100 novel mutations arriving in each person. Closely located mutations on the same DNA molecule are linked together forming haplotypes that are inherited as whole units and span over a considerable portion of a gene or several neighboring gene. An intense intermixture of millions of mutations occurs in every individual due to frequent meiotic recombinations during gametogenesis. Scientists and doctors are overwhelmed with this incredible amount of information revealed by new-generation sequencing techniques. Due to this complexity, we encountered significant challenges in deciphering genomic information and interpretation of genome-wide association studies.

The goal of this session is to discuss novel approaches and algorithms for processing of whole-genome SNP datasets in order to understand human health, history, and evolution.

Organizer: Alexei Fedorov, Ph.D, Department of Medicine, Health Science Campus, The University of Toledo (EEUU).

15. **SS14: Ambient Intelligence for Bioemotional Computing**

Emotions have a strong influence on our vital signs and on our behavior. Systems that take our emotions and vital signs into account can improve our quality of life. The World Health Organization (WHO) characterizes a healthy life first of all with the prevention of diseases and secondly, in the case of the presence of disease, with the ability to adapt and self-manage. Smart measurement of vital signs and of behavior can help to prevent diseases or to detect them before they become persistent. These signs are key to obtain individual data relevant to contribute to this understanding of healthy life.

The objective of this session is to present and discuss smart and unobtrusive methods to measure vital signs and capture emotions of the users and methods to process these data to improve their behavior and health.

Organizers: Prof. Dr. Natividad Martinez, Internet of Things Laboratory, Reutlingen University (Germany).

Prof. Dr. Juan Antonio Ortega, University of Seville (Spain).

Prof. Dr. Ralf Seepold, Ubiquitous Computing Lab, HTWG Konstanz (Germany).

In this edition of IWBBIO, we are honored to have the following invited speakers:

1. Prof. Xavier Estivill, Genomics and Disease group, Centre for Genomic Regulation (CRG), Barcelona (SPAIN).
2. Prof. Alfonso , Structural Computational Biology group, Spanish National Cancer Research Center (CNIO), Madrid (SPAIN).

3. Prof. Patrick Aloy, Structural Bioinformatics and Network Biology group, Institute for Research in Biomedicine (IRB), Barcelona (SPAIN).

It is important to note that for the sake of consistency and readability of the book, the presented papers are classified under 21 chapters. The organization of the papers is in two volumes arranged basically following the topics list included in the call for papers. The first volume (LNBI 9043), entitled "Advances on Computational Intelligence. Part I" is divided into seven main parts and includes the contributions on:

1. Bioinformatics for healthcare and diseases.
2. Biomedical Engineering.
3. Biomedical image analysis.
4. Biomedical signal analysis.
5. Computational genomics.
6. Computational proteomics.
7. Computational systems for modeling biological processes.

In the second volume (LNBI 9044), entitled "Advances on Computational Intelligence. Part II" is divided into 14 main parts and includes the contributions on:

1. E-Health.
2. Next generation sequencing and sequence analysis.
3. Quantitative and Systems Pharmacology.
4. Hidden Markov Model (HMM) for Biological Sequence Modeling.
5. Biological and bio-inspired dynamical systems for computational intelligence.
6. Advances in Computational Intelligence for Bioinformatics and Biomedicine.
7. Tools for Next Generation Sequencing data analysis.
8. Dynamics networks in system medicine.
9. Interdisciplinary puzzles of measurements in biological systems.
10. Biological Networks: Insight from interactions.
11. Toward an effective telemedicine: an interdisciplinary approach.
12. High-Performance Computing in Bioinformatics, Computational Biology, and Computational Chemistry.
13. Advances in Drug Discovery.
14. Ambient Intelligence for Bioemotional Computing.

This third edition of IWBBIO was organized by the Universidad de Granada together with the Spanish Chapter of the IEEE Computational Intelligence Society. We wish to thank our main sponsor BioMed Central, E-Health Business Development BULL (España) S.A., and the institutions Faculty of Science, Dept. Computer Architecture and Computer Technology and CITIC-UGR from the University of Granada for their support and grants. We also wish to thank the Editor-in-Chief of different international journals for their interest in editing special issues from the best papers of IWBBIO.

We would also like to express our gratitude to the members of the different committees for their support, collaboration, and good work. We especially thank the Local Committee, Program Committee, the Reviewers, and Special Session Organizers. Finally, we want to thank Springer, and especially Alfred Hoffman and Anna Kramer for their continuous support and cooperation.

April 2015

Francisco Ortuño
Ignacio Rojas

Organization

Program Committee

Jesus S. Aguilar	Universidad Pablo de Olavide, Spain
Carlos Alberola	Universidad de Valladolid, Spain
Hesham H. Ali	University of Nebraska at Omaha, USA
René Alquézar	Universitat Politécnica de Catalunya, Barcelona TECH, Spain
Rui Carlos Alves	University of Lleida, Spain
Eduardo Andrés León	Spanish National Cancer Center, Spain
Antonia Aránega	University of Granada, Spain
Saúl Ares	Spanish National Center for Biotechnology, Spain
Rubén Armañanzas	Universidad Politécnica de Madrid, Spain
Joel P. Arrais	University of Coimbra, Portugal
O. Bamidele Awojoyogbe	Federal University of Technology, Minna, Nigeria
Jaume Bacardit	University of Newcastle, Australia
Hazem Bahig	University of Haíl, Saudi Arabia
Pedro Ballester	Inserm, France
Oresti Baños	Kyung Hee University, Korea
Ugo Bastolla	Center of Molecular Biology "Severo Ochoa", Spain
Steffanny A. Bennett	University of Ottawa, Canada
Alfredo Benso	Politecnico di Torino, Italy
Concha Bielza	Universidad Politécnica de Madrid, Spain
Armando Blanco	University of Granada, Spain
Ignacio Blanquer	Universidad Politécnica de Valencia, Spain
Giorgio Buttazzo	Scuola Superiore Sant'Anna, Italy
Gabriel Caffarena	Universidad Politécnica de Madrid, Spain
Mario Cannataro	University Magna Graecia of Catanzaro, Italy
Carlos Cano	University of Granada, Spain
Rita Casadio	University of Bologna, Italy
Jose M. Cecilia	Universidad Católica San Antonio de Murcia, Spain
M. Gonzalo Claros	University of Málaga, Spain

Darrell Conklin	Universidad del País Vasco/Euskal Herriko Unibertsitatea, Spain
Clare Coveney	Nottingham Trent University, UK
Miguel Damas	University of Granada, Spain
Guillermo de La Calle	Universidad Politécnica de Madrid, Spain
Javier De Las Rivas	CSIC, CIC, USAL, Salamanca, Spain
Joaquin Dopazo	Centro de Investigación Principe Felipe Valencia, Spain
Hernán Dopazo	CIPF, Spain
Werner Dubitzky	University of Ulster, UK
Khaled El-Sayed	Modern University for Technology and Information, Egypt
Christian Esposito	ICAR-CNR, Italy
Afshin Fassihi	Universidad Católica San Antonio de Murcia, Spain
Jose Jesús Fernandez	University of Almeria, Spain
Jean-Fred Fontaine	Max Delbrück Center for Molecular Medicine, Germany
Xiaoyong Fu	Western Reserve University, USA
Razvan Ghinea	University of Granada, Sapin
Jesus Giraldo	Universitat Autònoma de Barcelona, Spain
Humberto Gonzalez	University of Basque Country, Spain
Daniel Gonzalez Peña	Bioinformatics and Evolutionary Computing, Spain
Concettina Guerra	Georgia Institute of Technology, USA
Christophe Guyeux	IUT de Belfort-Montbéliard, France
Michael Hackenberg	University of Granada, Spain
Luis Javier Herrera	University of Granada, Spain
Lynette Hirschman	MITRE Corporation, USA
Michelle Hussain	University of Salford, UK
Andy Jenkinson	European Bioinformatics Institute, UK
Craig E. Kapfer	KAUST, Saudi Arabia
Narsis Aftab Kiani	Karolinska Institute, Sweden
Ekaterina Kldiashvili	New Vision University/Georgian Telemedicine Union, Georgia
Tomas Koutny	University of West Bohemia, Czech Republic
Natalio Krasnogor	University of Newcastle, Australia
Marija Krstic-Demonacos	University of Salford, UK
Sajeesh Kumar	University of Tennessee, USA
Pedro Larrañaga	Universidad Politecnica de Madrid, Spain
Jose Luis Lavin	CIC bioGUNE, Spain
Rémi Longuespée	University of Liège, Belgium

Miguel Angel Lopez Gordo	Universidad de Cadiz, Spain
Ernesto Lowy	European Bioinformatics Institute, UK
Natividad Martinez	Reutlingen University, Germany
Francisco Martinez Alvarez	University of Seville, Spain
Marco Masseroli	Politechnical University of Milano, Italy
Roderik Melnik	Wilfrid Laurier University, Canada
Jordi Mestres	Universitat Pompeu Fabra, Spain
Derek Middleton	University of Liverpool, UK
Federico Moran	Complutense University of Madrid, Spain
Antonio Morreale	Repsol, Spain
Walter N. Moss	Yale University and Howard Hughes Medical Institute, New Haven, USA
Cristian R. Munteanu	University of Coruña, UDC, Spain
Enrique Muro	Johannes Gutenberg University, Institute of Molecular Biology, Mainz, Germany
Jorge A. Naranjo	New York University, Abu Dhabi, UAE
Isabel A. Nepomuceno	University of Seville, Spain
Michael Ng	Hong Kong Baptist University, China
Baldomero Oliva	Universitat Pompeu Fabra, Spain
Jose Luis Oliveira	University of Aveiro, IEETA, Portugal
Jose L. Oliver	University of Granada, Spain
Juan Antonio Ortega	University of Seville, Spain
Julio Ortega	University of Granada, Spain
Francisco Ortuno	University of Granada, Spain
Paolo Paradisi	National Research Council of Italy, Italy
Alejandro Pazos	University of Coruña, Spain
David Pelta	University of Granada, Spain
Alexandre Perera	Universitat Politècnica de Catalunya, Spain
Javier Perez Florido	Genomics and Bioinformatics Platform of Andalusia, Spain
María Del Mar Pérez Gómez	University of Granada, Spain
Violeta I. Pérez Nueno	Harmonic Pharma, France
Horacio Pérez Sánchez	Universidad Católica San Antonio de Murcia, Spain
Antonio Pinti	Valenciennes University, France
Alberto Policriti	Università di Udine, Italy
Héctor Pomares	University of Granada, Spain
Alberto Prieto	University of Granada, Spain
Carlos Puntonet	University of Granada, Spain

Omer F. Rana	Cardiff University, UK
Jairo Rocha	University of the Balearic Islands, Spain
Fernando Rojas	University of Granada, Spain
Ignacio Rojas	University of Granada, Spain
Maria Francesca Romano	Scuola Superiore Sant'Anna, Italy
Gregorio Rubio	Universitat Politècnica de València, Spain
Antonio Rueda	Genomics and Bioinformatics Platform of Andalusia, Spain
Michael Sadovsky	Siberian Federal University, Russia
Yvan Saeys	Ghent University, Belgium
Maria Jose Saez	University of Granada, Spain
José Salavert	European Bioinformatics Institute, UK
Carla Sancho Mestre	Universitat Politècnica de València, Spain
Vicky Schneider	The Genome Analysis Centre, UK
Jean-Marc Schwartz	University of Manchester, UK
Ralf Seepold	HTWG Konstanz, Germany
Jose Antonio Seoane	University of Bristol, UK
Istvan Simon	Research Centre for Natural Sciences, Hungary
Richard Sinnott	University of Glasgow, UK
Mohammad Soruri	University of Birjand, Iran
Prashanth Suravajhala	BioClues.org, India
Yoshiyuki Suzuki	Tokyo Metropolitan Institute of Medical Science, Tokyo
Li Teng	University of Iowa, USA
Pedro Tomas	INESC-ID, Portugal
Carolina Torres	University of Granada, Spain
Oswaldo Trelles	University of Málaga, Spain
Paolo Trunfio	University of Calabria, Italy
Olga Valenzuela	University of Granada, Spain
Alfredo Vellido	Universidad Politécnica de Cataluña, Spain
Julio Vera	University of Rostock, Germany
Renato Umeton	CytoSolve Inc., USA
Jan Urban	University of South Bohemia, Czech Republic

Additional Reviewers

Agapito, Giuseppe	Gonzalez-Abril, Luis
Alquezar, Rene	Ielpo, Nicola
Asencio Cortés, Gualberto	Julia-Sape, Margarida
Belanche, Lluís	König, Caroline Leonore
Calabrese, Barbara	
Caruso, Maria Vittoria	Milano, Marianna
Cárdenas, Martha Ivón	Mir, Arnau
Fernández-Montes, Alejandro	Mirto, Maria

Navarro, Carmen
Ortega-Martorell, Sandra
Politano, Gianfranco Michele Maria
Ribas, Vicent
Rychtarikova, Renata

Sarica, Alessia
Tosi, Alessandra
Vellido, Alfredo
Vilamala, Albert

Table of Contents – Part II

eHealth

Next Generation Sequencing and Sequence Analysis

Quantitative and Systems Pharmacology

Hidden Markov Model (HMM) for Biological Sequence Modeling

Advances in Computational Intelligence for Bioinformatics and Biomedicine

Tools for Next Generation Sequencing Data Analysis

Dynamics Networks in System Medicine

Interdisciplinary Puzzles of Measurements in Biological Systems

Biological Networks: Insight from Interactions

Towards an Effective Telemedicine: An Interdisciplinary Approach

High Performance Computing in Bioinformatics, Computational Biology and Computational Chemistry

Advances in Drug Discovery

Ambient Intelligence for Bioemotional Computing

Table of Contents – Part I

Bioinformatics for Healthcare and Diseases

Biomedical Engineering

Biomedical Image Analysis

Biomedical Signal Analysis

Computational Genomics

Computational Systems for Modelling Biological Processes

e-Health Informed Foreign Patient and Physician Communication: The Perspective of Informed Consent

Echo Huang[1], Shao-Fu Liao[1], and Shing-Lung Chen[2]

[1]Department of Information Management, National Kaohsiung First University of Science and Technology, No. 1, University Rd., Yenchao Dist., Kaohsiung City 82455, Taiwan (ROC)
echoh@nkfust.edu.tw, shaofu.liao@gmail.com
[2]Dept. of German
chensl@nkfust.edu.tw

Abstract. The rapidly growth of mobility population of exchange students, foreign workers, and tourists raises the communication challenge of foreign patient-physician communication in local hospitals (clinics). Language is not the only one skill during a medical visit, health literacy is another important skill used to understand the medical information delivered by doctors. Skills in understanding and applying information about health issues are critical to the process of patient-physician communication and may have substantial impact on health behaviors and health outcomes.

In recent years, the practice of foreigners using mobile apps to search map for location, access transportation information, and lookup translation, have increased steadily. However, the current mobile apps can't support foreigners for their sudden needs of hospital visits. Take into consideration of language barrier and low health literacy issue, an Internet health information-based patient-centered mobile system, is suggested to mediate the informed consent process of patient-physician communication. Thus, this research aims to enhance foreign visitors' capabilities in communication during exchange information with local foreign doctors by developing an effective patient-physician communication mobile system. Based on four case studies of patients' hospital visit experiences in Europe and Asia and their perception toward using the media-mediated communication. We concluded that 1) e-health informed foreign patients' language skill and health literacy were improved in compare with non-Internet-informed patients group; 2) e-health informed patients' states of anxiety were reduced; 3) their satisfaction of medical visit were improved; and 4) the process of informed consent was standardized into six parts (Symptom description, Disease judgment, Disease diagnosis, Examination, Treatment, and Healthcare education) for general foreign patient communications.

Keywords: Health literacy, language barrier, Informed Consent, e-Health system, foreign patient, Internet-informed patient.

F. Ortuño and I. Rojas (Eds.): IWBBIO 2015, Part II, LNCS 9044, pp. 1–11, 2015.
© Springer International Publishing Switzerland 2015

1   Introduction

Health communication consists of interpersonal or mass communication activities focused on improving the health of individuals and populations. Skills in understanding and applying information about health issues are critical to this process and may have substantial impact on health behaviors and health outcomes. The skills are so-called health literacy (HL). Health literacy (HL), refers to the capacity of individuals to access, understand, and use health information to make informed and appropriate health-related decisions, has been recognized as an important concept in patient education and disease management [1]. HL should be considered not only in terms of the characteristics of individuals, but also in terms of the interactional processes between individuals and their health and social environments. Improved HL may enhance the ability and motivation of individuals to find solutions to both personal and public health problems, and these skills could be used to address various health problems throughout life [2]. The process underpinning HL involves empowerment, one of the major goals of health communication.

The National Library of Medicine defines HL as "the degree to which individuals have the capacity to obtain, process, and understand basic health information and services needed to make appropriate health decisions." Nutbeam [6] proposed three levels of HL: functional, communicative and critical HL. *Functional literacy*, the basic level of reading and writing skills that let someone function effectively in everyday situations; *communicative literacy*, advanced skills that allow a person to extract information, derive meaning from different forms of communication, and apply new information to changing circumstances; and *critical literacy*, more advanced skills for critically analyzing information and using information to exert greater control over life events and situations.

Effective and timely communication between patients, physicians, nurses, pharmacists, and other healthcare professionals is vital to good healthcare. Current communication mechanisms, based largely on paper records and prescriptions, are old fashioned, inefficient, and unreliable. Recent research has found that the emergence of e-health, enhanced by advanced technology, has exerted substantial influence on people's health-related experiences [4]. The term 'e-health' generally refers to the use of the Internet as a health resource to improve the health status of patients [5]. Existing research has suggested that e-health has great potential to promote better health among patients. Essentially, e-health can help citizens facilitate and maintain healthier lifestyles. It could help patients increase health awareness and knowledge in general [7,8], learn about healthier life styles [9], and achieve a better understanding of health issues, diseases, and medical treatment [7,8]. The onset of e-health not only has influenced patients' health-related attitudes and behaviors, but it has also affected their roles in health care particularly health-related decision making [10,11]. Previous studies have suggested that e-health has heightened patients' sense of autonomy in their health care and has encouraged them to take a more active and responsible role in their medical decisions [10], [12]. Despite concerns about security and privacy, such e-health systems provide increased accuracy and efficiency, better communication among healthcare professionals, and reduced risk of prescription errors. As a result, e-health has at least partly contributed to a change in physician-patient interaction [10], [13].

An e-health communication system somehow can help to reduce the barriers of health literacy issue and encourage patients to address various health problem with confidence. E-health involves the delivery of customer service via web-based user accounts, social networks, mobile phone, and the Internet rather call center or facilities open to the public such as retail stores or service counters. Such digital services are increasingly demanded by customers, who are already using digital platforms to research and review products, as well as broadcast their service frustrations. The rewards of adopting e-healthcare are worth the effort, and virtually every customer-facing industry requiring extensive customer-relationship management-from cable operators and consumer electronic to healthcare and utilities, all can be benefit [3].

The use of Internet health information to supplement the informed consent process for patients has been discussed for years. Previous studies had tested the effect on the patients' state anxiety, comprehension, and satisfaction. As predicted, there was a significant positive correlation between comprehension and anxiety. The test group that viewed the e-Health scored significantly higher on the typical consent process and was rated significantly higher in comprehension by physicians than the comparison group. Thus, the goals of this research are to examine (a) the extent to which, and how, patients use e-health materials, and (b) how patients' beliefs about e-health and its effects on patients may be related to their decision-making of treatment during their medical visits.

2   Theoretical Framework: Media Mediated Communication

This study conceptualizes a mobile mediated patient-physician communication system. Physicians often serve as primary sources of patients' medical care and advice [11], [14] with a goal to help improve patients' overall well-being. The primary goal of media mediation is to empower a patient with health literacy skills by actively discussing e-health content with their physicians [4]. During the communication of e-health materials, physicians may endorse e-health by addressing the utility of e-health material, express their concerns about e-health [15], or strongly recommend patients use only certain e-health resources [16]. The result is an enhanced understanding of media messages that may help protect the patient from aversive media effects.

3   e-Health Communication System

Our prototype e-health communication system uses service-oriented architecture (SOA) to enforce basic software architecture principles and provide interoperability between different computing platforms and applications that communicate with each other. The devices accessing the patient and physician models can be desktop or server computers such as smartphones, tablet, and PDAs. To relieve the concerns of security and privacy, the system authenticates users and logs session information and only privileged users can view the data.

Fig. 1. Patient's smart phone displaying bilingual conversation and hyperlinks of medical terms

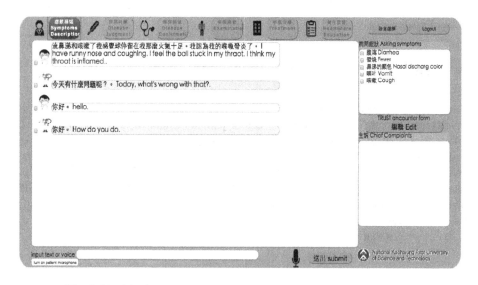

Fig. 2. Physician's screen displaying the six process of informed consent

3.1   Patient Module

The patient-side module supports three phases of a typical medical visit: before enter clinic room, during the process of patient-physician-communication, and after a hospital visit. In the first phase, a patient can edit the possible used sentences such as symptoms description, self-care treatment, allergy and personal medicines. During this period of time, a patient can click the function of 'frequently-mentioned symptoms' and 'how often' to logically organize his symptoms and conditions before talking with a doctor. In the second phase, the voice-to-text bilingual function, as depicted in Figure 1, can used to assist the possible misunderstanding of professional medical terms which seldom are used in our daily life and cause low health literacy patients frustrated and confused. By using the translation function and Internet hyperlinks of strange medical terms, a patient can negotiate with a doctor with greater confidence and better comprehension. In the third phase, a patient can access the electronic form of communication and consider the further healthcare decision.

3.2   Physician Module

In contrast with patient module, the physician module is mainly designed to standardize the process of informed consent. We categorize the flow of a medical visit into six parts: symptom description, disease judgment, disease confirmation, examination, treatment, and healthcare education, as depicted in Figure 2. A doctor is expected to control the flow of communication based on the theory of Informed Consent. Because the communication between a patient and a doctor is complicate, sometimes, very private, therefore, a media-mediated communication is much helpful to support the doctor leading the patient to share his/her situation logically and then give alternative advices for discussion to achieve shared-decision.

4   Research Method

Urquhart, et al. [17] suggests five guidelines for grounded theory studies in information systems, namely, constant comparison, iterative conceptualization, theoretical sampling, scaling up, and theoretical integration. In this study, we follow the theoretical sampling, this guideline stresses the importance of deciding on analytic grounds where to sample from next in the study. Theoretical sampling helps to ensure the comprehensive nature of the theory, and ensures that the developing theory is truly grounded in the data. We examined the foreign patients' perception toward their experiences of clinic visits. Based on the panel discussions, we then propose the system framework for further experimental study.

Study 1 Patient-Physician Communication Experiences Sharing

In Study 1, data collection took place in foreign students live in Vienna city of Austria and Kaohsiung city of Taiwan which host more than thousands of foreign students annually. Interviews were conducted with foreign students with medical visit experiences. Key interviewees were identified based on the level of their involvement in the medical visits. During the semi-structured interviews, questions on the rationales behind the conceptualization of the language barriers were first raised, followed by specific questions on how the interviewees attempted to identify and solve their health literacy issues. Lastly, the interviewees were asked to appraise the overall satisfactions of the medical visits. Information on the difficulties faced during the hospital visits, such as communication with healthcare givers and benefits attained from the hospitals, was also obtained.

Based on the suggestion of Informed Consent and patient interviews, as shown in Appendix, we divide the flow of a medical visit into six parts: symptom description, disease judgment, disease confirmation, examination, treatment, and healthcare education.

Study 2 Patient-Physician Communication System Design

As important, when Internet-informed patients feel they understand what are going to happen to themselves, their levels of state anxiety can be reduced significantly and satisfaction with the care they received from the health care professionals can be improved at the same time. In Study 2, we propose a media-mediated communication mechanism and test the effectiveness of a communication system presentation designed to standardize the informed consent.

Our prototype communication system uses service-oriented architecture (SOA) to enforce basic software architecture principles and provide interoperability between different computing platforms and applications that communicate with each other. The devices accessing the patient and physician models can be desktop or server computers such as smartphones, tablet, and wearable devices. To relieve the concerns of security and privacy, the system authenticates users and logs session information and only privileged users can view the data.

The experiment of system is shown in the Appendix. In addition, knowledge of local healthcare system, information-seeking behaviors, and self-efficacy were assessed for each patient through an interview.

5   Results and Discussion

As expected, patients' comprehension was inversely associated with anxiety with the help of an e-health system. When Internet-informed patients felt they understood what was going to happen to themselves, they reported lower levels of state anxiety and greater satisfaction with the care they received from the health care professionals. These findings are important to explain the importance of helping patients understand medical information. Taken together, our findings shed light on the importance of interventions that can improve patient health literacy. When patients understand medical information in a foreign local clinic, they feel less anxious and more likely to report greater satisfaction with medical care.

We then tested the effectiveness of an e-health system presentation designed to standardize the informed consent. As expected, foreign patients exposed to our prototype e-health system presentation scored significantly higher on a health literacy test than before using the system. The finding provides support for the improvement on the process of standardized informed consent by using multiple medical resources. For example, by combining reading visual types of Internet medical information and listening doctors' verbal explanation, we were able to standardize the consent process and improve patients understanding of complex medical information. Consultations with Internet-informed patients can thus involve a more patient-centered interaction. With the doctors' explanation of the Internet information which can reduce patient misinformation, avoid distress or patients' wrong self-diagnosis and detrimental self-treatment [19].

Patients admit that they do not know something details without using an e-health system and felt frustrated to discuss with a foreign doctor in a local clinic. Felt control and empowerment effects may also explain patients' perception toward using an e-health system. It appeared that patients generally felt that they knew a great deal about the procedure after the doctor delivered the informed consent. This finding is also corresponding with previous studies [20], Internet-informed patients usually report the improvement of their understanding and their ability to manage their health conditions.

References

1. Ishikawa, H., Takeuchi, T., Yano, E.: Measuring functional, communicative, and critical health literacy among diabetic patients. Diabetes Care 31(5), 874–880 (2008)
2. Ishikawa, H., Kiuchi, T.: Health literacy and health communication. BioPsychoSocial Medicine 4(18), 1–5 (2010)
3. Bianchi, R., Schiavotto, D., Svoboda, D.: Why companies should care about e-care. McKinsey&Company (2014), http://www.mckinsey.com/insights/marketing_sales/why_compani es_should_care_about_ecare (accessed)
4. Fujioka, Y., Stewart, E.: How do physicians discuss e-Health with patients? The relationship of physicians' e-Health beliefs to physician mediation styles. Health Communications 28, 317–328 (2013)
5. HIMSS SIG develops proposed e-health definition. HIMSS News 13(7), 12 (2003)
6. Nutbeam, D.: Health literacy as a public health goal: A challenge for contemporary health education and communication strategies into the 21st century. Health Promotion International 15(3), 259–267 (2006)
7. Bansil, P., Keenan, N.L., Zlot, A.I., Gilliland, J.C.: Health- related information on the web: Results from the HealthStyles sur- vey, 2002–2003. Preventing Chronic Disease: Public Health Research, Practice, and Policy 3(2), 1–10 (2006)
8. Cotten, S.R., Gupta, S.S.: Characteristics of online and offline health information seekers and factors that discriminate between them. Social Science & Medicine 59, 1795–1806 (2004)
9. Fox, S.: Pew Internet and American Life Project: The social life of health information (2011), http://pewinternet.org/reports/2011/social-life-of-health-in fo/summary-of-findings.Aspx, archived at http://www.Webcitation. Org/61imndlxs
10. Dedding, C., van Doorn, R., Winkler, L., Reis, R.: How will e-health affect patient participation in the clinic? A review of e-health studies and the current evidence for changes in the relationship between medical professionals and patients. Social Science & Medicine 72, 49–53 (2011)
11. Sommerhalder, K., Abraham, A., Zufferey, M., Barth, J., Abel, T.: Internet information and medical consultations: Experiences from patients' and physicians' perspectives. Patient Education and Counseling 77, 266–271 (2009)
12. Potter, S.J., McKinlay, J.B.: From a relationship to encounter: An examination of longitudinal and lateral dimensions in the doctor–patient relationship. Social Science & Medicine 61, 465–479 (2005)

13. McMullan, M.: Patients using the Internet to obtain health information: How this affects the patient–health professional relationship. Patient Education and Counseling 63, 24–28 (2006)

14. Meredith, K.L., Jeffe, D.B., Mundy, L.M., Fraser, V.J.: Sources influencing patients in their HIV medication decisions. Health Education & Behavior 28, 40–50 (2001)

15. Nathanson, A.I.: Parent and child perspectives on the presence and meaning of parental television mediation. Journal of Broadcasting & Electronic Media 45, 201–220 (2001)

16. Valkenburg, P.M., Kremar, M., Peeters, A.L., Marseille, N.M.: Developing a scale to assess three styles of television mediation: "Instructive mediation," "restrictive mediation," and "social coviewing. Journal of Broadcasting & Electronic Media 43, 52–66 (1999)

17. Ahmad, F., Hudak, P.L., Bercovith, K., Hollenberg, E., Levinson, W.: Are physicians ready for patients with Internet-based health information? Journal of Medical Internet Research 8, e22 (2006)

18. Akesson, K., Saveman, B.I., Nilsson, G.: Health care consumers' experiences of information communication technology—A summary of literature. International Journal of Medical Informatics 76, 633–645 (2007)

19. Urquhart, C., Lehmann, H., Myers, M.D.: Putting the theory back into grounded theory: Guidelines for grounded theory studies in information systems. Information System Journal 20, 357–381 (2010)

Appendix

Case I: Vision test
Interview date: November 2014
Place: An Ophthalmology clinic in Vienna, Austria

Table 1.

	Patient-Nurse-Doctor Communication
Registration (Patient-Nurse communication)	P: I have an appointment at 1.10p. Here is my e-card. Nurse A: Address please. Please fill the form on the counter. If you don't understand German, I'll help you later. P: I need your help for Q2 to Q7. Nurse A: Did you take any eye operation before? P: I did laser in 2006. Nurse A: How long you spent on computer a day? P: At least 8 hours. Nurse A: Do you feel headache? P: No. Nurse A: Do you pregnant now? P: No. Nurse A: Do you take medicine? P: No. Nurse A: Do you have allergy? P: No. Nurse A: Do you have the following disease? Diabetes, heart disease, family disease…etc. P: No.

Table 1. (*Continued*)

Pre-examination	Nurse A: Let me check your eyes. Please sit on the chair in front of this machine and put your forehead and chin on the bars. Please move to the next machine for eye pressure test. P: Thank you.
Syndrome description	Nurse B: Do you have any particular condition of your eyes? P: My right eye can view far distance objects clearly. Nurse B: Because this is your first visit, we better take a full examination including vision test, direction test, and reading distance test. After the examination, I'll use this eye drop on both of your eyes, after 10-15 minutes, my colleague will do the examination you just taken again. All the data collection done, Dr. will check up all the report and give you the advice. That's all for today. P: Thank you.
Disease judgment	(After the full examination) Nurse B: Your right eye is used for short-distance reading and your left eye is used for far-distance viewing therefore you're okay with no-eyeglass life. According to your examination, your sight is getting worse, so you definitely need to wear eyeglasses for better sight. That's my comment, please also discuss with Dr., let me drop eye drops into your eyes. Please use the cotton to clean eyelids. Please blink your eyes to make the eye drops active.
Disease confirmation	Doctor: You need to wear eyeglasses. P: Can I wear contact lens? Doctor: Let me check whether your eyes is suitable for contact lens or not.
Examination	Doctor: Put your chin on this bar. Ok, your right eye, look up, look down, right, and left. Ok, your left eye, up, down, right, and left. Ok, you can wear contact lens. Doctor: Your eye pressure is not okay, you might have coma risk, I need to take an advance examination, but insurance doesn't cover it. It costs you EUR80. Is any problem to you? Patient: Coma! Yes, I need to take further examination. Do I need to pay today? Doctor: Yes, at the front counter. Doctor: Please move to the chair next to the wall, put your chin on the bar. Please use your left eye to find the green X and watch it for seconds, next, right eye to find the green X and watch it for seconds. Doctor: Let me check your right eye again, you just blinked. Better now.
Treatment	Doctor: You can have your eyeglasses here or you can find an optician by yourself. Patient: I prefer to have one here. Doctor: I'll order your eyeglasses. Please come back in January for eye pressure test. Patient: Yes, thank you. I'll back in January. Can you give me a report of the advance examination? My private insurance can cover the expense. Doctor: Do you need a copy of the color printout of your coma risk analysis? Patient: Yes. Doctor: If you have any further questions, don't hesitate to ask.

Table 1. (*Continued*)

Healthcare Education	None
Registration	Nurse A: Here is your diagnosis report and prescription. Do you want to pay by bankomart? P: Yes. Nurse A: Your receipt. Do you want to make an appointment now or call me later? P: I'll call you later, no schedule on hand now.

Case II: Make an appointment for Mammography
Place: A clinic in Vienna, Austria
Interview date: November 2014
Barriers: Health literacy

Table 2. A comparison of media-mediated communication effect

Communication (10 minutes)	Media-mediated Communication (7 minutes)
Patient: Excuse me. I want to make an appointment of mammography for next week. Here is my invitation letter. Nurse A: Ok, when did you last have Menstruation? Patient: I don't understand. I just want to make an appointment X-ray of my breast for next week, not today. Nurse A: Yes. I can't translate. Please wait my colleague. Nurse B: Wen did you have menstruation? Bleeding…, woman…, before … Patient: What? Do you mean I need to take some blood test before take Mammography? Nurse B: No, let me find another word for you. (Then she looked up a Deutsch-English dictionary.) Period. Patient: Yes, I'm still have period regularly, the first week of each month. Nurse B: That's too long for now. Here your can find our phone number on this paper. Please make a phone call to us when the first day of period coming, then we can make arrangement of Mammography 8-12 days later of the first day. Patient: So, I'll call you when my first period day coming and check the available date for Mammography. It's very clear now. Thanks for your patient explanation.	Patient: Excuse me. I want to make an appointment of mammography for next week. Here is my invitation letter. Nurse A: Ok, when did you last have menstruation? Patient: I don't understand. I just want to make an appointment X-ray of my breast for next week, not today. Nurse A: Yes. I can't translate. Patient: Would you mind to speak to my mobile app, a tool, which can translate the difficult medical term to my language and help me to understand the meaning of it? Nurse A: menstruation. Patient: Period? Woman days? I still have period monthly. It's always coming at the first week of each month. Nurse A: Yes, we need to know your first period day next month, please call me to make an appointment when your next period coming. I'll arrange Mammography for you 8-12 days later the first day of your period. Patient: Ok, I'll call next month. Very helpful advice. Thank you very much.

Case II: Running nose and cough
Interview date: November 2014
Place: A clinic in Kaohsiung, Taiwan
Barrier: Language

Fig. 3a. Syndrome description

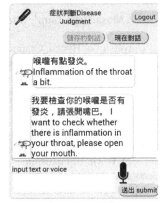

Fig. 3b. Disease judgment

Fig. 3c. Disease confirmation

Fig 3d. Examination

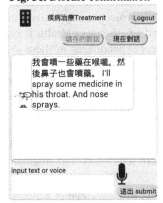

Fig. 3e. Treatment

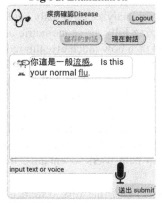

Fig. 3f. Healthcare education

Empirical Analysis of the Effect of eHealth on Medical Expenditures of Patients with Chronic Diseases

Masatsugu Tsuji[1], Sheikh Taher Abu[1,2], and Yusuke Kinai[1]

[1]University of Hyogo, Graduate School of Applied Informatics, Kobe, Japan
tsuji@ai.u-hyogo.ac.jp, ykinai@js8.so-net.ne.jp
[2]Jahangirnagar University, Department of Finance and Banking, Dhaka, Bangladesh
knipuljp@yahoo.co.jp

Abstract. This paper aims to evaluate empirically the effect of eHealth in Nishi-aizu Town, in Fukushima Prefecture, based on the mail survey to the residents and their receipt data of National Health Insurance from 2006 to 2007. Samples were divided into two groups: users and non-users, and the effects were analyzed in comparison of medical expenditures of two groups. Particularly what kind of diseases eHealth in this town is effective to reduce medical expenditures is focused. The targets are four chronic diseases such as heart diseases, high blood pressure, diabetes, and strokes. Panel data analysis is employed to estimate the monetary effect of eHealth in reduction of medical expenditures of users with these diseases. As a result, eHealth is verified to provide the positive effect to heart diseases, high blood pressure, and diabetes among four diseases. These results provide evidence for policy such as reimbursement from the medical insurance to eHealth.

Keywords: Heart diseases; high blood pressure, diabetes, strokes, medical expenditure, panel data analysis, Nishi-aizu Town, economic benefit.

1 Introduction

This paper analyzes rigorously evidences of an eHealth project in a rural town in Japan by utilizing fully an econometric analysis. What kind of diseases eHealth is effective for, and then how much eHealth reduces actual medical expenditures of these diseases, by examining Nishi-aizu Town, Fukushima Prefecture, Japan, as a case study. eHealth studied here is to monitor the health condition of the elderly at home by transmitting users' health-related data, such as blood pressure, ECG, and blood oxygen, to a remote medical institution via a telecommunications network. The system is equipped with a simple device that records an elderly person's condition or a patient's illness in graphs that are then used for diagnosis and consultation. Reports sent by the medical institution also help users to enhance their daily health consciousness and maintain good health.

It is obvious that this system is quite simple one, however, author's previous research ([1], [2], [3], [4], [5], for example) proved that this eHealth actually reduced medical expenditures or days for treatment of chronic diseases. This paper studies in

F. Ortuño and I. Rojas (Eds.): IWBBIO 2015, Part II, LNCS 9044, pp. 12–23, 2015.
© Springer International Publishing Switzerland 2015

more detail by focusing on chronic diseases such as heart disease, high blood pressure, diabetes, and strokes. There are quite few studies on economic evaluation of eHealth in Japan, and accordingly this analysis and results obtained here provide the rigorous economic foundation of eHealth in Japan as well as in the world ([6], [7]).

The paper consists of the following sections; Section 2 explains how we construct the data for a survey analysis, provides characteristics of the sample based on the data, and the method of analysis. Section 3 provides rigorous statistical methodology based on panel data analysis, and the results are presented in Section 4. Brief discussions and concluding remarks are stated in the final section.

2   eHealth of Nishi-Aizu Town

2.1   Background of eHealth

Nishi-aizu Town is located in the Northwest corner of Fukushima Prefecture, and is been an important point of transit to reach Niigata Prefecture and Aizu-wakamatsu, a nearby major city. The center of town is in a basin but the main area is surrounded by mountains, which cover 86% of the prefecture's area. The climate is severe in winter and summer, with lots of snow. The population is about 8,000; there are 3,000 households, and the percentage of the elderly ($\geq$ 65 years) was 42.0% in 2012. The main industry is agriculture, and rice is the main product.

As stated earlier, severe winter, especially heavy snow causes elderly people to lack physical exercise. In addition, due to a traditional diet of salty and protein-poor food, the town's death rate was 1.7 times higher than the national average during 1983-87, partly due to high rates of stomach cancer. The number of bedridden elderly people suffering from osteoporosis or arthritis is higher than the national average. In order to cope with these situations, the town office took initiative to establish a "total care system," which is referred to as the "Challenge to 100 Years Old," by unifying health, medical and welfare services. As a part of this project, e-Health was introduced in 1993.

In the town, there are three public clinics, named Nishi-aizu, Murooka and Shingo, which are operated by the National Insurance System, and two private clinics. The total number of medical doctors is four. One full-time physician is employed in the Murooka clinic, while the Shingo clinic has a part-time doctor dispatched from the other two clinics. There are private doctors, a surgeon and a neurologist, but both are more than 70 years old.

2.2   Implementation of e-Health

In order to prevent chronic diseases such as cerebral infarction and stroke, the town office introduced e-Health in 1994 which is Japan's longest-running e-Health. 300 peripheral devices called "*Urara*," manufactured by Nasa Corporation, were provided to residents who have symptoms of the above diseases. Each terminal is connected a host computer via PSTN (Public Switched Telephone Network), and health-related data of users, such as blood pressure, pulse, ECG, blood oxygen, weight and temperature are transmitted to a host computer. In 1996 and 1997, an additional 50 terminals each were purchased. These terminals use the CATV network for transmitting data. All costs of operating the system are paid by the town. In 2010, new peripheral device called "*Sauri*," was introduced in accordance with the network renovation of CATV for optical fiber. Currently all network were transformed to optical fiber.

The section in charge of e-Health is the town office's Department of Health and Welfare, which consists of seven public health nurses representing a much larger ratio than in other towns. They check the above health data transmitted by users and if these nurses observe unusual data, they ask medical doctors in clinics to see the patient in question. The health data of each user are summarized in a "Monthly Report," which is sent to a physician in charge. After a public health nurse adds their comments, the report is sent to the user. When the user sees a doctor, he/she is asked to bring the report with him/her.

e-Health is being operated as a part of the town's "Project for Promoting Total Care," and its essence lies in the close collaboration of health, medical and welfare activities. One important example of this collaboration are "Regional Care Meetings," which consist of doctors, nurses, public health nurses, staff of the town office, helpers of elderly people, and living advisers. The total number of participants in each meeting is over 20. Problems and treatments regarding a particular user, such as medical examinations, health advice, and care are discussed in detail. The health data of e-Health plays a role in this meeting. In Nishi-aizu Town, many such examples of exchanging information on residents can be found in the town office.

In addition, the town office organizes users' meetings five times a year in order to enhance motivation to use e-Health, and users exchange their experiences with using the system. These activities promote usage of the system. The introduction of e-Health is not the sole factor promoting regional healthcare; rather, it should establish a framework for the system to assist all related sections and personnel.

3   Date and Methodology

3.1   Sample Selection

As stated earlier, this paper examines the relationship between medical expenditures of Nishi-aizu's residents and eHealth. According to the Japanese medical insurance system, which is organized and operated by the Ministry of Welfare, Labor, and Health all people must be covered by one of several social health insurance systems. This paper focuses on people in Nishi-aizu who are covered by "National Health Insurance," since data on medical expenditures through this system are handled by local governments. National Health Insurance is not only for self-employed individuals such as farmers or owners and employees of small- and medium-sized firms, but also people who already retired.

One of the purposes of this paper is to compare days spent for treatment between two groups such as (i) users and (ii) non-users of eHealth from medical receipts of Nishi-aizu Town. Samples of two groups are selected according to the following way.

User Group. 412 users were selected from the list of registered users in the town according to the year they registered. The total number of users and that selected as the sample is shown Table 1. Then we send questionnaires to them and 311 replies were received.   Finally, after checking the replies, 199 replies remain as significant. The rate of significant reply is 38.05 percent.

Non-user Group. The study selected 450 residents who are covered by National Health Insurance out of total 3,528. Questionnaires were sent to 450 residents and we received 239 replies. Again by checking the replies, we had 209 significant replies. The rate of significant reply is 46.44 percent.

In sum, the total number of residents selected as the sample becomes 408. The research group is summarized in Table 1.

Table 1. Sampling Groups

	User	Non-user
Total	523	3528
Number of sent questionnaires	412	450
Number of respondents	311	239
Number of valid respondents	199	209
Rate of valid respondents	0.3805	0.4644

Next, their receipts from those stored in the town office were examined. The receipts of National Health Insurance of each month are kept at the town office, in which the data such as name and address of medical institution, birth date, name of disease, date of initial-visit, medicine, and score (amount) of medical treatment are described. In this paper, we use the following data: (i) name of resident, (ii) birth date, (iii) either regular outpatient treatment or hospitalized patient treatment, (iv) name(s) of major disease(s), (v) date of initial treatment, (vi) number of days spent for treatment, and (vii) score (amount) of medical treatment.

3.2   Characteristics of Data

As our data are the same as the previous research, here only essential aspects are presented. As for the detailed characteristics of data, refer to [1] [2] and [3].

This paper attempts to analyze what kind of diseases eHealth is effective for. First of all, we divided all diseases into twelve as shown Table 2, according to the advices of health professional and a public nurse in charge of the eHealth system there. Our interests lie in heart diseases, high blood pressure, diabetes, and strokes, since these diseases are categorized into chronic diseases which this paper and the eHealth system in Nishi-aizu place main targets.

Table 2 indicates diseases having treated within five years from 2002 to 2006 of both users and non-users. Obviously, these four diseases dominate high ratio. Our previous study ([3]) proved that this eHealth actually reduced the medical expenditures of a sum of these four diseases by approximately JPY15,688 (US$156.88). In particular, this paper attempts to estimate such monetary effect of each of chronic disease separately.

Table 2. Diseases Treated Within Five Years

	User	Non-user	Total
Heart diseases	44	23	67
High blood pressure	100	74	174
Diabetes	15	21	36
Strokes	14	10	24
Respiratory diseases	9	10	19
Cancer	8	3	11
Gastropathy	25	13	38
Lumbago, Arthritis	45	43	88
Ophthalmic diseases	57	46	103
Kidney diseases	3	1	4
Anal diseases	9	7	16
Others	19	7	26

3.3   Methodology

According to data obtained, we want to test hypotheses on medical expenditures related to chronic diseases. The objectives of the estimation consists of the following two questions: (i) how the experience of treatment of four chronic diseases provided effect to total medical expenditures, in other words, whether there is difference in medical expenditures among users and non-users due to chronic diseases; and (ii) how medical expenditures related to chronic diseases were different among two groups, user and non-user. In so doing, following equation is estimated which is aimed to show how medical expenditures including all diseases are determined by the selected variables. Let us refer the estimation model to explain the first question to as "estimation (1)," whereas the estimation model related to the second question to as "estimation (2)". The estimation model corresponding to estimation (1) is expressed in the following equation:

$$y_{it} = \alpha + X_{it}\beta + \sum_{j}^{12}\{disease_{it}^{j}(\gamma^{j} + User_{i}\delta^{j})\} + u_{it}$$
$$u_{it} = \lambda_{t} + v_{it}$$

(1)

where y_{it} denotes the medical expenditures of all disease of i-th subject at year t, X_{it} residents' characteristics such as sex, age, education, employment (dummy variable), the number of family living together, and income. $disease_{it}^{j}$ is a dummy variable which takes 1 if i-th resident have treated j-th disease in year t, and takes 0 otherwise. $User_{i}$ represents a user dummy variable. We utilized the panel data analysis with the one-way fixed effect model where λ_{t} denotes a year dummy variable, because the rule of calculating medical expenditures in the National Health Insurance was usually changed every two years. In addition, the individual effect, or dummy variables might cause serious multicollinearity with each characteristic, and accordingly only the time effect is considered. The difference in medical expenditures due to each disease between users and non-users can be identified by the coefficient of δ^{j}, which can be referred to as the cross effects. If coefficient δ^{j} is negatively (positively) significant, users' total medical expenditures affected by the j-th disease is statistically smaller (larger) than those of non-user. In sum, δ^j answers question (i).

Regarding estimation (2), equation (1) should be modified a bit. That is, medical expenditures of four diseases such as heart diseases, high blood pressure, diabetes, and strokes are taken as an explained variable, and independent variables are the same as equation (1). In this framework, how medical expenditures of four chronic diseases are different among users and non-users is clarified.

4   Results of Estimation

4.1   Result of Estimation (1): All Diseases

Individual Characteristics

The result of estimation (1) of all diseases is shown in Table 3. The detailed results on the variables of individual characteristics such as sex, age, education, employment, the number of family living together and income are summarized as follows. Among them, age, employment and the number of family living together are found to be significant.

- Medical expenditures increase around JPY2,223.7 (US$22.24) per year when they become one year older. It is natural that the older they become the larger medical expenditures.

- Medical expenditures of working group are lower than those not working by approximately JPY34,279.7 (US$342.80) per year. This is because working people might be healthier. However, because whether being sick or not could affect their working condition, this variable has to be treated as an endogenous. This problem should be taken care in the future study.

- The more the number of family living together, the more medical expenditures. It is difficult to see this relationship in the real life. However, this result coincides with our previous studies ([1], [2],[3], [4], [5], [6])

Diseases. Individual diseases are taken as dummy variables in equation (1), and they take 1 if the particular disease was treated, while take 0 otherwise. Thus, medical expenditures related to these variables can be interpreted as a part of total medical expenditures. According to Table 3, the proportion of each disease in the total medical expenditures is summarized as follows: heart diseases (20%); cancer (17%); diabetes (16%); and high blood pressure (10%). The shares of these diseases show higher ratios among diseases. In particular, the share of total expenditures related to four chronic diseases including heart diseases, high blood pressure, diabetes, and strokes is accounted for nearly 60% of total medical expenditures. This indicates that the reduction of expenditures of these diseases is indispensable for the reduction of total medical expenditures.

Cross Effect (User × Disease). By addition of the cross effect which consists of the dummy variables regarding diseases and user in estimation equation (1), it is possible to estimate the difference in medical expenditures of each chronic disease between users and non-users. According to the result shown in Table 3, heart diseases, high blood pressure, diabetes, and gastropathy show significantly negative coefficients, while strokes and kidney diseases significantly positive. By examining data further, however, one irregular amount was found in one resident's data related to kidney disease, and this is revealed to cause this result. Accordingly kidney diseases are ignored from this analysis. As for the former diseases, most of them are chronic diseases and users' expenditures are statistically lower than those of non-users. On the other hand, users' medical expenditure of strokes is higher than that of non-users. Since treatments or diagnosis of strokes require special medical techniques and skills, whether they have been using eHealth or not does not influence largely to the medical expenditures. eHealth is considered to be effective to prevent from being ill, but not suitable as means of medical treatment. In this sense, the town office distributes eHealth devices based on their health conditions such as having chronic diseases. This result seems to be consistent with previous studies such as [1], [2], [3], and [4].

In sum, eHealth in Nishi-aizu town is thus proved to have the positive effect to the reduction of medical expenditure due to chronic diseases such as heart diseases, high blood pressure, and diabetes. It is difficult, however, to calculate the exact monetary amounts reduced by this estimation. In order to calculate this, we have to estimate medical expenditures of particular diseases such as heart diseases, high blood pressure, diabetes and strokes by taking expenditures of these diseases as explained variables in the same framework of equation (1).

Table 3. Result of Estimation Ⅰ (All Diseases)

Variables	Characteristics, Diseases				Cross effect (User × Disease)			
	Coef.	S. E.	t value	p value	Coef.	S. E.	t value	p value
Sex	-747.67	1003.89	-0.74	0.457				
Age	222.37	62.19	3.58	0.000 ***				
Education	456.51	655.90	0.70	0.487				
Working	-3427.97	1041.30	-3.29	0.001 ***				
No. of family	1068.60	263.35	4.06	0.000 ***				
Income	-12.09	8.55	-1.42	0.157				
Heart diseases	20171.06	2823.36	7.14	0.000 ***	-6391.33	3569.24	-1.79	0.074 *
High blood pressure	10181.37	1375.35	7.40	0.000 ***	-3080.34	1803.75	-1.71	0.088 *
Diabetes	16166.76	2367.28	6.83	0.000 ***	-8837.79	3553.94	-2.49	0.013 **
Strokes	9254.46	3166.69	2.92	0.004 ***	9313.10	4287.74	2.17	0.030 **
Respiratory diseases	1668.48	2117.47	0.79	0.431 **	-692.95	3085.60	-0.22	0.822
Cancer	16843.44	2692.94	6.25	0.000 ***	-1165.97	3796.72	-0.31	0.759
Gastropathy	5257.63	1855.65	2.83	0.005 ***	-5701.81	2656.58	-2.15	0.032 ***
Lumbago, Arthritis	6193.50	1924.68	3.22	0.001 ***	478.86	2849.44	0.17	0.867
Ophthalmic diseases	6117.03	1772.52	3.45	0.001 ***	3358.96	2460.05	1.37	0.172
Kidney diseases	1697.60	4014.00	0.42	0.672 **	76684.07	6272.47	12.23	0.000 ***
Anal diseases	543.66	5249.44	0.10	0.918	-11084.02	10672.67	-1.04	0.299
Others	4793.56	1371.84	3.49	0.000 ***	2664.82	1988.11	1.34	0.180
Constant	-14600.09	4989.37	-2.93	0.003 ***				
Adjusted R^2	0.3675							
Number of Observation	1820							

Note: ***, **, and * indicate the significance level of 1, 5 and 10%, respectively.

4.2   Results of Estimation (2): Individual Chronic Diseases

This section attempts two types of estimation for each disease; with full sample and selected samples. The former indicates that the analysis covers all samples, while the latter covers only those who have been treated or examined this disease. The estimation model is same as the last section, based on the one-way fixed effect model with only time effect. We add user dummy variable as an explanatory variable, and it enables us to estimate whether there is difference between users and non-users in medical expenditures of each disease. The results of estimations are summarized in Table 4 for heart disease, Table 5 for high blood pressure, Table 6 for diabetes and Table 7 for strokes. Let us examine one by one in the next section.

Heart Diseases. The estimation of heart diseases in Table 4 shows that in the full sample case, Gender ($p<0.05$), Age ($p<0.01$), Income ($p<0.05$), and Recognizing Chronic Diseases ($p<0.01$) are found to be significant, where percentages in the parentheses stand for their significance levels. Although the user dummy variable is not significant in this full sample case, but in the selected sample case it is strongly in negative at the 1% significance level, which can be interpreted that users' medical expenditures of heart diseases are lower than those of non-users by approximately JPY39,080.9 (US$390.81) per year. The significance levels of some user characteristics become lower in the selected sample case than in the full sample case. This seems to be reasonable; since the number of observations is smaller in the latter than in the former and being heart disease might weaken the effect of aging or other factors to medical expenditures. As for estimated coefficients of user dummy, the same argument is applicable, that is, coefficient of the user dummy variable is larger in the selected sample case than in the full sample case.   The same holds for estimations of diabetes and strokes.

High Blood Pressure. Table 5 summarizes the result of estimation of high blood pressure. Sex ($p<0.01$), Age (($p<0.01$), Number of family (($p<0.01$), Income (($p<0.01$) and Recognizing chronic diseases (($p<0.01$) are significant. The user dummy variable has negative coefficient at the less than 10% significance level in the full sample case, and it amounts to JPY8,660.7 (US$86.60). Moreover, the difference is larger in the selected sample estimation, which amounts to JPY21,859.3 (US$218.60). In both cases, users' medical expenditures of high blood pressure are found to be lower than those of non-users.

Diabetes. The result of diabetes is shown in Table 8, which summarizes that Age ($p<0.1$), Education ($p<0.05$), Working ($p<0.05$), Number of family ($p<0.05$), and Recognizing chronic diseases ($p<0.05$) are significant. The coefficients of user dummy variables are negatively significant for both full and selected sample estimation, and the difference between two groups amounts to JPY8,784.5 (US$87.85) and JPY37,639.9 (US\$376.40), respectively. The main reason for Nishi-aizu Town to introduce eHealth was to manage diabetes. In this account, eHealth in Nishi-aizu is considered to be successful.

Table 4. Result of Estimation II (Heart Diseases)

Variables	Full Sample				Selected Sample (Only treated group)			
	Coef.	Std. Err.	t value	p value	Coef.	Std. Err.	t value	p value
Sex	572.58	286.49	2.00	0.046**	1136.15	1348.47	0.84	0.400
Age	45.34	17.36	2.61	0.009***	75.06	81.26	0.92	0.356
Education	90.45	185.25	0.49	0.625	194.34	1060.51	0.18	0.855
Working	-265.28	304.95	-0.87	0.384	149.64	1324.46	0.11	0.910
No. of family	80.37	76.63	1.05	0.294	988.26	353.48	2.80	0.006***
Income	-5.94	2.47	-2.40	0.016**	-24.19	11.64	-2.08	0.039**
Recognizing Chronic Diseases	835.36	285.06	2.93	0.003***	292.06	1364.25	0.21	0.831
User Dummy	-60.14	289.93	-0.21	0.836	-3908.09	1309.69	-2.98	0.003***
Constant	-2300.88	1414.66	-1.63	0.104	1399.02	7195.71	0.19	0.846
Adjusted R2	0.0164				0.0454			
Number of Observation	1545				315			

Note: ***, **, and * indicate the significance level of 1, 5 and 10%, respectively.

Table 5. Result of Estimation III (High Blood Pressure)

Variables	Full Sample					Selected Sample (Only treated group)				
	Coef.	S. E.	t	p		Coef.	S. E.	t	p	
Sex	2408.42	447.69	5.38	0.000	***	2229.11	681.87	3.27	0.001	***
Age	147.93	27.13	5.45	0.000	***	163.67	39.73	4.12	0.000	***
Education	-72.52	289.48	-0.25	0.802		-559.20	421.98	-1.33	0.185	
Working	-46.35	476.54	-0.10	0.923		903.69	704.09	1.28	0.200	
No. of family	421.12	119.74	3.52	0.000	***	577.38	176.52	3.27	0.001	***
Income	-10.39	3.86	-2.69	0.007	***	-13.03	6.76	-1.93	0.054	*
Recognizing Chronic Diseases	3696.00	445.45	8.30	0.000	***	3943.53	668.43	5.90	0.000	***
User Dummy	-866.07	453.07	-1.91	0.056	*	-2185.93	670.07	-3.26	0.001	***
Constant	-8428.52	2210.64	-3.81	0.000	***	-6688.60	3381.95	-1.98	0.048	**
Adjusted R2	0.0919					0.0764				
Number of Observation	1545					975				

Note: ***, **, and * indicate the significance level of 1, 5 and 10%, respectively.

Stroktes. Finally, the result of estimation of strokes is shown in Table 7. Again, Age (p<0.05), Number of family (p<0.1), Recognizing chronic diseases (p<0.05) are significant. Neither the coefficient of user dummy variable in the full sample case nor in the selected sample is significant. This coincides with the last estimation of all diseases shown in Table 3 which says there are no difference between users and non-users as for strokes. The reason is the same as explained previously.

Table 6. Result of Estimation IV (Diabetes)

	Full Sample				Selected Sample (Only treated group)			
	Coef.	S.. E.	t	p	Coef.	S. E.	t	p
Sex	-97.08	300.06	-0.32	0.746		-408.77	1776.25	-0.23	0.818
Age	31.18	18.18	1.71	0.087	*	61.11	108.29	0.56	0.573
Education	-413.88	194.02	-2.13	0.033	**	-1215.02	1266.35	-0.96	0.338
Working	722.28	319.40	2.26	0.024	**	2171.53	1987.25	1.09	0.276
No. of family	-193.26	80.26	-2.41	0.016	**	-1158.43	483.21	-2.40	0.017	**
Income	-2.74	2.59	-1.06	0.289		-9.64	16.19	-0.60	0.552
Recognizing Chronic Diseases	686.11	298.56	2.30	0.022	**	-649.29	1842.70	-0.35	0.725
User Dummy	-878.45	303.66	-2.89	0.004	***	-3763.99	1802.45	-2.09	0.038	**
Constant	101.37	1481.66	0.07	0.945		9386.29	8446.43	1.11	0.268
Adjusted R2	0.0167					0.0428			
Number of Observation	1545					245			

Note: ***, **, and * indicate the significance level of 1, 5 and 10%, respectively.

Table 7. Result of Estimation V (Strokes)

	Full Sample				Selected Sample (Only treated group)			
	Coef.	S. E.	t	p	Coef.	S. E.	t	p
Sex	-276.53	176.10	-1.57	0.117		-1201.41	1316.01	-0.91	0.362
Age	27.27	10.67	2.56	0.011	**	-84.73	78.68	-1.08	0.283
Education	56.36	113.87	0.49	0.621		328.28	979.60	0.34	0.738
Working	22.34	187.45	0.12	0.905		1849.92	1284.99	1.44	0.152
No. of family	84.10	47.10	1.79	0.074	*	-56.47	365.12	-0.15	0.877
Income	-1.04	1.52	-0.69	0.493		38.97	26.86	1.45	0.148
Recognizing Chronic Diseases	356.87	175.22	2.04	0.042	**	1264.20	1368.64	0.92	0.357
User Dummy	-22.80	178.22	-0.13	0.898		-1046.93	1451.75	-0.72	0.472
Constant	-1308.04	869.58	-1.50	0.133		10748.15	5942.92	1.81	0.072	*
Adjusted R2	0.0067					0.0141			
Number of Observation	1545					195			

Note: ***, **, and * indicate the significance level of 1, 5 and 10%, respectively.

According to the results obtained thus for, eHealth in Nishi-aizu town provides a positive effect especially to the diseases such as heart diseases, high blood pressure, and diabetes, and these effects are higher in particular for the people affected by those diseases. In other words, eHealth provides the greater effect to residents who have chronic diseases such as chronic diseases.

Table 8. Amount of Reduced Expenditures

	Heart diseases	High blood pressure
Full sample	not significant	8,660.7 yen* (US$86.23)
Selected sample	39,080.9 yen*** (US$434.2)	21,859.3 yen*** (US$242.9)

	Diabetes	Strokes
Full sample	8,784.5 yen*** (US$97.6)	not significant
Selected sample	37,639.9 yen** (US$418.2)	not significant

Note: ***, **, and * indicate the significance level of 1, 5 and 10%, respectively

5   Conclusion

By the rigorous regression analyses, the results obtained can be summarized in Table 8. These are amazing results, because town's original goals which were to manage chronic diseases such as high blood pressure and diabetes is proved to be successful by showing the medical expenditures of town were reduced by rigorous methods .

5.1   Success Factors of e-Health

The increase in medical expenditures is common phenomena all over the world. There are two measures to cope with this; the utilization of IT in medical area and prevention from being illness (or maintain health). The eHealth system can solve these issues. The results we obtained here provide the rigorous foundation of eHealth.

It is clear from authors' previous studies that eHealth is useful for consultation and maintaining the good health of the elderly and patients suffering from chronic diseases who are in stable condition. However, it is not effective at curing disease as there is no effect for strokes, estimated in this paper. It therefore has the psychological effect of providing a sense of relief to its users by the knowledge of being monitored by a medical institution 24-hours-a-day. This makes it difficult to estimate its benefits in concrete terms. This paper enables to provide concrete amounts for the reduction of users' medical expenditure of some chronic diseases. It is important to notice that solely introducing eHealth does not contribute to its success, but the fact that eHealth has been supported by eagerness of all related staff of the town office should be noticed. In Nishi-aizu town, there are six public nurses in charge of this system, and they always grasp health conditions of all users.

Moreover, it should be discussed e-Health projects in the UK and US which have the similar peripheral device and system ([6], [7]). According to our in-depth surveys of these projects, there are similarities and differences in these projects, but a common success factor lies in the enthusiasm of nurses, public or visiting nurses who participate in these projects to maintain health of the residents in the community. Further detailed study is required for factors of differences.

5.2   Economic Foundation of e-Health

Let us discuss on the economic foundation of the project. Nishi-aizu Town does not charge any fee to users. Other projects in the most of counties are the same. Neither this program nor those referenced in the UK and US charge user fees; rather, all are subsidized by the central as well as local government, as indeed are the UK demonstration programs, since they are national pilot projects. However, the ongoing sustainability of e-Health requires a new financial framework. [8] conducted a cost/benefit analysis of Nishi-aizu's e-Health and calculated a B/C (cost-benefit) ratio which was 0.25. The initial costs of the implementation, such as for host computers and peripheral devices, were borne by the central government, however, excluding them form analysis gave a B/C ratio for Town which bore only operational costs, which is 0.91. But this is not sustainable. One possibility for promoting e-Health is reimbursement using public medical insurance. The amount of reimbursement is based on economic effect of e-Health and must be obtained rigorous analysis. Most of countries are not recognized reimbursement for e-Health, which reason is simple; e-Health is not diagnosis but prevention of diseases. The simple e-Health system does reduce medical expenditures of users. The present paper provides important support for the development of evidence-based policies for the diffusion of e-Health.

References

1. Akematsu, Y., Tsuji, M.: Empirical Analysis of the Reduction of Medical Expenditures. Telemedicine and Telecare 15, 109–111 (2009)
2. Akematsu, Y., Tsuji, M.: An Empirical Approach to Estimating the Effect of eHealth on Medical Expenditure. Journal of Telemedicine and Telecare 16, 169–171 (2010)
3. Akematsu, Y., Tsuji, M.: Measuring the Effect of Telecare on Medical Expenditures without Bias Using the Propensity Score Matching Method. Telemedicine and e-Health 18, 743–747 (2012)
4. Akematsu, Y., Nitta, S., Morita, K., Tsuji, M.: Empirical analysis of the long-term effects of telecare use in Nishi-aizu Town, Fukushima Prefecture, Japan. Technology and Health Care 21, 173–182 (2013)
5. Akematsu, Y., Tsuji, M.: Does telecare reduce the number of treatment days? An empirical analysis in a Japanese town. Journal of Telemedicine and Telecare 19, 36–39 (2013)
6. Kent County Council, Promoting and Sustaining Independence in a Community Setting: A Study into the Management of People with Long Term Conditions (2010)
7. Darkins, A., Ryan, P., Kobb, R., Foster, L., Edmonson, E., Wakefield, B., Lancaster, A.E.: Care coordination/home telehealth: The systematic implementation of health informatics, home telehealth, and disease management to support the care of veteran patients with chronic conditions. Telemedicine and e-Health 4, 1118-1126 (2008)
8. Miyahara, S., Tsuji, M., Iizuka, C., Hasegawa, T., Taoka, F.: On the evaluation of economic benefits of Japanese telemedicine and factors for Its promotion. Journal of Telemedicine and eHealth 12, 291–299 (2007)

Impact of Health Apps in Health and Computer Science Publications. A Systematic Review from 2010 to 2014

Guillermo Molina-Recio[1], Laura García-Hernández[2,*], Antonio Castilla-Melero[1],
Juan M. Palomo-Romero[2], Rafael Molina-Luque[1], Antonio A. Sánchez-Muñoz[1],
Antonio Arauzo-Azofra[2], and Lorenzo Salas-Morera[2]

[1]Nursing Department. University of Córdoba, Spain
gmrsurf75@gmail.com, {antcasmel,rafael.moluq,samuaa}@gmail.com
[2]Area of Project Engineering. University of Córdoba, Spain
{ir1gahel,i52paroj,arauzo,lsalas}@uco.es

Abstract. Several studies have estimated the potential economic and social impact of the mHealth development. Considering the latest study by Institute for Healthcare Informatics, more than 40,000 Apps of health and medicine are offered in the Apple store. Adding those of Playstore and other platforms, the figure reaches 97,000. Thus, they have become the third-fastest growing category, only after games and utilities, and it has projected that its presence will grow about 23% annually over the next five years. This study aims to estimate the impact that the development of mHealth has had on the health and computer science field, through the study of publications in specific databases for each area which have been published since 2010 to 2014.

Keywords: mHealth, App, Health Sciences, Computer Sciences, Review.

1 Introduction

Several studies have estimated the potential economic and social impact of the mHealth development. mHealth is an abbreviation for mobile health, a term used for the practice of medicine and public health supported by mobile devices. According to WHO, nearly 90% of the world population could benefit from the opportunities offered by mobile technologies and with a relatively low cost. Considering the latest study by Institute for Healthcare Informatics (IMS) [1], more than 40,000 Apps of health and medicine are offered in the Apple store. Adding those of Playstore and other platforms, the figure reaches 97,000. Thus, they have become the third-fastest growing category, only after games and utilities, and it has projected that its presence will grow about 23% annually over the next five years [2].

Also, the IMS Institute indicates that 70% of health Apps is focused in general population, offering tools to reach and maintain wellness and improve physical activity. The remaining 30%, were designed to more concrete areas such as professionals or people affected by specific diseases.

Despite this situation, it is important to note that more than 50% of the available Apps received less than 500 downloads and only 5 of them comprise 15% of all those

F. Ortuño and I. Rojas (Eds.): IWBBIO 2015, Part II, LNCS 9044, pp. 24–34, 2015.

in the health category. The IMS attributed this situation to different causes, which include:

- Poor quality in many of them.
- The lack of guidance on the usefulness of the App.
- The lack of support from health professionals.

It is estimated that tools for monitoring chronic diseases will account for 65% of the global market for mHealth in 2017 [2].

This fact will represent revenue of 15,000 million dollars. The pathologies with a higher potential to increase business are, in order: diabetes and cardiovascular disease. They will also play an important role related to diagnostic services (they will reach 15% and will generate 3,400 million of dollars) and medical treatments (10% of the market and revenues of 2,300 million). By the other hand, it is estimated that business will increase from 4,500 million in 2013, to 23,000 million in 2017. Continents with largest market share are, in descending order, Europe and Asia (30%), United States of America and Canada (28%) [2].

However, we do not know if the Apps available to the population are based on scientific knowledge and therefore, it is difficult to assess the real impact of this spectacular development on the health of populations. On the other hand, we do not know how the spread of the phenomenon of Health 2.0 is reaching the scientific field (medical or computer), which should occur in parallel in order to offer products that positively affect the health of citizens.

Therefore, this study aims to estimate the impact that the development of mHealth has had on the health and computer science field, through the study of publications in specific databases for each area which have been published since 2010 to 2014.

2   Methods

A systematic review was conducted in two stages during November 2014. The first one was focused in locating papers available in databases from Health Sciences. After this step, we repeat the search but in the Computer Science field because we wanted to find the different penetration in each area. Both searches were bounded between 2010 and 2014. As recommended in the PRISMA Statement [4] for systematic reviews, we describe the search strategy and the number of papers located, discarded and finally selected for review. In the first stage, we have consulted the PubMed database. We used "mHealth" and "mHealth AND App" terms as search strategies. Finally 60 items were selected for reviewing (see Figure 1). We also consulted "Science Direct" and "Scopus" using "mHealth" but reducing the search to "Computer Science" area. Having initially located 335 publications, only 17 were chosen for our study (see Figure 2). Thus, a total of 77 items were reviewed. The impact factor of the journals that published the papers selected was consulted using the Journal Citation Report from Web of Science. Since this impact factor usually varies for each year, we took the corresponding to the year when the article was published.

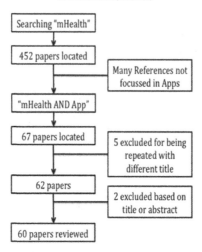

Fig. 1. Review and selection criteria of papers (health science)

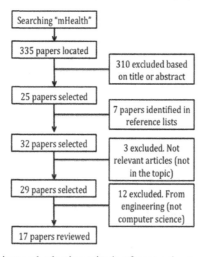

Fig. 2. Review and selection criteria of papers (computer science)

3   Results and Discussion

As it was mentioned above, the higher number of papers is located in health publications (over 75%), which, a priori, seems to show greater concern about developing research in mHealth. However, we cannot forget that there are much more scientific journals in this field and we should take into account that we find multidisciplinary teams in most of these papers, where programmers play a fundamental role. Another important fact which could explain this majority, is given by the concentration of articles in just two journals, Journal of Medical Internet Research, containing 41.6% of the published papers and International Journal of Medical Informatics, covering 7.8% articles. These two journals, are located in both areas of knowledge (Health and Computer Science). Thus, for example, Journal of Medical Internet Research is classified by Journal Citation Report (JCR) into several categories ("Health Care Sciences & Services" and "Medical Informatics") and the International Journal of Medical Informatics is listed by JCR into three categories, the two previously mentioned, as well as in "Computer Science, Information Systems" that clearly belongs to a non-health area. Another important result is the great impact factor of these publications. This way, in any category where they could be classified of these two journals belong to the first quartile. This may also explain that the average impact factor of the found publications was 3.1 ($\pm$ 1.8). Moreover, 63.3% of the articles are located in first quartile journals, 18.2% in the second and 7.8% in the third. Only 9.1% of the papers appear in not indexed journals. From our point of view, this fact highlights the high impact and newness that scientific work based on the use of mHealth technology represents for editors, allowing researchers to access to high impact publications.

However, the number of citations received of these articles is relatively small and widely dispersed. The sample shows an average of 6.35 ($\pm$ 13.56) references. Two papers highlight with 74[45] and 63[22] citations, facing many that have never been referenced. Interestingly, these two articles were published in 2011 and 2012 (respectively), when the exponential growth in the number of publications on mHealth began. Thus, we can find just 2 papers in 2011 (2.6% of total found) while we locate 45 in 2014 (58.4%). Similarly, only 4 articles were found in 2012 (5.2%) and 26 in 2013 (33.8%). In 2010 there were no relevant papers included in the search. This could be a sure sign that the scientific community is increasingly aware of the potential and the need to focus some research into this field.

Similarly, the mean number of citations received for years, shows a clear downward trend (F = 36.72, 3 df; p <0.001). As the years pass, the average is down from the 62 papers on average in 2011 till 1.53 in 2014. Temporal matters could explain this, ie, it stands to reason that more recent articles have had less time to be referenced, but it is a trend that is also seen in 2012 (with an average of 23 references) and 2013 (7.85). Post hoc tests found significant differences between every year (p$\leq$ 0.01).

Regarding the types of study (Figure 3), we found that original researches are more prevalent. This fact happens because most of the articles are focused on the user data collection and evaluation from Apps designed to assess adherence or modification of habits to healthier life styles (studies with larger sample sizes). Another percentage of the original researches are small pilot studies or clinical trials with smaller sample sizes. Reviews are also frequent, something justifiable given the exponential growth of publications. This situation force researchers to conduct periodically synthesis and evaluation of trends in the development of Apps and main findings, as well as the errors most frequently committed (especially as the methodological design is concerned). Another significant point was to find 3 papers from qualitative research. It is understandable if we comprehend that this type of studies is very helpful in trying to expose the different realities of the new or unknown phenomena which we have a very little information or experience. A good example could be the expected results of interactions between mHealth Apps and populations who want to improve their health. Furthermore, we were surprised because two of these articles had been published in Computer Science journals, where papers highly focused on applied and/or quantitative research are traditionally accepted.

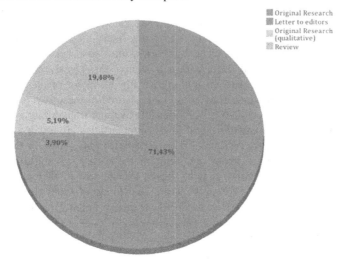

Fig. 3. Type of research

When we study comparatively these two types of journals (Health and Computer Science), placing the two journals mentioned above as Health journals, we find that the mean of the impact factor is almost double in health journals (p<0.01) (Table 1).

Table 1. Impact factor according to the type of journal

Type of Journal		N	Mean	Standard Deviation	Standard Mean error
Impact Factor	Health Science	59	3.448051	1.7437422	.2270159
	Computer Science	17	1.910882	1.6236743	.3937989
Independent Sample t-Test					

		Levene's test		t- Test		
		F	Sig.	t	gl	Sig (Bilateal)
Impact Factor	Equal variances are assumed	1.202	0.276	3.250	74	.002
	Equal variances are not assumed			3.382	27.562	0.002

4   Conclusions

- There are mHealth papers of all kinds (trials, analytical, descriptive, reviews, etc.) and its number grows at an exponential rate. This could show how technologies related to mHealth are reaching the scientific field. These technologies are reaching the population even faster. Therefore, mHealth is becoming an interesting research topic in both fields, health and computer sciences.
- These articles are published in high impact journals. We have found them in specific journals (focused on eHealth) and in generalist health and computing journals. Generalist journals have just begun to accept research based on the application of mHealth technologies. This highlights the growing importance of this topic.
- Classifying some journals as either Health or Computer Science has proved difficult. We understand this as a sign of surging interdisciplinary work in these fields, less common in the past.
- There are already studies that warn of: methodological deficits in some studies in mHealth, low accuracy and no reproducibility. This should compel publishers and researchers to be more stringent in the design of experiments and in the publication of results.

References

1. IMS Institute for Healthcare Informatics. Patient apps for improved healthcare: from novelty to mainstream (October 2013),
   `http://www.imshealth.com/deployedfiles/imshealth/Global/Conte nt/Corporate/IMS%20Health%20Institute/Reports/Patient_Apps/II HI_Patient_Apps_Report.pdf` (accessed July 22, 2014)
2. Deloitte Center for Health Solutions. How mobile technology is transforming health care (2012),
   `http://www2.deloitte.com/content/dam/Deloitte/ao/Documents/us _dchs_2012_mHealthHowMobileTechnologyisTransformingHealthcare 12062012.pdf` (accessed July 22, 2014)
3. Urrutia, G., Bonfill, X.: Declaración PRISMA: Una propuesta para mejorar la publicación de revisiones sistemáticas y metaanálisis. Medicina Clínica 135(11), 507–511 (2010)

Reviewed Articles

4. Abel, O., Shatunov, A., Jones, A.R., Andersen, P.M., Powell, J.F., Al-Chalabi, A.: Development of a Smartphone App for a Genetics Website: The Amyotrophic Lateral Sclerosis Online Genetics Database (ALSoD). JMIR mhealth and uhealth 1(2), e18 (2013)
5. Ahtinen, A., Mattila, E., Välkkynen, P., Kaipainen, K., Vanhala, T., Ermes, M., Lappalainen, R.: Mobile mental wellness training for stress management: Feasibility and design implications based on a one-month field study. JMIR mhealth and uhealth 1(2), e11 (2013)
6. Akter, S., D'Ambra, J., Ray, P.: Development and validation of an instrument to measure user perceived service quality of mHealth. Information & Management 50(4), 181–195 (2013)
7. Albrecht, U.V., Behrends, M., Matthies, H.K., von Jan, U.: Usage of Multilin-gual Mobile Translation Applications in Clinical Settings. JMIR mhealth and uhealth 1(1), e4 (2013)

8. Alnanih, R., Ormandjieva, O., Radhakrishnan, T.: Context-based and Rule-based Adaptation of Mobile User Interfaces in mHealth. Procedia Computer Science 21, 390–397 (2013)
9. Arnhold, M., Quade, M., Kirch, W.: Mobile Applications for Diabetics: A Sys-tematic Review and Expert-Based Usability Evaluation Considering the Special Require-ments of Diabetes Patients Age 50 Years or Older. Journal of Medical Internet Research 16(4) (2014)
10. Balsam, J., Bruck, H.A., Rasooly, A.: Capillary array waveguide amplified fluo-rescence detector for mHealth. Sensors and Actuators B: Chemical 186, 711–717 (2013)
11. Balsam, J., Rasooly, R., Bruck, H.A., Rasooly, A.: Thousand-fold fluorescent signal am-plification for mHealth diagnostics. Biosensors and Bioelectronics 51, 1–7 (2014)
12. Barwais, F.A., Cuddihy, T.F., Tomson, L.M.: Physical activity, sedentary be-havior and total wellness changes among sedentary adults: A 4-week randomized controlled trial. Health and Quality of Life Outcomes 11(1), 183 (2013)
13. Becker, S., Miron-Shatz, T., Schumacher, N., Krocza, J., Diamantidis, C., Albrecht, U.V.: mHealth 2.0: Experiences, Possibilities, and Perspectives. JMIR mHealth and uHealth 2(2), e24 (2014)
14. Bierbrier, R., Lo, V., Wu, R.C.: Evaluation of the accuracy of smartphone medical calcula-tion apps. Journal of Medical Internet Research 16(2) (2014)
15. BinDhim, N.F., McGeechan, K., Trevena, L.: Who Uses Smoking Cessation Apps? A Fea-sibility Study Across Three Countries via Smartphones. JMIR mhealth and uhealth 2(1), e4 (2014)
16. Boulos, M.N.K., Brewer, A.C., Karimkhani, C., Buller, D.B., Dellavalle, R.P.: Mobile medical and health apps: State of the art, concerns, regulatory control and certification. Online Journal of Public Health Informatics 5(3), 229 (2014)
17. Breland, J.Y., Yeh, V.M., Yu, J.: Adherence to evidence-based guidelines among diabetes self-management apps. Translational Behavioral Medicine 3(3), 277–286 (2013)
18. Breton, E.R., Fuemmeler, B.F., Abroms, L.C.: Weight loss—there is an app for that! But does it adhere to evidence-informed practices? Translational Behavioral Medicine 1(4), 523–529 (2011)
19. Bricker, J.B., Mull, K.E., Kientz, J.A., Vilardaga, R., Mercer, L.D., Akioka, K.J., Heffner, J.L.: Randomized, controlled pilot trial of a smartphone app for smoking cessation using acceptance and commitment therapy. Drug and Alcohol Dependence 143, 87–94 (2014)
20. Brooke, M.J., Thompson, B.M.: Food and Drug Administration Regulation of Diabetes-Related mHealth Technologies. Journal of Diabetes Science and Technology 7(2), 296–301 (2013)
21. Brown III, W., Yen, P.Y., Rojas, M., Schnall, R.: Assessment of the Health IT Usability Evaluation Model (Health-ITUEM) for evaluating mobile health (mHealth) tech-nology. Journal of Biomedical Informatics 46(6), 1080–1087 (2013)
22. Cafazzo, J.A., Casselman, M., Hamming, N., Katzman, D.K., Palmert, M.R.: Design of an mHealth app for the self-management of adolescent type 1 diabetes: A pilot study. Journal of Medical Internet Research 14(3), e70 (2012)
23. Carter, T., O'Neill, S., Johns, N., Brady, R.R.: Contemporary vascular smartphone medical applications. Annals of Vascular Surgery 27(6), 804–809 (2013)
24. Chen, L., Wang, W., Du, X., Rao, X., van Velthoven, M.H., Yang, R., Zhang, Y.: Effec-tiveness of a smart phone app on improving immunization of children in rural Sichuan Province, China: Study protocol for a paired cluster randomized controlled trial. BMC Public Health 14(1), 262 (2014)

25. Cornelius, C.T., Kotz, D.F.: Recognizing whether sensors are on the same body. Pervasive and Mobile Computing 8(6), 822–836 (2012)
26. Datta, A.K., Sumargo, A., Jackson, V., Dey, P.P.: mCHOIS: An application of mobile technology for childhood obesity surveillance. Procedia Computer Science 5, 653–660 (2011)
27. de la Vega, R., Miró, J.: mHealth: A Strategic Field without a Solid Scientific Soul. A Systematic Review of Pain-Related Apps. PloS One 9(7), e101312 (2014)
28. Dunford, E., Trevena, H., Goodsell, C., Ng, K.H., Webster, J., Millis, A., Neal, B.: FoodSwitch: A mobile phone app to enable consumers to make healthier food choices and crowdsourcing of national food composition data. JMIR mHealth and uHealth 2(3) (2014)
29. Eskenazi, B., Quirós-Alcalá, L., Lipsitt, J.M., Wu, L.D., Kruger, P., Ntimbane, T., Seto, E.: mSpray: A mobile phone technology to improve malaria control efforts and monitor human exposure to malaria control pesticides in Limpopo, South Africa. Environment International 68, 219–226 (2014)
30. Fiordelli, M., Diviani, N., Schulz, P.J.: Mapping mHealth research: A decade of evolution. Journal of Medical Internet Research 15(5) (2013)
31. Goldenberg, T., McDougal, S.J., Sullivan, P.S., Stekler, J.D., Stephenson, R.: Preferences for a Mobile HIV Prevention App for Men Who Have Sex With Men. JMIR mHealth and uHealth 2(4), e47 (2014)
32. Grindrod, K.A., Gates, A., Dolovich, L., Slavcev, R., Drimmie, R., Aghaei, B., Leat, S.J.: ClereMed: Lessons Learned From a Pilot Study of a Mobile Screening Tool to Identify and Support Adults Who Have Difficulty With Medication Labels. JMIR mHealth and uHealth 2(3) (2014)
33. Hao, W.R., Hsu, Y.H., Chen, K.C., Li, H.C., Iqbal, U., Nguyen, P.A., Jian, W.S.: LabPush: A Pilot Study of Providing Remote Clinics with Laboratory Results via Short Message Service (SMS) in Swaziland, Africa-A Qualitative study. Computer Methods and Programs in Biomedicine (2014)
34. Hilliard, M.E., Hahn, A., Ridge, A.K., Eakin, M.N., Riekert, K.A.: User Pref-erences and Design Recommendations for an mHealth App to Promote Cystic Fibrosis Self-Management. JMIR mHealth and uHealth 2(4), e44 (2014)
35. Hundert, A.S., Huguet, A., McGrath, P.J., Stinson, J.N., Wheaton, M.: Commercially Available Mobile Phone Headache Diary Apps: A Systematic Review. JMIR mHealth and uHealth 2(3) (2014)
36. Iwaya, L.H., Gomes, M.A.L., Simplício, M.A., Carvalho, T.C.M.B., Dominicini, C.K., Sakuragui, R.R., Håkansson, P.: Mobile health in emerging countries: A survey of research initiatives in Brazil. International Journal of Medical Informatics 82(5), 283–298 (2013)
37. Jibb, L.A., Stevens, B.J., Nathan, P.C., Seto, E., Cafazzo, J.A., Stinson, J.N.: A Smartphone-Based Pain Management App for Adolescents With Cancer: Establishing System Requirements and a Pain Care Algorithm Based on Literature Review, Interviews, and Consensus. JMIR Research Protocols 3(1) (2014)
38. King, C., Hall, J., Banda, M., Beard, J., Bird, J., Kazembe, P.: Electronic data capture in a rural African setting: evaluating experiences with different systems in Malawi. Global Health Action 7 (2014)
39. Kizakevich, P.N., Eckhoff, R., Weger, S., Weeks, A., Brown, J., Bryant, S., Spira, J.: A personal health information toolkit for health intervention research. Studies in Health Technology and Informatics 199, 35–39 (2013)
40. Klonoff, D.C.: The current status of mHealth for diabetes: Will it be the next big thing? Journal of Diabetes Science and Technology 7(3), 749–758 (2013)

41. Kuo, M.C., Lu, Y.C., Chang, P.: A newborn baby care support app and system for mHealth. In: NI 2012: Proceedings of the 11th International Congress on Nursing Informatics, vol. 2012. American Medical Informatics Association (2012)

42. Labrique, A., Vasudevan, L., Chang, L.W., Mehl, G.: H_pe for mHealth: More "y" or "o" on the horizon? International Journal of Medical Informatics 82(5), 467–469 (2013)

43. Lee, S.S.S., Xin, X., Lee, W.P., Sim, E.J., Tan, B., Bien, M.P.G., Thumboo, J.: The feasibility of using SMS as a health survey tool: An exploratory study in pa-tients with rheumatoid arthritis. International Journal of Medical Informatics 82(5), 427–434 (2013)

44. Lewis, T.L., Wyatt, J.C.: Mhealth and mobile medical apps: A framework to assess risk and promote safer use. Journal of Medical Internet Research 16(9) (2014)

45. Liu, C., Zhu, Q., Holroyd, K.A., Seng, E.K.: Status and trends of mobile-health applications for iOS devices: A developer's perspective. Journal of Systems and Software 84(11), 2022–2033 (2011)

46. Lopez, C., Ramirez, D.C., Valenzuela, J.I., Arguello, A., Saenz, J.P., Trujillo, S., Dominguez, C.: Sexual and Reproductive Health for Young Adults in Colombia: Teleconsultation Using Mobile Devices. JMIR mHealth and uHealth 2(3) (2014)

47. Lyons, E.J., Lewis, Z.H., Mayrsohn, B.G., Rowland, J.L.: Behavior change techniques implemented in electronic lifestyle activity monitors: A systematic content analysis. Journal of Medical Internet Research 16(8) (2014)

48. Mann, D.M., Kudesia, V., Reddy, S., Weng, M., Imler, D., Quintiliani, L.: Development of DASH Mobile: A mHealth Lifestyle Change Intervention for the Management of Hypertension. Studies in Health Technology and Informatics 192, 973–973 (2012)

49. Martínez-Pérez, B., de la Torre-Díez, I., López-Coronado, M.: Mobile health applications for the most prevalent conditions by the World Health Organization: Review and analysis. Journal of Medical Internet Research 15(6) (2013)

50. Martínez-Pérez, B., de la Torre-Díez, I., López-Coronado, M., Herreros-González, J.: Mobile Apps in Cardiology: Review. JMIR mhealth and uhealth 1(2), e15 (2013)

51. Martínez-Pérez, B., de la Torre-Díez, I., López-Coronado, M., Sainz-De-Abajo, B.: Comparison of Mobile Apps for the Leading Causes of Death Among Different Income Zones: A Review of the Literature and App Stores. JMIR mhealth and uhealth 2(1), e1 (2014)

52. Martínez-Pérez, B., de la Torre-Díez, I., López-Coronado, M., Sainz-de-Abajo, B., Robles, M., García-Gómez, J.M.: Mobile clinical decision support systems and applications: A literature and commercial review. Journal of Medical Systems 38(1), 1–10 (2014)

53. Masters, K.: Health professionals as mobile content creators: Teaching medical students to develop mHealth applications. Medical Teacher 36(10), 883–889 (2014)

54. Menezes Jr., J., Gusmão, C., Machiavelli, J.: A Proposal of Mobile System to Support Scenario-based Learning for Health Promotion. Procedia Technology 9, 1142–1148 (2013)

55. Mirkovic, J., Kaufman, D.R., Ruland, C.M.: Supporting cancer patients in illness management: Usability evaluation of a mobile app. JMIR mHealth and uHealth 2(3), e33 (2014) doi:10.2196/mhealth.3359

56. Mobasheri, M.H., Johnston, M., King, D., Leff, D., Thiruchelvam, P., Darzi, A.: Smartphone breast applications–What's the evidence? The Breast 23(5), 683–689 (2014)

57. Neto, O.B.L., Albuquerque, C.M., Albuquerque, J.O., Barbosa, C.S.: The Schisto Track: A System for Gathering and Monitoring Epidemiological Surveys by Con-necting Geographical Information Systems in Real Time. JMIR mhealth and uhealth 2(1), e10 (2014)

58. O'Malley, G., Dowdall, G., Burls, A., Perry, I.J., Curran, N.: Exploring the usability of a mobile app for adolescent obesity management. JMIR mHealth and uHealth 2(2) (2014)

59. Parmanto, B., Pramana, G., Yu, D.X., Fairman, A.D., Dicianno, B.E., McCue, M.P.: iM-Here: A novel mhealth system for supporting self-care in management of complex and chronic conditions. JMIR mhealth and uhealth 1(2), e10 (2013)
60. Pérez-Cruzado, D., Cuesta-Vargas, A.I.: Improving Adherence Physical Activity with a Smartphone Application Based on Adults with Intellectual Disabilities (APPCOID). BMC Public Health 13(1), 1173 (2013)
61. Ploderer, B., Smith, W., Pearce, J., Borland, R.: A Mobile App Offering Distrac-tions and Tips to Cope With Cigarette Craving: A Qualitative Study. JMIR mHealth and uHealth 2(2), e23 (2014)
62. Pulman, A., Taylor, J., Galvin, K., Masding, M.: Ideas and Enhancements Related to Mo-bile Applications to Support Type 1 Diabetes. JMIR mhealth and uhealth 1(2), e12 (2013)
63. Ribu, L., Holmen, H., Torbjørnsen, A., Wahl, A.K., Grøttland, A., Småstuen, M.C., Årsand, E.: Low-Intensity Self-Management Intervention for Persons With Type 2 Di-abetes Using a Mobile Phone-Based Diabetes Diary, With and Without Health Counseling and Motivational Interviewing: Protocol for a Randomized Controlled Trial. JMIR Re-search Protocols 2(2) (2013)
64. Sezgin, E., Yıldırım, S.Ö.: A Literature Review on Attitudes of Health Profes-sionals to-wards Health Information Systems: From e-Health to m-Health. Procedia Technology 16, 1317–1326 (2014)
65. Shishido, H.Y., Alves da Curze de, A.R., Eler, G.J.: mHealth Data Collector: An Applica-tion to Collect and Report Indicators for Assessment of Cardiometabolic Risk. Studies in Health Technology and Informatics 201, 425–432 (2013)
66. Silva, B.M., Rodrigues, J.J., Canelo, F., Lopes, I.C., Zhou, L.: A data encryption solution for mobile health apps in cooperation environments. Journal of Medical Internet Re-search 15(4) (2013)
67. Slaper, M.R., Conkol, K.: mHealth Tools for the Pediatric Patient-Centered Medical Home. Pediatric Annals 43(2), e39-43 (2014)
68. Sunyaev, A., Dehling, T., Taylor, P.L., Mandl, K.D.: Availability and quality of mobile health app privacy policies. Journal of the American Medical Informatics Association, amiajnl-2013 (2014)
69. Surka, S., Edirippulige, S., Steyn, K., Gaziano, T., Puoane, T., Levitt, N.: Evalu-ating the use of mobile phone technology to enhance cardiovascular disease screening by communi-ty health workers. International Journal of Medical Informatics 83(9), 648–654 (2014)
70. Tsui, I., Drexler, A., Stanton, A.L., Kageyama, J., Ngo, E., Straatsma, B.R.: Pilot Study Using Mobile Health to Coordinate the Diabetic Patient, Diabetologist, and Ophthalmolo-gist. Journal of Diabetes Science and Technology, 1932296814529637 (2014)
71. Turner-McGrievy, G.M., Tate, D.F.: Are we sure that Mobile Health is really mobile? An examination of mobile device use during two remotely-delivered weight loss interventions. International Journal of Medical Informatics 83(5), 313–319 (2014)
72. Turner-McGrievy, G.M., Tate, D.F.: Are we sure that Mobile Health is really mobile? An examination of mobile device use during two remotely-delivered weight loss interventions. International Journal of Medical Informatics 83(5), 313–319 (2014)
73. Turner-McGrievy, G.M., Beets, M.W., Moore, J.B., Kaczynski, A.T., Barr-Anderson, D.J., Tate, D.F.: Comparison of traditional versus mobile app self-monitoring of physical activi-ty and dietary intake among overweight adults participating in an mHealth weight loss program. Journal of the American Medical Informatics Association 20(3), 513–518 (2013)
74. van der Heijden, M., Lucas, P.J., Lijnse, B., Heijdra, Y.F., Schermer, T.R.: An autonom-ous mobile system for the management of COPD. Journal of Biomedical Informat-ics 46(3), 458–469 (2013)

75. van der Weegen, S., Verwey, R., Spreeuwenberg, M., Tange, H., van der Weijden, T., de Witte, L.: The development of a mobile monitoring and feedback tool to stimulate physical activity of people with a chronic disease in primary care: A user-centered design. JMIR mhealth and uhealth 1(2), e8 (2013)

76. van Drongelen, A., Boot, C.R., Hlobil, H., Twisk, J.W., Smid, T., van der Beek, A.J.: Evaluation of an mHealth intervention aiming to improve health-related behavior and sleep and reduce fatigue among airline pilots. Scandinavian Journal of Work, Environment & Health 40(6), 557–568 (2014)

77. Vriend, I., Coehoorn, I., Verhagen, E.: Implementation of an App-based neuromuscular training programme to prevent ankle sprains: A process evaluation using the RE-AIM Framework. British Journal of Sports Medicine, bjsports-2013 (2014)

78. Wang, A., An, N., Lu, X., Chen, H., Li, C., Levkoff, S.: A Classification Scheme for Analyzing Mobile Apps Used to Prevent and Manage Disease in Late Life. JMIR mhealth and uhealth 2(1), e6 (2014)

79. Yang, Y.T., Silverman, R.D.: Mobile health applications: the patchwork of legal and liability issues suggests strategies to improve oversight. Health Affairs 33(2), 222–227 (2014)

80. Zmily, A., Mowafi, Y., Mashal, E.: Study of the Usability of Spaced Retrieval Exercise Using Mobile Devices for Alzheimer's Disease Rehabilitation. JMIR mHealth and uHealth 2(3) (2014)

Automated Extraction of Food Intake Indicators from Continuous Meal Weight Measurements

Vasileios Papapanagiotou[1], Christos Diou[1], Billy Langlet[2], Ioannis Ioakimidis[2], and Anastasios Delopoulos[1]

[1] Aristotle University of Thessaloniki, Greece
{vassilis,diou}@mug.ee.auth.gr, adelo@eng.auth.gr
http://mug.ee.auth.gr
[2] Karolinska Institutet, Huddinge, Sweden
{billy.langlet,ioannis.ioakimidis}@ki.se

Abstract. Recent studies and clinical practice have shown that the extraction of detailed eating behaviour indicators is critical in identifying risk factors and/or treating obesity and eating disorders, such as anorexia and bulimia nervosa. A number of single meal analysis methods that have been successfully applied are based on the Mandometer, a weight scale that continuously measures the weight of food on a plate over the course of a meal. Experimental meal analysis is performed using the cumulative food intake curve, which is produced by the semi-automatic processing of the Mandometer weight measurements, in tandem with the video recordings of the eating session. Due to its complexity and the video recording dependence, this process is not suited to a clinical or a real-life setting.

In this work, we evaluate a method for automating the extraction of an accurate food intake curve, corrected for food additions during the meal and artificial weight fluctuations, using only the raw Mandometer output. Since the method requires no manual corrections or external video recordings it is appropriate for clinical or free-living use. Three algorithms are presented based on rules, greedy decisioning and exhaustive search, as well as evaluation methods of the Mandometer measurements. Experiments on a set of 114 meals collected from both normal and disordered eaters in a clinical environment illustrate the effectiveness of the proposed approach.

Keywords: Obesity, eating disorders, Mandometer, cumulative food intake curve.

1 Introduction

In 2008, more than 1.4 billion adults, 20 and older, were overweight. Of these over 200 million men and nearly 300 million women were obese[1]. Lifetime prevalence estimates of Anorexia Nervosa (AN), Bulimia Nervosa, and binge-eating disorder were 0.3%, 0.9%, and 1.6%, respectively [14]. Obesity (OB) is an important contributor to many diseases, such as cardiovascular disease and type II diabetes [15].

---

[1] http://www.who.int/mediacentre/factsheets/fs311/en/

F. Ortuño and I. Rojas (Eds.): IWBBIO 2015, Part II, LNCS 9044, pp. 35–46, 2015.
© Springer International Publishing Switzerland 2015

AN, on the other hand, is recognised as an important mortality cause in adolescent women, and even when treated is characterised by a high relapse rate [1].

Unfortunately, both for OB and Eating Disorders (EDs), the existing prevention and treatment guidelines are not effective, as they are often not evidence-based [11,16]. In EDs, less than 50% remission rates are often reported, accompanied by mortality rates are up to 25% [13]. Similarly, the predictions for the increase of OB internationally in the near future are bleak[2], pointing to the failure of current pharmacological interventions to halt the increase of OB in western societies [17]. Concerning EDs, the evidence for the effectiveness of psychopharmacological treatments is limited. In 5512 studies on traditional psychotherapy-based treatments, only 6 were found to fulfil scientific criteria, and only 2 appeared to be moderately effective [6], a pattern repeated in a more recent survey [7]. However, some attention has been recently given to monitoring and studying eating behaviour [12].

In 1996, a new treatment method was introduced, based on the Mandometer [4]. The Mandometer is a weighting scale, counting the progressive weight changes from a plate with food during a meal. According to [10], the cumulative Food Intake (FI) $w(t)$ of the subject can be modelled using a quadratic curve $w(t) = \alpha t^2 + \beta t + \gamma$, where α is the FI acceleration, β is the initial FI rate and $\gamma = 0$, since no food has been consumed at the beginning of a recording. In a study analysing the eating patterns of normal-weight individuals [18], two main patterns were observed: decelerated and linear eating. On the contrary, clinical populations with OB or EDs are characterised by linear eating [8]. For decelerated eaters ($\alpha < 0$), the initial FI rate β was typically higher than that of linear eaters ($\alpha \simeq 0$). However, the total meal FI does not differ significantly between normal-weight linear and decelerated eaters [18]. These findings combined, provide evidence for the relation of the linear eating style to the risk of developing disordered eating. Interestingly, the use of the Mandometer can train away the suspect behaviour [19].

In [2], a randomised controlled trial was presented, including 32 subjects, half of which received the Mandometer treatment. Out of the 16 subjects that received the Mandometer treatment, 14 recovered, compared to only 1 of the 16 subjects that did not receive it. Low relapse of 7% was also observed for another group of 83 subjects. In [3], results from $1,428$ subjects taken over 18 years show remission rates of 74%, out of which only 10% has relapsed.

In a laboratory environment, for the experimental analysis of eating behaviour, video recordings of the meals are combined with Mandometer data for the detailed description of a person's meal [9]. In the clinical practice, the raw Mandometer output is visually inspected by trained health professionals after the meal conclusion. Estimating FI in this manner requires a lot of manual labour and experience, since the recordings display weight fluctuations that do not always correspond to FI (e.g. food additions). Measurements from less controlled environments (for example screening at schools in the context of the

---

[2] http://www.sciencedaily.com/releases/2014/05/140509110711.htm

SPLENDID project[3]) suffer from even higher levels of noise and require additional manual effort for processing. Thus, both in the clinical practice and in real life use, a robust automatic method would greatly benefit the processing of the meal recordings. In this work we present three algorithms that automatically create the FI curve of a single meal based solely on the Mandometer recordings of that meal. We perform experiments on a large dataset and compare the efficency of each algorithm based on meaningful indicators, including the values of the coefficients α and β of the quadratic model, defined by clinical experts.

The rest of this paper is organised as follows. Section 2 presents the proposed algorithms in detail. Section 3 presents the experimental dataset that is used to evaluate the algorithms and the indicators that are extracted from the FI curves. Experimental results are presented in Section 4. Section 5 concludes the paper.

2 Methods

Creating the FI curve from the Mandometer recordings is based on a set of assumptions about the recording. The principle assumption is that any removed quantity of food from the plate, recorded as a decrease in the measured weight, is eaten by the subject; no amount of food is discarded from the plate in any manner. However, at the end of the meal, it is not required that all of the food has been consumed. Additionally, it is assumed that the beginning of the actual eating session does not necessarily start with the recording of the meal, but an arbitrary amount of time later. Similarly, the end of the recording can occur an arbitrary amount of time after the eating session is finished. Another assumption is that the subject can add additional food to the plate and continue the meal. This can happen more than once, at any point during the meal, and there is no limit in the minimum or maximum weight of the Food Addition (FA), except for the Mandometer sensitivity and maximum allowed weight. Finally, it is worth mentioning that removing a single spoonful of food from the plate usually registers more than a decrease on the recorded data; the applied pressure with the fork or spoon can register a temporary weight increase before the weight decrease (see Fig. 1).

2.1 Rule-Based Algorithm (RB)

Based on these assumptions, the first algorithm incorporates a set of decision rules in order to create the FI curve of a meal. The main components of the RB algorithm are presented in Algorithm 1, and the intermediate results are visually presented in Fig. 2.

At the pre-processing stage f_1, an opening morphological operator [5] is first applied in order to remove instances of weight increase due to the pressure applied on the plate with the fork and/or knife (see Fig. 1). Opening is the dilation of the erosion of a discrete signal. Given the sampling rate of 0.25 Hz

---

[3] http://splendid-program.eu/

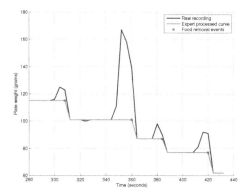

Fig. 1. Temporal weight increases in the recorded curve. Raw recording denoted with blue, and actual food weight removed with green. Red crosses denote the moments that bites occur.

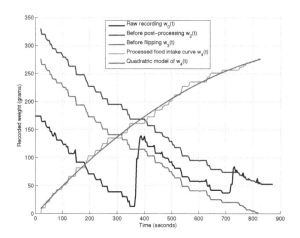

Fig. 2. Example of automated FI curve extraction. The black curve is the pre-processed recording w_1; blue w_2 is the result of restoring w_1 for 2 FAs. Red w_3 is the post-processed w_2, and green w_4 is the final FI curve. The quadratic model of $w_4(t)$ is the magenda curve. The α and β coefficients of the model are used to describe the eating behaviour of an individual across a meal.

Algorithm 1. Rule-based algorithm (RB)

Data: Mandometer raw meal recording $w_0(t)$
Result: FI curve $w_4(t)$

1 $w_1 = f_1(w_0; v)$;   // Pre-processing
2 $A = CFAD(w_1; w_{FA}, t_{FA}, w_{FAp1})$;   // FA detection
3 $w_2 = f_2(w_1; A)$;   // FA restoration
4 $w_3 = f_3(w_2; w_{\text{start}}, w_{\text{stop}})$;   // Post-processing
5 $w_4 = f_4(w_3)$;   // Flipping

---

of the Mandometer, a vector of ones of length 5 is selected as the structuring element v, as it was found to give best results.

Regarding the FAs, visual observation of the recorded signals shows that most of them cause a significant weight increase in relatively short time. Indeed, it is reasonable to assume that people do not repeatedly add and eat small amounts of food on their plates, but they rather only add larger quantities (compared to their average bite size) a few times only.

Based on this hypothesis, the FA detector

$$A = CFAD(w_1; w_{FA}, t_{FA}, w_{FApl}) \tag{1}$$

finds all meal segments (consecutive set of samples) during which weight increases at every sample, and a weight increase above a threshold w_{FA} occurs during a time period less than a threshold t_{FA}; approximately 20 grams in less than 12 seconds, to enable detection of small food additions. Both weight and time interval thresholds have been based on statistical analysis of the available recordings. In fact, minor decreases less than w_{FApl} (set at 3 grams) are actually allowed. Set A is the set of all the detected segments. Fig. 3 shows such an example where one such segment has been detected.

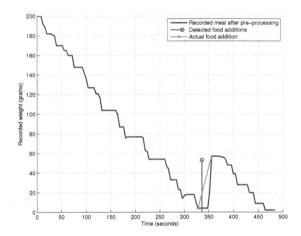

Fig. 3. A meal recording after the preprocessing step. The food additions detector has detected a single food addition of 53 grams at 336 seconds, which is accurate both in time of occurrence and added food weight according the manually annotated ground truth.

Once the FAs A have been detected, a restoration process follows that adjusts the curve, as if the entire food weight had been added to the plate before the beginning of the meal, as shown in Fig. 2. This allows information such as total FI to be easily extracted.

Accurate detection of FAs is detrimental to correctly constructing the food intake curve. In the example presented in Fig. 3, correct detection of the FA results in a meal of total weight of approximately 250 grams, whereas no detection of the FA will result in 200 grams of meal, and thus significant error.

The post-processing stage f_3 consists of four steps. First, the beginning of the meal is detected and earlier samples removed. To decide the amount of time elapsed between the start of the recording and the moment eating activity actually starts the following rule is employed. The longest interval starting from the first sample of the meal recording during which only small weight fluctuations are recorded is deleted. Secondly, the end of the meal is detected and remaining samples are removed, using a similar rule: the longest interval ending at the last sample of the recording during which small weight fluctuations are recorded is deleted. For these two steps, weight fluctuation in a segment is considered small if the difference of every sample from the first (or last) it is less than some w_{start} (or w_{stop} respectively). Thirdly, a smoothing operation removes any remaining weight increases by flattening them, and finally, the value of the last sample is subtracted from all samples. The resulting curve is strictly non-increasing. Total FI is equal to the value of the first sample.

Literally, the obtained curve is the "remaining food weight curve" of the plate. In order to obtain the FI curve of the subject the "flipping" stage f_4 is applied, where each sample is replaced with the difference between total FI and its value. This results in vertically flipping the curve, so that it represents the FI curve (which increases as time elapses). Fig. 2 shows such an example.

2.2 Greedy Quadratic Fitting (GQF)

The second algorithm aims at improving the robustness of the FA detection process. It is identical to the RB algorithm in all stages except for the $CFAD$ (Eq. (1)). Thus, only the new food detector is presented in this Section.

As described in Eq. (1), several parameters need to be selected. Furthermore, tuning the parameters still results in systematic error on specific cases. For example, a lower FA weight threshold w_{FA} allows the detection of smaller FAs, at the cost of increased number of falsely detected FAs. To deal with such problems, this algorithm uses a modified FAs detection step.

First, candidate food additions (CFAs) are detected using the same $CFAD$ of Eq. (1) of the RB algorithm, with low w_{FA} threshold. This results in high recall due to the lower threshold, and low precision since some temporal increases (see Fig. 1) are also detected as CFAs.

Given N CFAs, namely $x_i \in A_c$, the pre-processed curve w_1 is split into $N+1$ segments s_i. An iterative process starts that decides if each x_i is an actual food addition or not, from the oldest to the most recent, and restores the curve appropriately in the case of a positive decision. In order to decide about x_k, the following calculations are performed (note that this implies that all previous x_i, $i \in \{1, 2, 3, ..., k-1\}$ have been decided, and restoration has been performed on the processed recording for the x_is that have been positively decided as actual FAs).

The aim of this algorithm is to select the decision (about the current x_k) that least alters the so far trend of the meal curve. For this purpose, two segments

$$S_1 = \cup_{i=1}^{k} s_i \text{ and } S_2 = \cup_{i=1}^{k+1} s_i$$

are considered, representing the present and future trend of the meal. Both segments start from the first sample of the recording. Segment S_1 ends at the last sample before x_k, and segment S_2 ends at the last sample before x_{k+1} (or at the last sample of the recording if there are no CFAs after x_k).

Based on the quadratic model of [10], a quadratic curve $\alpha_1 t^2 + \beta_1 t + \gamma_1$ is fitted on S_1, yielding a set of coefficients $\boldsymbol{u}_1 = [\alpha_1 \, \beta_1 \, \gamma_1]$. The present trend of the meal is projected by evaluating the quadratic model on S_2. For the future of the meal, two cases exist: one where x_k is a FA and one where x_k is not a FA. Thus, two sets of coefficients $\boldsymbol{u}_{2+} = [\alpha_{2+} \, \beta_{2+} \, \gamma_{2+}]$ and $\boldsymbol{u}_{2-} = [\alpha_{2-} \, \beta_{2-} \, \gamma_{2-}]$ are produced on S_2, however for $\boldsymbol{u}_{2+}$ it is assumed that x_k is a FA and thus the curve is restored appropriately before the fitting, whereas for $\boldsymbol{u}_{2-}$ it is not restored.

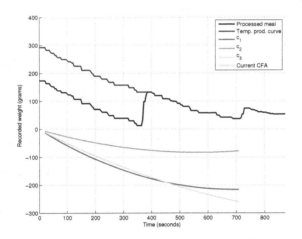

Fig. 4. Example of the GQF algorithm. The current processed data are denoted with black. The yellow area denotes the current CFA. The red curve c_1 represents the original trend of the meal curve. The green curve c_2 represents the trend of the meal as if the the current CFA was not an actual FA. The cyan curve c_3 represents the trend of the meal as if the current CFA was an actual food addition. The blue line denoted the temporarily produced curve.

Coefficients $\gamma_1, \gamma_{2+}, \gamma_{2-}$ are then set equal to 0, and three curves c_1, c_{2+} and c_{2-} are evaluated on S_2. Two mean squared error (MSE) values $\text{MSE}_+ = mse(c_1, c_{2+})$ and $\text{MSE}_- = mse(c_1, c_{2-})$ are calculated, where

$$mse(c_a, c_b) = \sum_{t=1}^{L} (c_a(t) - c_b(t))^2 \qquad (2)$$

and L is the number of samples of c_a (or c_b). The final decision of x_k is based on the comparison of the two MSEs. If $\text{MSE}_+ < \text{MSE}_-$ then x_k is decided to be an actual FA, and the so far processed curve is restored for this FA.

If $MSE_+ > MSE_-$, x_k is discarded. Finally, a new iteration begins, to decide x_{k+1} (as long as there are any remaining CFAs). Figure 4 shows an example of a single iteration of the algorithm.

2.3   Rule Based with Quadratic Fitting (RBQF)

The final algorithm aims at further improving the robustness of the FA detector stage. It is similar to GQF. However, instead of greedily deciding for each CFA, all possible combinations (subsets of the CFA set) are exhaustively examined and evaluated using a cost function. The combination/subset achieving the lowest score is selected for the final output. RBQF pseudocode is presented in Algorithm 2.

Algorithm 2. Rule-based with quadratic fitting algorithm (RBQF)

Data: Mandometer raw meal recording $w_0(t)$
Result: FI curve $w_4(t)$

1 $w_1 = f_1(w_0; s)$;   // Pre-processing
2 $A_c = CFAD(w_1; w_{FA}, t_{FA}, w_{FAp1})$;   // Candidate FA detection
3 **for** $A_i \in 2^{A_c}$;   // For all possible subsets of A_c
4 **do**
5 | $w_2 = f_2(w_1; A_i)$;   // Restore FAs of A_i
6 | $w_{3,i} = f_3(w_2; w_{start}, w_{stop})$;   // Post-processing
7 | $J(i) = cost(w_{3,i})$;   // Evaluate the cost function (Eq. (3))
8 **end**
9 $i^* = \arg\min J$;   // Select minimum cost case
10 $w_3 = w_{3,i^*}$;
11 $w_4 = f_4(w_3)$;   // Flipping

The cost function of each combination is based on the quadratic model of the FI curve of [10]. In particular, for iteration i, a quadratic curve $c_{3,i}$ is fitted on $w_{3,i}$, and evaluated on the duration of the meal. The cost is calculated as the MSE of $w_{3,i}$ and $c_{3,i}$ using Eq. (2), as

$$cost(w_{3,i}) = mse(w_{3,i}, c_{3,i}) \tag{3}$$

The rationale behind the cost function is that the FI curve is in general a smooth curve. Sharp weight increases and decreases are the results of non-eating related activity, such as food additions. Fig. 5 demonstrates two cases, one where an actual food addition is detected successfully, and one where an artifact that is detected as a CFA (and also falsely detected as an actual food addition by the first algorithm) is rejected.

3   Experimental Setup and Indicators

In order to evaluate the effectiveness of the proposed algorithms, experiments were performed on a dataset consisting of experimental eating behaviour data and clinical recordings of meals. The dataset consists of 114 meals that were recorded from

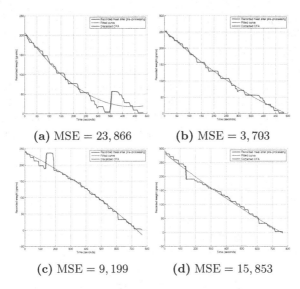

(a) MSE $= 23,866$ (b) MSE $= 3,703$

(c) MSE $= 9,199$ (d) MSE $= 15,853$

Fig. 5. Two examples of the $CFAD$ and cost function or RBQF. For Fig. 5a and 5b A_c contains one CFA at 350 seconds. In Fig. 5a $A_1 = \emptyset$ and so no restoration has been applied, resulting in a bad fit and high MSE. In Fig. 5b $A_2 = A_c$ and restoration has been applied, resulting in lower MSE. Similarly for Fig. 5c and 5d, A_c also contains one CFA, however this CFA is not an actual FA.

105 females and 9 males. Out of these 114 individuals, 45 have been characterised as *normal-weight* (mean age 22.8 years, mean Body Mass Index (BMI) $21.2\,kg/m^2$), 23 as *obese* (mean age 35.22 years, mean BMI $37.21\,kg/m^2$), and 46 as *anorexia nervosa cases* (mean age 21.7 years, mean BMI $17.4\,kg/m^2$). Thus, the dataset contains a variety of individuals regarding BMI and eating disorders.

For each of the 114 recorded meals, the raw Mandometer recordings were provided. The experimental meal data have been corrected by expert scientists in Karolinska Institute, with the help of video recordings of the meals, using the standard technique described in [9]. The clinical data (no video recordings were available) were manually corrected by clinical experts to the best of their ability. These FI curves have been used as the ground truth for the evaluation of the effectiveness of the algorithms, in the experiments presented in Section 4.

To evaluate the effectiveness of each algorithm, we apply each one of them on every meal of the dataset. Thus results in four FI curves per meal, including the ground truth curve provided by the clinical experts using manual annotation. Co-efficients α and β are extracted from every food intake curve.

Four meaningful classes are defined according to the value of coefficient α; $(-\infty, -0.004]$ for decelerated eaters C_1^α, $(-0.004, -0.0015]$ for semi-decelerated eaters C_2^α, $(-0.0015, 0.0015]$ for linear eaters C_3^α and $(0.0015, +\infty)$ for accelerated eaters C_4^α. Results for each algorithm are presented using confusion matrices, where the ground truth class is assigned based on the value of α extracted from the ground truth intake curve, and the predicted class is assigned based on the value of α of the output intake curve of each algorithm. Similarly, five class are defined for β using the intervals $(-\infty, 0.25]$ for C_1^β, $(0.25, 0.5]$ for C_2^β, $(0.5, 0.75]$ for C_3^β, $(0.75, 1]$ for C_4^β and $(1, +\infty)$ for C_5^β.

Furthermore, three additional meal indicators are extracted: total food intake in grams, total meal duration in seconds, and average bite size also in grams. Bites are detected as continuous weight drops (see Fig. 1). These indicators are evaluated using the average and standard deviation of the absolute error, which is the absolute value of the difference of the predicted value from the actual value of each characteristic.

4   Experimental Results

The confusion matrices regarding the classification based on α and β are presented in Table 1. The evaluation of the additional indicators is presented in Table 2.

The RB algorithm achieves moderate effectiveness both for α and β, as well as for the additional indicators. Both GQF and RBQF perform much better than RB however, and GQF achieves the highest precision lowest error values. It should be noted that for both α and β coefficients, the confusion matrices tend to have more zeros in cells further of the main diagonal, which is an indicator of *minor* confusion between neighbour classes.

Table 1. Confusion matrices and accuracy for α (top row) and β (bottom row) for each of the three algorithms

	Predicted			
	C_1^α	C_2^α	C_3^α	C_4^α
C_1^α	16	3	0	0
C_2^α	2	17	3	0
C_3^α	3	8	43	10
C_4^α	0	1	1	7

(a) RB (72.81%)

	Predicted			
	C_1^α	C_2^α	C_3^α	C_4^α
C_1^α	14	5	0	0
C_2^α	1	17	3	1
C_3^α	3	7	53	1
C_4^α	0	0	0	9

(b) GQF (81.58%)

	Predicted			
	C_1^α	C_2^α	C_3^α	C_4^α
C_1^α	15	3	0	1
C_2^α	2	17	3	0
C_3^α	2	6	50	5
C_4^α	0	0	2	7

(c) RBQF (78.76%)

	Predicted				
	C_1^β	C_2^β	C_3^β	C_4^β	C_5^β
C_1^β	18	9	1	1	1
C_2^β	2	22	2	0	0
C_3^β	0	2	21	2	0
C_4^β	0	0	1	14	3
C_5^β	0	0	0	4	11

(d) RB (75.44%)

	Predicted				
	C_1^β	C_2^β	C_3^β	C_4^β	C_5^β
C_1^β	27	2	1	0	0
C_2^β	0	25	1	0	0
C_3^β	2	2	19	2	0
C_4^β	0	0	1	13	4
C_5^β	0	0	0	4	11

(e) GQF (83.33%)

	Predicted				
	C_1^β	C_2^β	C_3^β	C_4^β	C_5^β
C_1^β	20	8	0	1	0
C_2^β	1	23	2	0	0
C_3^β	0	2	21	2	0
C_4^β	0	0	1	14	3
C_5^β	0	0	0	4	11

(f) RBQF (78.76%)

Table 2. Average (and standard deviation) of absolute error of each algorithm per indicator

	RB	GQF	RBQF
Total FI (gram)	53.3 (127.7)	25.7 (37.5)	30.9 (47.7)
Total meal duration (sec)	24.2 (50.1)	19.5 (26.9)	22.5 (46.9)
Average bite size (gram)	1.7 (3.5)	0.3 (0.5)	0.6 (1.5)

5 Conclusions

Processing of the meal curve, in order to produce the FI curve, is an essential part of the Mandometer treatment method, which has been shown to achieve high remission and very low relapse for the treatment of OB and EDs. Automatic processing significantly aids clinical practice by (i) reducing the the required manual labour, and (ii) removing the need for additional sensors such as monitoring video. It thus enables a faster and standardised method of processing, reduces the risk of misclassification due to manual error, and finally increases the accuracy in treatment. Furthermore, it allows for non-laboratory recording sessions, such as home-based application of the treatment, or screening tests at schools and other public environments.

In this work we have presented three algorithms that automatically construct the FI curve of a meal, given only the recording of the Mandometer. One using a set of rules to process the meal curve (RB), one greedy quadratic fitting algorithm (GQF) that is computationally light and allows for semi-real-time processing, achieving similar effectiveness with RBQF, and the RBQF that uses a quadrating fitting cost function to select the best possible interpretation of the meal curve.

Highly accurate automated processing of the meal curve is challenging, as indicated by the moderate effectiveness of the RB. Both GQF and RBQF algorithms presented in this work significantly improve effectiveness, however there is still room for improvement. Future work includes the development of more sophisticated FI curve extraction algorithms based on models of human eating.

Acknowledgment. The work leading to these results has received funding from the European Community's ICT Programme under Grant Agreement No. 610746, 01/10/2013 - 30/09/2016.

References

1. Arcelus, J., Mitchell, A.J., Wales, J., Nielsen, S.: Mortality rates in patients with anorexia nervosa and other eating disorders: A meta-analysis of 36 studies. Archives of General Psychiatry 68(7), 724–731 (2011)
2. Bergh, C., Brodin, U., Lindberg, G., Södersten, P.: Randomized controlled trial of a treatment for anorexia and bulimia nervosa. Proceedings of the National Academy of Sciences 99(14), 9486–9491 (2002)

3. Bergh, C., Callmar, M., Danemar, S., Hölcke, M., Isberg, S., Leon, M., Lindgren, J., Lundqvist, Å., Niinimaa, M., Olofsson, B., et al.: Effective treatment of eating disorders: Results at multiple sites. Behavioral Neuroscience 127(6), 878 (2013)
4. Bergh, C., Eklund, S., Eriksson, M., Lindberg, G., Södersten, P.: A new treatment of anorexia nervosa. The Lancet 348(9027), 611–612 (1996)
5. Gonzalez, R.C., Woods, R.E.: Digital image processing (2002)
6. Hay, P., Bacaltchuk, J., Claudino, A., Ben-Tovim, D., Yong, P.Y., et al.: Individual psychotherapy in the outpatient treatment of adults with anorexia nervosa. Cochrane Database Syst. Rev. 4 (2003)
7. Hay, P.P., Bacaltchuk, J., Stefano, S., Kashyap, P., et al.: Psychological treatments for bulimia nervosa and binging. Cochrane Database Syst. Rev. 4, CD000562 (2009)
8. Ioakimidis, I., Zandian, M., Bergh, C., Södersten, P.: A method for the control of eating rate: A potential intervention in eating disorders. Behavior Research Methods 41(3), 755–760 (2009)
9. Ioakimidis, I., Zandian, M., Ulbl, F., Ålund, C., Bergh, C., Södersten, P.: Food intake and chewing in women. Neurocomputing 84, 31–38 (2012)
10. Kissileff, H.R., Thornton, J., Becker, E.: A quadratic equation adequately describes the cumulative food intake curve in man. Appetite 3(3), 255–272 (1982)
11. Kushner, R.F., Ryan, D.H.: Assessment and lifestyle management of patients with obesity: Clinical recommendations from systematic reviews. JAMA 312(9), 943–952 (2014)
12. Sazonov, E.S., Fontana, J.M.: A sensor system for automatic detection of food intake through non-invasive monitoring of chewing. IEEE Sensors Journal 12(5), 1340–1348 (2012)
13. Steinhausen, H.-C.: The outcome of anorexia nervosa in the 20th century. American Journal of Psychiatry 159(8), 1284–1293 (2002)
14. Swanson, S.A., Crow, S.J., Le Grange, D., Swendsen, J., Merikangas, K.R.: Prevalence and correlates of eating disorders in adolescents: Results from the national comorbidity survey replication adolescent supplement. Archives of General Psychiatry 68(7), 714–723 (2011)
15. Weiss, R., Caprio, S.: The metabolic consequences of childhood obesity. Best Practice & Research Clinical Endocrinology & Metabolism 19(3), 405–419 (2005)
16. Wilson, G.T., Shafran, R.: Eating disorders guidelines from nice. The Lancet 365(9453), 79–81 (2005)
17. Yanovski, S.Z., Yanovski, J.A.: Long-term drug treatment for obesity: A systematic and clinical review. JAMA 311(1), 74–86 (2014)
18. Zandian, M., Ioakimidis, I., Bergh, C., Brodin, U., Södersten, P.: Decelerated and linear eaters: Effect of eating rate on food intake and satiety. Physiology & Behavior 96(2), 270–275 (2009)
19. Zandian, M., Ioakimidis, I., Bergh, C., Södersten, P.: Linear eaters turned decelerated: Reduction of a risk for disordered eating? Physiology & Behavior 96(4), 518–521 (2009)

System Development Ontology to Discover Lifestyle Patterns Associated with NCD

María J. Somodevilla, Maria Concepción Pérez de Celis, Ivo H. Pineda,
Luis E. Colmenares, and Ismael Mena

Facultad de Ciencias de la Computación
Benemérita Universidad Autónoma de Puebla
Puebla, México
{mariajsomodevilla,mcpcelish,ivopinedatorres,lecolme}@gmail.com,
ismael_mena85@hotmail.com

Abstract. The volume of biomedical spatial information available on line increases day by day, need to be exploited and shared by users in different knowledge areas. Although classical information retrieval techniques are efficient, they are not appropriate enough, and that is why in recent years the scientific community has focused on creating new semantic-based approaches to answer users' queries. In this article the problem of chronic noncommunicable diseases (NCDs) is addressed based on Semantic Web ontologies. NCDs, according WHO[1], cause 36 million deaths annually affecting any gender and age groups worldwide. Moreover, tobacco and alcohol addictions, physical inactivity and unhealthy diets represent major risk factors. Considering the seriousness of the problem globally, there are already underway government strategies to reduce risk factors and early detection and timely treatment. In particular, we propose an ontologies system for knowledge generation of patterns associated with lifestyle in NCDs. The system consists of the following ontologies: GeoOntoMex, HealthOntoMex, NutritionalOntoMex, PhysicalActivityOntoMex, OntoENTRiskFactors and OntoMedHistory. Also, the system of ontologies provides relevant information on time usage, food and tobacco-alcohol consumption and demographics aspects as lifestyle key dimensions. Furthermore, the system will facilitate the integration of data from different domains for decision making about lifestyle transformation.

Keywords: Ontology System, life style patterns, NCD.

1 Introduction

NCDs kill 36 million people each year and are classified as cardiovascular, cancer, respiratory and diabetes. Affect all age groups and all geographic regions. The risk factors associated with NCDs are unhealthy snuff consumption, physical inactivity, alcohol abuse and diets. These behaviors lead to four metabolic / physiological key changes that increase the risk of NCDs: hypertension, overweight / obesity,

---

[1] World Health Organization.

F. Ortuño and I. Rojas (Eds.): IWBBIO 2015, Part II, LNCS 9044, pp. 47–56, 2015.

hyperglycemia and hyperlipidemia. The exorbitant costs in the long term treatment of NCDs imply poverty and underdevelopment. WHO through the countries involved have developed strategies to reduce risk factors and early detection and timely treatment.

Ontologies have been developed to cover the areas of trade and discovery of knowledge about a particular domain. Approach is grounded in solving geospatial problems in the biomedical domain: a geospatial domain that makes inferences based on ontologies and generates information relevant to the user queries. Sometimes the use of multiple ontologies in one domain provides a more refined conceptualization, since concepts are collected from different sources.

The use of ontologies such as knowledge representation model supports applications that allow for specialized searches. In this project a system involving up to 6 ontologies is discussed. GeOntoMex [1], a spatial ontology of Mexico and HealthOntoMex[2], which is an ontology of health services offered in Mexico. The later is based on the taxonomy proposed by INEGI DENUES[2]. The 4 remaining ontologies are being developed and integrated into the system using the Protégé ontology manager. DLQuery[3] allows defining rules involving objects of different ontologies that make up the system. Proceeds from the application of these rules will discover new knowledge patterns associated with lifestyle. The rest of the paper is organized as follows. Section 2 discusses works related to integration, design and construction of ontologies. In Section 3 the proposed ontology is presented. Section 4 deals with the study case used to identify the lifestyle patterns followed by conclusions and references.

2   Related Work

Ontology integration can be achieved in three main ways: by merging ontologies, by mapping local ontologies to a global ontology, and by integrating local ontologies by means of semantic bridges that define mappings between the ontologies [4].

Ontology merging is suitable for use in traditional systems which are small or moderate in size and are fairly static, and where scalability is not a core requirement. In today's complex, large and dynamic systems, ontology merging can still be applied on groups of relevant ontologies in the system. In this approach, ontologies for different information sources and systems are merged into a new ontology which includes the source ontologies. This new ontology can then be used to support the activities of various applications, such as extracting, querying and reasoning about information. According to Choi and colleagues [5], some tools that support ontology merging include SMART, PROMPT/Anchor-PROMPT, OntoMorph, HICAL, CMS, FCA-Merge and Chimaera.

In ontology mapping, specific ontologies can be derived from global or 'reference' ontology. Ontology mapping in this case becomes much easier since concepts in different ontologies that need to be mapped are derived from the same ontology. There

---

[2] National Bureau of Economic Units of National Institute of Statistics and Geography.
[3] Description Logic.

currently exist initiatives to develop top-level ontologies which aim to define concepts that are generic to as many domains as possible. For example, the SUMO upper ontology aims to provide concepts that encompass all of the types of entities that exist in the universe. Some believe that the adoption of a single top-level ontology by all ontology developers would enable the controlled semantic integration of ontologies and consequently make possible some of the major goals of the Semantic Web. There are several ongoing research projects both in the development of top-level ontologies and in the application in large-scale semantic integration. Some of the achievements and difficulties faced in these directions are reported in [6].

In a third approach, the local ontologies are likely to be left unchanged, but are linked by semantic bridges (e.g., bridge axioms in first-order logic) that define the mappings between the ontologies. Querying and answering are carried out by the local servers in a cooperative manner. This approach is the most suitable for growing and highly dynamic systems (e.g., distributed agents and the Semantic Web). According to Choi et all[4], tools that support this type of integration include CTXMatch, GLUE, MAFRA, LOM, QOMm ONION, OKMS, OMEN and P2P ontology mapping.

According to Noy [7], ontology integration is typically carried out in three stages. First, ontology matching is performed to discover the correspondences that exist between concepts in the ontologies to be matched. Second, the ontology alignments derived from ontology matching are represented either as instances in an ontology, as bridging axioms in first-order logic which represent the transformation is required, or as views that describe mappings from a global ontology to local ontologies.

Finally, the ontology alignments that result are used for various integration tasks, including data transformation, query answering and web service composition. All of the ontology mapping methods presented in this section are infeasible to carry out without prior knowledge of semantic relationships between concepts in different ontologies.

Our currently work on ontology integration is based on a new approach of interaction between ontologies [16, 17] which is called ontology system, where a set of ontological modules with semantic relationships among them are defined. This process allows specifying domain ontologies separately and then integrating them into a new ontology, where rules are defined to generate new knowledge.

3   Methodology

The methodology used to develop our system of ontologies [16], is a combination of different design methodologies developed between 1990 and 2000. Among these methods we can cite MethoOntology[18], On-To-Knowledge[19] and NEON [20]. This new general methodology for the design and construction of ontologies is composed of the following steps: system design macro-level, single ontology design and integration and evaluation of ontologies. This design is characterized by its orientation to the domain, integration, modularity, incrementality, iteration and evaluation guided by principles of design. In figure 1 are represented by modules each stage comprising this methodology.

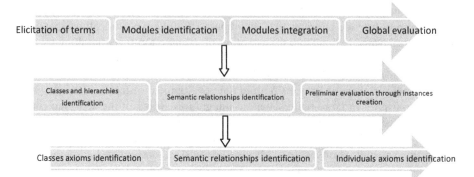

| Elicitation of terms | Modules identification | Modules integration | Global evaluation |

| Classes and hierarchies identification | Semantic relationships identification | Preliminar evaluation through instances creation |

| Classes axioms identification | Semantic relationships identification | Individuals axioms identification |

Fig. 1. General methodology for ontologies design and construction

4   Study Case

The case study deals with the identification of patterns of people lifestyle with a chronic noncommunicable disease. It argues that changing lifestyles from unhealthy to healthy decreases the risk of NCD or increase morbidity. A proposed solution to the problem is presented in Figure 2.

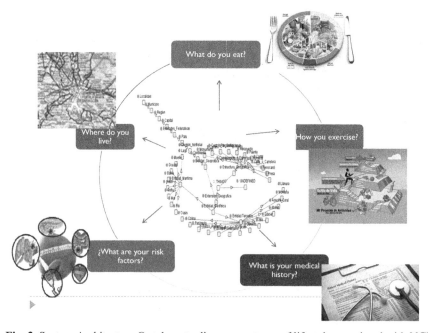

Fig. 2. System Architecture Ontology to discover patterns of lifestyle associated with NCD

4.1   Macrolevel Ontology Design

At this stage, elicitation of terms technique is applied based on competence questions. Below are some of these for our application:

- What are the <u>symptoms</u> a <u>person</u> with diabetes?
- How is the <u>nutrition</u> of a <u>person</u> suffering from <u>NCDs</u>?
- What is the <u>physical activity</u> of a <u>person</u> suffering from COPD[4]? What <u>regions</u> live with NCDs? What <u>re-gions</u> live <u>people</u> with NCDs?

It is identified nouns questions, which are those that are underlined, and will correspond to the classes of ontologies. Verbs are also identified, which indicate relationships between objects. In Figure 3, the macro-level ontology derived from this analysis is provided.

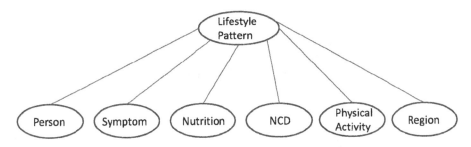

Fig. 3. Lifestyle Pattern: Macro Ontology System

4.2   Individual Ontology Design

The system consists of 6 ontologies. GeOntoMexb and HealhtOntoMex are already developed and the remaining are in development but already have enough classes and instances to the corresponding tests. Here are some of these ontologies.

GeOntoMex. Although existing ontologies as GeoOWL provide a good framework for spatial information in general, there has been no evaluation of the extent to which existing ontologies handle this information. In such ontologies spatial characteristics may be indicated by only a place name or a place name with some additional data (for example, the name of the place to which it belongs). Other ontologies provide abstraction schemes, i.e. spatial relations and organization of information. GeOntomex is a spatial ontology of Mexico developed at the Faculty of Computer Science of the Autonomous University of Puebla, as shown in figure 4.

---

[4] Chronic Obstructive Pulmonary Disease.

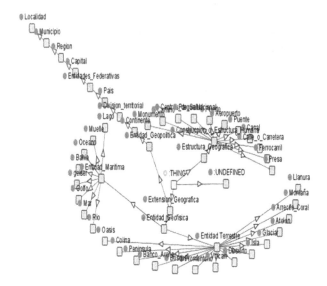

Fig. 4. GeOntoMex Ontology

```
hing
  OntoSalud
  ▼   Sector
          esPrivado
          esPublico
  ▼   Servicio
      ▼   Asistencia_Social
          ▼   Autoayuda
                  Alcoholicos
                  Otras_adicciones
          ▼   Capacitacion_para_Trabajo
                  Persona_Desempleada
                  Persona_Discapacitada
              Guarderia
          ▼   Orientacion_y_Trabajo_Social
                  Trabajo_social_para_la_niñez_y_la_juventud
          ▼   Servicio_Comunitario
                  Alimentacion
                  Emergencia
                  Refugio_Temporal
      ▼   Consulta_Externa
              Ambulancia
          ▼   Banco
                  Auxiliares
                  Organos
                  Sangre
          ▼   Centro
                  Centro_de_Enfermos_Mentales_y_Adictos
                  Centro_Planificacion_Familiar
          ▼   Consultorio
                  Consultorio_de_Audiologia
                  Consultorio_de_Nutriologos_Dietistas
                  Consultorio_de_Optometria
                  Consultorio_de_Psicologia
                  Consultorio_de_Quiropractica
                  Consultorio_Dental
                  Consultorio_Medico_Especializado
                  Consultorio_Medico_General
                  Consultorio_Ocupacional_Fisica_Lenguaje
              Enfermeria
          ▼   Laboratorio
                  Diagnostico
                  Medicos
      ▼   Hospital
              Especialidades_Medicas
              General
              Psiquiatrico
              Tratamiento_Adiccion
      ▼   Residencia
              Asilo
              Convalecientes_En_rehabilitacion_Incurables_Terminales
              Orfanato
              Retardo_Mental
              Trastorno_metal_y_Adicciones
```

Fig. 5. HealhtOntoMex Based on INEGIs

HealhtOntoMex. This project includes a variety of data related to general health services offered in Mexico, which include general hospitals, self-help groups, medical laboratories, childcare services, to specialized medicine, among others. These data were extracted from DENUE (National Statistical Directory of Economic Units) of INEGI over 170,000 instances. This taxonomy includes 54 classes for the treatment of various medical conditions. These instances are described by a set of attributes which are stored according to their latitude and longitude. Also, a relationship between a kind of "Service" and a "Public or Private Sector" is highlighted. This classification becomes important since a part of the population of Mexico has an opportunity to treat a disease in a government agency, which means, public and private institutions elsewhere. It is important to know this information, because the patient may choose to be assisted in a government agency where costs are reduced significantly so. However, there are services that are not offered in public institutions forcing the patient to attend to a private institution. Figure 5 shows its hierarchy of classes.

4.3   Ontologies Integration and Evaluation

At this stage the semantic relationships are verified for each ontology, i.e. taxonomic and relationships between objects. You need to ask what features are necessary and sufficient to define a concept. What if you have covered all the explicit relationship of competency questions? They should also identify characteristics of relations (functional, transitive, symmetric, reflexive). Figure 6 partially describes such integration.

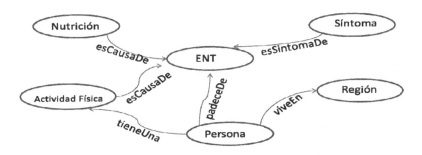

Fig. 6. Semantic relationships

Regarding the identification of type object properties relationships questions have to do with whether do you have covered all the explicit relationships of competency questions?. Figures 7 and 8 show the evaluation of rules that implement the logic descriptive competition questions.

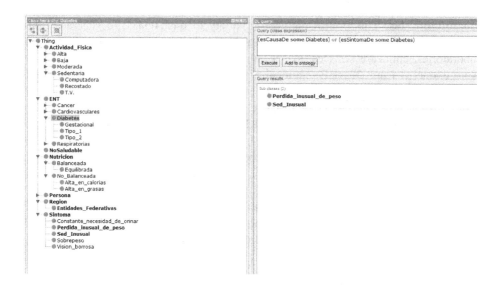

Fig. 7. Competence question: causes or symptoms of diabetes

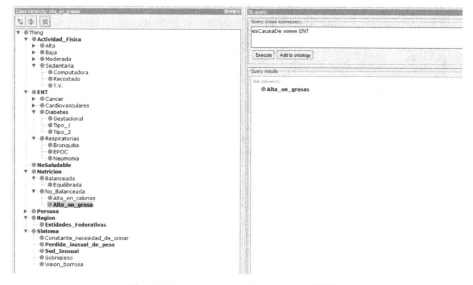

Fig. 8. Competence question: causes of NCD

5   Conclusions

A system of ontologies to support decision making related to the problem of chronic noncommunicable diseases has been submitted. Systems ontologies are a new efficient alternative to the problem of integration of ontologies, also exploits the wide range of semantic relationships established between individual ontologies system.

The lifestyle patterns are determined by multiple dimensions such as: people, symptoms, nutrition, NCD, physical activity and geographical region. Each of these dimensions has been modeled as an ontology and then be imported as a single macro ontology ontology representing lifestyles. A case study develops this subject following the methodology proposed and proved queries using descriptive logic is possible to cover the explicit relations questions of competence.

Acknowledgements. We thank the Vice-Rector for Research and Postgraduate Studies at the Autonomous University of Puebla and the National Council for Science and Technology for the support offered to perform this work.

References

1. Bentivogli, L., et al.: The seventh pascal recognizing textual entailment challenge. In: Proceedings of TAC (2011)
2. Achim, R., et al.: Mining the semantic web. Data Mining and Knowledge Discovery, 613–662 (2012)
3. Vandic, D., Van Dam, J.-W., Frasincar, F.: Faceted product search powered by the Semantic Web. Decision Support Systems, 425–437 (2012
4. Barry, C., Conboy, K., Lang, M., Wojtkowski, G., Wojtkowski, W.: Information Systems Development. Springer Editorial (2009) ISBN 978-0-387-68772-8
5. Gruber, T.R.: A translation approach to portable ontology specifications. Knowledge Acquisition, 199–220 (1993)
6. Zhang, Y., Vasconcelos, W., Sleeman, D.: Ontosearch: An ontology search engine. In: Research and Development in Intelligent Systems XXI, pp. 58–69. Springer, London (2005)
7. Pan, J.Z.: Edward Thomas, and Derek Sleeman: Ontosearch2: Searching and querying web ontologies. In: Proc. of WWW/Internet 2006, pp. 211–218 (2006)
8. d'Aquin, M., et al.: WATSON: A gateway for the semantic web (2007)
9. Jung, C.-T., Sun, C.-H., Yuan, M.: An ontology-enabled framework for a geospatial problem-solving environment. Computers, Environment and Urban Systems 38, 45–57 (2013)
10. Heydari, N., et al.: Ontology-based GIS web service for increasing semantic interoperability among organizations involving drilling in city of Tehran. In: 11th GSDI World Conference (GSDI 11), Spatial Data Infrastructure Convergence: Building SDI Bridges to Address Global Challenges, POSTER FORUM, Rotterdam, The Netherlands (2009)
11. Qinghai, C., Aiji, W., Chi, Z.: Design and Implement of East China Sea Economic Fisheries Resources Forewarning System. In: Wu, Y. (ed.) Software Engineering and Knowledge Engineering. AISC, vol. 114, pp. 1–10. Springer, Heidelberg (2012)
12. Priego, S., Pineda, H.: Geontomex: una ontología espacial de México para la desambiguación de topónimos. Research in Computing Science, Avances en Inteligencia Artificial 55, 103–112 (2012)
13. Tzitzikas, Y.: Faceted Taxonomy-Based Sources. In: Dynamic Taxonomies and Faceted Search, pp. 19–34. Springer, Heidelberg (2009)

14. Priego, B.: Metodología para la Resolución de Topónimos basada en Ontologías con Razonamiento Espacial. M.S. Thesis. BUAP-FCC (2012)
15. Parmar, S., Jain, S.K., Kaur, G., Kumar, A.: Ontology Construction for Polycystic Kidney Disease. In: Krishna, P.V., Babu, M.R., Ariwa, E., et al. (eds.) ObCom 2011, Part I. CCIS, vol. 269, pp. 245–254. Springer, Heidelberg (2012)
16. Bravo, M., Rodriguez, J., Pascual, J.: SDWS - Semantic Description of WebServices. International Journal of Web Services Research 11(2) (2014)
17. Bravo, M.: Smiilarity Measures for Web Service Composition Models. International Journal on Web Service Computing. Marzo 2014 5(1) (2014)
18. Fernández-López, M., Gómez-Pérez, A., Juristo, N.: Methontology: From ontological art towards ontological engineering. In: Proceedings of the Ontological Engineering AAAI 1997 Spring Symposium Series (1997)
19. Sure, Y., Staab, S., Studer, R.: On-to-knowledge methodology (OTKM). In: Handbook on Ontologies, pp. 117–132. Springer, Heidelberg (2004)
20. Gómez-Pérez, A., Suárez-Figueroa, M.C.: Scenarios for building ontology networks within the NeOn methodology. In: Proceedings of the Fifth International Conference on Knowledge Capture. ACM (2009)

USSD Technology a Low Cost Asset in Complementing Public Health Workers' Work Processes

Munyaradzi Zhou[1,2], Marlien Herselman[3], and Alfred Coleman[1]

[1]School of Computing, College of Science, Engineering and Technology,
University of South Africa, Florida Campus, Johannesburg, South Africa
[2]Midlands State University, Department of Computer Science and Information Systems,
Gweru, Zimbabwe
[3]CSIR, Meraka, Pretoria, South Africa

Abstract. Lowering costs and easy access to health information is important to public healthcare workers (PHWs) and patients who are both offline and online to improve equitable access to healthcare information. Harnessing mobile health (mHealth) improves the quality of and access to healthcare services in low income countries in which residents are remotely dispersed and have limitations in accessing the internet. Unstructured Supplementary Service Data (USSD) technology has been used to perform mobile money transactions through Ecocash, Telecash and Onewallet services but there is little use of the technology for clinical data repositories (CDRs). USSD codes facility is a cross-platform mobile handset support solution which allows the health services providers and patients to interact almost anywhere and at anytime. MHealth implementation through platforms such as EconetHealth merely focuses on health tips and there is need for emphasis on linking CDRs with USSD technology.

Keywords: mHealth·USSD technology·CDRs·EHRs·PHWs.

1 Introduction

USSD technology is a real time session based messaging service between the cell phones and an application server in the network. The service is a cross-platform handset support facility (basic feature phones and smartphones). The USSD service had mainly been used for mobile money transactions in both Zimbabwean rural and town communities. The Ministry of Health and Child care (MOHCC) can exploit the competence of USSD codes by providing PHWs a menu driven, interactive EHR patient profile checking and SMS notifications. EHR data from the healthcare facility level, district, provincial and national level can be integrated into Clinical Data Repositories (CDRs) for healthcare information access and decision making for real-time patient care [2]. Querying from the CDRs results in efficient access to data and informed patient care such as diagnosis and prescription practices.

Health applications need to be developed around USSD technology since it is not internet dependent. The internet has limitations such as poor network infrastructure,

F. Ortuño and I. Rojas (Eds.): IWBBIO 2015, Part II, LNCS 9044, pp. 57–64, 2015.

costly and limited access to other users. Even though, relatively low cost android smartphones such as G-Tide and Huawei are flourishing on the market the majority populous do not afford to connect to the internet 24/7. Registered pregnant women at clinics/hospitals must be registered to receive pregnancy related information and care from Ministry of Health and Child Care services (MOHCC). An Antenatal, perinatal and postpartum care platform using USSD technology is proposed which addresses limitations in information access to healthcare services (during the day, night or even holidays) and lack of information sharing (coordination and integration) among healthcare facilities. The platform is an integration of education and awareness programs, the helpline facility, and the remote data collection program.

2   USSD Technology and eHealth Projects in Zimbabwe

MHealth is a branch of eHealth (mainly supported by EHealth services/systems) which focuses on delivery of healthcare services using mobile devices such as cell phones and Personal Digital Assistants (PDAs), laptops and so on [15]. USSD communication protocol allows sending of USSD text between a mobile phone and programs running on the network. The USSD gateway application programming interface (API) allows easy integration of SMS services/functionality. USSD transactions can be initiated by either the network or the subscriber. USSD centre is totally open and can be integrated with any telecommunications system/device and the internet. The hub supports USSD codes execution from the home network at no cost during roaming.

Mobile penetration in Zimbabwe is approximately 60% which is more than half of the entire population [10]. The proliferation of mobile phones facilitates easy deployment of mHealth applications. Smartphone penetration stands at 15% [11] which implies that for the entire population; at most this proportion can access the internet using mobile phones [13].

Transactions using USSD codes are in the format *3digit# where the asterisk marks the beginning of the format string and the hash symbol marks the end. The format can have more than one asterisk marking the query levels to be performed. USSD software solutions have been developed for many applications such as mobile money and Facebook. A mobile money transaction using EcoCash for balance enquiry for example is in the form *151*200# and is then followed by menu commands until the user gets a transaction confirmation message using SMS. Facebook for USSD is accessed using the code *325# or *fbk# and allows the user to interact as if she/he is on the web. Interactive menu-based sessions can easily be navigated since almost all mobile phone users use USSD codes for airtime recharge and balance enquiry and mobile banking and, of course, the population is highly literate (approximately 92%). USSD platforms are cost-efficient to both the customer (no data charges applicable and can be used on basic feature phones) and mobile operators (no need for investment since the facility is supported by GSM networks), and is secure (requires login authentication).

Clients using the platform can freely access it or will be charged relatively low fees using methods such as pay-per-view. Since USSD codes can be accessed where there is a mobile operator network (without the internet), it can be accessed anytime and anywhere [8, 9]. The average mobile operator network coverage of the populated areas is above 86.67% with Telecel Zimbabwe having about 85%, Econet Wireless Zimbabwe above 90% and NetOne has network coverage above 85%. Network expansion programs can be rolled out to the unserviced communities and this can be facilitated by partnerships of mobile operators (being governed by Postal and Telecommunications Regulatory Authority of Zimbabwe (POTRAZ)) in sharing base station equipment [12].

In June 2014 the MOHCC officially launched the District Information Software Version 2 (DHIS-2) and the mHealth solution using mobile phones to report on HIV results for pregnant women in promotion of the Prevention of Mother to Child Transmission (PMTCT) program. The DHIS-2 supports integration of different existing information systems such as Tuberculosis (TB) and village health worker information systems. The SMS based mHealth solution for weekly disease surveillance has been rolled out in more than 75% of the nation's health facilities. DHIS-2 has been rolled out in 83 government owned health facilities including district, provincial and central hospitals. It supports collection, management and analysis of patient data. It also supports SMS based solutions such as capturing patient information, broadcasting SMSs to patients and Healthcare workers [6, 3].

DHIS-2's potential can be unlocked by integrating the platform with USSD technology by allowing healthcare personnel to register in their respective healthcare centres and to allow them to send information directly to the DHIS-2. Information in the DHIS-2 will be cleaned and analysed and then transmitted to the Clinical Data Repository which will be accessed through an interface of USSD codes. CDRs provide a cross-sectional view of patient information (from a variety of sources) which is easily accessible and facilitates pattern and trend analysis of patient information such as specific specimen results which results possibly in the improvement of illness diagnosis [1].

2.1   Mobile Money Transactions and mHealth Using USSD Technology

Mobile money transactions through Econet Wireless Zimbabwe's EcoCash service, Telecel Zimbabwe's TeleCash service and NetOne's Onewallet service are changing the landscape in which business transactions are carried out. The EcoCash mobile money platform offers services to receive and send money as well as pay bills using USSD codes. There are several products and services offered including buying airtime and data bundles, making payments to registered billers or merchants such as school fees, checking balance enquiry, banking services including linking your Eco-Cash account with the subscriber's bank if registered. The bank to wallet service allows a client to transfer money from the EcoCash wallet to the bank and vice-versa, check the account statement and balance enquiry on the bank account [4].

EconetHealth platform is an educational platform which focuses on general wellness health tips, disease outbreak alerts, women's health, and chronic diseases management such as diabetes through broadcasting SMSs (Bulk SMS) to its mobile subscribers. The prenatal and perinatal health tips are in the form of video tips on

YouTube. Health tips accessed via USSD technology are categorised into six, namely live well, and lose weight, pregnancy, healthy woman, stress and diabetes. Clients subscribe to access information on a specific topic for a daily fee of $0.05 which is 50% more costly than USSD sessions which can even be accessed at no cost [5]. MOHCC services must apply USSD code facility which it uses in integrating various health DSS applications. Deploying a USSD service typically requires collaboration with mobile network operators. The MOHCC must implement a costing model so that patients will access the platform at no cost or at a relatively low cost especially for interactive USSD menu-based sessions. The costing model should use optimal parameters in billing such as per subscriber request, per time interval and per session.

3   Method

A systematic review of literature was conducted concerning USSD technology, mobile subscriber penetration rate, mobile operator network coverage density, and perinatal care system PHWs' work processes. Content analysis was used to explore the cost and accessibility of USSD technology.

4   Discussion

PHWs' work processes can be improved through implementation of mHealth applications which support education and awareness, self-help USSD support portal facility, remote data collection and access. The prenatal and perinatal care system has been analysed and clinical processes which can be accomplished with USSD technology have been identified. Healthcare workers or the MOHCC can broadcast SMSs to all patients or specific patients across the mobile operator networks, thus targeting remote areas with few PHWs and have limited access to healthcare services. Healthcare system promotional programs including awareness and education are critical in transformation of the industry. EconetHealth tips already supports general perinatal care system health tips hence there is need to develop education and awareness tips which are patient-centric. Patients might be reminded about their pending visits thus promoting or encouraging a minimum of 4 ANC visits for pregnant women. In addition, they can be reminded about their drugs collection dates and their routine drug uptake. The postnatal care givers will be reminded about pending home visits to the mother and infant. Healthcare workers' sensitization to adhere to medical guidelines and protocols for maternal healthcare services can be done using SMS.

Helpline facilities are crucial as they aid patient self-service care, for example short codes are used to invoke attention from the healthcare workers or responsible authorities such as call back services or linking the patient to other healthcare services. In addition hotline facilities or dial up numbers for automated responses can be used which directs the user to appropriate assistance. Patients can easily be referred to other healthcare facilities since the platform must be an integration of private, public and faith-based healthcare facilities. PHWs can access specialist services such as remote diagnosis of a specimen which results in time and costs reduction. Mobile phones support geolocation services where the patients seeking care can easily locate the nearest healthcare facility using a USSD application linked to the geolocation finder. EconetHealth already has a facility to talk to trained physicians using the Dial-a-Doctor service.

MHealth supports remote data acquisition which can be in the form of a USSD interactive menu filling in questionnaires/survey (with validation checks) on related trimester information thus providing an effective facility to gather and transfer data quickly at no cost or at relatively low cost. This also bridges the gap for some patients who might not be able to visit their local clinics or any other health facility due to transport limitations (barriers), illness, etc. PHWs can easily collect data during home visits of mothers and infants during the postpartum period and during the first trimester period such as information relating to screening of racially specific conditions and identifying pregnant women's cultural beliefs and history.

The CDR will be developed from a data mart point of view since it focuses on pregnant women. Pregnant women who visit any healthcare facility should be registered to access the USSD technology platform irrespective of the mobile operator they're using. Conglomeration of all mobile network providers (Econet Wireless Zimbabwe, Telecel Zimbabwe and NetOne) for standardisation in accessing the service is essential. The technology is ideal since approximately 15% of the 60% of the entire population (who own a cell phone) which is less than one million residents) own a smartphone, thus the cross-platform facility bridges the gap for the 85% feature phone users [7]. Figure 1 overleaf describes the interaction of an EHR platform which is an eHealth platform incorporating a mHealth platform (USSD technology). The CDR retrieves information from the EHR which is optimised and then synchronised using an interface so as to connect to the USSD platform.

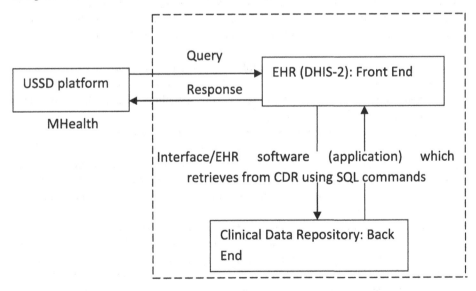

EHealth

Fig. 1. Clinical Data Repository synchronisation with USSD technology

The USSD platform which is accessed by the user interacts with the EHR (front-end using short codes) which sends a query message to the CDR (back-end) and upon retrieval of response, the response is sent back to the EHR using an interface and finally the front-end sends back the response to the user (USSD platform).

The annual growth in mobile penetration facilitates the adoption of mHealth initiatives by the government and private sector. The level of mHealth investment is moderate as the government with the aid of donors deployed the DHIS-2 solution from the DHIS-1 and the Weekly Disease Surveillance System (WDSS) and other platforms. Operational costs incurred while using WDSS such as sending SMSs is being funded by The Global Fund [14]. The limitation of DHIS-2 of being accessed over the internet is eliminated since real-time access to the network server is done using USSD codes. The shortage of desktop computers and laptops in the wards and erratic power cuts is eased since nurses and other healthcare personnel can easily access patients' information using their mobile handsets. Accounts creation for the platform must be decentralised to the healthcare facilities /centres with authorisation being done at national or provincial level depending on the duty station of the PHW.

5   Conclusions

POTRAZ has deployed a consultation paper to come up with legislation on base stations infrastructure sharing and must now focus on rolling out the program so that there will be mobile network coverage saturation. The USSD platform accesses information from the CDRs using the mHealth application interface for USSD codes. EHRs from different healthcare facilities must be centralised to the national EHR from which the CDRs are developed. Public, private and faith based healthcare institutions, mHealth developers, mobile operators and the MOHCC and other stakeholders must have a partnership which focuses on how to develop a framework which supports engagement of all major stakeholders to target almost the entire population in delivering competitive mHealth services at low cost. The platform must be piloted in Provincial Hospitals on short term basis while benchmarking performance and identifying opportunities for scaling up the project on long-term basis at national level. The framework must focus on policies such as subsidisation and exchanges e.g. tax incentives on mobile operators to reduce consumers' costs in accessing the USSD technology. In an attempt to reduce costs donor funding from Global Fund, UNDP and other donors must be solicited. The standardisation of USSD programming codes of the mHealth platform is important for easy integration of applications across networks and to the system users.

The parallel run duration of using the DHIS-2 platform with hard-copies needs to be trimmed through intervention strategies which facilitate early adoption of ICTs such as training healthcare personnel or giving incentives to those who fully implement the platform since it is a pre-requisite for the scaling of CDRs. PHWs will no longer capture patient demographics information during admissions or request for patient referral details from other facilities which results in less time in managing each patient case and executing tasks. The use of multiple registers is eliminated thus further reducing paper work as it will now be a matter of using USSD string to access the necessary information. Barriers and opportunities after implementation of the technology must be evaluated and addressed accordingly. Legislation, policy and compliance standards must be setup for uniformity in mHealth.

Android applications can act as a 'website' to access the EHRs and CDRs and these applications can be categorised as online and offline applications. Offline applications are considered essential since they are partly internet dependent, PHWs can accomplish their tasks even while offline and the application automatically synchronises with EHR whenever they connect to the internet. USSD technology integration with offline android applications will be essential since these applications will fill the gap of limited information access through USSD technology. In a nutshell, the use of USSD technology must result in improved productivity and efficiency in PHWs' work processes rather than the opposite and the limitations of computational power and battery power challenges are reduced which is commonly characterised with the use of installed applications and connecting to the internet. Offline android applications integration with USSD technology would be ideal to fully utilise the CDRs.

References

1. Ball, M.J., Douglas, J.V.: Performance Improvement Through Information Management. Springer Science & Business Media, p. 204 (1999)
2. Carter, J.H.: Electronic Health Records: A Guide for Clinicians and Administrators, p. 530. ACP Press (2008)
3. DHIS2, Collect, Manage, Visualize and Explore your Data | DHIS 2 (2015), https://www.dhis2.org/ (accessed January 21, 2015)
4. Econet Wireless Zimbabwe, EcoCash | Products and Services (2014a), https://www.econet.co.zw/ecocash/ecocash-products-and-services (accessed January 19, 2015)
5. Econet Wireless Zimbabwe, Econet Health Tips (2014b), https://www.econet.co.zw/econethealth/ (accessed January 13, 2015)
6. ICTedge, Zimbabwe Launches New Health Information System (HIS) | ICTedge (2014), http://www.ictedge.org/node/975 (accessed January 19, 2015)
7. IndexMundi, Zimbabwe Demographics Profile (2014), http://www.indexmundi.com/zimbabwe/demographics_profile.html (accessed January 16, 2015)
8. Information Resources Management Association (IRMA), Banking, Finance, and Accounting: Concepts, Methodologies, Tools, and Applications. IGI Global, p. 1593 (2014)
9. Ledgerwood, J., Earne, J., Nelson, C.: The New Microfinance Handbook: A Financial Market System Perspective, p. 530. World Bank Publications (2013)
10. TechZim, 60% of the Zimbabwean population connected on mobile: POTRAZ - Techzim (2014a), http://www.techzim.co.zw/2014/11/60-zimbabwean-population-connected-mobile-potraz/ (accessed January 13, 2015)
11. TechZim, Zimbabwe Internet: subscriptions 5.2 million, penetration 40% (2014b), http://www.techzim.co.zw/2014/01/zimbabwe-internet-statistics-5-2-million-subscriptions-40-penetration/ (accessed January 13, 2015)

12. The Financial Gazette, Telecel network coverage 85pc | The Financial Gazette – Zimbabwe News (2014a), `http://www.financialgazette.co.zw/telecel-network-coverage-85pc/` (accessed January 20, 2015)

13. The Financial Gazette, Zim smartphone penetration now over 15%, but what exactly is a smartphone? | The Financial Gazette – Zimbabwe News (2014b), `http://www.financialgazette.co.zw/zim-smartphone-penetration-now-over-15-but-what-exactly-is-a-smartphone/` (accessed January 13, 2015)

14. UNDP, Newsletter of The Global Fund Programme in Zimbabwe (2014)

15. Yañez, J.R.R.: Wikibook of Health Informatics. PediaPress (2011)

Energy Efficiency Study of Representative Microcontrollers for Wearable Electronics

Gregor Rebel, Francisco J. Estevez, and Peter Gloesekoetter

gregor@fh-muenster.de, {fjestevez,pgloesek}@ieee.org

Abstract. Wearable electronics are gaining commercial relevance nowadays. Applications like smartwatches and personal fitness trackers are already commercially available. Unfortunately, the operating time of most current devices is limited by small battery capacities. Many devices require to be charged every day. Human bodies can generate egnough power to operate electronic devices practically indefinitely. The challenge is to harvest this power and to use it as efficient as possible.

This paper bases on our previous paper [1] and our studies of energy harvesting on human bodies. It compares the amount of electrical energy being available from human body energy harvesting with the energy efficiency of certain representative microcontrollers. The paper gives a guideline to choose the best microcontroller for different mount points on human bodies according to their available, continuous power budget.

1 Introduction

Electronic devices being operated in or at the human body have to be lightweight and small to be accepted by the user. Powering wearable electronics directly from the human body can increase their usability. But harvesting energy from the human body itself is much more complex than using rechargeable batteries. This paper gives an overview of the state of the art in energy harvesting from human bodies. It lists different energy sources and their amount of available power. For each of these energy sources, some existing harvesting solutions are presented. For each harvesting solution, the amount of available power and its variation over time will be discussed.

Microcontrollers are well suited for wearable electronics with their limited built space and energy consumption. Their architecture is well tested and software development tools are widely available. The existence of prototype boards allows to start software development for a new product on the first day of the development phase. Certain microcontrollers can even be bought as soft IP and used as part of application specific circuits (ASICs) to built smaller and more energy efficient devices.

However, the task of finding the appropriate microcontroller for the corresponding design specifications is still difficult and time consuming. The datasheets provided by microcontroller manufacturers only show the amount of power being consumed at different operating voltages and clock frequencies. For a fair comparison, the processing power of each microcontroller architecture has to be taken

F. Ortuño and I. Rojas (Eds.): IWBBIO 2015, Part II, LNCS 9044, pp. 65–76, 2015.

into the equation. Different microcontrollers can draw the same power but provide different compute speeds. A microcontroller that gets the job done faster can be put into sleep mode earlier. So a fast architecture can have a lower average energy consumption than an architecture with a lower power draw, for a certain compute load. Putting a device to sleep for most of the time is a widely used technique to reduce the average power draw of ultra low energy applications.

Our selection of microcontrollers ranges from well known 8-bit to 32-bit architectures. In addition to our previous papers, we also added the newer STM32L0xx and STM32L1xx types. The list of tested devices is sorted by architecture and by name.

- 8-Bit Architectures:
  - Z8 Encore! XP F0822 from Zilog [2]
  - PIC18F46J50 from Microchip [3]
  - C8051F912 from Silicon Labs [4] with an 8051 core,
- 16-Bit Architectures:
  - MSP430G2553 from Texas Instruments [5] (16-bit RISC architecture),
- 32-Bit Architectures:
  - EFM32G210F128 from Energy micro, Silicon Labs [7] (ARM Cortex-M3),
  - STM32L100RCT6 from ST Microelectronics (ARM Cortex-M0),
  - STM32L053R8 from ST Microelectronics (ARM Cortex-M0).
  - STM32F051R8 from ST Microelectronics (ARM Cortex-M0),
  - STM32F107VCT6 from ST Microelectronics [6] (ARM Cortex-M3),
  - STM32F429ZIT6 from ST Microelectronics (ARM Cortex-M4)
  - STM32F302R8T6 from ST Microelectronics (ARM Cortex-M4).
  - STM32W108CB from ST Microelectronics (ARM Cortex-M3),

The aforementioned microcontrollers are compared by use of a common benchmark suite. The comparison is done in terms of power consumption, execution time and energy consumption from the microcontroller, excluding the board and the possible peripherals from this comparison. The benchmark suite has been choosen to represent different typical compute scenarios.

The rest of the paper is organized as follows. Section 2 presents an overview about eHealth applications and a state of art from energy harvesting on human body. In Section 3 we present and describe the benchmarks, their application on the microcontrollers and hardware test setup we adopted in order to measure the power and energy consumption of the microcontrollers, and in Section 4 we present and discuss the results of the measurements. Finally in Section 5 we conclude the paper.

Harvesting energy for electronic devices directly from the human body is a challenging task. The idea behind it is to provide electronics with a practically unlimited operational time. Wearable electronics could be put on and used like traditional clothes. Some of these devices still contain batteries as energy storage. And they might have to be recharged but in a greater interval than without using energy harvesting.

2   State of Art

The human body comprises quite a few potential power sources. Figure 1 shows an overview of the different available power sources which can be found on human bodies. In the below mentioned figure, brackets show the total amount of available power. The figure gives also the amount of harvestable energy by use of the harvesting technology.

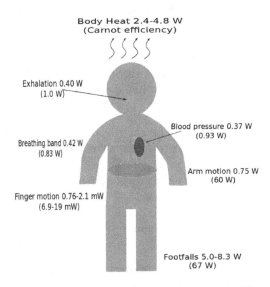

Fig. 1. Energy Sources on Human Body [9]

The amount of power available for energy harvesting must be much less than the total amount to avoid negative physical effects on the user [9]. Also, an up to date overview of human body applications, mainly eHealth applications, is presented.

2.1   Human Body Applications

Human Body applications are related to the interaction between human body and environment or the human body itself. They may use a Body Sensor Network (BSN) to implement a hierarchical communication structure. Smart Society as parameters measurement for contextual advertising or interaction with personal devices. On the other hand those Human Body applications can be related to the body itself, for example measuring or regulating internal parameters. This scope of application is known as eHealth, which is a very promising research area.

eHealth is a wide matter and some of the most interesting applications are:

- External monitoring
  These applications are based on analyze and transmit information from the human body. This information is being processed by external devices. A typical application are the physical activities trackers.
- Internal monitoring
  Internal monitoring applications are commonly part of another systems. Those systems measure different parameters (blood pressure, oxygen or glucose on blood, etc) and transmit this information to other external or internal systems as specifical medicine machines.
- Internal intervention
  An internal intervention makes reference to a small actuator located in the human body, in order to improve the performance of some human body functions. Some functions are the blood circulation, insulin distribution or hormone regulation.

One of the most important requisite for these applications is a sufficient hardware support, not only communications or sensors but also microcontrollers. Moreover, the power supply is another relevant study scope. The best approach to supply those devices is the energy directly scavenged from the human body, but there are some limitations and issues with those scavenging system, that makes this approach difficult to be developed. The next section gives an overview of the energy harvesting techniques on human bodies.

2.2   State of Art of Energy Harvesting on Human Bodies

Based on techniques and analysis of recent research results, this section gives a detailed overview of the different energy flows on human bodies and how to scavenge energy from them.

Several flows of energy can be found on the human body. Each flow is presented in a separate point focusing on the particular scavenging technique for it.

Chemical Energy. Food is the energy source of human bodies. It has nearly the same gravimetric energy density as gasoline and 100 times greater than batteries [10]. Human bodies store energy in fat cells distributed in various regions. Each gram of fat stores and equivalent of 37.7 kJ. An average person of 68 kg (150 lbs) with 15% body fat stores an approximate equivalent of 384 mega joule [9].

Energy from Liquid Fuels. Common liquid fuels are formic acid, ammonia or methanol. They show energy densities of 1.6, 5.2, and 5.6 kWh/kg. Liquid hydrocarbon fuels have gravimetric energy densities around 13 kWh/Kg. Typical electric generators based on these fuels provide a conversion efficiency of 25%. Electric energy generators powered from liquid fuel can provide energy densities of 3.25 kWh/Kg. For individual applications, the entire power system has to be evaluated including fuel, fuel storage delivery and power conversion. When complete power systems are taken into the equation, batteries remain a strong personal power option for many applications.

Mechanical Energy. Muscles in the human body convert food into mechanical work at efficiencies up to 25%. The usable mechanical output of human bodies can reach 100W for average persons and 200W for elite athletes [10]. Obviously only a small portion of this energy can be detoured to power an electric device without disturbing the human gait. For example, according to Starner and Paradiso, a maximum of 13W is available for energy harvesting for a 1 cm stroke [9].

Another kind of systems generate power when moved cyclically. Despesse et al. [11] research analyses a structure for electrostatic transduction with high electrical damping. This electrostatic transduction is designed to operate with low frequencies, typically less than 100 Hz. Other authors like Beeby et al [12] explain, for example, how an 18 cm^2 1cm volume device with a 0.104 kg inertial mass was electro discharge machined from tungsten and produced a scavenged power of 1052 μW for a vibration amplitude of 90 μm at 50 Hz. This represents a scavenged efficiency of 60% with the losses being accounted for by charge/discharge and transduction losses.

Thermal Energy. The human body produces waste heat in the range from 81W (sleeping) up to 1630 W (sprinting). Carnot efficiency limits the amount of power that can be harvested. Harvesting thermal energy always requires a temperature difference. The bigger the difference the more energy can be harvested. Assuming normal body temperature of 37°C, the Carnot efficiency is 5.5% for 20°C (difference of 17°K) and 3.2% for 27°C (difference of 10°K) room temperature. Using a Carnot heat engine to model the recoverable energy yields 3.7 – 6.4 W of power.

A design of a micro machined thermopile from 2007 shows output voltage of 13 mV/K/cm^2. Simulations showed an output power around 1.5 μW at 1V for a watch sized TEG placed on a human body [14].

3 Benchmarks on the Microcontrollers

A fair evaluation and comparison of different microcontrollers requires the existence of a common framework, a common set of tasks (benchmarks) to run on all of them. There are many published sets of benchmarks which are designed for testing microprocessors as well as the EEMBC [8] (Embedded Microprocessor Benchmark Consortium) benchmarks which aim at different kinds of application fields. For our work we focused on benchmarks designed for low power microcontrollers that are comercial available.

3.1 Benchmarks Overview

Instead of selecting exclusively one of these suites, we borrowed from each one the functionality most probably used in eHealth applications. Both of these benchmark suites have been tested on simulators before, but testing them on real microcontrollers presented us with one problem. The memory available on some of the microcontrollers is very limited compared to the needs of some of these benchmarks (such as the compression algorithms in the ImpBench). Therefore and due to the memory limitations, the set of benchmarks we assembled, consists of the following applications:

- Dijkstra: routing is a typical function needed in most networks.
- Mathematical operations: in order to stress the processor we include integer mathematical operations such as square root calculations.
- Checksum and CRC32: data integrity control is necessary in communications over unreliable connections.
- AES: security functionality is also vital in all communications.

3.2   Application of Benchmarks on the Microcontrollers

In order for the benchmarks to be applicable on the microcontrollers we made certain changes. Firstly in the case that a benchmark takes the input data from a file, we substituted the data file with a byte array stored in the ROM of the microcontroller. Secondly we simplified the benchmark functions where it was needed, in order to limit their memory needs. Table 1 presents the size and type of memories available on each microcontroller. From this table it becomes obvious that the 8-bit and 16-bit microcontrollers have very small memories which limit the programs that they can store and execute. With these changes we can fit the 4 benchmarks previously mentioned in any of the tested microcontrollers.

Table 1. All tested Microcontrollers

μC	Flash	RAM	Frequency Range	Voltage Range
Z8 Encore! XP F0822	8 KB	1 + 4 KB	5 - 18.4 MHz	1.8 - 3.6 V
PIC18F46J50	16 KB	768 bytes	0.032 - 48 MHz	2.0 - 3.6 V
C8051F912	16 KB	768 bytes	1.25 - 24.5 MHz	0.9 - 3.6 V
MSP430G2553	16 KB	512 bytes	0.125 - 16 MHz	1.8 - 3.6 V
EFM32G210F128	128 KB	16 KB	0.125 - 32 MHz	1.98 - 3.8 V
STM32F051R8	64 KB	8 KB	8 - 64 MHz	2.0 - 3.6 V
STM32F107VCT6	256 KB	64 KB	8 - 72 MHz	2.0 - 3.6 V
STM32F302R8T6	64 KB	16 KB	8 - 64 MHz	2.0 - 3.6 V
STM32F429ZIT6	2048 KB	256 KB	8 - 168 MHz	1.8 - 3.6 V
STM32L053R8	64 KB	8 KB	0.128 - 48 MHz	1.65 - 3.6 V
STM32L100RCT6	156 KB	16 KB	0.65 - 48 MHz	1.65 - 3.6 V
STM32W108CB	128 KB	8 KB	8 - 168 MHz	2.1 - 3.6 V

The maximum and minimum frequencies listed here are the limits of the prototype boards being available for the benchmark. Some boards (e.g. for STM32F302R8T6) did not provide an external crystal and were only tested with their internal RC oscillator.

3.3 Hardware Test Setup

In this section we describe the methodology we followed in order to measure the energy consumption of the microcontrollers. The test setup includes one RMS amperometer and one voltmeter, with the amperometer we measure the current that drives the microcontroller and with the voltmeter the operating voltage of the microcontrollers. According to the datasheets, each microcontroller can operate at a range of voltages. In order to get the minimum possible energy consumption measurements, we use the minimum operating voltage for each microcontroller.

Subsequently we use the measured current and voltage to calculate the power consumption. The energy consumption calculation is made according to Equation 1. The calculation of the Clock Cycles has been described in Section 3. The results of these measurements are presented in the following section.

$$Energy = Power \times Execution\ Time = Power \times Clock\ Cycles \times Clock\ Period \qquad (1)$$

4 Performance Analysis

In this section we present the results we acquired through the measurement procedure described in the previous section. For every microcontroller we show the results for a variety of operating frequencies and different frequency sources, where possible. The Z8 Encore! XP microcontroller worked correctly with external oscillators at frequency values varying from 5MHz to 18.4MHz. We tested the PIC with its internal oscillator (INTOSC) from 32kHz to 8MHz, and with the high speed oscillator (HS) from 8MHz to 48MHz. We operated the C8051 with its low power internal oscillator (INTOSC) from 1.25MHz to 20MHz, and with its programmable precision internal oscillator (LP) from 3.062kHz to 24.5MHz. We configured the MSP430 to operate with the internal digitally controlled oscillator from 125kHz to 16MHz. We operated the EFM32 in 2 modes: high frequency RC oscillator (HFRCO) from 7MHz to 28MHz and high frequency crystal oscillator (HFXO) from 2MHz to 32MHz. We tested the STM32F0, STM32F3, STM32L0 and STM32L1 operated with their internal oscillator, in a range from 65KHz to 48MHz, STM32F1 and STM32F4 with an external oscillator in the frequency range from 8MHz to 72MHz, and 8MHz to 168MHz, respectively. Finally the STM32W at the nominal frequency 12MHz of the High-frequency internal RC oscillator. The STM32F4 provides the widest frequency range of all tested microcontollers followed by the PIC.

The Figures 2 and 3 present the power and energy consumption respectively, on each of the microcontrollers for every frequency and mode of operation that we tested. The illustrated power consumption is the power that a microcontroller consumes while in active mode and it has been measured through the procedure described in Section 3.3. The illustrated energy consumption is the energy that the microcontrollers consume for a single execution of the benchmarks presented in Section 3.1 and it is calculated according to Equation 1 described previously. The x and y axes on both graphs are logarithmic for visual purposes.

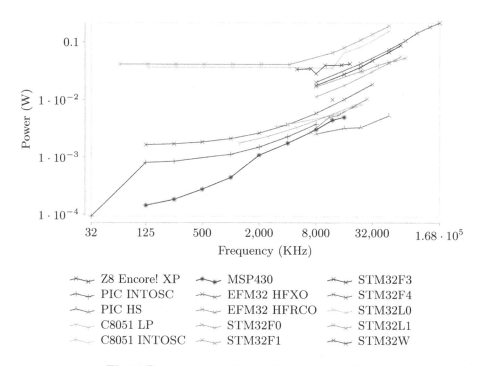

Fig. 2. Power consumption in the microcontrollers

As we see in Figure 2, the power consumption appears to increase almost linearly within this logarithmic representation.

Furthermore we notice that the 8- and 16-bit architectures have generally smaller power consumption compared to the 32-bit ones. The Z8 Encore! XP is a significant exception in this observation. It stands out to draw more power per megahertz than any other microcontroller in our test. This is astounding because its 8-bit architecture is expected to consume less than 32-bit architectures. The microcontrollers with lowest power consumption in our test prove to be the MSP430 and the PIC. These consume 143 μW at 125 kHz and 92.7 μW at 32 kHz respectively.

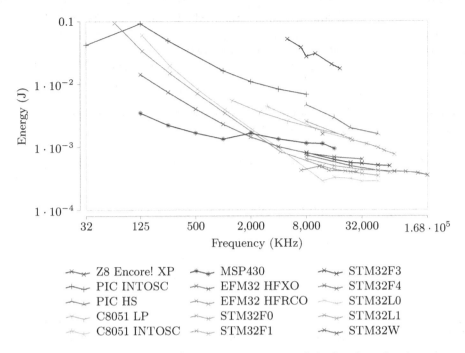

Fig. 3. Energy consumption for one execution of the benchmark suite

Figure 3 shows that the energy efficiency increases with the operating frequency for all tested architectures if we don't consider the idle time after the benchmark execution. So the gain of computing power is greater than the increase of power consumption. It is usual that the manufacturers of microcontrollers optimize each device for its maximum frequency, the graph reflects that fact. Moreover, the curves of some architectures (EFM32 HFXO, PIC INTOSC) show a saturation effect at higher frequencies. So the gain of computing performance decreases at higher frequencies. The reason may be internal limiting factors. We also observe that the 32-bit architectures are generally more energy efficient than the 8-bit ones. 32-bit microcontollers can compute more data per clock cycle than 8-bit derivatives. We can also observe that the Z8 Encore! XP is the least energy efficient microcontroller, followed by the PIC. Even though the PIC microcontroller shows an overall low power consumption, due to its very small instruction set. The old 8-bit instruction set is much slower in executing the benchmarks than its competitors. At low frequencies the PIC consumes even more energy than the Z8 Encore! XP. Less energy demanding are the C8051 and the two STM32L microcontrollers. The C8051 is much faster than the other 8-bit microcontrollers. This makes it the most energy efficient 8-bit microcontroller in our test. The STM32W microcontroller on the other hand is slower and more power hungry than any other 32-bit microcontroller in our test and thus less energy efficient. At the low frequencies the MSP430 microcontroller is quite energy efficient thanks to its remarkably low power consumption and relatively fast execution. This microcontroller is a good choice for applications that do not require the raw computing performance of 32 bit architectures. The STM32F series has been designed for high computing performance. At its maximum clock frequency the STM32F is even more energy efficient than the MSP430. The STM32L series has been designed for ultra low energy consumption and high computing performance. At its high frequency, both STM32L microcontrollers are the most energy efficient devices in our test. Moreover the STM32F0 and STM32F4 have a very good results. Each one provides the best efficiency in a different range of clock frequency. The STM32L0 rules the competition in the range from 12 to 48 MHz and the the STM32F0 is best in class from 48 to 72 MHz. If even higher computing power is required, the STM32F4 can provide it at the same efficiency as the other two.

Figure 3 also shows some surprising behaviours. The PIC INTOSC, Z8 Encore! XP, MSP430 and EFM32 HFRCO curves show peaks where the microcontrollers are either less power efficient or more power efficient than expected from the general trend. For example the PIC INTOSC at 125 kHz becomes much more energy hungry than for all other frequencies. The Z8 Encore! XP shows the opposite behaviour as at 8 MHz it is more energy efficient than at 7 MHz and 10 MHz. The MSP430 shows a declining energy consumption but at 2 MHz the energy consumption rises again. The reasons may be due to efficiency variations of internal DC-DC converters. The STM32F and STM32L families show a typical declining curve that gets more even for higher frequencies. These architectures provide their highest computing efficiency at highest operating frequency. The STM32L1 is more efficient at higher frequencies than the STM32F4. For frequencies lower than 4 Mhz, the STM32F4 is better. The most efficient microcontroller from ST STM32L0 is also the best of all tested microcontrollers. Its optimal operating frequency was found at 12. The EFM32 HFRCO shows an almost constant energy consumption for all the frequency values that we tested it. It is only beaten by the STM32L0 at frequencies above or equal 12 MHz.

Table 2 compares the amount of benchmark runs that can be computed by the different microcontrollers being powered from energy harvesting on the human body. Each column shows the estimation for a certain mount position. The variations come from different human activities as shown by [13]. Each microcontroller is assumed to run at its most efficient frequency. The more power being available and the more efficient a microcontroller computes, the more benchmarks can be run per hour (B/h). This table can be used as a decision matrix to choose the right microcontroller for a certain application. Some human energy sources provide enough power to operate all microcontrollers, while other require more efficient and eventually more expensive microcontrollers. The STM32F4 and STM32L1 share the same power efficiency, though the F4 can compute a result much faster. This can be an application requirement.

Table 2. Estimation of Benchmark execution time (Benchmark per hour or B/h) for the selected microcontrollers

μC	Clock	Hip	Shank	Foot Instep	Wrist
	(kHz)	1.40 - 22.89 μW	0.02 - 28.74 μW	0.39 - 28.44 μW	0.04 - 3.08 μW
Z8 Encore!	18432	0.29 - 5 B/h	0.1 - 6 B/h	0.1 - 6 B/h	0.1 - 0.6 B/h
PIC18F	48000	3 - 52 B/h	0.1 - 65 B/h	0.9 - 64 B/h	0.1 - 7 B/h
C8051	24500	4 - 64 B/h	0.1 - 80 B/h	1 - 79 B/h	0.1 - 9 B/h
MSP430	16000	5 - 88 B/h	0.1 - 110 B/h	1 - 109 B/h	0.2 - 12 B/h
EFM32	28000	13 - 205 B/h	0.2 - 258 B/h	3 - 254 B/h	0.4 - 28 B/h
STM32F0	64000	13 - 2210 B/h	0.2 - 263 B/h	4 - 260.34 B/h	0.4 - 28 B/h
STM32F1	72000	7 - 108 B/h	0.1 - 136 B/h	2 - 134.18 B/h	0.2 - 15 B/h
STM32F3	64000	10 - 164 B/h	0.1 - 207 B/h	3 - 204 B/h	0.3 - 22 B/h
STM32F4	168000	14 - 233 B/h	0.2 - 293 B/h	4 - 290 B/h	0.4 - 31 B/h
STM32L0	32000	18 - 291 B/h	0.3 - 365 B/h	5 - 361 B/h	0.5 - 40 B/h
STM32L1	48000	14 - 237 B/h	0.2 - 298 B/h	4 - 295 B/h	0.4 - 32 B/h
STM32W1	12000	3 - 51 B/h	0.1 - 65 B/h	0.9 - 64 B/h	0.1 - 7 B/h

5   Conclusion

Powering electronics via energy harvesting from the human body means to save energy wherever possible. Selecting the right microcontroller for an embedded system is the key to build a suitable solution. Lower frequency and smaller busses mean less power draw. But the computing power shrinks too. If an 8- or 16-bit microcontroller does not fit the computing requirements, 32-bit architectures may provide it. Most of these high performance microcontrollers are most computing efficient at their highest operating frequency. At low clock frequencies, they provide less computing power than the smaller architectures. For a high performance human body application this means to put the microcontroller to sleep as often as possible. Every energy saving feature must be used. A software developer must implement a more complex power state control than for older 8- and 16- bit architectures. The benefit are smaller reaction times due to the much higher compute performance of modern 32-bit architectures.

Our tests showed that the PIC (8-bit) and the MSP430 (16-bit) consume least power. Unlike the power consumption, the energy consumption per benchmark run decreases as the operating frequency increases. At low frequencies the MSP430 shows a higher energy efficiency than STM32F1 and STM32W1. This makes it a good alternative if its low computing power is sufficient for the intended application. However the STM32L0 seems to be currently the best choice for bodily devices. It is a good combination of high computing power, high computing efficiency and low power consumption. The EFM32 comes very close to the STM32L0 and is more suitable at lower frequencies.

References

1. Tsekoura, I., Rebel, G., Gloesekoetter, P., Berekovic, M.: An evaluation of energy efficient microcontrollers. In: ReCoSoC (2014)
2. Zilog, http://www.zilog.com/
3. Microchip, http://www.microchip.com/
4. Silicon Labs, http://www.silabs.com/
5. Texas Instruments, http://www.ti.com/
6. ST Microelectronics, http://www.st.com/
7. Energy micro, Silicon Labs, http://www.energymicro.com/
8. Embedded Microprocessor Benchmark Consortium, http://www.eembc.org/index.php
9. Starner, T., Paradiso, J.A.: Human Generated Power for Mobile Electronics GVU Center, College of Computing Responsive Environments Group, Media Laboratory Georgia Tech MIT Atlanta, Responsive Environments Group, Media Laboratory MIT Cambridge (2004), http://resenv.media.mit.edu/pubs/books/Starner-Paradiso-CRC.1.452.pdf
10. Maxwell, J., Naing, V., Li, Q.: Biomechanical Energy Harvesting Locomotion Lab. Simon Fraser University, Burnaby, BC, V5A 1S6, Canada (2009)
11. Despesse, G., Jager, T., Chaillout, J., Leger, J., Vassilev, A., Basrour, S., Chalot, B.: Fabrication and characterisation of high damping electrostatic micro devices for vibration energy scavenging. In: Proc. Design, Test, Integration and Packaging of MEMS and MOEMS, pp. 386–390 (2005)
12. Beeby, S. P., Tudor, M. J., and White, N. M.: Energy harvesting vibration sources for microsystems applications Meas. Sci. Technol., 17, R175–R195 (2006)
13. Pollak, M., Mateu, L., Spies, P.: Step-Up Dc-Dc-Converter With Coupled Inductors For Low Input Voltages Thermogenera- TORS. Power Efficient Systems Department, Fraunhofer IIS, Nuremberg
14. Wang, Z., Leonov, V., Fiorini, P., Van Hoof, C.: Micromachined Thermopiles for Energy Scavenging on Human Body. In: Transducers & Eurosensors, B-3000, Leuven, Belgium. Katholieke Universiteit Leuven (2007)

Analysis of Inter-rater Reliability of the Mammography Assessment after Image Processing

Kemal Turhan[1], Burçin Kurt[1], Sibel Kul[2], and Aslı Yazağan[3]

[1] Department of Biostatistics and Medical Informatics,
Karadeniz Technical University, Trabzon, Turkey
[2] Department of Radiology, Karadeniz Technical University, Trabzon, Turkey
[3] Department of Computer Technologies, Recep Tayyip Erdoğan University, Rize, Turkey
{kturhan.tr,burcinnkurt,ayazagan}@gmail.com, sibel_ozy@yahoo.com

Abstract. The aim of this study is to assess whether image processing causes information lost on mammography images. In the study, 50 mammogram from MIAS database (open database of mammograms) are selected: 20 images include mass, 20 images include calcification and 10 normal images. Selected images are read and marked by radiologists. The same radiologists read the enhanced version of images after three months later. In order to assess the consistency, inter-rater reliability statistics are used. Results indicate that image processing on mammography images especially images without calcification does not affect radiologists' evaluation consistency. Also, it is indicated that images including calcifications reduce evaluation consistency of the radiologists and it is decided to use other image processing methods for images with calcifications.

Keywords: image processing, mammography, radiology, interpretation, consistency, reliability.

1 Introduction

Breast density is an important and widely accepted risk factor for breast cancer and interpreting mammography image is significantly important for diagnosis [1]. However, radiographic interpretations often rely on some degree of subjective interpretation by observers [14]. There are inter and intra-observer variability exists in any measurements made on medical images [13].

In the literature, it is widely used to compare assessments of radiologists with and without image processing of medical images using different image processing techniques [2,3,4,5]. Krupinski et al. [2] measured the diagnostic performance in reading of computed chest radiographs with and without the use of image processing by means of receiving operating characteristic analysis and found that the effect of image processing does not greatly influence diagnostic performance in chest radiography. Lund et al [6] investigated reader performance and image quality on standard film, computer film and computer monitor radiography viewing formats in the evaluation of skeletal extremity trauma. ROC is used to evaluate the observers' performance.

F. Ortuño and I. Rojas (eds.): IWBBIO 2015, Part II, LNCS 9044, pp. 77–84, 2015.

Feng et al. [3] evaluated radiologists' ability to detect subtle nodules by the use of standard chest radiographs alone and compared with bone suppression imaging used together with standard radiographs. ROC curves, with and without localization, were obtained for the observers' performance. Rockette et al. [4] conducted a multireader observer-performance study on poster anterior chest images. Images are rated by three observers as to the difficulty of determining the presence or absence of each abnormality when the true diagnosis was known and when it was not known. Zanca et al. [7] compared two methods for assessment of image-processing algorithms in digital mammography. Four radiologists assessed the mammograms for the detection of microcalcifications. After 8 months, the same images were interpreted by the same radiologists. ROC analysis is used to measure the ability of an observer to detect and correctly interpret microcalcifications in mammograms.

There are also studies to analyze inter-observer agreement in the interpretation of medical images [8, 9, 16, 17].

In Chen's study [8], two different images were evaluated by two radiologists. Agreement on images was analyzed using intra-class correlation coefficient statistics and inter-observer variability between the two radiologist evaluators was evaluated using kappa statistics. In Inah's [9] study, radiographs were subjectively evaluated by two radiographers and radiologists and coefficient of variability was used to check intra-reader consistency while agreement between the raters was determined by Cohen Kappa statistic. Skaane et al. [16] analyzed inter-observer agreement in the interpretation of palpable noncalcified breast masses by means of mammography. The images were analyzed independently and without knowledge of the final diagnosis. The inter-observer variation was analyzed by means of observed agreement, kappa statistics. Hopstaken et al. [17] assessed inter-observer variation in the interpretation of chest radiographs of individuals with pneumonia versus those without pneumonia and used Kappa statistics.

This is a preliminary work of an ongoing project to develop a decision support system for mammography results (MDSS). MDSS has three basic steps: image enhancement, mass detection and classification of detected mass as malign and benign. MDSS is expected to classify detected mass automatically so it is important not losing diagnostic information after image processing for system performance. The objective of this study is to test whether image enhancement step causes data loss that affects radiologists' decision. Inconsistency of radiologists' evaluation is considered as an indication of diagnostic information loss. Thus, radiologists' evaluation consistency is analyzed.

2   Materials and Methods

2.1   Collection of Mammography Images

The study is a part of a computer-aided breast cancer diagnosis system. In the study, database of the Mammographic Image Analysis Society (MIAS) was used. MIAS is the most easily accessed and the most commonly used database for digital mammography images [18]. Images in the database also include radiologist's truth-markings on the location of any abnormalities that may be present. In our study, 50 mammogram from MIAS database were selected: 20 images include tumors, 20 images include calcification and 10 normal images.

2.2   Evaluation of Images by Radiologists

We worked with 2 female and 3 male radiologists who have different working experience. Selected images were screened and marked by the radiologists before and after image enhancement process. Image sets were read by radiologists after three months interval between two sessions in order to reduce the influence of learning effects on the results. In order to assess the consistency of radiologist's evaluation, inter-rater reliability statistics was used.

First assessment was conducted before the image enhancement process. Selected images were evaluated and marked on the scale 1-6 by five radiologists. The mark indicates the abnormality level of the suspicious region according to the radiologist's estimate of the probability of disease, or confidence level; e.g., $1 \equiv$ less that 2% probability of disease, $2 \equiv 3$ to 5%, probability of disease, $3 \equiv 6 - 30\%$ probability, $4 \equiv 31$ to 70%, $5 \equiv 71$ to 94% probability and $6 \equiv$ greater than 95% probability.

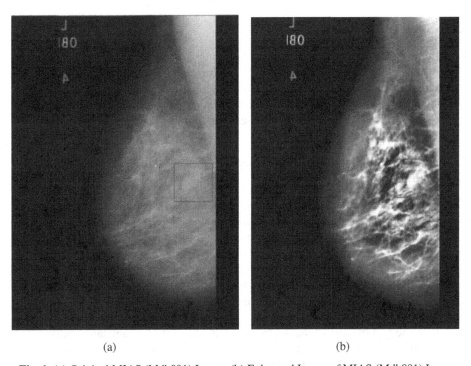

(a)   (b)

Fig. 1. (a) Original MIAS (Mdb091) Image, (b) Enhanced Image of MIAS (Mdb091) Image

For the second assessment, selected images were processed by image processing techniques such as Median Filtering, Wavelet Based Thresholding, CLAHE, Anizotopik Diffusion Filtering and they were presented to the same radiologists after three months for evaluation.

Figure 1 (a) shows original MIAS mammography image and Figure 1 (b) shows the enhanced image using the image enhancement techniques. In both images, originally labeled mass is marked in red square. It is clear that images are visually different. Evaluation results of the radiologists in two sessions are shown in Table 1 for these images. All 50 mammography images were evaluated by the same way.

Table 1. Radiologists' evaluation on a MIAS image (Mdb091)

Radiologists	PreProcessing Eval.	PostProcessing Eval.
1	mass	mass
2	normal	mass
3	normal	mass
4	normal	normal
5	mass	mass

2.3   Data Transformation

Evaluation results from all radiologists before and after image enhancement were combined for each image. Abnormality counts for each image were multiplied by the maximum probability of each marked label and then a cumulative probability value was obtained for each image. Before and after image enhancement, cumulative probability scores for each radiologist's decision were obtained and inter-class correlation coefficients were calculated.

Cumulative probability score = $\sum$ Abnormality counts of the label* maximum probability of the label

An example of evaluation table for one image is shown in Table 2. dr1_pre and dr1_post values represent the number of suspicious regions identified by one radiologist before and after machine learning process on mammography. Columns represent the degree of the probability of being malign of those regions. Cumulative probability score is calculated by the product of maximum degree of the probability of being malign of the region and the number of suspicious region. In the example, radiologist identified four abnormalities before the image enhancement; two of them had malignancy probability of 2%, one of them had maximum malignancy probability of 30% and other one had maximum malignancy probability of 94%. Cumulative probability score of this evaluation (ppre) was calculated as 1.28. In the same way, cumulative probability score after image enhancement (ppost) was calculated as 1.04.

Table 2. An Example Evaluation for one image

Scale	1	2	3	4	5	6
Probability	2%	3 to 5%	6 to 30%	31 to 70%	71 to 94%	>95%
dr1_pre	2		1		1	
dr1_post	2		1	1		

ppre = (0,02*2) + (0,30*1) + (0,94*1) = 1,28
ppost = (0,02*2) + (0,30*1) + (0,70*1) = 1,04

It is expected that these two cumulative probability scores, ppre and ppost, are as close as possible each other. The closeness of these two values indicates the consistency of the radiologist's decision on that image.

2.4   Observer Consistency

Categorical evaluation data were transformed continuous data by the calculation of cumulative probability score. For this reason, Cronbach's Alpha can be used to evaluate radiologist reading consistency [10]. There is no best way to choose one statistics over others because of normality, interval data or continuous data measurement type so that besides Cronbach's Alpha, Sperman Correlation, Pearson's R and Kappa statistics were also calculated to analyze variability.

3   Results

Consistency can be defined as to find the same result on repetitive occasions in different times. In this study, instead of assessing radiologists' diagnosis performance, the degree to which the same radiologists give consistent results on the same mammography images was assessed using Cronbach's Alpha and other statistics.

Presented values in Table 3 and Table 4 are designed to show consistency statistics of five radiologists. Between before and after readings of each tree radiologists had high consistency especially according to Cronbach's Alpha. However, the reading of the other two radiologists showed a moderate consistency. Inconsistency might depend on other factors such as experience, age, sex, psychological situation etc. [7]. This can be statistically analyzed as a future work.

Table 3. Consistency scores without calcification images

Radiologists	Cronbach's Alpha	Spearman Correlation	Pearson's R	Kappa
2	0.925	0.870	0.870	0.334
1	0.884	0.788	0.817	0.346
5	0.706	0.785	0.340	0.340
3	0.627	0.591	0.511	0.207
4	0.438	0.436	0.322	0.130

Table 3 shows the consistency statistics on evaluated images without calcification. According to tables, it can be concluded that mass detection performance consistency of radiologists was more consistent than calcification detection.

Table 4. Consistency scores of all Images

Radiologists	Cronbach's Alpha	Spearman Correlation	Pearson's R	Kappa
1	0,837	0,698	0,749	0,182
2	0,787	0,669	0,268	0,268
3	0,382	0,361	0,283	0,130
5	0,363	0,550	0,486	0,224
4	0,273	0,278	0,188	0,152

Our results indicate that image enhancement on mammography images especially images without calcification does not affect radiologists' evaluation consistency significantly. Also, it is indicated that calcification images reduces the agreement of each radiologists between before and after image enhancement. Thus, it is decided to use other image processing methods for calcification images [11].

4   Discussion

There are many studies to measure the diagnosis quality of radiologists after image processing but it is very difficult to infer a general judgment that image processing improves or reduces radiologists' diagnostic quality [2,3,4,5]. There are studies comparing image processing techniques e.g. Zanca et al. compared some algorithms in their study [7]. An image processing technique might give good results on some type of images as indication of diagnostic quality improvement of radiologists, but the same technique might not work on another type of image. For example, an image processing algorithm that gives satisfactory results on chest radiography might give poor result on mammogram image. Diagnostic performance studies depend on radiological problem domain and image processing techniques that used in the study. Thus, wide and long term projects should be planned for generic usage that helps radiologists' interpretation of images.

MDSS is expected to classify mammography abnormalities automatically so it is important not losing diagnostic information after image enhancement for accurate classification [11,12]. Findings support that image enhancement process on mammography images does not affect evaluation consistency of radiologists especially for images without calcification. These findings obtained by using image enhancements techniques mentioned in this study on mammography images. Similar findings were obtained using different algorithms and different radiological domains in other studies [16, 17].

5   Conclusion

This study focuses on the consistency of radiologists' evaluation instead of their diagnosis performance before and after image enhancement. Image enhancement techniques can create a quite significant visual difference between original and processed image. The main problem in this context is whether this visual difference hides any valuable information for the detection of abnormalities. Because, it is important not losing diagnostic information that might affect diagnostic decision of radiologists after image enhancement. Our findings emphasize that, for accurate diagnosis for automated classification systems, information should not be lost during image processing by doing this type of research.

Acknowledgement. The study is a part of a computer-aided breast cancer diagnosis system which is supported as a SANTEZ project by Republic of Turkey Science, Technology and Industry Ministry and AKGÜN Computer Programs and Services Industry Tic. Ltd. Şti.

References

1. Boyd, N.F., et al.: Breast tissue composition and susceptibility to breast cancer. PubMed, 1224-1237 (2010)
2. Krupinski, E.A., et al.: Influence of image processing on chest radiograph interpretation and decision changes. Academic Radiology 79–85 (1998)
3. Li, F., et al.: Improved Detection of Subtle Lung Nodules by Use of Chest Radiographs With Bone Suppression Imaging: Receiver Operating Characteristic Analysis With and Without Localization. Cardiopulmonary Imaging (2011)
4. Rockette, H.E., et al.: Selection of subtle cases for observer-performance studies: The importance of knowing the true diagnosis. Academic Radiology, 86–92 (1998)
5. Vicas, C., Lupsor, M., Socaciu, M., Nedevschi, S., Badea, R.: Influence of Expert-Dependent Variability over the Performance of Noninvasive Fibrosis Assessment in Patients with Chronic Hepatitis C byMeans of Texture Analysis. Computational and Mathematical Methods in Medicine (2012)
6. Lund, P.J., Krupinski, E.A., Pereles, S., Mockbee, B.: Comparison of conventional and computed radiography: Assessment of image quality and reader performance in skeletal extremity trauma. Academic Radiology, 570–576 (1997)
7. Zanca, F., et al.: Comparison of visual grading and free-response ROC analyses for assessment of image-processing algorithms in digital mammography.The British Journal of Radiology, 1233–1241 (2012)
8. Chen, C.-Y., et al.: Utility of the Iodine Overlay Technique and Virtual Nonenhanced Images for the Preoperative T Staging of Colorectal Cancer by Dual-Energy CT with Tin Filter Technology. Plos (2014)
9. Inah, G.B., Akintomide, A.O., Edim, U.U., Nzotta, C., Egbe, N.O.: A study of pelvic radiography image quality in a Nigerian teaching hospital based on the Commission of European Communities (CEC) criteria. South African Radiographers (2013)
10. Karen, D.: Multon, Encyclopedia of Research Design.: SAGE Reference Online (2010)

11. Kurt, B., Nabiev, V., Turhan, K.: A Novel Automatic Suspicious Mass Regions Identification Using Havrda & Charvat Entropy And Otsu'S N Thresholding. Computer Methods and Programs in Biomedicine, 349–360 (2014)
12. Kurt, B., Nabiev, V., Turhan, K.: Contrast Enhancement and Breast Segmentation of Mammograms. In: 2nd World Conference on Information Technology, Antalya, pp. 26–31 (2011)
13. Sampat, M.P., et al.: Measuring Intra- and Inter-Observer Agreement in Identifying and Localizing Structures in Medical Images. Image Processing, 81–84 (2006)
14. Viera, A.J., Garrett, J.M.: Understanding interobserver agreement: The kappa statistic. Family Medicine, 360–363 (2005)
15. Trochim, W.M.K.: Research Methods Knowledge Base (January 2015), http://www.socialresearchmethods.net/kb/reltypes.php
16. Skaane, P., Engedal, K., Skjennald, A.: Interobserver Variation in the Interpretation of Breast Imaging Comparison of mammography, ultrasonography, and both combined in the interpretation of palpable noncalcified breast masses. Acta Radiologica, 497–502 (1997)
17. Hopstaken, R.M., Witbraad, J., van Engelshoven, J.M., Dinant, G.J.: Inter-observer variation in the interpretation of chest radiographs for pneumonia in community-acquired lower respiratory tract infections. Clinical Radiology, 743–752 (August 2004)
18. Grgic, M., Delac, K.: Mammographic Image Analysis Homepage (January 2015), http://www.mammoimage.org/databases/

A LOD-Based Service for Extracting Linked Open Emergency Healthcare Data

Mikaela Poulymenopoulou[1], Flora Malamateniou[1], and George Vassilacopoulos[1,2]

[1]Department of Digital Systems, University of Piraeus, Piraeus, Greece
[2]New York University, New York, NY, USA

Abstract. The linked open data (LOD) initiative – an initiative taken by governments around the world to open up and link the vast repositories of data they hold across agencies and departments – features particular potential in the health care sector. The real value of linked open data comes from its interpretation, analysis and linking up which, in the healthcare sector, is expected to result in improved quality of care and lower healthcare costs. In particular, emergency healthcare quality is expected to improve by making healthcare data, which is related to emergency healthcare, available to authorized users at the point of care (suitably anonymized for security reasons) and by providing researchers with access to large volumes of data. In addition, the analysis of emergency healthcare LOD can provide insights on a variety of factors contributing to emergency medical services (EMS) usage and to EMS failures so that to formulate sustained emergency healthcare policies and enable effective and efficient decision making that results in improving emergency case morbidity and mortality indices. This paper addresses the general problem of LOD usage in emergency healthcare delivery and describes a LOD-based cloud service that seeks to automatically export appropriate emergency healthcare data of interest from a variety of sources, semantically annotate this data and enriching it through the creation of links with other, relevant, data. To this end the service is designed to interact with EMS information systems, electronic medical records (EMRs) and personal health records (PHRs).

Keywords: LOD, open data, emergency healthcare, emergency medical services.

1 Introduction

Across the western world, the ageing populations, the increased rates of non-communicable or chronic diseases like diabetes, and annually outbreaks like influenza have resulted in a substantial growth on requests for healthcare services including emergency medical services (EMSs) [1-3]. The latter are often considered some of the most important gateways to healthcare systems, consisting of ambulance agencies and hospital emergency units that perform a variety of interrelated activities (administrative, paramedical and medical) from the time of a request (call) for an ambulance till the time of patient's exit from the emergency department (ED) of a hospital [2].

The increased emergency facility usage results in EMS failure to efficiently and timely manage emergency cases with high impact to emergency patient morbidity

F. Ortuño and I. Rojas (Eds.): IWBBIO 2015, Part II, LNCS 9044, pp. 85–92, 2015.

and mortality. This holds especially in critical accidents or emergencies that require appropriate treatment at the place of incident and en route to a hospital and timely transfer to the most appropriate healthcare unit.

The systematic collection, analysis and interpretation of big datasets related to emergency healthcare delivery from multiple sources helps in understanding patient factors associated with high utilization of EMSs so that to enable healthcare policy makers in making appropriate decision for optimizing emergency healthcare utilization, improving emergency healthcare outcomes and containing costs [1,3-5]. In particular, the use of dispersed emergency healthcare data retrieved from a variety of sources like EMS information systems, hospital electronic medical records (EMRs) and personal health records (PHRs) enables the identification of a sufficient number of subjects with specific characteristics (e.g. obese people that have frequent/non-frequent EMS incidents due to a specific disease related to obesity) required for an analysis scenario. The aggregation of this data according to which only the counts of data having specific characteristics are considered, instead of using record raw level data, helps in preserving patient anonymity and its analysis can help in gaining knowledge of the relationships between pathological process (e.g. diseases or adverse events) and risk factors that lead to increased EMS usage [6]. Then, the derived knowledge can drive health policy in preventing unnecessary EMS usage (e.g. patients that can be treated in the local community), in better managing emergency cases (e.g. use the most appropriate emergency healthcare protocol, transfer cases to the most appropriate and available healthcare facility) and in efficiently utilizing emergency healthcare resources. Despite the importance of EMS resource utilization to the healthcare system, very little health analytics effort has been devoted to the area of gaining insights about the cause of EMS use (or abuse) and about the impact of EMS utilization to healthcare delivery effectiveness and efficiency.

Open Data is an emergent trend which consists of freely available data usually from public organizations that can be used in various analysis scenarios to provide valuable information and knowledge for enhanced decision-making [7]. Recently, the Linked Open Data (LOD) initiative has also emerged as a new semantic web paradigm related to a set of best practice standards in publishing and linking heterogeneous data. LOD principles can be applied to both the representation and management of data stemming from multiple sources in order to tackle interoperability problems [7-9]. Examples of recent efforts on incorporating LOD principles in healthcare data are the Linked Life Data and the Linked Open Drug Data [10,11].

The LOD initiative presents a very powerful approach to integrating heterogeneous data, although its true potential can only be realized when LOD is combined with "real" patient data from existing information systems (e.g. EMRs or EMS information systems). However, in practice, due to several privacy, security, ethical and policy issues, few patient datasets are available online and those mostly concern administrative, as opposed to medical, data [4,5,10,11]. On the other hand, the creation of an emergency healthcare LOD set drawn from a variety of sources (suitably anonymised to ensure data privacy) and its linkage with other LOD available can assist in providing more complete and accurate healthcare data for analysis using appropriate health analytics techniques , such as machine learning and natural language processing.

This paper addresses the general problem of LOD usage in emergency healthcare delivery and describes a LOD-based cloud service that interacts with multiple data sources in order to generate an emergency healthcare LOD set drawn from a variety of emergency healthcare sources. In particular, this service aims at making data not only accessible, but also readable, comprehensible and usable for research as well as linked with other relevant LOD sets on the cloud.

The basic motivation for this research stems from our involvement in a research project concerning the use of health analytics in the emergency healthcare context to derive insight for the factors contributing to increased EMS usage and its impact on hospital-based healthcare delivery so that to enable fact-based decision making for improved emergency healthcare quality and for better emergency healthcare resource management. The stringent needs of using large scale emergency medical data that is dispersed into multiple, diverse, data sources which are usually closely guarded and monitored for unauthorized access within institutional firewall boundaries, motivated this work and provided some of the background supportive information for the development of the LOD-based cloud service.

2   Methods

2.1   LOD Standards

Linked data is a method to publish structured data by using standard web technologies, to connect related data and make them accessible on the Web. According to LOD principles, HTTP uniform resource identifiers (URIs) are used for identifying data items, the RDF for describing data and links for describing the relationships among relevant data. Other standards used in LOD applications include Resource Description Framework Schema (RDFS) for describing RDF vocabularies and the SPARQL Protocol and RDF Query Language (SPARQL) for querying RDF graphs [6-8].

Lately, there is wide adoption of the LOD best practices by governments that has lead to the extension of the web into an unbound, global data space with connected data stemming from multiple sources and diverse domains. This is named "web of data" and is considered as opening up new opportunities for both domain-specific and cross-domain data analytic applications. Moreover, the data analytics results can be continually refined and provide better answers as new data or new data sources appear on the web [5,9,12].

2.2   The LOD-Based Service

The proposed LOD-based cloud service is considered to interact with multiple sources which are mainly divided into the following categories: a) EMS information systems and hospital EMRs that usually store their data in relational or document-based repositories and b) PHRs that usually provide an Application Programming Interface (API) or services to enable extracting patient data of interest in a standard format like the Continuity of Care Document (CCD).

A high level view of the LOD-based cloud service is show in Figure 1. It is seen that the main modules comprising the service are: data retrieval, data storage, data translation, data mapping and data linkage.

The data retrieval module consists of client applications that make calls to the services provided by the data sources and/or by services created in order to connect and retrieve aggregated emergency healthcare data from the designated data sources according to specific characteristics as opposed to raw level data, in order to preserve patient anonymity. Basically, the retrieved data sets are considered to exist in two formats: SQL data views and XML documents.

The data storage module consists of a cross-platform data repository that enables the storage and management of SQL, XML and RDF data and also supports SPARQL queries over data. Hence, the retrieved data, by the data retrieval module, is stored according to its format to the data repository as follows: SQL data views are stored in SQL like database tables and XML documents are stored in XML databases.

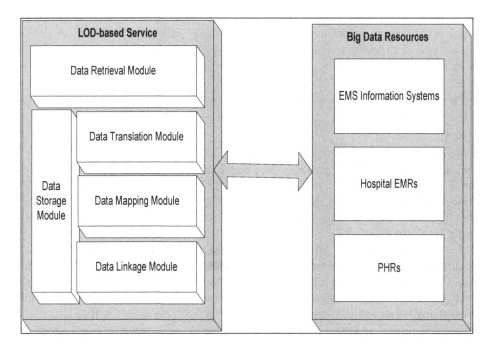

Fig. 1. A LOD-based cloud service

The data translation module consists of shared common vocabularies and ontologies used in LOD to represent data and express relationships among data and new emergency medicine domain ontologies created to include concepts specific to emergency healthcare, like emergency medical protocols, medical procedures, observations and chief complaints coded according to standard medical terminologies like International Classification of Diseases (ICD) and SNOMED which are linked with the standard ontologies used in LOD. Examples of standard LOD vocabularies and ontologies used are RDF, RDFS and Web Ontology Language (OWL). Moreover, the ICD ontology has been used to classify diseases and other health problems including signs and symptoms and the Experimental Factor Ontology (EFO) ontology has been used to combine parts of several biological ontologies such as anatomy and disease compounds.

The data mapping module maps the SQL data and the XML documents, of the data repository, into RDF graphs using direct mapping or mapping documents. The direct mapping takes as input the data and generates a standard RDF graph while mapping documents provide guidance for the creation of the desired RDF graphs according to the mappings defined among the input data and the resulting RDF. Those mappings are created using the ontologies provided by the ontology module. The resulted RDF graphs are stored into the data repository of the data storage module in the form of RDF triples.

The data linkage module interlinks RDF documents of the data storage module and other datasets, which are already in the LOD cloud. In particular, similar link types are identified in the RDF documents and suitable matching links are found in external datasets enabling all those to be linked through the use of RDF links.

The LOD-based service is considered to be either used on demand by the connected data sources or be executed in pre-specified intervals for making freely available on the web new data of emergency healthcare interest.

2.3   Security Issues

Despite the technical and interoperability challenges faced in discovering and accessing scattered medical data there also many ethical, legal and security constraints. In particular, in big data analysis and LOD-based systems (like the one presented in this paper) that require access to multiple data sets from dispersed data sources, a variety of security issues need be considered. These include the regulations that protect patient data and prevent patient re-identification by any means, the healthcare providers security policies, the agreements based on consent forms and patient data sharing preferences as imposed by the patient-oriented paradigm [6,10]. In big data analysis and LOD initiatives, it is widely suggested to aggregate patient data in order to ensure non-identification and anonymization while the original data remain safe from any modifications [6]. However, even with the use of aggregate data, the healthcare provider security policies and patient sharing preferences should be considered.

In the LOD-based service described in this paper, aggregate emergency healthcare data is retrieved from multiple big data sources subject to the constraints imposed by the security policies enforced by those sources. For example, access requests to aggregate EMR data of a hospital are subject to the constraints set by the access control rules of the particular hospital's security policy. However, access requests to patient data existing in PHR services (following the patient-oriented paradigm) are subject to patient sharing preferences. Therefore, is considered that big data sources connected to the LOD-based service should define access control rules of the form which patient data (objects) is allowed to be accessed by the LOD-based service (subject) and under what circumstances. For example, a healthcare provider access control rule may imply that demographic patient data such as name, address and social security number can't be shared with the LOD-based service for preserving patient anonymity while another access control rule may imply that patient age, body mass index, regular exercise, vegetable consumption, health problems and emergency incidents information can be shared with the LOD-based service.

With regard to access control, the emergency healthcare LOD (in RDF format) that is created by the LOD-based service and stored into the cloud, there is a need for developing a security module to control access to RDF graphs by authorized subjects. However, a description of the access control mechanism devised for this purpose base on the attribute-based access control (ABAC) general policy is beyond the scope of this paper but will be described elsewhere.

3   Results

The LOD-based cloud service described is able to integrate heterogeneous, in both format and semantics, emergency healthcare data from multiple data sources into a data repository, maps this data into RDF format, stores it in the data repository in the form of RDF triples and then links it to other relevant datasets in the LOD cloud.

The service is currently under development using a laboratory-based cloud infrastructure and open-source tools such as the Protégé OWL editor for ontology engineering, the cross-platform OpenLink Virtuoso Universal Server as the data repository of the data storage module, the embedded mapping language of Virtuoso for creating the mapping documents of the data mapping module and the Silk link discovery framework that provides a language for specifying the types of RDF links that should be discovered and linked with other datasets in the LOD cloud.

Currently, the LOD-based cloud service is connected to a PHR system called PINCLOUD, which is a cloud-based service that integrates patient health and social care information from various sources such as the patient, non-healthcare providers, home care systems (that store medical information transmitted from Internet connected medical devices to the patient) and multiple healthcare information systems (e.g. EMS information systems, primary care systems, hospital EMR systems). In the future, the LOD-based cloud service is planned to be connected to multiple EMS information systems, to hospital EMRs and PHR systems.

The PINCLOUD PHR provides services that can be set to retrieving patient data according to user defined criteria. These services have been altered to support the retrieval of aggregate data from multiple data sources. Moreover, client applications of the data retrieval module have been created to call the PINCLOUD PHR web services that result in retrieving relevant aggregate emergency healthcare data sets in the form of XML documents. As an example, consider the retrieval of aggregate data related to patients who are obese, have a "heart disease", belong to age group 50-60 and have multiple emergency encounters (more than 3 the last 3 years) with critical severity. The analyses of such datasets are expected to reflect the relative burden incurred on the EMS system by various case categorizations.

The retrieved aggregate datasets in the form of XML documents are stored to the data storage module as XML documents. To this end, mapping documents have been created where mappings among the retrieved XML documents fields and the resulting RDF graphs are defined using emergency medicine ontologies and other LOD vocabularies and ontologies (RDF, OWL) of the ontology module. In particular, the emergency medicine domain (EMS-DO) ontology proposed in [13] has been used as a basis and adapted to the particular needs of the LOD-based service. The resulting RDF triples are also stored to the data repository of the data storage module.

At this stage of development, links have been identified and created to connect emergency healthcare LOD with a subset of DBpedia relevant entities. DBpedia is a well-known entity source, which covers various fields, including biomedicine [11]. Given its broad use range, only a related dataset of relevant entities have been selected from DBpedia by finding entities having a common category or meaning with the entities existing in the resulting RDFs.

The LOD-based service is executed on demand for retrieving new emergency healthcare relevant data from the PINCLOUD PHR to be made available on the cloud as LOD.

4   Conclusion

LOD of emergency healthcare interest that originates from multiple data sources can provide increased opportunities for more robust population-based observational and prevention studies and therefore to lead to improved emergency healthcare quality of service to patients and better use of emergency healthcare resources [1-4]. Recent semantic tools have eased the conversion of data into the form of RDF triples. However, appropriate data selection is critical and considerable work is needed to ensure data consistency and validity across sources, platforms and systems. To this end, a special purpose LOD-based cloud service is presented that attempts to provide rich and meaningful emergency healthcare related datasets that is converted to a linked data cloud. Other ongoing efforts in healthcare and life science domain is the Linked Open Drug Data, LinkedCT, Open Biomedical Ontologies and the World Wide Web Consortium's (W3C) Health Care and Life Sciences working group [5,8-11].

The degree of advancement that LOD technology can bring in emergency healthcare data analysis is difficult to measure. From the technological perspective, the advantage of LOD is that it makes data discoverable by search engines and machine understandable and therefore it is tackled the current healthcare interoperability problem [8]. Currently, LOD is still in its infancy and therefore immature as compared to other database technologies however certainly is a very promising candidate for use in healthcare big data analytics scenarios, like the one presented in this paper in the field of emergency healthcare. Our intention is to test this service for its performance with data from different data sources and then to try using the resulting LOD for answering emergency healthcare research questions like which are the main clinical characteristics of the majority of patients requesting ambulance transfer to a hospital. Moreover, we intend to extend LOD-based service functionality for including online medical informatics publications like PubMed as data sources with emergency healthcare relevant data that will be also made available in the LOD cloud.

References

1. Abe, T., Ishimatsu, S., Tokuda, Y.: Descriptive analysis of patients' EMS use related to severity in Tokyo: A population-based observational study. Plos One 8(3) (2013)
2. Poulymenopoulou, M., Malamateniou, F., Vassilacopoulos, G.: Emergency healthcare process automation using mobile computing and cloud services. Journal of Medical Systems 36(5), 3233–3241 (2012)
3. Knowlton, A., Weir, B., Hughes, B., Southerland, H., Schultz, C., Sarpatwari, R., Wissow, L., Links, J., Fields, J., McWilliams, J., Gaasch, W.: Patient demographic and health factors associated with frequent use of emergency medical services in a midsized city. Academic Emergency Medicine 20(11), 1101–1111 (2013)
4. Ortmann, J., Limbu, M., Wang, D., Kauppinen, T.: Crowdsourcing linked open data for disaster management. In: Terra Cognita 2011 Workshop at the ISWC (2011)
5. Chiolero, A., Santschi, V., Paccaud, F.: Public health surveillance with electronic medical records: At risk of surveillance bias and overdiagnosis. European Journal of Public Health 23(3), 350–351 (2013)
6. Kamateri, E., Kalampokis, E., Tambouris, E., Tarampanis, K.: The linked medical data access control framework. Journal of Biomedical Informatics 50, 213–225 (2014)
7. Janssen, M., Estevez, E., Janowski, T.: Interoperability in big, open, and linked data – organizational maturity, capabilities, and data portfolios. IEEE Computer Society (2014)
8. Heath, T., Bizer, C.: Linked Data: Evolving the Web into a Global Data Space (1st edition). Synthesis Lectures on the Semantic Web: Theory and Technology 1(1), 1–136 (2011)
9. Tilahum, B., Kauppinen, T., Kebler, C., Fritz, F.: Design and development of a linked open data-based health information representation and visualization system: Potentials and preliminary evaluation. Journal of Medical Internet Research (JMIR) Medical Informatics 2(2) (2014)
10. Pathak, J., Kiefer, R., Chute, C.: Using linked data for mining drug-drug interactions in electronic health records. Studies in Healthcare Technology and Information 192, 682–686 (2013)
11. Tian, L., Zhang, W., Bikakis, A., Wang, H., Yu, Y., Ni, Y., Cao, F.: MeDetect: A LOD-based system for collective entity annotation in biomedicine. In: Proceedings of the International Conferences on Web Intelligence and Intelligent Agent Technology, pp. 233–240. IEEE Computer Society (2013)
12. Ferguson, M.: Architecting a big data platform for analytics. Intelligent Business Strategies (2012),
    http://www.ndm.net/datawarehouse/pdf/Netezza%20-
    %20Architecting%20A%20Big%20Data%20Platform%20for%20Analytics
    .pdf
13. Poulymenopoulou, M., Malamateniou, F., Vassilacopoulos, G.: Document management mechanism for holistic emergency care. International Journal of Healthcare Information Systems and Informatics 9(2) (2014)

Development of an Auditory Cueing System to Assist Gait in Patients with Parkinson's Disease

Vânia Guimarães[1], Rui Castro[1], Ana Barros[1], João Cevada[1], Àngels Bayés[2], Sheila García[2], and Berta Mestre[2]

[1] Fraunhofer Portugal AICOS, Porto, Portugal
[2] Parkinson's Disease Unit, Teknon Medical Center, Barcelona, Spain

Abstract. Patients with Parkinson's Disease often experience motor symptoms that compromise their ability to walk independently and safely. One of the key problems is the inability to generate sufficient step length, which is typically compensated by an increase in stepping frequency. In this work, a system providing real-time auditory stimuli through a headset connected to a smartphone is developed and tested. Stimuli are provided when certain episodes are identified so as to modify speed and amplitude of movements. In this study, the feasibility of the system in stimulating gait using self-adaptive cueing rhythms is investigated and system's usefulness and acceptance are evaluated. Experimental results suggest that better gait patterns can be stimulated when individuals follow sounds whose rate is close to their natural step rate. Results also suggest that the system would be readily accepted by patients, provided that it can help them in real time during their daily activities.

Keywords: Parkinson's Disease, Motor Symptoms, Auditory Cueing, Gait, Rhythm, Smartphone, Headset.

1 Introduction

Parkinson's disease (PD) is a progressive neurological disorder characterized by specific motor impairments, which include tremor, rigidity, bradykinesia (slowness of movement) and postural instability [1,2].

One of the key symptoms of PD is the diminished ability in walking. Typically, people with PD have problems in maintaining gait rhythm, resulting in an abnormal gait pattern with increased variability, short steps, slow walking speed and increased step rate [1,2,3]. With disease progression, episodic gait disturbances such as freezing of gait (FOG) and shuffling steps can occur [1,4].

Despite the success of pharmacological therapies in ameliorating some symptoms of PD, gait deficits can be resistant to medication and over time become one of the most incapacitating symptoms [5]. Stimulation in the form of sounds, referred to as *auditory cueing*, is known to have an immediate effect on gait, with improvements in walking speed, step length and cadence [1,3,6]. Auditory stimuli enhance gait by inducing a sense of rhythm and a stable coupling between footfalls and the beat [3,6]. Auditory cueing may also facilitate gait initiation and minimize the occurrence of episodic problems such as FOG [1].

F. Ortuño and I. Rojas (Eds.): IWBBIO 2015, Part II, LNCS 9044, pp. 93–104, 2015.

The Auditory Cueing System (ACS) is part of REMPARK, a Personal Health System with detection, response and treatment capabilities for the management of PD [7]. REMPARK enables the real-time extraction of spatio-temporal features of gait, as well as the identification of motor symptoms. ACS aims to provide auditory stimuli in real time, whenever a motor symptom is detected. Stimuli last until the individual's walking behaviour is corrected. The smartphone is responsible for generating cueing sounds, controlling their rhythm according to the inputs it receives and streaming them to a headset over a Bluetooth connection. The technology chosen may allow individuals to use the system in both clinical and social settings, with an immediate effect on gait and a minimal disruption of their daily routine.

In this work, the ACS is developed and its functionalities demonstrated isolated from the other REMPARK system's components, such as sensors and server. The system is tested with a group of individuals with PD to reach conclusions about its feasibility in stimulating gait. After a brief description of related research and background, testing methodologies are described. Results are then presented and discussed.

2 Background

The beneficial effect of auditory cueing on gait performance in PD has been widely documented, including improvements in step length [1,8,9], walking speed [1,9], variability [6,10], and even in dual-task [1,8].

Sounds act as temporal cues, informing about movement timing [3]. Their rhythm induces a sense of beat, as an internal clock, that helps regulate pace in walking [3]. Therefore, to improve gait quality, cueing rhythm needs to be carefully adjusted to each individual, in each situation.

In the study by Hausdorff, J. M. (2009) [10], when auditory stimuli was administered at a rate either equal to the individuals' baseline step rate or 10% higher, stride length and gait speed increased. Variability, however, did not improve significantly in response to auditory cueing set to the individual's baseline cadence. At a 10% higher rate, variability decreased significantly [10]. According to Ledger, S. et al. (2008) [1], an increase in auditory cueing frequency by 10% to 20% of individuals' natural step rate has an immediate effect on walking speed and stride length [1]. However, these effects have been reported inconsistent in people with PD who freeze [8].

While increased cueing rates may in general affect non-freezers' gait positively, evidence suggests that these rates affect freezers' walk negatively [1,8,9,11]. On the basis of the present results, it is recommended to lower the cueing rate settings for freezers, whereas for non-freezers an increase of up to 10% may have potential therapeutic effect [11]. According to Nanhoe-Mahabier, W. et al. (2012) [8], decreased cueing rates may benefit both freezers and non-freezers [8]. Moreover, providing auditory stimuli at the moment of a frozen gait has been shown to help patients to overcome FOG episodes [5,13].

According to Roerdink, M. (2011) [6], gait is coupled best to cueing rates near the preferred cadence, with a reduction in variability [6]. Also, auditory-motor coordination is more stable when both footfalls are paced [12]. Combinations of sounds can be used to pace differently right and left steps, which may also be helpful for patients [6].

Nieuwboer, A. *et al.* (2007) [14] showed that the effects of cueing can be maintained in the absence of cues after training, indicating that some degree of motor learning can be preserved in PD [14]. Training with auditory cueing has also been shown to diminish the occurrence of FOG episodes in PD [8,14].

Nevertheless, the technology to date has not been practically viable for use outside clinics. Previous approaches have several limitations, including the need of triggering cueing manually, thus being limited to training sessions where continuous auditory stimuli is applied [13]. To overcome these limitations, a system which automatically triggers stimuli at the time of the detection of a gait disorder is developed. The system is evaluated to conclude about the effectiveness of a self-adaptive cueing rate in stimulating gait when problems are detected.

3 Methods

3.1 Participants

After giving their written informed consent, twelve people diagnosed with PD, 7 men, 5 women, average age 71.2 ± 1.4 years, participated in the study (Table 1). Only subjects who were able to walk independently were considered for participation.

Table 1. Participants Characteristics ($n = 12$)

Characteristics	Values [†]
Number of subjects	12
Age (years)	71.2 ± 1.4
Gender (men/women)	7/5
Height (cm)	166.8 ± 11.1
MMSE	28.6 ± 1.3
Hoehn and Yahr	2.4 ± 0.3
Disease Duration (years)	9.4 ± 5.6
UPDRS (part III)	19.3 ± 8.8
Freezers/Non-freezers	5/7
Use devices to help walking / don't use	3/9

[†] Values are $\bar{x} \pm STD$, except where indicated; $\bar{x}$, average; STD, standard deviation; $MMSE$, Mini-Mental State Examination; $UPDRS$, Unified Parkinson's Disease Rating Scale.

3.2 System Setup

Auditory stimuli were administered through a headset connected to a smartphone, which was responsible for generating sounds and controlling their rhythm according to the inputs it received, i.e. walking rhythm, quality of walking and symptoms.

At this stage, and in order to simulate the inputs the system would receive automatically from REMPARK sensors, another smartphone connected via Bluetooth to the main unit was used. An assistant was responsible for touching an interface button each time the individual took a step, enabling the calculation of cadence. The main unit was carried by the therapist and was used not only to control stimuli, but also to input information about quality of walking and symptoms. Therapists could also adjust cueing rhythm in real time (Fig. 1).

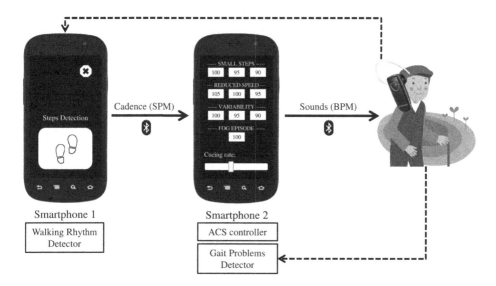

Fig. 1. System setup and tasks (*SPM*–Steps per minute, *BPM*–Beats per minute)

To make the system adaptive, different gait impairment levels were considered and each level was by default associated to cueing rhythms that were equal (i.e. 100%), higher (i.e. >100%) or lower (i.e. <100%) than the average stepping rate measured when no symptoms were present (Table 2). In general, lower rhythms were given (progressing from the left to the right in Table 2) when impairment severity was higher. These values were chosen based on the literature, considering the reported effects of cueing in each specific gait parameter.

Several types of sounds were available to be administered to patients, including metronome sounds, musical beats, clapping and verbal cueing (i.e. "one-two-one-two"). Combinations of sounds were also included so that right and left steps could be paced differently. Sounds started playing whenever a symptom was detected and stopped playing when symptoms were overcome.

Table 2. Default values for cueing rate percentages

Event	Default cueing rate percentages		
	Level 1	Level 2	Level 3
Small Steps	100	95	90
Reduced Speed	105	100	95
Variability	100	95	90
FOG Episodes	100		

3.3 Procedures

Non-cued and cued trials using rhythmic auditory cues were considered in this study. Walking parameters were measured in both non-cued and cued conditions, so that effects on gait could be evaluated. Patients were offered the opportunity to hear different types of sounds, select one and adjust its volume according to their preferences.

Two Android devices, Samsung Nexus S, were used. To deliver the sounds, the headset Samsung HM3500 was used.

Firstly, a 20-m walking test (Fig. 2) was performed to get reference values for cadence, walking speed and step length. Patients were instructed to walk at their comfortable pace and measurements of test duration and number of steps were taken. The test was completed twice and mean values of walking parameters calculated.

Fig. 2. 20-m walking test

Then, a rhythmic sound fixed at 10% below the preferred stepping frequency was administered. The test was repeated, but this time in the presence of auditory cues. Mean values of gait parameters in cued condition were computed.

In the last test, patients were asked to walk at will, under the supervision of the therapist, and using the ACS. Outdoor and indoor conditions were captured (Fig. 3) and data was collected, regarding activations and deactivations of cueing, applied cueing rhythm, walking rhythm, type of impairment detected and the adjustments of rhythm performed by the therapist in real time.

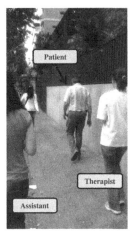

Fig. 3. Free walking test

Patients were then asked to fill in a questionnaire to evaluate their perceptions and satisfaction with the system (see Table 3). For the questions to which patients had to provide a self-assessment, the Visual Analogue Scale (VAS) was used [15]. Additionally, patients were given the opportunity to provide additional comments.

Table 3. Questionnaire applied to participants

Nr	Question	Rate
1	How do you grade your current walking performance? (Note: before hearing the sounds)	0 (worst walking) to 10 (best walking)
2	Is the volume of sounds adequate?	yes/no
3	Do you feel comfortable with the sounds provided, or do they annoy you?	0 (very uncomfortable) to 10 (very comfortable)
4	How do you grade your walking performance when using the ACS?	0 (worst walking) to 10 (best walking)
5	Do you feel comfortable wearing the headset?	0 (very uncomfortable) to 10 (very comfortable)
6	Would you be willing to wear the system during your everyday life?	yes/no

3.4   Analysis

A statistical analysis was performed to test the hypothesis that there is a statistically significant difference in walking parameters from baseline to cued conditions. The efficacy of each rhythm in overcoming gait problems was also assessed and questionnaire results analysed. Differences between normally distributed variables were assessed with Student's T test (with 95% confidence, $p>0.05$) in Microsoft Excel 2010®.

4   Results

4.1   Non-Cued vs. Cued Walking Tests

A T-test for paired samples was used to test the hypothesis that there is a statistically significant difference in walking parameters from non-cued to cued (10% below natural step rate) conditions. The null hypothesis defined was that the means of the two conditions are equal, i.e. no effects of cueing on gait can be observed.

Table 4. Descriptive data for non-cued and cued walking tests, $^*p>0.05$ ($\bar{x}$ – average, STD – standard deviation)

Non-cued and cued walking tests				
Parameter	Condition	$\bar{x} \pm STD$	Statistic Value	P-value
Walking Speed (m/s)	Non-cued walking	1.02 ± 0.21	$t = 0.71$	$p = 0.49^*$
	Cued walking	1.01 ± 0.21		
Step Length (m)	Non-cued walking	0.54 ± 0.12	$t = -0.33$	$p = 0.75^*$
	Cued walking	0.54 ± 0.14		
Cadence (steps/min)	Non-cued walking	115.20 ± 12.44	$t = 1.10$	$p = 0.29^*$
	Cued walking	112.93 ± 14.38		
Rhythm (steps or beats per min)	Applied	103.43 ± 10.81	$t = 27.21$	$p<0.01$
	Measured	112.93 ± 14.38		

According to the results presented in Table 4, no significant differences in walking speed, step length and cadence ($p>0.05$) were detected. Expectedly, a significant difference between applied and measured rhythms ($p<0.01$) was observed, which means that patients did not follow the established rhythm.

4.2   Free Walking Results

Symptoms, cueing rhythm and successful rhythms (i.e. those that enabled the person to overcome a specific gait problem) were analysed to inspect whether each cueing modality and settings were appropriate to each patient in each specific situation.

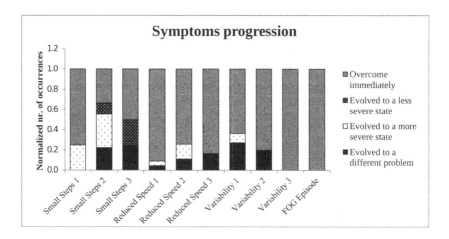

Fig. 4. Symptoms progression

A total of 102 events were captured in this study, including 55 events of reduced speed, 22 events of variability and 21 events of small steps. Only four FOG episodes were registered, so their results were not analysed.

In the majority of the cases, gait problems were overcome after cueing had started. However, it was frequent to observe a symptom evolving to another symptom or to another state of the same symptom (categorized in terms of severity from 1 - least severe state - to 3 - worst state of the symptom), as presented in Fig. 4. The four most frequent patterns were:

- Small Steps-2 to Small Steps-3 (33% of Small Steps-2 occurrences)
- Small Steps-1 to Small Steps-3 (25% of Small Steps-1 occurrences)
- Small Steps-2 to Reduced Speed-2 (22% of Small Steps-2 occurrences)
- Variability-2 to Reduced Speed-1 (20% of Variability-2 occurrences)

A therapist was required to change cueing rhythm sometimes. The most frequent symptoms to which a change in rate was applied were "Variability-2", "Small Steps-2" and "Reduced Speed-2". Fig. 5 presents the average rate of changes in rhythm applied by the therapist. In this graph, positive values represent an increase in cueing rhythm, whereas negative values represent a decrease in rhythm. As can be observed, all changes aimed at increasing the rate that was being applied when a symptom was detected (Fig. 5).

The relation between the successful cueing rhythms applied and the baseline cadence (measured before starting to use cueing) was also analysed. A T-test for paired samples was used to test whether the difference between cueing rhythm and baseline cadence was significantly different from zero. The null hypothesis defined was that the means of the two conditions were equal. According to Table 5, where results are presented, the majority of participants (7 out of 12) overcame their gait problems (regardless of their origin) when no significant differences between cueing rhythm and baseline cadence were observed ($p > 0.05$).

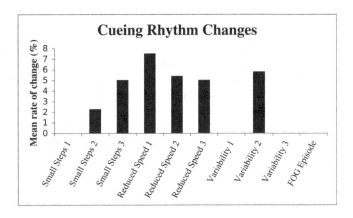

Fig. 5. Average percentage of cueing rhythm change set by the therapist

Table 5. Relation between baseline cadence and successful cueing rhythms applied, *$p>0.05$ ($\bar{x}$ – average, STD – standard deviation, in beats or steps per minute)

Patient ID	Parameter	$\bar{x} \pm STD$	Statistic Value	P-value
1	Cueing rhythm	102.50 ± 9.38	$t = -3.51$	$p<0.01$
	Baseline cadence	114.13 ± 0.00		
2	Cueing rhythm	102.00 ± 3.46	$t = -6.61$	$p<0.01$
	Baseline cadence	112.24 ± 0.00		
3	Cueing rhythm	111.40 ± 6.54	$t = -1.51$	$p = 0.21^*$
	Baseline cadence	115.81 ± 0.00		
4	Cueing rhythm	136.25 ± 2.75	$t = -4.79$	$p = 0.02$
	Baseline cadence	142.85 ± 0.00		
5	Cueing rhythm	108.33 ± 9.99	$t = -2.26$	$p = 0.05^*$
	Baseline cadence	115.87 ± 0.00		
6	Cueing rhythm	102.60 ± 13.32	$t = 0.53$	$p = 0.63^*$
	Baseline cadence	99.46 ± 0.00		
7	Cueing rhythm	117.86 ± 7.56	$t = -4.96$	$p<0.01$
	Baseline cadence	132.03 ± 0.00		
8	Cueing rhythm	122.50 ± 7.94	$t = 0.44$	$p = 0.69^*$
	Baseline cadence	120.75 ± 0.00		
9	Cueing rhythm	89.00 ± 5.94	$t = -3.53$	$p = 0.04$
	Baseline cadence	99.48 ± 0.00		
10	Cueing rhythm	108.50 ± 7.78	$t = -1.13$	$p = 0.46^*$
	Baseline cadence	114.73 ± 0.00		
11	Cueing rhythm	102.50 ± 5.74	$t = -2.05$	$p = 0.13^*$
	Baseline cadence	108.39 ± 0.00		
12	Cueing rhythm	106.00 ± 5.66	$t = -0.18$	$p = 0.89^*$
	Baseline cadence	106.70 ± 0.00		

4.3 Questionnaires

According to Table 6, the majority of participants (11 out of 12) found the volume of sounds adequate and felt comfortable wearing the headset (7.6 ± 1.9). Also, they felt comfortable with the sounds provided (6.9 ± 1.9). Hence, all patients would be willing to use the system during their everyday life. The majority (8/12) rated with a higher score their walking performance when they were walking with the help of cueing sounds in comparison with their perceived walking performance when walking without cueing. However, no significant statistical differences in perceived walking performance were found ($t = -2.31$, $p = 0.06$).

Table 6. Answers provided by participants ($\overline{x}$ – mean, STD – standard deviation)[†]

Question	Answers
How do you grade your current walking performance?	5.7 ± 1.8
Is the volume of sounds adequate? (yes/no)	11/1
Do you feel comfortable with the sounds provided, or do they annoy you?	6.9 ± 1.9
How do you grade your walking performance when using the ACS?	6.5 ± 1.8
Do you feel comfortable wearing the headset?	7.6 ± 1.9
Would you be willing to wear the system during your everyday life? (yes/no)	12/0

[†] Values are $\overline{x} \pm STD$, except where indicated.

5 Discussion

In general, patients felt comfortable with the sounds provided, even in outdoor conditions, where volume could have been an issue. Despite being not statistically significant, a tendency to rate with a higher score the cued walking against non-cued walking performance could be observed. As such, all participants would be willing to use the system during their everyday life, considering that the system would help them in real time during their daily activities (Table 6).

Contrary to what one might expect, when a 90%-fixed cueing rhythm was applied patients did not synchronize their steps with the rate of sounds ($t = 27.21$, $p<0.01$). Also, changes in walking speed, cadence and step length were not statistically significant ($p>0.05$), which means that patients did not change their walking pattern when cueing at this fixed-rate was added (Table 4).

In free walking tests, cueing rate settings benefiting rates lower than natural stepping rate were worse for the patients participating in the study, and frequently led to the occurrence of reduced speed and shorter steps. Also, generally, the therapist interacted with the system to increase the default cueing rates defined as lower than baseline, as is the case of "Small Steps-2", "Small Steps-3" and "Variability-2". The majority of participants were able to overcome their gait problems (regardless of the type of problem detected) when the cueing rhythm applied was close to their natural stepping rate (see section 4.2).

Results suggest that cueing rhythm needs to be individualized to the person, so that it will stimulate the best walking pattern possible, which is achieved when their natural, desirable, step rate is followed. When a 90%-fixed cueing rhythm was applied at the beginning, patients were walking with a good pattern, and therefore, they did not feel the need to follow the rhythm. Also, the rate applied was not adequate, since it was lower than their natural rhythm, the rhythm they were capable of following still with an adequate length of steps and speed.

These findings differ from the results presented in other studies, where different than baseline rhythms were used to improve gait. However, these studies were performed in different conditions, which may partially explain the different conclusions. Some studies, for example, base their results on the use of a treadmill (as in [6], [8] and [12]), where a fixed speed is imposed and maintained, something that cannot be applied when someone walks on the ground. Others, on the other side, use cueing mainly as training (e.g. [1] and [14]) and they just evaluate gait quality before and after intervention. One must have in mind, however, the limitations encountered in this study, including the small sample size and the limited accuracy of the instruments used to measure walking parameters. Also, identification and categorization of symptoms severity is made manually based on observation, which introduces some subjectivity.

6   Conclusion

In this work, a system providing real-time auditory stimuli through a headset connected to a smartphone was developed and tested. This study shows that stimuli can be applied when certain episodes are identified to modify speed and amplitude of movements. The system was proven effective in stimulating gait whenever a symptom was detected in case sounds rate was adjusted to the individuals' natural step rate associated to the their best walking pattern. All participants would accept using the system during their everyday life, considering that it would help them in real time during their daily activities.

Further tests would be required to assess the efficacy of the system in stimulating gait in real life conditions, including an evaluation of its effects on gait when FOG episodes are detected. To that purpose, ACS needs to be integrated with REMPARK movement sensors capable of detecting walking rhythm and symptoms autonomously. Further investigations are required to assess quality of life of people using the system continually during their daily life.

Acknowledgments. The authors would like to acknowledge all patients who participated in this study for their time and effort. This work has been performed in the framework of the FP7 project REMPARK ICT-287677, which is funded by the European Community. The authors would like to acknowledge the contributions of their colleagues from REMPARK Consortium.

References

1. Ledger, S., Galvin, R., Lynch, D., Stokes, E.K.: A randomised controlled trial evaluating the effect of an individual auditory cueing device on freezing and gait speed in people with Parkinson's disease. BMC Neurol. 8 (2008)
2. Casamassima, F., Ferrari, A., Milosevic, B., Ginis, P., Farella, E., Rocchi, L.: A Wearable System for Gait Training in Subjects with Parkinson's Disease. Sensors 14(4), 6229–6246 (2014)
3. Nombela, C., Hughes, L., Owen, A., Grahn, J.: Into the groove: Can rhythm influence Parkinson's Disease? Neurosci. Biobehav. Rev. 37(10), 2564–2570 (2013)
4. Giladi, N., Shabtai, H., Rozenberg, E., Shabtai, E.: Gait festination in Parkinson's disease. Parkinsonism Relat. Disord. 7(2), 135–138 (2001)
5. Mazilu, S., Blanke, U., Hardegger, M., Tröster, G., Gazit, E., Hausdorff, J.M.: Gait Assist: A Daily-Life Support and Training System for Parkinson's Disease Patients with Freezing of Gait. In: Proceedings of the SIGCHI Conference on Human Factors in Computing Systems, pp. 2531–2540 (2014)
6. Roerdink, M., Bank, P.J., Peper, C., Beek, P.J.: Walking to the beat of different drums: Practical implications for the use of acoustic rhythms in gait rehabilitation. Gait & Posture 33(4), 690–694 (2011)
7. Personal Health Device for the Remote and Autonomous Management of Parkinson's Disease (REMPARK), http://www.rempark.eu/
8. Nanhoe-Mahabier, W., Delval, A., Snijders, A.H., Weerdesteyn, V., Overeem, S., Bloem, B.R.: The possible price of auditory cueing: Influence on obstacle avoidance in Parkinson's disease. Mov. Disord. 27(4), 574–578 (2012)
9. Picelli, A., Camin, M., Tinazzi, M., Vangelista, A., Cosentino, A., Fiaschi, A., Smania, N.: Three-dimensional motion analysis of the effects of auditory cueing on gait pattern in patients with Parkinson's disease: A preliminary investigation. Neurol Sci. 31(4), 423–430 (2010)
10. Hausdorff, J.M.: Gait dynamics in Parkinson's disease: Common and distinct behavior among stride length, gait variability, and fractal-like scaling. Chaos 19(2), 1–14 (2009)
11. Willems, A.M., Nieuwboer, A., Chavret, F., Desloovere, K., Dom, R., Rochester, L., Jones, D., Kwakkel, G., Wegen, E.: The use of rhythmic auditory cues to influence gait in patients with Parkinson's disease, the differential effect for freezers and non-freezers, an explorative study. Disabil Rehabil 28(11), 721–728 (2006)
12. Roerdink, M., Lamoth, C.J., van Kordelaar, J., Elich, P., Konijnenbelt, M., Kwakkel, G., Beek, P.J.: Rhythm perturbations in acoustically paced treadmill walking after stroke. Neurorehabil Neural Repair 23(7), 668–678 (2009)
13. Bächlin, M., Plotnik, M., Roggen, D., Giladi, N., Hausdorff, J.M., Tröster, G.: A wearable system to assist walking of Parkinson's disease patients. Methods Inf. Med. 49(1), 88–95 (2010)
14. Nieuwboer, A., Kwakkel, G., Rochester, L., Jones, D., Wegen, E., Willems, A.M., Chavret, F., Hetherington, V., Baker, K., Lim, I.: Cueing training in the home improves gait-related mobility in Parkinson's disease: The RESCUE trial. J. Neurol. Neurosurg. Psychiatry 78(2), 134–140 (2007)
15. Wewers, M.E., Lowe, N.K.: A critical review of visual analogue scales in the measurement of clinical phenomena. Res. Nurs. Health 13(4), 227–236 (1990)

Linear Accelerator Bunkers: Shielding Verification

Khaled Sayed Ahmed[1] and Shereen M. El-Metwally[2]

[1]Department of Bio-Electronics, Modern University for Technology and Information,
Cairo, Egypt
khaled.sayed@k-space.org
[2]Faculty of Engineering, Cairo University, Giza, Egypt
shereen.elmetwally@yahoo.com

Abstract. Many designs of radiation therapy rooms miss the correct shielding requirements due to improperly selected equipment or incorrectly chosen design. These designs may have been established based on the most common devices in the market at design time or based on previous work or experiences. Furthermore, the device upgrading from low energy to high energy or from cobalt tele-therapy gamma-ray units to linear accelerator may result in an imperfect shielding. Through implementation, problems may be faced to keep the required safety levels, furthermore, to satisfy the needed protection. In this paper, we illustrate through selected case studies, some of the problems faced and how it could be overcome in order to reach the required protection levels. This study includes seven bunkers at different places whose designs prevented to establish a linear accelerator due to room areas, room surrounding spaces or type of existing shielding: wall or layers thickness and material. Three rooms were predesigned for a cobalt tele-therapy system with an overall wall thickness of 70 cm; two rooms were inadequately designed with irregular walls of 1 m thickness, and the other two rooms were designed for low energy devices and are to be upgraded to high energy, in-addition to different inadequate conditions surrounding these bunkers. Combined solutions were used to overcome the faced problems. The presented solutions incorporate using other shielding materials than concrete as lead, borated polyethylene, gamma-600 together with changing and/ or controlling the obliquity factor (changing the incident angle of radiation). The existing shielding may be also be modified by adding an internal/external layer of lead. Moreover, the position of the device inside the room was re-allocated with a certain angle to the incident rays in order to reduce leakage radiation. The resulting design solutions were validated via the atomic energy commission at those countries.

Keywords: Acceptable level, shielding, radiation protection, adequate space.

1   Introduction

Cancer treatment is a complete removal of tumor from the patient's body. This goal is mainly achieved using three treatment methods: 1) surgery (involving direct resection of the tumor), 2) chemotherapy (use of antineoplastic drugs), and 3) radiation therapy

F. Ortuño and I. Rojas (Eds.): IWBBIO 2015, Part II, LNCS 9044, pp. 105–113, 2015.
© Springer International Publishing Switzerland 2015

(use of ionizing radiation to kill tumor cells). Radiation therapy may be used radically (one treatment modality alone) or adjuvant (combination of modalities) depending on the type and location of the tumor, its grade (how aggressive it is), its stage (how advanced it is), and the general state of the patient. Worldwide, radiation is used to treat about 50% of all cancer patients [1]. This radiation may be described by a percentage depth dose (PDD) which depends on the beam type (photon or charged particle), beam energy (higher energy beams are more penetrating), and type of the absorbing material (water, soft tissue, bone, etc). Photon beams for use in external beam radiation therapy are produced using either superficial or ortho-voltage X-ray machines (10 kV to 100 kV and 100 kV to 500 kV, respectively), cobalt tele-therapy gamma-ray units (1.25 MeV) or linear accelerators (2 MV to 25 MV) which is the most critical device based on its radiated dose. Superficial and ortho-voltage X rays are used to treat lesions on or close to the patient's skin, where the external photon beam is delivered from just one direction. Higher energy beams, used to deliver the dose to deeper lesions, are delivered over a range of angles. Accordingly, tele-therapy units and linear accelerators (linacs) are typically arranged with the radiation source on a gantry that can rotate around the patient (iso-centric setup) [2].

Radio therapy is one of the most medical procedures having a double-edged sword; its effectiveness in treating cancer and several other diseases and potential for radio-iatrogenesis, in the form of short- or late-terms both to the patient and to the therapist [3]. In order to limit the exposure of individuals and society to radiation, such radiation exposure must be measurable and its biological effects quantifiable. Radiation dosimetry is the science of measuring radiation exposure, while radiobiology is the science of understanding the biological effects of radiation. In health physics, a number of dosimetric quantities and units, encompassing aspects of both dosimetry and radiobiology, are defined [4, 5]. These quantities include: absorbed dose, equivalent dose, effective dose, and collective effective dose, which are defined as follows:

— Absorbed dose
It is simply a measure of the energy of ionizing radiation absorbed per unit mass of absorbing material. The unit of absorbed dose is the Gray (Gy).

— Equivalent dose
Since some radiations are biologically more effective (more dangerous) than others, the International Commission on Radiological Protection (ICRP) defined the quantity equivalent dose which, for a particular tissue, is the absorbed dose multiplied by a radiation weighting factor. The unit of equivalent dose is the Sievert (Sv).

— Effective dose
The quantity effective dose was defined by the ICRP to account for the variation in radiation sensitivity among the tissues and organs of the body. The effective dose is defined as the sum of the equivalent doses to exposed tissues and organs multiplied by the appropriate tissue weighting factors. The Sievert (Sv) is also the unit of effective dose.

— Collective effective dose

In order to compare the effective doses between exposed population groups, the ICRP introduced the quantity collective effective dose. The collective effective dose is defined as the product of the average effective dose to an exposed population and the number of persons exposed. The unit of collective effective dose is the Person-Sievert.

In order to measure the collective effective dose and to determine the critical margins of patient safety, there are three tenets underlying radiation protection: distance, time, and shielding [5], as described below:

— Distance

The distance from the source should be maximized. The inverse-square law governs the fall-off in dose as a function of source distance, e.g., a doubling of the distance will reduce the exposure level by a factor of 4.

— Time

The duration of an exposure should be minimized, since the accumulated exposure increases linearly as a function of time.

— Shielding

The amount of shielding around the source should be maximized. Radiation beams will be attenuated exponentially based on used the shielding materials and its thickness.

Because the linear accelerator is the highest energy generator, the purpose of radiation shielding is to reduce the equivalent dose from this device to a point outside the bunker to a certain target dose limit or constraint set by national standards. The standard site plan for linear accelerator bunkers is illustrated in Fig.1. The basic shielding material is concrete.

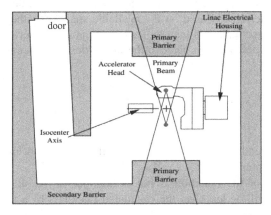

Fig. 1. Plan view of a typical radiation therapy treatment room (linear accelerator)

Radiation generated by linear accelerators is mainly divided into primary and secondary components, the latter being divided into scatter and leakage radiation [6-9]. Primary radiation is that radiation used in patients treatment within a formed treatment field size of 40 x 40 cm^2 in most common devices, and can be directed into primary barrier.

For primary barriers, a required barrier transmission factor, B_p, is given by:

$$B_p = \frac{Pd^2}{WUT} \tag{1}$$

where, P is the weekly target dose limit derived from the annual limit appropriate for the type of space protected by the barrier (Sv/wk), d is the distance from the target to the point of measurement (m), W is the workload (Gy/wk) defined as the average number of radiated monitor units (MU) used per week in patients having treatment over the course of a year, U is the usage factor, and T is the occupancy factor.

On the other hand, secondary radiation components represent indirect rays which can be reflected on a secondary barrier. Scatter radiation is that radiation generated by the patient or by the primary beam (attenuated by the patient) striking a primary barrier while Leakage radiation refers to x-rays generated in the head from interactions of the primary electrons in the target, flattening filter collimator jaws and other surroundings. These types of radiations typically set limits for an uncontrolled or public area, and a controlled area staffed by radiation workers. Some other factors may have an impact on determining the required shielding. These include neutrons generated from the interactions of useful X-rays (at energies above 8 MeV), which should be covered through shielding calculations together with the maze area and the thickness of the paraffin wax of the room door. In addition to the usage of IMRT (intensity modulated radiation therapy) which results in radiation doses about 2 to 4 times higher than for conventional treatments, also the usage of a multi-leaf collimator (MLC).

The basic concept of radiation shielding is to limit radiation exposures to members of public and to satisfy an acceptable level to employees. The thickness of shielding materials can be determined based on the available spaces together with the type of used equipment. These materials may be lead, steel and / or interlocking bricks. The barrier thickness, s, necessary to achieve the target dose rate is calculated using the following formulas:

$$n = log\left(\frac{1}{B}\right),$$

$$s = TVL_1 + (n-1)TVL_e \tag{2}$$

where, TVL is the Tenth Value Layer, defined as the thickness of material required to allow 10% transmission.

When the angle of the radiation is not orthogonal to the shielding barrier, the required thickness will be less than the calculated thickness by a factor that depends on the angle of incidence. This factor is known as the obliquity factor [10, 11]. It varies as $cos(\alpha)$, where α is the angle between the incident ray and the normal to the shielding wall. Thus, the effective thickness of a barrier, T, is related to the actual thickness, s, by:

$$T = \frac{s}{cos(\alpha)} \tag{3}$$

This relationship is certainly valid for all energies and angles of incidence up to 45° for concrete barriers.

2   Data and Methodology

2.1   Case I: Cobalt System to Linear Accelerator Conversion

In this case, three rooms A, B, and C have been established for cobalt system since 8 years. The rooms are typically designed as: an overall wall thickness of 70 cm, internal rooms' space of 7.6 m x 7 m, and the shielding material used was concrete. Rooms A and B are adjacent with a double thickness in between (140cm) as illustrated in Fig. 2 (a). It was required to upgrade these three rooms to be suitable for linear accelerators establishment at 6 MV. The purpose of radiation shielding is to reduce the effective equivalent dose from a linear accelerator to a point outside the room to a sufficiently low level. This level is determined as per case, but, in general, 0.02 mSv/week for public or uncontrolled area. Often, a higher level is chosen for areas restricted from public access (controlled areas) and occupied only by workers; this limit is 0.1 mSv/ week [7]. This problem was tackled by first studying the surrounding conditions to inspect the possibility of the shielding expansion from outside the room. As the rooms (A and B), and room C were located in separate buildings, the rooms' internal spaces were minimized to 6.8 m x 6.2 m to satisfy the needs of most common devices. Upon calculating the shielding requirements of a 6 MV linear accelerator, the required thicknesses of concrete for the primary and secondary barriers were 2.4 m and 1.6 m, respectively. It can be noted that the maximum thickness to be added is 40 cm overall without compromising the operation of the linear accelerator such as the couch rotation. Either lead or steel can be used for shielding rooms A and B in order to minimize the required barriers' thickness. Therefore, lead barriers with thicknesses of 6 cm and 12 cm were used as secondary and primary barriers, respectively. These reduced- thickness lead barriers are equivalent to concrete barriers of 40 cm and 80 cm thickness, respectively. In room C, concrete was added from outside the room to be more cost-saving.

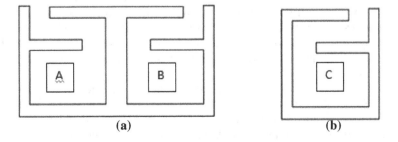

Fig. 2. The three cobalt rooms before modifications

2.2   Case II: Two Rooms with Irregular Walls of 1 m Thickness

In this case, two rooms have an inadequate irregularly-shaped design, i.e., the room has more than 4 sides, as illustrated in Fig. 3. Concrete was the used material for cost reduction. The internal clear spaces were 7.8 m x 7.8 for the two rooms. It was re-quired to re-shield these rooms to meet the shielding requirements for two linear acce-lerators of 6 MV, where the needed internal spaces are 7 m x 6.8 m. This shielding design problem was managed by adding 1 m of concrete from one side and 80 cm from the other side. The thicknesses required for the primary and secondary barriers are 2.4 m and 1.6 m, respectively. Due to the common wall of 2 m thickness between the two rooms, a 60 cm can be added to this wall. The other sides (outer sides) can be extended to 1.5 m. Also, by taking into account the obliquity factor with a maximum angle of 45° (*the maximum angle which reflects the minimum shielding*) the required thicknesses could be reduced to 1.15 m and 1.45 m, instead of 1.6 m and 2.4 cm, respectively.

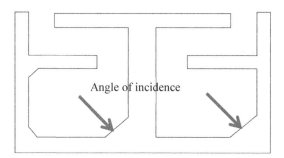

Fig. 3. Two rooms with irregular walls of 1 m thickness

2.3   Case III: Two Rooms Design Upgrading from Low Energy to High Energy Devices

In this case, two rooms were designed for low dose equipment of 4 MV and 6 MV as shown in Fig. 4. The primary barrier thickness was 2.4 m, the secondary barrier thickness was 1.6 cm, and the connected parts of the two adjacent rooms had 3.4 m thickness. Concrete was the used material for cost reduction. The internal clear spaces were 8 m x 8 m, and 7 m x 7 m for the first and second rooms, respectively. Room A is located adjacent to a PET-CT room, while room B is located towards the street (outside direction). It is required to re-shield these rooms to meet the shielding requirements, where the needed spaces for the two high energy devices (15 MV, 18 MV) to be installed are 7 m x 6.8 m.

This situation is common as there is a frequent need to re-shield rooms that hold a low single energy machine as 4 and 6 MV in order to accommodate a dual energy machine with maximum photon energy of 15 or 18 MV. Re-shielding is required because of the difference in the TVL of concrete for the low and high energies. The TVL_1 and TVL_e of concrete are 0.35 m and 0.35 m at 6 MV, and 0.47 m and 0.43 m at 18 MV.

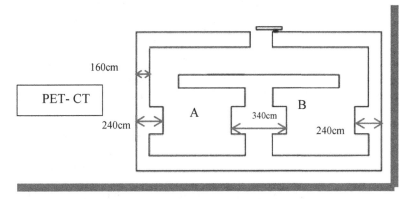

Fig. 4. Two rooms for low energy devices located at the hospital corner

The difficulties encountered to make this change depend mainly upon the space within the room and also the surrounding areas. Here, it was noted that room A has a sufficient space, so either poured concrete or concrete blocks can be used for the additional shielding. A 30 cm concrete layer can be added to the overall room walls (primary and secondary barriers).

In room B, the situation is very different. This is a frequent situation where there is a little extra space inside the room for added shielding without compromising the operation of the linear accelerator, such as the couch rotation. Either lead or steel can be used. Both lead and steel require structural support, but lead has the advantage that its TVL is half that of steel. However, it should be noted that steel has the advantage of a lower photo-neutron production. The first step in the design tackling was to calculate how much additional lead is required to meet the regulatory requirements and to put it from the room outer side. Here, a 6 cm lead was added from outside the room (right side) keeping the left side as it is. In other cases, if the required lead is greater than a certain threshold, so interlocking bricks can be used. The shielding for both rooms A and B is shown in Fig. 5. Concrete has been used (grey color) from the left side of room A, lead (red color) has been used at room B from the inner side of the secondary barrier and external side of the primary barrier.

Also, as a high energy linear accelerator (>10 MV) is involved, a complete neutron survey was be carried out. The effect of lining high-energy linear accelerator mazes with neutron was examined. Some neutron-moderating materials are often used in order to reduce scattered neutron dose at the accelerator room door. These materials may be polyethylene alone, polyethylene combined with flex-boron panels. Much greater reductions in both neutron and gamma ray dose can be obtained by incorporating polyethylene and boron into either internal or external maze doors, which is used in our solution. Our results support the conclusion that neutrons directly incident on the maze from the accelerator contribute little to the neutron dose at the door, and that the majority of neutron dose is due to scattered and thermal neutrons.

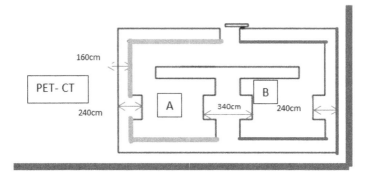

Fig. 5. Two rooms for low energy devices after upgrading to high energy devices. Concrete, lead, and hospital wall are shown in grey, red and blue colors, respectively.

3   Discussion and Conclusion

The main objective of radiation shielding is to limit radiation exposures to members of public and to satisfy an acceptable level to employees. In this paper, technical information and recommendations related to the design (redesign/ modification) and installation of structural shielding for radiation therapy facilities have been presented. The case studies presented in this paper aimed to establish a target dose-rate at a certain point behind a barrier, together with calculating the barriers thickness necessary to achieve shielding requirements. The barriers' material and their thicknesses have been determined based on the available spaces together with the type of the used equipment. These materials may be lead, steel and / or interlocking bricks. The obliquity factor has been also used in some cases to modify the angle of incidence. The resulting design solutions were validated via the national atomic energy commission. The presented solutions in this paper may be employed in other similar conditions based on the available spaces and location of the rooms. If a high energy linear accelerator is involved, a complete neutron survey must be carried out. Recent shielding designs, Radiation Protection Design Guidelines for 0.1-100 MeV Particle Accelerator Facilities, and Neutron Contamination from Medical Electron Accelerators have to refer to NCRP Reports 51, 79, and 151, respectively [7, 8, 9].

References

1. IAEA: IAEA TECDOC Series 1040 (1998)
2. Podgorsak, E.B.: Radiation Physics for Medical Physicists. Springer (2006)
3. Hall, E.J., Giaccia, A.J.: Radiobiology for the Radiologist. Lippincott Williams and Wilkins (2006)
4. ICRP Publication 90: Annals of the ICRP (2003)
5. ICRP Publication 103: Annals of the ICRP (2007)

6. NCRP report #49, National Council on Radiation Protection and Measurements. Structural shielding design and evaluation for medical use of x-rays and gamma rays of energies up to 10 MV. Bethesda, MD: NCRP; NCRP Report No. 49 (1976)

7. NCRP report #51, National Council on Radiation Protection and Measurements. Radiation protection guidelines for 0.1-100 MeV particle accelerator facilities. Bethesda, MD: NCRP; NCRP Report No. 51 (1977)

8. NCRP report #79, National Council on Radiation Protection and Measurements. Neutron contamination from medical electron accelerators. Bethesda, MD: NCRP; NCRP Report No. 79 (1984)

9. NCRP report #151, National Council on Radiation Protection and Measurements. Structural shielding design and evaluation for megavoltage x-and gamma-ray radiotherapy facilities. Bethesda, MD: NCRP; NCRP Report No. 151 (2005)

10. Biggs, P.J.: Primary Beam Widths in Ceiling Shielding for Megavoltage Linear Accelerators. Radiation Protection Management 19(4) (2002)

11. Biggs, P.J.: Linear Accelerator Shielding: Thirty Years Beyond NCRP 49

New IT Tools and Methods Improving Lifestyle of Young People

Alexandru Serban[1], Mihaela Crisan-Vida[1], Maria-Corina Serban[2], and Lacramioara Stoicu-Tivadar[1]

[1]University Politehnica Timisoara, Romania,
Department Automation and Applied Informatics, Timisoara, Romania
{alex.serban81,mihaela.vida}@yahoo.com,
lacramioara.stoicu-tivadar@aut.upt.ro
[2]University of Medicine and Pharmacy "Victor Babes "Timisoara, Romania,
Department of Functional Sciences, Timisoara, Romania
dr.corinaserban@yahoo.com

Abstract. Internet addictions, stress, different types of dependencies and poor diets have generated detrimental effects on the educational, relationship and work-related characteristics of young generation. The aim of this study was to develop a multidimensional **cloud-based application** which combines different aspects of lifestyle. This tool is used by students from two universities from Romania, University of Medicine and Pharmacy Victor Babes Timisoara and University Politehnica Timisoara. The student answer to questions regarding age, gender, ethnicity, location, height, weight, dependency of Internet, level of stress, consumption of alcohol, cigarettes, coffee and energy drinks, sleep schedule, daily diet, and mealtimes. The answers of students will be saved in a cloud database and the application will calculate useful lifestyle parameters. The data is anonymized and secured and the HL7 standard ensures high connectivity to other medical applications. Using **emotion-oriented computing** and **lifestyle questions,** the application can monitor and limit different addiction habits, improving lifestyle.

Keywords: User-computer interface, Cloud-computing, lifestyle, internet addiction, emotion computing.

1   Introduction

It is a reality that today internet plays an important role in our lives. The users have accessibility to an incredible quantity of information and tend to be more comfortable in handling their own lifestyle. The development of the internet offer new opportunities for creation of different applications and solutions which can improve the quality of lifestyle by means of user engagement and shared decision making [1]. Moreover, modern lifestyle applications are more complex in terms of cloud computing technology and user experience design (UX).

Cloud computing, as characterized by NIST (National Institute of Standards and Technology) facilitates on request access to the network in order to share computing resources (e.g. servers, networks, applications, services, or storage). It can be released

F. Ortuño and I. Rojas (Eds.): IWBBIO 2015, Part II, LNCS 9044, pp. 114–122, 2015.
© Springer International Publishing Switzerland 2015

and created promptly without service provider connection, but with minimum management [2]. Cloud computing technology has five essential properties: extensive accessibility to the network (might be utilized from PCs, or smartphones interconnected to the Internet), on-demand services (end users can use web services and connect to a website anytime), elasticity (permits customers costs up or down as needed and remarkable flexibleness for scaling systems) and measured services (allows monitoring and recording of resources in a pay – per – use manner) and pooling resources (customers might share the resources of computing with different clients, so these resources might be dynamically reallocated and hosted anyplace), [3]. Cloud architecture includes virtualization and service oriented architecture (SOA), offering a lot of benefits such as: language (neutral integration like XML), reutilization of components (after creation, an application can be reused without rewriting code), organizational agility (after building blocks of software considering user specification allows quickly recombination and integration) and usage of existing systems (enable integration between new and old systems components) [4]. PaaS, SaaS, and IaaS are specific cloud service models. PaaS, such as Microsoft Azure, Google's Apps Engine, Force platform, and Salesforce.com relates to various solutions functioning on a cloud to offer platform computing for users. SaaS, including Gmail, Google Docs, Online Payroll and Salesforce.com, represents applications running on a remote cloud system proposed by the cloud supplier as solutions which can be utilized using Internet. IaaS, comprising Flexiscale and Amazon's EC2, designates hardware equipment functioning on a cloud offered by service suppliers and utilized by users on desire [5]. By all described, Microsoft Azure is a cloud computing platform and infrastructure created and provided by Microsoft for building, deploying and managing applications and services through a global network of data centers managed by Microsoft. It supports multiple programming languages, tools and architectures. Microsoft Azure provides mobile services, in this way greatly facilitating communication between mobile application on different platforms (e.g. Android and Windows 8) to have the possibility to send and receive information from the cloud [6].

User Experience (UX) is described as a result of the appearance, system performance, interactive behavior, efficiency, and helpful abilities of an interactive system, both hardware and software. UX includes interaction methods and design, information architecture or visual design, becoming an established field of research in Design. Furthermore, User Experience Design short called UXD or UED defines a process of enhancing user satisfaction by increasing the level of interaction of a user with a product or service provided. UXD can be nominated as a process of interaction with interface, graphics, animation, motion graphics or physical design [7].

It is already known that interconnected UX online applications designed on cloud-based platforms facilitate faster data processing and interpretation than old technologies. Furthermore, new developed cloud-based applications have been gaining a growing number of users in online environment. The Centers for Medicare and Medical Services started to use internet as a communication platform, collecting information from users in an electronic format [8]. Furthermore, all existing web-based questionnaires are depended of self-reported evaluation, a useful and cost-effective approach for obtain data through surveys. The answers obtained from users through self-reported checks are susceptible to bias, being based on cognitive ability to remember different personal information. The questions used in lifestyle surveys comprise information about physical activities (type, regularity, period, intensity) and dietary habits (the varieties and quantity of food consumed) of users [9]. Nowadays, the questionnaires are essential options used in epidemiologic studies and clinical surveys. From the medical point of view, this online lifestyle cloud-based application comprise a lot of parameters related to the extensive evaluation of ideographic attitudes and behaviors related to health of a person such as amount of sleep, different types of alcohol or cigarettes dependencies, physical activity behavior and diet characteristics of young users, a population at great risk for problematic health behaviors.

Amount of sleep is an important indicator of health and well-being in students. Adequate sleep, defined as 6–8 hours per night regularly, is a critical factor in health-related behaviors. The sleep length of time and daytime sleepiness are considered to be independent factors related to psychosocial parameters and function as independent predictors of health and overall performance. Furthermore, inadequate sleep might be a screening indicator for an unhealthy lifestyle and poor health status [10]. Many physiologic studies showed that sleep deprivation might have an impact on weight as a result of effects on physical activity, appetite, and thermoregulation [11].

Internet addiction is recognized by abnormal or badly handled desires, preoccupations or behaviors concerning Internet use which produce distress or impairment. Nowadays were described 3 subtypes of Internet addiction: socializing or social networking consisting of text/email messaging, intensive gaming-gambling and sexual preoccupations (cybersex). Internet addicts could use for prolonged intervals the Internet, concentrate nearly completely on the Internet instead of broader life events and separate themselves from the rest of forms of interpersonal contact [12]. Because leisure time is crucial for well-being and health, and cell phone use has been linked with mental health, the relationship between the use of cell phone and leisure need to be more effective comprehended. The cell phones permit users to browse the internet, to be involved in popular online social networks like Twitter and Facebook, Hi5, to create and share personal photos and videos, to play many online and offline games or to stream movies and live sports nearly anytime and anyplace [13]. Since Facebook turn into probably the most utilized site on the Internet, *Facebook addiction* was predictable and seems to be directly associated with sleep habits. Moreover, it was observed that environmental and psychological factors may contribute to food choices young adults make.

Poor diet can lead to co morbidities like diabetes and obesity according to American Heart Association, 2014. Furthermore, young users have diets rich in fat, sugar, and sodium intake and regularly drink sugar-sweetened beverages increasing the risk for chronic diseases. Higher levels of stress in students may increase the risk of poor eating habits at night as a means of maladaptive coping. *Night eating syndrome* (NES) is influencing lifestyle behavior of young users. The diagnostic criteria for NES describe an important caloric intake in the evening and/or night-time, characterized by consuming at least 25% of food intake, and /or experiencing at least two episodes of nocturnal eating per week [14]. A lifestyle education tool should be focused on both behavioral and dietary factors. Users might be advised to reduce saturated fats and replace saturated fats with unsaturated fat and low trans-fats and to choose healthier behaviors and promote better dietary choices.

Since the use of the Internet for extracting, transforming, and disseminating information was realized only by means of papers, books, pamphlets, and instruction materials, it was a challenge to develop a lifestyle UX cloud-based application based on online questionnaires, for data collecting and interpretation, which is able to advise users in terms of lifestyle changes.

The aim of this study was to develop a UXD cloud-based application based on questions about physical activities and dietary habits of users, being able to advise users in terms of lifestyle changes. From the medical point of view, the application comprise multidimensional aspects of users: age, gender, ethnicity, location, height, weight, amount of sleep, dependence of cell phones, of social media games and Internet, indirect assessment of frustration and stress and consumption of alcohol, cigarettes, coffee and energy drinks. The application comprises also questions regarding daily diet, favorite types of food and mealtimes.

2 Materials and Methods

This paper presents a new multidimensional cloud-based application designed for online users. Each user has the possibility to register a new account in order to use this lifestyle online tool. A large database is created comprising answers related to ethnicity, age, race, gender, socio-economic status, leisure-time physical activity, anthropometric parameters, cigarettes-smoking, alcohol consumption, caffeine consumption, internet addiction, video game addiction, phone addiction and food diet. This online tool creates a detailed lifestyle profile of the registered user. The data collection is realized through multiple questionnaires using various lifestyle and dietary questions. This online application is capable to process obtained data from users: validation (verification if supplied data is clean, correct and useful), summarization (reducing details to specific points), classifying (separates data into multiple categories) and reporting (listing computed information data). After collecting data, this online tool is capable to offer different lifestyle advices for users.

The application comprises three sets of questions. The first set of questions will collect data regarding age, location, gender, ethnicity, height and weight. Based on this first set of data, the tool calculates automatically the body mass index of the user. The second set of questions collects data regarding addiction pattern of internet use, internet addiction, video gaming disorder, consumption of alcohol, smartphone overuse, cigarettes, coffee and other drinks. The third questionnaire collects information regarding daily diet, fruit and vegetables consumption, favorite food and mealtimes. Based on the last two sets of questionnaires, the online tool automatically calculates risk factor of diseases that predispose the interviewed user. After this step, the user receives helpful advices on how to reduce the risk for obesity, diabetes mellitus through diet and lifestyle changes. All information is saved in a database that can be used for generating statistics and comparisons between certain dietary behaviors from different users interviewed. The user can always update the data entered into the application with new data; in this way he will be able to improve his nutritional lifestyle.

2.1   The Design Architecture

This multidimensional application can be accessed via an internet browser from laptops, mobile phones or tablets. The online tool uses the latest trends in user experience design (UXD). It has a **responsive design**, which means it provides an optimal viewing experience. This feature will allow design adaptation to different resolutions on any device from where is accessed (Figure 1). The application will auto adjust the layout and content for the device from which is accessed, ensuring that the user has a richer viewing experience. The adjustments for resolution are realized by adding media queries and scripts for resolution scale and orientation ranges. The online tool has many integrated multimedia elements, allowing the interaction between user and application, increasing the grade of UXD.

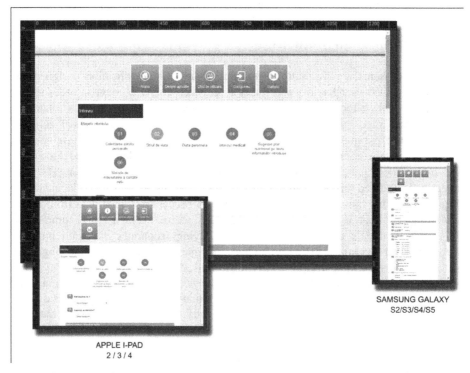

Fig. 1. The interface scalability simulation of the online tool on different devices

Technically, the application is divided in three parts:
1. User friendly interface - even by the users who are not very experienced with the modern online tools.
2. The application database – where all the input data is stored

3. The engine module – where all the entered data is processed into information and displayed to the user trough interface. With the help of modular design, the application can always be developed and enriched using new elements, in terms of architecture and in terms of UX, without expensive development.

2.2   The System Architecture

This online cloud-based multidimensional lifestyle application is in trend with the latest technology in the field of information architecture and design. It has a dynamically generated interface/content. The application's information is stored in a database on an internet server. The application is developed with latest technology in *server side dynamic pages,* where the web page construction is controlled by an application server processing server-side scripts.

In server-side scripting, the parameters determine how the construction of each new web page continues, integrating the arranging of additional client-side developing. This will help users to find the required information more rapidly. It is already known that an advantage of dynamic web pages is that, the server can process large quantity of information to multiple online users in the same time, without any risk of losing the resources of the servers. The main advantage of the dynamic-content is the fasting loading through a web server, because all the information content is stored in a single database instead of multiple files. Because the application has a modular design, every improvement can be quickly performed and without damaging the rest of the modules in the tool. Generated observed errors are reported in an XML file, to be easily identified and corrected, allowing future improvements of the online application. There are two kinds of errors that can be encountered in such kind of application: **development errors** – mostly encountered in the development stage of the application or when is tested, before launching online, and **user's errors** – that appears when a user makes a mistake. The application has a special dedicated module for determining the errors encountered when the application is used. The error module saves the errors encountered through the navigation in a XML file on the web server. This file plays a great role in detecting bugs, errors and the deficiencies of the application, allowing easy identifying and solving them.

Different standards can be used to transmit data between different units. The standard used for this application is HL7 Clinical Document Architecture, a document markup standard that specifies the structure and semantics of "clinical documents" for the purpose of data exchange and it is a complete information object which can include text, images, sounds, and other multimedia data and could be any of the following: discharge summary, referral, clinical summary, history/physical examination, diagnostic report, prescription, or public health report [15].

The application is connected to a database in Windows Azure cloud, where it can access the data in real time. Figure 2 presents the database in Windows Azure.

Currently the application is uploaded online on a web server and is tested by students from University Politehnica (UPT) Timisoara and University of "Medicine and Pharmacy" Timisoara (UMFT). In this phase, this online tool has a special feedback questionnaire in order to see the student's opinion regarding the application in terms of user interface. In this manner, some undiscovered errors or bugs can be reported. After this stage, the application will be improved in accordance with the student's feedback. In the next phase the results will include a comparison of the lifestyle of the students from both universities. In the test group there are included 5 groups of students from UPT, around 110 students and around 100 from UMFT. They register and use the application. The intention is to compare the feedback of IT and non IT users, and also medical and nonmedical users. The timeline of the process is: 3 month (October-December 2014) for testing at UPT and UMFT, 2 month processing the data (Jan-Feb 2015), 2 month redesign based on results of processing (Mar-Apr 2015). This will end the stage of design.

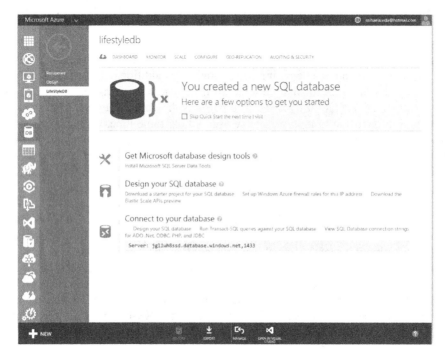

Fig. 2. DB on Windows Azure

Based on the new tool the test users group will be diversified and the study of lifestyle behavior reports and statistics will be provided during the second half of the year. Having a dynamic architecture and a user-centered design, this online tool is actually a web 2.0 generation application. The modular system allows easy further improvement of the application. The development and integrated multimedia elements will help the user to access information more quickly, in this way increasing its UXD. Furthermore, the application is based on a optimization system for the most advanced browsers. The novelty for this type of applications is its adaptability on any mobile device connected to the Internet. The development of the application is based on "responsive design" using HTML/HTML 5 and CSS3 scripting language, a system that is increasingly used in modern websites. It has a fluid layout using a flexible grid, and a range of the website according to browser's full size and graphics.

3   Conclusions

Considering the fact that there is a high demand for research studies to accurate determine the dietary habits of individuals, the suggested cloud-based application presented in this research intends to provide a personalized solution for evaluating physical activities and dietary habits assessment among different types of user. The application is a useful tool for improving the life-style of users and promoting healthy eating attitudes.

Acknowledgements. "This work was partially supported by the strategic grant POSDRU/159/1.5/S/137070 (2014) of the Ministry of National Education, Romania, co-financed by the European Social Fund – Investing in People, within the Sectorial Operational Programme Human Resources Development 2007-2013"

References

1. Weinhold, I., Gastaldi, L.: From Shared Decision Making to Patient Engagement in Health Care Processes: The Role of Digital Technologies. In: Challenges and Opportunities in Health Care Management, pp. 185–196. Springer (2015)
2. Mell, P., Grance, T.: The NIST definition of cloud computing (2011)
3. Abolfazli, S., Sanaei, Z., Sanaei, M., Shojafar, M., Gani, A.: Mobile cloud computing: The-state-of-the-art, challenges, and future research (2015)
4. Raines, G.: Cloud Computing and SOA. Systems Engineering at MITRE, p. 20 (2009)
5. Hashem, I.A.T., Yaqoob, I., Anuar, N.B., Mokhtar, S., Gani, A., Khan, S.U.: The rise of "big data" on cloud computing: Review and open research issues. Information Systems 47, 98–115 (2015)
6. Calder, B., Wang, J., Ogus, A., Nilakantan, N., Skjolsvold, A., McKelvie, S., Xu, Y., Srivastav, S., Wu, J., Simitci, H.: Windows Azure Storage: A highly available cloud storage service with strong consistency. In: Proceedings of the Twenty-Third ACM Symposium on Operating Systems Principles, pp. 143–157. ACM (Year)
7. Pucillo, F., Cascini, G.: A framework for user experience, needs and affordances. Design Studies 35, 160–179 (2014)
8. Steffen, M.W., Murad, M.H., Hays, J.T., Newcomb, R.D., Molella, R.G., Cha, S.S., Hagen, P.T.: Self-report of tobacco use status: Comparison of paper-based questionnaire, online questionnaire, and direct face-to-face interview–implications for meaningful use. Population Health Management 17, 185–189 (2014)
9. Dinzeo, T.J., Thayasivam, U., Sledjeski, E.M.: The development of the lifestyle and habits questionnaire-brief version: Relationship to quality of life and stress in college students. Prevention Science 15, 103–114 (2014)
10. Chen, M.-Y., Wang, E.K., Jeng, Y.-J.: Adequate sleep among adolescents is positively associated with health status and health-related behaviors. BMC Public Health 6, 59 (2006)

11. Patel, S.R., Hu, F.B.: Short sleep duration and weight gain: A systematic review. Obesity 16, 643–653 (2008)
12. Weinstein, A., Dorani, D., Elhadif, R., Bukovza, Y., Yarmulnik, A., Dannon, P.: Internet addiction is associated with social anxiety in young adults. Annals of Clinical Psychiatry 27, 3 (2015)
13. Lepp, A., Li, J., Barkley, J.E., Salehi-Esfahani, S.: Exploring the relationships between college students' cell phone use, personality and leisure. Computers in Human Behavior 43, 210–219 (2015)
14. Allison, K.C., Lundgren, J.D., O'Reardon, J.P., Geliebter, A., Gluck, M.E., Vinai, P., Mitchell, J.E., Schenck, C.H., Howell, M.J., Crow, S.J.: Proposed diagnostic criteria for night eating syndrome. International Journal of Eating Disorders 43, 241–247 (2010)
15. Kawamoto, K., Honey, A., Rubin, K.: The HL7-OMG healthcare services specification project: Motivation, methodology, and deliverables for enabling a semantically interoperable service-oriented architecture for healthcare. Journal of the American Medical Informatics Association 16, 874–881 (2009)

XTENS - A JSON-Based Digital Repository
for Biomedical Data Management

Massimiliano Izzo[1,2,*], Gabriele Arnulfo[1,3], Maria Carla Piastra[1],
Valentina Tedone[1], Luigi Varesio[2], and Marco Massimo Fato[1]

[1]Department of Computer Science Bioengineering Robotics and Systems Engineering,
University of Genoa, Viale Causa 13, 16145 Genoa, Italy
[2]Laboratory of Molecular Biology, Giannina Gaslini Institute, Largo Gaslini 5, 16147 Genoa,
Italy
[3]Neuroscience Center, P.O. Box 56, FI-00014 University of Helsinki, Finland
massimorgon@gmail.com

Abstract. Biomedical Science poses unique challenges in data management.
Heterogeneous information - such as clinical records, biological specimens,
imaging and genomic data, different technology-associated formats - must be
collected and integrated to provide a unified overview of each patient. International scale research collaborations involve different disciplines (Medi-
cine/Biology, Engineering/IT, Physics,...). Extensive metadata is required to
maximize information sharing among the partners. To properly tackle these issues, we have developed XTENS, a data repository built on a flexible and extensible JSON-based data model. The JSON data model is conceived to achieve
maximal flexibility, to allow adaptive metadata management, and to perceive
metadata as a dynamical process of scientific communication rather than an en-
during product fixed in time. XTENS is integrated with iRODS, a data grid
software that allows distributed storage, metadata file annotation and advanced
policies for data curation. We have adopted the platform for a functional connectomics multicentric project where heterogeneous data sources (radiological
images, electroencephalography signals) must be integrated and analysed to
compute connectivity maps of the brain. To this end, we have tested the repository prototype allowing the external programs to interact with XTENS using a
service-oriented REST interface. We demonstrated XTENS usefulness because
we could input heterogeneous data, run the required processing tool and store
the process output.

1 Background

Data Management in Biomedical Science presents peculiar challenges. Research
projects in the field are constantly moving towards international collaboration with
participants coming from a great variety of disciplines (medicine, biology, engineer-
ing/IT, etc.). In such a scenario, extensive metadata are required to improve the accessibility of the shared information, be it clinical records, biological specimens'

---

* Corresponding author.

F. Ortuño and I. Rojas (Eds.): IWBBIO 2015, Part II, LNCS 9044, pp. 123–130, 2015.

(i.e. samples) characteristics, or the output of high-throughput analyses. Many data repositories have been proposed in the field. Some of them - such as SimBioMS [1] and openBIS [2] - are more focused on molecular biology and genomics, and provide support for sample management; others - like XNAT [3], COINS [4] or LORIS [5] - are more oriented towards Neuroscience/Neuroimaging, and are equipped with tools for automatic metadata extraction from common radiological formats (DICOM/NIFTI) and quality controls on the uploaded images. A common trait of all these systems is that, in order to create novel data types - such as a hitherto unsupported genomic assay or a novel imaging format - the system must be reconfigured by an operator with sufficient informatics skills, able to modify SQL relations or XML schemas. As a consequence, shared data is often described with a somewhat fixed and limited set of metadata, out of the control of the researchers themselves, with strong drawbacks on the quality and completeness of the shared information, and the output of biomedical collaborative projects. Recent studies in Neural [7] and Social Sciences [8] promote the view of metadata as a fluid, dynamical process rather than a fixed product; therefore we feel that modern repository should comply this requirement, providing adaptive metadata management and configuration tools to maximize information sharing and understanding in multidisciplinary, international collaborations.

Here we describe a novel implementation of XTENS, a data repository based on a JSON metadata model [9], as a proper tool to solve data management issues in Biomedical Sciences and specifically in Functional Connectomics studies. Previously, XTENS was developed with Java technology and had been successfully used both in Neuroscience [10] and in integrated biobanking management [11]. In the next paragraphs, we describe the novel JSON-compliant XTENS architecture and its underlying data model. Then we will illustrate a use case scenario - a Stereo-ElectroEncephaloGraphy (SEEG) collaborative project where our group was providing support for data management and image/signal processing.

2   Results

2.1   The Repository

In XTENS metadata model, each Data instance is characterized by its Data Type. The Data Types provide a hierarchical JSON schema that works as a 'template' for the Data instance. The schema is composed by (i) a header with name, a brief description and the version of the data type, and (ii) a body, that contains the full set of metadata properties (i.e. attributes) usually collected in metadata groups. Each property is characterized by a name, a primitive type (i.e. number, string, date, boolean...), an (optional) ontology term, value options and units, a set of validation properties and a sensitivity flag. Each Data instance contains a metadata field in JSON format, where all the properties defined in the Data Type schema are stored as a name-value-unit triples. Subjects and Samples are treated as specialized versions of the data instance class, containing additional fields, methods and relationships. This paradigm ensures security, anonymisation of personal details, and dedicated biobanking management. The new XTENS implementation supports management of samples from different biobanks, and describes biobanks according to the MIABIS specifications of the

BBMRI consortium [11]. An outline of XTENS data model is shown in Figure 1. Figure 2 provides a visual example of the JSON schema. XTENS source code is available on Github [12]

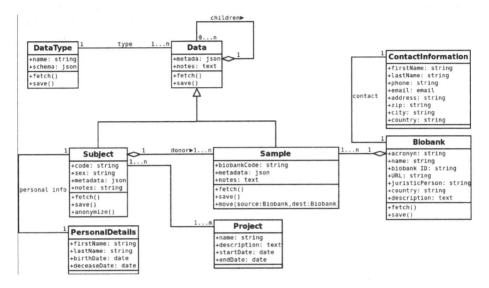

Fig. 1. XTENS data model. Each Data instance is characterized by its Data Type, and the Data Type schema provides the structure to build up the metadata JSON object. Subject and Sample are specialized classes of Data. Personal Details are managed in a separate class to allow easy anonymisation and satisfy privacy requirements.

The model has been implemented on the novel XTENS repository using a modular structure, separating the concerns between the client interface (i.e. front-end) and the back-end. The latter consists of a web application running on a Node.js server, a PostgreSQL database, and a distributed file system based on the iRODS data grid software [13]. The adoption of iRODS makes XTENS the first data management platform relying on a distributed file storage. The web application is written in JavaScript to provide full compliance with the JSON metadata model, and exposes a RESTful interface to allow a common access specification for the XTENS front-end and for external applications. We have chosen PostgreSQL as database management system because it supports JSON as a native type. The upcoming 9.4 version will introduce a binary JSON format (JSONB) that improves dramatically the query speed on retrieval. iRODS is equipped as well with a REST API (*irods-rest*) that allows direct and transparent upload/download from the client interface. When a new Data Instance is created using the front-end web form, the user can upload one or more associated files. Uploaded files are temporary stored in a 'landing' collection, and moved to their permanent location after the Data form is submitted. The association between the Data instance and Data file location in then stored in the database. A dynamical query interface based on the Composite design pattern allows users to perform queries on every Data Type previously defined. Data Type schemas can be updated with new properties using the graphical interface; new and updated data types can be used immediately for create-retrieve-delete-update (CRUD) operations with no additional programming, compilation or restart steps. The user (with administrative authorization) is given full control to describe its data and experiments, without having to resort every time to an IT expert. New Data Types can be created with minimal effort, to improve data sharing in different scenarios and to promote the concept of metadata as a dynamical, ephemeral and fluid process.

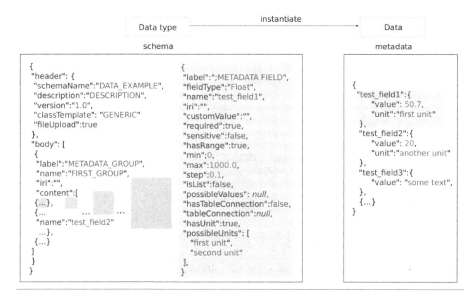

Fig. 2. Outline of the JSON metadata schema used in XTENS. Each Data Type has a schema composed by a header and a body. The latter is an array of Metadata Groups. Each contains one or more Metadata Field or Loops. Metadata Loops are not shown for sake of simplicity. A Metadata Field (shown in the second column) represents the leaf of the model and is described by a set of property that specify its primitive type, name, an optional International Resource Identifier (IRI) for linked data support, a set of validation properties to determine if it is required, if it stores sensitive/personal information, if (in case of numeric type) the value must fall in a determined range, if the value must be selected from a list of controlled terms, and if it has a measure unit. The "metadata" column of a Data entity stores the instances of each metadata field as a name-value-unit triple.

2.2 SEEG Use Case

We have set up a first XTENS 2.0 prototype to manage imaging data - computed tomography (CT), magnetic resonance imaging (MRI) and Stereo-ElectroEncephaloGraphy (SEEG) - data in a collaborative project involving three centres: Niguarda Hospital in Milan (providing data), the Neuroscience Centre at the University of Helsinki (developing methods), and the Department of informatics, Bioengineering, Robotics and System Engineering (DIBRIS) at the University of Genoa (Data storage and management). The aim of the project is to exploit recent advancements in functional and effective connectomics to tentatively define biomarkers for focal epilepsy. Functional (and effective) connectomics studies describe how different brain regions interact with each other and how modification of such functional (or effective) couplings is directly linked to neurological pathologies. In this context, it is crucial to have access to high quality tools to store, analyse, and retrieve multimodal datasets in order also to comply with national and international laws that rules the sharing of medical information and patient details.

Details about the methods used in data preprocessing and analyses can be found elsewhere [14] [15], but we briefly summarize the peculiar steps that interact with the XTENS platform. SEEG is a highly invasive techniques to record neural activity that is routinely used in clinical application aimed at localizing seizure onset zones in patients with drug-resistant focal epilepsy undergoing presurgical evaluation [16]. Despite the sparsity of SEEG implants, recently we showed that it can successfully be used in the context of functional connectomics studies fully exploiting its potential. We estimated that ~100 patients are required to reach a 85% coverage of all possible interactions in a 250 anatomical parcels atlas. The entire analyses can be divided in two domains: structural and functional. Each domain is characterized by different data, methods and analyses outputs. The structural domain deals with anatomical data and is composed of a post-implant CT (postCT) scans that show the electrode in their final locations and pre-implant MRI (preMRI) that contain the information about individual brain anatomy. We provided to the physicians a set of medical image processing tools that are specifically designed to deal with SEEG implants to (i) localize each contact in both individual and common geometrical spaces and (ii) to assign to each contact its neuronal source on a probabilistic reference atlas (Destrieux,). The functional domain deals with signal processing techniques aimed at quantifying the degree of synchrony between brain regions and at characterizing the so called functional connectome. We developed several tools (i) to correctly estimate phase differences and (ii) to assess statistical significance of the observed phase couplings.

XTENS successfully manages data describing both domains and provides client-side services for physicians to submit data and retrieve analyses results. We have installed XTENS on a cluster of Linux Servers (Ubuntu 12.04 LTS) located at the Department of Informatics, Bioengineering, Robotics and System Engineering (DIBRIS), Genoa, Italy. Details of the installation are shown in Figure 2. We currently have defined the following Data Types: Patient, Preimplant_MRI, Postimplant_CT, Fiducial_List, SEEG_Implant, SEEG_Data, and Adjacency_Matrix. SEEG_Implant data instances are the output of the segmentation process operated on Postimplant_CT using Fiducial_List metadata as reference. On the other hand, Adjacency_Matrix is the data describing brain region phase couplings in individual patient geometry estimated using SEEG_Data.

Postimplant_CT, Fiducial_List, and SEEG_Data are directly uploaded by the physician on the XTENS repository. We have developed a Node.js package, called *xpr-seeg*, to provide a web interface with the segmentation tool running on a separate server. The user fires through the XTENS client interface. In turn, XTENS sends a POST request to xpr-seeg forwarding all the required information about the two data instances. Xpr-seeg executes a bash script that retrieves the required files from iRODS and runs the segmentation algorithm. Once the procedure is done, the computed SEEG Implant is stored on a file. A novel data instance of SEEG_Implant is composed by xpr-seeg and saved in XTENS through a POST request. In a similar way, SEEG_Data are downloaded by operators, manually investigated to rejected artefactual channels (i.e., non physiological data) and analyzed to build the Adjacency_matrix. Here, xpr-seeg provides the tool to correctly upload the analysed data to its data parent (i.e., Patient) in the XTENS repository.

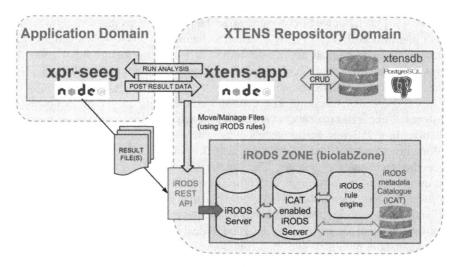

Fig. 3. XTENS setup for the Stereo-EEG collaborative project. XTENS communicates with xpr-seeg using REST. After the analysis (e.g. segmentation) is run by xpr-seeg, the results' file is stored in iRODS and a new Data instance is saved on XTENS.

3   Discussion

We have developed a novel data management and service providing platform that is a valid alternative to more established technological solutions, because it presents a number of advantages. First, the whole system is structured on JSON format, and does not require lengthy and cumbersome data transformation or binding often required when dealing with XML format or entity-attribute-value (EAV) paradigms. Secondly, the system is conceived to be user-friendly, easily configurable by non-IT people. This is a major advantage in biomedical science which is hampered by the difficulties in dealing with complex informatics systems. For instance, XNAT, among the most popular data repositories for Neuroscience projects, provides data schema in XML format. To modify data schema expressed in XML it is necessary to have basic/advanced knowledge of the syntax itself. Furthermore, after data schema changes the database needs to be updated, XNAT application redeployed and security XNAT setup must also be performed. In XTENS, the users can create new Data Types and setup the security and authorization levels using an intuitive graphical interface.

These characteristics make XTENS suitable for projects in labs that cannot afford a system administrator. XTENS only requires the initial configuration and deploy while new data types can be added at runtime.

The previous version of XTENS was based on Java Servlet technology (running on a Tomcat server), and was backed by a MYSQL database. We moved to Node.js and PostgreSQL to provide an environment more compliant to our JSON model. The new version is suitable to test the scaling capabilities of Node.js in conjunction with PostgreSQL JSONB format when managing large amounts of metadata. We are planning to run some stress tests on the system, before adopting it in production, and to tune the database for improving the performances. Moreover, the extensibility and simple management of the presented platform make it the optimal tool in connectomics studies. In this evolving and challenging scientific context, exchanging knowledge between scientific operators and multimodal data approaches can sensibly increase the interpretability of the results and the applicability of the method itself to the clinical context. In focal epilepsy studies, the complexity of the pathology itself and difficulties in the localization process can benefit from modern distributed technologies such as the one suggested in the present work.

4   Conclusions

We developed a novel data repository, with a highly configurable JSON-based data model, for research collaborations in Biomedical Science. We have tested the repository prototype in an ongoing SEEG project where external programs interacted with the repository using a service-oriented REST interface. We demonstrated its usefulness in Computational Neuroscience because we could (i) input CT/MRI and SEEG data, (ii) run the required processing tool and (iii) output the phase couplings co-localized on both individual and common anatomical spaces.

References

1. Krestyaninova, M., Zarins, A., Viksna, J., Kurbatova, N., Rucevskis, P., Neogi, S.G., Gostev, M., Perheentupa, T., Knuuttila, J., Barrett, A., et al.: A System for Information Management in BioMedical Studies–SIMBioMS. Bioinformatics 25, 2768–2769 (2009)
2. Bauch, A., Adamczyk, I., Buczek, P., Elmer, F.J., Enimanev, K., Glyzewski, P., Kohler, M., Pylak, T., Quandt, A., Ramakrishnan, C., et al.: openBIS: A flexible framework for managing and analyzing complex data in biology research. BMC Bioinformatics 12, 468 (2012)
3. Herrick, R., McKay, M., Olsen, T., et al.: Data dictionary services in XNAT and the Human Connectome Project. Front. Neuroinform. 8, 65 (2014)
4. Scott, A., Courtney, W., Wood, D., et al.: COINS: An Innovative Informatics and Neuroimaging Tool Suite Built for Large Heterogeneous Datasets. Front. Neuroinform. 5, 33 (2011)
5. Das, S., Zijdenbos, A.P., Harlap, J., Vins, D., Evans, A.C.: LORIS: A web-based data management system for multi-center studies. Front. Neuroinform. 5, 37 (2011)

6. Neu, S.C., Crawford, K.L., Toga, A.W.: Practical management of heterogeneous neuroimaging metadata by global neuroimaging data repositories. Front. Neuroinform. 6, 8 (2012)
7. Edwards, P.N., Mayernik, M.S., Batcheller, A.L., Bowker, G.C., Borgman, C.L.: Science friction: data, metadata, and collaboration. Soc. Stud. Sci. 41, 667–690 (2011)
8. JSON, http://www.json.org/
9. Corradi, L., Porro, I., Schenone, A., Momeni, P., Ferrari, R., Nobili, F., Ferrara, M., Arnulfo, G., Fato, M.M.: A repository based on a dynamically extensible data model supporting multidisciplinary research in neuroscience. BMC Med. Inform. Decis. Mak. 12, 115 (2012)
10. Izzo, M., Mortola, F., Arnulfo, G., Fato, M., Varesio, L.: A digital repository with an extensible data model for biobanking and genomic analysis management. BMC Genomics 15 (2014)
11. Norlin, L., Fransson, M., Eriksson, M., Merino-Martinez, R., Anderberg, M., Kurtovic, S., Litton, J.: A Minimum Data Set for Sharing Biobank Samples, Information, and Data: MIABIS. Biopreservation and Biobanking 10 (2012)
12. XTENS source code, https://github.com/biolab-unige/xtens-app
13. iRODS, http://irods.org/
14. Arnulfo, G., Hirvonen, J., Nobili, L., Palva, S., Palva, J.M.: Phase and Amplitude correlations in resting state activity in human stereoEEG recordings. NeuroImage (submitted)
15. Arnulfo, G., Narizzano, M., Cardinale, F., Fato, M.M., Palva, J.M.: Automatic Segmentation of Deeep intracerebral Electrodes in Computed Tomography Scans. BMC Bioinformatics (submitted)
16. Cardinale, F., Cossu, M., Castana, L., Casaceli, G., Schiariti, M.P., Miserocchi, A., et al.: Stereoelectroencephalography: surgical methodology, safety, and stereotactic application accuracy in 500 procedures. Neurosurgery 72(3), 353–366 (2013)

An Innovative Platform for Person-Centric Health and Wellness Support

Oresti Banos[1], Muhammad Bilal Amin[1], Wajahat Ali Khan[1],
Muhammad Afzel[1], Mahmood Ahmad[1], Maqbool Ali[1],
Taqdir Ali[1], Rahman Ali[1], Muhammad Bilal[1], Manhyung Han[1],
Jamil Hussain[1], Maqbool Hussain[1], Shujaat Hussain[1], Tae Ho Hur[1],
Jae Hun Bang[1], Thien Huynh-The[1], Muhammad Idris[1], Dong Wook Kang[1],
Sang Beom Park[1], Hameed Siddiqui[1], Le-Ba Vui[1], Muhammad Fahim[2],
Asad Masood Khattak[3], Byeong Ho Kang[4], and Sungyoung Lee[1],[*]

[1] Department of Computer Engineering, Kyung Hee University, Korea
{oresti,mbilalamin,wajahat.alikhan,muhammad.afzal,rayemahmood,maqbool.ali,
taqdir.ali,rahmanali,bilalrizvi,smiley,jamil,maqbool.hussain,
shujaat.hussain,hth,jhb,thienht,idris,dwkang,sbp,siddiqi,
lebavui,sylee}@oslab.khu.ac.kr
[2] Department of Computer Engineering, Istanbul Sabahattin Zaim University, Turkey
muhammad.fahim@izu.edu.tr
[3] College of Technological Innovation, Zayed University, UAE
asad.khattak@zu.ac.ae
[4] School of Computing and Information Systems, University of Tasmania, Australia
Byeong.Kang@utas.edu.au

Abstract. Modern digital technologies are paving the path to a revolutionary new concept of health and wellness care. Nowadays, many new solutions are being released and put at the reach of most consumers for promoting their health and wellness self-management. However, most of these applications are of very limited use, arguable accuracy and scarce interoperability with other similar systems. Accordingly, frameworks that may orchestrate, and intelligently leverage, all the data, information and knowledge generated through these systems are particularly required. This work introduces Mining Minds, an innovative framework that builds on some of the most prominent modern digital technologies, such as Big Data, Cloud Computing, and Internet of Things, to enable the provision of personalized healthcare and wellness support. This paper aims at describing the efficient and rational combination and interoperation of these technologies, as well as their integration with current and future personalized health and wellness services and business.

Keywords: Human behavior, Context-awareness, Big data, Big information, Big knowledge, Cloud computing, Quantified self, Digital health, Health devices, Social networks, User interface, User experience, Knowledge bases, Personalized recommendations.

---

[*] Author to whom correspondence should be addressed.

F. Ortuño and I. Rojas (Eds.): IWBBIO 2015, Part II, LNCS 9044, pp. 131–140, 2015.

1   Introduction

A drastic change in the delivery of health and wellness services is envisioned for the forthcoming years. The current global socio-economic situation, with cuts in government spending, an increasing population of pensioners and a growing unemployment rate, has particularly fostered the need of more efficient health and wellness care models. These new models particularly build on the concepts of proactivity and prevention, which in simple words refer to avoiding as much as possible the need of care. In fact, it is well-known that most prevalent diseases are partly caused by lifestyle choices that people make during their daily living. Therefore, bringing these lifestyle diseases under control may have a great impact on healthcare and assistance spending, and certainly on people's health itself. To that end, empowerment, encouragement and engagement of people in their personal health care and wellbeing is especially required.

In this context, information and communication technologies appear as the main driver of change to support people empowerment, encouragement and engagement. Actually, an increasing number of applications and systems for personalized healthcare and wellness management have been developed during the last years. These solutions, mainly oriented to fitness purposes, are used for detecting very primitive user routines and behaviors, and are also utilized for providing track of progresses and simple motivational instructions. Withings Pulse [4], Jawbone Up [2] and Fitbit Flex [1] are some examples of instrumented bracelets and wristbands accompanied by mobile apps capable of providing some basic recommendations based on the measured taken steps and slept hours. More prominent health and wellness systems have been provided at the research level, yet they are fundamentally prototypes. Examples of these systems are [9] for detecting cardiovascular illnesses, [6] for alerting on physical conditions or [10] for tracking changes in physiological responses of patients with chronic diseases. These solutions have a very limited application scope, lack of interoperability with other similar systems and rarely personalize to the particular user needs and preferences. Therefore, to neatly support all health and wellness aspects of each particular user, comprehensive frameworks capable of tackling more complex and realistic scenarios are required. Although a few attempts have been very recently provided in this regard [8,7,5], most of them lack essential requirements of a person-centric framework for health and wellness support.

In this work we present Mining Minds [3], an innovative framework that builds on the core concepts of the digital health and wellness paradigm to enable the provision of personalized healthcare and wellness support. Mining Minds is further devised to intelligently exploit digital health and wellness data to generate new businesses and services, which are unquestionably called upon to change the actual healthcare and wellness panorama. The rest of the paper is organized as follows. The essential requirements devised for a framework supporting personalized healthcare and wellness services are shown in Section 2. Section 3 thoroughly describes the proposed Mining Minds Framework. A potential business model and service scenario that may be supported through Mining Minds is presented in Section 4. Finally, main conclusions and future directions are shown in Section 5.

2   Requirements of a Person-Centric Digital Health and Wellness Framework

People health and wellness states can be represented through data of a very diverse nature, such as physical -sensory-, logical -personal profile and interests-, social -human cyber relations- and clinical -medical- data. Accordingly, one of the most important challenges posed to digital health and wellness systems refers to the intelligent and comprehensive collection, processing and organization of these data. For data collection, several modern systems such as wearable self-quantifiers, ambient sensors, SNS or advanced clinical systems, are increasingly available. Thus, a certain level of abstraction from heterogeneous resources is required to make their utilization transparent to the user. Moreover, data types are of a very diverse nature, ranging from structured - e.g., electronic health records -, semi-structured - e.g., multimedia - or unstructured - e.g., SNS -. Thus, an important requirement of person-centered digital health and wellness frameworks is to be capable of dealing with this new dimension of heterogeneous data. Not only data variety constitutes a key factor to be considered, but also data volume and velocity. Massive amounts of health and wellness-related data are generated over time on and around the subject at different paces. Therefore, these frameworks need to provide procedures to support the storage and real-time processing of such amounts of data. Similarly, mechanisms for load balancement and scalability are utterly required when dealing with several potential users and data collection mechanisms.

Determining a person's health and wellness state is a very challenging task that require more than simply collecting and persisting personal data. Thus, automatic intelligent mechanisms to process person-centered data, and extract interpretable information, are needed. Moreover, insights should be also gained not only from individual users but from the collectivity. To do so, advanced techniques to process people's health and wellness information in an anonymized form are particularly required. These insights can be leveraged by health and wellness care systems to extend, adapt and evolve the knowledge provided by human domain experts.

Mechanisms such as alerts, recommendations or guidelines, generally known as service enablers, are particularly used to catalyze both information and knowledge to be delivered in a human-understandable way to users and stakeholders in general. Not only should person-centered digital health and wellness frameworks provide these enablers, but also support mechanisms to customize them to each particular person needs and demands, for example, by mapping user needs to the best possible recommendations or personalizing the explanation of these recommendations.

Another important requirement of these frameworks refer to the presentation of health and wellness outcomes to the end-user. Since users of these systems may play a different role, the information needs to be presented in the most convenient way given their interest and expertise. For example, comprehensive user interfaces need to be prepared for clinical professionals, while more simplified and appealing presentations may be required by average users. Moreover, the analysis of the user interaction with the system is seen to be of great value. Users perceptions of system aspects such as utility and ease of use need to be fed back into the framework in order to provide the most personalized possible experience, as well as to help identify potential inconsistencies in the operation of the system.

Finally, all the aforementioned requirements need to be neatly accommodated to user security and privacy principles. Person-centered digital health and wellness frameworks deal with sensitive information, thus it is of utmost importance to adequate privacy, security, protection and risk management measures to all the processes involved in the treatment of this information.

3   Mining Minds Platform

In the light of the requirements presented in the previous section, a novel person-centric digital health and wellness framework is proposed here. Hereafter referred to as "Mining Minds", this framework consists of a collection of innovative services, tools, and techniques, working collaboratively to investigate on human's daily life data, generated from heterogeneous resources, for personalized health and wellness support. Mining Minds philosophy revolves around the concepts of data, information and service curation, which refer to the adequation, adaptation and evolution of both contents and mechanisms used for the provision of high quality health and wellness services. Motivated by these concepts, a multilayer architecture is particularly devised for Mining Minds. The architecture, depicted in Figure 1, is composed by three main layers, respectively, Data Curation Layer (DCL), Information Curation Layer (ICL) and Service Curation Layer (SCL), and an additional one, Supporting Layer (SL), to ensure the suitable operation of the former ones. In the following, the Mining Minds architecture layers are described.

3.1   Data Curation Layer

Data Curation Layer, DCL, is in charge of processing and persisting the data obtained from the Sensor Layer, which abstractly defines the possible sources of user health and wellness data, i.e., SNS, questionnaires, wearable biomedical devices or ambient intelligence systems, among others. The DCL is composed of Data Curation, Data Representation and Mapping, and Big Data Persistence components. Data Curation is responsible for the acquisition, labeling and analysis of the data obtained from the diverse sources, in both real-time and offline manners, as generic streams. The format of the acquired streams is based on the source devices, thus their specifications are hosted by device registry of the Data Curation component. To classify the data streams by device and usage,

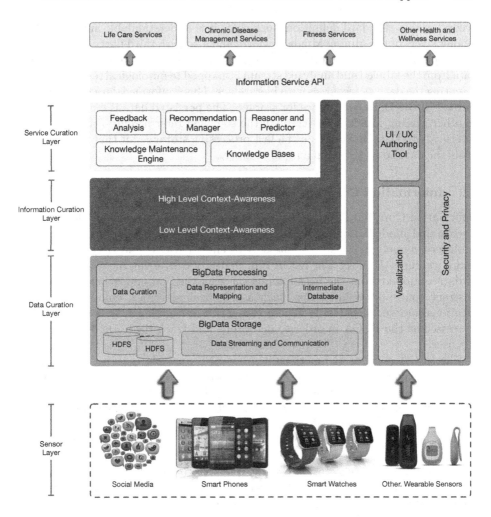

Fig. 1. Mining Minds Framework Architecture

the Data Curation component provides real-time data labeling, which converts the unstructured data into semi-structured format. As the volume of the data collected is large and type of this data is heterogeneous, the possibility of data noise and redundancy is high; therefore, the labeled data stream is forwarded for analysis where several data analysis filters are executed to ensure the reliability of data, keeping its comprehensiveness preserved. Apart from analysis of real-time data, the Data Curation also ensures the reliability of already preserved data with its provenance features. These features are executed as filters over offline data batch processes.

After analysis, data streams are forwarded to the Data Representation and Mapping component. The role of Data Representation and Mapping is to conform data according to a standard definition; such that, it is understandable and shareable among layers of the Mining Minds platform and also with third party components and applications. The conformance definition is based upon an ontology, where data from the labeled and analyzed stream is mapped to ontological resources, representing the data as resources with hierarchies. This conformed data model is forwarded to Big Data persistence for storage. The persisted data is made available to other Mining Minds layers through the so-called Intermediate Database. The Intermediate Database consists of a fast processing storage unit that temporarily hosts the data to be served in a rapid manner.

3.2   Information Curation Layer

Information Curation Layer, ICL, represents the Mining Minds core for the inference and modeling of the user context and behavior. It is composed by two sublayers, namely, Low Level Context Awareness (LLCA) and High Level Context Awareness (HLCA). The LLCA is in charge of converting the wide-spectrum of data obtained from the user interaction with the real and cyber-world, into abstract concepts or categories, such as physical activities, emotional states, locations and social patterns. These categories are intelligently combined and processed at the HLCA in order to determine and track the normal behavior of the user.

The LLCA consists of functionalities for SNS analysis, activity recognition, emotion recognition and location detection. The SNS analysis relates to the processing of the data generated by the user during their interactions in regular social networks such as 'Facebook' or 'Twitter'. This comprises from posts or tweets generated by the user themselves, user mentions, user traces and even global social trends, in the form of both text and multimedia data. From here, personal and general people interests, needs, conducts and states may be determined. The activity recognition process refers to the identification of primitive physical actions performed by the user, such as, 'standing', 'walking' or 'jogging'. This process may build on several sensing modalities, as they happen to be available to the user. Examples of these modalities are wearable inertial sensors, video and audio. The emotion recognition process is defined to infer user emotional states, such as 'happiness' or 'anxiety', by using sensor data similar to the aforementioned, as well as more sophisticated sources exploring human physiological variations and responses. In order to determine the user situation, it is of extreme importance to track the user ambulation. This is the role of the location detection functionality, which essentially builds on the data collected through indoor and outdoor positioning sensors, such as video and GPS, to specify the exact location or direction of the user. The information generated on top of the LLCA is unified and delivered to the DCL, in order to make it accessible to not only the HLCA, but other Mining Minds components and applications.

The diverse categories identified through the LLCA are used by the HLCA to define a more comprehensive representation of the user context and behavior. Two main functionalities are considered to that end. The first one corresponds to the context awareness and modeling, which enables the interpretation and representation of the user context. The modeling of the context is performed through ontologies, which have been adopted in the past as a unified conceptual backbone for modeling, representing and inferring context, while its interpretation is done through a rule-based reasoning process. Thus for example, based on the actual time (e.g., midday), location (e.g., restaurant) and inferred activities (e.g., sitting), this functionality can determine the precise user context (e.g., lunch). The context awareness and modeling is also used to populate the LifeLog Repository. This repository is used to store the contexts determined for the person during the use of the Mining Minds system. This information can be served to other Mining Minds components and applications, although it is primarily devised as input to the second essential HLCA functionality, so-called behavior modeling and analysis, which is devised to identify the user behavior patterns and routines. For example, if it has been identified that the user normally goes for lunch during a specific time span in work days, it can be determined as a personal behavior pattern or routine of this particular user. LLCA and HLCA information is regularly stored in the Intermediate Database, making it accessible to the SCL and other potential parties.

3.3   Service Curation Layer

Service Curation Layer, SCL, provides the means to transform the data and information generated by DCL and ICL into actual services. To do so, SCL supports automatic and expert-based knowledge creation and maintenance, personalized recommendations and predictions, and users feedback analysis. The knowledge creation capability is activated either by the domain expert or knowledge engineer, by using data driven, knowledge driven or hybrid approaches. The created knowledge, which is persisted in the Knowledge Bases of SCL, has various levels of granularity, which range from abstract or general to personalized or user-specific knowledge. The knowledge managed by SCL is used to generate personalized health and wellness recommendations. First, the Reasoner component uses the abstract level knowledge for generating general recommendations, that are further personalized by the Recommendation Manager. Then, the Recommendation Manager makes use of the personalized knowledge, which encodes user preferences and contextual information. Once the personalized recommendations are delivered to the user, feedback can be obtained from their acceptance - i.e., recommendation is followed - or rejection - i.e., recommendation is not followed -. This feedback is analyzed through the Feedback Analysis component, which converts it into information interpretable by the Knowledge Maintenance Engine component. This valuable information is then used by the Knowledge Maintenance Engine to update and evolve the user-based knowledge, in order to ensure a more personalized and adequate health and wellness support.

3.4   Supporting Layer

The role of the Supporting Layer, SL, is to enrich the overall Mining Minds functionalities through advanced visualization, interactive and personalized UI/UX and adequate procedures to ensure privacy and security in all aspects. The main role of the Visualization component is to adjust the style of the information delivered to the users based on their expertise and role. On the one hand, for example, average users may receive certain recommendations related to their daily life activities in the form of comprehensive textual or audiovisual instructions. More complex analytics may be displayed to human experts in relation to users health and wellness data, information and knowledge.

UI/UX is a major supportive component aimed to engage the end-user with the Mining Minds system in an intuitive fashion. Considering user preferences, habits, attitude and mood, the UI/UX component enables the end-user applications interface to adapt accordingly. This adaptation is required to fine tune the human-computer interaction experience with respect to font size, color, theme, or audio levels, among other characteristics.

Considering the sensitivity and associated concerns of the collected personalized information, the Mining Minds system need to assure and exhibit adequate privacy and security, not only at a storage level, but also during processing and delivery of services. Mining Minds employs state of the art existing cryptographic primitives along with indigenous protocols to exhibit more control over possible states of data. For secure storage, AES standard is considered, whereas for oblivious processing homomorphic encryption and private matching is used. Taking into account the intensive data flow between end-users applications and systems and the Mining Mind platform, randomization techniques are considered. These procedures ensure a high entropy for minimal leakage of information. For sharing personalized information and recommendations with the users, an authorized model ensures the legitimate disclosure. Slow processing of information is an effect caused by the encryption; however, to assist partial swiftness to Mining Minds, sensitive and non-sensitive information is decoupled where required.

4   Business Model and Service Scenario

A potential business model for Mining Minds consists of enhancing the relationship between health insurance companies and their customers. A healthier customer is beneficial for an insurance company as it can help reduce medical and assistance costs, ultimately resulting in higher profits. Likewise, customers can benefit from managing their health and wellness by improving their health conditions and also receiving rewards in the form of cheaper health insurances, lower co-pays, deductibles and out-of-pocket health expenses.

The management of people's health and wellness through Mining Minds requires defining diverse service scenarios. These services should particularly relate to the user's daily life activities, covering those aspects of their lifestyle that may have a direct or indirect impact on their health and wellness status. One of the simplest but at the same time most challenging case application scenario refers to the weight management of a person. The determinants of abnormal weight have been deeply explored in the past; however, practical mechanisms to encourage and guide users to lose or gain weight are very primitive and of limited success. Mining Minds aims at tackling this problem from an holistic perspective, supporting diverse key services necessary for an efficient weight management, such as healthy diet menu management, restaurant recommendations, convenient food store suggestion and exercise encouragement, among others. These services are not only seen to serve as pillars to empower users and promote a healthy weight management, but also open a new branch of potential third party businesses. Other envisioned case study scenarios that are in the scope of Mining Minds include, but are not limited to, health management of chronic disease patients, anti-aging habits promotion, pregnancy management and infant care assessment.

5   Conclusions

This work introduced Mining Minds, an innovative digital health framework for personalized healthcare and wellness support. The framework has been neatly designed taking into account crucial requirements of technologies and applications of the digital health and wellness domain. As a result, a multilayered architecture defined to provide the necessary functionality to enable a broad range of services for personalized healthcare and wellness has been presented. The proposed architecture, being the result of both technical and business-oriented research, could enable a new marketplace and the creation of a new business ecosystem around healthcare, wellness and other related domains.

This paper also showed the feasibility of the Mining Minds concept as well as an initial realization of the key architectural components. Future work includes the enhancement of the existing components as well as an evaluation of the presented architecture and its services on a large scale testbed, which is currently under construction.

Acknowledgments. This research was partially funded by the Korean Ministry of Science, ICT & Future Planning (MSIP) as part of the ICT R&D Program 2013. This work was also supported by the Industrial Core Technology Development Program, funded by the Korean Ministry of Trade, Industry and Energy (MOTIE), under grant number #10049079.

References

1. Fitbit Flex (2014), http://www.fitbit.com/flex (accessed: October 22, 2014)
2. Jawbone Up (2014), https://jawbone.com/up (accessed: October 22, 2014)
3. Mining Minds Project (2014), http://www.miningminds.re.kr/
4. Withings Pulse (2014), http://www.withings.com/es/withings-pulse.html (accessed: October 22, 2014)
5. Banos, O., Garcia, R., Holgado-Terriza, J.A., Damas, M., Pomares, H., Rojas, I., Saez, A., Villalonga, C.: mHealthDroid: A novel framework for agile development of mobile health applications. In: Pecchia, L., Chen, L.L., Nugent, C., Bravo, J. (eds.) IWAAL 2014. LNCS, vol. 8868, pp. 91–98. Springer, Heidelberg (2014)

6. Banos, O., Villalonga, C., Damas, M., Gloesekoetter, P., Pomares, H., Rojas, I.: Phys-iodroid: Combining wearable health sensors and mobile devices for a ubiquitous, con-tinuous, and personal monitoring. The Scientific World Journal 2014(490824), 1–11 (2014)

7. Fortino, G., Giannantonio, R., Gravina, R., Kuryloski, P., Jafari, R.: Enabling ef-fective programming and flexible management of efficient body sensor network ap-plications. IEEE Transactions on Human-Machine Systems 43(1), 115–133 (2013)

8. Gaggioli, A., Pioggia, G., Tartarisco, G., Baldus, G., Corda, D., Cipresso, P., Riva, G.: A mobile data collection platform for mental health research. Personal Ubiqui-tous Comput. 17(2), 241–251 (2013)

9. Oresko, J.J., Jin, Z., Cheng, J., Huang, S., Sun, Y., Duschl, H., Cheng, A.C.: A wearable smartphone-based platform for real-time cardiovascular disease detection via electrocardiogram processing. IEEE Transactions on Information Technology in Biomedicine 14(3), 734–740 (2010)

10. Patel, S., Mancinelli, C., Healey, J., Moy, M., Bonato, P.: Using wearable sensors to monitor physical activities of patients with copd: A comparison of classifier performance. In: Proceedings of 6th International Workshop on Wearable and Im-plantable Body Sensor Networks, Washington, DC, USA, pp. 234–239 (2009)

An Ontology for Dynamic Sensor Selection in Wearable Activity Recognition

Claudia Villalonga[1], Oresti Banos[2], Hector Pomares[1], and Ignacio Rojas[1]

[1] Research Center for Information and Communications Technologies of the
University of Granada (CITIC-UGR), C/Periodista Rafael Gomez Montero 2,
Granada, Spain
cvillalonga@correo.ugr.es, {hector,irojas}@ugr.es
[2] Department of Computer Engineering, Kyung Hee University, Korea
oresti@oslab.khu.ac.kr

Abstract. A strong effort has been made during the last years in the autonomous and automatic recognition of human activities by using wearable sensor systems. However, the vast majority of proposed solutions are designed for ideal scenarios, where the sensors are pre-defined, well-known and steady. Such systems are of little application in real-world settings, in which the sensors are subject to changes that may lead to a partial or total malfunctioning of the recognition system. This work presents an innovative use of ontologies in activity recognition to support the intelligent and dynamic selection of the best replacement for a given shifted or anomalous wearable sensor. Concretely, an upper ontology describing wearable sensors and their main properties, such as measured magnitude, location and internal characteristics is presented. Moreover, a domain ontology particularly defined to neatly and unequivocally represent the exact placement of the sensor on the human body is presented. These ontological models are particularly aimed at making possible the use of standard wearable activity recognition in data-driven approaches.

Keywords: Ontologies, Activity Recognition, Wearable sensors, Sensor selection, Sensor placement, Human anatomy.

1 Introduction

In the recent years, an enormous interest has awaken in the human physical self-quantification. Particularly devoted to health and wellness improvement, the personal self-tracking and evaluation of people's wellbeing is flourishing as a key business in which hundreds of applications and systems are increasingly available at the reach of most consumers. Most of these systems build on mobile and portable sensor devices that are carried on, or directly worn, by their users. Generally named "wearables", these devices are capable of measuring important physical and physiological human characteristics such as body motion or vital signs, which are principally used to quantify physical activity patterns [7,11] as well as to determine abnormal vital conditions [15,14,9].

F. Ortuño and I. Rojas (Eds.): IWBBIO 2015, Part II, LNCS 9044, pp. 141–152, 2015.

By far, most of the effort in the personal self-quantification has been devoted to the analysis of human behavior by using wearable systems, also known as wearable activity recognition. Many solutions have been provided to that end, and although accurate systems are available, most of them are designed to work in closed environments, where the sensors are pre-defined, well-known and steady. However, real-world scenarios do not fulfill these conditions, since sensors might suffer from diverse type of anomalies, such as failures [6] or deployment changes [8]. Realistic dynamic sensor setups pose important challenges to the practical use of wearable activity recognition systems, which translate into specific requirements to ensure seamless recognition capabilities. One of the most important requirements refers to the support of anomalous sensor replacement to maintain the recognition systems operation properly. In order to enable sensor replacement functionalities in an activity recognition system, mechanisms to abstract the selection of the most adequate sensors are needed. To that end, a comprehensive and interoperable description of the available sensors is required, so that the best ones could be selected to replace the anomalous ones.

Although technical characteristics may be extracted from data or spec sheets, more practical definitions such sensor location or availability are required for an accurate sensor selection at runtime. Accordingly, models that may integrate these heterogeneous sensor descriptions are required. In this work, the use of ontologies is proposed to neatly and comprehensively describe the wearable sensors available to the user. Concretely, this work aims at defining ontologies to support the intelligent and dynamic selection of the best replacement wearable sensor in case an anomalous one is determined. To the best of the authors' knowledge, this is the first time that ontologies are used in this regard, which goes beyond the state-of-the-art utilization of these models to detect activities in a knowledge-based recognition approach. On the contrary, it can be said that, ontologies are used here to enhance the machine learning activity recognition used in data-driven approaches. The rest of the paper is as follows. In Section 2 an overview on the use of ontologies in activity recognition is provided. The key motivations for the use of ontologies in dynamic sensor selection is presented in Section 3. Section 4 thoroughly describes the ontology proposed for the sensor selection problem. Finally, main remarks and conclusions are provided in Section 5.

2   Related Work

The use of ontologies in activity aware systems is principally focused on the application of knowledge-based recognition techniques. In these approaches, the activities are described through ontologies and recognized using reasoning and inference methods. For example, Bae [4] presents an ontology-based smart home system that discovers and monitors activities of the daily living. Nguyen et al. [13] also propose an ontological approach using the outputs of binary sensors to detect office activities. A similar use of the ontologies is made by Cheng et al. [10] to both represent and reason activities based on the analysis of the user interaction with smart objects in pervasive environments.

Previous approaches rely on binary or very simple sensors to detect primitives or atomic activities, which are described in an ontological model and used for ontological reasoning to detect high level activities. However, they do no exploit the potential of data-driven approaches in activity recognition, where the sensor data is analyzed using machine learning techniques to detect patterns matching known activities. Therefore, and in order to move one step forward, knowledge-driven approaches have been combined with data-driven approaches to recognize activities. For example, BakhshandehAbkenar and Loke [5] define a hybrid model using machine learning techniques applied to body motion data and reasoning based on the ontological representation of the activities. Riboni and Bettini [16,17] utilize ontological reasoning to recognize complex activities based on simple actions, which are detected via supervised learning algorithms building on data from wearable sensors and mobile devices.

3   Motivation for the Use of Ontologies for Sensor Selection

In order to provide interoperability, heterogeneous sensors used in wearable activity recognition systems should be abstracted from the actual underlying network infrastructure. This of utmost importance to be able to replace a sensor suffering from anomalies with another one which could provide the activity recognition system with the same sensing functionality.

A semantic description is needed to define the wearable sensor capabilities; not only the information the sensor measures and its intrinsic characteristics, but also its location on the human body. In case the anomalous sensor selection and replacement were done by human users, it would be sufficient to describe the sensor with a number of keywords or tags. However, free-text tags are insufficient for any machine-based interaction, where the selection and replacement of anomalous sensors have to be executed by a machine. In this case, the syntax and semantics of the sensor description need to be clearly defined.

In the sensor description, the semantics could be implemented using different representations. For example a language with implicit semantics like XML or an ontology language that formally describes the semantics. XML descriptions do not provide the full potential for machines to acquire and interpret the emerging semantics from data, therefore the semantic meaning of the data has to be previously agreed between machines. Conversely, an ontology-based data representation solves these problems and enables efficient selection for heterogeneous sensors. The drawbacks of ontologies are the overhead in their representation and the complexity of defining the models. However, the interpretation of the semantics out of the data is a great advantage that overcomes these disadvantages. For all this reasons, ontologies one of the best options to capture the semantics in the sensor description.

Moreover, one of the properties of a formal structure like an ontology is the interoperability. Therefore, using ontologies the sensor descriptions provided for sensors of different vendors are sufficiently rich to be automatically interpreted by the activity recognition system to apply methods to select a replacing sensor. This work proposes an ontology to describe heterogeneous wearable sensors and which supports the replacement of anomalous sensors in activity recognition systems.

4 The Sensor Selection for Real-World Wearable Activity Recognition Ontology

An ontology to describe heterogeneous wearable sensors and which will enable the selection replacement of anomalous sensors in activity recognition systems is presented in this work. This ontology named Sensor Selection for Real-World Wearable Activity Recognition Ontology (SS4RWWAR Ontology) needs to have two main characteristics: extensibility and evolvability. These refer to the ability of the SS4RWWAR Ontology to support the description of new sensors not envisioned at design time and used in new application domains. Extensibility and evolvability require that the ontology is designed to assure that the mechanisms to select the best sensors for replacement are still valid and do not need to be re-implemented when new sensors are added and new concepts are included to the ontology.

The SS4RWWAR Ontology needs to be defined as an upper ontology which defines the basic common concepts and several plugable domain ontologies which inherit from the concepts in the upper ontology. New concepts, that could be required in future activity recognition applications, are defined in domain ontologies that extend these models. Extending the SS4RWWAR Ontology in a distributed fashion by generating the new concepts for the sensor descriptions in a decentralized manner could be achieved in the future using an approach based on Linked Data [2]. Moreover, in order to allow extensibility, existing ontologies have to be reused if possible, for example for the definition of the sensing magnitudes, units or body locations.

4.1 SS4RWWAR Upper Ontology

The SS4RWWAR Upper Ontology specifies the sensor description and includes the list of magnitudes that can be measured by the sensor, the location where the sensor is placed, the sensor internal characteristics and a human readable description of the sensor. The *WearableSensor* class is the main concept of the SS4RWWAR Upper Ontology and an instance of this class is the actual sensor description. In this work we use the well-known ontology language OWL2 as encoding for the sensor descriptions because of its expressiveness. The graphical representation of the SS4RWWAR Upper Ontology with all its classes and properties is shown in Fig. 1.

Magnitudes measured by the wearable sensors need to be clearly specified in order to support the definition of heterogeneous sensor modalities used in activity recognition. In the SS4RWWAR Upper Ontology the magnitudes are represented by the *Magnitude* class. In order to link the *Magnitude* class to the

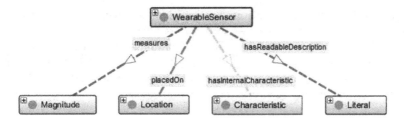

Fig. 1. SS4RWWAR Upper Ontology representing the description of Wearable Sensors

sensor description the *measures* property has been defined. This object property has as domain the *WearableSensor* class and as range the *Magnitude* class. The *Magnitude* class has to be further specified in a domain ontology in order to describe the details of each magnitude. At this moment, the ontology only defines three types of magnitudes measured by Inertial Measurement Units (IMU). The three subclasses of the *Magnitude* class are the *Acceleration* class representing the measurement of accelerometers, the *TurnRate* class representing the measurement of gyroscopes, and the *MagneticFieldOrientation* class representing the measurements of magnetometers. In the future we plan to provide a complete ontology with the most important sensor modalities which allows the description of the most common sensors.

Wearable sensors location affects the performance of the activity recognition systems. In order to allow sensor replacement, the locations of the sensors need to be well specified. In the SS4RWWAR Upper Ontology, the position where the wearable sensor is placed is described by the *Location* class. The link to the sensor description is established through the *placedOn* property which has as domain the *WearableSensor* class and as range the *Location* class. Since wearable sensors are placed on the human body, the actual location of the sensor will be a body part. In order to describe the human body parts and use them as sensor locations, the *HumanBodyPart* class, which is a subclass of the *Location* class, has been defined in the SS4RWWAR Human Body Ontology. This ontology, described in Section 4.2, is one of the main contributions of this work and is the key to support the selection of replacement sensors placed on closed by body locations.

Wearable sensors from different vendors have different characteristics, for example sensor dynamic range, bias, or offset, which have to be properly described in the ontology. The *Characteristic* class is used to describe these internal sensor characteristics. The link between the characteristics and the actual sensor description is done via the *hasInternalCharacteristic* property, which has as domain the *WearableSensor* class and as range the *Characteristic* class. The *Characteristic* class needs to be further specified in the future in order to comprehensively describe all the sensor characteristics.

The sensor description may contain some human readable information about the sensor. Examples of these descriptions could be "SHIMMER 3", "Fitbit Flex" or any other name that could identify the sensor. The property *hasReadableDescription* is used to link the human readable text, represented by the class *rdfs:Literal*, to the sensor description. The property *hasReadableDescription* has as domain the *WearableSensor* class and as range the *rdfs:Literal* class.

The SS4RWWAR Upper Ontology is quite simple since all the potential will be derived of the domain ontologies, like the SS4RWWAR Body Ontology presented in the forthcoming section. As any other ontology, the SS4RWWAR Upper Ontology is subject to any future extensions and revisions.

4.2   SS4RWWAR Human Body Ontology

Wearable sensors are placed on the human body, they are located on concrete body parts. In order to represent human body parts the SS4RWWAR Body Ontology has been defined. The possibility of using available ontologies to describe the human body parts has been analized. A candidate ontology was the Foundational Model of Anatomy ontology (FMA) [1], one of the most complete knowledge source for bioinformatics which represents the phenotypic structure of the human body. Another candidate was the Uber anatomy ontology (Uberon) [12,3], an anatomy ontology that integrates any type of animal species. These ontologies are too extensive for the purpose of this work since the location of the sensors does not require the definition of the internal organs, neural network, skeletal system or musculature. In fact, the FMA ontology is composed of over 75.000 classes and the Uberon of over 10.000 classes, which makes them too complex for reasoning on the selection of best wearable sensors. For these reasons, a new body ontology describing only the body locations where sensors can be worn has been created in this work. This ontology is based on the lessons learned from studying the well-known anatomical ontologies.

The main class of the SS4RWWAR Body Ontology is the *HumanBodyPart* and represents each one of the body parts (see Fig. 2). The main division of the body is done in four parts: head, trunk, upper limbs and lower limbs. Therefore, four classes are defined as subclasses of the *HumanBodyPart*: the *Head*, the *Trunk*, the *UpperLimb* and the *LowerLimb*. Moreover, each of the main body parts can be further partitioned in subdivisions, which are also parts of the human body and therefore subclasses of the *HumanBodyPart* class. The *HeadSubdivision* class has been specified to define the subdivisions of the head: face and scalp. The *TrunkSubdivision* has been specified to define the subdivisions of the trunk: thorax, abdomen and back. The *UpperLimbSubdivision* class has been specified to define the subdivisions of the upper limbs: shoulder, arm, elbow, forearm, wrist, and hand. The *LowerLimbSubdivision* class has been specified to define the subdivisions of the lower limbs: hip, thigh, knee, leg, ankle, and foot.

In order to set the links between the each of the main body parts and their corresponding subdivisions, the *hasPart* object property has been defined, as well as its inverse property the *partOf* property which relates the subdivisions to their containing main body part (see Fig. 2). The link between the *HeadSubdivision* class and the *Head* class is created by using the *partOf* property and defining the *HeadSubdivision* as a subclass of the axiom *partOf some Head*. Similarly, the inverse property *hasPart* links the *Head*class to the *HeadSubdivision* class. In the same way, these properties are used to establish the relations between the

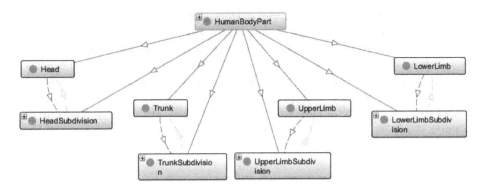

Fig. 2. Top part of the SS4RWWAR Body Ontology defining the body and its four main parts - head, trunk, upper limb and lower limb - represented by the *HumanBodyPart*, the *Head*, the *Trunk*, the *UpperLimb* and the *LowerLimb* classes, and their subdivisions represented by the *HeadSubdivision*, the *TrunkSubdivision*, the *UpperLimbSubdivision* and the *LowerLimbSubdivision* classes. The continuous purple arrows represent the *has subclass* property, which links the *HumanBodyPart* class with its eight subclasses. The dashed brown arrows represent the *hasPart* property, which relates the main body parts to their corresponding subdivisions. The dashed yellow arrows represent the *partOf* property, which relates the subdivisions with the main body parts.

rest of body parts. The link between the *TrunkSubdivision* class and the *Trunk* class is created by using the *partOf* property and defining the *TrunkSubdivision* as a subclass of the axiom *partOf some Trunk*, and the inverse property *hasPart* links the *Trunk* class to the *TrunkSubdivision* class. The link between the *Upper-LimbSubdivision* class and the *UpperLimb* class is created by using the *partOf* property and defining the *UpperLimbSubdivision* as a subclass of the axiom *partOf some UpperLimb*, and the inverse property *hasPart* links the *UpperLimb* class to the *UpperLimbSubdivision* class. The link between the *LowerLimbSub-division* class and the *LowerLimb* class is created by using the *partOf* property and defining the *LowerLimbSubdivision* as a subclass of the axiom *partOf some LowerLimb*, and the inverse property *hasPart* links the *LowerLimb* class to the *LowerLimbSubdivision* class.

Not only are the different body parts subdivided in a hierarchical manner, they are also connected to other parts. Several object properties have been defined in the SS4RWWAR Body Ontology to describe the connections between body parts. The top property is the *connectedTo* property and has eight sub-properties which define the connections of body parts according to the standard human directional terminology: superior or inferior, anterior or posterior, medial or lateral, proximal or distal. The *superiorlyConnectedTo* property relates a body part with another which is located towards the top of the body or human head. Its inverse, the *inferiorlyConnectedTo* property relates a body part with another which is located towards the bottom of the body or feet. The *anteriorly-ConnectedTo* property relates a body part with another which is located towards the front of the body. Its inverse, the *posteriorlyConnectedTo* property relates a body part with another which is located towards the back of the body. The *laterallyConnectedTo* property relates a body part with another which is located towards the lateral of the body. Its inverse, the *mediallyConnectedTo* property relates a body part with another which is located towards the middle of the body. The *proximallyConnectedTo* property relates a body part with another which is located towards the main mass of the body. Its inverse, the *distallyConnectedTo* property relates a body part with another which is located more distantly of the main mass of the body.

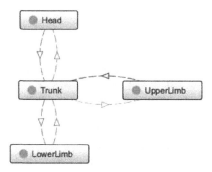

Fig. 3. Representation of the connections between the main body parts - head, trunk, upper limb and lower limb -. The dashed red arrows represent the *superiorlyConnectedTo* property, which relates the *Trunk* class to the *Head* class, and the *LowerLimb* class to the *Trunk* class. The dashed blue arrows represent the *inferiorlyConnectedTo* property, which relates the *Head* class to the *Trunk* class, and the *Trunk* class to the *LowerLimb* class. The dashed green arrow represents the *laterallyConnectedTo* property, which relates the *Trunk* class to the *UpperLimb* class. The dashed purple arrow represents the *mediallyConnectedTo* property, which relates the *UpperLimb* class to the *Trunk* class.

The connections between the main body parts can be established through the eight subproperties of the *connectedTo* property as shown in Fig. 3. Since the head is the top of the body and has located the trunk below, the connection between the *Head* class and the *Trunk* class is created by using the *inferiorlyConnectedTo* property and defining the *Head* as a subclass of the axiom *inferiorlyConnectedTo some Trunk*. Inversely, the connection between the *Trunk* class and the *Head* class is created by using the *superiorlyConnectedTo* property and defining the *Trunk* as a subclass of the axiom *superiorlyConnectedTo some Head*. The same reasoning applies to the connection between the trunk and the lower limbs, since the trunk is on top of the lower limbs. Thus, the connection between the *Trunk* class and the *LowerLimb* class is created by using the *inferiorlyConnectedTo* property and defining the *Trunk* as a subclass of the axiom *inferiorlyConnectedTo some LowerLimb*. Inversely, the connection between the *LowerLimb* class and the *Trunk* class is created by using the *superiorlyConnectedTo* property and defining the *LowerLimb* as a subclass of the axiom *superiorlyConnectedTo some Trunk*. Finally, the trunk is in the middle of the body and the upper limbs are in a lateral position from the trunk. Thus, the connection between the *Trunk* class and the *UpperLimb* class is created by using the *lateralyConnectedTo* property and defining the *Trunk* as a subclass of the axiom *lateralyConnectedTo some UpperLimb*. Inversely, the connection between the *UpperLimb* class and the *Trunk* class is created by using the *mediallyConnectedTo* property and defining the *UpperLimb* as a subclass of the axiom *mediallyConnectedTo some Trunk*.

In order to complete the the SS4RWWAR Body Ontology definition, the subdivisions of the main body parts need to be specified and the connections between these subdivisions need to be established. Fig. 4 shows the classes and properties related to the body subdivisions.

The *HeadSubdivision* class (see Fig. 4(a)) has two subclasses, the *Face* and the *Scalp*, which inherit from the *HeadSubdivision* class being a subclass of the axiom *partOf some Head*. The face is the anterior part of the head and the scalp the posterior part of it. Thus, the connection between the *Face* class and the *Scalp* class is created by using the *posteriorlyConnectedTo* property and defining the *Face* as a subclass of the axiom *posteriorlyConnectedTo some Scalp*. Inversely, the connection between the *Scalp* class and the *Face* class is created by using the *anteriorlyConnectedTo* property and defining the *Scalp* as a subclass of the axiom *anteriorlyConnectedTo some Face*.

The *TrunkSubdivision* class (see Fig. 4(b)) has three subclasses, the *Thorax*, the *Abdomen* and the *Back*, which inherit from the *TrunkSubdivision* class being a subclass of the axiom *partOf some Trunk*. The thorax and the abdomen conform the anterior part of the trunk and the back the posterior part of it. Thus, the connection between the *Thorax* class and the *Back* class is created by using the *posteriorlyConnectedTo* property and defining the *Thorax* as a subclass of the axiom *posteriorlyConnectedTo some Back*. Similarly, the connection between the *Abdomen* class and the *Back* class are created by using the *posteriorlyConnectedTo* property and defining the *Abdomen* as a subclass of the axiom *posteriorlyConnectedTo some Back*. Inversely, the connection between the *Back* class and the *Thorax* class is created by using the *anteriorlyConnectedTo* property and defining the *Back* as a subclass of the axiom *anteriorlyConnectedTo some Thorax*. Also the connection between the *Back* class and the *Abdomen* class is created by using the *anteriorlyConnectedTo* property and defining the *Back* as a subclass of the axiom *anteriorlyConnectedTo some Abdomen*. Moreover, the thorax is located on top of the abdomen in the anterior of the trunk. Thus, the connection between the *Thorax* class and the *Abdomen* class is created by using the *inferiorlyConnectedTo* property and defining the *Thorax* as a subclass of the axiom *inferiorlyConnectedTo some Abdomen*. Inversely, the connection between the *Abdomen* class and the *Thorax* class is created by using the *superiorlyConnectedTo* property and defining the *Abdomen* as a subclass of the axiom *superiorlyConnectedTo some Thorax*.

The *UpperLimbSubdivision* class (see Fig. 4(c)) has six subclasses, the *Shoulder*, the *Arm*, the *Elbow*, the *Forearm*, the *Wrist* and the *Hand*, which inherit

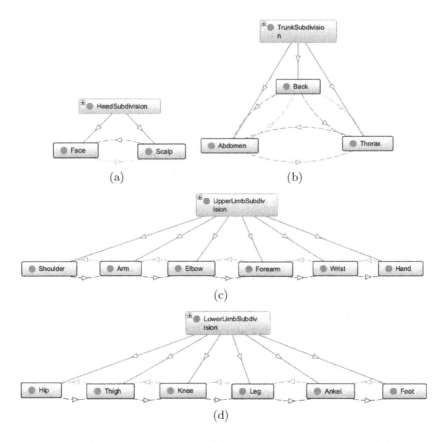

Fig. 4. SS4RWWAR Body Ontology for (a) the *HeadSubdivision* class, (b) the *Trunk-Subdivision* class, (c) the *UpperLimbSubdivision* class, and (d) the *LowerLimbSubdivision* class. The continuous purple arrows represent the *has subclass* property. The dashed red arrows represent the *superiorlyConnectedTo* property. The dashed blue arrows represent the *inferiorlyConnectedTo* property. The dashed dark gray arrows represent the *anteriorlyConnectedTo* property. The dashed light gray arrows represent the *posteriorlyConnectedTo* property. The dashed orange arrows represent the *proximallyConnectedTo* property. The dashed green arrows represent the *distallyConnectedTo* property.

from the *UpperLimbSubdivision* class being a subclass of the axiom *partOf some UpperLimb*. The shoulder is connected to the arm, the arm to the elbow, the elbow to the forearm, the forearm to the wrist, and the wrist to the hand. From these upper limb subdivisions, the hand is the most distant from the trunk, which is the main mass of the body, and the shoulder is the closest to it. The connections between upper limb subdivisions are created by using the *distallyConnectedTo* property and defining the *Shoulder* as a subclass of the axiom *distallyConnectedTo some Arm*, the *Arm* as a subclass of the axiom *distallyConnectedTo some Elbow*, the *Elbow* as a subclass of the axiom *distallyConnectedTo some Forearm*, the *Forearm* as a subclass of the axiom *distallyConnectedTo some Wrist*, and the *Wrist* as a subclass of the axiom *distallyConnectedTo some Hand*. The inverse property *proximallyConnectedTo* is used to create the inverse connections by defining the *Hand* as a subclass of the axiom *proximallyConnectedTo some Wrist*, the *Wrist* as a subclass of the axiom *proximallyConnectedTo some Forearm*, the *Forearm* as a subclass of the axiom *proximallyConnectedTo some Elbow*, the *Elbow* as a subclass of the axiom *proximallyConnectedTo some Arm*, and the *Arm* as a subclass of the axiom *proximallyConnectedTo some Shoulder*.

The *LowerLimbSubdivision* class (see Fig. 4(d)) has six subclasses, the *Hip*, the *Thigh*, the *Knee*, the *Leg*, the *Ankle* and the *Foot*, which inherit from the *LowerLimbSubdivision* class being a subclass of the axiom *partOf some LowerLimb*. The hip is connected to the thigh, the thigh to the knee, the knee to the leg, the leg to the ankle, and the ankle to the foot. From these lower limb subdivisions, the foot is the most distant from the trunk, which is the main mass of the body, and the hip is the closest to it. The connections between lower limb subdivisions are created by using the *distallyConnectedTo* property and defining the *Hip* as a subclass of the axiom *distallyConnectedTo some Thigh*, the *Thigh* as a subclass of the axiom *distallyConnectedTo some Knee*, the *Knee* as a subclass of the axiom *distallyConnectedTo some Leg*, the *Leg* as a subclass of the axiom *distallyConnectedTo some Ankle*, and the *Ankle* as a subclass of the axiom *distallyConnectedTo some Foot*. The inverse property *proximallyConnectedTo* is used to create the inverse connections by defining the *Foot* as a subclass of the axiom *proximallyConnectedTo some Ankle*, the *Ankle* as a subclass of the axiom *proximallyConnectedTo some Leg*, the *Leg* as a subclass of the axiom *proximallyConnectedTo some Knee*, the *Knee* as a subclass of the axiom *proximallyConnectedTo some Thigh*, and the *Thigh* as a subclass of the axiom *proximallyConnectedTo some Hip*.

5   Conclusions

Human physical self-quantification systems for health and wellness improvement build on mobile and portable sensor devices. Body-worn sensor devices are subject to changes that may prevent the correct functioning of wearable activity recognition systems. Accordingly, mechanisms to support the selection of adequate sensor replacements are required in real-world scenarios. In this work, a novel use of ontologies for dynamic sensor selection has been presented. The ontological model is composed by an upper ontology describing wearable sensors and their main properties, as well as a supportive domain ontology particularly defined to neatly and unequivocally represent the exact placement of the sensor on the human body. Next steps of this work include the extension of the presented models towards the magnitude and sensor characteristics domains, as well as the application of ontological reasoning techniques to automate the selection of the most adequate sensors.

References

1. Foundational Model of Anatomy ontology (FMA),
   http://sig.biostr.washington.edu/projects/fm/AboutFM.html
2. Linked data, http://linkeddata.org/
3. Uber anatomy ontology (Uberon), http://uberon.github.io/
4. Bae, I.-H.: An ontology-based approach to adl recognition in smart homes. Future Generation Computer Systems 33, 32–41 (2014)
5. BakhshandehAbkenar, A., Loke, S.W.: Myactivity: Cloud-hosted continuous activity recognition using ontology-based stream reasoning. In: 2014 2nd IEEE International Conference on Mobile Cloud Computing, Services, and Engineering, pp. 117–126 (2014)
6. Banos, O., Damas, M., Guillen, A., Herrera, L.-J., Pomares, H., Rojas, I., Villalonga, C.: Multi-sensor fusion based on asymmetric decision weighting for robust activity recognition. Neural Processing Letters, 1–22 (2014)
7. Banos, O., Damas, M., Pomares, H., Prieto, A., Rojas, I.: Daily living activity recognition based on statistical feature quality group selection. Expert Systems with Applications 39(9), 8013–8021 (2012)
8. Banos, O., Toth, M.A., Damas, M., Pomares, H., Rojas, I.: Dealing with the effects of sensor displacement in wearable activity recognition. Sensors 14(6), 9995–10023 (2014)
9. Banos, O., Villalonga, C., Damas, M., Gloesekoetter, P., Pomares, H., Rojas, I.: Physiodroid: Combining wearable health sensors and mobile devices for a ubiquitous, continuous, and personal monitoring. The Scientific World Journal 2014(490824), 1–11 (2014)
10. Chen, L., Nugent, C.D., Okeyo, G.: An ontology-based hybrid approach to activity modeling for smart homes. IEEE T. Human-Machine Systems 44(1), 92–105 (2014)
11. Mannini, A., Intille, S.S., Rosenberger, M., Sabatini, A.M., Haskell, W.: Activity recognition using a single accelerometer placed at the wrist or ankle. Medicine and Science in Sports and Exercise 45(11), 2193–2203 (2013)
12. Mungall, C., Torniai, C., Gkoutos, G., Lewis, S., Haendel, M.: Uberon, an integrative multi-species anatomy ontology. Genome Biology 13(1), R5 (2012)
13. Nguyen, T.A., Raspitzu, A., Aiello, M.: Ontology-based office activity recognition with applications for energy savings. Journal of Ambient Intelligence and Humanized Computing 5(5), 667–681 (2014)
14. Oresko, J.J., Jin, Z., Cheng, J., Huang, S., Sun, Y., Duschl, H., Cheng, A.C.: A wearable smartphone-based platform for real-time cardiovascular disease detection via electrocardiogram processing. IEEE Transactions on Information Technology in Biomedicine 14(3), 734–740 (2010)
15. Patel, S., Mancinelli, C., Healey, J., Moy, M., Bonato, P.: Using wearable sensors to monitor physical activities of patients with copd: A comparison of classifier performance. In: Proceedings of 6th International Workshop on Wearable and Implantable Body Sensor Networks, Washington, DC, USA, pp. 234–239 (2009)
16. Riboni, D., Bettini, C.: Cosar: Hybrid reasoning for context-aware activity recognition. Personal Ubiquitous Computing 15(3), 271–289 (2011)
17. Riboni, D., Bettini, C.: Owl 2 modeling and reasoning with complex human activities. Pervasive Mobile Computing 7(3), 379–395 (2011)

Proposal for Interoperability Standards Applications in the Health Sector

Alejandro Paolo Daza[1], Brayan S. Reyes Daza[1], and Octavio J. Salcedo Parra[2]

[1] Universidad Distrital Francisco José de Caldas, Bogotá D.C., Colombia
{internetinteligente,bsreyesd}@correo.udistrita.edu.co
[2] Universidad Nacional de Colombia, Bogotá D.C., Colombia
osalcedo@udistrital.edu.co

Abstract. The technological framework of the health sector is complex and problematic; each actor has created islands of information, where the exchange of information with other actors is the exception, where this provides a sea of technological heterogeneity and information systems providers with different stages of development even inside institutions. This scheme is not very effective in coordinating policies and implements concrete joint projects to improve the global search. In this scenario the efforts are not coordinated, processes are duplicated, yield is lost and increase costs. This paper presents the results of a technology proposal that allows solving this problematic.

Keywords: CDA, HL7, HCEU, Interoperation, PIX, SOA, XDS.

1 Introduction

Current technological developments have made available a number of opportunities for access to information historically unprecedented, at the same time of this evolution the current health models are changing. Today there are applications to manage within the institutions providing health services the clinical information of patients, the major problem of this technology in the health sector today has been the course of history of how it has developed and the current stage of the existing information systems in the sector. The technological framework of the health sector is complex and problematic because every institution providing health services, funding, insurance and regulatory as well as various governments agencies, universities and other actors in the health system, have created islands of information in themselves, where the exchange of information with other parties is the exception (Acevedo-Bernal, 2011), where the present provides a sea of technological heterogeneity and information systems providers with different stages of development even inside institutions. This scheme is not very effective in coordinating policies and implement concrete joint projects to search the global improve. In this scenario the efforts are not coordinated, processes are duplicated, yield is lost and increase costs. Then, regulators entities and government have a partial and belated vision what happens in institutions, unable to evaluate the implementation and impact of the projects. Finally, this makes the planning of new health policies and adds noise to the decision-taking at the level of the

F. Ortuño and I. Rojas (Eds.): IWBBIO 2015, Part II, LNCS 9044, pp. 153–164, 2015.

regulator sector. This problematic presented to the rector entities of the sector is add the problematic of attention processes of whose patients who are duplicated information in each of the institutions that have lent them service.

In addition to the current problematic the Congress of the Republic of Colombia through Law 1438 of 2011 decreed the obligatory application of the Unique Health Record before December 31, 2013 by placing a "dead line" to the solution of the current problem (CONGRESS OF THE REPUBLIC, 2011). So this law pretend the strengthen of the General System of Social Security in Health establishing as one of its axes application of Unique Clinical Story ensuring portability of benefits or provision anywhere in the country thanks to the availability of timely and accurate clinical information. This paper presents the results of a technological proposal for the implementation of health standards that permit interoperability in industry without throw away the efforts jointly by the governing body of the health sector for the Capital District (SDS) and by the entities health providers of Network Attached via a service-oriented architecture that provides the ability to integrate all efforts previously made and provides the evolutionary capacity of the HCEU platform.

2   Research Methodology

Based on the proposed objectives as a result of the operation model for the system of Electronic Health Record Unified (HCEU) it must be determined at the application level the overall architecture of the system, to determine this architecture is used the projective research methodology to guide the process of developing of the architectural system proposal. In this methodological process by reviewing reports on states of management of clinical information about patients of the health system of the Capital District previously developed by the Ministry of Health, existing the problematic of the impossibility of sharing by a timely way the clinical information regarding the medical history of the patients given the high degree of heterogeneity and various states of development or absence of the information systems of the Provider Entities Health Network Attached Capital District (University District, 2012). Based on this study the Ministry of Health raised a number of objectives aimed at interoperability of the different existing information systems, the refinement of these objectives allowed the selection of standards for sharing clinical information based on international experiences such as the United States, Spain, Argentina or Mexico (HL7-Mexico, 2012) (DGIS, 2012) (DGIS-NOM, 2012) (HL7-Spain, 2012). The standard selected for the system because its different advantages and great adaptability was HL7, with the implementation of IHE profiles for application and review of architectural styles was selected because the service-oriented architecture advantages. How is applied the service-oriented architecture using HL7 is the result of this research.

3   HL7

HL7 (Health Leven Seven) is not a standard, it is a set of standards that allow us to specify between some domains messaging management for transmission and communication of clinical information between information systems as the strongest domain of HL7. The complete set of standards allows us to exchange, share, integrate and retrieve health concerning electronic information. For these purposes HL7 defined how information is packaged and communicated between applications, setting the language, structure and types of data required for the uncoupled system integration.

4   Research Results

4.1   General Architecture of the HCEU System

The design of the architecture supports the integration of existing applications on each attached network entity, created to support the mission of the District Health System, by building a technological solution based on interoperability, and fully available, which has a lower impact in each one of the consumers applications of the HCEU services. As part of the construction process of the overall system architecture, system actors directly involved in the information flow of the system, such as different health professionals Network Attached and the SDS. The system professional will be responsible for consulting and feed the system with clinical information from patient through the handed down existing systems in each state and that consume services offered by the system to carry out the processes impacted for the HCEU according by model operation document (University District, 2012). The SDS will be responsible for the administration of HCEU, and consulting transactional information that will feed the analysis process indicators and the management, control and monitoring of the work done by the different entities of the network attached using a web application designed to satisfy the specific requirements of information of the SDS in their job of regulator entity. Access to interoperability platform HCEU will take place through the exchange of secure messages that allow the exchange of information through channels whose implement appropriate measures to protect the identity of patients and clinical character information policies. HCEU is designed as a service platform responsible for mediating access to information incorporating service-orientation as a paradigm, with the objective to respond to changing information requirements, in an agile, extensible and scalable manner that will allow the achievement of strategic objectives that motivated the project and which are specific to service-oriented (Susanti, 2011) (Xiaoqing, 2009), allowing the SDS and each network entities attached rely on the platform to optimize the performance of its work, benefiting from tangible and direct way the different actors.

The architecture of the HCEU proposed the disposition of an Inventory of Services for each context of information involved in the different business processes expressed in system implementation, PIX Service Inventory, XDS Service Inventory, CDA Service Inventory and the HCEU Service Inventory. In order to give way to logical solutions independently governed which permit reach the different strategic objectives that motivate SDS automation and centralization of the processes involved in the delivery of health services and regulation, surveillance and control required. In satisfaction of the security policies, the disposition of the platform HCEU, exposed through a secure channel the HCEU Inventory Services, which will be responsible for providing consumers applications registered in the system, the different capacities which abstract logic of the consultation process of the clinical history of the patients, and feeding repositories of medical and demographic information of patients and allowing the SDS to perform system configuration, in addition to building reports and indicators of management of each of the entities belonging to the District Network Attached, other services Inventories, designed to support the interaction of the services exposed by HCEU platforms and specific information will only be available to be accessed by orchestration and mediation services defined by the HCEU inventory services, responsible for validating access to each platform according to the access profile enable to the health professional or official SDS required to use the platform interoperability across existing applications and which have been adapted to support this kind of transactions.

The CDA Repository requires an inventory of services which allows consult medical records of patients, exposing only the own abilities from consolidation and retrieval of information within the context of specific information and documents covering the clinical Network Attached patients. Similarly, an inventory of services for the interaction with the XDS, document and headers repository designed for clinical documents exchange in accordance with standard HL7, and an inventory of specific services it is defined according to identification context and access to patient demographic information PIX. Each service inventory arising from the process of analysis and design HCEU has been modeled using Service Oriented Design Patterns Architecture enabling which permit better automating processes Outpatient, Admissions, Hospitalization, surgery and Emergency through a technological solution focused on obtaining the expected benefits of service orientation, and intrinsic interoperability required by the project, looking a low impact over each existing applications in each local HIS entity providing health services. Additionally, it is important the HCEU positioning as a resource of SDS technology and generally the District Health System, which from the beginning is projected to extend its capabilities, and its use and re-use in changing processes and evolutionary focusing on providing a health service highest quality to each of the beneficiaries.

4.2   HCEU Services

The inventory of services HCEU defined taking into account the security role, mediation, audit, and analysis of information that must support. In general, and to reach each of these roles business entities are identified that subsequently will determine candidates to HCEU inventory services regardless of their role within it, demonstrating the importance of meeting the needs of information in accordance with the processes and specific tasks required to perform the different functions identified for achieving the strategic objectives of the SDS by implementing the system. Business entities: user, application, role, authorization, log, alert, alert type, entity, administer entity of benefit plan, patient, patient type, pathology, clinical document, type of clinical document, municipality and department. Business tasks: authentication, time synchronization, alerting, log, report access, report clinical events, reporting and transactional history. Business processes: extern consult, hospital admissions, and emergency surgery. Once identified the different candidate services, we proceed to classify them according to the service model. Due to the convergence of functionality required for different consumers of the features of the inventory services and to better understand the design of it, the business inventory is divided consistent with the functional contexts in which it is group different HCEU inventory services.

Authentication. The functional context of authentication supports the identification, authentication, authorization and assignation for the right role to each user for the players in the system.

- Authentication: task service responsible for the identification, authentication, authorization and assignment of user role in the system task. Is delegated the responsibility of administering users, roles, profiles, applications and the relationship between themselves, and the access of allocation profiles for different HCEU components, once the user is authenticated successfully.
- User: entity service which abstract a system user. It will offer the capacities for the user management of the HCEU.
- Application: Entity service which abstracts the application through HCEU services are consumed. It will provide capacities for the manage applications that actually can access to the system functionality
- Role: entity service which abstracts the roles that determine the access profiles which has a specific user group. It will offer the capacities for the configured roles management of the HCEU.
- Authorization: entity service that abstracts the permits of a system user based on their role. It will provides capacities for permissions management for HCEU users.
- Entity: service that abstracts health service entity that belongs to the SDS attached network. It will offer the capacities for the configured roles management of the HCEU.
- Transaction Log: utility Service that abstracts the functionality of the System Event Log. The responsibility is delegated to carry the access log and system administration.

Transaction Reports. The functional context of transactional reports is responsible for report generation of the access by system users and the reporting of transactions carried out by the different consumers of the HCEU services.

- Access Report: Task services responsible for generating the report system access according to different filter criteria and information classification.
- Transaction Report: Task service responsible for generating report transactions executed on the system by different users and consuming applications of the HCEU services according to different criteria of information classification.
- Log: Entity service that abstracts the system event log. It offers capacities for the query repository where the Event Log is stored.
- Transaction Log: Utility service that abstracts the functionality of the System Event Log. The responsibility is delegated to bring the access log, reports generation and system administration.

Specific Reports. Functional context that includes the generation of specific reports as the clinical event report required for HCEU.

- Report of Clinical Episodes: Task service that generates reports on possible interest peaks alarms health events, such as acute respiratory disease, these reports should present unidentified information and allow aggregate this information according to the demographic information that can be gotten then as unidentified records.
- Clinical episode: entity service that abstract diferent clinical episodes configured for the HCEU.
- Entity: entity service that abstracts health service entity that belongs to the SDS attached network. It will offer the capacities for the configured roles management of the HCEU.
- Managing Entity of the Benefit Plan: Entity service that abstracts an Administrative Entity of the benefit plan of a patient. It will provide capacities for managing the entities configured for HCEU.
- Transaction Log: Utility service that abstracts the functionality of the System Event Log. The responsibility is delegated to bring the access log, reports generation and system administration.

Functional. Product from the provision of services, a Network Attached Entity requires consult and feed within the processes involved, the patient history information and documents, plus access to demographic information of the recipients of their services. In this way, HCEU must provide mechanisms and tools to reach this requirement, which implies the need for a set of task services responsible for providing access to information whenever required. According to the general architecture of the application, HCEU will have a mediating role between data repositories where is the patient demographic information, clinical episodes information and clinical documents related with the clinical records of the patient. To allow quick and efficient way to meet these system requirements, 3 new services are postulated to play a role of orchestration within the system, consuming HCEU own resources and those of the various information repositories related to the mission of the SDS and the health institutions in the district.

- Clinical history: Orchestration controller service to coordinate the interaction between entity services of the HCEU inventory services and consume the Services Inventories of the CDA Services, the XDS and the PIX to satisfy the information requirements relating to Medical Records of the patient.
- Clinical document: Orchestration controller service to coordinate the interaction between entity services of the HCEU inventory services and consume the Services Inventories of the CDA Services, the XDS and the PIX to satisfy the information requirements relating to consultation and information load of Clinical Documents of the beneficed Patients of services offered by the different entities of the attached network.
- Clinical Episode: Orchestration controller service to coordinate the interaction between entity services of the HCEU inventory services and consume the Services Inventories of the CDA Services to reach the information requirements relating to

consultation and information load of Clinical Documents of the beneficed Patients of services offered by the different entities of the attached network.

- Clinical Episode: As entity service which abstracts the diferents clinical episodes configured for the HCEU.
- Entity: entity service which abstracts an entity health service that belongs to the attached network SDS. It will provide capacities for managing the entities configured for HCEU.
- Managing Entity of the Benefits Plan: Entity service that abstracts a Managing Entity of the Benefits Plan of a patient. It will provide capacities for managing the entities configured for HCEU.
- Document Type: entity Service that abstracts Clinical Document Types set for HCEU. It offers capacities to manage the identified types of clinical documents for HCEU taking as basis the CDA structure for: consultation notes, background notes, progress notes, delivery notes, procedure notes, report discharge notes, continuity notes in patient management and unstructured documents notes.
- Patient Type: entity service which abstract beneficiaries' types configured for the HCEU. It offers capacities to manage the types of patients.
- Patient: entity service which abstracts the HCEU information patient, according to the definition given by the PIX.
- Clinical Document: service entity which abstracts a clinical document.
- Transaction Log: Utility Service that abstracts the functionality of the System Event Log. Is delegated the responsibility to bring the access log, reporting and system administration.

Provided Services by the District Entities. At the HCEU process level, services that allow to perform different functionalities required for the system according to the flow of information involving the consultation of the medical history, consultation and updating of clinical documents are identified, and consultation and income information on the care episode in each of the processes linked to the system, such as: outpatient, inpatient admissions, and emergency surgery. These processes allow postulate 4 new services oriented at coordinating HCEU specific functionality according to the own business conditions of the context of each process, involving different functionality in each business entities that are required.

- External consult: task service responsible for coordinating the interaction between HCEU services representing business entities and other services task oriented healthcare processes in which the patient, for the type of pathology requires no attention immediately. It will provide the skills to argue the need for an appointment, consult the patient demographic information and consult the patient's medical history, in addition to load information about the episode of care.
- Emergency: task service responsible for coordinating the interaction between HCEU services representing business entities and other services task oriented healthcare processes by the type of patient has a disease that require immediate attention. It will provide the capacities relating to the entry of the incidence in the HCEU, initial appraisal, consulting demographic and patient history, admission

procedures for immediate care or resuscitation, the application of new services and entering information on the process.

- Hospitalization: task service responsible for coordinating the participation and interaction among entity HCEU services that represent business entities and other task services, oriented to processes relating to hospitalization service task. It will provide the skills to argue the need for hospitalization, consultation of patient demographics and medical history and income information on the process HCEU.
- Surgery: task service responsible of coordination of the interaction between entities HCEU services representing business entities and other task services, oriented to inherent processes to the surgica l service. It will provide the skills to argue the need for a surgical act in HCEU, consultation of patient demographic information and medical history, and to the entry of the procedures performed in HCEU product of the service in each of its phases.

4.3   CDA Services

The CDAs repository is the repository of information product of the care clinical episodes of patients, which is stored using the HL7 Templates (HL7-CDA, 2012). The persistent structure of CDA proposes a set of Business Entities that interact to provide detailed information from medical records based on HL7. HCEU uses CDA Repository to shape the details of the clinical information of the beneficiaries of the health system that integrate the different entities of the Attached Network. For the conceptual definition of the service inventory CDA, it have taken into account the major business entities, without ruling out the participation of new entities according to the required depending on the functional demands that are considered in the development of HCEU. As evidenced in the HCEU Process Model (District University, 2012), the main business entities taking part in the processes of CDA are:

- Act: presence attend or administrative acts in a health sector organization.
- Participation: represent the application or execution scenery in a care act.
- Entity: tangible entities (physics) that participate in the health system.
- Role: defines the role and the entities participation.
- ActRelationShip: establish relations between two care acts.
- RoleLink: establish the relation between two roles.

Consistent, CDA inventory services in its definition should incorporate at least the corresponding entity services, additional of task services that allow for query functionalities required by HCEU expressed through their inventory of services.

Task Services. The inventory of CDA services must provide tools for searching information concerning procedures for the care of patients, in order to respond swiftly to the information requirements of each consumer HCEU services. On this it is based the following task services are identified, that through the configuration of different compositions of entity services encapsulating the process of storage and retrieval of clinical documents.

- Clinical Episodes Consultation: Service responsible for coordinating and setting up the interactions between entity services representing processes and business entities of the CDA for medical records consultation, product of the provision of health services in the different task entities that are part of the attached network.
- Clinical Documents entrance: Task service responsible for coordinating and setting up the interactions between entity services representing processes and business CDA entities for loading and updating medical records in compliance with the HL7 Standard task.

Entity Services

- Act: entity service that abstracts the presence care or administrative acts in a organization of the health sector.
- Participation: entity service that abstracts the application or execution scenery of an attention act.
- Entity: entity service that abstracts the tangible entities (physics which participe in the health system.
- Role: entity service that abstracts the role and the entities participation.
- ActRelationShip: entity service that abstracts the relations between two care acts.
- RoleLink: entity service that establish relation between two roles.

Utility Services

- Transactional Log: utility service which allows the persistence and consult of transactions done inside CDA component.
- Notification: utility service responsible of consumers' notification of the resources of CDA component of new information responsibility, or changes done to configuration and component data.

4.4   XDS Services

The XDS is the index references of repository of clinical documents relevant to medical records of patients who operates directly with the CDA for the exchange of clinical information and details of the clinical activity product of health service delivery in each of the entities of the attached SDS network and permit sharing these documents between different institutions and HCEU (IHE-XDS, 2012). According to HCEU requirements, and how it is developed in the HCEU operation model document, are identified three operations that are relevant to the relationship between actors and transactions for HCEU: registration of documents, extracting and managing documents headers. These operations are consistent with the processes of the ESE and mission processes of SDS, because the creator and consumer of documents are the ESE, and SDS has a role of agent contents. The services identified in the conceptual model of service inventory together have the responsibility to provide access to records of medical information, playing specific roles for configuration management and information consolidation.

Task Services. Services that correspond to this service layer must configure service compositions expressing consultation processes and loads of information on clinical documents that are generated and require between processes you want to automate the HCEU.

- Documents Consultation: Task service responsible for setting up and consolidating the extraction process of Clinical Documents of a patient according to different filter criteria and query task.
- Documents income: Task service responsible for feeding information repository of XSD with clinical information encoded, simple texts and images in alignment with the requirements defined for the use of CDA.

Entity Service

- Clinical Document: entity service that correspond to the abstraction of a Clinical Document Stored in XDS Repository. It will provide capacities for consultation and admission of clinical patient documents.
- Document Type: entity service that abstracts the document types supported by the XDS repository. It provides capacities for management and configuration of the different document types supported by the XDS repository.
- Headline: entity service that abstract headers of a clinical document type configured for the system. It will provide capacities for managing headers documents existing types and configured for the platform.

Utility Services

- Logger: utility service that allow persistence and query transactions carried out within the CDA component.
- Notification: utility service responsible for notification to consumers of resources component of the CDA on the availability of new information or changes to the configuration and component data utility.

4.5   PIX Services

The PIX profile (Patient Identifier Cross Referencing), Cross-Reference patient Identifier provides cross references of the patient identifiers from multiple Patient ID domains. These patient identifiers are used by identity consumer systems to correlate information about a single patient from sources who know the patient by different identifiers (IHE-PIX, 2012). The profile compatible supports collation of multiple patient identification patient identification domains is done through:

- The transmission of patient identity information from a source of identity to the Administrator of cross-references of patient identifier.
- Provides the ability to access the list (s) of identification patients with cross references, either via a query / response or via update notification.

The inventory of PIX services should provide services that allow interaction with the profile and information repository for the query and update demographic information and identification of a patient responding to requests from individual consumers to HCEU services meeting the information requirements of each of the applications that make use of HCEU for automation of the processes of care to beneficiaries of health services of different health institutions of attached network.

Task Services. Patient Identification: task service responsible for coordinating the process of identifying patients administrator cross reference identification task. It will provide capabilities to identify and validate the identity of a patient based on the information provided, looking into the different sources of Demographic configured for the system.

Entity Services

- Patient: Entity Service abstracted demographic and body of a patient information. It will provide capabilities for updating patient demographic information.
- Demographic Source: entity service responsible for managing sources of demographic configured for the system. It provides capabilities for creating, modifying and updating the reference to the sources of information.

Utility Services

- Update Notification: utility service responsible for generating alerts consumers of resources PIX inventory services, concerning the availability of new information on the sources of income information and changes made on the demographics of a patient.

5   Conclusions

The application of set of standards defined by HL7 in conjunction with the profiles defined by IHE deployed on an architectural service-oriented style empowers the SDS and health entities of Capital District integrating their services on an integrated platform clinical information (HCEU) that will allow the patient care in the Capital District with the guarantee of the availability of clinical information in a timely manner to optimize the health care episodes thereof.   Using HL7 allows the District and SDS supported by integrated services architecture integrate on a scalable form the various entities that make up the framework of the Capital District Health, and even the extension of the national model by integrating health networks substantially improve the nation's health. The resulting health model of the Capital District supported by HL7 standards group, IHE profiles and service-oriented architectural style is a viable, scalable and evolutionary model that allows the determination of health sector policies that facilitate the promotion and prevention management of the morbidity - mortality due to the availability of information for episodes of care of patients in the sector.

References

1. Bernal O., Forero J.C.: Information Systems in the health sector in Colombia (2011), http://rev_gerenc_polit_salud.javeriana.edu.co/vol10_n_21/estudios_1.pdf (recovered)
2. IHE. Cross-enterprise document sharing (2012), http://wiki.ihe.net/index.php?title=Cross-Enterprise_Document_Sharing (recovered)
3. IHE. Patient identifier cross-referencing (2012), http://wiki.ihe.net/index.php?title=Patient_Identifier_Cross-Referencing (recovered)
4. Health Level Seven International. CDA release 2 (2012), http://www.hl7.org/implement/standards/product_brief.cfm?product_id=7 (recovered)
5. Health Level Seven International. Introduction to hl7 standards (2012), http://www.hl7.org/implement/standards/index.cfm?ref=nav (recovered)
6. Xiaoqing, F.L.: Design of soa based web service systems using qfd for satisfaction of quality of service requirements. In: 2009 IEEE International Conference on Web Services, pp. 567–574 (2009)
7. CONGRESO DE LA REPUBLICA. Ley 1438 de 2011, (2011), http://www.secretariasenado.gov.co/senado/basedoc/ley/2011/ley_1438_2011.html (recovered)
8. Universidad Distrital SDS. Modelo de Operación HCEU. Universidad Distrital, Reading, MA (2012)
9. Susanti, F., Sembiring, J.: The mapping of interconnected soa governance and itil v3.0. In: 2011 International Conference on Electrical Engineering and Informatics, pp. 1–5 (2011)
10. Capítulo HL7 México. Estándares (2012), http://hl7.org.mx/estandares (recovered)
11. Dirección General de Información en Salud. Intercambio De Información En Salud (2012), http://www.dgis.salud.gob.mx/intercambio/index.html (recuperado)
12. Dirección General de Información en Salud. Norma Oficial Mexicana NOM-024-SSA3-2012 (2012), http://www.dgis.salud.gob.mx/intercambio/nom024.html (recovered)
13. Healt Level Seven Spain. Casos de Éxito (2012), http://www.hl7spain.org/exitos.html (recovered)

An Event-Driven Architecture for Biomedical Data Integration and Interoperability

Pedro Lopes and José Luís Oliveira

Universidade de Aveiro, Campus Universitário de Santiago 3810-193 Aveiro, Portugal
{pedrolopes,jlo}@ua.pt

Abstract. Connecting data and services is nowadays an essential skill for life sciences researchers. Handling data from unrelated sources using problem-specific software or labor-intensive tools are common tasks. Despite the latest advances, integration and interoperability developments still involve primitive interactions and manual triggers. On the one hand, available tools cover specific niches, ignoring more generic problems. On the other hand, overly complex tools with excessive features dominate the market. In this proposal we introduce a cloud-based architecture to simplify real-time integration and interoperability for biomedical data. We support our strategy in an event-driven service-oriented architecture, where new data are pushed from any content source, through an intelligent proxy, for delivery to heterogeneous endpoints. This enables a passive integration approach, providing developers with a streamlined solution for deploying integrative biomedical applications.

Keywords: Distributed/Internet based software engineering tools and techniques, interoperability, data translation, data warehouse and repository, bioinformatics.

1   Introduction

The biomedical knowledge landscape, with ever growing datasets, increasingly demands for hardware and software progress [1]. The true value behind existing raw data lies in its adequate exploitation. However, data by itself are no longer valuable if there are no additional services for their better exploration [2].

Life sciences datasets are challenging: the structure is heterogeneous, they originate from distinct places, they adopt multiple standards, the content is stored in different formats and their meaning is always changing [3]. From next generation sequencing hardware [4] to the growing availability of biomedical sensors, exploring these data is an on-going endeavour [5]. Whether through the manual aggregation of data, or the use of specific software, retrieving and integrating data is an everyday requirement in life sciences' research work.

Recently, biomedical data integration and interoperability is focused on cloud-based service-oriented architectures strategies [6-8]. Cloud-based approaches are useful where the computational requirements are the main bottleneck. For instance, executing intensive analysis algorithms on next-generation sequencing data. Furthermore, using

F. Ortuño and I. Rojas (Eds.): IWBBIO 2015, Part II, LNCS 9044, pp. 165–174, 2015.

cloud-based technologies also simplifies common researchers' tasks, as they do not need to install complex application stacks to perform relatively simple integration operations.

In addition to cloud-based operations, workflow management tools are also responsible for notable advances in biomedical data interoperability. The ability to create comprehensive workflows eased researchers' work [9]. Nowadays, connecting multiple services and data sources is a trivial task. Yet, tools such as Yabi [10], Galaxy or Taverna [11] lack automation strategies.

Although the adoption of cloud-based and workflow management tools is widespread, there is room for improving their operability, namely through the inclusion of event-driven strategies. These will promote process automation and reduce the burden on researchers and developers alike. Hence, in this work we introduce an architecture to streamline data integration and interoperability. Our objective is to enable the automated event-driven analysis of data, empowering the creation of state-of-the-art integration and interoperability applications.

Traditional integration approaches, in use by warehouses, rely on batch, off-line Extract-Transform-Load (ETL) processes. These are manually triggered on regular intervals of downtime, which can range from weeks to months up to years. However, in the life sciences domain, the demand for fresh data cannot be ignored. Hence, we need to deploy new strategies that are dynamic [12], reactive [13] and event-driven [14, 15]. Modern platforms must act intelligently, i.e., in real-time and autonomously, to new events in integration ecosystems.

Our approach starts with an intelligent event handler, to where any content owner or producer can push newly generated data. When these events occur, the system initiates an Extract-Transform-Load (ETL) process, analysing received data and, when data are new, delivering it to a preconfigured set of endpoints. The delivery process enables interacting with files, databases, emails or URLs. As a result, there are endless combinations for customisable integration tasks, connecting events' data with actions configured in templates. Among others, the framework empowers live data integration and heterogeneous many-to-many interoperability.

This architecture is targeted at bioinformatics developers. The goal is to enable the creation of new client applications on top of a barebones framework. These cloud-based applications can feature event-driven, publish/subscribe, Integration-as-a-Service architectures.

Biomedical data integration and interoperability is the cornerstone of modern life sciences research projects. In this proposal, we bring forward an innovative architecture to quickly deploy automated and real-time integration workflows.

2   Methods

2.1   Background

This proposal introduces an architecture that can act as the foundation for distinct systems in event-driven environments. Traditional service-oriented architectures follow a request-response interaction model [16]. Although this basic operation principle sustains many systems, lack of support for responses to events is a major drawback [17].

Event-driven architectures adopt a message-based approach to decouple service providers from consumers [18]. Sending and receiving messages, the events, control the conditional execution of tasks within the system. This architectural strategy is used in both standalone applications and workflow environments [19]. Hence, event-driven architectures are used for direct responses to various events and for coordination with business process integration in ubiquitous scenarios [20, 21].

Our framework enables this kind of reactive integration. An intelligent event handler was designed to process incoming events, with a central server acting as a message broker and router. Events are received, via push mechanisms, in a publicly open service endpoint.

This event-driven strategy also enables publish-subscribe software [22]. The basic principle behind this strategy revolves around a dynamic endpoint, the publisher, which transmits data to a dynamic set of subscribers [23]. Publish/subscribe architectures decouple communicating clients and complement event-driven architectures with the introduction of notifications. These can be a specialized form of events and ensure that we can actively push the desired data to subscribers in the shortest possible time.

Translating this basic principle to our architecture, anyone can be a remote publisher, pushing event data into our system to interact with specific external endpoints, the subscribers. This architecture features the required tools that will allow applications to harvest the full potential of this paradigm in a seamless way. Likewise, we can perform asynchronous push-based communication, broadcasting event data to any number of assorted destinations.

Adopting the proposed work, data owners now have the tools to deliver custom notifications when new data are generated. This further advances the state of the art, namely on the life sciences [24], becoming vital when we consider the amount of legacy systems used to store data and the availability (or lack thereof) of interoperability tools to connect these systems. Event-driven architectures are being used more often in healthcare and bioinformatics [25, 26]. However, existing systems are *ad hoc* solutions to specific challenges. In opposition, our approach brings a general architecture for building any kind of event-driven integration application.

2.2 Architecture

Fig. 1 details the framework's architecture, including its basic components, which are described next.

Origin resources are the external entities that own the data. They are responsible for sending events into the system. The Event Handler processes these events, starting the integration process.

The internal data store is the system database for storing all application data. The platform implementation uses PostgreSQL and a Redis cache for faster change detection during event processing.

The core engine is the main application controller, registering and proxying everything. With all components deployed independently, the core engine acts as the glue keeping all pieces together, and as a flow manager, ensuring that all operations are performed smoothly, from event management to final delivery. Moreover, the core engine also enables the framework's API, empowering the various platform web services.

The delivery engine, as implied by its naming, performs the data delivery to each endpoint. Destination resources are the templates' objects to which the delivery engine will connect for final delivery. Initially, there are templates for interacting with files (via user-based workspace or Dropbox), SQL databases (MySQL or PostgreSQL), emails or HTTP-based web service calls.

A log engine stores summary information for all actions and flows. Each log entry contains the minimal set of information required to re-enact specific transactions or errors. This includes timestamps, origin/destination and the messages sent. For improved tracking and analysis, the backend uses the Sentry platform.

2.3   Workflow

Fig. 2 showcases a sequence diagram featuring the basic system interactions, and highlights the event management and the data delivery ones. These are also described in detail next.

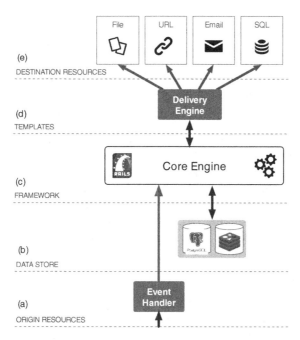

Fig. 1. System architecture, highlighting the different system layers. (a) external Original Resources generate events for processing; (b) the internal Data store uses a relational database (PostgreSQL) to store data and an object cache (Redis) for improved performance; (c) the framework engine is implemented in Ruby, with the Rails framework, and controls the entire application and its API; (d) the Delivery Engine applies event data to Templates and executes the final delivery; (e) the external Destination Re-sources receive the data from the system.

Event Management

The event handler component manages all data input into the system. Event data messages arrive at a public service endpoint, containing a unique identifier for each configured integration task. Data are submitted via HTTP POST to a unique address, generated for each integration.

When new data arrives, the system searches the internal data store for the integration settings. If the integration is fully configured, including data extraction definitions and matching delivery templates, the integration process starts.

To ensure a reliable stream of changed data, the architecture features a set of content change detection services. These ensure that only fresh content, which has not been processed before, enters the system.

Content change detection adopts an atomic data storage approach. That is, we can configure how to extract specific data elements from the event data, and each of these is independently stored. This change detection process occurs in four steps:

- The change detection service loads integration metadata according to the event identification.
- The change detection service extracts the data for analysis from the event message.
- The detector service validates the retrieved data in an internal cache. The cache acts as the atomic data storage component, storing each data element uniquely.
- If the retrieved data are not in the cache, i.e., data are fresh, the event handler triggers the integration delivery and stores the new data in the cache. When the data has been seen, the event handler stops the integration process.

The cache verification process can use two distinct data elements. On the one hand, we can configure the cache property for each event specifying what variable change detection will use. This should be set to track unique data properties, namely identifiers. On the other hand, if there are no data elements that can identify integrated data unequivocally, the architecture creates and stores an MD5 hash of the data elements' content. As changing content results in a new hash, we can detect which events are new without compromising the system performance.

Data Delivery

From an integration perspective, this architecture is a basic intelligent ETL proxy. With it, we can simplify the transformation and loading process associated with the aggregation of data from distinct heterogeneous systems.

Extracting data from events is based on selectors. Selectors are key-value pairs, mapping a unique variable name, the key, with an expression to extract data elements, the value. Where keys can be generic strings, values are specific to each data format.

Delivery templates perform the transform and load process. Their configuration can have several variables, named with the *%{<variable name>}* expression. These are identified at runtime and should match the selector keys configured in the agent.

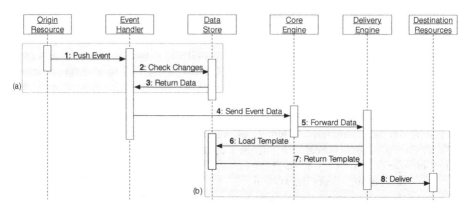

Fig. 2. Integration sequence diagram. In addition to the listed steps, all actions are logged internally for auditing, error tracking and performance analysis. The two main sequences are a) event management, which starts the integration process and includes the content change detection; b) data delivery, which finishes the integration process and includes the template loading and execution.

For instance, an integration task that delivers data to a SQL database has an event configured with one or more data selectors. During the transform and load process, the delivery engine replaces SQL template variables (in the INSERT INTO... statement) with values obtained for each selector.

Besides transforming static data for variables, templates can call internal functions or execute custom transformation code. Functions, named within *${<function name>}*, provide quick access to programmatic operations.

For example, *${datetime}* outputs the time of delivery in the template. More complex operations are written in Ruby code. The *${code(<ruby script>)}* function enables wri-ing simple scripts that are evaluated during the delivery. This endows templates with a generic, simple and flexible strategy for performing complex data transformations. Furthermore, it enables conditional transformations, solving equations or manipulating strings, among many other operations.

3   Discussion

This architecture adds true value to the scientific data integration landscape, through three key outcomes:

1. Enabling automated real-time data integration based on modern event-driven web hooks strategies.

2. Improving data delivery to heterogeneous destinations using a comprehensive template-based approach that allows transmitting and transforming data.

3. Facilitating integration and interoperability, as we can use this architecture to empower multiple event-driven platforms, from publish/subscribe to cloud-based integration-as-a-service.

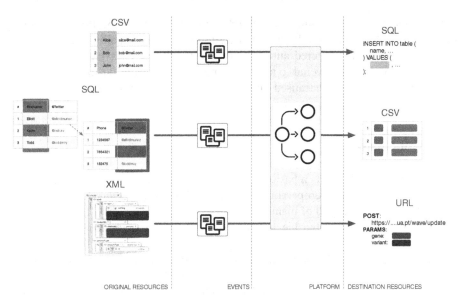

Fig. 3. Data integration from original to destination resources. Data from Original Resources can be easily transformed and published to URL requests, files, SQL queries or email, using delivery templates.

The clear benefits for data integration are highlighted in Fig. 3, where we show how data owners can enable the distribution of their data to multiple heterogeneous endpoints.

These integration perspectives come to fruition in several bioinformatics scenarios. For instance, we can use this architecture to automate the integration of human mutation data. The collection of unique mutations associated with specific locus is already available in several platforms. Diseasecard [27], WAVe [28] or Cafe Variome [29] aggregate data from distributed locus-specific databases (LSDBs) and make it available through web interfaces. Despite their quality, these systems' integration is limited by various constraints: the adopted integration pipelines - extract and curate data, enrich datasets, deliver results – only creates a time-based snapshot of the available mutation data.

LOVD provides a turnkey solution to launch new LSDBs, with web and database management interfaces [30]. These features make LOVD the *de facto* standard for LSDBs, with more than 2 million unique variants stored throughout 78 distinct installations. With little effort, LOVD can be updated to include a simple notification system capable of sending message to a new application built with our architecture.

Hence, when curators submit new variants to LOVD, a message with event data will be sent to the system for distribution to any configured endpoint. One of these endpoints can be WAVe. In this case, the integration started when the system receives the new event will deliver data directly to WAVe's database (using a SQL template) or to a URL-based service (using a URL route template connecting to COEUS' services [31]).

This simple scenario highlights the architecture's three key benefits: autonomy, delivery flexibility, and overall better integration and interoperability. First, the data integration process is triggered autonomously, without user intervention. Next, data for new mutations are delivered to any number of heterogeneous destinations. At last, variome datasets in the centralised application are always up-to-date with the latest discovered variants.

4   Conclusions

Access to biomedical data is a basic research commodity. This is due to the evolution of data integration and interoperability strategies. Yet, despite their quality, existing approaches need some kind of manual management effort. Currently, connecting data and services is a labor-intensive user-controlled endeavor, producing time-limited solutions.

In this research work we introduce an architecture to simplify the biomedical data integration process. We do this by enhancing the way data are shared and integrated. With a unique event-driven approach, data import and export processes are triggered passively, in a push-based environment. In summary, when owners produce new content, they notify a central server, sending the newly generated for distribution. In his turn, the central server analyses and delivers the new data to any preconfigured endpoint. This workflow's automation is a new approach to the field and stimulates reactive and event-driven integration tasks. The proposed architecture ensures live data integration, maintaining data up-to-dateness with minimal changes to the original content sources. Moreover, by enabling the delivery of data to miscellaneous endpoints, we promote interoperability.

This article introduces an architecture created for bioinformatics developers, bridging the gap between heterogeneous data and services through an autonomous event-driven integration and interoperability layer, where new researcher-oriented applications can be built.

Acknowledgments. The research leading to these results has received funding from the European Community (FP7/2007-2013) under ref. no. 305444 – the RD-Connect project, and from the QREN "MaisCentro" program, ref. CENTRO-07-ST24-FEDER-00203 – the CloudThinking project.

References

1. Sascha, S., Kurtz, S.: A New Efficient Data Structure for Storage and Retrieval of Multiple Biosequences. IEEE/ACM Transactions on Computational Biology and Bioinformatics 9, 345–357 (2012)
2. Thiam Yui, C., Liang, L.J., Jik Soon, W., Husain, W.: A Survey on Data Integration in Bioinformatics. In: Abd Manaf, A., Sahibuddin, S., Ahmad, R., Mohd Daud, S., El-Qawasmeh, E. (eds.) ICIEIS 2011, Part IV. CCIS, vol. 254, pp. 16–28. Springer, Heidelberg (2011)

3. Darmont, J., Boussaid, O., Ralaivao, J.-C., Aouiche, K.: An architecture framework for complex data warehouses. arXiv preprint arXiv:0707.1534 (2007)
4. http://www.genome.gov/sequencingcosts
5. Goble, C., Stevens, R.: State of the nation in data integration for bioinformatics. Journal of Biomedical Informatics 41, 687–693 (2008)
6. Dudley, J.T., Butte, A.J.: Reproducible in silico research in the era of cloud computing. Nature Biotechnology 28, 1181 (2010)
7. Schönherr, S., Forer, L., Weißensteiner, H., Kronenberg, F., Specht, G., Kloss-Brandstätter, A.: Cloudgene: A graphical execution platform for MapReduce programs on private and public clouds. BMC Bioinformatics 13, 200 (2012)
8. Sansone, S.-A., Rocca-Serra, P., Field, D., Maguire, E., Taylor, C., Hofmann, O., Fang, H., Neumann, S., Tong, W., Amaral-Zettler, L.: Toward interoperable bioscience data. Nature Genetics 44, 121–126 (2012)
9. Jamil, H.M.: Designing Integrated Computational Biology Pipelines Visually. IEEE/ACM Transactions on Computational Biology and Bioinformatics 10, 605–618 (2013)
10. Hunter, A., Macgregor, A.B., Szabo, T.O., Wellington, C.A., Bellgard, M.I.: Yabi: An online research environment for grid, high performance and cloud computing. Source Code for Biology and Medicine 7(1) (2012)
11. Abouelhoda, M., Issa, S.A., Ghanem, M.: Tavaxy: Integrating Taverna and Galaxy workflows with cloud computing support. BMC Bioinformatics 13, 77 (2012)
12. Salem, R., Boussaïd, O., Darmont, J.: Active XML-based Web data integration. Inf. Syst. Front. 15, 371–398 (2013)
13. Tank, D.M.: Reducing ETL Load Times by a New Data Integration Approach for Real-time Business Intelligence. International Journal of Engineering Innovation and Research 1, 1–5 (2012)
14. Naeem, M.A., Dobbie, G., Webber, G.: An Event-Based Near Real-Time Data Integration Architecture. In: 2008 12th Enterprise Distributed Object Computing Conference Workshops, pp. 401–404 (2008)
15. Mouttham, A., Peyton, L., Eze, B., Saddik, A.E.: Event-Driven Data Integration for Personal Health Monitoring. Journal of Emerging Technologies in Web Intelligence 1(2) (2009), Special Issue: E-health Interoperability (2009)
16. Papazoglou, M., Heuvel, W.-J.: Service oriented architectures: Approaches, technologies and research issues. The VLDB Journal 16, 389–415 (2007)
17. Kong, J., Jung, J.Y., Park, J.: Event-driven service coordination for business process integration in ubiquitous enterprises. Computers and Industrial Engineering 57, 14–26 (2009)
18. Niblett, P., Graham, S.: Events and service-oriented architecture: The oasis web services notification specification. IBM Systems Journal 44, 869–886 (2005)
19. Zhao, Z., Paschke, A.: Event-Driven Scientific Workflow Execution. In: La Rosa, M., Soffer, P. (eds.) BPM Workshops 2012. LNBIP, vol. 132, pp. 390–401. Springer, Heidelberg (2013)
20. Overbeek, S., Janssen, M., van Bommel, P.: Designing, formalizing, and evaluating a flexible architecture for integrated service delivery: Combining event-driven and service-oriented architectures. Service Oriented Computing and Applications 6, 167–188 (2012)
21. Frank, J.H., Zeng, L.: On Event-Driven Business Integration. In: 2013 IEEE 10th International Conference on e-Business Engineering (ICEBE), pp. 82–89. IEEE (Year)
22. Fotiou, N., Trossen, D., Polyzos, G.C.: Illustrating a publish-subscribe internet architecture. Telecommunication Systems 51, 233–245 (2012)

23. Eugster, P.T., Felber, P.A., Guerraoui, R., Kermarrec, A.-M.: The many faces of publish/subscribe. ACM Computing Surveys (CSUR) 35, 114–131 (2003)
24. Linlin, L., Shizhong, Y.: XPath-Based Filter for Publish/Subscribe in Healthcare Environments. In: 2012 IEEE 12th International Conference on Computer and Information Technology (CIT), pp. 1092–1096 (2012)
25. Yu, H.Q., Zhao, X., Zhen, X., Dong, F., Liu, E., Clapworthy, G.: Healthcare-Event driven semantic knowledge extraction with hybrid data repository. In: 2014 Fourth International Conference on Innovative Computing Technology (INTECH), pp. 13–18. IEEE (2014)
26. Llambias, G., Ruggia, R.: A middleware-based platform for the integration of Bioinformatic services. In: 2014 XL Latin American Computing Conference (CLEI), pp. 1–12. IEEE (2014)
27. Lopes, P., Oliveira, J.L.: An innovative portal for rare genetic diseases research: The semantic Diseasecard. Journal of Biomedical Informatics 46, 1108–1115 (2013)
28. Lopes, P., Dalgleish, R., Oliveira, J.L.: WAVe: web analysis of the variome. Human Mutation 32, 729–734 (2011)
29. Lancaster, O., Hastings, R., Dalgleish, R., Atlan, D., Thorisson, G., Free, R., Webb, A., Brookes, A.: Cafe Variome-gene mutation data clearinghouse. Journal of Medical Genetic BMJ, S40–S40 (Year)
30. Fokkema, I.F., Taschner, P.E., Schaafsma, G.C., Celli, J., Laros, J.F., den Dunnen, J.T.: LOVD v. 2.0: The next generation in gene variant databases. Human Mutation 32, 557–563 (2011)
31. Lopes, P., Oliveira, J.L.: COEUS: "Semantic web in a box" for biomedical applications. Journal of Biomedical Semantics 3, 1–19 (2012)

Local Search for Multiobjective Multiple Sequence Alignment

Maryam Abbasi[1], Luís Paquete[1], and Francisco B. Pereira[1,2]

[1] CISUC, Department of Informatics Engineering, University of Coimbra, Portugal
{maryam,paquete}@dei.uc.pt
[2] Polytechnic Institute of Coimbra, Portugal
xico@dei.uc.pt

Abstract. Recently, there has been a growing interest on the multiobjective formulation of optimization problems that arise in bioinformatics, such as sequence alignment. In this work, we consider the multiobjective multiple sequence alignment, with the goal of maximizing the substitution score and minimizing the number of indels. We introduce several local search approaches for this problem. Several neighborhood definitions and perturbations are presented and discussed. The local search algorithms are tested experimentally on a wide range of instances.

Keywords: Multiple sequence alignment, Local search, Multiobjective optimization.

1 Introduction

Most approaches to multiple sequence alignment are based on a weighted sum formulation, such as the classical dynamic programming approach [7]. This formulation is not "natural" for the problem, given the conflicting nature of reducing the number of indels and increasing substitution score. By considering the multiobjective formulation of multiple sequence alignment, the practicioner is provided a set of optimal alignments, *the Pareto optimal alignment set*, which represents the trade-off between indels and substitution score. This set contains not only all the optima of a weighted sum formulation, but also many other alignments that are not possible to find by the weighted sum aproach. Each of these alignments can be seen as a potential explanation of the relationship between the sequences; see the application of this concept in pairwise sequence alignment for the analysis of phylogenetic trees in Abbasi et al. [1].

Several works have been extending algorithms for pairwise sequence alignment for a multiobjective setting [1,3,9,10,11,13]; see also an extensive review about bioinformatics problems recast as multiobjective optimization problems in Handl et al. [5]. In this work, we introduce heuristic methods for solving the multiobjective sequence alignment problems (MMSA) for an arbitrary number of sequences. This work extends the formulation given in Abbasi et al. [1]. A *sum-of-pairs* score vector function is considered: maximization of the substitution score, given by a substitution matrix, and minimization of indels.

F. Ortuño and I. Rojas (Eds.): IWBBIO 2015, Part II, LNCS 9044, pp. 175–182, 2015.

2 Multiple Sequence Alignment

In multiple sequence alignment (MSA), a natural extension of pairwise sequence alignment, homologous characters among a set of sequences are aligned together in columns. Multiple alignment can be formally defined analogously to the alignment of two sequences as follows [2]:

Definition 1. Multiple sequence alignment. Let $A_1 = (a_{1,1}, \ldots, a_{1,n_1})$, $\ldots$, $A_m = (a_{m,1}, \ldots, a_{m,n_m})$ be m strings over an alphabet Σ. Let $'-' \notin \Sigma$ be an indel symbol and let $\Sigma' = \Sigma \cup \{-\}$. Let $h : (\Sigma')^* \mapsto \Sigma^*$ be a homomorphism defined by $h(a) = a$ for all $a \in \Sigma$, and $h(-) = \lambda$. A multiple sequence alignment ϕ of $(A_1, \ldots, A_m)$ is a m-tuple of $B_1 = (b_{1,1}, \ldots, b_{1,\ell}), \ldots, B_m = (b_{m,1}, \ldots, b_{m,\ell})$ of strings of length $\ell \geq \max\{A_i | 1 \leq i \leq m\}$ over the alphabet Σ', such that the following conditions are satisfied:

i) $|B_1| = |B_2| = \ldots = |B_m|$
ii) $h(B_i) = A_i$ for all $i \in \{1, \ldots, m\}$
iii) For all $j \in \{1, \ldots, \ell\}$ there exists an $i \in \{1, \ldots, m\}$ such that $b_{i,j} \neq '-'$.

A way of evaluating the quality of an alignment ϕ is to score its columns by the *sum-of-pairs* function. The score of a column $j = 1, \ldots, \ell$ is defined as:

$$S(j) = \sum_{1 \leq i < k \leq m} sc(b_{i,j}, b_{k,j})$$

where the score $sc(b_{i,j}, b_{k,j})$, for $b_{i,j}, b_{k,j} \in \Sigma$, comes from a substitution matrix used for scoring pairwise sequence alignments (such as PAM and BLOSUM). Indels are scored by defining $sc('-', \cdot) = sc(\cdot, '-') = W_d$, where W_d is the weight of an indel, and $sc('-', '-') = 0$. The score for alignment ϕ is computed as

$$SP(\phi) = \sum_{j=1}^{\ell} S(j)$$

3 Multiobjective Multiple Sequence Alignment

In most real life problems, objectives conflict very often with each other, and optimizing a particular solution with respect to a single objective can result in poor solutions with respect to the remaining objectives. A reasonable approach to a multiobjective problem is to find a set of solutions, each of which cannot be improved in one objective without deteriorating at least one of the others [4].

Traditionally, sequence alignment is conducted with respect to a single objective function, which depends on determining the weights for matches, mismatches, insertions and deletions. Moreover, there is often a considerable disagreement about how to specify fixed values for these parameters in the most commonly used alignment software packages. MMSA overcomes the problem of defining weights. Furthermore, it provides different alignments that may give more information to the practitioners [10].

Let ϕ be an alignment of m sequences $(A_1, \ldots, A_m)$, as defined in the previous section. We define the following two score functions

$$S_s(j) = \sum_{1 \leq i < k \leq m} s(b_{i,j}, b_{k,j}) \quad \text{and} \quad S_d(j) = \sum_{1 \leq i < k \leq m} d(b_{i,j}, b_{k,j})$$

where the score $s(b_{i,j}, b_{k,j})$, for $b_{i,j}, b_{k,j} \in \Sigma$ is obtained from a substitution matrix and $d(b_{i,j}, b_{k,j})$, for $b_{i,j}, b_{k,j} \in \Sigma'$, is 1 if either $b_{i,j} =' -'$ or $b_{k,j} =' -'$, and 0, otherwise. The multiobjective score sum-of-pairs for alignment ϕ is

$$BSP(\phi) = \left(BSP_s(\phi) = \sum_{j=1}^{\ell} S_s(j), BSP_d(\phi) = \sum_{j=1}^{\ell} S_d(j) \right)$$

Given two alignments ϕ and ϕ', $BSP(\phi) \geq BSP(\phi')$ (ϕ dominates ϕ') if and only if it holds that $BSP_s(\phi) \geq BSP_s(\phi')$ and $BSP_d(\phi) \leq BSP_d(\phi')$, with at least one strict inequality. An alignment ϕ^* is *Pareto optimal* if there exists no other alignment ϕ such that $BSP(\phi) \geq BSP(\phi^*)$. The set of all Pareto optimal alignments is called *Pareto optimal alignment set*. The image of a Pareto optimal alignment in the score space is a *non-dominated score* and the set of all non-dominated scores is called *non-dominated score set*.

In this article, the goal is to solve the problem of finding a Pareto optimal alignment set for a given set of sequences. However, if the number of sequences is larger than two, the problem becomes NP-hard. Therefore, we consider local search methods that find an approximation to the Pareto optimal alignment set in a reasonable amount of time.

4   Pareto Local Search

A local search algorithm starts from a feasible solution and searches *locally* for better neighbors to replace the current one. This neighborhood search is repeated until no improvement is found anymore and the algorithm stops in a local optimum. In the context of MMSA, let Φ denote the set of feasible alignments and let $N \mapsto 2^\Phi$ be a neighborhood function that associates a set of feasible alignments $N(\phi)$ to every feasible alignment ϕ. An alignment ϕ is a Pareto local optimum if there exists no alignment ϕ' in $N(\phi)$ such that $BSP(\phi') \geq BSP(\phi)$.

Pareto local search (PLS) [8] is a generic local search framework for multi-objective optimization problems that follows the principle above by keeping the best alignments into a special data structure, known as the *archive*. Each neighboring alignment that is nondominated with respect to the alignments in the archive, is added to it. The algorithm "naturally" terminates once there is no neighbor of any alignment in the archive that is not dominated (a local optimal set). See [8] for more details about this approach.

In the following we describe some options that were considered to apply PLS for MMSA: neighborhood function and starting alignment.

Neighborhood. We consider a *k-block* neighborhood for this problem, which consists of exchanging a substring of at most k characters in Σ with indels in a *gap*, that is, a contiguous sequence of indels. In the following we establish the conditions for two alignments to be k-block neighbors. Let $A_1 = (a_{1,1}, \ldots, a_{1,n_1})$, $\ldots$, $A_m = (a_{m,1}, \ldots, a_{m,n_m})$ denote m strings and let ϕ_B and ϕ_C be two alignments of A, where $\phi_B = B_1, \ldots, B_m$ and $\phi_C = C_1, \ldots, C_m$. The alignments ϕ_B and ϕ_C are k-block neighbors if and only if the following conditions hold:

i) $|B_i| = |C_i|$, for $i = 1, \ldots, m$;
ii) Let $J = \{j \mid j \in \{1, \ldots, m\} : B_j \neq C_j\}$; then $|J| = 1$;
iii) For $j \in J$ let $\pi^i_{B_j}$ and $\pi^i_{C_j}$ denote the position of $a_{j,i} \in A_j$ in string B_j and C_j, respectively. Let $I_{B_j} = \{\pi^i_{B_j} \mid \pi^i_{B_j} \neq \pi^i_{C_j}, i = 1, \ldots, |A_j|\}$ and $I_{C_j} = \{\pi^i_{C_j} \mid \pi^i_{B_j} \neq \pi^i_{C_j}, i = 1, \ldots, |A_j|\}$. Then, $\ell = |I_{B_j}| = |I_{C_j}| \leq k$.
iv) Let $I_{B_j} = \{I_{B_j,1}, \ldots, I_{B_j,\ell}\}$ and $I_{C_j} = \{I_{C_j,1}, \ldots, I_{C_j,\ell}\}$; then $I_{B_j,i+1} - I_{B_j,i} = I_{C_j,i+1} - I_{C_j,i} = 1$, for $i = 1, \ldots, \ell - 1$.

Conditions i) and ii) state that both alignments should have the same size, and differ only in the j-th string, respectively. In addition, condition iii) and iv) state that at most k characters from string A_j do not occupy the same position in both alignments, and that those characters are contiguous. For illustration purpose, consider the following alignment:

```
D-GGF-
-D-FGL
```

We list all possible 2-block neighbors as follows:

```
-DGGF- DG-GF- DGG-F- D-GG-F D-G-GF D-GGF- D-GGF- D-GGF- D-GGF-
-D-FGL -D-FGL -D-FGL -D-FGL -D-FGL D--FGL --DFGL -DF-GL -DFG-L
```

It is possible to visit all k-block neighbors of an alignment ϕ in a straightforward way. Given the first string of an alignment ϕ, consider the leftmost substring of size one that contains only a character and an indel to its right. Then, exchange it with every indel in its right and stop when no indel is found. In case the substring has also indels to its left, repeat the same exchange procedure. If it is still possible, increase the size of the substring by one and repeat the same moves to its right and to its left; repeat the overall procedure until reaching a substring of size k. Then, consider the next substring of size one to the right and repeat the same procedure as above. Note that each move generates a new k-block neighbor of alignment ϕ. In order to maintain feasibility, the columns that contain only indels are deleted from the alignment.

Starting Alignment. The starting solution may have a strong impact in the overall performance of local search. We considered the two following possibilities for PLS:

i) Rand: A random feasible alignment, which is obtained by inserting indels randomly into the strings, except in the largest one. In principle, this approach generates a feasible alignment with a small number of indels.

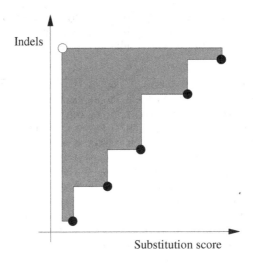

Fig. 1. Illustration of the hypervolume indicator

ii) `Clust`: A feasible alignment obtained from program Clustal Omega [12] (available at `http://www.clustal.org/omega`). It is expected that starting from "good" alignment allows the local search to find high quality alignments.

5   Experimental Analysis

We peformed several experiments with sequences obtained from the benchmark database BAliBASE [14]. In particular, we chose all the sequences of set RV11, which contains 38 equidistant families with sequence identity less than 20%.

We investigated the effect of the k-block neighborhood with respect to different k values and the effect of the two starting alignments on the performance of local search. For a given instance, let `Min` denote the length of the smallest string; we consider $k \in \{\texttt{Min}, \texttt{Min}/2, \texttt{Min}/4, \texttt{Min}/8, \texttt{Min}/16\}$. PLS terminates once it is not possible to find nondominated neighboring alignments or the time limit of 5 minutes of CPU-time is reached. Since PLS is a stochastic approach, we ran it ten times for each instance and recorded the contents of the archive for each run, namely the *approximate set*, once the termination criterion was met.

In order to evaluate the quality of each approximate set, we computed its hypervolume indicator value. This indicator measures the area that is dominated by the approximate set, bounded by a reference point [15]; see an in-depth discussion about the hypervolume indicator in relation with other performance assessment methods in Zitzler et al. [16]. Figure 1 illustrates the hypervolume indicator (shaded area) for a given approximate set (black points) and a given reference point (white point). We chose, as the reference point for each instance, the minimum substitution score minus one and the maximum number of indels plus one that were obtained from the runs of all local search variants. We merged

Table 1. Percentage with respect to the reference hypervolume indicator value for local search for several k-block neighborhood sizes and starting solutions (`Clus` and `Rand`) for each benchmark instance. The results are averaged over 10 runs. See text for more details.

id	m	Min	Max	$k = \mathrm{Min}$ Clust	Rand	$k = \mathrm{Min}/2$ Clust	Rand	$k = \mathrm{Min}/4$ Clust	Rand	$k = \mathrm{Min}/8$ Clust	Rand	$k = \mathrm{Min}/16$ Clust	Rand
22	4	63	205	0.81	0.92	0.70	0.87	0.64	**0.93**	0.48	0.89	0.41	0.76
25	4	64	103	**0.92**	0.66	0.81	0.68	0.67	0.65	0.57	0.67	0.47	0.61
29	4	81	138	0.89	0.74	**0.93**	0.71	**0.93**	0.74	0.87	0.69	0.74	0.70
1	4	83	91	**0.95**	0.71	0.70	0.70	0.68	0.66	0.68	0.52	0.63	0.41
9	4	97	337	**0.90**	0.67	0.70	0.67	0.42	0.65	0.27	0.65	0.14	0.58
21	4	102	139	**0.92**	0.53	0.86	0.57	0.78	0.55	0.70	0.56	0.59	0.54
8	4	104	540	**0.91**	0.81	**0.91**	0.81	0.78	0.81	0.50	0.82	0.34	0.75
17	4	247	264	**0.96**	0.38	0.95	0.31	0.89	0.32	0.86	0.30	0.75	0.24
15	4	297	327	0.96	0.50	**0.97**	0.48	0.94	0.57	0.92	0.53	0.92	0.49
12	4	320	397	0.96	0.41	**0.97**	0.38	0.95	0.47	0.90	0.39	0.85	0.38
24	4	372	465	**0.90**	0.52	**0.90**	0.52	0.78	0.53	0.71	0.54	0.61	0.54
4	4	390	456	0.92	0.36	**0.93**	0.37	**0.93**	0.36	0.85	0.35	0.77	0.36
3	4	414	516	**0.94**	0.45	0.91	0.45	0.90	0.43	0.87	0.45	0.84	0.45
10	4	490	492	**0.95**	0.03	0.91	0.03	0.78	0.03	0.70	0.01	0.64	0.01
13	5	51	101	**0.87**	0.75	**0.87**	0.75	0.76	0.75	0.73	0.71	0.58	0.68
35	5	71	138	0.90	0.77	**0.91**	0.77	**0.91**	0.79	0.80	0.80	0.62	0.72
11	5	160	242	0.94	0.54	**0.95**	0.54	0.92	0.52	0.80	0.50	0.70	0.53
37	5	335	1192	0.71	0.38	0.77	0.46	0.83	0.55	**0.85**	0.63	**0.85**	0.72
14	6	502	634	0.93	0.37	0.96	0.35	0.97	0.46	**0.98**	0.36	0.97	0.48
26	7	76	906	0.74	0.56	0.77	0.73	0.82	0.86	0.76	0.88	0.62	**0.96**
27	7	175	432	0.83	0.48	0.89	0.58	0.95	0.66	**0.96**	0.72	0.92	0.69
23	7	231	407	0.85	0.58	0.90	0.66	0.94	0.71	**0.97**	0.67	0.93	0.78
2	8	52	193	**0.93**	0.86	0.89	0.85	0.82	0.86	0.72	0.85	0.66	0.83
6	8	186	283	0.82	0.33	0.88	0.33	0.87	0.34	**0.89**	0.34	0.79	0.35
32	8	226	403	0.77	0.46	0.83	0.51	0.86	0.58	0.90	0.60	**0.91**	0.63
38	8	261	614	0.82	0.23	0.85	0.28	0.93	0.36	0.95	0.48	**0.96**	0.55
36	8	298	436	0.77	0.41	0.83	0.46	0.87	0.47	**0.92**	0.52	**0.92**	0.51
16	8	316	729	0.77	0.29	0.83	0.35	0.87	0.42	0.91	0.55	**0.96**	0.64
34	8	401	729	0.80	0.31	0.83	0.35	0.87	0.41	0.92	0.53	**0.94**	0.63
20	9	201	237	0.91	0.32	0.93	0.33	**0.94**	0.33	0.91	0.30	0.87	0.30
7	9	385	457	0.78	0.40	0.84	0.42	0.89	0.40	**0.94**	0.44	**0.94**	0.43
28	10	93	211	**0.92**	0.56	**0.92**	0.60	0.86	0.59	0.83	0.60	0.76	0.54
19	10	299	396	0.74	0.36	0.80	0.39	0.85	0.41	**0.92**	0.43	0.91	0.45
33	11	85	239	0.80	0.62	0.85	0.68	**0.87**	0.70	0.84	0.73	0.77	0.71
31	11	300	611	0.75	0.22	0.79	0.26	0.86	0.31	0.89	0.40	**0.97**	0.47
5	14	329	465	0.63	0.24	0.67	0.26	0.70	0.32	0.78	0.37	**0.85**	0.40
30	14	236	392	0.75	0.26	0.79	0.31	0.83	0.36	0.90	0.43	**0.95**	0.48
18	14	418	750	0.75	0.10	0.77	0.14	0.81	0.17	0.85	0.21	**0.88**	0.27

all approximate sets produced from all runs for a given instance, extracted the nondominated scores and computed the hypervolume indicator value of the resulting set, namely, the *reference hypervolume indicator value*, which is then used as a reference value to evaluate the relative performance of each approach. Then, for each approximate set, we computed its hypervolume indicator value and the percentage of this value with respect to the reference hypervolume indicator value; the larger the value, the better is the approximate set in terms of this indicator.

Table 1 reports the results obtained, averaged over 10 runs, for the five k values and the two starting alignments. Column id corresponds to the instance id from set RV11 (BB1100*.tfa, where * denotes the id); column m gives the number of sequences; columns Min and Max correspond to the length of the smallest and the largest sequence, respectively. The instances are lexicographically ordered in terms of the number of sequences and the length of the smallest sequence. The values in bold correspond to the best result obtained for each instance. The results clearly indicated the best performance of the starting alignment Clust. However, the best k-block value is strongly dependent of sequences sizes and number of sequences. In principle, for the given cut-off time, a small (large) k value gives better performance on larger (smaller) sequences and larger (smaller) number of sequences.

6   Conclusions and Further Work

The multiobjective formulation of the classical sequence alignment problem is of particular interest for an in-depth analysis of the evolutionary relationship between species. A similar concept has been applied to the analysis of phylogenetic trees of primates [1], which gave two possible explanations for the relationship between the species considered. The framework is based on the biobjective pairwise alignment for each pair of sequences, which produces several sets of alignments. The current approach allows to simplify the biobjective framework mentioned above for any number of sequences, by producing a single set of alignments.

The local search introduced in this article is able to provide high quality alignments in a reasonable amount of time. However, there are still ways of improving it further. For instance, in some cases, the local search gets easily stuck in a local optimum. In that case, it makes sense to perturb the set of alignments in the archive and restart the local search, such as done by iterated local search [6].

Acknowledgement. M. Abbasi thanks the Fundação para a Ciência e Tecnologia (FCT), Portugal, for the Ph.D. Grant SFRH/BD/91451/2012. This work was partially supported by iCIS (CENTRO-07-ST24-FEDER-002003).

References

1. Abbasi, M., Paquete, L., Liefooghe, A., Pinheiro, M., Matias, P.: Improvements on bicriteria pairwise sequence alignment: Algorithms and applications. Bioinformatics 29(8), 996–1003 (2013)

2. Bockenhauer, H.J., Bongartz, D.: Algorithmic Aspects of Bioinformatics. Springer (2007)
3. DeRonne, K.W., Karypis, G.: Pareto optimal pairwise sequence alignment. IEEE/ACM Transactions on Computational Biology and Bioinformatics 10(2), 481–493 (2013)
4. Erhgott, M.: Multicriteria optimization. Springer (2005)
5. Handl, J., Kell, D.B., Knowles, J.D.: Multiobjective optimization in bioinformatics and computational biology. IEEE/ACM Transactions on Computational Biology and Bioinformatics 4(2), 279–292 (2007)
6. Lourenço, H.R., Martin, O., Stützle, T.: Iterated local search. In: Handbook of Metaheuristics. International Series in Operations Research & Management Science, vol. 57, p. 321. Kluwer Academic Publishers (2013)
7. Needleman, S.B., Wunsch, C.D.: A general method applicable to the search for similarities in the amino acid sequence of two proteins. Journal of Molecular Biology 48, 443–453 (1970)
8. Paquete, L., Schiavinotto, T., Stützle, T.: On local optima in multiobjective combinatorial optimization problems. Annals of Operations Research 156(1), 83–97 (2007)
9. Paquete, L., Matias, P., Abbasi, M., Pinheiro, M.: MOSAL: Software tools for multiobjective sequence alignment. Source Code for Biology and Medicine, 9(2) (2014)
10. Roytberg, M.A., Semionenkov, M.N., Tabolina, O.I.: Pareto-optimal alignment of biological sequences. Biophysics 44(4), 565–577 (1999)
11. Schnattinger, T., Schöning, U., Kestler, H.: Structural RNA alignment by multiobjective optimization. Bioinformatics 29(13), 1607–1613 (2013)
12. Sievers, F., Wilm, A., Dineen, D.G., Gibson, T.J., Karplus, K., Li, W., Lopez, R., McWilliam, H., Remmert, M., Söding, J., Thompson, J.D., Higgins, D.G.: Fast, scalable generation of high-quality protein multiple sequence alignments using Clustal Omega. Molecular Systems Biology 7, 539 (2011)
13. Taneda, A.: Multi-objective pairwise RNA sequence alignment. Bioinformatics 26(19), 2383–2390 (2010)
14. Thompson, J.D., Koehl, P., Ripp, R., Poch, O.: BAliBASE 3.0: Latest developments of the multiple sequence alignment benchmark. Proteins 61(1), 127–136 (2005)
15. Zitzler, E., Thiele, L.: Multiobjective optimization using evolutionary algorithms - A comparative case study. In: Eiben, A.E., Bäck, T., Schoenauer, M., Schwefel, H.-P. (eds.) PPSN 1998. LNCS, vol. 1498, pp. 292–301. Springer, Heidelberg (1998)
16. Zitzler, E., Thiele, L., Laumanns, M., Fonseca, C.M., Grunert da Fonseca, V.: Performance assessment of multiobjective optimizers: An analysis and review. IEEE Transactions on Evolutionary Computation 7(2), 117–132 (2003)

Alignment Free Frequency Based Distance Measures for Promoter Sequence Comparison

Kouser[1,*], Lalitha Rangarajan[1,*], Darshan S. Chandrashekar[2,3],
K. Acharya Kshitish[2,4], and Emin Mary Abraham[1]

[1] DoS in Computer Science, Mysore, India
[2] Institute of Bioinformatics and Applied Biotechnology (IBAB), Biotech Park, Electronic City,
Bengaluru (Bangalore) - 560100, Karnataka State, India
[3] Research Scholar at Manipal University, Manipal – 576104, Karnataka State, India
[4] Shodhaka Life Sciences Pvt. Ltd., IBAB, Biotech Park,
Bengaluru (Bangalore) - 560100, Karnataka State, India
kauserdt@gmail.com

Abstract. With the massive amount of biological sequence data being generated by current technologies there is an urgent need to come out with faster sequence comparison methods. Most of the existing sequence comparison methods are alignment based which are proven to be very computationally complex when compared to the alignment free methods. In this paper, we have proposed alignment free methods for analysis of promoter sequences. Promoter sequences play a crucial role in gene regulation. After extracting the promoter sequence, matrices of motif frequency with position information (Position Specific Motif Matrix (PSMM)) is constructed, this is further taken for promoter analysis. These designed Frequency Based (FD) algorithms are tested on three different promoter datasets obtained from NCBI and UCSC repositories. The results show high similarity values for promoters with similar functionality and low values otherwise.

Keywords: Alignment free, Sequence comparison, Frequency distribution, Promoter sequences, Distance measure.

1 Introduction

Emergence of large scale sequencing projects and upsurge in accumulation of gene expression data make it increasingly important to understand the hidden information from DNA sequences. Also it is important to combine in vivo validation with computational approaches to perceive transcriptional regulatory networks better [1]. Computational differentiation of gene regulation networks has become a major task in the post genome-sequencing era. Promoters are the parts of the genome believed to be responsible for most transcriptional control [2]; its structure is characterized by the organization of TFBs within the promoter which is specific to gene groups [3].

---

* Corresponding authors.

F. Ortuño and I. Rojas (Eds.): IWBBIO 2015, Part II, LNCS 9044, pp. 183–193, 2015.
© Springer International Publishing Switzerland 2015

Despite the importance promoters have, the techniques to identify and analyze them are less advanced than the methods for coding regions. This is because of the diverse nature of the promoters and even well-known TFBs are not always conserved across all promoters [1].

Conventional sequence analysis methods rely mostly on pair wise or multiple sequence alignments [4] which are computationally intensive when compared to the alignment free methods. Mutual information based [5] methods which outperform conventional distance based phylogeny methods, Bayesian network based Dragon Promoter Mapper [DPM] is used to classify promoters to its target groups. There are several popular tools for multiple sequence alignment namely MUSCLE [6], T-Coffee, MAFFT, CLUSTAL W etc. A detailed comparison of these tools can be found in [6].

The alignment free methods are not affected by genome re arrangements, less dependent on sequence substitutions and help readily assess the degree of similarity in sequences. However these methods are less accurate when compared to the alignment based techniques and have limited use in Phylogenomics. There are several alignment free methods that are already available and are predicted to become more and more important as more sequence data becomes available [5]. In the literature, alignment free methods are available from as early as 1986 based on first and second order Markov Chain [7], Suffix tree and L-words frequency [8], Spaced-word frequency methods [9] and Relative frequencies of dual nucleotides [10]. Many methods relating to k-mer/word frequency, substring, information theory and graphical representation are also available. Some software available for the methods described above include Kmacs, Spaced words, AGP, Alfy, WNV typer, d2Tools etc. A detailed review of the distance measures for biological sequences is done in [11, 12] which are very useful in Biological Sequence Analysis (BSA). It is proved that the dissimilarity values observed by using distance measures, based on word frequencies are directly related to the ones requiring sequence alignment [13].

However, most of the methods described above are nucleotide comparison based which are only useful in analysis of the coding regions. Promoter sequences are a part of the non-coding regions of the genes and nucleotide wise comparison is not suited for their analysis. It is reported that promoter functionality is based on motif conservation unlike nucleotide conservation found in the coding regions. There have been a few attempts in the recent past on motif based promoter sequence comparison methods [14, 15, 16, 17]. In this paper we devise motif/TFBs based comparison methods to understand the underlying similarity in the promoter sequences. We also present a comparative analysis with Cumulative Frequency Distribution (CDF) based method described in literature [18]. The general outline of the steps involved in our model is as described in the Fig.1.

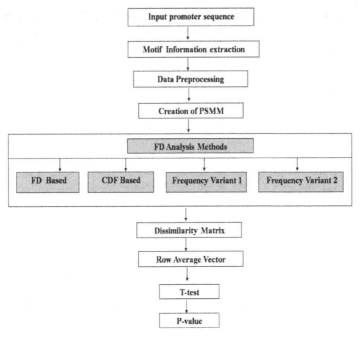

Fig. 1. Overall schema of the method

2   Method

In this section we describe generation of PSMM, the working details of our model and the proposed algorithms.

2.1   Generation of PSMM

The PSMM of promoter sequence are created using the following steps. At first, we obtain the nucleotides in the region behind the start of a gene till the end of the gene behind it for forward strands and in front of start of the gene till the end of the next gene for backward strands from the NCBI database. This constitutes the entire promoter of that particular gene. Then, we spot the first 'TATA box' around initiator codon (ATG) of the gene and then take 500bps from that TATA for our experiments. Later, these sequences are submitted to the 'TFSEARCH' [21] tool with a default threshold of 85% (specified in tool) to obtain the details of all the motifs with their position and score. This process is carried out for all the organisms and enzymes.

The PSMM matrix is generated by considering all the unique motifs identified by the TFSEARCH tool and are greater than 5bp in length. The matrix has 10 columns representing positions 1-50, 51-100, up to ......451-500 of the considered 500 bps. The number of rows depends on the number of unique motifs present in that particular promoter. We build a matrix where the multiple occurrences of a motif are counted. In case the motif is spread over two different ranges then we put '1' in the cell/cells where at least 40% of the motif is present. Every promoter has different number of motifs resulting in varying number of rows. Hence before we submit the matrix to our algorithm we do initial preprocessing by normalizing the number of rows by simply taking union of all the motifs from different promoters. The data flow in the generation of PSMM and subsequently the analysis proposed and the modules involved are shown in Fig.2.

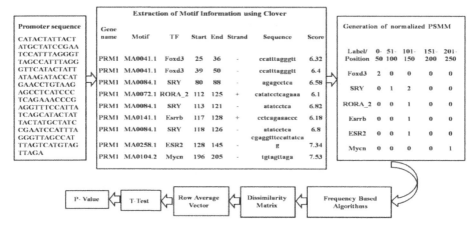

Fig. 2. Architecture of the method

2.2   Frequency Distribution Analysis Methods

We have devised two methods based on the frequency distribution of the motifs in a specified band 'N' and range 'R' and its corresponding   generated PSSM. For example the user's interest may be in the range 100 to 500 (R) and in the band of 50 (N) nucleotides in this range.

2.2.1   Frequency Distribution Based Approach

In the frequency distribution algorithm we calculate the row sum corresponding to each motif which keeps track of the frequency of each motif in a promoter. This is taken as a frequency feature vector and a dissimilarity value between two promoters is calculated by taking the sum of the squared differences (SSDF) between the two vectors.

Algorithm: FD Based.

Input: PSMMs of promoters.
Step 1: Calculate FD of motifs for each promoter.
Step 2: Calculate SSDF between each unique pair of promoters.
Output DM

2.2.2 Frequency Variant 1

In the Frequency variant 1 algorithm we calculate the column sum corresponding to each band of 50nt which keeps track of the frequency of all motifs in a particular band of promoter's PSMM. This is taken as a frequency feature vector and a dissimilarity value between two promoters is calculated by taking the sum of the squared differences (SSDF) between the two vectors.

Algorithm: Frequency Variant 1.
　　Input PSMMs of promoters.
　　Step 1: Calculate column sum of each band of size N.
　　Step 2: Calculate SSDF between each unique pair of promoters.
　　Output DM

2.2.3 Frequency Variant 2

In the frequency variant 2 algorithm we calculate the average of the output values of the previous two methods.

Algorithm: Frequency Variant 2.
　　Input PSMMs of promoters.
　　Step 1: Calculate average of SSDF of FD Based and Frequency variant 1 methods
　　Output DM

2.2.4 Cumulative Frequency Distribution

This method is described in detail in [17]. In this algorithm all the steps remain same as described in the FD based method. The only difference is that the frequency is cumulatively calculated.

Algorithm: Cumulative Frequency Distribution.
　　Input PSMMs of promoters.
　　Step 1: Calculate CFD of motifs for each promoter.
　　Step 2: Calculate SSDCF between each unique pair of promoter.
　　Output DM.

3   Experimental Analysis

The description of the data sets used for experiment and the results are discussed in this section.

3.1   Description of the Datasets

Dataset 1 is the promoter sequences consisting of pyruvate kinase gene of one prokaryote and nine eukaryotes (listed in title of Table .1). Dataset 2 comprises of promoters of all the 10 enzymes of glycolysis pathway of Homosapiens (human). Entire promoter sequence is extracted for each promoter of both datasets from NCBI database. The sequences are of varying lengths up to 35,000bps. For analysis we have considered 500bps for each of the promoter sequence and then used TFSEARCH tool to obtain the motif details and later created the PSMMs from the same. This results in matrices of size 73x10 for dataset 1 and 79x10 for dataset 2. The matrix is a record of the presence (entry in the cell is 1) or absence (entry in the cell is 0) of a motif in various positions 1-50, 51-100,101-150...451-500 (10 bands) of the 500bps considered. For dataset 3, among 176 genes listed in the supplementary notes of [19], we found that only 124 are known functional genes and extracted promoter sequences for these genes using UCSC chromosomal sequences, BioMart annotations and a PERL program. Similarly, two background sets of promoters of genes, which are not differentially expressed, are also obtained. Using CLOVER tool, JASPAR matrices were scanned to obtain the motif information of the promoter as shown in Fig. 2.

3.2   Results and Discussion

The results from all the 4 methods discussed reveal higher similarity in organisms belonging to the same family in dataset 1 as shown in Tables 1 and 2. However there are some results that go against our expectation and these are highlighted in the Tables 1 and 2. You may observe that promoters of a few closely related organisms (Human, Monkey; Monkey, Chimpanzee) show high dissimilarity and also some distantly related organisms (Fruit fly, Dog; Cat, Rat) show high similarity. Similarly few unexpected results are highlighted in Table 2. However using frequency variant 2 such problems are comparatively minimized (Refer Table.3).Dissimilarity in Fig.6 are high. This is not unusual as frequencies are cumulated and dissimilarity calculated. Generally the differences are better seen in this method. In dataset 2, we observe some underlying similarity existing between all the 10 enzymes involved in the glycolysis pathway of the human promoters which paves way for more analysis of promoters of organisms in different metabolic pathways. All our methods behave almost the same as seen in Figures 3, 4, 5 and 6. In Figure.7 we have separately plotted the values of 1 enzyme (HK1 hexokinase 1) with rest of the 9 enzymes of the glycolysis pathway. The analysis results on dataset 3 are presented in Tables 4, 5, 6, and 7. These results are obtained by performing T-test between dissimilarity matrices of two sets of promoters and p-value less than 0.05 signifies the success of the methods in differentiating the gene promoters between the two sets. The complexity of the discussed methods is linear when compared to the nonlinear complexity of alignment based methods. The implementation is done using MATLAB and it takes roughly 180seconds for analysis of 100 promoter sequences on a processor speed less than 2GHz and RAM memory of 4GB.

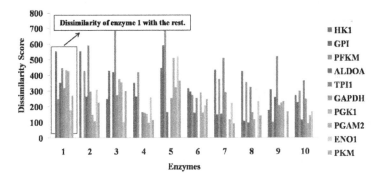

Fig. 3. Frequency method on dataset 2

(Enzymes : 1.HK1 hexokinase 1, 2.GPI glucose-6-phosphate isomerase , 3.PFKMphosphofructokinase, 4. ALDOA aldolase A, fructose-bisphosphate, 5.TPI1 triose-phosphate isomerase 1, 6.GAPDH glyceraldehyde-3-phosphate dehydrogenase, 7.PGK1 phos-phoglycerate kinase 1 , 8.PGAM2 phosphoglycerate mutase 2 (muscle) ,9.ENO1 enolase 1, (alpha) and 10.PKM pyruvate kinase (muscle))

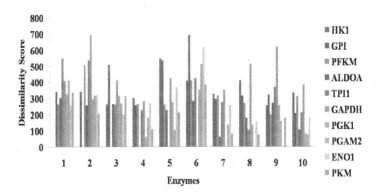

Fig. 4. Frequency variant 1 method on dataset 2

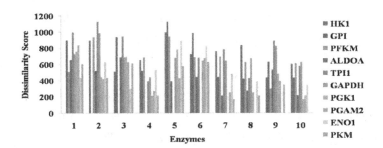

Fig. 5. Frequency variant 2 method on dataset 2

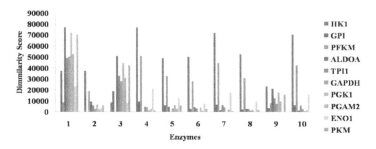

Fig. 6. Cumulative Frequency method on dataset 2

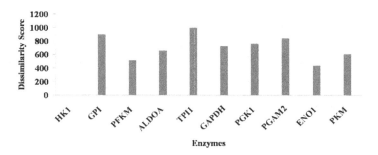

Fig. 7. Frequency variant 2 method on dataset 2 (Dissimilarity of enzyme 1 with the rest)

Table 1. Dissimilarity matrix of frequency variant 1 on dataset 1 (Organisms: 1. Homo sapien (human) , 2. Gorilla gorilla (gorilla) , 3.Macaca mulatta (rhesus monkey), 4. Bos taurus (cattle), 5.Felis catus (cat), 6.Pan troglodytes (chimpanzee) 7. Canis lupus (dog), 8.Rattus norvegicus (rat), 9.Drosophila melanogaster (fruit fly) and 10. Pseudomonas aeruginosa (bacterium))

	1	2	3	4	5	6	7	8	9	10
1	0	299	1629	903	1411	1378	1439	451	1531	1346
2	299	0	890	256	568	581	604	266	796	587
3	1629	890	0	400	500	283	488	1608	306	629
4	903	256	400	0	300	213	256	820	256	369
5	1411	568	500	300	0	267	198	1302	480	125
6	1378	581	283	213	267	0	169	1263	101	298
7	1439	604	488	256	198	169	0	1310	262	173
8	451	266	1608	820	1302	1263	1310	0	1606	1275
9	1531	796	306	256	480	101	262	1606	0	497
10	1346	587	629	369	125	298	173	1275	497	0

Table 2. Dissimilarity matrix frequency based method on dataset 1

	1	2	3	4	5	6	7	8	9	10
1	0	311	331	721	1177	1154	1043	139	1003	950
2	311	0	274	218	558	599	452	96	798	371
3	331	274	0	306	462	435	410	158	634	325
4	721	218	306	0	382	369	404	362	852	259
5	1177	558	462	382	0	65	112	572	560	109
6	1154	599	435	369	65	0	185	577	555	164
7	1043	452	410	404	112	185	0	484	628	81
8	139	96	158	362	572	577	484	0	604	417
9	1003	798	634	852	560	555	628	604	0	607
10	950	371	325	259	109	164	81	417	607	0

Table 3. Dissimilarity matrix of frequency variant 2 method on dataset 2 (Enzymes : 1.HK1 hexokinase 1, 2.GPI glucose-6-phosphate isomerase , 3.PFKM phosphofructokinase, 4. ALDOA aldolase A, fructose-bisphosphate, 5.TPI1 triosephosphate isomerase 1, 6.GAPDH glyceraldehyde-3-phosphate dehydrogenase, 7.PGK1 phosphoglycerate kinase 1 , 8.PGAM2 phosphoglycerate mutase 2 (muscle) ,9.ENO1 enolase 1, (alpha) and 10.PKM pyruvate kinase (muscle))

	1	2	3	4	5	6	7	8	9	10
1	0	449	256	328	499	363	381	420	217	303
2	449	0	470	261	564	495	222	212	315	218
3	256	470	0	344	474	343	348	314	151	307
4	328	261	344	0	196	223	108	138	266	111
5	499	564	474	196	0	340	394	214	446	290
6	363	495	343	223	340	0	323	339	414	316
7	381	222	348	108	394	323	0	128	242	86
8	420	212	314	138	214	339	128	0	196	109
9	217	315	151	266	446	414	242	196	0	174
10	303	218	307	111	290	316	86	109	174	0

Table 4. FD Based Analysis on Dataset 3

	Background1	Background2
Test	0.1946	0.1572
Background1		0.8298
Background2		

Table 5. CDF Based Analysis on Dataset 3

	Background1	Background2
Test	0.0156	0.0075
Background1		0.1980
Background2		

Table 6. Frequency Variant 1 Based Analysis on Dataset 3

	Background1	Background2
Test	0.2843	0.0458
Background 1		0.0425
Background 2		

Table 7. Frequency Variant 2 Based Analysis on Dataset 3

	Background1	Background2
Test	0.2671	0.0549
Background1		0.0425
Background2		

4   Conclusion

It is shown that the structural similarity between the genomic sequences correspond to their underlying functional similarity. The need for good similarity measures has gained importance since the possibility of sequencing of whole genome has grown, giving rise to the question of discovering common features between these biological sequences belonging to different species reflecting common functionality and evolutionary process. The methods discussed are successful in identifying the structural similarity of the promoters based on its motif information and hence aid biologists to make inferences about the functionality of the promoter sequences. We need to make a note that not all co-regulated gene promoters share a common motif, because some of the identified genes in a given cluster might be secondary response genes. Also, the combinatorial nature of TFs result in same motif being found in promoter regions of genes that are not co-regulated[1]. Though the methods discussed are all fast or linear time complex, the results suggest that the alignment free methods need better testing for scalability and robustness.

References

1. Qui, P.: Recent Advances in Computational Promoter Analysis in Understanding the Transcriptional Regulatory Network. Biochem. Biophys. Res. Commun. 309(3), 495–501 (2003)

2. Ohler, U., Niemann, H.: Identification and Analysis of Eukaryotic Promoters: Recent Computational Approaches. Trends Genet. 17(2), 56–60 (2001)

3. Chowdhary, R., Lam Tan, S., Ali, R.A., Boerlage, B., Wong, L., Bajic, B.V.: Dragon Promoter Mapper (DPM): A Bayesian Framework for Modelling Promoter Structures. Bioinformatics. 22(18), 2310–2312 (2006)

4. Felsenstein, J.: Inferring Phylogenies. Am. J. Hum. Genet. 74(5), 1074. Sinauer Associates, Sunderland M A (2004)

5. Penner, O., Grassberger, P., Paczuski, M.: Sequence Alignment, Mutual Information, and Dissimilarity Measures for Constructing Phylogenies. PLoS One. 6(1), (2011)

6. Edgar, R.C.: MUSCLE: Multiple Sequence Alignment with High Accuracy and High Throughput. Nucleic Acids Res. 32(5), 1792–1797 (2004)

7. Blaisdell, B.E.: A Measure of Similarity of Sets of Sequences not Requiring Sequence Alignment. Proc. Natl. Acad. Sci. 83(14), 5155–5159 (1986)

8. Soares, I., Goios, A., Amorim, A.: Sequence Comparison Alignment-Free Approach Based on Suffix Tree and L-Words Frequency. The Scientific World Journal 2012, (2012)

9. Leimeister, C.A., Boden, M., Horwege, S., Linder, S., Morgenstern, B.: Fast Alignment-Free Sequence Comparison Using Spaced-Word Frequencies. Bioinformatics. 30(14), 1991–1999 (2014)

10. Luo, J., Li, R., Zeng, Q.: A Novel Method for Sequence Similarity Analysis Based on the Relative Frequency of Dual Nucleotides. MATCH Commun. Math. Comput. Chem. 59, 653–659 (2008)

11. Mantaci, S., Restivo, A., Sciortino, M.: Distance Measures for Biological Sequences: Some Recent Approaches. Internat. J. Approx. Reason. 47, 109–124 (2008)

12. Vinga, S., Almeida, J.: Alignment-Free Sequence Comparison-A Review. Bioinformatics. 19(4), 513–523 (2003)

13. Blaisdell, B.E.: Effectiveness of Measures Requiring and not Requiring Prior Sequence Alignment for Estimating the Dissimilarities of natural Sequences. J. Mol. Evol. 29, 526–537 (1989)

14. Hu, J., Liang, X., Zhao, H., Chen, D.: The Analysis of Similarity for Promoter Sequence Structures in Yeast Genes. In: 5th BMEI, 919–922 (2012)

15. Meera, A., Rangarajan, L.: Comparison of Promoter Sequences Based on Inter-Motif Distance. IJSSCI 3(3), 57–68 (2011)

16. Meera, A., Rangarajan, L., Bhat, S.: Computational Approach Towards Finding Evolutionary Distance and Gene Order Using Promoter Sequences of Central Metabolic Pathway. Interdiscip. Sci. 3(1), 43–49 (2011)

17. Rangarajan, L.: Similarity Analysis of Position Specific Motif Matrices using Lacunarity for Promoter Sequences. In: ICONIAAC 2014, Article. 37. ACM, New York (2014)

18. Meera, A., Rangarajan, L., Shilpa, N.: New distance Measure for Sequence Comaprison Using Cumulative Frequency Distribution. IJCA. 19(2), 13–18 (2011)

19. Huang, A.C., Hu, L., Kauffman, S.A., Zhang, W., Shmulevich, I.: Using Cell Fate Attractors to Uncover Transcriptional Regulation of HL60 Neutrophil Differentiation. BMC Syst. Biol. 20(3), (2009)

20. Levenshtein, V.I.: Binary Codes Capable of Correcting Deletions, Insertions, and Reversals. Cybern. Control Theory 10(8), 707–710 (1966)

21. http://www.cbrc.jp/research/db/TFSEARCH.html

Energy-Efficient Architecture for DP Local Sequence Alignment: Exploiting ILP and DLP*

Miguel Tairum Cruz, Pedro Tomás, and Nuno Roma

INESC-ID, Instituto Superior Técnico, Universidade de Lisboa, Portugal
{miguel.tairum,pedro.tomas,nuno.roma}@inesc-id.pt

Abstract. Typical approaches to solve Dynamic Programming algorithms explore data level parallelism by relying on specialized vector instructions. However, the fully-parallelizable scheme is often not compliant with the memory organization of general purpose processors, leading to a less optimal parallelism exploitation, with worse performance. The proposed processor architecture overcomes this issue by relying on a data stream loader and a set of especially designed instructions. Furthermore, to make it compliant with the strict power and energy limitations of embedded systems, it maximizes resource utilization by exploiting both data and instruction level parallelism, by statically scheduling a bundle of instructions to several vector execution units. To evaluate the proposed architecture, we compare it with two embedded processors, an ARM Cortex-A9 and an application-specific processor with SIMD extensions, using two benchmarks from the sequence alignment domain, namely Smith-Waterman and Viterbi. The obtained results show that the proposed architecture achieves up to 5.16x and 2.19x better performance-energy efficiency and up to 5.44x and 1.25x better energy efficiency than the ARM Cortex-A9 and the dedicated processor, respectively.

Keywords: Low-Power Architecture, DLP, ILP, VLIW, Dynamic Programming, Sequence Alignment.

1 Introduction

Sequence alignment applications frequently make use of Dynamic Programming (DP) algorithms to extract information from large databases [1]. As an example, the Smith-Waterman (SW) algorithm [2] is widely adopted for local sequence alignment, computing the alignment score by filling up a scoring matrix, where each cell presents a vertical, horizontal and diagonal dependency with its neighbors. Similarly, the Viterbi algorithm [3], used to find the most probable state sequence in a Hidden Markov Model (HMM), can also be represented in a matrix form, resulting in similar computations steps and dependencies as the SW algorithm. Accordingly, Data Level Parallelism (DLP) can be naturally exploited in these algorithms by using Single Instruction Multiple Data (SIMD) extensions (often present in most processors), as long as the computation is performed along the matrix anti-diagonal.

---

* This work was partially supported by national funds through Fundação para a Ciência e a Tecnologia (FCT), under projects: Threads (PTDC/EEAELC/117329/2010) and project UID/CEC/50021/2013.

F. Ortuño and I. Rojas (Eds.): IWBBIO 2015, Part II, LNCS 9044, pp. 194–206, 2015.

Due to the gradual adoption of embedded systems in these type of applications (e.g. portable biochips [4]), efficient solutions are highly required, not only to guarantee the performance results required by these applications, but also to comply with strict power and energy consumption constraints. Typical processing solutions rely on General Purpose Processors (GPPs) [5,6] and dedicated hardware [7], that make use of SIMD extensions to exploit the DLP. On the other hand, the amount of memory that is required to accommodate the dependencies between states on DP algorithms is often very large, requiring the implementation of techniques to cache and reuse results from previous state computations, in order to quickly retrieve them without redundant computations. Furthermore, the previously referred anti-diagonal parallelism usually requires non-adjacent memory accesses, leading to a large performance impact in GPP implementations. To overcome this issue, GPP implementations often adopt a different processing pattern that better complies with a traditional GPP memory organization, such as vertical or horizontal simultaneous processing pattern, resulting in a better performance at the cost of introducing additional *lazy* loops to the algorithm computations. Dedicated implementations can approach this problem with special memory organization or special data management mechanisms, often resulting in better performance results [7]. However, these implementations only focus on a particular algorithm, disregarding any support for a broader class or different types of algorithms, thus limiting the range of application support.

In this paper, a new low-power programmable processor capable of supporting different DP algorithms is proposed, focusing on the performance extraction and optimization of sequence alignment algorithms, namely the SW and the Viterbi algorithms. The objective is not only to provide a high-performance processor for DP algorithms, but also to provide an energy-efficient solution suitable for low-power embedded environments. To attain the aimed performance levels with a low-power consumption, the architecture exploits: *i)* both DLP and Instruction Level Parallelism (ILP), supporting concurrent data computation in several independent and parallel execution units, with the minimal hardware requirements to accommodate it; and *ii)* a concurrent memory access mechanism, supported on a Data Stream Unit (DSU) that accesses the memory in parallel with the algorithm computations, thus maximizing the performance obtained with the most parallel processing scheme (e.g. the anti-diagonal parallelism).

This paper is organized as follows. After a brief overview of the studied DP algorithms and corresponding parallelism, in section 2, the proposed architecture is detailed in section 3. Section 4 briefly describes the prototyping and scalability evaluation of the proposed architecture, followed, in section 5, by a performance and energy efficiency evaluation of the architecture, against different state-of-the-art architectures. Finally, the conclusions are drawn in section 6.

2   Dynamic Programming Algorithms

DP algorithms are often represented in matrix form, where each cell corresponds to a sub-problem that depends on the adjacent cells (sub-problem dependencies).

This results in a final matrix where the value of the last cell can only be determined after all the previous cells have been computed (optimal substructure property). In a 2D matrix, each cell frequently presents horizontal, vertical and diagonal dependencies, allowing for DLP extraction along the anti-diagonal of the matrix. Furthermore, since all sub-problems are similar, they will require the exact same loop to be computed, allowing for an ILP exploitation by computing different steps of the loop at the same time. Popular sequence alignment algorithms such as the SW and the Viterbi algorithms, manifest the referred proprieties of DP algorithms and thus will be considered as two particular and independent case-studies.

2.1   Smith-Waterman

The SW algorithm computes the optimal local alignment between two sequences by considering a predefined substitution score matrix and a gap penalty function [2]. Gotoh [8] subsequently improved such algorithm definition by using an affine gap penalty model, allowing multiple sized gap penalties.

Given an arbitrary pair composed by a query and a reference sequences, together with a substitution score matrix and a gap initialization and extension penalties, the score matrix can be computed by solving three recursive relations, which correspond to a diagonal, horizontal and vertical dependencies, thus resulting in an anti-diagonal processing direction to extract the maximum parallelism. After the score matrix is filled, a traceback runs over such matrix, returning the local alignment.

2.2   HMM Viterbi

The Viterbi algorithm [3] is a DP algorithm that finds the most likely sequence path of hidden states in a HMM for a given sequence of observed outputs. A HMM consists in a stochastic model, where the future states of a process depend only on the present state and not on the complete sequence of states that preceded it. Furthemore, some (or all) states are hidden from the observer, with only the sequence of outputs generated by the model visible to the observer.

HMMs can be used to model alignment profiles, thus permitting the Viterbi algorithm to solve sequence alignment problems, similarly to the SW algorithm. These profiles model a family of sequences by highlighting the common features of them. They are usually generated by an initial multiple alignment, followed by a probabilistic breakdown of the elements present in each position.

A Profile HMM contains three different main interconnected states: Match States (M), that represent each column of the profile sequence; Insertion States (I), that represent the gaps in the alignment, and Deletion States (D), that represent the portions of the profile not matched by the sequence. Additional special states are also included, to support multiple local alignments. These states consist of flanking states, which delimit the sub-regions of the local alignment (i.e., they separate an aligned region from an unmatched region, in the sequence). When using Profile HMMs, the Viterbi algorithm becomes very similar to the

SW algorithm, with diagonal, horizontal and vertical recursions, resulting in the same anti-diagonal parallel processing pattern.

2.3   Parallelism Exploitation

The anti-diagonal processing pattern that is present in DP algorithms raises two problems to its exploitation: harder memory organization / access and a large amount of resource hardware requirements with a reduced utilization. While the former can be solved by implementing specialized memory access units to gather cell values in non-adjacent memory positions, the latter requires exploiting a different type of parallelism, namely ILP. Since the operations over a vector of elements along the anti-diagonal are independent, different parallel instructions can simultaneously compute different operations. This not only increases the potential for additional parallelism, but also reduces the hardware requirements, specially the number of FUs. The use of ILP is also supported by the common steps in a DP algorithm. Usually, these steps consist in dependency loads, followed by cell computations, and are finalized with the results storing. Hence, by assigning these different steps of the algorithms to the different elements in the vector, not only ensures ILP while maintaining data coherence (given the independence between the elements), but also improves the FUs utilization ratio.

In accordance, the proposed architecture reduces the hardware control by exploiting static ILP alongside DLP. This is achieved by issuing instructions in bundles, composed of several different parallel instructions, each operating over a vector of independent elements (DLP) (see Fig. 1(a)) in different and independent execution units (ILP). This way, instead of using a single large vector computing the same instruction as it is typical in vector architectures, the architecture has several smaller SIMD vectors, each computing their own instruction, in a similar fashion to a Very Long Instruction Word (VLIW) architecture. Parallelizing cells of a DP algorithm in different steps of the algorithm can, however, lead to data races between the cells if two conditions are not met: all cells currently being processed must be independent; and the cells that are being processed in advance must never have dependencies to the results of the other cells simultaneously being executed. The first condition is easily solved in the presented algorithms by following an anti-diagonal processing pattern. The impact caused by the irregular memory accesses referred by this parallelization scheme will be mitigated by an additional unit that performs the memory access operations concurrently to the main algorithm execution - a DSU. The second condition requires that the units operating over the down-left cells of the matrix are in advance regarding the units computing the cells at the top-right section (see Fig. 1(b)), due to the existing data dependencies.

3   Proposed Architecture

To exploit both DLP and ILP, as proposed in the previous section, the proposed architecture (see Fig. 2(a)) is composed of a bundle of vector execution units

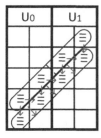
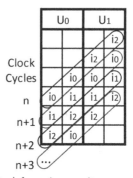
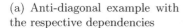

(a) Anti-diagonal example with   (b) Advancing units example.
the respective dependencies   Unit 1 has a 1-instruction delay.

Fig. 1. Anti-diagonal DLP (left) and the adopted ILP based on an advancing units scheme (right). Both figures illustrate an example of two execution units with 2-cell vectors. 3 instructions (1 clock cycle each) are required to compute each cell: i_0, i_1 and i_2. The dependencies are represented by the arrows.

(each with an independent register bank) and a DSU (to allow autonomous data-transfers from the main shared memory). The architecture's instruction word thus consists in a bundle of several smaller SIMD instructions packed in a VLIW macro word: one for each execution units and one for the DSU. The architecture is also characterized by a pipeline structure (see Fig. 2(b)) with 4 pipeline stages, namely FETCH, DECODE, EXECUTE and WRITE-BACK stages, with data forwarding mechanisms to minimize the number of stalls due to data dependencies.

The architecture includes support for both *scalar* and *vector* functional units, a large main shared memory and a smaller local fast memory. All these blocks can be accessed by all the execution units, provided that no structural hazards occur. The DSU can only communicate with the main memory. If conflicts arise when accessing registers or FUs, a stall mechanism is implemented in a priority list manner, where the processor is stalled until all conflicts have been resolved (taking the instructions more clock cycles to compute).

The execution units, registers, memories and FUs share the same maximum vector width, with support for different word widths. This design paradigm allows for different accuracy compromises, improving algorithm performance if higher accuracy is not required (using more but smaller SIMD words), while still supporting problems that require a higher level of precision (using fewer but larger SIMD words).

The architecture can also be easily scaled in two distinct ways: by increasing the length of each execution unit and thus increasing the vector length (DLP); and by increasing the number of execution units, and thus increasing the number of parallel instructions (ILP).

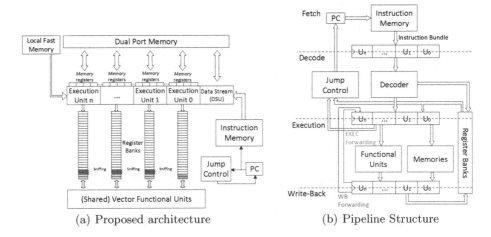

(a) Proposed architecture   (b) Pipeline Structure

Fig. 2. Proposed architecture scheme and respective pipeline structure

3.1   Architectural Units

As referred before, the architecture presents two distinct memories (in addition to the instructions memory): a main shared RAM memory and a local fast memory. The former can be accessed by the DSU and the execution units to store and read values, while the latter is a small memory that can only be read by the execution units, storing constant values required by DP algorithms. The existence of two memories minimizes the delay that is introduced by concurrent memory accesses and also promotes a better data organization, by separating the constant data values of the algorithms from the changing intermediary results. These memories present an access latency of 2 clock cycles: one cycle to index the correct address, and another cycle to load the value in the previous indexed address. In fact, there are two instructions to compute these two steps of a memory load, enabling a parallel usage of a memory load instruction between two units: one indexing an address, and the other effectively loading a data word. This allows achieving a throughput of 1 clock cycle with a latency of 2 clock cycles, when loading a value from memory.

The FUs in the architecture are shared by all the execution units. However, since the number of available FUs is limited, the execution units should not issue more operations of a given FU type than those available on the architecture, in order to maximize the processing performance. The FUs implemented in the architecture consist of SIMD Sum, Maximum, Shift, Logic (AND, OR, XOR) and Comparison units, with support for easily adding new FUs.

Each execution unit in the architecture has its own private register bank and a separate small set of 4 shared *memory* registers, allowing data sharing between units without memory accesses. These *memory* registers can eliminate the redundant memory loads introduced by some DP algorithms, whenever several cells (and therefore, execution units) share the same dependencies. Furthermore, these

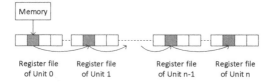

Fig. 3. Register Window mechanism: a value is loaded from memory to a register while the previous register value is shifted between adjacent units

registers will also be used by the DSU to communicate with the main memory (hence the *memory* name tag), storing and loading values from it in parallel to the algorithm computations, thus reducing the memory access latency impact. To further reduce the impact of memory accesses, the *memory* registers can also be used in a register window mechanism. This mechanism allows a memory load operation to fetch a data value to a *memory* register in one of the units on the periphery, while shifting the previous stored value on that register to the adjacent units (at the same time), as depicted in figure 3. This will update the selected *memory* registers in all units, enabling the pre-load of data values required by future iterations of the algorithms, such as sequence elements that are computed sequentially in all units.

Additionally, there is also a small subset of registers in the execution units' register banks, where a sniffing mechanism is implemented. This mechanism mirrors the sniffed registers to the register bank of the adjacent unit, effectively storing the register value in the adjacent unit. It is thus possible to share data values between adjacent units (which are very common in DP algorithms) without resorting to the main shared memory, at the same time that the *memory* registers are busy reading or storing values.

3.2   Instruction Set

While the DSU only performs memory access and shift operations over the *memory* registers (enabling the use of the referred register window mechanism), the execution units compute common arithmetic, logic, memory and control instructions. Additionally, optimized instructions for the targeted algorithms are also present, such as the `MAXMOV` instruction (useful for the SW algorithm), that not only does a maximum operation, but also adjusts some registers in the background to support the affine gap model without requiring additional clock cycles. The instruction set also presents modifiers to the implemented instructions. These modifiers can change the operand source of an instruction (e.g. use of immediate values) or add background operations to the instruction, enabling some algorithmic optimization. An example of the latter consists in a broadcast sum operation, where up to 3 parallel sums are performed in only 1 clock cycle, enabling the computation of multiple dependencies in DP algorithms.

3.3   Interface

In order to allow an efficient interconnection of the proposed architecture with different processing platforms, it was incorporated an interfacing structure aiming FPGA-based implementations. Such interfacing structure is based on the Advanced Microcontroller Bus Architecture (AMBA), structured according to its Advanced eXtensible Interface (AXI), allowing an easy interconnection between a GPP and the proposed architecture, which acts as an accelerator core.

The only communication requirement between the GPP element and the proposed architecture consists in a writing access directly to the three memories present in the proposed architecture, namely: *i)* the instructions memory; *ii)* the local fast memory; *iii)* and the RAM memory. Regarding the instructions memory, a control unit, outside the proposed architecture, controls the instructions flow from memory, with the instructions bundles being transferred in parallel to the architecture core. This control unit is further simplified when accounting that DP algorithms usually consist in one main loop that is repeated several times.

The two remaining memories require writing access from inside the proposed architecture, in addition to write access from the GPP. This way, there will be multiplexers at the entrance of the writing ports in these memories, where a control unit (external to the proposed architecture) decides which source will write to the memories. Therefore, the proposed architecture must stay idle when large sequences of data are being transferred to the memories (specially to the main memory). For the targeted sequence alignment algorithms, the idle time can be minimized by aligning a reference sequence to multiple query sequences, reducing the transfer times only to the query sequence transfers.

4   Prototyping

The proposed architecture was prototyped in a Zynq SoC 7100 FPGA [9]. The implemented configuration of the proposed architecture is composed of 1 DSU and 4 32-bits execution units, each using vectorial instructions to process multiple cells in parallel, resulting in a 128-bit wide VLIW. The word width was defined at 8 bits for the SW algorithm implementation and 16 bits for the Viterbi algorithm (due to higher precision requirements), resulting in 16 cells and 8 cells being computed in parallel, respectively. The register banks and memories share the same word widths.

The resources occupied in the FPGA (post place-&-route) can be seen in figure 4, accompanied by the operating frequency, power and scalability results.

4.1   Scalability

The impact on the hardware resources, frequency and power, introduced by both DLP and ILP scalability, is depicted in figure 4. The DLP scalability was evaluated by increasing the vector length from 32 bits to 40 bits, while the ILP

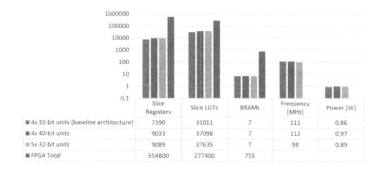

Fig. 4. Hardware, frequency and power scalability for the proposed architecture by increasing the DLP (40-bit vectors) and by increasing the ILP (5 execution units)

scalability was achieved by increasing the number of execution units to 5. In both cases, the number of slice registers and LUTs increased approximately by 18% and 16%, respectively. The number of BRAMs remained the same, since it only changes with a greater variation in the vector width. Regarding the operating frequency and power consumption, the DLP scalability test resulted in approximately the same frequency as the baseline architecture, with an increase of 12% in the power consumption. The ILP scalability achieved a frequency decrease of 12%, with a power consumption increase of 4%.

5   Evaluation

To evaluate the proposed architecture, different metrics were used to measure both the performance and energy efficiency. The performance evaluation was measured in Cell Updates per Second (CUPS) and the energy efficiency was measured in Cell Updates per Joule (CUPJ). Additionally, the architecture was also evaluated regarding its performance-energy efficiency, measured in Cell Updates per Joule-Second (CUPJS).

5.1   Reference State-of-the-Art Architectures

To better assess the proposed architecture, it was compared with two state-of-the-art architectures representing distinct low-power domains: *i)* a mobile low-power GPP (ARM Cortex-A9) operating at 533MHz; and *ii)* a dedicated programmable ASIP (Bioblaze [7]) operating at 158MHz and implemented in the same Zynq FPGA (for a fair comparison). Additionally, the proposed architecture was also compared with an intel i7 3820 GPP, operating at 3.6GHz, for a high-performance reference evaluation (although this processor does not comply with the typical power constraints of embedded biochips). All these architectures make use of 128-bit SIMD extensions (ARM's NEON, Intel's SSE, and Bioblaze ISA), ensuring a fair comparison against the proposed architecture. The Bioblaze was used to evaluate only the SW algorithm, since it was developed with the objective of accelerating such particular algorithm (although being programmable), while the remaining architectures were evaluated with both algorithms.

5.2 Algorithm Implementations

Both the SW and Viterbi algorithm implementations follow the anti-diagonal processing pattern, which is particularly efficiently exploited in the proposed architecture. The ILP is explored by having the left-most computing instructions in advance to the right-most cells. Hence, the SW main loop is composed of 5 instructions per execution unit (namely comparisons, sums and maximum operations), to process a complete vector of cells. The Viterbi algorithm requires a total of 23 instructions for an execution unit to compute the cells in its vector. These instructions mainly consist in simple sum and maximum operations, with additional loads and stores in the outer loop to account for the special states.

In the proposed architecture, the main memory is pre-loaded with the reference and query sequence, while both memories (main and local fast memories) are pre-loaded with all the necessary constants and cost/score values required by the evaluated algorithms. Therefore, only the algorithm steps are accounted for in the performed evaluations. Accurate clock cycle measurements of the required time to execute each biological sequence analysis in the proposed platform were obtained using Xilinx ISim.

For the remaining evaluated architectures, the algorithms follow the state-of-art implementations of Farrar [5] and HMMER [6], for the SW and Viterbi algorithms, respectively. In these implementations, it is used a processing flow along the query sequence (vertical), leading to the existence of additional lazy loops in the computations.

In the Bioblaze, the clock cycle measurements were achieved by using Modelsim SE 10.0b [7]. In the ARM Cortex-A9 and Intel Core i7, the system timing functions were used to determine the total execution time of the DNA sequence alignment. To improve the measurement accuracy, several repetitions of the same alignment were done and the obtained values were subsequently divided by the number of repetitions and the processor clock frequency.

5.3 Smith-Waterman Evaluation

To benchmark the SW algorithm, a DNA dataset composed of several reference sequences (ranging from 128 to 16384 elements) and a set of query sequences with length ranging from 20 to 2276 elements was used. The reference sequences correspond to twenty indexed regions of the Homo sapiens breast cancer susceptibility gene 1 (BRCA1gene) (NC_000017.11). The query sequences were obtained from a set of 22 biomarkers for diagnosing breast cancer (DI183511.1 to DI183532.1) and a fragment, with 68 base pairs, of the BRCA1 gene with a mutation related to the presence of a Serous Papillary Adenocarcinoma (S78558.1).

The evaluation results for the SW algorithm can be observed in Fig. 5, together with the scalability results introduced in section 4.1.

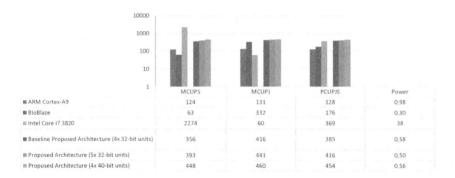

	MCUPS	MCUPJ	PCUPJS	Power
■ ARM Cortex-A9	124	131	128	0,98
■ BioBlaze	63	332	176	0,30
■ Intel Core i7 3820	2274	60	369	38
■ Baseline Proposed Architecture (4x 32-bit units)	356	416	385	0,58
■ Proposed Architecture (5x 32-bit units)	393	441	416	0,50
■ Proposed Architecture (4x 40-bit units)	448	460	454	0,56

Fig. 5. SW algorithm performance and energy-efficiency results for all evaluated architectures

When comparing with the low power architectures, namely the ARM Cortex-A9 and the Bioblaze, the proposed architecture achieves a speedup of 2.86x and 5.65x, respectively. Although the frequency of the ARM processor is 4.8x higher than the frequency of the proposed architecture, the fact that the algorithm implementation does not present lazy loops in the proposed architecture results in a better throughput and therefore in a better performance. Regarding the energy efficiency, the proposed architecture achieved a better efficiency than the ARM Cortex-A9 (3.18x) and the Bioblaze (1.25x), with the latter having a lower power consumption, thus demonstrating that the proposed architecture enables a more efficient implementation. Finally, the obtained performance-energy efficiency results show that the proposed architecture offers a significant gain over the ARM (3.02x) and the Bioblaze (2.19x).

As it was expected, in the high-performance domain the proposed architecture achieved a lower performance (0.16x) and a higher energy efficiency (6.95x) than the Intel i7, due the disparity in operating frequencies and power consumptions. Regarding the performance-energy efficiency, the proposed architecture managed to achieve better results, than the Intel i7, proving again that by efficiently exploiting the available parallelism it is possible to compensate for the difference in operating frequency.

Regarding the scalability results, the DLP scalability achieved superior performance and energy efficiency results when compared to the ILP scalability. In fact, the DLP scalability achieves a speedup of 1.26x and gains of 1.11x and 1.18x (for the performance, energy efficiency and performance-energy efficiency metrics, respectively) in comparison to the baseline architecture. As it was previously seen in section 4.1, on top of that, the DLP scalability also resulted in slightly less hardware resources and a higher operating frequency than the ILP scalability (only losing to the power consumption), proving to be the best scalability option given the amount of cell parallelism added.

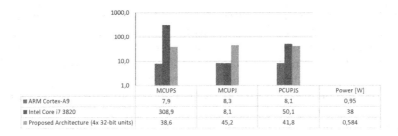

	MCUPS	MCUPJ	PCUPJS	Power [W]
■ ARM Cortex-A9	7,9	8,3	8,1	0,95
■ Intel Core i7 3820	308,9	8,1	50,1	38
■ Proposed Architecture (4x 32-bit units)	38,6	45,2	41,8	0,584

Fig. 6. Viterbi algorithm performance and energy-efficiency results for all evaluated architectures

5.4 Viterbi Evaluation

To evaluate Viterbi's algorithm implementation, a sample of 28 HMMs from the Dfam database of *Homo Sapiens* DNA [10] was used. The adopted model lengths vary from 60 to 3000, increasing by a step of roughly 100 model states, and were created by the HMMER3.1b1 tool [6]. Query sequences (generated by the HMMER tool [6]) ranging from 20 to 10000 symbols were used to evaluate the alignment against all the above reference sequences.

The results for the performance and energy efficiency metrics for the Viterbi algorithm can be observed in Fig. 6. Similarly to the SW results, the proposed architecture achieved a better performance (4.89x), energy efficiency (5.44x) and performance-energy efficiency (5.16x) than the ARM Cortex-A9. When comparing to Intel i7, the results for the performance and energy-efficiency were also similar to those obtained for the SW algorithm, with the proposed architecture achieving a lower performance (0.13x) and a higher energy efficiency (5.56x). Regarding the performance-energy efficiency, and contrary to the SW evaluation, the proposed architecture achieved a worse result than the Intel i7 (0.83x). This result shows that the higher energy efficiency of the proposed architecture, together with the optimized algorithm implementation, are not enough to surpass the higher performance of the Intel i7, even with a less efficient algorithm implementation. However, due to its higher power consumption, the Intel i7 architecture cannot be seen as a viable option to the targeted low-power embedded systems domain.

6 Conclusion

The proposed architecture exploits DLP and ILP to provide a high performance platform to DP algorithms, offering low-power consumption and high energy efficiency, to comply to the strict requisites of embedded systems (e.g. biochips).

Furthermore, the proposed architecture is scalable in two distinct fronts: at a DLP level, by increasing the vector lengths, and at an ILP level, by increasing the number of execution units. In both cases, it was shown speedup and energy efficiency gains against the baseline architecture, which demonstrates its potential scalability. The architecture also provides an extended algorithmic support when compared with traditional dedicated processors, by implementing a instruction set with optimized instructions and modifiers. Furthermore, the proposed architecture template permits the addition of new instructions and FUs, allowing further algorithmic optimization or support.

In face of the conducted evaluations, the proposed architecture can target low power systems without showing any significant loss against commonly used GPP alternatives and dedicated architectures. According to the obtained results, it presents better performance and energy efficiency than all the low-power architectures. In terms of performance-energy efficiency, the proposed architecture achieved gains of up to 5.16x against the ARM Cortex-A9 and 2.19x against the dedicated Bioblaze. For a high-performance reference, the proposed architecture also managed to obtain a better performance-energy efficiency than the Intel i7 3820 processor.

References

1. Benson, D.A., Clark, K., Karsch-Mizrachi, I., Lipman, D.J., Ostell, J., Sayers, E.W.: Genbank. Nucleic Acids Research (2014)
2. Smith, T.F., Waterman, M.S.: Identification of Common Molecular Subsequences. Journal of Molecular Biology 147(1), 195–197 (1981)
3. Viterbi, A.: Error Bounds for Convolutional Codes and an Asymptotically Optimum Decoding Algorithm. IEEE Transactions on Information Theory 13(2), 260–269 (1967)
4. Cardoso, F., Costa, T., Germano, J., Cardoso, S., Borme, J., Gaspar, J., Fernandes, J., Piedade, M., Freitas, P.: Integration of Magnetoresistive Biochips on a CMOS Circuit. IEEE Transactions on Magnetics 48(11), 3784–3787 (2012)
5. Farrar, M.: Striped Smith–Waterman Speeds Database Searches Six Times Over Other SIMD Implementations. Bioinformatics 23(2), 156–161 (2007)
6. Eddy, S.R.: Profile Hidden Markov Models. Bioinformatics 14(9), 755–763 (1998)
7. Neves, N., Sebastião, N., Matos, D., Tomás, P., Flores, P., Roma, N.: Multicore SIMD ASIP for Next-Generation Sequencing and Alignment Biochip Platforms. In: IEEE Transactions on Very Large Scale Integration (VLSI) Systems (in press) (2015)
8. Gotoh, O.: An improved Algorithm for Matching Biological Sequences. Journal of Molecular Biology 162(3), 705–708 (1982)
9. Xilinx.: Xilinx DS190 Zynq-7000 All Programmable SoC Overview (2013), http://www.xilinx.com/support/documentation/data_sheets/ds190-Zynq-7000-Overview.pdf (last accessed on December 10, 2014)
10. Finn, R.D., Bateman, A., Clements, J., Coggill, P., Eberhardt, R.Y., Eddy, S.R., Heger, A., Hetherington, K., Holm, L., Mistry, J., et al.: Pfam: The Protein Families Database. Nucleic Acids Research, gkt1223 (2014)

Hierarchical Assembly of Pools

Riccardo Vicedomini[1], Francesco Vezzi[3], Simone Scalabrin[2], Lars Arvestad[4],
and Alberto Policriti[1]

[1] Dept. of Mathematics and Computer Science, University of Udine, Udine, Italy
riccardo.vicedomini@uniud.it
[2] IGA, Institute of Applied Genomics, Udine, Italy
[3] Science for Life Laboratory, Department of Biochemistry and Biophysics,
Stockholm University, Stockholm, Sweden
[4] Science for Life Laboratory, Swedish e-Science Research Centre, Dept. of Computer
Science and Numerical Analysis, Stockholm University, 17121, Sweden

Abstract. This study introduces a method to address the problem of
building a draft *de novo* assembly of complex genomes when a collection
of well-assembled long-insert pools is available. Sequencing and assembling a collection of such pools reduces the complexity of the assembly
and has been proven to be a viable strategy in order to carry out downstream analyses in recent sequencing projects. In this work we depict a
framework to tackle this problem: we propose a novel fingerprinting technique to speed up overlap detection and we describe a merging technique
based on the well established string graph structure in order to carry out
the reconciliation step. Finally, we show some preliminary results on simulated data sets based on the human chromosome 14 obtained with an
early implementation of a tool we called Hierarchical Assemblies Merger.

Keywords: assembly reconciliation, hierarchical assembly, pool sequencing, next-generation-sequencing.

1 Introduction

In the last decade sequencing costs have been continuously dropping down with
the advent of the so called Next Generation Sequencing (NGS).

Nevertheless, these cost-effective technologies – based on a high coverage of
very short reads – created new challenges for the *de novo* assembly problem,
especially for large and complex genomes [1]. Indeed, resolving repeats longer
than read length is often unfeasible, particularly in repeat-abundant data sets.
Several algorithms have been proposed to increase assembly's contiguity and
correctness, though, the quality of the reconstructed sequences is often unsatis-
factory for downstream analyses.

A recent approach, seeking for a trade-off between sequencing costs and assembly's quality, consists in sequencing long-insert DNA fragments (*e.g.*, fosmids
and BAC clones) in pools using NGS technologies. In this way, since a pool represents just a small subset of the entire genome, the complexity of assembly

F. Ortuño and I. Rojas (Eds.): IWBBIO 2015, Part II, LNCS 9044, pp. 207–218, 2015.

highly decreases. In particular, compared to the canonical whole genome shotgun (WGS) sequencing, a higher quality is expected for an assembled pool in terms of sampling, repeats resolution, and allele reconstruction [1]. Also, pools introduces helpful constraints that can be exploited during reconciliation.

To the best of our knowledge, when facing either a large or a complex genome, where pool sequencing proved to be a viable and promising approach [2,3], a standard methodology which systematically handles these kinds of data sets has not been proposed yet. Available assembly reconciliation tools such as GAA [4], GAM-NGS [5], and Mix [6], in fact, have not been designed for this kind of task. In particular, the first two are based on a master-slave approach and they are able to reconcile just two WGS assemblies at a time. The third one, instead, while being able to accept multiple assemblies as input, had been specifically thought for small bacterial genomes. Moreover, its negligence in dealing with mis-joins makes it unsuitable for the job.

For these reasons we propose a hierarchical scheme, along with a first implementation, whose goal is to build a draft assembly from multiple sets of independently assembled pools. Our idea relies on the effectiveness of the methods used to tackle two main sub-problems: overlap detection and merging of input sequences. For the first one, we depict a fingerprinting-based solution which let us to quickly carry out the task. We describe then a possible data structure and heuristics in order to deal with the second one. Some preliminary results are reported, obtained using a prototype tool we named *Hierarchical Assemblies Merger* (HAM). We want also to remark that this work could also be easily adapted to take advantage of new sequencing technologies, which produce long reads and are affected by high error rates.

2   Methods

Let $\mathcal{P} = \{\mathcal{P}_1, \ldots, \mathcal{P}_m\}$ be a collection of assembled pools (in the following the adjective "assembled" may be omitted). Each $\mathcal{P}_i$ is the result of the *de novo* assembly of multiple long-insert fragments and it is supposed to be obtained by a state-of-the-art *de novo* assembler, using a high-coverage set $\mathcal{R}_i$ of short reads.

In order to guide the reconciliation, we are going to make use of two assumptions about pools. First, a pool covers a small percentage of the genome. Thus, two contigs $C_1, C_2 \in \mathcal{P}_i$ most likely *do not* belong to the same genomic *locus* (or at least we expect this would occur rarely), unless they represent the same insert that could not be entirely assembled. Second, a contig cannot be longer than the maximum size of the sequenced fragments (say, approximately, 40 Kbp for a fosmid and 100 Kbp for a BAC clone).

Our hierarchical assembly scheme can be naturally depicted as a binary tree $\mathcal{T}$ which recursively decompose the problem. $\mathcal{T}$ has m leaves and minimum height (*i.e.*, $\mathcal{O}(\log m)$). More precisely, the i-th leaf is labeled $\mathcal{P}_i$, while an internal node corresponds to the result of the reconciliation of two or more pools and it is labeled $\mathcal{L}_1 \oplus \mathcal{L}_2$, where $\mathcal{L}_1$ and $\mathcal{L}_2$ identify the "partial" assemblies of its children and $\oplus$ is the merging operation. The root, thus, consists in the final assembly of the collection $\mathcal{P}$. $\mathcal{T}$ is also partitioned "vertically" with respect to the depth of its vertices. More precisely, we propose to consider three major classes of nodes, namely $\mathcal{A}_h$, $\mathcal{A}_m$, and $\mathcal{A}_l$, where different strategies could be applied (see Fig. 1).

$\mathcal{A}_h$ comprises nodes with highest depth. They correspond to the reconciliation of assemblies which still cover a quite small part of the genome. Thus, in this scenario we can afford to use a less sophisticated (even quadratic) algorithm for the overlap detection and a simple merging algorithm based on minimum length and identity. However, the more a node is closer to the root the more an unsophisticated/greedy method would be computationally expensive and error-prone.

For this reason we introduce $\mathcal{A}_m$ and $\mathcal{A}_l$, whose reconciliations are thought to be done with more efficient techniques. In the following sections we define an original method for $\mathcal{A}_m$ (valid also for $\mathcal{A}_l$) based on the construction of smaller objects – fingerprints – to be used in place of the entire contigs for overlap detection. We then formulate an open problem whose solution could improve the performances for $\mathcal{A}_l$. Finally, we present an algorithm based on the String Graph [7] used to carry out the actual merge.

In conclusion, we can summarize our hierarchical assembly approach in:

1. a pool pre-processing step;
2. several reconciliation steps consisting in a depth-based strategy for overlap detection and merging (see Fig. 1).

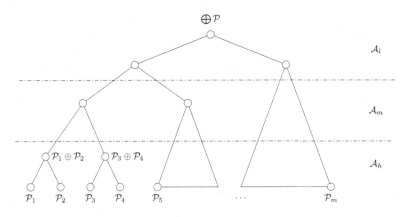

Fig. 1. Hierarchical assembly of pools guided by a binary tree $\mathcal{T}$ with reconciliation strategies depending on node depth (*i.e.*, regions $\mathcal{A}_h$, $\mathcal{A}_m$, $\mathcal{A}_l$)

2.1 Pool Pre-processing

This phase is thought to be performed on leaves of $\mathcal{T}$. The goal is to filter poorly assembled pools, exploiting features (*e.g.*, paired reads mapping, k-mer analyses) which may spot errors in input contigs. However, we do not want to discard these sequences completely but we can try to integrate them in a later stage (as soon as a reliable draft assembly has been achieved).

An intuitive pre-processing procedure to find putative mis-assemblies consists in using a mapping of $\mathcal{R}_i$ against $\mathcal{P}_i$ in order to break contigs in regions showing a low physical coverage (*i.e.*, uncovered by paired-read inserts). This idea has already been exploited in tools like REAPR [8] and FRC^{bam} [9] – which also take advantage of other "bad-mapping" evidences – for correction and evaluation purposes, respectively.

Finally, a length threshold can be applied to keep only long-enough contigs. Since we expect good quality assemblies for most of the pools, for instance, we can set it to a fairly high value (*e.g.*, 5 Kbp).

2.2   Overlap Detection

For this task we use that pools are unlikely to contain overlapping fragments. We can formulate the constraint that any pair of contigs from the same pool are distinct.

A simple approach to solve the problem could be performing an all-against-all alignment among the pools to be merged. We then keep overlaps above user-defined length and identity thresholds. Indeed, using this approach could be computationally demanding for large numbers of sequences and could be afford-able only for merges in $\mathcal{A}_h$.

In order to perform the task in $\mathcal{A}_m$ and $\mathcal{A}_l$ one may think to use overlap detection algorithms of state-of-the-art assemblers. However, to the best of our knowledge, the presence of significant indels is not kept into account by them. More precisely, NGS assemblers usually expect *short good-quality* reads as input and not *long pre-assembled contigs*. Sanger-read assemblers, instead, were not designed to work agilely on very large data sets. For this particular task, we want to find an effective method to detect overlaps while still being able to handle errors. The idea we are currently exploring consists in replacing each contig C with a k-mer-based *fingerprint* $\mathcal{F}(C)$ when seeking for sequences C' that are likely to overlap with C. This idea should help in quickly finding putative approximate overlaps between contigs.

Non-deterministic Fingerprints for Overlap Detection. Let S be an assembled sequence, $|S|$ be its length, and let k_i be the k-mer starting at position i in S, where $i \in [1, |S| - k]$. More precisely, k_i is the sub-sequence $S[i, i + k - 1]$. We want to compute a significantly smaller structure $\mathcal{F}(S)$ to be used in place of S which shall allow us to find overlaps with high probability. In particular, the idea is to build a collection $\mathcal{F}(\mathcal{P}) = \{\mathcal{F}(\mathcal{P}_1), \ldots, \mathcal{F}(\mathcal{P}_m)\}$, where $\mathcal{F}(\mathcal{P}_i) = \{\mathcal{F}(S) \mid S \in \mathcal{P}_i\}$, and exploit it to speed up the overlap detection. Fingerprint is built by wisely picking an ordered list of its k-mers. We also assume the global frequency $f(t)$ to be available for every k-mer t in the data set. The fingerprint computation problem is defined as follows.

Problem. Let S be a sequence, T_{freq} be a k-mer frequency threshold, and T_{gap} be a maximum distance threshold. We define n_{HF} as the number of k-mers $k_i \in S$ for which $f(k_i) > T_{freq}$ and n_{LF} as the number of k-mers $k_i \in S$ for which $f(k_i) \le T_{freq}$. The problem is to seek for an ordered list of k-mers $\mathcal{F}(S) = \langle k_{i_1}, \ldots, k_{i_z} \rangle$ where $i_j \in \{1, \ldots, |S| - k + 1\}$, $i_1 < i_2 < \ldots < i_z$, and such that the following constraints are fulfilled:

- $i_{j+1} - i_j \le T_{gap}$ for $j = 1, \ldots, z-1$ (*i.e.*, two consecutive k-mers are not too far apart in S);
- n_{HF} is minimum;
- n_{LF} is minimum among those lists which minimize n_{HF}.

In other words, the problem is building a fingerprint which fulfills the gap constraint and minimizes the pair (n_{HF}, n_{LF}) in the lexicographic order.

The rationale behind this approach is to take as many low-frequency k-mers as possible while assuring that long sub-sequences are not left "uncovered". Keep in mind that these k-mers will be aligned and minimizing the number of highly frequent k-mers improves the performance.

Fingerprint Construction. An approximate solution can be found linearly with two scans of the ordered list of S's k-mers. In the first one, we just pick those which are boundaries of *low-frequency regions* (*i.e.*, maximal sub-sequences comprising exclusively k-mers t such that $f(t) \le T_{freq}$). In the second one, whenever two subsequent k-mers in $\mathcal{F}(S)$ violate the gap constraint, a minimal list of k-mers is added between them in order to fulfill the gap threshold. This solution, while not being optimal, has the advantage of providing the minimum number of high-frequency k-mers. It is also pretty straightforward and does not take much computational effort (see Figure 2).

Fig. 2. Example of fingerprint construction. Low-frequency regions are filled with a darker tint. First, marked k-mers are added to $\mathcal{F}(C)$. Second, remaining k-mers are chosen in order to fulfill the gap constraints (*i.e.*, $i_{j+1} - i_j < T_{gap}$). The output fingerprint is then $\mathcal{F}(C) = \langle k_{i_1}, \ldots, k_{i_z} \rangle$.

Fingerprint-based Overlaps Detection. After the fingerprints are built, we map each k-mer in $\mathcal{F}(S)$ against the set of sequences S' for which we want to detect overlaps. Taking into account the distances between the mapped k-mers and their mapping order allows us to reduce the number of false positives (*i.e.*, sequences which do not overlap). S' can be indexed either using a db-Hash [10] or an FM-index. The purpose of this mapping, however, is just to identify putative overlaps while reducing at the same time the number of exhaustive alignment computations (*i.e.*, performed using a banded Smith-Waterman algorithm).

First, we introduce a parameter c_{min}, which is the minimum number of shared k-mers required to check whether two contigs overlap. This parameter should be chosen in order to guarantee we are able to find true overlaps with high probability and a low false positive rate.

Second, we take into account the distances and the order of the mapped k-mer. Let A be a contig sharing at least c_{min} k-mers with $\mathcal{F}(C)$ and let $\mathcal{M}_A$ be a list of pairs (k_{i_j}, w), also referred as *hits*, such that $k_{i_j} \in \mathcal{F}(C)$ and w is the position where k_{i_j} occurs in A. After sorting $\mathcal{M}_A$ according to j (*i.e.*, the index of k_{i_j} in $\mathcal{F}(C)$), we seek for a long enough interval I of hits with the following constraints:

1. the k-mers of two consecutive hits reflect the order in $\mathcal{F}(C)$;
2. the actual number of k-mers in $\mathcal{M}_A$ between two *uniquely* mapped k-mers differs at most by 3 from the expected number (*i.e.*, the one in $\mathcal{F}(C)$);
3. all k-mers should be mapped with the same orientation.

Since I corresponds to a region in both A and C, we can think of it as an approximate overlap and we want to choose the one which minimizes the sum of the left and right tips of A and C (these tips are accounted with respect to I). An interval fulfilling these constraints is then extended from both ends considering also non-uniquely mapped k-mers (see Figure 3). This putative overlap is then assessed using Smith-Waterman algorithm and retained only if its length is greater than T_L and the identity exceed id_{min}, where T_L and id_{min} are two user-defined thresholds.

This approach might be seen as a speed up of GAM-NGS's blocks construction [5]. In GAM-NGS's work a set of reads had to be mapped on every assembly and the idea had not been generalized to efficiently handle more than two assembly. In this work, instead, we just consider a subset of k-mers to quickly build an approximate layout of pools' sequences.

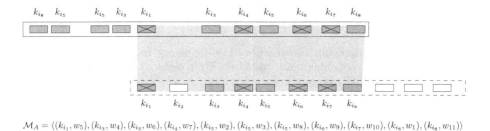

$$\mathcal{M}_A = \langle (\underline{k_{i_1}, w_5}), (k_{i_3}, w_4), (k_{i_3}, w_6), (\underline{k_{i_4}, w_7}), (k_{i_5}, w_2), (k_{i_5}, w_3), (k_{i_5}, w_8), (\underline{k_{i_6}, w_9}), (\underline{k_{i_7}, w_{10}}), (k_{i_8}, w_1), (k_{i_8}, w_{11}) \rangle$$

Fig. 3. Example of overlap detection. Colored k-mers are those found in the target sequence: crossed ones identify unique hits (underlined in $\mathcal{M}_A$), while the others are mapped in multiple position. Blank k-mers, instead, are absent in the sequence. The first interval computed is $[k_{i_1}, k_{i_7}]$, which is then extended with k_{i_8}. The approximate overlap reported consists of the sequences $C[i_1, i_8 + k - 1]$ and $A[w_5, w_{11} + k - 1]$.

Deterministic Fingerprints for Overlap Detection. When dealing with the reconciliation of large assemblies, both in terms of number and length of the contigs, a computationally efficient overlap detection strategy is extremely important. In our hierarchical framework, if we were able to build fingerprints in a deterministic way we could further improve overlap detection in the most critical part of $\mathcal{T}$ (*i.e.* $\mathcal{A}_l$), leading us to a fingerprint-against-fingerprint comparison. This construction, however, is not trivially achievable and, for the time being, we define it as the following open problem.

Problem. Given a set of sequences, is it possible to build a *small* k-mer based fingerprint such that, whenever two sequences C_1 and C_2 overlap, their fingerprint also "overlap", that is, they share most of the same k-mers in the overlapping region?

2.3 A Merging Strategy

A reasonable structure to reconcile assemblies would certainly be a "variant" of the String Graph (SG) introduced by Myers in [7]. This small variant consists in considering approximate overlaps and will be referred in the following as Overlap Graph (OG). Moreover, the SG can be seen as a *bidirected graph*, which is a graph where a directed head is attached to both ends of an edge. There are four kinds of edges and they correspond to the different types of overlaps [7]. Thus, we can define the in-degree of a vertex as the number of incident heads that point inwards and the out-degree as the number of incident heads that point outwards with respect to the vertex.

The SG construction is done as follows: a vertex is put for each contigs, while edges link overlapping pairs; transitive edges are removed; simple paths are collapsed in unitigs. The main idea of the SG is to represent sequences as vertices and prefix-suffix overlaps as edges. We expect that input sequences (*i.e.*, assembled pools) might contain different kind of errors: simple mismatches, insertions/deletions and misjoins. Thus, before connecting two vertices (contigs), we check whether their approximate overlap exceeds a user-defined identity threshold. The alignment is then retained only if the "overhangs" introduced by the overlap's intervals are short enough (*e.g.*, less than 100 bp). Moreover, fully included contigs are discarded and not represented as vertices. However, they will be used to compute the sequence coverage of the graph node in which they are included.

Graph Simplification. The OG is simplified removing transitive edges using Myers's algorithm [7] and unambiguous paths (*i.e.*, non-branching paths) can be collapsed into single nodes. At this point, we can take care of some graph topologies which may arise due to putative misjoins and small indels: bubbles and cut vertices adjacent to bifurcations.

Bubbles are characterized by two (or more) paths starting and ending at the same nodes. They are usually caused by small errors (*e.g.*, insertions/deletions) or biological variants. We can simply rely on algorithms used by state-of-the-art assemblers to take care of these structures. For instance, we may perform a linear visit of the graph (*e.g.,* depth-first) and when a bubble is found, if the identity of the branches is above a certain threshold, we retain only the path which is better covered and remove the other ones. Otherwise, if branches diverge too much, we do not remove any of them.

When an input sequence contains a mis-join such as a relocation (*i.e.,* regions far apart within the same chromosome are spliced together) or a translocation (*i.e.,* regions belonging to different chromosomes are joined) we can witness long almost-unambiguous paths which are connected by a single vertex (the contig containing the mis-join) which causes a bifurcation in each path. These vertices can bee seen as *cut vertices* (*i.e.,* their removal increases by one the number of connected components of the graph) with in and out degrees equal to 1 and adjacent to nodes characterized by a bifurcation.

Finally, an additional constraint could be exploited to identify consistent (or inconsistent) paths: giving the assumption that pool inserts have been sampled uniformly and independently, we expect contigs belonging to the same pool to be found far apart in a path and close enough contigs (*i.e.,* below the insert-size) are likely to represent the assembly of a single insert.

Consensus Sequence. Due to the use of approximate alignments to compute overlaps, we decided to output the sequence for each remaining vertex as follows. Each vertex corresponds to a simple unambiguous path in the former OG. Thus, we simply start from the first contigs in the path and we extend with the following one (see Figure 4). A better approach might be weighting the edges (possibly considering also transitive ones) and providing the best path according to a certain function.

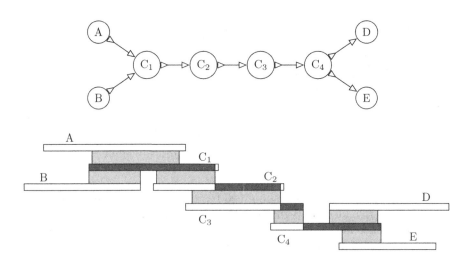

Fig. 4. Example of OG. A possible consensus sequence is colored with a darker tint

3   Results

Two simulated data sets on the 88 Mbp-long Human chromosome 14 have been taken into account. All the results are based on an early implementation of a tool we named Hierarchical Assemblies Merger (HAM) which performs the overlap detection and the merge strategies in a single step. The first one is carried out with the non-deterministic fingerprint-against-sequence alignment depicted in Section 2.2, while the second one takes advantage of the OG and the heuristics presented in Section 2.3. All the experiments were run on the same machine using 8 threads and setting the following HAM's parameters: $T_{freq} = 20$, $T_{gap} = 50$bp, $c_{min} = 8$, $T_L = 1$Kbp, $id_{min} = 0.95$.

The first data set (*ExactHG14*) has been built sampling a $4X$ coverage of 40Kbp-long inserts randomly assigned to 123 pools. In this way, each pool covers roughly the 3% of the reference genome. Then, inserts in pools have been randomly sheared to generate, on average, 19Kbp-long *exact* contigs.

The time required by BLAST to produce an all-against-all alignment took approximately 1 hour and 14 minutes. HAM's overlap detection, instead, took 18 minutes distributed as in Table 1.

Table 1. Time required to compute overlaps

sub-step	time
indexing	8m 22s
k-mer freq. computation	3m 24s
fingerprint construction	34s
fingerprint alignment + SW	5m 40s
whole procedure	**18m 0s**

Consider also that, if the data set is fixed, the index can be built once and quickly loaded from the secondary memory in successive experiments. The indexing and the alignments were performed with BWA's implementation of the FM-index and the Smith-Waterman algorithm, while the k-mer frequencies computation was carried out with an efficient hash table implementation.

Since we are considering an exact data set, we set the minimum identity of 100% for BLAST to find all the overlaps grater than 1 Kbp. The same threshold for sequence identity has been chosen for HAM as well. We were able to obtain a very high sensitivity while keeping a low number of false positives (see Table 2).

The second data set (*SimHG14*), consists of a $8X$ coverage of 40Kbp-long inserts randomly assigned to 353 pools (each one containing approximately 50 elements and representing the 2.27% of the reference genome). We chose the insert size to follow a normal distribution with mean 40Kbp and standard deviation 5Kbp.

For each pool, we simulated a $42X$-coverage PE-read library of Illumina fragments with (500 ± 25) bp-long insert-size and 100 bp-long reads. For this task

Table 2. BLAST overlaps, HAM overlaps, True Positives (TP), False negatives (FN), False Negatives (FN), and Sensitivity in HAM's overlap detection

BLAST	HAM	TP	FP	FN	Sensitivity
69,950	69,506	69,252	254	698	99%

we chose the tool pIRS [11]. Then, in order to obtain good-quality assemblies, pools have been independently assembled using two state-of-the-art *de novo* assemblers: ABySS (version 1.5.2) and MaSuRCA (version 2.3.1). These assembled pools will allow us to test HAM on a more realistic (yet still small) data set with respect to *ExactHG14*.

In a real scenario, a reference genome is usually not available. We decided, however, to validate the input assemblies using GAGE's validation script [12] against the available references for each pool. In this way, we can better assess the performances of our tool, especially when mis-joins are concerned. The choice of the assembler has been pretty clear. ABySS returned very contiguous assemblies with good quality metrics, while MaSuRCA returned results with also good quality metrics but they were more fragmented and presented on average a higher percentage of missing reference sequence.

ABySS has been first executed with mostly default parameters using a k-mer size of 71bp ($k = 71$), a higher maximum bubble length ($b = 1,000,000$), and a higher threshold for the unitig size required to build contigs ($s = 500$). A second run has been carried out with two additional parameters: a lower minimum alignment length of a read ($l = 1$ instead $l = k$) and a higher minimum sequence identity for a bubble ($p = 0.95$ instead of $p = 0.9$). Both executions achieved similar results as inversions and relocations are concerned. The second one, however, led to a significantly higher number of translocations (*i.e.,* rearrangements moving sequences between different inserts). Therefore, the first one has been selected.

We then run HAM with the previously mentioned parameters. The overlap detection performance is shown in Table 3. Using GAGE's validation script we finally computed assembly correctness and contiguity statistics. As shown in Table 4, HAM was able to reconstruct most of the genome with a low number of mis-joins and with good contiguity statistics.

Table 3. BLAST overlaps, HAM overlaps, True Positives (TP), False negatives (FN), False Negatives (FN), and Sensitivity in HAM's overlap detection on *SimHG14*

BLAST	HAM	TP	FP	FN	Sensitivity
143,591	145,404	140,876	4,528	2,715	98.11%

Table 4. GAGE statistics on human chromosome 14. For each assembler we report the number of contigs (Ctg), the NG50, assembly's size, the length of reference's regions which cannot be found in the assembly, the length of assembly's regions that cannot be found in the reference, the percentage of duplicated and compressed regions in the assembly, the number of SNPs, indels shorter than 5 bp and indels greater (or equal) than 5 bp. Finally, the number of mis-joins is the sum of inversions (parts of contigs reversed with respect to the reference genome) and relocations. Percentages refers to the ungapped reference genome size.

Ctg num	NG50 (kb)	Assembly size (%)	Unaligned reference (%)	Unaligned assembly (%)	Duplication (%)	Compression (%)
453	293	96.59	2.85	0.03	0.03	1.18

SNPs	Indels < 5 bp	Indels ≥ 5 bp	Mis-joins	Invertions	Relocations
4339	542	230	20	2	18

4   Conclusions and Future Work

This study introduces a novel approach in order to build a draft *de novo* assembly of complex genomes when a collection of well-assembled long-insert pools is available. Moreover, sequencing and assembling a collection of such pools has been proven to be a viable strategy for improving downstream analyses in several sequencing projects (*e.g.*, the Norway spruce and the Pacific oyster genomes). The main advantage is that pool assemblies are less likely hindered by repeats and allelic differences.

In order to exploit these kinds of data sets properly, we designed a strategy to perform reconciliation in a hierarchical manner. In particular, we proposed a method based on fingerprints to carry out the overlap detection, while we relied on the String Graph to merge assemblies.

We showed that in practice the fingerprint-against-sequence comparison allows us to retain most of the significant overlaps, while producing a low number of false positives. While still being a proof of concept, we were able to obtain promising preliminary results on two relatively small data sets based on the human chromosome 14.

In the future, while improving HAM's implementation to reflect precisely the scheme depicted in this work, our intent is to devise additional heuristics based on the particular input data set. Finally, we want to improve the fingerprint construction for the overlap detection, making it deterministic (*i.e.*, if two sequence overlap, then their fingerprints should contain the same k-mers in the overlapping region).

References

1. Alexeyenko, A., Nystedt, B., Vezzi, F., Sherwood, E., Ye, R., Knudsen, B., Simonsen, M., Turner, B., de Jong, P., Wu, C.C., Lundeberg, J.: Efficient de novo assembly of large and complex genomes by massively parallel sequencing of fosmid pools. BMC Genomics 15(1), 439 (2014)
2. Nystedt, B., Street, N., Wetterbom, A., Zuccolo, A., Lin, Y., Scofield, D., Vezzi, F., Delhomme, N., Giacomello, S., Alexeyenko, A., Vicedomini, R., Sahlin, K., Sherwood, E., Elfstrand, M., Gramzow, L., Holmberg, K., Hallman, J., Keech, O., Klasson, L., Koriabine, M., Kucukoglu, M., Kaller, M., Luthman, J., Lysholm, F., Niittyla, T., Olson, A., Rilakovic, N., Ritland, C., Rossello, J., Sena, J.: The norway spruce genome sequence and conifer genome evolution. Nature 497, 579–584 (2013)
3. Zhang, G., Fang, X., Guo, X., Li, L., Luo, R., Xu, F., Yang, P., Zhang, L., Wang, X., Qi, H., Xiong, Z., Que, H., Xie, Y., Holland, P., Paps, J., Zhu, Y., Wu, F., Chen, Y., Wang, J., Peng, C., Meng, J., Yang, L., Liu, J., Wen, B., Zhang, N., Huang, Z., Zhu, Q., Feng, Y., Mount, A., Hedgecock, D., Xu, Z., Liu, Y., Domazet-Loso, T., Du, Y., Sun, X., Zhang, S., Liu, B., Cheng, P., Jiang, X., Li, J., Fan, D., Wang, W., Fu, W., Wang, T., Wang, B., Zhang, J., Peng, Z., Li, Y., Li, N., Wang, J., Chen, M., He, Y., Tan, F., Song, X., Zheng, Q., Huang, R., Yang, H., Du, X., Chen, L., Yang, M., Gaffney, P., Wang, S., Luo, L., She, Z., Ming, Y., Huang, W., Zhang, S., Huang, B., Zhang, Y., Qu, T., Ni, P., Miao, G., Wang, J., Wang, Q., Steinberg, C., Wang, H., Li, N., Qian, L., Zhang, G., Li, Y., Yang, H., Liu, X., Wang, J., Yin, Y., Wang, J.: The oyster genome reveals stress adaptation and complexity of shell formation. Nature 490, 49–54 (2012)
4. Yao, G., Ye, L., Gao, H., Minx, P., Warren, W.C., Weinstock, G.M.: Graph accordance of next-generation sequence assemblies. Bioinformatics (2011)
5. Vicedomini, R., Vezzi, F., Scalabrin, S., Arvestad, L., Policriti, A.: Gam-ngs: genomic assemblies merger for next generation sequencing. BMC Bioinformatics 14(suppl. 7), S6 (2013)
6. Soueidan, H., Maurier, F., Groppi, A., Sirand-Pugnet, P., Tardy, F., Citti, C., Dupuy, V., Nikolski, M.: Finishing bacterial genome assemblies with mix. BMC Bioinformatics 14(suppl. 15), S16 (2013)
7. Myers, E.W.: The fragment assembly string graph. Bioinformatics 21(suppl. 2), ii79–ii85 (2005)
8. Hunt, M., Kikuchi, T., Sanders, M., Newbold, C., Berriman, M., Otto, T.: Reapr: a universal tool for genome assembly evaluation. Genome Biology 14(5), R47 (2013)
9. Vezzi, F., Narzisi, G., Mishra, B.: Reevaluating assembly evaluations with feature response curves: Gage and assemblathons. PLoS One 7(12), e52210 (2012)
10. Policriti, A., Prezza, N.: Hashing and indexing: Succinct data structures and smoothed analysis. In: Ahn, H.-K., Shin, C.-S. (eds.) ISAAC 2014. LNCS, vol. 8889, pp. 155–166. Springer, Heidelberg (2014)
11. Hu, X., Yuan, J., Shi, Y., Lu, J., Liu, B., Li, Z., Chen, Y., Mu, D., Zhang, H., Li, N., Yue, Z., Bai, F., Li, H., Fan, W.: pirs: Profile-based illumina pair-end reads simulator. Bioinformatics 28(11), 1533–1535 (2012)
12. Salzberg, S.L., Phillippy, A.M., Zimin, A.V., Puiu, D., Magoc, T., Koren, S., Treangen, T., Schatz, M.C.: Delcher, a.L., Roberts, M., Marcais, G., Pop, M., Yorke, J.A.: GAGE: A critical evaluation of genome assemblies and assembly algorithms. Genome Research (December 2011)

SimpLiSMS: A Simple, Lightweight and Fast Approach for Structured Motifs Searching[*]

Ali Alatabbi[1], Shuhana Azmin[2], Md. Kawser Habib[2], Costas S. Iliopoulos[1], and M. Sohel Rahman[1,2,**]

[1] Department of Informatics, King's College London
[2] AℓEDA Group, Department of CSE, BUET, Dhaka-1000, Bangladesh

Abstract. A Structured Motif refers to a sequence of simple motifs with distance constraints. We present SimpLiSMS, a simple, lightweight and fast algorithm for searching structured motifs. SimpLiSMS does not use any sophisticated data structure, which makes it simple and lightweight. Our experiments show excellent performance of SimpLiSMS. Furthermore, we introduce a parallel version of SimpLiSMS which runs even faster.

1 Introduction

A *Structured Motif* (alternatively, a *structured pattern*) is defined by a list of simple (as opposed to structured) sub-patterns (or seeds) separated by variable length gaps defined by a list of intervals [7, 13]. In other words, a structured motif imposes a sort of variable constraint on the relative distances among its sub-patterns: between two consecutive sub-patterns, a structured motif allows any gap within the minimum and maximum limit provided as part of the definition. It is also referred to as "compound patterns" in [7,13].

Structured motifs find interesting and useful applications in computational biology and bioinformatics. For example, the PROSITE database [10, 18] supports searching for structured motifs. Different application scenarios for structured motif pop up during different biological experiments. This is especially useful in the identification of conserved features in a set of DNA or protein sequences. Readers are kindly referred to [13–15] for further motivations.

The problem of structured motif search has received significant attention in the literature. The simplest approach is to solve this problem using a regular expression matching (REM) algorithm. But as has been argued by [4], such approach cannot be efficient unless some special care is taken in the translation of the problem to REM. Among theoretically efficient algorithms for this problem using REM, the recent work by [5] is notable. [14, 15] presented a very fast and practical implementation of an REM algorithm to solve the problem exploiting bit parallelism. However, there algorithm becomes less efficient as gaps get longer [4]. Also, bit operations are more costly when they have

---

[*] Part of this research has been supported by an INSPIRE Strategic Partnership Award, administered by the British Council, Bangladesh for the project titled "Advances in Algorithms for Next Generation Biological Sequences".

[**] Commonwealth Academic Fellow, funded by the UK government. Currently on a Sabbatical Leave from BUET.

F. Ortuño and I. Rojas (Eds.): IWBBIO 2015, Part II, LNCS 9044, pp. 219–230, 2015.

to be performed on several computer words instead of one. An alternative approach, suggested independently by [13] and [17], is to design algorithms in two phases. In the first phase, the occurrences of the seeds of the structured motif are computed and in the second, the intervals are considered to identify the occurrences of the complete structured motif. In what follows, the algorithms handling this problem using the 2-phase approach will be referred to as 2-ϕ-algorithms. Additionally, [13] performed extensive experiments to examine the usefulness of this approach. In the implementation of the algorithm of [13], suffix trees were used in the first phase to compute the occurrences of the seeds. One of the problem of this approach is the huge space requirement due to the summation of the number of occurrences of all the seeds, many of which ultimately may turn out to be useless in the context of the actual occurrences of the complete structured motif. Apart from the above we are aware of two more works namely, SMO-TIF [19] and a follow up work on SMOTIF in [9] as a conference paper. Notably the work of [9] almost resembles the 2-ϕ-algorithms mentioned above and it uses suffix tree as the index and hence suffers the same problem suffered by [13] as mentioned above. Unfortunately most of the prior works including [13, 17] were not cited in [9].

In this article, we present SimpLiSMS (pronounced "Simply SMS"), a simple, lightweight and fast algorithm for searching structured motifs. SimpLiSMS exploits an idea of a search context (to be defined, shortly) and combines the two phases of 2-ϕ-algorithms into one. It identifies the occurrences of the seeds, mostly through a character by character matching, rather than a seed by seed matching and thereby refrains from using a heavy-weight data structure like a suffix tree. As a result, not only that SimpLiSMS is lightweight, as it turns out, it is also extremely fast in practice. Moreover, SimLiSMS lends itself easily to a parallel implementation which makes the searching even faster.

2   Preliminaries

A string or sequence is a succession of zero or more symbols from an alphabet Σ of cardinality σ. The empty string is the empty sequence (i.e., of zero length) and is denoted by ε. The set of all strings over the alphabet Σ including ε, is denoted by Σ^*. The set of all non-empty strings over the alphabet Σ is denoted by Σ^+. So, $\Sigma^* = \Sigma^+ \cup \varepsilon$. A string or sequence x of length n is represented by $x[1..n]$. The i-th symbol of x is denoted by $x[i]$. The length of a string x is denoted by $|x|$. A string w is a factor of x if $x = uwv$ for $u, v \in \Sigma^*$. It is a prefix of x if u is empty and a suffix of x if v is empty. We denote by $x[i..j]$ the factor of x that starts at position i and ends at position j.

In this work, unless otherwise specified, the underlying alphabet is assumed to be the DNA alphabet, i.e., $\Sigma = \{A, C, G, T\}$.

Definition 1. *A structured motif can be defined as a pair* $(\mathcal{S}, \mathcal{G})$, *where* $\mathcal{S} = (s_1, .., s_k)$ *is a sequence of seeds (i.e., patterns) and* $\mathcal{G} = ([a_1, b_1], .., [a_{k-1}, b_{k-1}])$, *with* $a_i, b_i \in \mathbb{Z}$ *and* $a_i \leq b_i$ *for* $1 \leq i < k$ *is a sequence of closed intervals characterizing the gaps between the consecutive seeds. So, in an occurrence of a structured motif, the distance between two consecutive seeds* s_{k-1} *and* s_k *must be within the close interval* $[a_{k-1}, b_{k-1}]$. *A structured motif* $\mathcal{M}$ *is usually expressed in the following form:*

$$\mathcal{M} = s_1\ [a_1, b_1]\ s_2\ [a_2, b_2]\ \ldots\ s_{k-1}\ [a_{k-1}, b_{k-1}]\ s_k. \tag{1}$$

Problem 1. (Structured Motif Search)
We are given a sequence x and a structured motif $M = (S, G)$. We need to find the occurrences of M in x.

We exploit the idea of a search context which is defined as follows.

Definition 2. *(Search Context). Given a sequence x, a search context is the smallest factor of x that starts with an occurrence of the first seed s_1 and has the maximum length equal to $\sum_{i=1}^{k-1}(|s_i| + b_i) + |s_k|$.*

We use the following notations, with respect to a search context. We use $\mathcal{L}_{min}(\mathcal{L}_{max})$ to denote the minimum (maximum) possible length of a search context. More formally, we have the following:

$$\mathcal{L}_{min} = \sum_{k=1}^{k=i-1} |s_i| + a_i + |s_k|$$

$$\mathcal{L}_{max} = \sum_{k=1}^{k=i-1} |s_i| + b_i + |s_k|.$$

We maintain a data structure called Map which is defined as follows.

Definition 3. *(Map). Given a structured motif M, assume that $y = s_1 s_2 .. s_k$ and let $|y| = \ell$. Then $Map[1..\ell]$ is an array of length ℓ where $Map[i], 1 \leq i \leq \ell$ is the pair (A, B) as defined below:*

$$Map[i].A = \begin{cases} i; & \text{if } i = 0 \\ Map[i-1].A + a_{k'} + 1; & \text{if } i = \sum_{j=1}^{k'} |s_j|, \text{for some } k' < k. \\ Map[i-1].A + 1; & \text{otherwise} \end{cases}$$

$$Map[i].B = \begin{cases} i; & \text{if } i = 0 \\ Map[i-1].B + b_{k'} + 1; & \text{if } i = \sum_{j=1}^{k'} |s_j|, \text{for some } k' < k. \\ Map[i-1].B + 1; & \text{otherwise} \end{cases}$$

In other words, given a position i, the range $[Map[i].A, Map[i].B]$ gives us the valid positions for $M[i]$. Example 1 computes the Map for $M = CATA[1, 3]TACA[0, 2]GGG$.

Example 1. Given the pattern (structured motif) $M = CATA[1,3]TACA[0,2]GGG$, our algorithm will construct the Map as follows:

$$M = CATA[1, 3]TACA[0, 2]GGG$$

i	0	1	2	3	4	5	6	7	8	9	10
y	C	A	T	A	T	A	C	A	G	G	G
Map	[0,0]	[1,1]	[2,2]	[3,3]	[5,7]	[6,8]	[7,9]	[8,10]	[9,13]	[10,14]	[11,15]

3   Methods

As has been mentioned by [17], there could be an explosive number of occurrences of a structured motif especially because different occurrences of it can exist having exactly the same start and end positions. This happens because of the so called 'elasticity' of the gaps, i.e., variable length gaps. To avoid reporting such explosive number of occurrences we exploit the concept of a search context. Given a sequence x, a search context is the smallest factor of x that starts with an occurrence of the first seed s_1 and has length equal to $\sum_{i=1}^{k-1}(|s_i| + b_i) + |s_k|$. In other words, the longest possible pattern corresponding to the structured motif can (barely) fit in the search context. SimpLiSMS takes a simple approach. It identifies all possible search contexts in x and consider each of those one after another. In each search context it checks whether there is indeed an occurrence of the structured motif in it and if yes, it identifies and report the start position of that structured motif. Notably, the start position of the search context is the start position of the structured motif that exists in it. Then it moves to the next search context and so on. Since a search context is defined by the occurrence of the first seed, SimpLiSMS uses an exact pattern matching algorithm (e.g., the famous KMP algorithm for exact matching [12]) to identify all the occurrences of the first seed and thereby identify all the search contexts. To facilitate the search process within a search context, it makes use of a data structure called Map that keeps track of the valid positions for a particular character in the structured motif with respect to its preceding character. Map is constructed as a preprocessing step before the actual search in a search context can begin.

A parallel version of the SimpLiSMS, referred to as SimpLiSMS-P, is also implemented as follows. The sequence x is first divided into f overlapping subsequences $\{x_1, x_2, ..., x_f\}$ of length ℓ each, except possibly for the last one, x_f, which may have a lesser length. The overlapping is done so that no search context is missed due to the cutting of the sequence into subsequences. Each subsequence $x_i, 1 \le i \le f$ is then handled as a separate thread or process in a multi-processor/multi-threaded machine.

4   SimpLiSMS Algorithm outline

Recall that the input of our problem is a sequence x and a structured motif M. An outline of the *SimpLiSMS* algorithm is as follows.

1. PREPROCESSING PHASE

   Step 1. [Preprocessing the sequence x]: SimpLiSMS segments the sequence x into overlapping factors of length ℓ, called ℓ-factors, where $\ell > \mathcal{L}_{max}$, such that each consecutive ℓ-factors overlap with each other by $\mathcal{L}_{max}$. The overlap is to cover all possible search contexts in x.

   Step 2. [Preprocessing the pattern M]: In this step, SimpLiSMS constructs Map for the structured motif M.

   Step 3. [Preprocessing for exact string matching]: SimpLiSMS uses an exact string matching algorithm, e.g., the KMP algorithm, to identify the start positions of the first seed s_1. For the KMP algorithm, it needs to compute the so called failure function table π for the first seed s_1 of the structured motif

M. We have also implemented SimpLiSMS with Boyer-Moore (BM) algorithm [6]. For BM algorithm, it needs to build the *bad character shift array* and *good suffix shift array*. Such preprocessing is done during this step.

2. **PATTERN MATCHING PHASE**

   Step 1. The algorithm searches the ℓ-factors obtained in Step 1.1. For each ℓ-factor the algorithm finds α, the list of starting positions of all occurrences of the first seed s_1 of the structured motif M within the given ℓ-factor.

   Step 2. For each match of s_1 we calculate the boundaries of the search context ($\mathcal{L}_{min}$ and $\mathcal{L}_{max}$) relative to the position of s_1 in the sequence x.

   Step 3. Now SimpLiSMS determines whether there exist at least one occurrence of the structured motif M in the current search context. The algorithm performs a guided search for M in the current search context SC with the help of Map computed during the pre-processing. Suppose the start position of the current search context is p. Recall that, $y = s_1 s_2 .. s_k$ and the occurrence of s_1 coincides with the start of the search context. So, SimpLiSMS starts checking for a valid match from $y[|s_1| + 1]$. Now, suppose we have valid matches up to $x[i_1] = y[j], j > |s_1|$. Now we are going to check $y[j + 1]$. Suppose we have $x[i_2] = y[j + 1]$. Then, SimpLiSMS only needs to check whether $i_2 - p + 1 \in [Map[i_2].A, Map[i_2].B]$. If yes, then we continue to check $y[j + 1]$. Otherwise, we need to start re-checking from $y[|s_1| + 1]$ all over again.

   Step 4. If M is found in the current searching context, then SimpLiSMS reports the start position of the search context and proceed to the next search context.

5 Handling Degenerate Strings

A degenerate string (also referred to as the indeterminate string in the literature) is a sequence $x = x[1..n]$, where $x[i] \subseteq \Sigma$ for all i. A position of a degenerate string may match more than one elements from the alphabet Σ; such a position is said to have a *non-solid* symbol (also called a character class). If in a position we have only one element of Σ, then we refer to this position as *solid*. The length of a degenerate string is defined in the same way as it is for a regular string: a degenerate string x has length n, when x has n positions, where each position can be either solid or non-solid. We represent non-solid positions using [..] and solid positions omitting [..].

For degenerate strings the definition of a matching relation is extended as follows. A degenerate character (or character class) s_1 is said to match another degenerate character s_2, if and only if $s_1 \cap s_2 \neq \varepsilon$.

Problem 2. (Degenerate Structured Motif Search)
Given a degenerate sequence x and a structured motif M, compute all starting positions of M in x.

Note that there could be $2^\sigma - 1$ possible subsets of Σ and hence there are in total as many degenerate characters including the σ solid characters. For example, for DNA alphabet there are 4 letters, namely A, C, G and T. Hence there could be 15 degenerate characters including the 4 solid characters. In order to efficiently match degenerate characters, we represent each degenerate character as a sequence of $|\Sigma|$ bits. We maintain an array of bit masks $\mathcal{U}$ of length $2^\sigma - 1$ in a way such that the bit mask of any given degenerate symbol (solid or non-solid) can be accessed in constant time. For each character of the DNA alphabet, i.e., $\{A, C, G, T\}$ the conversion to the corresponding 4 bit mask is done as follows: $\mathcal{U}(A) = 0001; \mathcal{U}(C) = 0010; \mathcal{U}(G) = 0100; \mathcal{U}(T) = 1000$. Then, the bit mask of each non-solid symbol s can be computed as follows. Suppose s contains the characters $a_{i_1}, a_{i_2}, ..., a_{i_k}$ where $k \leq 4$ and $a_{i_j} \in \Sigma, 1 \leq j \leq k$. Then the bit mask of s, namely, $\mathcal{U}(s) = \mathcal{U}(a_{i_1})$ OR $\mathcal{U}(a_{i_2})$ OR ... OR $\mathcal{U}(a_{1_k})$. Clearly, given two degenerate characters s_1 and s_2, now we can say that s_1 and s_2 (degenerate) match if and only if $\mathcal{U}(s_1)$ AND $\mathcal{U}(s_2) > 0$.

Degenerate characters or character classes can be found in biological sequences and are ubiquitous in PROSITE database [10, 18]. This is why, search in PROSITE database supports the use of character classes. Now, SimpLiSMS can be extended for degenerate structured motif searching by simply plugging in an appropriate function called *isEquivalent* for checking the degenerate matching as described above.

6   Results

We have evaluated the performance of SimpLiSMS through extensive experiments. We have used 4 sequences, namely, a sequence taken from the *Homo sapiens* genome (size: 256,053,182 bytes), the *Arabidopsis thaliana* DNA sequence (size: 321,118,972 bytes), the *Oryza sativa* DNA sequence (size: 634,849,961 bytes) and finally a randomly generated sequence (size: 104,8576,000 bytes). We have followed the experimental strategy of [13]. A set of 1000 structured motifs were randomly generated by randomly choosing, for each one, the number $k \in [3, 8]$ of simple motifs, the length $\ell \in [5, 10]$ of each motif and $k - 1$ intervals of $[0, 100]$ as variable length gaps. The experiments were run on a Windows Server 2008 R2 64-bit Operating System, with Intel(R) Core(TM) i7 2600 processor @ 3.40GHz having an installed memory (RAM) of 8.00 GB. We have implemented SimpLiSMS in *C#* language using Visual Studio 2010. To compare the performance we have also implemented the 2-ϕ-algorithm of [17] and [13] using the Aho-Corasick pattern matching machine of [1] to implement the first phase (i.e., search phase). Due to the huge space requirement of suffix tree we did not use the suffix tree in the search phase of the 2-ϕ-algorithm. We also slightly modify Phase 1 of the 2-ϕ-algorithm to incorporate the concept of a search context to ensure a level-playing ground with SimpLiSMS.

We do not compare SimpLiSMS with the work of [14, 15] because the length of the structured motif we use in our experiments are larger than computer words, for which their algorithm is reported to be quite slow. This is why their algorithm was not considered in the experimentation of [13] as well.

Although originally we devised SimpLiSMS using KMP algorithm, we also implemented a variant where KMP algorithm was replaced by the famous Boyer-Moore algorithm [6]. Our motivation for this comes from the fact that despite much better theoretical running time, the Boyer-Moore algorithm outperforms KMP algorithm in practice. And indeed as will be reported shortly, the performance of SimpLiSMS with Boyer-Moore algorithm performs better in most cases. Table 1 describes the different

Table 1. Different Variants/Implementations of SimpLiSMS

Name	Searching Algorithm (for S_1)	Sequential/Parallel
SimpLiSMS-KMP-S	KMP	Sequential
SimpLiSMS-KMP-P	KMP	Parallel
SimpLiSMS-BM-S	Boyer-Moore	Sequential
SimpLiSMS-BM-P	Boyer-Moore	Parallel

implementations of our algorithm. The results are illustrated in Figures 1 through 10. These figures basically present two different types of comparisons. Since the size of the structured motif largely depends on the gap length, in Figures 1, 3, 5 and 9 the time required to compute the occurrences of the set of structured motifs are reported against the gap lengths in those. On the other hand, the number of occurrences also affect the search time significantly. Hence, in Figures 2, 4, 6 and 10, the time vs. number of occurrences are plotted. In particular, in Figures 1 and 2, we present the comparison among the 2-ϕ-algorithm, SimpLiSMS-KMP-S and SimpLiSMS-KMP-P. On the other hand, in Figures 3 and 4, we present the comparison among the 2-ϕ-algorithm, SimpLiSMS-BM-S and SimpLiSMS-BM-P. Finally, in Figures 5 and 6 we put SimpLiSMS-KMP-S and SimpLiSMS-BM-S against each other. From the figures, performance superiority of SimpLiSMS over the 2-ϕ algorithm is clearly evident. It is also clear that SimpLiSMS-P runs even faster as expected. Also, in most cases, SimpLiSMS-BM outperforms SimpLiSMS-KMP.

At this point a brief discussion on SimpLiSMS-P is in order. We note that SimPliSMS-P can be configured depending of the length of the input sequence and the machine resources (RAM size). For example to run a search on a sequence of length 600MB, we configure SimpLiSMS to segment the sequence into 75 smaller segments each of length 8MB. So the queue contains 75 threads and we set the concurrent thread limit to 25 (the number of threads running at anytime).

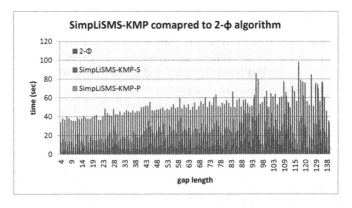

Fig. 1. Comparison of SimpLiSMS-KMP-S, SimpLiSMS-KMP-P and 2-ϕ-algorithm (time vs. gaps length)

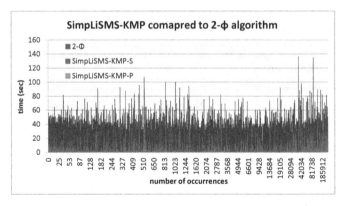

Fig. 2. Comparison of SimpLiSMS-KMP-S, SimpLiSMS-KMP-P and 2-φ-algorithm (time vs. number of occurrences)

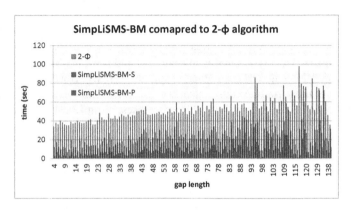

Fig. 3. Comparison of SimpLiSMS-BM-S, SimpLiSMS-BM-P and 2-φ-algorithm (time vs. gaps length)

We have also made an attempt to compare SimpLiSMS with SMOTIF [19]. However the implementation available for SMOTIF turned out to be a bit problematic and was crashing for long motifs. As a result we could not run the experiments for longer gap length. However, as is evident from Figures 7 and 8, the run-time of SMOTIF remains almost invariant with regards to the changing gap length or the number of occurrences. And clearly, the performance of SimpLiSMS is better than SMOTIF.

We also have considered degenerate motifs. As a second experiment, we compared the performances of SimpLiSMS-BM, SimpLiSMS-KPM and the 2-φ algorithm by processing the same data-set used in the first experiment to search for a set of 1,000 degenerate structured motifs over the IUPAC alphabet, randomly generated by randomly choosing, for each model, the length $\ell \in [5, 10]$ of each motif and $k - 1$ intervals of $[0, 100]$ as variable length gaps. The results, averaged over 10 trials, are illustrated in Figures 9 and 10. As expected, the performance superiority of SimpLiSMS over 2-φ algorithm is clearly evident from the figures.

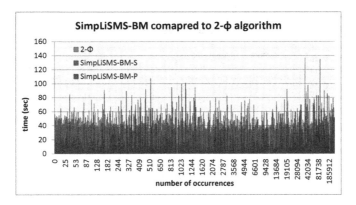

Fig. 4. Comparison of SimpLiSMS-BM-S, SimpLiSMS-BM-P and 2-ϕ-algorithm (time vs. number of occurrences)

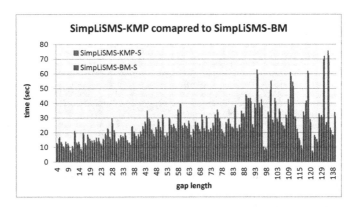

Fig. 5. Comparison of SimpLiSMS-BM-S and SimpLiSMS-KMP-S (time vs. gaps length)

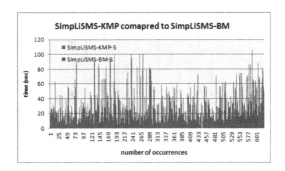

Fig. 6. Comparison of SimpLiSMS-BM-S and SimpLiSMS-KMP-S (time vs. number of occurrences)

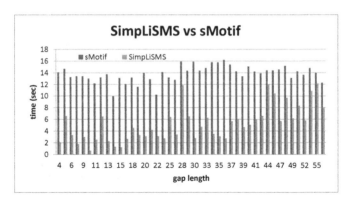

Fig. 7. Comparison of SimpLiSMS and sMotif (time vs. gaps length)

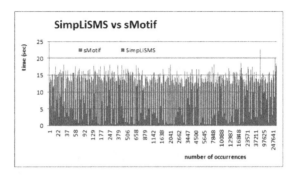

Fig. 8. Comparison of SimpLiSMS and sMotif (time vs. number of occurrences)

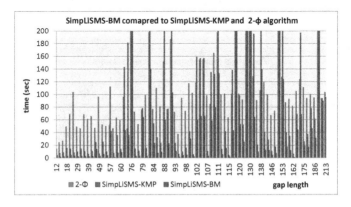

Fig. 9. Comparison of SimpLiSMS-BM, SimpLiSMS-KMP and 2-ϕ-algorithm for degenerate structured motifs (time vs. gaps length)

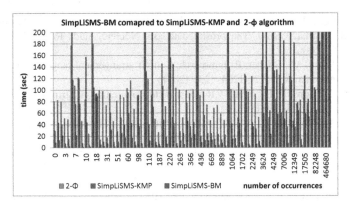

Fig. 10. Comparison of SimpLiSMS-BM, SimpLiSMS-KMP and 2-φ-algorithm for degenerate structured motifs (time vs. number of occurrences)

7 Conclusion

In this article we have presented SimpLiSMS, a simple and lightweight algorithm for structured motif searching that runs extremely fast in practice. We have also implemented SimpLiSMS-P, a parallel version of SimpLiSMS that runs even faster. We augmented our algorithm with the capability to enable search for degenerate motifs.

References

1. Aho, A.V., Corasick, M.J.: Efficient string matching: an aid to bibliographic search. Communications of the ACM 18(6), 333–340 (1975)
2. Bailey, T.L., Bodén, M., Buske, F.A., Frith, M.C., Grant, C.E., Clementi, L., Ren, J., Li, W.W., Noble, W.S.: MEME SUITE: tools for motif discovery and searching. Nucleic Acids Research 37(Web-Server-Issue), 202–208 (2009)
3. Bailey, T.L., Williams, N., Misleh, C., Li, W.W.: MEME: discovering and analyzing DNA and protein sequence motifs. Nucleic Acids Research 34(Web-Server-Issue), 369–373 (2006)
4. Bille, P., Gortz, I.L., Vildhoj, H.W., Wind, D.K.: String matching with variable length gaps. Theor. Comput. Sci. 443, 25–34 (2012)
5. Bille, P., Thorup, M.: Regular expression matching with multi–strings and intervals. In: Charikar, M. (ed.) ACM–SIAM Symp. on Discrete Algorithms, pp. 1297–1308. SIAM (2010)
6. Boyer, R.S., Moore, J.S.: A fast string searching algorithm. Communications of the ACM 20(10), 762–772 (1977)
7. Crochemore, M., Sagot, M.-F.: 1. motifs in sequences. In: Compact Handbook of Computational Biology, p. 47 (2004)
8. Grundy, W.N., Bailey, T.L., Elkan, C., Baker, M.E.: Meta-meme: motif-based hidden markov models of protein families. Computer Applications in the Biosciences 13(4), 397–406 (1997)
9. Halachev, M., Shiri, N.: Fast structured motif search in DNA sequences. In: Elloumi, M., Küng, J., Linial, M., Murphy, R.F., Schneider, K., Toma, C. (eds.) BIRD 2008. CCIS, vol. 13, pp. 58–73. Springer, Heidelberg (2008)

10. Hulo, N., Bairoch, A., Bulliard, V., Cerutti, L., De Castro, E., Langendijk-Genevaux, P.S., Pagni, M., Sigrist, C.J.A.: The prosite database. Nucleic Acids Research 34(suppl. 1), D227–D230 (2006)

11. Junier, T., Pagni, M., Bucher, P.: mmsearch: a motif arrangement language and search program. Bioinformatics 17(12), 1234–1235 (2001)

12. Knuth, D.E., Morris, J.H., Pratt, V.R.: Fast pattern matching in strings. SIAM Journal of Computing 6(2), 323–350 (1977)

13. Morgante, M., Policriti, A., Vitacolonna, N., Zuccolo, A.: Structured motifs search. Journal of Computational Biology 12(8), 1065–1082 (2005)

14. Navarro, G., Raffinot, M.: Fast and simple character classes and bounded gaps patternmatching, with application to protein searching. In: RECOMB, pp. 231–240 (2001)

15. Navarro, G., Raffinot, M.: Fast and simple character classes and bounded gaps pattern matching, with applications to protein searching. Journal of Computational Biology 10(6), 903–923 (2003)

16. Pissis, S.P.: Motex-ii: structured motif extraction from large-scale datasets. BMC Bioinformatics 15, 235 (2014)

17. Rahman, M.S., Iliopoulos, C.S., Lee, I., Mohamed, M., Smyth, W.F.: Finding patterns with variable length gaps or don't cares. In: Chen, D.Z., Lee, D.T. (eds.) COCOON 2006. LNCS, vol. 4112, pp. 146–155. Springer, Heidelberg (2006)

18. Sigrist, C.J.A., de Castro, E., Cerutti, L., Cuche, B.A., Hulo, N., Bridge, A., Bougueleret, L., Xenarios, I.: New and continuing developments at prosite. Nucleic Acids Research 41(D1), D344–D347 (2013)

19. Zhang, Y., Zaki, M.J.: SMOTIF: efficient structured pattern and profile motif search. Algorithms for Molecular Biology, 1 (2006)

Successes and Pitfalls in Scoring Molecular Interactions

Heloisa S. Muniz and Alessandro S. Nascimento[*]

Instituto de Física de São Carlos. Av. Trabalhador Sãocarlense,
400. Centro. São Carlos, SP, Brazil 13566-590
asnascimento@ifsc.usp.br

Abstract. The appropriate evaluation of molecular interactions is still an important challenge in molecular modeling field. The difficulties for scoring those interactions become clear when the performance of the so-called 'ligand-based' methods is compared with the 'structure-based' methods. Although more information is provided for the latter, the former very often performs better in recovering actual binders from a set of ligands and decoys. Here, we compare some results of different scoring functions as implemented and tested in our hybrid-docking algorithm named LiBELa. The results show that the parameters devised from force fields such as AMBER provide a very good estimate of the polar and van der Waals contribution for binding. When properly set up, soft-docking, i.e., a smooth potential, can, at best, reproduce the results obtained with force fields, but hardly outperforms the former. Finally, the results obtained in our group and from other groups clearly indicate that an adequate potential for modeling the solvent effect is still a goal to be achieved. At best, the current empirical solvation models used in docking algorithms can lead to improvement of enrichment in ROC curves for a fraction (50% or less) of the current and gold-standard test sets. In conclusion, the scoring of molecular interactions at an atomic level is a promising field with many important advances achieved but also with a number of open issues.

1   Introduction

The evaluation of molecular interaction is central to the 'structure-based' approach in drug design. Its relevance to the field emerged early in the seventies when Beddell and coworkers designed three compounds based on the crystal structure of haemoglobin, the first protein structure solved by Max Perutz and coworkers [1]. During the time of this writing, roughly 80% of the protein structures deposited in the PDB [2] contain a ligand, revealing that a large set of structural models are now available to improve the current interaction models and for the proposal of new models.

Despite this large volume of structural data, the evaluation of interaction models based on the structural data is still challenging [3]. The situation is even harder when a rapid evaluation of the binding (potential) energy is ought to be used as the criteria for selecting compounds during screening trials, which is the typical scenario in

---

[*] Corresponding author.

F. Ortuño and I. Rojas (Eds.): IWBBIO 2015, Part II, LNCS 9044, pp. 231–237, 2015.
© Springer International Publishing Switzerland 2015

docking campaigns. In this context, many simplifications are made, including the approximation of a rigid receptor, the usage of empirical solvation models and limited ligand flexibility, in order to achieve the computational efficiency necessary for a virtual screening.

In recent years, many 'ligand-based' methodologies were proposed where ligand similarities were used for the discovery of new binders [4-7] and, curiously, most of these methods outperform the 'structure-based' methods such as docking in their ability to recover true binders among decoy molecules that preserve similar physicochemical properties when compared to the true binders set. This apparent paradox reveal that, although more relevant information is provided in 'structure-based' methods, the ability to correctly score ligand interactions and the ruggedness of the binding energy landscape [8, 9] impose important limitations to the structure-based methods in drug design.

Here, we evaluated a combination of ligand-based approach in a structure-based modeling tool called LiBELa. The results obtained show a hybrid approach is efficient in distinguishing actual binders from decoy molecules and was found to be more effective than the well-established structure-based tool DOCK6.

2   Methods

2.1   Interaction Model

In LiBELa, the docking procedure is divided in two stages. In the first stage, a set of ligand conformers is generate by the genetic algorithm, as implemented in the Open-Babel API [10]. Each ligand conformer is then overlaid in a reference ligand (already posed in the protein active site) using the matching in volume and charge distribution as parameters. This ligand-based superposition is carried out using our previously validated tool *MolShaCS* [7]. The best-scored conformers (based on degree of overlap or on the initial interacting energy) are selected for the second stage, where an optimization in the Cartesian space is done using the ligand-receptor interaction energy as the objective function. The interaction energy is evaluated as:

$$E_{bind} = E_{ele} + E_{VDW} + E_{desol} = \sum_{i \neq j} \frac{q_i q_j}{D d_{ij}} + \sum_{i \neq j} \frac{A_{ij}}{d_{ij}^{12}} - \frac{B_{ij}}{d_{ij}^6} + (S_i X_j + S_j X_i) \quad (1)$$

where i and j account for ligand and receptor atoms, respectively, q is the atomic charge, d_{ij} is the interatomic distance, D is the dielectric constant, A and B are Lennard-Jones parameters defined in AMBER force field as $A_{ij} = \epsilon_{ij}(r_i + r_j)^{12}$ and $B_{ij} = 2\epsilon_{ij}(r_i + r_j)^6$, where ε is the well depth and $r_0 = r_i + r_j$ is the typical distance. Assuming, for instance, that $A_{ij} = \sqrt{A_{ii}}\sqrt{A_{jj}}$ and $B_{ij} = \sqrt{B_{ii}}\sqrt{B_{jj}}$, the van der Waals terms can be pre-computed in grids equally spaced in a cube encompassing the active site [11, 12].

We used an empirical solvation function in LiBELa that is a linear function of the square of its atomic charge ($S_i = \alpha q_i^2 + \beta$). This linear function is then multiplied bu the volume of solvent displaced by the interacting agent with a Gaussian envelope function ($X_j = f_j exp^{\left(\frac{-r^2}{2\sigma^2}\right)}/\sigma^3$). The volume f_j is computed using AMBER force field radii and σ is a constant (σ=3.5Å). This solvation model has two main advantages in the context of molecular docking. First, it is a very simple model, although retaining some phenomenological rationale [13], and, second, it can also be used in a decomposable way, allowing pre-computation of receptor terms in grids [11]. Unless otherwise stated, the parameters used in this work were α=0.4 kcal/(mol.e^2) and β=-0.005 kcal/mol.

We also tested a smooth version of the binding energy given by:

$$E_{bind} = \sum_{i \neq j} \frac{q_i q_j}{D(d_{ij} + \delta_{ele})} + \sum_{i \neq j} \frac{A_{ij}}{(d_{ij}^6 + \delta_{VDW}^6)^2} - \frac{B_{ij}}{(d_{ij}^6 + \delta_{VDW}^6)} + (S_i X_j + S_j X_i) \quad (2)$$

In equation (2), energies are computed in a way similar to the original force field energy given in equation (1), except for the smoothing term δ in the denominator of VDW and electrostatic terms.

The parameters for van der Waals terms were taken from AMBER force field (FF99SB and GAFF). The atomic charges for receptors were also taken from AMBER FF while ligand atomic charges were used as provided or computed using AM1-BCC method as implemented in ANTECHAMBER [14].

The optimization of molecular overlay and binding energy are accomplished by the Augmented Lagrangian and Dividing Rectangles algorithms, respectively [15, 16]. The latter has the advantage of the larger radius of search, being classified as a global optimization algorithm. These optimization algorithms were used as implemented in the NLOPT library [17].

2.2 Ligand Pose Analysis

The ability of LiBELa to recover experimental (crystallographic) ligand poses was evaluated with the test set SB2012 [18]. In this case, ligands were docked on their own receptors and the root mean square deviation of the ligand atoms after binding energy optimization were evaluated using the crystallographic pose as the reference. The protein binding mode test set [19], containing 144 experimental complexes and available in UCSF DOCK website was used for binding pose prediction analysis. In this stage, the RMSD computed for all atoms (hydrogen atoms included) was used as a measure of accuracy.

2.3 Enrichment Analysis

The benchmark sets DUD [20] and a subset DUD-E [21] including the targets CP2C9, CXCR4, GRIK1, MK10, XIAP, MCR, THB, HIVINT, KITH, PUR2, LKHA4,

PPARD and DYR were used in order to test LiBELa's ability to recover actual binders among decoy molecules. In all cases, ligand and decoys were docked against their own receptor and the enrichment was evaluated in semi-log ROC curves using the adjusted logAUC parameter [22]. The targets were prepared using UCSF Chimera DockPrep tool [23].

3   Results and Discussion

First of all the hybrid approach implemented in LiBELa was evaluated on over 1,000 crystallographic complexes as available in the dataset SB2012, provided by Rizzo's group. After ligand superposition, the initial conformation was energy-optimized in the Cartesian space in a global optimizer. The search radius was limited to a maximal translation of 12.0Å in each direction and full rotation around Euler's angles. For this test, the binding energies were computed directly, i.e., with no pre-computation in grids. An analysis of the RMSD of achieved ligand against experimental poses revealed an average RMSD of 1.79Å with median in 0.52Å and standard deviation in 3.14Å (N=1030). For the sake of comparison, a pure FF docking was also performed, i.e., neglecting any desolvation term, and resulted in an average RMSD of 1.05Å with median in 0.51Å and standard deviation of 1.8Å, indicating that these models are effective in reproducing experimental binding conformations.

For the smoothed potential, an average RMSD of 0.87Å with median in 0.52Å and standard deviation of 1.27Å was found, slightly lower than the values observed for force field energy evaluation. In the absence of a solvation term, the smoothed potential resulted in an average RMSD of 2.52Å with median in 0.62Å and standard deviation of 4.35Å. In this analysis the parameters δ_{VDW} and δ_{ele} were set to 1.5Å and 1.0Å, respectively, according to the results of preliminary tests. These results suggest that a smoothed potential partially account for 'receptor flexibility' allowing some accommodation upon ligand binding, what would explain the slightly better RMSD values observed. On the other hand, the Stouten-Verkhivker (SV) solvation model seems to be very well parametrized to be used together with smoothed potentials, as originally proposed [9], given the reasonable increase in the RMSD values observed when the smoothed potential was used without the solvation model.

Since the hybrid docking starts with a ligand-based superposition, it was expected that a low average RMSD would be found. This test shows, however, that the energy evaluation and its implementation using AMBER force field parameters are good enough to reproduce experimental binding poses even with an increased search radius as available in the global optimization process.

We also set out a more stringent test where LiBELa's ability to reproduce binding poses in cross-docking experiments was evaluated. For this purpose, we took the protein binding mode test set [19], as available in UCSF DOCK website e filtered for targets with more than one binder. The final set consisted of 14 complexes different (non-native) complexes. An average RMSD of 3.8Å (median 2.0Å) was achieved for this non-native set.

We then evaluated the enrichment in ROC curves when using our hybrid model on validated benchmarks such as DUD, or DUD-E. For the original DUD, using a typical AMBER force field energetic analysis coupled with the empirical solvation term, an logAUC of 15.2 was observed (median 14.9 and standard deviation 14.2) after averaging over 37 different targets. Curiously, the same evaluation for a pure FF energy model, i.e., without a solvation term, resulted in an increase in the observed enrichment to 20.0. The same behavior is observed when the AUC rather than the adjusted logAUC is taken as a parameter (64.6% and 70.2% in the presence and absence of solvation term, respectively). The best sets of smoothing parameters resulted in enrichments similar to those observed in non-smoothed potential. An average logAUC of 15.5 (median 15.1) was observed in the presence of a solvation term while in the absence of a solvation term an average logAUC of 13.7 (median in 14.3) was observed (Table 1).

For DUDE, the AMBER force field energy model resulted in average logAUC of 7.2 and 8.1 in the presence and absence of the solvation term, respectively (median 5.4 and 5.5). The smooth energy model resulted in average logAUC of 7.0 and 8.5 (median 6.1 and 5.6, respectively).

Table 1. Enrichment values (logAUC) observed for DUD and DUDE for AMBER FF energy model and smooth AMBER FF model in the presence and absence of a solvation term

	DUD	DUDE
AMBER FF + Solvation	15.2 (14.9)	7.2 (5.5)
AMBER FF	20.0 (20.1)	8.1 (5.4)
Smooth AMBER FF + Solvation	15.5 (15.1)	7.0 (6.1)
Smooth AMBER FF	20.7 (20.5)	8.5 (5.6)

A number of interesting conclusions can be devised from the results observed with LiBELa. First, the hybrid docking seems to be an effective way to direct structure-based methods. This can be observed by the reasonably good binding poses, even in cross docking experiments, and by the good overall enrichment values in retrospective docking. The same sets were used in 'classical' structure-based tools, such as DOCK6 and the results observed in LiBELa were significantly better than those observed with DOCK. It is interesting to point out that DOCK also uses AMBER FF energy model in its grid score.

Second, smoothed potentials, in principle, could be effective as interaction model since they somewhat implicitly account for some receptor flexibility by allowing local bad contacts. In addition, these models in principle could reduce the ruggedness and increase the convergence towards the global minimum. Despite the theoretical advantages, the implementation of smooth potential in LiBELa, at best could reproduce the results found with AMBER FF model. Therefore, our results suggest that there is no general advantage in using such potentials.

Third, the effectiveness of the SV empirical solvation model is intriguing. Looking at the enrichment observed in AMBER FF energy model, the solvation model seems to worse the overall enrichment in about 25%, suggesting that the model is ineffective. In the other hand, for a couple of specific targets, the solvation model was shown to be effective (COX1 and PDGFRB in DUD, KITH and PUR2 in DUDE, for example). Since the model was conceived for a smooth potential, it is highly probable that the parameters used are not directly applicable here. This hypothesis is supported by the observation that the solvation term increased the enrichment in the smooth AMBER potential. The results from our best energy model (FF) however do advocate against the model as proposed (SV) as an efficient model for docking purposes.

In conclusion, our experience with LiBELa show that simple energy models such as molecular mechanics force field-based models can be very effective when coupled to global optimization methods. Additionally, the ligand-based sampling of initial conformations seem to be a rapid and efficient way of driving the docking to correct conformations. The analysis of our empirical solvation model suggests that better fast and efficient solvation models are still necessary to bring interaction models closer to actual binding free energies. LiBELa can be considered a minimalistic docking tool, in the sense that very simple concepts are put together on its algorithm. The results provided here indicate that it might be possible to take good findings out of simple ideas. But not simpler.

References

[1] Beddell, C., Goodford, P., Norrington, F., Wilkinson, S., Wootton, R.: Compounds Designed to Fit a Site of Known Structure in Human Hemoglobin. British Journal of Pharmacology 57, 201–209 (1976)
[2] Berman, H., Battistuz, T., Bhat, T., Bluhm, W., Bourne, P., Burkhardt, K., Iype, L., Jain, S., Fagan, P., Marvin, J., Padilla, D., Ravichandran, V., Schneider, B., Thanki, N., Weissig, H., Westbrook, J., Zardecki, C.: The Protein Data Bank. Acta Crystallographica Section D-Biological Crystallography 58, 899–907 (2002)
[3] Leach, A.R.: Molecular modelling: principles and applications, 2nd edn. Prentice Hall, Harlow (2001)
[4] Good, A.C., Hodgkin, E.E., Richards, W.G.: Similarity Screening of Molecular-Data Sets. Journal of Computer-Aided Molecular Design 6, 513–520 (1992)
[5] Armstrong, M.S., Morris, G.M., Finn, P.W., Sharma, R., Moretti, L., Cooper, R.I., Richards, W.G.: ElectroShape: fast molecular similarity calculations incorporating shape, chirality and electrostatics. J. Comput. Aided Mol. Des. 24, 789–801 (2010)
[6] Armstrong, M.S., Finn, P.W., Morris, G.M., Richards, W.G.: Improving the accuracy of ultrafast ligand-based screening: incorporating lipophilicity into ElectroShape as an extra dimension. J. Comput. Aided Mol. Des. 25, 785–790 (2011)
[7] Vaz de Lima, L.A.C., Nascimento, A.S.: MolShaCS: A free and open source tool for ligand similarity identification based on Gaussian descriptors. European Journal of Medicinal Chemistry 59, 296–303 (2013)
[8] Verkhivker, G., Bouzida, D., Gehlhaar, D., Rejto, P., Freer, S., Rose, P.: Complexity and simplicity of ligand-macromolecule interactions: the energy landscape perspective. Current Opinion in Structural Biology 12, 197–203 (2002)

[9] Verkhivker, G.M., Rejto, P.A., Bouzida, D., Arthurs, S., Colson, A.B., Freer, S.T., Gehlhaar, D.K., Larson, V., Luty, B.A., Marrone, T., Rose, P.W.: Towards understanding the mechanisms of molecular recognition by computer simulations of ligand-protein interactions. Journal of Molecular Recognition 12 (1999)

[10] O'Boyle, N.M., Banck, M., James, C.A., Morley, C., Vandermeersch, T., Hutchison, G.R.: Open Babel: An open chemical toolbox. J. Cheminform. 3, 33 (2011)

[11] Luty, B.A., Wasserman, Z.R., Stouten, P.F.W., Hodge, C.N., Zacharias, M., McCammon, J.A.: A Molecular Mechanics Grid Method For Evaluation of Ligand-Receptor Interactions. Journal of Computational Chemistry 16, 454–464 (1995)

[12] Meng, E.C., Shoichet, B.K., Kuntz, I.D.: Automated Docking with Grid-Based Energy Evaluation. Journal of Computational Chemistry 13, 505–524 (1992)

[13] Stouten, P.F.W., Frommel, C., Nakamura, H., Sander, C.: An Effective Solvation Term Based on Atomic Occupancies For Use in Protein Simulations. Molecular Simulation 10, 97–120 (1993)

[14] Wang, J., Wang, W., Kollman, P.A., Case, D.A.: Automatic atom type and bond type perception in molecular mechanical calculations. J. Mol. Graph. Model. 25, 247–260 (2006)

[15] Andreani, R., Birgin, E.G., Martinez, J.M., Schuverdt, M.L.: On Augmented Lagrangian Methods with General Lower-Level Constraints. Siam Journal on Optimization 18, 1286–1309 (2007)

[16] Jones, D.R., Perttunen, C.D., Stuckman, B.E.: Lipschitzian optimization without the Lipschitz constant. Journal of Optimization Theory and Applications 79, 157–181 (1993)

[17] Jhonson, S.G.: The NLopt nonlinear-optimization package, The NLopt nonlinear-optimization package, http://ab-initio.mit.edu/nlopt

[18] Mukherjee, S., Balius, T., Rizzo, R.: Docking Validation Resources: Protein Family and Ligand Flexibility Experiments. Journal of Chemical Information and Modeling 50, 1986–2000 (2010)

[19] Moustakas, D., Lang, P., Pegg, S., Pettersen, E., Kuntz, I., Brooijmans, N., Rizzo, R.: Development and validation of a modular, extensible docking program: DOCK 5. Journal of Computer-Aided Molecular Design 20, 601–619 (2006)

[20] Huang, N., Shoichet, B.K., Irwin, J.J.: Benchmarking sets for molecular docking. J. Med. Chem. 49, 6789–6801 (2006)

[21] Mysinger, M., Carchia, M., Irwin, J., Shoichet, B.: Directory of Useful Decoys, Enhanced (DUD-E): Better Ligands and Decoys for Better Benchmarking. Journal of Medicinal Chemistry 55, 6582–6594 (2012)

[22] Mysinger, M.M., Shoichet, B.K.: Rapid Context-Dependent Ligand Desolvation in Molecular Docking. Journal of Chemical Information and Modeling 50, 1561–1573 (2010)

[23] Pettersen, E.F., Goddard, T.D., Huang, C.C., Couch, G.S., Greenblatt, D.M., Meng, E.C., Ferrin, T.E.: UCSF chimera - A visualization system for exploratory research and analysis. Journal of Computational Chemistry 25, 1605–1612 (2004)

The Use of Random Forest to Predict Binding Affinity in Docking

Hongjian Li[1], Kwong-Sak Leung[1], Man-Hon Wong[1], and Pedro J. Ballester[2]

[1] Department of Computer Science and Engineering, Chinese University of Hong
Kong, Sha Tin, New Territories, Hong Kong
[2] Cancer Research Center of Marseille, INSERM U1068, F-13009 Marseille, France,
Institut Paoli-Calmettes, F-13009 Marseille, France, Aix-Marseille Université,
F-13284 Marseille, France, and CNRS UMR7258, F-13009 Marseille, France
pedro.ballester@inserm.fr

Abstract. Docking is a structure-based computational tool that can be
used to predict the strength with which a small ligand molecule binds
to a macromolecular target. Such binding affinity prediction is crucial
to design molecules that bind more tightly to a target and thus are
more likely to provide the most efficacious modulation of the target's
biochemical function. Despite intense research over the years, improving
this type of predictive accuracy has proven to be a very challenging task
for any class of method.

New scoring functions based on non-parametric machine-learning
regression models, which are able to exploit effectively much larger volumes of experimental data and circumvent the need for a predetermined functional form, have become the most accurate to predict binding
affinity of diverse protein-ligand complexes. In this focused review, we
describe the inception and further development of RF-Score, the first
machine-learning scoring function to achieve a substantial improvement
over classical scoring functions at binding affinity prediction. RF-Score
employs Random Forest (RF) regression to relate a structural description of the complex with its binding affinity. This overview will cover
adequate benchmarking practices, studies exploring optimal intermolecular features, further improvements and RF-Score software availability
including a user-friendly docking webserver and a standalone software for
rescoring docked poses. Some work has also been made on the application
of RF-Score to the related problem of virtual screening. Comprehensive
retrospective virtual screening studies of RF-based scoring functions constitute now one of the next research steps.

Keywords: Molecular docking, scoring functions, random forest, chemical informatics, structural bioinformatics.

1 Introduction

Molecular docking is a key computational method in structure-based drug design,
which has several important applications. First, given a X-ray crystal structure of

F. Ortuño and I. Rojas (Eds.): IWBBIO 2015, Part II, LNCS 9044, pp. 238–247, 2015.
© Springer International Publishing Switzerland 2015

a protein-ligand complex (e.g. Figure 1), docking can identify 3D conformations of a known ligand that are conformationally close to the co-crystallised conformation (native pose identification). Second, docking can discriminate between known binders and non-binders with the goal of finding previously unknown binders in large databases of candidate molecules (virtual screening). Third, docking can predict the binding strength of known binders and their derivatives in order to identify more potent binders against the target (increasing potency) or less potent binders against an off-target (increasing selectivity).

Operationally, docking predicts the position, orientation and conformation of a molecule when docked to the target's binding site (pose generation), as well as how tightly the docked pose of such putative ligand binds to the target (scoring). State-of-the-art docking software such as AutoDock Vina [19] and its faster implementation idock [13] perform reasonably well in pose generation, achieving a redocking success rate of more than 50% [11] on native pose identification benchmarks. However, the single most critical limitation of docking is the low accuracy of the scoring functions that estimate the binding strength of a ligand. This binding affinity can thereafter be used to prioritize the molecules predicted to bind strongly for subsequent biological assays.

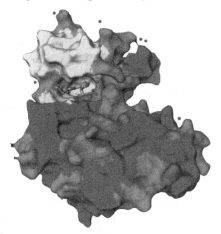

Fig. 1. Example of the X-ray crystal structure of a protein-ligand complex (PDB ID: 3NUX). The molecular surface of the protein is colored by secondary structure with 0.9 opacity to indicate the underlying secondary structure. The ligand molecule is pictured in stick representation colored by atom type. This figure was created by iview [12], an interactive WebGL visualizer freely available at http://istar.cse.cuhk.edu.hk/iview/. iview does not require Java plugins, yet supports macromolecular surface construction and virtual reality effects.

Classical scoring functions [9] assume a functional form relating a description of the crystal protein-ligand complex to its binding affinity, usually through Multiple Linear Regression (MLR). Non-parametric machine-learning scoring functions, on the other hand, circumvent the need of modelling assumptions implicit in functional forms, and have recently been demonstrated [3,4] to introduce a substantial increase in the accuracy of scoring functions for binding affinity prediction.

In this focused review, we describe the inception and further development of RF-Score [3], the first machine-learning scoring function to achieve a large improvement over classical scoring functions [9] at binding affinity prediction. The next sections cover adequate benchmarking practices, studies exploring optimal intermolecular features, further improvements and RF-Score software availability including a user-friendly docking webserver and a standalone executable for rescoring docked poses. Much less work has been done so far on the application of RF-Score to the related problem of virtual screening, as it will be explained in the last section.

2   Random Forest (RF) Scoring Functions

RF-Score [3], the first scoring function using Random Forest (RF) [8] as the regression model, was found to outperform a range of widely-used classical scoring functions by a large margin. RF-Score employs Random Forest (RF) regression [8] to relate a structural description of the complex at atomic level with its binding affinity. RF-Score features are elemental occurrence counts of a set of protein-ligand atom pairs in a complex. To calculate these features, atom types were selected so as to generate features that are as dense as possible, while considering all the heavy atoms commonly observed in PDB complexes (C, N, O, F, P, S, Cl, Br, I). As the number of protein-ligand contacts is constant for a particular complex, the more atom types are considered, the sparser the resulting features will be. Therefore, a minimal set of atom types is selected by considering atomic number only. Furthermore, a smaller set of interaction features has the additional advantage of leading to computationally faster scoring functions.

RF-Score features are defined as the occurrence count of intermolecular contacts between elemental atom types i and j, as shown in equations 1 and 2, where d_{kl} is the Euclidean distance between the kth protein atom of type j and the lth ligand atom of type i calculated from a structure; K_j is the total number of protein atoms of type j ($\#\{j\} = 9$) and L_i is the total number of ligand atoms of type i ($\#\{i\} = 4$) in the considered complex; $\mathcal{H}$ is the Heaviside step function that counts contacts within a d_{cutoff} neighbourhood. For example, $x_{7,8}$ is the number of occurrences of protein oxygen atoms hypothetically interacting with ligand nitrogen atoms within a chosen neighbourhood. This representation led to a total of 36 features. Therefore, each complex was characterized by a vector with 36 integer-valued features. Full details on RF-Score features are available at [3,1].

$$x_{ij} = \sum_{k=1}^{K_j} \sum_{l=1}^{L_i} \mathcal{H}(d_{cutoff} - d_{kl}) \qquad (1)$$

$$\mathbf{x} = \{x_{ij}\} \in N^{36} \qquad (2)$$

A second version of RF-Score, RF-Score-2 [5] , arised from a study intended to investigate the impact of the chemical description of the protein-ligand complex on the predictive power of the resulting scoring function. Strikingly, it was found that a more precise description did not generally lead to a more accurate prediction of binding affinity. However, RF-Score-2 was built on 123 features and the software implementing it was not sufficiently user-friendly. To address these issues, a recent paper [15] introduced RF-Score-3, which uses a more compact feature vector without sacrificing performance. That is, a total of 42 features: the 36 RF-Score features [3] in addition to the 6 AutoDock Vina features [19]. Furthermore, an easy-to-use executable implementing this re-scoring function has been recently released and it is now available online (http://istar.cse.cuhk.edu.hk/rf-score-3.tgz and http://crcm.marseille.inserm.fr/fileadmin/rf-score-3.tgz).

3   Experimental Setup

The PDBbind benchmarks [9,17] are arguably the most widely used for prediction of binding affinity of diverse complexes. Using a third-party and well-defined benchmark, where many scoring functions had been tested previously, has the advantage of minimizing the risk of using a benchmark complementary to a particular scoring function. This risk is further minimized by using two different, pre-established benchmarks. This section describes the benchmarks that have been used so far to validate RF-Score.

3.1   The PDBbind v2007 Benchmark

Based on the 2007 version of the PDBbind database, it contains a particularly diverse collection of protein-ligand complexes from a systematic mining of the entire Protein Data Bank [7,6]. This procedure led to a refined set of 1300 protein-ligand complexes along with their binding affinities. The PDBbind benchmark essentially consists of testing the predictions of scoring functions on the 2007 core set, which comprises 195 diverse complexes with measured binding affinities spanning more than 12 orders of magnitude, while training on the remaining 1105 complexes in the refined set. In this way, a set of protein-ligand complexes with measured binding affinity can be processed to give two non-overlapping data sets, where each complex is represented by its feature vector $\mathbf{x}_i$ and its binding affinity y_i, which includes both pKd and pKi measurements, henceforth referred to as just pKd for simplicity:

$$D_{train} = \{y_i, \mathbf{x}_i\}_{i=1}^{1105} \tag{3}$$

$$D_{test} = \{y_i, \mathbf{x}_i\}_{i=1106}^{1300} \tag{4}$$

$$y = pK_d = -\log_{10} K_d \tag{5}$$

3.2 The 2013 Blind Benchmark

A new benchmark mimicking a blind test has been recently proposed [15] to provide a more realistic performance estimation than that provided by the PDBbind benchmark, where higher performance is to be expected due to the protocol that generates this benchmark [4]. The new test set comprises all the structures in the 2013 release of the PDBbind refined set that were not already in the 2012 release, i.e. the new protein-ligand complexes added in 2013, whereas the 2012 refined set is used for training. This is hence conducted as a blind test in that only data available until 2012 is used to build the scoring function that predicts the binding affinities of 2013 complexes as if these had not been measured yet. Therefore, the 2897 complexes in the PDBbind v2012 refined set were used as a training set, whereas the 382 new complexes included in the PDBbind v2013 refined set were used as the non-overlapping test set.

3.3 Performance Measures

Predictive performance is quantified through commonly-used metrics [9], including standard deviation SD in linear correlation, Pearson correlation coefficient Rp and Spearman correlation coefficient Rs between the measured and predicted binding affinities of the test set. The SD metric is essentially the residual standard error (RSE) metric used in some other studies [16]. In some applications, the ultimate goal of a scoring function is to predict an absolute value of binding affinity as close to the measured value as possible. Hence the root mean square error RMSE between measured and predicted binding affinities without coupling a linear correlation constitutes a more realistic metric. In all cases, lower values in RMSE and SD and higher values in Rp and Rs denote better predictive performance.

Equations 6, 7, 8 and 9 describe the mathematical expressions of the four metrics. Given a scoring function f and the features $\mathbf{x}_i$ characterizing the ith complex out of n complexes in the test set, $p_i = f(\mathbf{x}_i)$ is the predicted binding affinity, $\{\hat{p}_i\}_{i=1}^n$ are the fitted values from the linear model between $\{y_i\}_{i=1}^n$ and $\{p_i\}_{i=1}^n$ on the test set, whereas $\{y_i^r\}_{i=1}^n$ and $\{p_i^r\}_{i=1}^n$ are the rankings of $\{y_i\}_{i=1}^n$ and $\{p_i\}_{i=1}^n$, respectively.

$$RMSE = \sqrt{\frac{1}{n}\sum_{i=1}^n (p_i - y_i)^2} \tag{6}$$

$$SD = \sqrt{\frac{1}{n-2}\sum_{i=1}^n (\hat{p}_i - y_i)^2} \tag{7}$$

$$R_p = \frac{n\sum_{i=1}^n p_i y_i - \sum_{i=1}^n p_i \sum_{i=1}^n y_i}{\sqrt{(n\sum_{i=1}^n (p_i)^2 - (\sum_{i=1}^n p_i)^2)(n\sum_{i=1}^n (y_i)^2 - (\sum_{i=1}^n y_i)^2)}} \tag{8}$$

$$R_s = \frac{n\sum_{i=1}^n p_i^r y_i^r - \sum_{i=1}^n p_i^r \sum_{i=1}^n y_i^r}{\sqrt{(n\sum_{i=1}^n (p_i^r)^2 - (\sum_{i=1}^n p_i^r)^2)(n\sum_{i=1}^n (y_i^r)^2 - (\sum_{i=1}^n y_i^r)^2)}} \tag{9}$$

4   Results and Discussion

4.1   Comparing the Performance of Scoring Functions

Table 1 compares the performance of 23 scoring functions on the same test set. It is worth noting that the top 4 are all RF-based scoring functions. These machine-learning scoring functions outperform a range of widely-used classical scoring functions (in the table, those below SVR-Score) by a large margin. An area for further work is investigating the performance of RF-Score on complexes whose proteins belong to the same family (i.e. on different protein families) and more broadly to design family-specific RF-based scoring functions. However, because the test set contains different protein families, the larger the improvement of RF-Score over a given classical scoring function on this diverse test set, the more protein families in the test set will be also better predicted by RF-Score.

Table 1. Predictive performance of 23 scoring functions evaluated on PDBbind v2007 benchmark

Scoring function	Rp	Rs	SD
RF-Score-3	0.803	0.798	1.42
RF-Score-2	0.803	0.797	1.54
SFCscoreRF	0.779	0.788	1.56
RF-Score	0.774	0.762	1.59
ID-Score	0.753	0.779	1.63
SVR-Score	0.726	0.739	1.70
Cyscore	0.660	0.687	1.79
X-Score::HMScore	0.644	0.705	1.83
DrugScoreCSD	0.569	0.627	1.96
SYBYL::ChemScore	0.555	0.585	1.98
DS::PLP1	0.545	0.588	2.00
GOLD::ASP	0.534	0.577	2.02
SYBYL::G-Score	0.492	0.536	2.08
DS::LUDI3	0.487	0.478	2.09
DS::LigScore2	0.464	0.507	2.12
GlideScore-XP	0.457	0.435	2.14
DS::PMF	0.445	0.448	2.14
GOLD::ChemScore	0.441	0.452	2.15
SYBYL::D-Score	0.392	0.447	2.19
DS::Jain	0.316	0.346	2.24
GOLD::GoldScore	0.295	0.322	2.29
SYBYL::PMF-Score	0.268	0.273	2.29
SYBYL::F-Score	0.216	0.243	2.35

4.2   Comparing RF-Score-3 with Vina

As seen in Table 1, RF-Score-3 constitutes a remarkable improvement over classical scoring functions at binding affinity prediction. Figure 2 shows a head-to-head comparison on this problem with the widely-used Vina software, which is one of the top performers among classical scoring functions (compare the upper left plot of this figure with performances in Table 1). An emerging trend is that one can take any classical scoring function, substitute its regression model by RF and achieve a more accurate machine-learning scoring function. This has been already demonstrated in the case of Vina [15] and Cyscore [14].

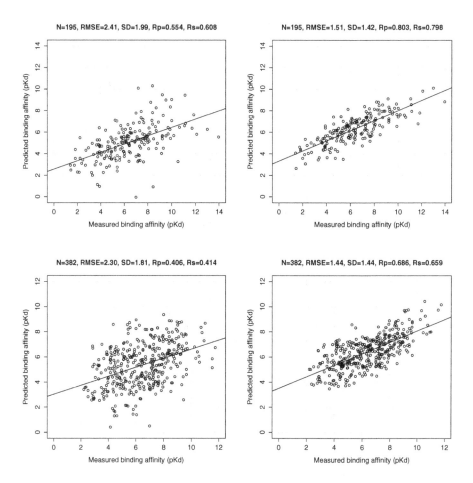

Fig. 2. Predictive performance of AutoDock Vina (left) and RF-Score-3 (right) on the PDBbind v2007 benchmark (top) and the PDBbind v2013 blind benchmark (bottom)

4.3   Performance Improvement with Training Set Size

In a recent study [15], the blind benchmark was employed to analyse how test set performance increases with training set size. In the case of a classical scoring function, its linear regression model could not assimilate data effectively beyond a few hundred training complexes and thus its test set performance remained flat regardless of training with more data. In contrast, performance of RF models on the same test continued to increase even after having assimilated 3, 000 training complexes. The latter was observed in those RF models that used exactly the same set of features as Vina. This is an important result because it shows that, without this machine learning approach, more data would go wasted without producing any further improvement. The interested reader can find the full set of experiments in [15].

4.4   RF for Virtual Screening

Previous sections have given an overview of the use of RF for binding affinity prediction. This machine learning approach can also be taken to address the related problem of virtual screening. The first version of RF-Score [3] has already been used [2] to discover a large number of innovative binders of antibacterial targets in a prospective study. Furthermore, it has now been incorporated [11] into a large-scale docking tool for prospective virtual screening (http://istar.cse.cuhk.edu.hk/idock/). Despite this preliminary work, the performance of RF-based scoring functions on virtual screening is yet to be benchmarked across a range of protein targets. However, RF classifiers [18], which are not scoring functions because they cannot be applied to predict binding affinity, have already been found to outperform classical scoring functions on retrospective virtual screening benchmarks. Moreover, scoring functions based on Support Vector Machines [10] have also demonstrated substantial improvements over classical scoring functions on these benchmarks. Therefore, future research on a RF-based scoring function tailored to this related problem is very promising.

4.5   Conclusions and Future Prospects

RF-Score has been rigorously demonstrated to be on average much more accurate than classical scoring functions at predicting binding affinity of protein-ligand complexes. Furthermore, we have seen that this performance gap will broaden as more structural and interaction data are made available, as RF exploits data more effectively than MLR. There are now a number of very promising extensions of this initial work on RF-based scoring functions. Firstly, RF-Score has so far focused on re-scoring generic protein-ligand complexes. The next step would be to study how well complexes from a particular protein family are predicted by current RF-Score versions as well as the best way to generate family-specific versions of RF-Score which may be more accurate than those trained on complexes of any type. Secondly, the design of RF-based scoring functions for the related problems of virtual screening and pose generation is yet to be investigated, while being very promising. Lastly, pose generation error is generally believed to have a major negative impact on binding affinity prediction. So far, prediction has been made on co-crystallised ligands, i.e. with no pose generation error at all.

Therefore, studies intended to evaluate the impact that pose generation error really has on binding affinity predition as well as machine-learning strategies to compensate for any of these errors are needed.

Acknowledgements. This work has been carried out thanks to the support of the A*MIDEX grant (n° ANR-11-IDEX-0001-02) funded by the French Government «Investissements d'Avenir» program, the Direct Grant from the Chinese University of Hong Kong and the GRF Grant (Project Reference 414413) from the Research Grants Council of Hong Kong SAR.

References

1. Ballester, P.J.: Machine Learning Scoring Functions Based on Random Forest and Support Vector Regression. In: Shibuya, T., Kashima, H., Sese, J., Ahmad, S. (eds.) PRIB 2012. LNCS, vol. 7632, pp. 14–25. Springer, Heidelberg (2012)
2. Ballester, P.J., Mangold, M., Howard, N.I., Robinson, R.L.M., Abell, C., Blumberger, J., Mitchell, J.B.O.: Hierarchical virtual screening for the discovery of new molecular scaffolds in antibacterial hit identification. Journal of The Royal Society Interface 9(77), 3196–3207 (2012)
3. Ballester, P.J., Mitchell, J.B.O.: A machine learning approach to predicting protein-ligand binding affinity with applications to molecular docking. Bioinformatics 26(9), 1169–1175 (2010)
4. Ballester, P.J., Mitchell, J.B.O.: Comments on "Leave-Cluster-Out Cross-Validation Is Appropriate for Scoring Functions Derived from Diverse Protein Data Sets": Significance for the Validation of Scoring Functions. Journal of Chemical Information and Modeling 51(8), 1739–1741 (2011)
5. Ballester, P.J., Schreyer, A., Blundell, T.L.: Does a More Precise Chemical Description of Protein-Ligand Complexes Lead to More Accurate Prediction of Binding Affinity? Journal of Chemical Information and Modeling 54(3), 944–955 (2014)
6. Berman, H., Henrick, K., Nakamura, H.: Announcing the worldwide Protein Data Bank. Nature Structural & Molecular Biology 10(12), 980–980 (2003)
7. Berman, H.M., Westbrook, J., Feng, Z., Gilliland, G., Bhat, T.N., Weissig, H., Shindyalov, I.N., Bourne, P.E.: The Protein Data Bank. Nucleic Acids Research 28(1), 235–242 (2000)
8. Breiman, L.: Random Forests. Machine Learning 45(1), 5–32 (2001)
9. Cheng, T., Li, X., Li, Y., Liu, Z., Wang, R.: Comparative Assessment of Scoring Functions on a Diverse Test Set. Journal of Chemical Information and Modeling 49(4), 1079–1093 (2009)
10. Ding, B., Wang, J., Li, N., Wang, W.: Characterization of Small Molecule Binding. I. Accurate Identification of Strong Inhibitors in Virtual Screening. Journal of Chemical Information and Modeling 53(8), 114–122 (2013)
11. Li, H., Leung, K.S., Ballester, P.J., Wong, M.H.: istar: A Web Platform for Large-Scale Protein-Ligand Docking. PLoS ONE 9(1), e85678 (2014)
12. Li, H., Leung, K.S., Nakane, T., Wong, M.H.: iview: an interactive WebGL visualizer for protein-ligand complex. BMC Bioinformatics 15(1), 56 (2014)
13. Li, H., Leung, K.S., Wong, M.H.: idock: A multithreaded virtual screening tool for flexible ligand docking. In: Proceedings of the 2012 IEEE Symposium on Computational Intelligence in Bioinformatics and Computational Biology (CIBCB), pp. 77–84 (2012)

14. Li, H., Leung, K.S., Wong, M.H., Ballester, P.J.: Substituting random forest for multiple linear regression improves binding affinity prediction of scoring functions: Cyscore as a case study. BMC Bioinformatics 15(1), 291 (2014)
15. Li, H., Leung, K.S., Wong, M.H., Ballester, P.J.: Improving AutoDock Vina using Random Forest: the growing accuracy of binding affinity prediction by the effective exploitation of larger data sets. Molecular Informatics (2015) doi:10.1002/minf.201400132
16. Li, L., Wang, B., Meroueh, S.O.: Support Vector Regression Scoring of Receptor-Ligand Complexes for Rank-Ordering and Virtual Screening of Chemical Libraries. Journal of Chemical Information and Modeling 51(9), 2132–2138 (2011)
17. Li, Y., Liu, Z., Li, J., Han, L., Liu, J., Zhao, Z., Wang, R.: Comparative Assessment of Scoring Functions on an Updated Benchmark: 1. Compilation of the Test Set. Journal of Chemical Information and Modeling 54(6), 1700–1716 (2014)
18. Sato, T., Honma, T., Yokoyama, S.: Combining Machine Learning and Pharmacophore-Based Interaction Fingerprint for in Silico Screening. Journal of Chemical Information and Modeling 50(1), 170–185 (2010)
19. Trott, O., Olson, A.J.: AutoDock Vina: Improving the speed and accuracy of docking with a new scoring function, efficient optimization, and multithreading. Journal of Computational Chemistry 31(2), 455–461 (2010)

Strong Inhomogeneity in Triplet Distribution Alongside a Genome

Michael Sadovsky and Xenia Nikitina

Institute of computational modelling of SB RAS,
Akademgorodok, 660036 Krasnoyarsk, Russia
msad@icm.krasn.ru,
ksklokova@yandex.ru
http://icm.krasn.ru

Abstract. The distribution of triplets alongside a genome is studied. We explored the distribution to the nearest neighbour, that is the pattern where two triplets are fixed, and the distance is determined from the former to the latter so that the second triplet takes place nowhere inside the observed gap surrounded with the couple of the given triplets. The distribution differs strongly, for different organisms. Yeast and bacteria seem to have rather smooth pattern, while mammalia and other higher eukaryotes exhibit very complex patterns with long-range correlations in the triplet distribution.

Keywords: Order, periodicity, inhomogeneity, track, longest gap.

1 Introduction

Finite symbol sequence (from a finite alphabet) is a typical mathematical object. Eventually, it presents naturally in genetic matter of any living being; namely, as DNA sequence. Further we shall consider the finite symbol sequences of (various) length N (here N is the number of symbols in a sequence), from four-letter alphabet $\aleph = \{\mathsf{A}, \mathsf{C}, \mathsf{G}, \mathsf{T}\}$. We stipulate the occurrence of neither other symbols, nor blank spaces in a sequence.

An identification and search of structures in DNA sequence is the core issue of bioinformatics, biophysics and related scientific fields, including programming and information theory. Structures observed within a sequence define an order and provide easier understanding of functional roles of a sequence or its fragments. Reciprocally, a new function (or a connection between function and structure, or taxonomy) might be discovered through a search for new patterns in symbol sequences corresponding to DNA moluculae. There are issues of an order mentioned above: the former is a structuredness revealed within a sequence that yields further (and deeper) understanding of the functions and other related properties of a sequence itself, the latter is a kind of classification observed over a set of sequences (say, population genomics, etc.).

A number of structure has been found in DNA sequences; a lot would be found more. Basically, a structure to be found in a sequence could be reduced to the

F. Ortuño and I. Rojas (Eds.): IWBBIO 2015, Part II, LNCS 9044, pp. 248–255, 2015.
© Springer International Publishing Switzerland 2015

study of the probability density function of an occurrence of some strings within the sequence (i. e. frequency dictionary). Another issue is a structure determined by the mutual interlocation of those (short) strings to be found in a sequence. Among all such patterns, there is one looking very simple; meanwhile, it is still heavily understudied. This is the mutual distribution of triplets alongside a DNA sequence. Surprisingly, being very easy in terms of technical implementation, this study does not draw any attention of researchers. Surprisingly, very few attention has been paid to the study of such long range structuredness in DNA sequences. A model of l-tipple distribution to be expected at DNA sequence was proposed in [1]; giving a promising approach to study a long-range correlations, the authors of this paper nonetheless kept themselves within a study of a structure a frequency dictionary. Also, a short range correlations in DNA sequences were studied in a numerous papers, here we mention very few of them (see, e. g., [2,3]). Paper [2] provides an abundant experimental data accompanied with a relevant theoretical analysis of the correlations to be found in DNA sequences, while again no long-range correlations are taken into consideration. Finally, the paper [4] should be mentioned as a good basis for theoretical analysis of the problem. Everywhere below we shall concentrate on the distribution of the words of the length $q = 3$ (so called triplets, in biology); the method provided at this paper is (formally) feasible for a set of the string of any length, while a study of the distribution functions for arbitrary stings falls beyond the scope of this paper.

2   Distribution to the Nearest Neighbour

Yet, the distribution of triplets alongside a sequence may bring a lot new issues to biologists; same is true for the mutual distribution of the triplets. To begin with, we should fix the patterns of correlations to be studied within symbol sequences. Here two types of a structure could be expected: the former is related to a "static" distribution, and the latter to a "dynamic" one. Let fix two triplets, $\omega_1 = \nu_1\nu_2\nu_3$ and $\omega_2 = \mu_1\mu_2\mu_3$. The key question of the paper is what is a character of the mutual interlocation of the triplets. Here two different structures could be defined.

The first one consists in a study of the frequency distribution function

$$f_{\langle \omega_1, \omega_2 \rangle}(l)$$

of the mutual interlocation of these two triplets at the fixed distance l; this approach is pretty close to a Fourier analysis of a symbol sequence. Indeed, seeking for an increased (or, vice versa, decreased) frequency of the occurrence of some specific couple of triplets at the given distance each other, one can surely say the "hidden" periodicity takes place, or at least a structure resembling very much a periodical occurrence of two triplets.

Another approach is to seek for the "gaps" in a sequence; these are the (rather long) series of symbols met in a sequence under consideration, remarkable for some specific feature. A subsequence observed within the given symbol sequence that does not contain some specific word ω is called *gap*. Two given triplets $\omega_1 = \nu_1\nu_2\nu_3$ and $\omega_2 = \mu_1\mu_2\mu_3$ surrounding a gap that does not embed $\omega_2 = \mu_1\mu_2\mu_3$ make *bridge*. Consider a fragment of sequence

...AAAAAAAAAACCCCCCCCCCCCCCCGGGGGGGGGTTTTTACACACACAGG...

Then the string

...AAA CCCCCCCCCCCCCCGGGGGGGGGTTTTTACACACA CAG...

is the bridge determined by the triplets AAA and CAG, respectively, while the string

AAC CCCCCCCCCCCCCCGGGGGGGGGTTTTTA CAC

is the similar bridge determined by the triplets AAC and CAC, respectively. The very beginning results for the bridge search in some genetic entities are present in paper [5].

Figure 1 illustrates the distribution to the nearest neighbour observed over 22nd chromosome of *P. troglodytes*. The figure just demonstrates possible complexity and unexpectedness of a pattern that might be observed over a sequence. Surely, a figure like that one shown in Fig. 1 depends on the sequence under

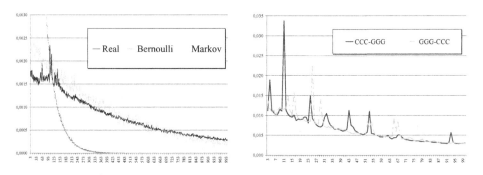

Fig. 1. (left) The distribution for AAA ⇔ CGA couple of triplets; take a look over the scale. (right) Short-range pattern of the distribution of the couples CCC ⇔ GGG and GGG ⇔ CCC.

consideration, and the specific couple of triplets. We shall not discuss this point further here; the most intriguing point in such distributions is that it decays extremely slowly, in comparison to any Markov model of the sequence [6], thus yielding a kind of "heavy tails" distributions.

2.1 The Longest Bridges

The main purpose of this paper is the introduction of the structure that is a limit case of the nearest neighbour distribution described above. That latter is the study of the longest gaps observed within a sequence. Few words should be said towards the terminology of the study. We calculated the longest gap length $l^*(\omega_1, \omega_2)$ for all 4096 couples of triplets ω_1 and ω_2. The bridges (that are the longest gaps enclosed within two given triplets). Both the length $l^*(\omega_1, \omega_2)$ of the bridge, and the location $n^*_{\omega_1}$ of the starting triplet ω_1 were determined; the observed data have been written in text file. The location mentioned above is the number of nucleotide giving the start to the bridge. Some of the bridges (while very rare) occurred in two copies; all these copies we also kept in the file.

Thus, any sequence was ascribed with 4096 lines corresponding to a triplet couple each. Definitely, the specific pattern of the list depends rather heavily both on the length of a sequence under consideration, and the taxonomy of the bearer of a sequence. Actually, we tried to figure out some general differences in the patterns of the bridges distribution over a sequence, in dependence on the taxonomy of the sequence bearer. In such capacity, one has to figure out the way to eliminate the impact of the length difference of the different sequences; in other words, one needs a normalization to avoid the impact of the sequence length difference.

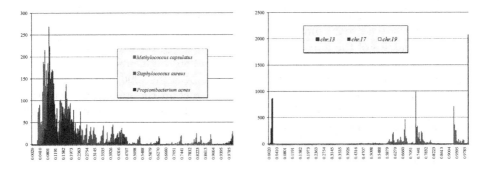

Fig. 2. Histograms of the figures (2) distribution for *Bacteria* (left) and *Callithrix jacchus* (right). The number of intervals was 256.

To implement the normalization, one has to introduce some adequate model of a sequence; unfortunately, one hardly could expect to figure out some model [6]. Nonetheless, for random non-correlated sequence, the expected length of a bridge goes down exponentially; thus, we used the following normalization

$$r^*(\omega_1, \omega_2) = \frac{l^*(\omega_1, \omega_2)}{\ln N}, \tag{1}$$

where N is the length of a sequence.

3   Results

To begin with, let consider the behaviour of the figure $l^*(\omega_1, \omega_2)$, for some examples. Fig. 2 shows the histograms of the distribution of all 4096 couples of triplets over the normalized figures of $l^*(\omega_1, \omega_2)$. To do that, the figures $l^*(\omega_1, \omega_2)$ have been changed for

$$\tilde{l}(\omega_1, \omega_2) = \frac{l^*(\omega_1, \omega_2)}{\max_{\Omega \ni (\omega_1, \omega_2)} \{l^*(\omega_1, \omega_2)\}}, \tag{2}$$

where maximum is sought over the set of couples (ω_1, ω_2), for each sequence, independently. Thus, for each sequence the figures (2) lie at the interval $[0, 1] \ni \tilde{l}(\omega_1, \omega_2)$. This normalization provides an adequate visualization of the data of $l^*(\omega_1, \omega_2)$, for various species and/or chromosomes. Fig. 2 shows the patterns of (2) for three bacterial species (left part of the figure), and for three chromosomes of monkey *Callithrix jacchus* (right part of the figure).

Another important question is whether those gaps and bridges are located alongside a sequence rather homogeneously, or they tend to concentrate in some specific regions of DNA. Figure 3 addresses this question, to some extent. The figure shows the histogram of the embedments alongside a sequence: namely, each embedment was determined through the relevant number of a start9ing nucleotide. Then the distances (expressed in nucleotide numbers) were normalized to 1, by dividing the numbers on the sequence length. Thus, we obtained the relative figures independent on the sequence length; here any embedment is indexed with the real number $s \in [0, 1]$. Let us remind that there are (at least) 4096 entries of the gaps, in a sequence.

Then the data massive has been divided into 256 fragments of equal length. We have chosen this number of fragments just because 4096 and 256 are the powers of 2. Then the number of embedments met in the given fragment has been calculated, and the result is shown in Fig. 3. The same procedure has been implemented to get Fig. 2.

4   Discussion

A new type of structuredness is determined by the mutual distribution of triplets alongside a sequence. Surprisingly, very few attention has been paid to this structure, while it is easily detected. The pattern of distribution of a specific couple of triplet definitely depends on the couple: even within the same sequence the patterns for different couples may vary significantly. Besides, the taxonomy of a sequence bearer impact heavily the patterns. Apparently, higher clades exhibit simpler patterns of those distributions (see Figs. 2 and 3 for illustration).

The paper presents a novel approach to identify a structuredness in DNA sequences, for the specific constraint: we studied the limit case of the distribution shown in Fig. 1. Namely, we explored the longest gaps ever observed in a sequence. Usually, the longest gap presents in a single copy; several copies may take place, while the probability seems to decay significantly with the growth of the sequence; yet, we have no any evident proof for this statement. Nonetheless, the observations allow to say that the distribution of the longest gaps is very for from any physically based one: it exhibits so called "heavy tails". More strictly, the expected probability to meet a gap of the length l is several orders less in comparison to the observed one.

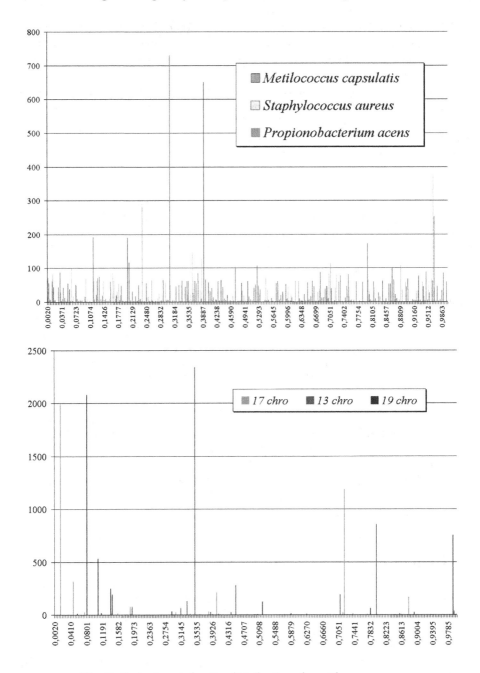

Fig. 3. Embedment density distribution alongside a sequence

We've examined $\sim 2 \times 10^2$ genomes of the organisms ranking from bacteria to primates. General observation is that bacterial genomes exhibit rather diverse and complicated pattern of the longest gaps distribution, in comparison to higher ranked organisms. Another fact is that all the genomes have the stretches inside the DNA sequences that are free from some specific triplets, and the length of such stretches is unexpectedly great. Apparently, bacterial genomes have much more diverse ensemble of the gaps, in comparison to eukaryotic organisms, as expressed in the terms of length of the gaps. Indeed, the histograms shown in Fig. 2 evident diversity of the gaps in terms of their lengths, for bacterial genomes, while that one observed for monkey genome seems to have the sited within a genome containing highly concentrated starting points of the longest gaps. Table 1 illustrates this feature.

Table 1. Some statistical characters of the gaps of bacterial genomes vs. monkey genome. N is the genome length; $l^*(\text{max})$ is the longest gap; $l^*(\text{min})$ is the shortest one; $\overline{l^*}$ σ are average length, and the standard deviation, respectively; sum is the total length of all the gaps; ratio is the ratio of the total length to the genome length.

species	N	$l^*(\text{max})$	$l^*(\text{min})$	$\overline{l^*}$	σ	sum	ratio
S. aureus	2902619	7822	260	1662.45	1273.28	6809390	2.35
M. capsulatus	3304561	9727	289	1650.25	1922.16	6759444	2.05
P. acnes	2560265	6734	335	1238.97	1026.13	5074824	1.98
chr.17	74750902	26887	11039	18710.08	1924.00	76636489	1.03
chr.13	117903854	1002671	10192	513823.00	491258.50	2104618996	17.85
chr.19	49578535	21163	9295	17482.41	2605.94	71607967	1.44

The couples yielding the results shown in Table 1 are the following (the first one corresponds to the longest one¡ and the second one corresponds to the shortest one): S. aureus — GGG – GGG, CCC – AAT; P. acnes — TAC – CTA, TAA – CGA; and M. capsulatus — TTA – TAA, ATC – GCC. Similar couples for monkey are the following: chr.17 — GCG – GCG, CGT – TGG; chr.13 — ACG – ACG, CGG – CAG; and chr.19 — GCG – CGG, CGC – TGG.

5   Conclusion

The proposed approach to the analysis of nucleotide sequences identifies a new structuredness in these latter. The structure is determined by the mutual location of triplets considered with respect to the "nearest neighbour" location. This latter is the distribution of triplets so that for a given couple of triplets there is no other triplet coinciding to the second one somewhere inside the site bounded by these two given triplets. Surprisingly, the distribution of such longest gaps, for each couple, shows a heavy tail distribution, not particular for natural systems. Extremely non-random character of a distribution of the gaps over a sequence allows to assume that this pattern is related to some biologically charged issues encoded in sequences.

References

1. Chen, Y.-H., Nyeo, S.-L., Yeh, C.-Y.: Model for the distribution of k-mers in DNA sequence. Phys. Rev. E 72, 011908 (2005)
2. Baldazzi, V., Bradde, S., Cocco, S., Marinari, E., Monasson, R.: sequences from mechanical unzipping data: the large-bandwidth case. Phys. Rev. E 75, 011904 (2007)
3. Gonzalez, D.L., Giannerini, S., Rosa, R.: Strong short-range correlations and dichotomic codon classes in coding DNA sequences. Phys. Rev. E 78, 051918 (2008)
4. Allegrini, P., Barbi, M., Grigolini, P., West, B.J.: Dynamical model for DNA sequences. Phys. Rev. E 52, 5281–5296 (1995)
5. Bushmelev, E.Y., Mirkes, E.M., Sadovsky, M.G.: On the structures revealed from symbol sequences. Journal of Siberian Federal University. Mathematics & Physics 5(4), 507–514 (2012)
6. Sadovsky, M.G., Nikitina, X.A.: Very Low Ergodicity of Real Genomes. Journal of Siberian Federal University. Mathematics & Physics 7(4), 530–532 (2014)

Predicting Sub-cellular Location of Proteins Based on Hierarchical Clustering and Hidden Markov Models

Jorge Alberto Jaramillo-Garzón[1,2], Jacobo Castro-Ceballos[1],
and Germán Castellanos-Dominguez[1]

[1] Universidad Nacional de Colombia, Sede Manizales, Colombia
[2] Institute Tecnológico Metropolitano, Medellín, Colombia
{jcastroc,cgcastellanosd}@unal.edu.co, jorgejaramillo@itm.edu.co

Abstract. Sub-cellular localization prediction is an important step for inferring protein functions. Several strategies have been developed in the recent years to solve this problem, from alignment-based solutions to feature-based solutions. However, under some identity thesholds, these kind of approaches fail to detect homologous sequences, achieving predictions with low specificity and sensitivity. Here, a novel methodology is proposed for classifying proteins with low identity levels. This approach implements a simple, yet powerful assumption that employs hierarchical clustering and hidden Markov models, obtaining high performance on the prediction of four different sub-cellular localizations.

1 Introduction

The information derived from sequenced genomes has grown with an exponential behaviour, and the number of protein sequences with missing annotation increases rapidly. Therefore, the functional annotation of proteins has become a theme of great importance in molecular biology. Nonetheless, this task posses a big challenge due to the huge amount of available data.

The localization of a given protein can indicate how and what kind of cellular environments the proteins interact, and thus, it can help to elucidate its function [1]. A commonly used experimental method for determining the localization of a given protein is to fuse the sequence encoding a green fluorescent protein (GFP) to one end of the gene sequence for the query protein, and then use its intrinsic fluorescence to monitor where the protein is in the cell [2]. However, as this approach must be focused on specific proteins, it turns very expensive and time consuming, especially when considering the current size of unannotated protein sequences [3].

Several computational predictors of protein sub-cellular localizations have been proposed in the past few years. The most common methods among biologists are the alignment-based methods, which consist on searching query proteins against public databases of annotated proteins by using local alignment search tools such as BLAST or PSI-BLAST [4]. These methods, however, tend to attain low sensitivity for databases of proteins with low identity levels, due to the inability of the method for identifying homologous proteins at significant E-values [5].

A second category of methods are based on machine learning strategies. In this kind of methods, a set of numerical features from protein sequences is computed and a classifier is trained to label query protein sequences according to one of the classes from the

F. Ortuño and I. Rojas (Eds.): IWBBIO 2015, Part II, LNCS 9044, pp. 256–263, 2015.

training dataset. Among the classifiers used in this kind of methods are support vector machines [6–8, 3], neural networks [9], or ensembles of multiple specialized classifiers [10–13].

A more biologically driven alternative are the subsequence-based methods, which explore the fact that the functionality of proteins is mainly due to functional domains that may reside in different portions of the proteins. These methods employ stochastic models for describing protein families. Large collections of protein families and domains can be found in public databases [14] and methods based on Hidden Markov Models (HMM) can efficiently represent family profiles [15]. However, both machine learning based methods and subsequece-based methods have shown low sensitivities for categories with high diversity among its samples, as it is the case for several subcellular localizations. Since these methods try to represent the whole category by a single trained model, potentially useful information is necessarily discarded and a big amount of false-negatives appear.

Considering those precedents, this work proposes a subsequence-based methodology, that aims to improve the prediction of sub-cellular localizations with low identity levels, by using HMM models. The proposed methodology assumes that there is not a single cluster of samples belonging to a given category and that each cluster may have its own distinctive profiles. The proposed methodology is compared with two other subsequence-based methods in the literature PfamFeat [15] and Plant-mPloc [16] for the prediction of four sub-cellular localizations in a set of *Embryophyta* proteins. The results show that implementing this simple, yet powerful assumption, the classification model is enhaced and the sensitivity of the system rises, thus increasing the overall performance of the system.

2 Background

The proposed methodology involves two software packages that have been extensively used in the literature. First sequences are clustered together with the CD-HIT software package [17] and then, each cluster is modelled into an HMM profile using the HMMer software [18]. These packages will be described in the present section.

Hierarchical Clustering of Protein Sequences Using CD-HIT: CD-HIT [17] uses a method based on greedy incremental clustering for detecting clusters of similar sequences in the data. Briefly, sequences are first sorted in order of decreasing length and the longest one becomes the representative of the first cluster. Then, each remaining sequence is compared to the representatives of existing clusters. If the similarity with any representative is above a given threshold, it is grouped into that cluster. Otherwise, a new cluster is defined with that sequence as the representative. The process continues until all sequences have been assigned to a cluster.

This algorithm has been extensively used for a large variety of applications ranging from non-redundant dataset creation [19], protein family classifications [20], metagenomics annotation [21], among others.

HMM-Based Modelling of Sequence Clusters Using HMMER: HMMs are stochastic model, which assume the system can be modelled as a Markov process, with unknown parameters. In discrete cases, these parameters are represented by a set of Q states $\theta \in \mathbb{R}^Q$ that has to be computed by an underlying optimization process. Each state is associated to one of K possible observation values. The model is then composed by three parameters: an initial state probability π with elements $\{\pi(\theta_i) \in \mathbb{R}[0,1]\}$, that describes the distribution over the initial state set; a transition matrix $A \in \mathbb{R}^{Q \times Q}$ where each $a_{ij} \in \mathbb{R}^+, i, j \in [1, Q]$ represents the transition probability from state i to state j; and an observation matrix $B \in \mathbb{R}^{Q \times K}$ with the elements $\{b_{i,k} \in [0,1]\}$ representing the probability of each observed symbol $k \in [1, K]$, given that the system remains at state i [22].

The HMMER software package [18] uses the (MSV) score for target sequence x. it is a log likelihood ratio score of singles optimal (maximally likely) alignment: the ratio of the probability of the optimal alignment Ω for x given the MSV model $\mathcal{M}$ and the probability of the sequence given a null hypothesis model $\mathcal{R}$:

$$S^{MSV}(x) = \log_2 \frac{Prob(x, \Omega|\mathcal{M})}{Prob(x|\mathcal{R})} \tag{1}$$

For a query of length m positions, the MSV Profile has Km match emission parameters (where K is the alphabet size, 4 nucleotide of 20 amino acids), plus $m + 8$ additional state transition parameters involving the flanking and N, B, E, C and J states that account for non-homologous residues. Other state transitions in the original profile are ignored, which means implicitly in the original profile are ignored, which means implicitly treating match-match transitions as 1.0.

The null model R is assumed to be an HMM with a single state R emitting residues a with background frequencies $f(a)$ (i.e. a standard i.i.d null model: independent, identically distributed residues), with a geometric length distribution specified by a transition parameter t_{RR}.

The mK positions-specific match scores $\sigma_k(a)$ are precomputed as log-odss ratios for a residues a emitted from match state M_k with emission probability $e_k(a)$, compared to the null model background frequencies f_a:

$$\sigma_k(a) = \log_2 = e_k(a)/f_a \tag{2}$$

These match scores (as well as the emission probabilities and background frequencies) are the same as in the original profile. The only state transition parameters in the MSV model are those that control target sequences length modelling, the uniform local alignment fragment length distribution, and the number of hits to the core homology model per target sequence [23]. These too are identical to the parametrization of the original profile.

3   Experimental Set-Up

The workflow of the experimental set-up has five main components: the input *database* that is labeled with the corresponding sub-cellular localizations; the *preprocessing* stage, where sequences are grouped into clusters with the CD-HIT software; the alignment stage, where a sequence alignment is obtained from the sequences in each cluster using the Clustal Omega software package [24]; and the *HMM construction* stage, where the HMMER software package is used to generate profiles from each alignment. Finally a statistical *validation* is performed in order to test the performance of the designed predictor and obtain performance measures. Figure 1 shows the experimental set-up workflow.

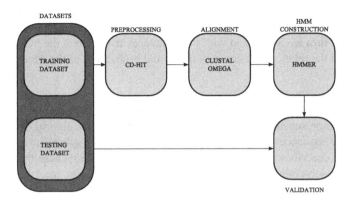

Fig. 1. Scheme of the baseline classifiers

3.1 Database

The dataset used in this work is a subset of the database published in [19]. It is conformed from all the available *Embryophyta* proteins at UniProtKB/Swiss-Prot database ([25] file version: 10/01/2013), with at least one annotation in the Cellular Component Ontology according to *Gene Ontology Annotation* (GOA) project ([26], file version: 7/01/2013). Proteins with unknown evidence of existence or resulting from computational predictions were discarded.

This dataset is composed by 2643 proteins, associated to four Gene Ontology terms, as shown in Table Table 1.

Table 1. Number of protein sequences associated to each sub-cellular localization

Localization	GO ID	Samples
Cytoplasm	GO_0005737	572
Cell Wall	GO_0005618	412
Extreacellular Region	GO_000556	374
Membrane	GO_0016020	1285

Preprocessing: In order to identify clusters of sequences with similar primary structures, the CD-HIT software package is used. The software searches for sequences with identities under a predefined cut-off. This identity cut-off was set at 30% in order to test the accuracy of the proposed methodology at a low identity level. CD-HIT performs a hierarchical clustering and retrieves the clustered proteins sequences. Then, a multiple sequence alignment is performed with the Clustal Omega software package in order to obtain an alignment model for being used as input of HMMER.

HMM Training: With the aligned protein sequences grouped into clusters, the HMMs can be generated from the alignments. The obtained profiles are associated with the GO terms which in turn are associated with the sequences from which the models came, in order to to find the relationship between them, and probability of the dataset belonging to any model.

Validation: Three performance measures are used to analyse the generalization capability of the predictor: sensitivity (s_n) describes the capacity of the algorithm to recognize as positives the sequences that are indeed associated to a given sub-cellular component; specificiy (s_p) describes how the algorithm is able to reject sequences that are not associated to the sub-cellular component; and the geometric mean (g_m) between those measures as a global performance measure.

$$s_n = \frac{n_{TP}}{n_{TP} + n_{FN}} \qquad s_p = \frac{n_{TN}}{n_{TN} + n_{FP}}$$

$$g_m = \sqrt{\frac{n_{TP} n_{TN}}{(n_{TP} + n_{FN})(n_{TN} + n_{FP})}}$$

where n_{TP}, n_{TN}, n_{FP} and n_{FN} are true positive, true negative, false positive, and false negative, respectively.

4   Results and Discussion

Figure 2 show the results obtained for the four sub-cellular components. It is important to note how *Cytoplasm* and *Membrane*, which are subcellular localizations containing protein sequences with a high variety, reached very high performance results. In contrast, *Extracellular region* and *Cell Wall*, which are more specific categiories, attained high specificities but remained with poor sensitivities.

In order to compare the results of the proposed methodology against other proposed strategies, Table 2 contrasts the results with the ones reported by the PfamFeat algorithm [15]. Also, since the database used for the experiments contains only *Embryophyta* (land plants) proteins, the Plant-mPloc [16] server was also tested for comparison purposes.

As can be observed on Table 2, the proposed methodology outperforms the results of PfamFeat on three out of four sub-cellular localizations, and the results from Plant-mPloc in all cases. The most important result relies on the fact that the specificity of the system was significantively improved for all the subcellular localizations, while keeping an acceptable sensitivity. Although Plamnt-mPloc obtained the highest values of sensitivity for three localizations, it also obtained the lowest specificities, thus achieving poor global performances.

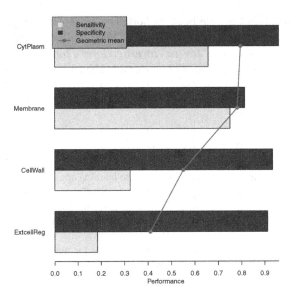

Fig. 2. Main results obtained for each sub-cellular localization

Table 2. Comparison of methods

Class	HMM			Pfam			Plant-mPloc		
	$s_p\%$	$s_n\%$	$g_m\%$	$s_p\%$	$s_n\%$	$g_m\%$	$s_p\%$	$s_n\%$	$g_m\%$
Cytoplasm	**96.15**	**65.75**	**79.51**	72.5	37.7	52.28	1.43	41.70	7.74
Cell Wall	**93.47**	32.43	**55.06**	88.01	19.7	41.63	2.32	**88.08**	14.31
Extcell. Region	**91.30**	18.46	41.05	78.5	40.7	**56.52**	1.25	77.73	9.88
Membrane	**81.33**	75.03	**78.10**	72.2	37.1	51.75	0.00	**77.30**	0.00

This results demonstrate that the inclusion of the hierarchical clustering stage, in conjunction with the HMM profile representations, provides a methodology that efficiently describes diversity of protein sequences associated to each localization, even when sequences have low identity levels among them.

5   Conclusions and Future Work

This paper presented a novel methodology for the prediction of sub-cellular localization of proteins. It proved to have better overall performances than other similar methods recently proposed in the literature, enhacing the sensitivity of the predictor for categories with high diversity. As a future work, it is important to test the methodology with the other GO ontologies (Molecular Function and Biological Process). Also it would be interesting to test several other clustering algorithms and semi-supervised strategies in order to improve even further the classification performances.

Acknowledgment. This work has been supported by the research project *"Metodología de clasificación multi-etiqueta mediante técnicas de optimización aplicado a la clasificación funcional de proteínas"* (*"Jóvenes Investigadores e Innovadores 2012"* program by COLCIENCIAS), and 20101007497 Universidad Nacional de Colombia.

References

1. Chou, K.-C., Shen, H.-B.: Cell-ploc: a package of web servers for predicting subcellular localization of proteins in various organisms. Nature Protocols 3(2), 153–162 (2008)
2. Baldi, P., Brunak, S.: Bioinformatics: the machine learning approach. The MIT Press (2001)
3. Jaramillo-Garzón, J., Perera-Lluna, A., Castellanos-Domiínguez, C.: Predictability of protein subcellular locations by pattern recognition techniques. In: 2010 Annual International Conference of the IEEE Engineering in Medicine and Biology Society (EMBC), pp. 5512–5515. IEEE (2010)
4. Conesa, A., Götz, S.: Blast2go: A comprehensive suite for functional analysis in plant genomics. International Journal of Plant Genomics 2008 (2008)
5. Hawkins, T., Chitale, M., Luban, S., Kihara, D.: PFP: Automated prediction of gene ontology functional annotations with confidence scores using protein sequence data. Proteins 74(3), 566–582 (2009)
6. Yu, C., Lin, C., Hwang, J.: Predicting subcellular localization of proteins for gram-negative bacteria by support vector machines based on n-peptide compositions. Protein Science 13(5), 1402–1406 (2004)
7. Shi, J., Zhang, S., Pan, Q., Cheng, Y., Xie, J.: Prediction of protein subcellular localization by support vector machines using multi-scale energy and pseudo amino acid composition. Amino Acids 33(1), 69–74 (2007)
8. Nanni, L., Lumini, A.: An ensemble of support vector machines for predicting the membrane protein type directly from the amino acid sequence. Amino Acids 35(3), 573–580 (2008)
9. Ma, J., Liu, W., Gu, H.: Predicting protein subcellular locations for Gram-negative bacteria using neural networks ensemble. In: Proceedings of the 6th Annual IEEE Conference on Computational Intelligence in Bioinformatics and Computational Biology, pp. 114–120. The Institute of Electrical and Electronics Engineers Inc. (2009)
10. Shen, Y., Burger, G.: 'Unite and conquer': enhanced prediction of protein subcellular localization by integrating multiple specialized tools. BMC Bioinformatics 8(1), 420 (2007)
11. Shen, H., Yang, J., Chou, K.: Euk-PLoc: an ensemble classifier for large-scale eukaryotic protein subcellular location prediction. Amino Acids 33(1), 57–67 (2007)
12. Niu, B., Jin, Y., Feng, K., Lu, W., Cai, Y., Li, G.: Using adaboost for the prediction of subcellular location of prokaryotic and eukaryotic proteins. Molecular Diversity 12(1), 41–45 (2008)
13. Khan, A., Majid, A., Choi, T.: Predicting protein subcellular location: exploiting amino acid based sequence of feature spaces and fusion of diverse classifiers. Amino Acids 38(1), 347–350 (2010)
14. Punta, M., Coggill, P.C., Eberhardt, R.Y., Mistry, J., Tate, J., Boursnell, C., Pang, N., Forslund, K., Ceric, G., Clements, J., et al.: The pfam protein families database. Nucleic Acids Research 40(D1), D290–D301 (2012)
15. Arango-Argoty, G., Ruiz-Munoz, J., Jaramillo-Garzon, J., Castellanos-Dominguez, C.: An adaptation of pfam profiles to predict protein sub-cellular localization in gram positive bacteria. In: 2012 Annual International Conference of the IEEE Engineering in Medicine and Biology Society (EMBC), pp. 5554–5557. IEEE (2012)

16. Chou, K.-C., Shen, H.-B.: Plant-mploc: a top-down strategy to augment the power for predicting plant protein subcellular localization. PloS One 5(6), e11335 (2010)
17. Fu, L., Niu, B., Zhu, Z., Wu, S., Li, W.: Cd-hit: accelerated for clustering the next-generation sequencing data. Bioinformatics 28(23), 3150–3152 (2012)
18. Finn, R.D., Clements, J., Eddy, S.R.: Hmmer web server: interactive sequence similarity searching. Nucleic Acids Research 39(suppl. 2), W29–W37 (2011)
19. Jaramillo-Garzón, J.A., Gallardo-Chacón, J.J., Castellanos-Domínguez, C.G., Perera-Lluna, A.: Predictability of gene ontology slim-terms from primary structure information in embryophyta plant proteins. BMC Bioinformatics 14(1), 68 (2013)
20. Yooseph, S., Li, W., Sutton, G.: Gene identification and protein classification in microbial metagenomic sequence data via incremental clustering. BMC Bioinformatics 9(1), 182 (2008)
21. Sun, S., Chen, J., Li, W., Altintas, I., Lin, A., Peltier, S., Stocks, K., Allen, E.E., Ellisman, M., Grethe, J., et al.: Community cyberinfrastructure for advanced microbial ecology research and analysis: the camera resource. Nucleic Acids Research 39(suppl. 1), D546–D551 (2011)
22. Rabiner, L.: A tutorial on hidden markov models and selected applications in speech recognition. Proceedings of the IEEE 77(2), 257–286 (1989)
23. Freyhult, E.K., Bollback, J.P., Gardner, P.P.: Exploring genomic dark matter: a critical assessment of the performance of homology search methods on noncoding rna. Genome Research 17(1), 117–125 (2007)
24. Sievers, F., Wilm, A., Dineen, D., Gibson, T.J., Karplus, K., Li, W., Lopez, R., McWilliam, H., Remmert, M., Söding, J., et al.: Fast, scalable generation of high-quality protein multiple sequence alignments using clustal omega. Molecular Systems Biology 7(1) (2011)
25. Jain, E., Bairoch, A., Duvaud, S., Phan, I., Redaschi, N., Suzek, B., Martin, M., McGarvey, P., Gasteiger, E.: Infrastructure for the life sciences: design and implementation of the UniProt website. BMC Bioinformatics 10(1), 136 (2009)
26. Barrell, D., Dimmer, E., Huntley, R.P., Binns, D., O'Donovan, C., Apweiler, R.: The goa database in 2009-an integrated gene ontology annotation resource. Nucleic Acids Research 37(suppl. 1), D396–D403 (2009)

Risk Quantification of Multigenic Conditions for SNP Array Based Direct-to-Consumer Genomic Services

Svetlana Bojić[1] and Stefan Mandić-Rajčević[2]

[1]Faculty of Biology, University of Belgrade, Serbia
ceca.bojic@gmail.com
[2]International Centre for Rural Health, Department of Health Sciences, University of Milan, Italy
stefan.mandicrajcevic@gmail.com

Abstract. Genome wide association studies (GWAS) are typically designed as case-control studies, collecting thousands of sick and healthy individuals, genotyping hundreds of thousands of SNPs, and documenting the SNPs which are more abundant in one group or the other. Direct-to-consumer genetic testing has opened the possibility for a regular person to receive data about his/her genotype, but the validity of risk assessment procedures and the final genetic risk estimate have been questioned. Many authors have discussed the advantage of use of the asymptotic Bayes factor (ABF) to measure the strength of SNP/trait associations, over the use of p-values. We propose a ABF based heuristic to filter-our and select SNP/trait associations to be used in multigenic risk assessment.

A raw genotype result from the 23andMe web service was merged with the GWAS catalog, and SNP/trait associations were filtered and selected using the R programming language together with free and publicly available databases.

From the initial 3195 SNP/trait associations, only 425 remained after the initial filters on descent, replicated findings, qualitative trait and availability of the number of cases and controls in the study. Selecting only one SNP/trait association from repeated studies and studies done with proxy SNPs left us with 377 SNP/trait associations available for multigenic risk assessment. After excluding the associations with unsatisfying ABF, only 300 SNP/trait associations remain for the multigenic risk assessment.

Whatever the link between SNP/trait associations and final DTC multigenic risk assessment for a given trait is, the final value of a risk score is heavily influenced by the number, as well as strength of evidence for individual SNP/trait pairs that are used for calculation. The ABF provides an unambiguous and simple criterion for ranking and including SNP/trait associations in multigenic risk assessment.

Keywords: GWAS, multigenic risk assessment, Bayes factor, SNP selection, direct to consumer.

1   Introduction

Genome wide association studies (GWAS) test for association between a disease or a quantitative trait and multiple genetic markers. These associations fall under the

F. Ortuño and I. Rojas (Eds.): IWBBIO 2015, Part II, LNCS 9044, pp. 264–275, 2015.
© Springer International Publishing Switzerland 2015

"common disease – common variant" hypothesis. The hypothesis suggests that the occurrence of common complex diseases is influenced by a moderate number of (interacting) disease alleles, called "casual variants". The most common type of markers are single nucleotide polymorphisms (SNPs). Nowadays, the SNP array technology offers the possibility to determine the state of even more than one million SNPs in one experiment. A typical GWAS is designed as a case-control study, collecting thousands of sick and healthy individuals, genotyping hundred thousand SNPs, and documenting the SNP-genotypes which are more abundant in one group or the other. They have been increasingly popular since almost a decade ago, but have been criticized, among other things, for lack of reproducibility or unclear utility for clinicians [1]. Nevertheless, GWAS have successfully identified many genetic variants contributing to the susceptibility for complex diseases.

To this day thousands of SNPs have been flagged as associated with hundreds of diseases, but the ability to predict one's disease status based solely on SNPs fails short of what would be expected, having in mind the high heritability of these diseases. One explanation could be that many phenotypes might be defined by a large group of SNPs with tiny effects, and present day GWAS are underpowered to detect them. For instance, an author estimated the overall number of SNPs which affect height to be 93,000 [2]. The naïve idea that it is enough to use the SNPs flagged as significantly associated with diseases or traits in GWAS has lead to low predictive value of risk assessments done in this way, as a large number of SNPs remain outside that scope.

Direct-to-consumer (DTC) genetic testing applies to the situations where genetic tests are marketed directly to the consumer via television, print advertisements or the Internet, as opposed to being ordered through healthcare providers such as physicians or genetic counselors. The upside of DTC genetic testing is that its growing market might promote awareness of genetic diseases and even more importantly allow patients/consumers to have a more proactive role in their health care, change their lifestyle, and organize their life better. There are, however, significant risks, as consumers might be misled by the results of invalid tests, or take important decisions regarding their health based on incomplete or misunderstood information [1].

Risk assessment is often defined as the determination of quantitative or qualitative value of risk related to a concrete situation and a recognized threat, and is also an essential part of genetic testing and counseling. It must be calculated as accurately as possible to enable the clinician and the patient to make sound health-related decisions. The calculation of genetic risk should incorporate all available information at a particular point in time. It should be considered an ongoing process of analysis of estimates [3]. As for DTC risk assessment, two trends dominate the industry [4]. First, practiced for instance by the web services Promethease, Interpretome, or LiveWello limits itself to listing SNP-wise (allele or genotype) based risk, as extracted from dbSNP database [5]. This information might be enriched with ClinVar entry providing the basis for dividing those SNP-wise risks into "good news" and "bad news" [6], with unquestionable scientific background, but yet not coherent enough to direct potential health related intervention. Higher level of DTC risk assessment implies combining SNP-wise evaluated risks into some form of multigenic risk score for a given trait. This has proven to be tricky, because we are far from knowing how these

SNPs interact shaping the influence on any phenotype. The three leading companies (23andMe, deCODE, Navigenics) use both absolute lifetime risk and relative risk compared to the general population, as obtained from odds ratios for individual SNPs under multiplicative effect assumption. Both approaches require high level of transparency regarding the exact method used, since there are numerous sources of potential bias (starting with inappropriate specification of disease prevalence in the control population, to unclear assumptions about prior odds), not to mention the high variability of the risk estimates even when well defined workflow is strictly followed [7]. These sources of compromised validity have been recognized by the Federal Drug and Administration agency, which resulted in the (in)famous ban of the 23andMe health reports [8].

In the frequentist inference, an association of a marker with a disease is evaluated based on whether or not the p-value of an association test is less than a significance level. The markers that have significant p-values are tested for replication in subsequent studies. A typical significance level for GWA studies is determined by Bonferroni procedure, and it lies somewhere around 5×10^{-7}, although other levels have emerged and have been evaluated in the literature [9, 10]. However, this significance level is used in practice independently of the sample size of studies, the effect size of associated markers, and the power to detect association [10]. For many common diseases, the effect size in studies is small, having an odds ratio in the range of 1.1-1.5. Therefore, markers with true associations can have p-values greater than the significance level, or even be ranked far away from the top, if based solely on p-values [11].

In Bayesian hypothesis testing, Bayes factor (BF) is often used and reported to measure the strength of association. The BF is a ratio of the probabilities of the observed data under the alternative H_1 and the null H_0 hypotheses, in which all the parameters are treated as random and averaged out with respect to their prior distributions [12, 13]. Many authors have discussed the advantage of BF over p-values for genetic association studies [14-16]. It measures the strength of association by integrating the significance of association with the power to detect it, while the p-value only measures the significance of association [15, 16]. When the sample size is large enough (thousands of cases and thousands of controls), large BF values ($\log_{10}BF > 5$) strongly support observed small p-values and therefore the association. On the other hand, small BF values ($\log_{10}BF < 0$) can be used to exclude the markers with small p-values as false positives with high confidence [15]. In addition, to compare results across studies, the BF is a better measure that the p-value, since it integrates both the significance of association and the power (sample size and effect size), as demonstrated by simulating different studies with different sample sizes and genetic effects, showing that different studies had different BFs, while the p-values remained the same [15]. The more widespread use of the Bayes factor has been hampered by the need for prior distributions to be specified for all of the unknown parameters in the model, and the need to evaluate multidimensional integrals, a complex computational task. Still, Wakefield [14, 17] has proposed the asymptotic Bayes factor (ABF) which avoids each of these requirements.

Whatever is the link between SNP/trait associations and final DTC multigenic risk assessment for a given trait, the final value of any multigenic risk score is heavily influenced by the number, as well as strength of evidence for individual SNP/trait pairs that are used for calculation. This is where ABF based inference can provide unambiguous and simple criterion for the ranking of SNP/trait associations, which embeds all the relevant information at the same time. In the present work we demonstrate the usefulness of ABF values in selecting SNP/trait associations for DTC multigenic risk assessment.

2   Methodology

Our subject was a European 30-year old Caucasian man, from the Balkan peninsula. He had received his raw genotype data from the 23andMe web service [18] and expressed interest in gaining knowledge regarding his own genetic risks.

2.1   Importing Data and Associating SNPs with GWAS

Raw data were a tab delimited text file, downloaded from the 23andMe website, with four columns: rsid (SNP marker ID), chromosome, genomic position and (unphased) genotype.

The data.frame with raw genotypes was merged with data from the National Human Genome Research Institute (NHGRI) GWAS catalog [19] using the *gwascat R* package [20]. This way we had several additional columns, describing all the GWAS that have implied some kind of trait association with our SNPs.

The rows with the "strongest risk allele" entry from a GWAS study matching any of the genotyped alleles for a particular SNP were retained for the future analysis, representing the body of evidence for this subject's genotype/trait risk estimation.

2.2   Filtering SNP-Trait Associations

After reducing our *data.frame* to only the SNPs with "risk" alleles, we applied several filters to account for the subject's origin and to rule out studies which had insufficient information by our criteria. Filters were created using regular expressions, defined as sequences of characters which form a search pattern, implemented through the *stringr R* package [21].

First we filtered out studies which were not done on European population, then studies which were not replicated. Finally we excluded studies which had no odds ratios (OR) and/or confidence intervals (CI).
Before calculating the ABF, number of cases and controls in each study were extracted using regular expressions.

2.3   Asymptotic Bayes Factor Calculation

Prior probability of H_1 was set to be equal for all the SNPs, dependent only on the number of SNPs investigated in the particular study. The logic behind such approach was that, for given trait, only about 100-1000 SNPs are truly expected to be associated, so the probability that a given SNP (on a given platform) is one of them approximates to 500/number of SNPs printed on the platform. That gave us prior probabilities in the range of 10^{-4} to 10^{-5}, which was in accordance with previous studies [22].

Asymptotic Bayes factor was calculated for each SNP/trait association, according to the formula:

$$ABF_0 = \sqrt{\frac{V+W}{V}} \, exp \left(-\frac{z^2 W}{2(V+W)} \right) \tag{1}$$

where V is the estimated variance of the parameter θ, W is the variance of the prior probability of null hypothesis, $z^2 = \hat{\theta}^2/V$ is the usual Wald statistic, and $exp(\theta)$ is estimated odds ratio [14]. Subscript "0" is there for this form of ABF summarizes the evidence for/against null hypothesis, whilst its reciprocal value $(1/ABF_0)$ then supports the working hypothesis - the evidence for the association. The crucial property of the relationship between the power of the study (represented with V of the estimated parameter θ) and evidence pro H_1 is that, providing all other elements, except V, of the equation (1) are fixed, the evidence for H_1 would grow as the power decreases, but only until it reaches its maximum at $V = W/(z^2 - 1)$. After that, ABF for H_1 would decline, since the power is not sufficient to provide strong evidence. This contrasts strongly with the behavior of the p-values, where very small departures from H_0 would produce small p-values when the power is high [14].

Parameters of prior variances were set as advised by Kraft and Evangelou [23, 24]. The usage of ABF guaranteed comparability of the "amount of evidence" for each SNP/trait association across studies, and allowed for the filtering procedures that followed [16]. We assumed autosomal-dominant model of inheritance at this point for the sake of simplicity.

2.4   Redundancy Check

SNP/trait associations were checked for redundancy, and repeated SNP/trait associations, as well as the trait association with SNPs that are in LD (defined as $R^2 > 0.8$) and/or physically closer than 500bp were discarded, keeping only the SNP/trait pairs with the highest ABF within their "redundancy cluster". We used SNAP web based tool [25] for identification of those "proxy" SNPs, and the search was conducted over SNPs in 1000GenomesPilot1 data.

Overall, the data analysis was done using R language and environment [26], with additional packages from the Comprehensive R Archive Network (CRAN) and Bioconductor [27].

3   Results

3.1   Filtering the SNPs for Multigenic Risk Assessment

The initial raw genotype file consisted of approximately 700,000 SNPs. Of that number, around 600,000 passed the genotype quality control. Merging with the NHGRI

GWAS database yielded 9172 SNP/trait associations. In the next step we excluded all the SNPs where the subject did not have at least one risk allele, leading to 3195 SNP/trait associations.

When we filtered out non-European sample studies, non-replicated findings, and studies not reporting OR and CI, we were left with 438 SNP/trait associations. *Figure 1* shows a flow-chart of the filters applied and the number of SNP/trait associations left after each filter was applied.

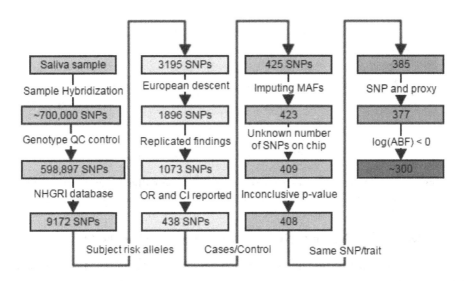

Fig. 1. Flow-chart of the filtering and selection process of SNPs

In the next step, missing MAFs were imputed where possible from the dbSNP database [5], and entries without reported/imputable MAFs were excluded. Rows with no inferable number of SNPs on the chip used, and samples with inconclusive p-value from the original studies were left out.

3.2   Asymptotic Bayes Factor

ABF was calculated for the remaining 408 SNP/trait associations. *Figure 2* shows the Manhattan plot of genomic coordinates of SNP/trait associations along the X-axis and the base 10 logarithm of the respective ABF on the Y-axis, together with the proposed cut-off values for $\log_{10}(ABF)$ of 5 (dashed line). Twelve percent of SNP/trait associations had $\log_{10}(ABF)$ values above cut-off, meaning there really is strong evidence to support those associations. Sixty seven percent had $\log_{10}(ABF)$ between 5 and 0, meaning there is enough evidence that those SNP/trait associations should be taken into consideration for the multigenic risk assessment. It is interesting to note that as many as 21% of SNP/trait associations fell below the second cut-off level of 0,

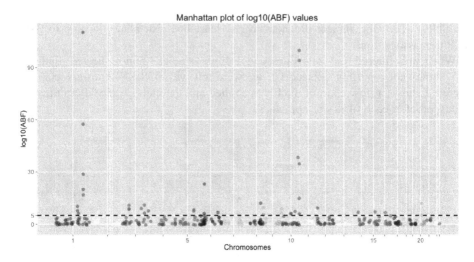

Fig. 2. Manhattan plot of $\log_{10}$(ABF) values

offering no firm evidence for those associations. Still, for illustration purposes these "unsupported" SNP/trait associations were not excluded from further analysis.

3.3   Removing Redundant SNP/Trait Association Based on the ABF

It is a common practice to explore indicated SNP/trait associations in several independent GWA studies, as to pursue verification through replication. However, when constructing the multigenic risk assessment profile only one estimation of OR and CI per SNP should be used in the calculation. A criteria should be defined to consistently select the "correct" OR and CI. *Table 1* reports 23 SNP/trait associations, each targeted by 2 studies, as well as the estimated odds ratios. The OR from the SNP/trait association with highest ABF from every pair was kept for the future analysis and bolded in the table. In this step the initial data.frame was reduced by another 23 rows.

Some studies explore the SNP/trait association using a proxy SNP instead of the original. A proxy SNP is usually defined as a SNP in LD with the original SNP (R2 > 0.8). *Table 2* reports original SNPs and their proxies for the same trait, along with the their respective odds ratios and confidence intervals. The OR and CI of the SNP/trait association selected using the ABF to remain in the analysis is bolded in the table.

Based on the previous steps, we have reduced the initial data.frame from 408 to 378 SNP/trait associations using the ABF. Having in mind that another 21% of SNP/trait association would not pass the strength of evidence criteria considering their ABF, our final data.frame would consist of only around 300 rows. These SNP/trait associations should pose a high enough reliability to be used downstream in any established form of multigenic risk assessment.

Table 1. Selected ORs and CI based on the ABF from pairs of studies targeting the same SNP/trait association

rsID	Trait	OR[CI] - 1	OR [CI] - 2
rs10737680	Age-related macular degeneration	3.11[2.76-3.51]	**2.43[2.39-2.47]**
rs2075650	Age-related macular degeneration	1.37[1.22-1.54]	**1.23[1.13-1.34]**
rs429608	Age-related macular degeneration	2.16[1.84-2.53]	**1.74[1.68-1.79]**
rs9621532	Age-related macular degeneration	**1.61[1.37-1.89]**	1.41[1.27-1.57]
rs2294008	Bladder cancer	**1.15[1.10-1.20]**	1.13[1.09-1.17]
rs4973768	Breast cancer	1.14[1.09-1.19]	**1.10[1.08-1.12]**
rs6651252	Crohn's disease	1.23[1.17-1.30]	**1.19[1.13-1.25]**
rs7702331	Crohn's disease	**1.12[1.07-1.17]**	1.09[1.05-1.13]
rs12700667	Endometriosis	**1.20[1.13-1.27]**	1.18[1.11-1.25]
rs6457327	Follicular lymphoma	1.69[1.43-2.00]	**1.47[1.27-1.72]**
rs9298506	Intracranial aneurysm	1.35[1.22-1.49]	**1.28[1.20-1.38]**
rs6719884	Myasthenia gravis	1.35[1.19-1.53]	**1.35[1.19-1.52]**
rs7078160	Orofacial clefts	**1.38[1.21-1.58]**	1.36[1.21-1.53]
rs7590268	Orofacial clefts	1.42[1.23-1.64]	**1.42[1.26-1.59]**
rs8001641	Orofacial clefts	**1.35[1.14-1.61]**	1.31[1.13-1.51]
rs11782652	Ovarian cancer	**1.24[1.16-1.33]**	1.19[1.12-1.26]
rs3018362	Paget's disease	1.52[1.36-1.70]	**1.45[1.34-1.56]**
rs11672691	Prostate cancer	**1.11[1.02-1.20]**	1.08[1.04-1.12]
rs10488631	Systemic sclerosis	**1.50[1.35-1.67]**	1.35[1.20-1.51]
rs7583877	Type 1 diabetes nephropathy	1.29[1.18-1.40]	**1.29[1.17-1.42]**
rs11739663	Ulcerative colitis	**1.15[1.09-1.21]**	1.07[1.03-1.12]
rs4728142	Ulcerative colitis	**1.10[1.07-1.14]**	1.07[1.03-1.11]
rs6017342	Ulcerative colitis	**1.23[1.19-1.27]**	1.20[1.15-1.26]

Based on the previous steps, we have reduced the initial *data.frame* from 408 to 378 SNP/trait associations using the ABF. Having in mind that another 21% of SNP/trait association would not pass the strength of evidence criteria considering their ABF, our final *data.frame* would consist of only around 300 rows. These SNP/trait associations should pose a high enough reliability to be used downstream in any established form of multigenic risk assessment.

Table 2. Selected ORs and CIs based on the ABF from paris of studies targeting the SNP/trait associations in LD

SNP-proxy pairs	R^2	Trait	OR[CI] - 1	OR[CI] - 2
rs6666258 rs13376333	1	Atrial fibrillation	1.18[1.13-1.23]	**1.52[1.40-1.64]**
rs646776 rs599839	0.90	Coronary heart disease	1.14[1.09-1.19]	**1.11[1.08-1.15]**
rs10490924 rs3793917	0.95	Age-related macular degeneration	**3.67[3.33-4.05]**	3.4[2.94-3.94]
rs10801555 rs1061147	1	Age-related macular degeneration	**2.33[2.08-2.63]**	1.4[1.32-1.48]
rs1329424 rs1061147	0.93	Age-related macular degeneration	**1.88[1.68-2.10]**	1.4[1.32-1.48]
rs1219648 rs2981579	0.97	Breast cancer	1.32[1.22-1.42]	**1.27[1.24-1.29]**
rs10801555 rs1329424	0.93	Age-related macular degeneration	2.33[2.08-2.63]	**1.88[1.68-2.10]**
rs4474514 rs995030	0.86	Testicular cancer	**3.07[2.29-4.13]**	2.26[1.95-2.61]

4   Discussion

Our work focuses on the possibility of performing multigenic risk assessment using the raw genotype data from a DTC genomic service, in this case the 23andMe web service [18], and utilizing only the publicly and freely available databases and data analysis tools. We present an overview of the advantages of using the ABF over *p*-values for the selection of "significant" SNP/trait associations, and we illustrate its multipurposeness in developing the heuristic for including only the meaningful SNP/trait associations in multigenic risk assessment.

The flow-chart (see *Figure 1*) depicts how the number of useful SNP/trait associations rapidly reduces when using just the basic filters on descent, insist on replicated findings, and select only SNP/trait associations for qualitative traits (reporting ORs and CIs), even before checking for redundancies.

It is necessary to have in mind that all the SNP/trait associations reported in GWAS had been selected based on *p*-values only, and had passed stringent *p*-value criteria. Nevertheless, the use of *p*-values only has been characterized as problematic by many authors [14, 15], and an alternative has been offered: using $\log_{10}(ABF)$ and two cut-off criteria for the strength of evidence: $\log_{10}(ABF) > 5$ and $\log_{10}(ABF) > 0$ [15]. In *Figure 2* we demonstrate how the SNP/trait associations would rank under the ABF criteria, showing that more than 20% of SNP/trait associations would not satisfy the minimum criteria of $\log10(ABF) > 0$, even though their *p*-values were significant enough for reporting in GWAS. It could be claimed that these SNP/trait associations cannot be trusted enough to enter a multigenic risk score. Indeed, all the studies that had $\log_{10}(ABF) < 0$ were seriously overpowered, and the use of evidence from them in downstream analysis would be misleading, since, as argued above, such conclusion would not account for the adverse effect the high power has on the strength of evidence for the true association [14].

In the following step we examined the common situation where the same SNP/trait association is examined in different studies, each of these studies proposing different OR estimate. As shown in *Table 1*, when selecting which study to take into account for multigenic risk assessment, the answer is often not the one with larger OR. That is paradoxal, but only at first glance, since we should remember that ABF is influenced more by the precision of the estimate of OR, not as much by OR itself [17]. The situation is similar when selecting between studies where the original and a proxy SNP/trait association has been explored, as shown in *Table 2*. An important benefit in the method we use is the prevention of an overestimation of the "true" odds ratio.

The filtering and selection procedure (see *Figure 1*) has "approved" only around 300 SNP/trait association for the further downstream use in multigenic risk assessment. We must take into consideration the fact that, at this point, only SNPs with sufficiently low p-values are indexed in the NHGRI database. The threshold *p*-value has been decided by each study's investigators, meaning the SNP/trait associations available to us had already been pre-filtered. Since a standard procedure for the selection of a threshold value might not be well defined [9, 10], there is an immense loss of information, and this might prove critical for the downstream procedures that are relying on the large initial set of SNPs to work on, such as building predictive models [28].

Furthermore, the Authors are aware that the decision to work only with GWAS studies which had the string "European" in the population description field is questionable. Our subject was from the Balkans, and as pointed out [29], that might imply quite different referent minor allele frequencies, LD patterns as well as disease prevalence and corrupt validity of risk assessment. Still, facing the scarcity of studies from the Balkan peninsula, we compromised on this.

Another issue was that is that the only output readily available to us, in terms of effect sizes, was the allelic odds ratio. Since risk ratios are considered more intuitive and precise [30], and can be approximated with odds ratios only under the "rare disease" assumption, we would suggest the NHGRI database report also the risk ratios, as well as the number of cases and controls with a specific risk allele, which would reduce the potential error due to data mining. Until it becomes common practice, we propose the use of an available R package *orsk* [31] for RR estimation from incomplete GWAS based data, with perhaps additional MAF based constraints when choosing the optimal solution for RR.

Even with a well established method for the filtering and selection of SNP/trait associations supported by enough evidence to be included into a multigenic risk score, a question remains on how a trait is defined, as different GWAS can explore the association of SNPs with a "general" trait, or just a characteristic (a "subtrait") falling under the general trait in question. A prospective direction of improvement in conveying the risk information might be MeSH term enrichment analysis of the genotype profile as a whole, and some tools have already been developed for this aim [32, 33]. In this scenario, the proper selection of SNPs for which we are positive there is an association with a trait plays and even more crucial role.

In conclusion, the price of DTC genomic services has gone from several thousand down to several hundreds of dollars in the last 10 years, dropping even bellow 100$ in the past few years [1, 18], and it is easy to see the use of having a genetic report of a patient at your doctor's disposal. We have demonstrated the utility of the ABF in developing the criteria for selecting SNP/trait associations supported by strong enough evidence for the downstream multigenic risk assessment. Future steps to improve this process include the update of the information reported in the NHGRI database, using risk ratios instead of odds ratios, and MeSH term enrichment. An accurate genetic report is crucial to correctly interpret genetic information, but also to influence a range of health-promoting behaviors.

References

1. Mitchell, J.A., Fun, J., McCray, A.T.: Design of Genetics Home Reference: a new NLM consumer health resource. Journal of the American Medical Informatics Association 11(6), 439–447 (2004)
2. Goldstein, D.B.: Common genetic variation and human traits. N. Engl. J. Med. 360(17), 1696–1698 (2009)
3. Baptista, P.V.: Principles in genetic risk assessment. Ther. Clin. Risk Manag. 1(1), 15–20 (2005)
4. Regaldo, A.: How a Wiki Is Keeping Direct-to-Consumer Genetics Alive (2014), http://www.technologyreview.com/featuredstory/531461/how-a-wiki-is-keeping-direct-to-consumer-genetics-alive/
5. Sherry, S.T., et al.: dbSNP: the NCBI database of genetic variation. Nucleic Acids Research 29(1), 308–311 (2001)
6. Landrum, M.J., et al.: ClinVar: public archive of relationships among sequence variation and human phenotype. Nucleic Acids Research, gkt1113 (2013)
7. Yang, Q., et al.: Using lifetime risk estimates in personal genomic profiles: estimation of uncertainty. The American Journal of Human Genetics 85(6), 786–800 (2009)
8. Szoka, B. FDA Just Banned 23andMe's DNA Testing Kits, and Users Are Fighting Back (2013), http://www.huffingtonpost.com/berin-szoka/fda-just-banned-23andmes-_b_4339182.html
9. Jannot, A.-S., Ehret, G., Perneger, T.: P<5*10-8 has emerged as a standard of statistical significance for genome-wide association studies. Journal of Clinical Epidemiology (2015)
10. Zheng, G., Yuan, A., Jeffries, N.: Hybrid Bayes factors for genome-wide association studies when a robust test is used. Computational Statistics & Data Analysis 55(9), 2698–2711 (2011)
11. Zaykin, D.V., Zhivotovsky, L.A.: Ranks of Genuine Associations in Whole-Genome Scans. Genetics 171(2), 813–823 (2005)
12. Wang, L., et al.: Bayes Factor Based on a Maximum Statistic for Case-Control Genetic Association Studies. Journal of Agricultural, Biological, and Environmental Statistics 17(4), 568–582 (2012)
13. Kass, R.E., Raftery, A.E.: Bayes Factors. Journal of the American Statistical Association 90(430), 773–795 (1995)

14. Wakefield, J.: Bayes factors for genome-wide association studies: comparison with P-values. Genet. Epidemiol. 33(1), 79–86 (2009)
15. Sawcer, S.: Bayes factors in complex genetics. Eur. J. Hum. Genet. 18(7), 746–750 (2010)
16. Stephens, M., Balding, D.J.: Bayesian statistical methods for genetic association studies. Nat. Rev. Genet. 10(10), 681–690 (2009)
17. Wakefield, J.: A Bayesian measure of the probability of false discovery in genetic epidemiology studies. The American Journal of Human Genetics 81(2), 208–227 (2007)
18. 23andMe. 23andMe Web Service (2014), https://www.23andme.com/
19. Welter, D., et al.: The NHGRI GWAS Catalog, a curated resource of SNP-trait associations. Nucleic Acids Research 42(D1), D1001–D1006 (2014)
20. Carey, V.: gwascat: structuring and querying the NHGRI GWAS catalog (2013)
21. Wickham, H.: stringr: modern, consistent string processing. The R Journal 2(2), 38–40 (2010)
22. Burton, P.R., et al.: Genome-wide association study of 14,000 cases of seven common diseases and 3,000 shared controls. Nature 447(7145), 661–678 (2007)
23. Evangelou, E., Ioannidis, J.P.: Meta-analysis methods for genome-wide association studies and beyond. Nat. Rev. Genet. 14(6), 379–389 (2013)
24. Kraft, P., Zeggini, E., Ioannidis, J.P.: Replication in genome-wide association studies. Statistical Science: A Review Journal of the Institute of Mathematical Statistics 24(4), 561 (2009)
25. Johnson, A.D., et al.: SNAP: a web-based tool for identification and annotation of proxy SNPs using HapMap. Bioinformatics 24(24), 2938–2939 (2008)
26. Team, R.C.: R: A Language and Environment for Statistical Computing. R Foundation for Statistical Computing, Vienna, Austria (2012) ISBN 3-900051-07-0
27. Gentleman, R.C., et al.: Bioconductor: open software development for computational biology and bioinformatics. Genome Biology 5(10), R80 (2004)
28. Wei, Z., et al.: From disease association to risk assessment: an optimistic view from genome-wide association studies on type 1 diabetes. PLoS Genet. 5(10), e1000678 (2009)
29. Barrett, J.: Why prediction is a risky business (2010), http://genomesunzipped.org/2010/08/why-prediction-is-a-risky-business.php
30. Cummings, P.: The relative merits of risk ratios and odds ratios. Arch. Pediatr. Adolesc. Med. 163(5), 438–445 (2009)
31. Wang, Z.: Converting Odds Ratio to Relative Risk in Cohort Studies with Partial Data Information. Journal of Statistical Software 55(5) (2013)
32. Nikaido, I., Tsuyuzaki, K., Morota, G.: meshr: Tools for conducting enrichment analysis of MeSH. R package version 1.2.4
33. Tsuyuzaki, K., et al.: How to use MeSH-related Packages (2014)

Computational Inference in Systems Biology

Benn Macdonald and Dirk Husmeier

University of Glasgow, Department of Mathematics and Statistics,
Scotland, G12 8QW
b.macdonald.1@research.gla.ac.uk,
Dirk.Husmeier@glasgow.ac.uk

Abstract. Parameter inference in mathematical models of biological
pathways, expressed as coupled ordinary differential equations (ODEs),
is a challenging problem. The computational costs associated with repeatedly solving the ODEs are often high. Aimed at reducing this cost,
new concepts using gradient matching have been proposed. This paper
combines current adaptive gradient matching approaches, using Gaussian
processes, with a parallel tempering scheme, and conducts a comparative
evaluation with current methods used for parameter inference in ODEs.

Keywords: Parameter Inference, Ordinary Differential Equations, Gradient Matching, Parallel Tempering, Gaussian Processes.

1 Introduction

Ordinary differential equations (ODEs) have many applications in systems biology. Conventional inference methods involve numerically integrating the system
of ODEs, to calculate the likelihood as part of an iterative optimisation or sampling procedure. However, the computational costs involved with numerically
solving the ODEs are large. To reduce the computational complexity, several
authors have adopted an approach based on gradient matching (e.g. [1] and
[15]). The idea is based on the following two-step procedure. In a first, preliminary smoothing step, the time series data are interpolated; in a second step, the
kinetic parameters $\boldsymbol{\theta}$ of the ODEs are optimised so as to minimise some metric
measuring the difference between the slopes of the tangents to the interpolants,
and the $\boldsymbol{\theta}$-dependent time derivative from the ODEs. In this way, the ODEs never
have to be solved explicitly, and the typically unknown initial conditions are effectively profiled over. A disadvantage of this two-step scheme is that the results
of parameter inference critically hinge on the quality of the initial interpolant.
A better approach, first suggested in [13], is to regularise the interpolants by the
ODEs themselves. Dondelinger et al. [4] applied this idea to the nonparametric
Bayesian approach of [1], using Gaussian Processes (GPs), and demonstrated
that it substantially improves the accuracy of parameter inference and robustness with respect to noise. As opposed to [13], all smoothness hyperparameters
are consistently inferred in the framework of nonparametric Bayesian statistics,
dispensing with the need to adopt heuristics and approximations.

F. Ortuño and I. Rojas (Eds.): IWBBIO 2015, Part II, LNCS 9044, pp. 276–288, 2015.

We extend the work of [4] in two respects. Firstly, we combine adaptive gradient matching using GPs with a parallel tempering scheme for the gradient mismatch parameter. This is conceptually different from the inference paradigm of the mismatch parameter that [4] employs. If the ODEs provide the correct mathematical description of the system, ideally there should be no difference between the interpolant gradients and those predicted from the ODEs. In practise, however, forcing the gradients to be equal is likely to cause parameter inference techniques to converge to a local optimum of the likelihood. A parallel tempering scheme is the natural way to deal with such local optima, as opposed to inferring the degree of mismatch, since different tempering levels correspond to different strengths of penalising the mismatch between the gradients. Since our modelling process is created using a products of experts approach (see section 2), parallel tempering should work well on penalising the mismatch. A parallel tempering scheme was explored by Campbell & Steele [3], however, their approach uses a different methodological paradigm, and thus the results are not directly comparable to the GP approach in [4]. In this paper, we present for the first time, a comparative evaluation of parallel tempering versus inference in the context of gradient matching for the same modelling framework, i.e. without any confounding influence from the model choice. Secondly, we compare the method of Bayesian inference with Gaussian Processes with a variety of other methodological paradigms, within the specific context of adaptive gradient matching, which is highly relevant to current computational systems biology.

2   Methodology

Consider a set of T arbitrary time points $t_1 < ... < t_T$, and a set of noisy observations $\mathbf{Y} = (\mathbf{y}(t_1), ..., \mathbf{y}(t_T))$, where $\mathbf{y}(t) = \mathbf{x}(t) + \boldsymbol{\epsilon}(t) + \boldsymbol{\mu}$, $N = \dim(\mathbf{x}(t))$, $\mathbf{X} = (\mathbf{x}(t_1), ..., \mathbf{x}(t_T))$. The signals of the system are described by ordinary differential equations (ODEs), of the form

$$\mathbf{x}' = \frac{d\mathbf{x}(t)}{dt} = \mathbf{f}(\mathbf{x}(t) + \boldsymbol{\mu}, \boldsymbol{\theta}, t); \qquad \mathbf{x}(t_1) = \mathbf{x}_1 \qquad (1)$$

where $\boldsymbol{\theta}$ is a parameter vector of length r, $\boldsymbol{\mu}$ is a vector of integration constants, for simplicity set as the sample mean, and $\boldsymbol{\epsilon}$ is multivariate Gaussian noise, $\boldsymbol{\epsilon} \sim N(\mathbf{0}, \sigma_n^2 \mathbf{I})$. Then,

$$p(\mathbf{Y}|\mathbf{X}, \boldsymbol{\sigma}) = \prod_n \prod_t N(y_n(t)|x_n(t) + \mu_n, \sigma_n) \qquad (2)$$

and the matrices $\mathbf{X}$ and $\mathbf{Y}$ are of dimension N by T. Now let $\mathbf{x}_n$ and $\mathbf{y}_n$ be T-dimensional column vectors containing the nth row of $\mathbf{X}$ and $\mathbf{Y}$. Following [1], we place a Gaussian process (GP) prior on $\mathbf{x}_n$,

$$p(\mathbf{x}_n|\boldsymbol{\phi}) = N(\mathbf{x}_n|\mathbf{0}, \mathbf{C}_{\phi_n}) \qquad (3)$$

where $\mathbf{C}_{\phi_n}$ is a positive definite matrix of covariance functions with hyperparameters ϕ_n. Assuming additive Gaussian noise, with state-specific error variance σ_n^2, we get $p(\mathbf{y}_n|\mathbf{x}_n, \sigma_n) = N(\mathbf{y}_n|\mathbf{x}_n, \sigma_n^2 \mathbf{I})$, and

$$p(\mathbf{y}_n|\boldsymbol{\phi}_n, \sigma_n) = \int p(\mathbf{y}_n|\mathbf{x}_n, \sigma_n)p(\mathbf{x}_n|\boldsymbol{\phi}_n)d\mathbf{x}_n = N(\mathbf{y}_n|\mathbf{0}, \mathbf{C}_{\phi_n} + \sigma_n^2 \mathbf{I}) \quad (4)$$

The conditional distribution for the state derivatives is then

$$p(\mathbf{x_n}'|\mathbf{x_n}, \boldsymbol{\phi_n}) = N(\mathbf{m}_n, \mathbf{K}_n) \quad (5)$$

as the derivative of a GP is itself a GP, provided the kernel is differentiable [16], [1]. For closed form solutions to $\mathbf{m}_n$ and $\mathbf{K}_n$, see Rasmussen & Williams [10]. Assuming additive Gaussian noise with a state-specific error variance γ_n, from (1) we get

$$p(\mathbf{x}'_n|\mathbf{X}, \boldsymbol{\theta}, \gamma_n) = N(\mathbf{f}_n(\mathbf{X} + \boldsymbol{\mu}, \boldsymbol{\theta}), \gamma_n \mathbf{I}) \quad (6)$$

Dondelinger et al. [4] link the interpolant in (5) with the ODE model in (6) using a products of experts approach, obtaining a joint distribution

$$p(\mathbf{X}', \mathbf{X}, \boldsymbol{\theta}, \boldsymbol{\phi}, \boldsymbol{\gamma}) = p(\boldsymbol{\theta})p(\boldsymbol{\phi})p(\boldsymbol{\gamma}) \prod_n p(\mathbf{x}'_n|\mathbf{X}, \boldsymbol{\theta}, \boldsymbol{\phi}, \gamma_n)p(\mathbf{x}_n|\boldsymbol{\phi}_n) \quad (7)$$

They show that you can marginalise over the derivatives to get a closed form solution to

$$p(\mathbf{X}, \boldsymbol{\theta}, \boldsymbol{\phi}, \boldsymbol{\gamma}) = \int p(\mathbf{X}', \mathbf{X}, \boldsymbol{\theta}, \boldsymbol{\phi}, \boldsymbol{\gamma})d\mathbf{X}' \quad (8)$$

Using (2) and (8), our full joint distribution becomes

$$p(\mathbf{Y}, \mathbf{X}, \boldsymbol{\theta}, \boldsymbol{\phi}, \boldsymbol{\gamma}, \boldsymbol{\sigma}) = p(\mathbf{Y}|\mathbf{X}, \boldsymbol{\sigma})p(\mathbf{X}|\boldsymbol{\theta}, \boldsymbol{\phi}, \boldsymbol{\gamma})p(\boldsymbol{\theta})p(\boldsymbol{\phi})p(\boldsymbol{\gamma})p(\boldsymbol{\sigma}) \quad (9)$$

where Dondelinger et al. [4] show

$$p(\mathbf{X}|\boldsymbol{\theta}, \boldsymbol{\phi}, \boldsymbol{\gamma}) = \frac{1}{\prod_n |2\pi(\mathbf{K}_n + \gamma_n \mathbf{I})|^{\frac{1}{2}}} \exp[-\frac{1}{2}\sum_n$$
$$(\mathbf{x}_n^T \mathbf{C}_{\phi_n} \mathbf{x}_n + (\mathbf{f}_n - \mathbf{m}_n)^T (\mathbf{K}_n + \gamma_n \mathbf{I})^{-1}(\mathbf{f}_n - \mathbf{m}_n))] \quad (10)$$

$p(\mathbf{Y}|\mathbf{X}, \boldsymbol{\sigma})$ is described in equation (2) and $p(\boldsymbol{\theta}), p(\boldsymbol{\phi}), p(\boldsymbol{\gamma}), p(\boldsymbol{\sigma})$ are the priors over the respective parameters. We use the same MCMC sampling scheme as [4], which uses the whitening approach of [20], to efficiently sample in the joint space of GP hyperparameters $\boldsymbol{\phi}$ and latent variables $\mathbf{X}$.

Parallel Tempering: The gradient matching approach is based on the intrinsic slack parameter γ (see equation (6)), which theoretically should be $\gamma = 0$, since this corresponds to no mismatch between the gradients. In practise, it is allowed to take on larger values, $\gamma > 0$, to prevent the inference scheme from getting stuck in sub-optimal states. However, rather than inferring γ like a model parameter, as carried out in [4], γ should be gradually set to $\gamma \to 0$. To this end we combine our gradient matching with Gaussian processes with the tempering approach in [3] (for details on tempering: [17], [18]) and temper this parameter to zero. Consider a series of "temperatures", $0 = \beta_1 < \ldots < \beta_M = 1$ and a power posterior distribution of our ODE parameters [14]

$$p_{\beta_i}(\boldsymbol{\theta}_i|\mathbf{y}) \propto p(\boldsymbol{\theta}_i)p(\mathbf{y}|\boldsymbol{\theta}_i)^{\beta_i} \quad (11)$$

(11) reduces to the prior for $\beta_i = 0$, and becomes the posterior when $\beta_i = 1$, with $0 < \beta_i < 1$ creating a distribution between our prior and posterior. The M approximations are used as the target densities of M parallel MCMC chains [3]. At each MCMC step, each chain independently performs a Metropolis-Hastings step to update $\boldsymbol{\theta}_i$. Also at each MCMC step, two chains are randomly selected, and a proposal to exchange parameters is made, with acceptance probability: $p_{swap} = min(1, \frac{p_{\beta_j}(\boldsymbol{\theta}_i|\mathbf{y})p_{\beta_i}(\boldsymbol{\theta}_j|\mathbf{y})}{p_{\beta_i}(\boldsymbol{\theta}_i|\mathbf{y})p_{\beta_j}(\boldsymbol{\theta}_j|\mathbf{y})})$. We now choose values of γ and fix them in place, associated with each β_i, for different chains, such that chains closer to the prior allow the gradients from the interpolant to have more freedom to deviate from those predicted by the ODEs, chains closer to the posterior to more closely match the gradients, and for β_M, we wish that the mismatch is approximately zero. Since γ corresponds to the variance of our state-specific error variance (see (6)), as $\gamma \to 0$, we have less mismatch between the gradients, and as γ gets larger, the gradients have more freedom to deviate from one another. Hence, we temper γ towards zero. Then each chain has a β_i for tempering of the power posterior and a γ_i for the gradient mismatch. For schedules, see table 3.

3   Alternative Methods for Comparison

We have carried out a comparative evaluation of the proposed scheme with various state-of-the-art gradient matching methods. These methods are based on different statistical modelling and inference paradigms: non-parametric Bayesian statistics with Gaussian processes without tempering, penalised regression splines, splines-based smooth functional tempering, and penalised likelihood based on reproducing kernel Hilbert spaces. Since many methods and settings are used in this paper for comparison purposes, for ease of reading, abbreviations are used. Table 1 is a reference for those methods and an overview of the methods is given below.

Table 1. Abbreviations of the methods used throughout this paper

Abbreviation	Method	Reference
C&S	Tempered mismatch parameter using splines-based smooth functional tempering (SFT)	Campbell & Steele [3]
GON	Reproducing kernel Hilbert space and penalised likelihood	González et al. [11]
INF	Inference of the gradient mismatch parameter using GPs	Dondelinger et al. [4]
LB2	Tempered mismatch parameter using GPs in Log Base 2 increments	Our method
LB10	Tempered mismatch parameter using GPs in Log Base 10 increments	Our method
RAM	Penalised splines & 2nd derivative penalty	Ramsay et al. [13]

INF [4]: This method conducts parameter inference using adaptive gradient matching and Gaussian processes. The penalty mismatch parameter γ is inferred rather than tempered. **GON** [11]: Parameter inference is conducted in a non-Bayesian fashion, implementing a reproducing kernel Hilbert space (RKHS) and penalised likelihood approach. Comparisons between RKHS and GPs have been previously explored (for example, see [10], [19]) conceptually, and in this paper we analyse

Table 2. Particular settings of Campbell & Steele's [3] method

Abbreviation	Definition	Details
10C	10 Chains	When comparing our methods, it was of interest to see how the performance depended on the number or parallel MCMC chains, as originally the authors used 4 chains.
Obs20	20 Observations	Originally, the authors use 401 observations. We reduced this to a dataset size more usual with these types of experiments to observe the dependency of the methods on the amount of data.
15K	15 Knots	The method in C&S uses B-splines interpolation. We changed the original tuning parameters from the author's paper to observe the sensitivity of the parameter estimation by these tuning parameters.
P3	Polynomial order 3 (Cubic Spline)	The original polynomial order is 5 and again, we wanted to observe the sensitivity of the parameter estimation by these tuning parameters.

Table 3. Ranges of the penalty parameter γ for LB2 and LB10

Method	Chains	Range of Penalty γ	Method	Chains	Range of Penalty γ
LB2	4	[1 , 0.125]	LB10	4	[1 , 0.001]
LB2	10	[1 , 0.00195]	LB10	10	[1 , $1e^{-9}$]

this empirically in the specific context of gradient matching. The RKHS gradient matching method in [11] involves linearising the ODEs and obtaining a gradient using a differencing operator. **RAM** [13]: This method uses a non-Bayesian optimisation process for parameter inference. They use 2 penalties: the 2nd derivative of the interpolant (penalising by the curvature of the interpolant to avoid overfitting) and the difference between the gradients (using penalised splines). **C&S** [3]: Parameter inference is carried out using adaptive gradient matching and tempering of the mismatch parameter. The choice of interpolation scheme is B-splines. Table 2 outlines particular settings with some of the methods in table 1. The ranges of the penalty parameter for γ, for LB2 and LB10 methods are given in table 3. The increments are equidistant on the log scale. The M β_is from 0 to 1 are set, by taking a series of equidistant M values and raising them to the power 5 [14].

4   Data

Fitz-Hugh Nagumo ([5], [8]): These equations model the voltage potential across the cell membrane of the axon of giant squid neurons. There are two "species": Voltage (V) and Recovery variable (R), and 3 parameters; α, β and ψ. Species in [ ] denote the time-dependent concentration for that species:

$$\dot{V} = \psi([V] - \frac{[V]^3}{3} + [R]); \quad \dot{R} = -\frac{1}{\psi}([V] - \alpha + \beta * [R]) \tag{12}$$

Protein Signalling Transduction Pathway [12] : These equations model protein signalling transduction pathways [12] in a signal transduction cascade, where the free parameters are kinetic parameters governing how quickly the proteins ("species") convert to one another. There are 5 "species" (S, dS, R, RS, Rpp) and 6 parameters ($k_1, k_2, k_3, k_4, V, K_m$). The system describes the phosphorylation of a protein, $R \rightarrow Rpp$ (equation (17)), catalysed by an enzyme S, via an active protein complex (RS, equation (16)), where the enzyme is subject to degradation ($S \rightarrow dS$, equation (14)). The chemical kinetics are described by a combination of mass action kinetics (equations (13), (14) and (16)) and Michaelis-Menten kinetics (equations (15) and (17)). Species in [ ] denote the time-dependent concentration for that species:

$$\dot{S} = -k_1 * [S] - k_2 * [S] * [R] + k_3 * [RS] \tag{13}$$

$$\dot{dS} = k_1 * [S] \tag{14}$$

$$\dot{R} = -k_2 * [S] * [R] + k_3 * [RS] + \frac{V * [Rpp]}{K_m + [Rpp]} \tag{15}$$

$$\dot{RS} = -k_2 * [S] * [R] - k_3 * [RS] - k_4 * [RS] \tag{16}$$

$$\dot{Rpp} = k_4 * [RS] - \frac{V * [Rpp]}{K_m + [Rpp]} \tag{17}$$

5   Simulation

We have compared the proposed GP tempering scheme with the alternative methods summarised in Section 3. For those methods for which we were unable to obtain the software from the authors ([11] and [13]), we compared our results directly with the results from the original publications. To this end, we generated test data in the same way as described by the authors and used them for the evaluation of our method. For methods for which we did receive the authors' software ([3] and [4]), we repeated the evaluation twice, first on data equivalent to those used in the original publications, and again on new data generated with different (more realistic) parameter settings. For comparisons with other Bayesian methods, we used the authors' specifications for the priors on the ODE parameters. For comparisons with non-Bayesian methods, we applied our method with the parameter prior from [3], since the ODE model was the same. Our software is available upon request.

Reproducing Kernel Hilbert Space Method [11]: They tested their method on the Fitz-Hugh Nagumo data (see section 4) with the following parameters: $\alpha = 0.2; \beta = 0.2$ and $\psi = 3$. Starting from initial values of $(-1, -1)$ for the two "species", the authors generated 50 timepoints over the time course $[0, 20]$, producing 2 periods, with iid Gaussian noise (sd $= 0.1$) added. 50 independent data sets were generated in this way.

Penalised Splines & 2nd Derivative Penalty Method [13]: This method was included in the study by [11], and we have used the results from their paper. For comparison, our method was applied in the same way as for the comparison with [11].

Tempered Mismatch Parameter Using Splines-Based Smooth Functional Tempering [3]: They tested their method on the Fitz-Hugh Nagumo system with the following parameter settings: $\alpha = 0.2$, $\beta = 0.2$ and $\psi = 3$, starting from initial values of $(-1, 1)$ for the two "species". 401 observations were simulated over the time course $[0, 20]$ (producing 2 periods) and Gaussian noise was added with sd $\{0.5, 0.4\}$ to each respective "species". In inferring the ODE parameters with their model, the authors chose the following settings: splines of polynomial order 5 with 301 knots; four parallel tempering chains associated with gradient mismatch parameters $\{10,100,1000,10000\}$; parameter prior distributions for the ODE parameters: $\alpha \sim N(0, 0.4^2)$, $\beta \sim N(0, 0.4^2)$ and $\psi \sim \chi_2^2$.

In addition to comparing our method with the results the authors had obtained with their settings, we made the following modifications to test the robustness of their procedure with respect to these (rather arbitrary) choices. We reduced the number of observations from 401 to 20 over the time course $[0, 10]$ (producing 1 period) to reflect more closely the amount of data typically available from current systems biology projects. For these smaller data sets, we reduced the number of knots for the splines to 15 (keeping the same proportionality of knots to data points as before), and we tried a different polynomial order: 3 instead of 5. Due to the high computational costs of their method (roughly $1\frac{1}{2}$ weeks for a run), we could only repeat the MCMC simulations on 3 independent data sets. The respective posterior samples were combined, to approximately marginalise over data sets and thereby remove their potential particularities. For a fair comparison, we mimicked the authors' tempering scheme and only applied our method with 4 rather than 10 chains.

Inference of the Gradient Mismatch Parameter Using GPs and Adaptive Gradient Matching [4]: We applied the method in the same way as described in [4], using the authors' software and selecting the same kernels and parameter/hyperparameter priors as for the method proposed in the present paper. Data were generated from the protein signal transduction pathway described in Section 4. We applied our methods to data simulated from the same system, with the same settings as in [4]; ODE parameters: $(k_1 = 0.07, k_2 = 0.6, k_3 = 0.05, k_4 = 0.3, V = 0.017, K_m = 0.3)$; initial values of the species: $(S = 1, dS = 0, R = 1, RS = 0, Rpp = 0)$; 15 time points covering one period, $\{0, 1, 2, 4, 5, 7, 10, 15, 20, 30, 40, 50, 60, 80, 100\}$. Following [4], we added multiplicative iid Gaussian noise of standard deviation $\{0.1\}$ to all observations. For Bayesian inference, we chose the same gamma prior on the ODE parameters as used in [4], namely $\Gamma(4, 0.5)$. For the GP, we used the same kernel; see below for details. In addition to this ODE system, we also applied this method to the set-ups previously described for the Fitz-Hugh Nagumo model.

Choice of Kernel: For the GP, we need to choose a suitable kernel, which defines a prior distribution in function space. Two kernels were considered in our study (to match the authors' set-ups in [4]), the radial basis function (RBF) kernel

$$k(t_i, t_j) = \sigma^2_{RBF} exp(-\frac{(t_i - t_j)^2}{2l^2}) \tag{18}$$

with hyperparameters σ^2_{RBF} and l^2, and the sigmoid variance kernel

$$k(t_i, t_j) = \sigma^2_{sig} arcsin \frac{a + (bt_i t_j)}{\sqrt{(a + (bt_i t_i) + 1)(a + (bt_j t_j) + 1)}} \tag{19}$$

with hyperparameters σ^2_{sig}, a and b [10].

To choose initial values for the hyperparameters, we fit a standard GP regression model (i.e. without the ODE part) using maximum likelihood. We then inspect the interpolant to decide whether it adequately represents our prior knowledge. For the data generated from the Fitz-Hugh Nagumo model, we found that the RBF kernel provides a good fit to the data. For the protein signalling transduction pathway, we found that the non-stationary nature of the data is not represented properly with the RBF kernel, which is stationary [10], in confirmation of the findings in [4]. Following [4], we tried the sigmoid variance kernel, which is non-stationary [10] and we found this provided a considerably improved fit to the data. **Other settings:** Finally, the values for our variance mismatch parameter of the gradients, γ, needs to be configured. Log_2 and Log_{10} increments were used (with an initial start at 1), since studies that indicate reasonable values for our technique are limited (see [1], [14]). All parameters were initialised with a random draw from the respective priors (apart from GON, which did not use priors. For details of their technique, see [11]).

6   Results

Reproducing Kernel Hilbert Space [11] and **Penalised Splines & 2nd Derivative Penalty Methods** [13]: For this configuration, to judge the performance of our methods, we used the same concept as in GON to examine our results. For each parameter, the absolute value of the difference between an estimator and the true parameter ($|\hat{\theta}_i - \theta_i|$) was computed and the distribution across the datasets were examined. For the LB2, LB10 and INF methods, the median of the sampled parameters was used as an estimator, since it is a robust estimator. Looking at figure 1 left, the LB2, LB10 and INF methods, do as well as the GON method, for 2 parameters (INF doing slightly worse for ψ) and outperform it for 1 parameter. All methods outperform the RAM method.

Tempered Mismatch Parameter Using Splines-Based Smooth Functional Tempering [3]: The C&S method shows good performance over all parameters in the one case where the number of observations is 401, the number of knots is 301 and the polynomial order is 3 (cubic spline), since the bulk of the distributions of the sampled parameters surround the true parameters in figures 1 right and 2 right and are close to the true parameter in figure 2 left. However, these settings require a great deal of "hand-tuning" or time expensive cross-validation and would be very difficult to set when using real data. The sensitivity of the splines based method can be seen in the other settings, where the results deteriorate. It

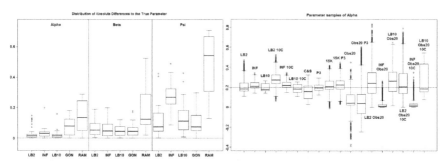

Fig. 1. Left: Boxplots of absolute differences to the true parameter over 50 datasets. The three sections from left to right represent the parameters α, β and ψ from equations (12). Within each section, the boxplots from left to right are: LB2 method, INF method, LB10 method, GON's method (boxplot reconstructed from [11]) and RAM's method (boxplot reconstructed from [11]). **Right:** Distributions of sampled Alphas from equation (12) over 3 datasets. From left to right: LB2, INF, LB10, LB2 10C, INF 10C, LB10 10C, C&S, C&S P3, C&S 15K, C&S 15K P3, C&S Obs20, C&S Obs20 P3, LB2 Obs20, INF Obs20, LB10 Obs20, LB2 Obs20 10C, INF Obs20 10C and LB10 Obs20 10C. The solid line is the true parameter. For definitions, see tables 1 and 2.

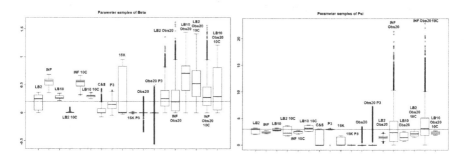

Fig. 2. Left: Distributions of sampled Betas from equation (12) over 3 datasets. From Left to right: LB2, INF, LB10, LB2 10C, INF 10C, LB10 10C, C&S, C&S P3, C&S 15K, C&S 15K P3, C&S Obs20, C&S Obs20 P3, LB2 Obs20, INF Obs20, LB10 Obs20, LB2 Obs20 10C, INF Obs20 10C and LB10 Obs20 10C. The solid line is the true parameter. **Right:** Distributions of sampled Psis from equations (12) over 3 datasets. From Left to right: LB2, INF, LB10, LB2 10C, INF 10C, LB10 10C, C&S, C&S P3, C&S 15K, C&S 15K P3, C&S Obs20, C&S Obs20 P3, LB2 Obs20, INF Obs20, LB10 Obs20, LB2 Obs20 10C, INF Obs20 10C and LB10 Obs20 10C. The solid line is the true parameter. For definitions, see tables 1 and 2.

is also important to note that when the dataset size was reduced, the cubic spline performed very badly. This inconsistency makes these methods very difficult to apply in practise. The LB2, LB10 and INF methods consistently outperform the C&S method with distributions overlapping or being closer to the true parameters. On the set-up with 20 observations, for 4 chains and 10 chains, the INF method produced largely different estimates over the datasets, as depicted by the wide boxplots and long tails. The long tails in all these distributions are due to the combination of estimates from different datasets.

Inference of the Gradient Mismatch Parameter Using GPs, Adaptive Method [4]: In order to see how our tempering method performs in comparison to the INF method, we can examine the results from the protein signalling transduction pathway (see section 4), as well as comparing how each method did in the previous set-ups. The distributions of parameter estimates minus the true values for the protein signalling transduction pathway are shown in figure 3 left. The author's code was unable to converge properly for some of the datasets, so in order to present a clear indication of the methods' performance, we show the distributions across the dataset that showed the median parameter estimation, as determined by root mean square of the parameter samples.

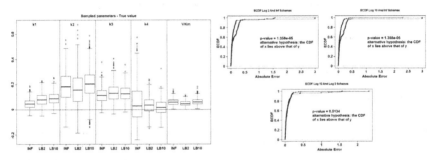

Fig. 3. Left: Average performance of the parameter inference for the INF, LB2 and LB10 methods. The distributions are of the sampled parameters from equations (13)-(17) minus the true value. The horizontal line shows zero difference. **Right:** ECDFs of the absolute errors of the parameter estimation. Top left - ECDFs for LB10 and INF, Top right - ECDFs for LB2 and INF and Bottom left - ECDFs for LB10 and LB2. Included are the p-values for 2-sample, 1-sided Kolmogorov-Smirnov tests. For definitions, see tables 1 and 2.

By examining figure 3 left, we can see that for each parameter, the distributions are close to the true values and so the methods are performing reasonably. Overall, there does not appear to be a significant difference between the INF, LB2 and LB10 methods for this model.

For the set-up in [11] and [13]: Figure 3 right shows the Expected Cumulative Distribution Functions (ECDFs) of the absolute errors of the parameter samples for the tempering and inference schemes. P-values for 2-sample, 1-sided Kolmogorov-Smirnov tests. If a distribution's ECDF is significantly higher than anothers, this constitutes better parameter estimation, since the distributions are of the average error. A higher curve means that more values are located in the lower range of absolute error.

Figure 3 right shows that both the LB2 and LB10 methods outperform the INF method, shown by p-values of less than the standard significance level of 0.05. Therefore we conclude that the CDFs for LB2 and LB10 are significantly higher than those for INF. Since we are dealing with absolute error, this means that the parameter estimates from the LB2 and LB10 methods are closer to the true parameters than the INF method. The LB2 and LB10 method show no significant difference to each other.

For the set-up in [3]: The LB2 and LB10 methods do well over all the parameters and dataset sizes, with most of the mass of the distributions surrounding or being situated close to the true parameters. The LB2 does better than the LB10 for 4 parallel chains (distributions overlapping the true parameter for all three parameters) and the LB10 outperforms the LB2 for 10 parallel chains (distribution overlapping true parameter in figure 1 right, being closer to the true parameter in figure 2 left and narrower and more symmetric around the true parameter in figure 2 right). The INF method's bulk of parameter sample distributions are located close to the true parameters for all dataset sizes. However, the method produces less uncertainty at the expense of bias. When reducing the dataset to 20 observations, for 4 and 10 chains, the inference deteriorates and is outperformed by the LB2 and LB10 methods. This could be due to the parallel tempering scheme constraining the mismatch between the gradients in chains closer to the posterior, allowing for better estimates of the parameters.

7 Discussion

We have proposed a modification of a recently proposed gradient matching approach for systems biology (INF), and we have carried out a comparative evaluation of various state-of-the-art gradient matching methods. These methods are based on different statistical modelling and inference paradigms: non-parametric Bayesian statistics with Gaussian processes (INF, LB2, LB10), penalised regression splines (RAM), splines-based smooth functional tempering (C&S), and penalised likelihood based on reproducing kernel Hilbert spaces (GON). We have also compared the antagonistic paradigms of Bayesian inference (INF) versus parallel tempering (LB2, LB10) of slack parameters in the specific context of adaptive gradient matching. The GON method produces estimates that are close the true parameters in terms of absolute uncertainty. This however, was for the case with small observational noise (Gaussian iid noise sd = 0.1) and it would be interesting to see how the parameter estimation accuracy is affected by the increase of noise. The C&S method does well only in the one case, where the number of observations are very high (higher than what would be expected in these types of experiments) and the tuning parameters are finely adjusted (which in practise is very difficult and time-consuming). When the number of observations was reduced, all settings for this method deteriorated significantly. It is important also to note that the settings that we found to be optimal were slightly different than in the original paper, which highlights the sensitivity and unreliability in the splines based method. The INF method shows a reasonable performance in terms of consistently producing results close to the true parameters, across all the set-ups we have examined. However, this technique's decrease in uncertainty is at the expense of bias. The LB2 and LB10 methods show the best performance across the set-ups. The parameter accuracy is unbiased across the different ODE models and the different settings of those models. The parallel tempering seems to be quite robust, performing similarly across the various set-ups. We have explored four different schedules for the parallel tempering scheme (as shown in table 3). Overall, the performance of parallel tempering has been found to be reasonably robust with respect to a variation of the schedule.

References

1. Calderhead, B., Girolami, M.A., Lawrence, N.D.: Accelerating Bayesian inference over non-linear differential equations with Gaussian processes. Neural Information Processing Systems (NIPS), 22 (2008)
2. Calderhead, B.: A study of Population MCMC for estimating Bayes Factors over nonlinear ODE models. University of Glasgow (2008)
3. Campbell, D., Steele, R.J.: Smooth functional tempering for nonlinear differential equation models. Stat. Comput. 22, 429–443 (2012)
4. Dondelinger, F., Filippone, M., Rogers, S., Husmeier, D.: ODE parameter inference using adaptive gradient matching with Gaussian processes. In: The 16th Internat. Conf. on Artificial Intelligence and Statistics (AISTATS). JMLR, vol. 31, pp. 216–228 (2013)
5. FitzHugh, R.: Impulses and physiological states in models of nerve membrane. Biophys. J. 1, 445–466 (1961)
6. Lawrence, N.D., Girolami, M., Rattray, M., Sanguinetti, G.: Learning and Inference in Computational Systems Biology. MIT Press, Cambridge (2010)
7. Lotka, A.: The growth of mixed populations: two species competing for a common food supply. Journal of the Washington Academy of Sciences 22, 461–469 (1932)
8. Nagumo, J.S., Arimoto, S., Yoshizawa, S.: An active pulse transmission line simulating a nerve axon. Proc. Inst. Radio Eng. 50, 2061–2070 (1962)
9. Pokhilko, A., Fernandez, A.P., Edwards, K.D., Southern, M.M., Halliday, K.J., Millar, A.J.: The clock gene circuit in Arabidopsis includes a repressilator with additional feedback loops. Molecular Systems Biology 8, 574 (2012)
10. Rasmussen, C.E., Williams, C.K.I.: Gaussian Processes for Machine Learning. The MIT Press (2006)
11. González, J., Vujačić, I., Wit, E.: Inferring latent gene regulatory network kinetics. Statistical Applications in Genetics and Molecular Biology 12(1), 109–127 (2013)
12. Vyshemirsky, V., Girolami, M.A.: Bayesian ranking of biochemical system models. Bioinformatics 24(6), 833–839 (2008)
13. Ramsay, J.O., Hooker, G., Campbell, D., Cao, J.: Parameter estimation for differential equations: a generalized smoothing approach. J. R. Statist., 741–796 (2007)
14. Friel, N., Pettitt, A.N.: Marginal likelihood estimation via power posteriors. J. Royal Statist. Soc.: Series B (Statistical Methodology) 70, 589–607 (2008)
15. Liang, H., Wu, H.: Parameter Estimation for Differential Equation Models Using a Framework of Measurement Error in Regression Models. J. Am. Stat. Assoc., 1570–1583 (December 2008)
16. Solak, E., Murray-Smith, R., Leithead, W.E., Leith, D.J., Rasmussen, C.E.: Derivative observations in Gaussian Process Models of Dynamic Systems. Advances in Neural Information Processing Systems, 9–14 (2003)

17. Mohamed, L., Calderhead, B., Filippone, M., Christie, M., Girolami, M.: Population MCMC methods for history matching and uncertainty quantification. Comput Geosci., 423–436 (2012)
18. Calderhead, B., Girolami, M.: Estimating Bayes Factors via thermodynamic integration and population MCMC. Comp. Stat. & Data Analysis. 53, 4028–4045 (2009)
19. Murphy, K.P.: Machine Learning. A Probabilistic Perspective. MIT Press (2012)
20. Murray, I., Adams, R.: Slice sampling covariance hyperparameters of latent Gaussian models. Advances in Neural Information Processing Systems (NIPS) 23 (2010)

A Price We Pay for Inexact Dimensionality Reduction

Sarunas Raudys[1], Vytautas Valaitis[1], Zidrina Pabarskaite[1], and Gene Biziuleviciene[1,2]

[1] Vilnius University, Faculty of Mathematics and Informatics, Lithuania
[2] State Research Institute, Centre for Innovative Medicine, Vilnius, Lithuania

Abstract. In biometrical and biomedical pattern classification tasks one faces high dimensional data. Feature selection or feature extraction is necessary. Accuracy of both procedures depends on the data size. An increase in classification error caused by employment of sample based K-class linear discriminant analysis for dimensionality reduction was considered both analytically and by simulations. We derived analytical expression for expected classification error by applying statistical analysis. It was shown theoretically that with an increase in the sample size, classification error of $(K-1)$-dimensional data decreases at first, however, later it starts increasing. The maximum is reached when the size of K class training sets, n, approaches dimensionality, p. When $n > p$, classification error decreases permanently. The peaking effect for real world biomedical and biometric data sets is demonstrated. We show that regularisation of the within-class scattering can reduce or even extinguish the peaking effect.

Keywords: dimensionality reduction, complexity, sample size, biometrics, linear discriminant analysis.

1 Introduction

One nasty problem pretty often encountered in biometrics and bioinformatics data mining is the overabundance of features or attributes. Over and over again the number of features, p, is much larger than a training set size, N_1, N_2, ..., N_K (K denotes the number of pattern classes). Few examples of this type of problem are: genomics, microarray, and hyper-spectral biomedical images analysis, speech and face recognition. In such situations one needs either to select the most important features or to perform feature extraction or to use very simple classification rules.

Among the first methods employed in pattern classification the linear discriminant analysis (LDA) [1] remains the most popular one. While seeking the best linear mapping of p-D (dimensional) vectors into the $(K-1)$-D space, this method tries to maximize a scattering between K sample mean vectors of the classes. At the same time, the algorithm fixes a within-class scattering of training vectors and keeps it constant (see Fig. 1a; for details see Section 2). Fig. 1a shows three class 30-D (dimensional) chromosome data (to be discussed latter) mapped into a 2-D space by this method. We see three clusters (diverse chromosomes, the pattern classes) tinted in diverse colors. Radial lines between mean vectors (marked as squares) and vectors of each solitary class portray scattering of training vectors within one class. Fig. 1a shows perfect

F. Ortuño and I. Rojas (Eds.): IWBBIO 2015, Part II, LNCS 9044, pp. 289–300, 2015.

separation of the classes. In this example we had $N=17$ training vectors within one pattern class and $p=30$ geometrical features of the chromosomes. Small size of the training data explains reasons of *apparent* good separation of the classes. Thus, if parameters of LDA data mapping rule are estimated from the training set, accuracy of feature extraction declines. Fig. 1b shows much worse 2-D separation of the 483+483+483 test vectors when for LDA training we used 17 vectors from each class. If all, 500+500+500, vectors available in our research were used for developing the rule to map the data from 30-D space into the 2-D space, reliability of the LDA rule would be much higher (Fig. 1c). Thus, Fig. 1a demonstrates too optimistic result which is not confirmed by the test.

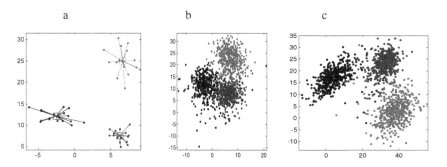

Fig. 1. Mapping of 30-dimensional chromosome data into two dimensional space

Generally, the LDA performs well when data is Gaussian, the classes share common covariance matrix (CM) Σ, the pattern classes are well separated, and parameters of the feature distribution are known. Any decline from these assumptions worsens the LDA. For practitioners it is important to know a cost we have to pay when the data is transformed into the low dimensional space inexactly. In the finite sample size situation we obtain a decline in data mapping accuracy (Fig.1b).

Small sample issues in feature extractions algorithms remain non-investigated up to now. For that reason, a number of improvements have been considered while developing multitude of novel approaches (see [2-12] and references therein).

Analysis of the data mapping papers mentioned advocates that new dimensionality reduction methods outperform existing ones in the experiments reported in the papers. Unfortunately, only minority of researchers paid attention to the sample size/dimensionality issues while reporting their results. Development of the new methods is helpful. For a better understanding of small samples problems one needs to analyze feature extraction accuracy analytically. It is the main objective of the present paper.

The quality of the feature extraction methods can be evaluated *by the increase in classification error rate evaluated after dimensionality reduction*. We will follow this idea.

In pattern recognition, relationships between the classifiers' complexity, dimensionality, training sample size and classification error were considered for a number of classifiers [13, 14, 15]. It was demonstrated that in two class pattern classification tasks an increase in classification error depends on the dimensionality sample size ratio, p/N, if only mean vectors of the classes are used to make classification rule. Additional estimation of the p variances, σ_1, σ_2, ..., σ_p, does not affect classification error asymptotically, if $p \to \infty$, $N \to \infty$, $p/N \to constant$ if it is supposed the σ_1, σ_2, ..., σ_p, are the same for all pattern classes. If we refuse from latter assumption, then an increase in classification error depends on the ratio $2p/N$. In addition, when we estimate common for both classes CM, we have a supplementary term, $2N/(2N-p)$, that affects an increase in the classification error [13]. In mean square regression we have a term $N/(N-p)$ that affects an increase in prediction error. In financial investments portfolio management, we have two terms to analyze. One of them, ratio p/N, is responsible for the decline of sample based portfolio profit due to the inaccurate estimation of p-dimensional vector of returns. Another term, $N/(N-p)$, is responsible for inaccurate estimation of $p \times p$ - D (dimensional) matrix of correlations between p components of the portfolio. Asymptotically, when $(p \to \infty$, $N \to \infty$, $p/N \to constant$, estimation of p variances does not worsen the portfolio profit [16].

In feature selection analysis it was shown both theoretically and experimentally that while ranking the features according to sample based hold-out classification error estimates, the best feature (or the feature subset) is not the first one in the ranking row. On average, the best features can be the 10-th, the 100-th or even the 1000-th in the row. To use this theoretical result for finding the interval of the best features in the row one needs to know the distribution function of true classification errors of the features or feature subsets [17]. Nevertheless, knowledge of this fact can be useful for practitioners.

Unfortunately, we do not have similar results for feature extraction algorithms used in pattern recognition tasks. Present paper analyzes sample size influence on accuracy of standard LDA used in multi-category situations. Accuracy is measured in terms of an increase in classification error due to imperfect feature extraction.

2   LDA, a Case When Sample Size Exceeds Dimensionality

In multi-category case, standard LDA maximizes a scattering between K sample mean vectors, $\bar{x}_i$ ($i = 1, 2, \ldots , K$) of K pattern classes. At the same time, the algorithm fixes a within-class scattering and keeps it constant. Actually, one maximizes ratio [1]

$$W = \underset{w}{\arg\max} \left(\left| w' s_b w \right| / \left| w' s_w w \right| \right), \tag{1}$$

where W stands for $p \times (K-1)$ - D (dimensional) transformation matrix. Here from the p-D original feature vector-column X, one obtains $(K-1)$ – dimensional vector $Y = X'W$,

$S_b = (K-1)^{-1}\sum_{i=1}^{K}(\overline{X}_i - \overline{X})(\overline{X}_i - \overline{X})'$ stands for the "between classes sample covariance matrix", $\overline{X}$ stands for a mean vector of K class training data, $\overline{X}_i$ stands for the i-th class sample estimate of the mean vector,

$$S_w = (K(n-1))^{-1}\sum_{i=1}^{K}\sum_{j=1}^{N_i}(\overline{X}_i - X_{ij})(\overline{X}_i - X_{ij})', \tag{2}$$

stands for the within-classes scatter (covariance) matrix, X_{ij} represents j-th training vector-column of i-th pattern class, and sign " $'$ " denotes transposition.

The $p\times(K-1)$-D transformation weight matrix W is composed of $(K-1)$ eigenvectors (columns) of matrix $T = S_w^{-1} S_b$, corresponding to $(K-1)$ largest eigen-values of matrix T. Instead of matrix inversion often one uses an alternative calculation of matrix T, the matrix division of S_w into S_b. After the dimensionality reduction, $Y = X'W$, the $(K-1)$-dimensional vectors can be classified by means of any linear or non-linear classification rule. To investigate the finite data size effects for estimation of data transformation matrix W (matrices S_w and S_b, actually) we will consider Gaussian data model, where all classes share the same covariance matrix, Σ. For the simplicity sake, we assume that sample sizes of all classes are the same, i.e. $N_1 = N_2 = \ldots, N_K = N$. Therefore, a number of training vectors used to estimate matrix S_w, $n_K = KN$, depends on the number of classes. The number n_K is important since in small sample situations, where $n = n_K - K < p$, this matrix becomes singular and we cannot invert it. Singular matrix case will be considered in Section 3. In Section 2 we consider an increase in classification errors, when $n>p$.

In multi-class LDA, actually we perform classification by means of K discriminant functions, $g_i(X)$, where we calculate

$$g_i(X) = (X - \tfrac{1}{2}\overline{X}_i)'S_w^{-1}\overline{X}_i \tag{3}$$

and select solution with maximal $g_i(X)$. If we consider two pattern classes then we perform classification according to a sign of the discriminant function (DF)

$$g_{ij}(X) = g_i(X) - g_j(X) = X'S_w^{-1}(\overline{X}_i - \overline{X}_j) - \tfrac{1}{2}(\overline{X}_i + \overline{X}_j)'S_w^{-1}(\overline{X}_i - \overline{X}_j). \tag{4}$$

It is the sample based standard Fisher linear discriminant function (DF) [1]. With an increase in dimensionality, distribution function of DF (4) approaches Gaussian. In such case, one can calculate its expected classification error

$$EP_n = \tfrac{1}{2}\text{Prob}(g_{ij}(X)>0|X\in \pi_i)+ \tfrac{1}{2}\text{Prob}(g_{ij}(X)\leq 0|X\in \pi_j)=$$

$$\tfrac{1}{2}\Phi(Eg_{ij}(X)/\text{std}(g_{ij}(X)\,0|X\in \pi_i)) + \tfrac{1}{2}\Phi(-Eg_{ij}(X)/\text{std}(g_{ij}(X)|X\in \pi_j)), \tag{5}$$

where E and std stands as mean and standard deviation respectively. Following the previous postulation, $N_2=N_1=N$, we assumed that prior probabilities of the classes are equal as well. After employment the methods of multivariate analysis and rather tedious algebra, very simple asymptotic equation was derived [18, 19].

$$EP_{ij} \approx \Phi\left(-\frac{1}{2}\delta_{ij}((1+\frac{2p}{N\delta_{ij}^2}) \times \frac{2N-p}{p})^{-1/2} \right) \tag{6}$$

In Equation (6), δ_{ij} stands for Mahalanobis distance between classes, π_i and π_j. This distance determines asymptotic probability of misclassification when $p = constant$, and sample size $N \rightarrow \infty$: $P_\infty = \Phi(-\frac{1}{2}\delta_{ij})$. In Eq. (6), $\Phi(\cdot)$ stands for standard Gaussian (zero mean and unity variance) cumulative distribution function. While deriving Eq. (6) it was postulated n > p. Therefore, our assumption that p is large requires that sample size, $2N$, must be large as well. Subsequent simulation study showed that accuracy of Eq. (6) was high. It outperformed six other simple asymptotic formula derived to evaluate the expected classification error [20].

In analysis of K class LDA, we use $n_K = KN$ vectors to evaluate the within-class scatter matrix, S_w. Therefore, asymptotic Eq. (6) converts into

$$EP_{ij}^K \approx \Phi\left(-\frac{1}{2}\delta_{ij}\left((1+\frac{2p}{N\delta_{ij}^2}) \times \frac{K(N-1)-p}{p} \right)^{-1/2} \right) \tag{7}$$

Serious difficulties arise in analytical investigation of the classification error rates in multi-class pattern recognition tasks. A simplified equation

$$P_K = \sum_{i=1,}^{K} \sum_{j=1,j\neq i}^{K} q_i P_{ij} , \tag{8}$$

is expressed by K prior probabilities, $q_1, \ldots , q_K$, of K classes and $K\times(K-1)$ pair-wise conditional classification error rates, EP_{ij}^K. This expression is correct if the i-th class can be confused only with one solitary "non-self" class. To obtain exact equation one needs integrating in $(K-1)$-dimensional space. A practical solution can be suggested: a) to consider several the most overlapping pairs of the classes, b) to use Eq. (7) for calculating EP_{ij}^K and a relative increase in the classification error rate afterwards.

3   LDA, a Case When Dimensionality Exceeds Sample Size

In situations, where rank $n = n_K - K$ of $p\times p$ - D matrix S_w becomes singular, we cannot use standard LDA for feature extraction directly. A popular approach is substitute matrix S_w^{-1} by Moore-Penrose inverse matrix S_w^+ or apply matrix division of S_w into S_b. The pseudo-inverse is easier to use in analysis. It can be explained easily by utilizing singular value decomposition

$$S_w = TDT', \tag{9}$$

where rows, $t_1, t_1, ..., t_p$, of $p \times p$ dimensional matrix T are orthogonal, i.e. $t_i t_j' = 1$ if $j = i$, and $t_i t_j' = 0$ otherwise (when $j \neq i$). Diagonal matrix D is composed of eigen-values of the learning sample based matrix S_w. Then $T' S_w T = D$. Let us rank the diagonal elements of matrix D. Consequently, we can assume: $d_1 \geq d_2 = ... \geq d_p$.

If $n < p$, the last $p - n_K + K$ eigen-values are equal to zero, i.e. $D = \begin{pmatrix} D_1 & 0 \\ 0 & 0 \end{pmatrix}$, where D_1 is $n \times n$ dimensional diagonal matrix with elements $d_1, d_2, ..., d_n$. Then

$$S_w^+ = T_1 \begin{pmatrix} D_1^{-1} & 0 \\ 0 & 0 \end{pmatrix} T_1', \tag{10}$$

where T_1 is $n \times p$ dimensional upper part of matrix $T = [T_1', T_2']'$.

Use of pseudo-inverse of the CM in the two category case was already discussed in the pattern recognition literature [21-25]. It was noticed that reconstruction of the sample based covariance matrix by its inverse is worst when $n \to p$ [25]. Similar results are obtained by simulation of training and testing processes of the Fisher linear DF [21, 25]. Attempts were made to derive asymptotic analytical formula for genera-lization error if the pseudo-inverse S_w^+ is used [26-28]. Below we will use double asymptotic approach ($p \to \infty$, and $N \to \infty$) to analyze generalization error.

To simplify understanding of the behavior of the generalization error, with a small loss in generality, the analytical expressions will be derived for the two class case, when $\Sigma = I_p$ ($p \times p$ - dimensional identity matrix) and use the previous assumptions: prior probabilities of the classes $q_i = q_j = \frac{1}{2}$, $N_j = N_i = N$.

We will denote the p-D mean vectors of two classes μ_i and μ_j, $\mu_i - \mu_j = \mu$. Under Gaussian assumptions mentioned, vectors $X - \frac{1}{2}(\overline{X}_i + \overline{X}_j) = \frac{1}{2}\mu + Z$, and $\overline{X}_i - \overline{X}_j = \mu + U$ are independent [29]. Their distributions are: $U \sim N(0, I \times 2/N)$ and $Z \sim N(0, I_n \times (1 + 1/(2N)))$.

Above, we assumed the CM is common for both pattern classes. For that reason, we evaluate only the classification error $P_{1/2}$ where vector X belongs to the first (i-th) class. The calculation of $P_{2/1}$ is similar. Moreover, vectors Z, U and matrix S are independent as well. Thus, discriminant function (DF) (4) can be rewritten as

$$g_{ij} = (\frac{1}{2}\mu_{ij} + Z)' S^{-1}(\mu_{ij} + U)$$

$$= (\frac{1}{2}\mu_{ij} + Z)' [T_1' T_2'][T_1' T_2']' S^{-1} [T_1' T_2'][T_1' T_2']' (\mu_{ij} + U). \tag{11}$$

We analyze situation where $p > n$, and instead of inverse S^{-1} one uses pseudo-inverse, S^+ defined in Eq. (10). Consequently, DF (11) can be rewritten as

$$g_{ij} = (\frac{1}{2}\mu_{ij}^n + Z^n)' D_1^{-1}(\mu_{ij}^n + U^n), \tag{12}$$

where

$$\boldsymbol{\mu}_{ij}^n = (\mu_{ij}^1, \mu_{ij}^2, \dots, \mu_{ij}^n)' = T_1 \boldsymbol{\mu}_{ij},$$

$$Z^n = (z_{ij}^1, z_{ij}^2, \dots, z_{ij}^n)' = T_1 Z, U^n = (y_{ij}^1, y_{ij}^2, \dots, y_{ij}^n)' = T_1 U,$$

D_1 was defined in Eq. (10), and upper indexes, n, remind that dimensionality of vectors (marked by n and printed in bold) is n.

The $n \times p$-dimensional matrix D_1 is orthogonal and independent of Z and U [29] According to our assumption that $\Sigma = I_p$, if dimensionality, p, is constant and sample size $N \to \infty$, then $D_1 \to I_p$. In finite sample case, especially when $p > n$, the diagonal components of D_1 are random variables. Denote, mean value of random variable $1/d_s$ is E_d and variance is V_d. Ranking of randomly distributed elements $d_1, d_2, \dots, d_n$ shows very small absolute values of components of vector $\boldsymbol{\mu}_{ij}^n = \left(\mu_{ij}^1, \mu_{ij}^2, \dots, \mu_{ij}^n\right)$. Therefore, we make an assumption that components μ_{ij}^s ($s = 1, 2, \dots, n$) are zero mean Gaussian variables with variance δ_{ij}^2/p. In such case, expected squared Mahalanobis distance for n components would be

$$E\left(\sum_{s=1}^{n}(\mu_{ij}^s)^2\right) = n/p \times \delta_{ij}^2. \tag{13}$$

To obtain an equation to calculate expected generalization error we will follow methodology of multidimensional analysis discussed in Section 2 (see Eq. (5)), where both, p and N are large and simultaneously tend to infinity with keeping ratio p/N and Mahalanobis distance invariable. In such a case, distribution of discriminant function, $g_{ij}(X, (\overline{X}_i, \overline{X}_j, S^+))$, approaches Gaussian distribution with following mean and variance

$$Eg_{ij} = (\tfrac{1}{2}\boldsymbol{\mu}_{ij}^n + Z^n)' D_1^{-1}(\boldsymbol{\mu}_{ij}^n + Y^n) = E\sum_{s=1}^{n}\left(\tfrac{1}{2}(\mu_{ij}^s)^2 + \tfrac{1}{2}\mu_{ij}^s y_{ij}^s + \mu_{ij}^s z_{ij}^s + z_{ij}^s y_{ij}^s\right) \times E_d =$$

$$= \tfrac{1}{2}\sum_{s=1}^{n}(\mu_{ij}^s)^2 \times E_d = \tfrac{1}{2}\delta_{ij}^2 \frac{n}{p} \times E_d, \tag{14}$$

$$Vg_{ij} = E\sum_{s=1}^{n}\left(\tfrac{1}{4}(\mu_{ij}^s)^4 + \tfrac{1}{4}(\mu_{ij}^s)^2(y_{ij}^s)^2 + (\mu_{ij}^s)^2(z_{ij}^s)^2 + (z_{ij}^s)^2(y_{ij}^s)^2\right) \times (E_d^2 + V_d) +$$

$$+ (E_d^2 + V_d) + \tfrac{1}{4}\sum_{r \neq s}^{n}(\mu_{ij}^s)^2(\mu_{ij}^r)^2 E_d^2 - (Eg_{ij})^2. \tag{15}$$

Expectations of random variables μ_{ij}^s, z_{ij}^s, and y_{ij}^s are zero. Therefore

$$E(\mu_{ij}^s)^2 = \delta_{ij}^2/p,\ E(\mu_{ij}^s)^4 = 3\delta_{ij}^4/p^2,\ E(z_{ij}^s)^2 = 1 + \frac{1}{2N},\ \text{and}\ E(y_{ij}^s)^2 = \frac{2}{N},$$

$$Vg_{ij} = \left(\delta_{ij}^4\frac{3n}{4p^2} + \delta_{ij}^2\frac{n}{p}\frac{1}{2N} + \delta_{ij}^2\frac{n}{p}(1+\frac{1}{2N}) + (1+\frac{1}{2N})\frac{2n}{N}\right)\times(E_d^2 + V_d) - \delta_{ij}^4\frac{n}{4p^2}E_d^2$$

$$= E_d^2\frac{n}{p}\delta_{ij}^2\left(\delta_{ij}^2\frac{3}{4p} + (1+\frac{1}{N}) + \frac{1}{\delta_{ij}^2}(1+\frac{1}{2N})\frac{2p}{N}\right)\times(1+V_d/E_d^2) - \delta_{ij}^2\frac{1}{4p}\right). \qquad (16)$$

Then expected PMC $EP_{ij} = \mathrm{Prob}(g_{ij} < 0 \mid X \in \pi_i) = \Phi(-\tfrac{1}{2}\,Eg_{ij}/\sqrt{Vg_{ij}})$.

$$EP_{ij} \approx \Phi\left(\frac{-\delta_{ij}}{2}\sqrt{\frac{n}{p}}\left((1+\frac{1}{N})(1+\frac{2p}{\delta_{ij}^2}) + \frac{p}{N^2\delta_{ij}^2} + \frac{3}{4p}\delta_{ij}^2)\times(1+\frac{V_d}{E_d^2}) - \frac{1}{4p}\delta_{ij}^2\right)^{-1/2}\right). \qquad (17)$$

To avoid extinction of the peaking when $n\to\infty$, $p\to\infty$, while derivation of Eq. (17) we did not reject small terms that tend to zero when only n, or p are approaching the infinity. In Fig. 2a we display expected probability of misclassification calculated according to Eq. (17), for 50-D Gaussian data model with δ_{ij}= 4.65. Term V_d/E_d^2 concerning eigen-values of the sample CM is necessary for calculations. It cannot be expressed explicitly. Thus, for N = 4, 5, ... , 48, it was evaluated numerically. For larger sample sizes $n=2N-2>p$, Eq. (7) was used.

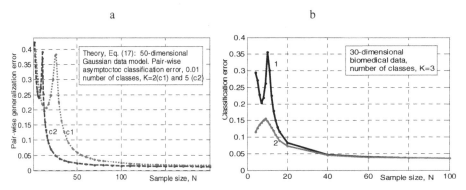

Fig. 2. Generalization error of LDA based data mapping into $(K$-1)-D space versus sample size: a – theoretical calculation (curve c1 – with K=2; curve c2 – with K=5), b – simulations with three class 30-D chromosome data: 1 – without regularization, 2 – with regularization.

Curve c1 in Fig. 2a is calculated for a single pair of the pattern classes. It shows peaking effect very clearly. Calculations performed for variety of Gaussian model parameters showed that for situations $n < p$, the smallest generalization error can be obtained when $n \approx p/2$. This fact hints that *all pairs of the classes* exhibit similar peaking behaviour. Thus, one can hope that the peaking effect will be observed in multiple class situations. Curve c2 in Fig.2a demonstrates situation where we have multiple classes. In this case, for estimation of the mean vectors we used N vectors of each pattern class. For estimation of the CM we used $K{\times}N$ vectors. The rank of sample estimate of the covariance matrix is $n=K{\times}(N\text{-}1)$ in such situation. Therefore, for $K=5$ we observe peaking effect much earlier, $N{\approx}p/K=10$. For sample sizes $n<p$, the minimal generalization error we obtain at $N{\approx}p/(2K)=5$.

4   Simulation Experiments

We cannot obtain exact analytical formula for calculation of K class classification error. To verify peaking behavior that follow from theory, we performed experiments with multi-class real world biomedicine and biometrics pattern recognition tasks in multi-category case. In Fig. 2b we have analogous to the Fig. 2a graphs, however, here we display experimentally evaluated classification error obtained while classifying $K=3$ class 30-D chromosome data. This biomedicine data set was chosen due to a fact that here we had 500 vectors in each pattern classes. Thus, we had enough data to estimate classification error in the mapped, $(K\text{-}1) = 2$-D, space.

In series of our experimental evaluations, we used randomly selected $N = 4, 5, 6,$ ..., 60, 80, or 100 vectors to train LDA based feature extraction rules. After dimensionality reduction, we obtained 500 2D vectors from each of three pattern classes. To evaluate the performance of dimensionality reduction we used two fold *cross-validation* method to estimate the asymptotic classification error rates in the 2-D space [1]. Here 250 randomly selected vectors, A_{250}, were used for training the classifier (3). Remaining, B_{250}, vectors were used for testing. Afterwards, the data sets, A_{250}, and B_{250}, were interchanged. In parallel, for the both sets, A_{250}, and B_{250}, we preformed *resubstitution* error rate estimations [1]. For training and testing we use the same data in the resubstitution method. Average of four error estimates approximately characterizes *asymptotic classification error* (see Section 6.3.1.3 in [30]). To increase the accuracy of empirical evaluations, the experiments were repeated 1000 times.

Curve 1 in Fig. 2b confirms conclusions obtained in theoretical calculations and reported in the considerations associated with Fig. 2a. In situations where $K{\times}N<p$, one possible ways to increase feature reduction accuracy is to use only $N = p/(2K)$ training vectors per class. Much better alternative is the regularization of the "within-class covariance matrix", S_w. Regularization of the CM is a popular mean to soften small sample problems in data mining procedures, including LDA [10, 30]. Curve 2 in Fig. 2b confirms usefulness of regularization.

To analyze value of regularization in high-dimensional tasks we performed experiments with two data sets used in the biometrics (face recognition). To obtain sufficient amount of data for training and for validation of the feature extraction algorithms, we calculated K mean vectors of the classes and covariance matrices in p-dimensional original space. To avoid zero eigenvalues we increased the smallest eigen-values of the covariance matrix. Finally we used these parameters to generate

10,000 700-D vectors per class. This synthetic data was used in experiments performed similarly as these reported above.

In Fig. 3 we present results obtained with $K=15$, $p=700$-D Yale data (http://cvc.yale.edu/projects/yalefaces/yalefaces.html), 1000 repetitions of the experiments. In Fig. 3a we display generalization errors versus sample size, N, for four different levels, λ, of regularization $S_{wregularixed} = S_w \times (1-\lambda) + \text{diagonal}(S_w) \times \lambda$. Fig. 3b demonstrates: the peaking effect is observed for diverse separation of the pattern classes. To obtain the graphs we increased/decreased proportionally the distances between the mean vectors of the classes, however, we kept CM, $S_{wregular}$ fixed.

a   b

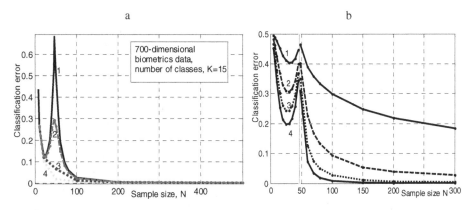

Fig. 3. Averages of generalization error of LDA based 700-D Gaussian data mapping into 14-D space versus sample size, N ; a – an influence of regularization, λ, of matrix S_w: curve 1 – no regularization $(\lambda=0)$, 2 - $\lambda=0.001$, 3 - $\lambda=0.01$, 4 $\lambda=0.1$, 2); b- no regularization, however, asymptotic classification errors we modeled were different: 1 - $P_\infty= 0.1$ (curve 1), 0.01 (c2), 0.001 (c3), 0.0001 (c4).

If $n<p$, we obtained minimum of generalization error rate when $N \approx \frac{1}{2} \, p/K$. Conclusions concerning peaking in dimensionality reduction are valid for variety of biomedical image analysis tasks where original dimensionality of images is very high. Correctly selected regularization parameter, λ, allows avoiding the negative peaking effect. Means of finding of the most suitable value of parameter λ constitute topics for future research.

5   Conclusions

In many multi-class real world biomedicine and bioinformatics data analysis tasks we deal with high-dimensional data and comparative small number of training vectors. Traditional way to avoid small sample size problems is dimensionality reduction prior to employment of pattern classification algorithms. We considered an increase in classification error that was caused by employment of sample based K- class linear discriminant analysis for dimensionality reduction. For this purpose we derived analytical expression for expected classification error by applying techniques of the multivariate statistical analysis and double asymptotic approach. Theoretical analysis shows that with an increase in sample size, the classification error in the mapped, $(K\text{-}1)$-dimensional data decreases at first, however, after $N \approx p/(2K)$ it starts increasing. If $n < p$, we have the maximum when the size of K class training sets, $K \times N$, approaches dimensionality, p. Further increase in the number of training vectors diminishes classification error permanently.

We demonstrated the peaking effect for real world biomedical and biometric data sets and confirmed that regularisation of within-class scatter, S_w, can reduce or even extinguish this effect. Analytical calculations show that the increase in classification error rate is notable for many multi-class data analysis tasks we deal with in typical real world biomedicine and bioinformatics problems. Therefore, the small sample effects that arise in dimensionality reduction should be taken into account.

Acknowledgements. This research was funded by grant MIP 057/2013 from Research Council of Lithuania.

References

1. Fukunaga, K.: Introduction to Statistical Pattern Recognition, 2nd edn. Academic Press, New York (1990)
2. Jaiswal, A., Kumar, N., Agrawal, R.K.: A Hybrid of principal component analysis and partial least squares for face recognition across pose. In: Alvarez, L., Mejail, M., Gomez, L., Jacobo, J. (eds.) CIARP 2012. LNCS, vol. 7441, pp. 67–73. Springer, Heidelberg (2012)
3. Morris, J.S., Coombes, K.R., Koomen, J., Baggerly, K.A., Kobayashi, R.: Feature extraction and quantification for mass spectrometry in biomedical applications using the mean spectrum. Bioinformatics 21(9), 1764–1775 (2005)
4. Kou, B.C., Chang, K.Y.: Feature extraction for sample size classification problem. IEEE Transactions on Geoscience and Remote Sensing 45, 756–764 (2007)
5. Kumar, N., Jaiswal, A., Agrawal, R.K.: Performance evaluation of subspace methods to tackle small sample size problem in face recognition. In: Proceedings of the International Conference on Advances in Computing, Communications and Informatics, pp. 938–944. ACM, New York (2012)
6. Lee, C., Landgrebe, D.A.: Feature extraction based on decision boundaries. IEEE Transactions Pattern Analysis and Machine Intelligence 15, 388–400 (1993)
7. Pang, Y., Wang, S., Yuan, Y.: Learning regularized LDA by clustering. IEEE Trans. Neural Networks & Learning Systems 25(12), 2191–2201 (2014)
8. Raudys, S.: Taxonomy of classifiers based on dissimilarity features. In: Singh, S., Singh, M., Apte, C., Perner, P. (eds.) ICAPR 2005. LNCS, vol. 3686, pp. 136–145. Springer, Heidelberg (2005)
9. Sun, S., Wang, H., Jiang, Z., Fang, Y., Tao, T.: Segmentation-based heart sound feature ex-traction combined with classifier models for a VSD diagnosis system. Expert Systems with Applications 41(4), Part 2, 1769–1780 (2014)
10. Yang, W., Wu, H.: Regularized complete linear discriminant analysis. Neurocomputing 137(5), 185–191 (2014)
11. Yang, W., Wang, Z., Sun, C.: A collaborative representation based projections method for feature extraction. Pattern Recognition 48, 20–27 (2015)

12. Zhang, T., Fang, B., Tang, Y.Y., Shang, Z., Xu, B.: Generalized discriminant analysis: a matrix exponential approach. IEEE Trans. on Systems, Man and Cyb. Pt.2: Cybernetics 40(1), 186–197 (2010)
13. Raudys, S., Young, D.: Results in statistical discriminant analysis: A review of the former Soviet Union literature. J. of Multivariate Analysis 89(1), 1–35 (2004)
14. Amari, S., Fujita, N., Shinomoto, S.: Four types of learning curves. Neural Computation 4, 605–618 (1992)
15. Vapnik, V.N.: The Nature of Statistical Learning Theory. Springer, Berlin (1995)
16. Raudys, S.: Portfolio of automated trading systems: Complexity and learning set size issues. IEEE Trans. on Neural Networks and Learning Systems 24(3), 448–459 (2013)
17. Raudys, S.: Feature over-selection. In: Yeung, D.-Y., Kwok, J.T., Fred, A., Roli, F., de Ridder, D. (eds.) SSPR 2006 and SPR 2006. LNCS, vol. 4109, pp. 622–631. Springer, Heidelberg (2006)
18. Deev, A.D.: Asymptotic expansions for distributions of statistics W, M, W in discriminant analysis. In: Statistical Methods of Classification, vol, 31, pp. 6–57. Moscow University Press, Moscow (1972) (in Russian)
19. Raudys, S.: On the amount of a priori information in designing the classification algorithm. Engineering Cybernetics N4, 168–174 (1972) (in Russian)
20. Wyman, F., Young, D., Turner, D.: A comparison of asymptotic error rate expansions for the sample linear discriminant function. Pattern Recognition 23, 775–783 (1990)
21. Duin, R.P.W.: Small sample size generalization. In: Borgefors, G. (ed.) Proceedings of the 9th Scandinavian Conference on Image Analysis, vol. 2, pp. 957–964 (1995)
22. Krzanowski, W., Jonathan, P., McCarthy, W., Thomas, M.: Discriminant analysis with singular covariance matrices: Methods and applications to spectroscopic data. Applied Statistics 44, 101–115 (1995)
23. Raudys, S., Duin, R.P.W.: On expected classification error of the Fisher classifier with pseudoinverse covariance matrix. Pattern Recognition Letters 19, 385–392 (1998)
24. Schafer, J., Strimmer, K.: An empirical Bayes approach to inferring large scale gene association networks. Bioinformatics 21, 754–764 (2005)
25. Hoyle, D.C.: Accuracy of pseudo-inverse covariance learning—A random matrix theory analysis. IEEE Trans. Pattern Analysis and Machine Intelligence 33, 1470–1481 (2011)
26. Yamada, T., Hyodo, M., Seo, T.: The asymptotic approximation of EPMC for linear discriminant rules using a Moore-Penrose inverse matrix in high dimension. Communications in Statistics - Theory and Methods 42(18), 3329–3338 (2013)
27. Kubokawa, T., Hyodo, M., Srivastava, M.S.: Asymptotic expansion and estimation of EPMC for linear classification rules in high dimension. J. Multivariate Analysis 115, 496–515 (2013)
28. Srivastava, M.S., Katayama, S., Kano, Y.: A two sample test in high dimensional data. J. Multivariate Analysis 114, 349–358 (2013)
29. Anderson, T.W.: An Introduction to Multivariate Statistical Analysis. Willey, NY (1958)
30. Raudys, S.: Statistical and Neural Classifiers: An Integrated Approach to Design. Springer, London (2001)

An Ensemble of Cooperative Parallel Metaheuristics for Gene Selection in Cancer Classification

Anouar Boucheham, Mohamed Batouche, and Souham Meshoul

MISC Laboratory, Computer Science Department, College of NTIC,
Constantine University 2, 25000 Constantine, Algeria
{anouar.boucheham,mohamed.batouche,souham.meshoul}
@univ-constantine2.dz

Abstract. Biomarker discovery becomes the bottle-neck of personalized medicine and has gained increasing interest from various research fields recently. Nevertheless, producing robust and accurate signatures is a crucial problem in biomarker discovery and relies heavily on the used feature selection algorithms. Feature selection is a preprocessing step which plays a crucial role in omics data analysis to improve learning. The accumulating evidence suggests that ensemble methods and swarm intelligence are two growing solutions for improving feature selection algorithms. In this paper, we propose a two stages approach to identify a predefined number of biomarkers from gene expression data. It is designed as a wrapper-based ensemble method; each part of the ensemble is performed through cooperative parallel meta-heuristics and a filter-based mechanism. Experiments from twelve DNA microarray datasets have shown that our approach competes with and even outperforms recent state-of-the-art methods in terms of accuracy and robustness. Also, biological interpretation shows that our approach selects highly informative genes for cancer diagnosis.

Keywords: Personalized medicine, Biomarker identification, Cancer classification, Gene expression analysis, Data mining, Ensemble feature selection.

1   Introduction

In less than 3 decades, discoveries in genomics are irrevocably changing the face of biomedicine impacting diagnosis, disease activity monitoring and response to therapy prediction [1]. This revolution is the result of the development and the use of high throughput technologies for the treatment of genetic diseases, particularly Cancer which is the first cause of mortality worldwide. Expanded understanding of the genome has led to personalized medicine which can through genetic consultations achieve an individualized therapy for each patient. One of the most productive areas in personalized medicine is biomarker signature discovery [2]. It represents the key mechanism in current biomedical research and, more slowly, clinical medicine [3].

DNA microarrays technologies have made the monitoring of the expression pattern of thousands of genes simultaneously easier, as a result a huge amount of gene expression data has been produced during last decade. Gene expression data are primarily

F. Ortuño and I. Rojas (Eds.): IWBBIO 2015, Part II, LNCS 9044, pp. 301–312, 2015.

characterized by high dimensionality and relatively small sample sizes [4]. Harnessing the full potential of omics data to understand diseases and improve health requires the use of machine learning techniques. The direct application of these methods on high-dimensional data is usually ineffective. In addition, the data sets contain a large number of irrelevant, redundant and noisy genes [5]. Therefore, a smaller set of genes contains useful biological interpretations that helps achieving a high accuracy for cancer diagnosis. Moreover, dealing with a large amount of features affects not only the performance of prediction, but also the computational time of classifiers [6].

The identification of the most informative genes as well as eliminating the redundant and irrelevant ones is an NP-hard problem, where researchers try to identify the smallest subset of genes that can still achieve good prediction [7]. As a consequence, gene selection is often performed with machine learning and dimension reduction methods. Accordingly, biomarker discovery may be considered as a feature selection problem and cancer diagnosis as a supervised classification problem where each class is the phenotype of a specific cancer [8].

There has been a number of feature selection methodologies developed for biomarker discovery. Depending on how they combine search of selected subset with the building of a classification model, we can broadly group them into six categories [9, 7 and 10]: filter methods [11], wrapper methods [12], embedded methods, ensemble methods [13], hybrid methods [14] and unsupervised methods [15].

Filter methods are simple and fast, and they are independent of any classification model. A common drawback of filter methods is that they ignore the interaction with the classifier and take no notice of feature dependencies [9]. In contrast, wrapper methods tend to obtain better performance. They are very computationally intensive and have the risk of over-fitting due to high dimensionality of data. Advantages of wrapper approaches include the interaction between feature subset search and classification model, and the ability to take into account feature dependencies [14]. Therefore, filter methods are usually the best choice to obtain simply a reduced set of features, when the number of features is very large, and wrapper methods are the best option to achieve better classification performance [16].

In this work, we focus on ensemble methods, which have been widely applied in bioinformatics to improve the robustness of feature selection techniques [17]. The main improvement offered by ensemble methods is their ability in dealing with the curse-of-dimensionality in gene expression data. In most reviewed studies, ensemble methods proposed to handle feature selection for biomarker discovery are generally filter-based ensembles. To the best of our knowledge, there is no work which has proposed a wrapper-based ensemble method in this field, except, Ghorai Santanu, et al. [18] proposed a genetic algorithm-based simultaneous feature selection scheme to train a number of nonparallel plane proximal classifier (NPPC) in multiple subspaces.

The current study presents a novel wrapper-based ensemble feature selection method called ECPM-FS, to identify a predefined number of biomarkers from gene expression data. As ensemble systems use a two-step procedure to make decisions, the proposed approach is designed as an ensemble of cooperative parallel metaheuristics (ECPM-FS) performed in two levels. In order to do that, we first adapt the generalized islands model (GIM) to feature selection problem. GIM is a recently developed framework for cooperative parallel population-based metaheuristics in the field of optimization [19]. The generation of the different part of the ensemble is based on the cooperative parallel metaheuristics framework, which is a hybrid wrapper/filter feature selection algorithm proposed in our previous work [20]. At this stage, we replaced random initialization in traditional metaheuristics by a filter-based initialization mechanism. Furthermore, in order to repair individuals which have irregular number of selected genes, we propose an enhanced filter-based repair mechanism which can also improve the selection of the most informative genes. After a predefined number of iterations and migrations, each GIM selects a subset of best features.

The next step is the aggregation of all selected subsets. It is performed through one GIM using as input a reduced set of features (genes). The latter represents the union of all selected features within the different parts of the ensemble, in the previous step. Thereby, the ECPM-FS allows selecting more robust and accurate subset of features, especially when using wrapper method to aggregate different subsets of the ensemble.

The rest of this paper is organized as follows. Section 2 describes the proposed ECPM-FS method for biomarker discovery. Experimental results are discussed in Section 3 and finally conclusions and future work are drawn in Section 4.

2   Proposed Approach   ECPM-FS

2.1   General Framework of the Proposed Approach

In this section, we provide a detailed description of the proposed approach for biomarker discovery from gene expression data. The proposed ECPM-FS approach is designed as an ensemble method which performs feature selection in two steps.

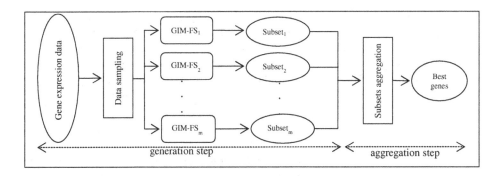

Fig. 1. Flowchart describing the general framework of ECPM-FS

As shown in Fig. 1, we first construct the different parts of the ensemble {subset$_1$, subset$_2$ … subset$_m$} followed by a consensus step to select the final subset of best features. As the construction of the ensemble's subsets is the most important step, we use a hybrid filter/wrapper mechanism to perform selection [20]. The latter allows taking into account the power of each gene separately through using filter within the selection process. Also, it considers the relationship between genes (strength of group) by studying the different combinations of genes through the exploration of the solution space by wrapper method. Furthermore, the diversity within the ensemble's parts is guaranteed, since they are based on stochastic and random methods.

To ensure a better exploration of the solution space, the selection within each part of the ensemble is based on cooperative parallel optimization (CPO). The cooperation of several nature inspired metaheuristics represents an effective solution to avoid premature convergence while speeding up the search process [21]. Hence, we employ different population based metaheuristics such as Particle Swarm Optimization (PSO), Ant Colony Optimization (ACO) and Genetic algorithm (GA) which are deployed in parallel with solution migration mechanism [12].

Izzo et al proposed the Generalized Island Model (GIM) which is a new framework to implement CPO [19]. The latter employs different optimization algorithms on different islands with the introduction of a migration operator. According to Izzo et al, the GIM can speed up computation time significantly and further increase the accuracy of prediction with respect to homogeneous case in several optimization problems. Recently, Boucheham et al. was adapted the GIM framework into a hybrid wrapper/filter method to perform feature selection with a predefined number of selected features [20]. It constitutes each part of the proposed ensemble with a specific configuration.

Once the construction of all parts of the ensemble is attained, the first step is completed by the selection of the best subset of features over all islands from each GIM-FS. These subsets are aggregated in the second step through a wrapper-based consensus function, as shown in Fig. 1.

2.2   Generation Step

We describe in this section the parallel model of cooperative metaheuristics for feature selection which constitutes each part of the ECPM-FS method. It is based on the generalized island model along with an adjustment of its parameters to the context of feature selection [19]. To reach the goal of functional diversity we use inside one GIM-FS different configuration of population-based metaheuristics (see Fig. 2) [20]. The selection process is started by the initialization of populations being the initial solutions in each island.

A. L. Gutiérrez et al have shown that the initialization strategies used inside the metaheuristics perform differently in different kinds of problems with high dimensional search spaces [22]. Thus, the initial positions of the populations determine the convergence of metaheuristics. Accordingly, we propose a new initialization technique which is based on an ensemble of filters [13]. All selected subsets ($S_1, S_2... S_N$) through filters ($F_1, F_2... F_N$) within the ensemble represent the initial population of each island as shown in Fig. 2. Therefore, to construct the different sets of the ensemble of filters based on sub-samples already generated. We choose one of the most popular and successful filters which is Information Gain (IG), since it is simple, fast and meaningful for an appropriate ensemble method [23]. In order to ensure the data diversity among different populations, we have used data partitioning with overlap which allows creating many reduced datasets for all selectors within the ensemble of filters. Dataset perturbation involves generating sub-samples by removing instances from the original datasets randomly [13].

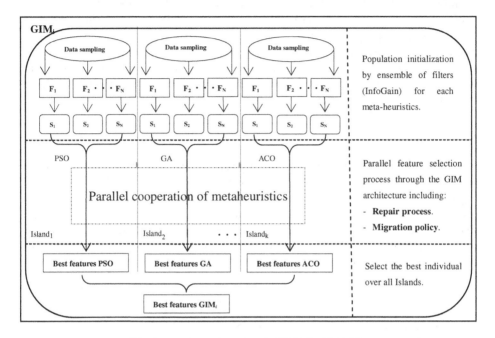

Fig. 2. Parallel model of each GIM-FS (CPM-FS)

Moreover, the migration of solutions is done in different intervals with an asynchronous communication between islands according to fully connected topology and an elitist policy. The appropriate choice of the migration mechanism prevents optimization algorithms for getting stuck into local optima. As a consequence, we can achieve good results and reduce the number of iterations and the size of populations at the same time. On the other hand, the objective function used to guide search is only the accuracy value given by the SVM classifier using 5-cross-validation procedure, since the number of selected features is predefined. Once a termination criterion is met in all islands, we select the subset with the best fitness over all GIM-FS.

Repair Process. In order to apply the different metaheuristics (PSO, GA, ACO...) for feature selection with a predefined number of selected features n, certain adjustments should be performed when updating solutions of all metaheuristics. Noting that Solutions are represented as a binary vector of length D (D is the total number of features). The value of the ith gene is set to 1 or 0 to indicate whether the ith feature is selected or not respectively. The application of the update operators of each metaheuristics on solutions leads to the overflowing of the number of selected features in individuals, it can be less or greater than the desired value [20].

To overcome this problem, we propose integrating a repair process that will be in-troduced for each new solution that does not satisfy this requirement. The proposed repair process is based on information given by the filter Information Gain (IG). Therefore, two cases will appear. The first one is when the number of selected fea-tures in each individual is greater than n. In this case, some selected features must be removed. For this purpose, we use the IG filter to rank all selected features and elimi-nate the least ranked ones. The second case is when the number of selected features is less than n. In this case, we add to the subset the best ranked features from the whole set excluding the already selected ones. Unlike to our previous work where we have add randomly selected features in this case [20].

2.3   Consensus Step

In analogy with ensemble methods, the aggregation of the generated subsets in the previous step is a key part in our ECPM-FS method. The main objective consists of aggregating M subsets of features to obtain one final subset containing the most rep-resentative features (genes). The use of consensus functions which are based on the study of each gene separately as counting the most frequent features is not efficient in our case. Alternatively, the best consensus function in the issue of biomarker discov-ery must be based on the strength of the group of genes which work together to attain a common objective.

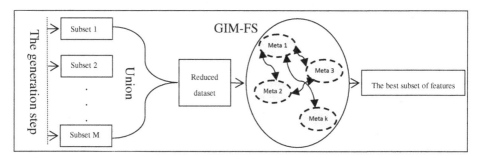

Fig. 3. Consensus process

In order to achieve this goal, we propose a wrapper-based consensus function to aggregate the different subsets of the ensemble which represents the novelty of our work. The aggregation is done by performing a GIM-FS starting from a reduced set of features. The latter represents the union of features inside all selected subsets in the previous step, as shown in Fig. 3. The same framework of the aforementioned GIM-FS is used to establish the consensus among the ensemble's parts, and selects subset of best features.

Finally, the selected biomarkers are employed to construct a classification model that helps taking appropriate decisions regarding treatments. This can provide patients with better treatment or response to therapy, especially when the disease has been identified at its early time.

3   Experiments and Discussions

In this section, we present the empirical study conducted in order to assess the performance of the proposed approach and compare it against other state of the art methods from the literature. Our study focuses on supervised feature selection, since several DNA microarray datasets have class values which are useful for prediction.

We have implemented our method using MATLAB®'s Parallel Computing Toolbox (PCT). To test the effectiveness of the proposed ECPM-FS, twelve DNA microarray datasets are used [12]. Table 1 summarizes the details of these datasets.

Table 1. Characteristics of the microarray datasets used in this study

Dataset	#Features	#Classes	#Samples
9_tumors	5726	9	60
11_Tumors	12,533	11	174
Prostate_Tumor	10,509	2	102
Colon	2,000	2	62
Leukemia	7129	2	72
Ovarian	15,154	2	253
Leukemia1	5327	3	72
Leukemia2	11,225	3	72
DLBCL	5469	2	77
SRBCT	2308	4	83
Brain_Tumor1	5920	5	90
Brain_Tumor2	10,367	4	50

3.1   Experimental Setting

The proposed EPCM-FS consists of an ensemble of four GIM-FS each of which is composed of three islands. Each island performs selection through a modified population-based metaheuristic as mentioned in section 2. To this aim, we have used three well known metaheuristics namely PSO [24], GA [24] and ACO [25]. Experimental settings of the different GIM-FS parameters are given in Table 2.

Table 2. Approach Calibration

Selection policy	*Elitism strategy*
Recombination policy	*Replacing worst solutions*
Topology	*Bidirectional Fully connected*
Communication	*Asynchronous initiated by the source*
Maximum number of iterations	*250*
Migration interval	*50*
Fitness function	*SVM classifier with 5-fold cross validation*
Swarm/Population size	*30*
c1 , c2 of PSO	*2*
Probability of crossover	*0.8*
Probability of mutation	*0.05*

3.2   Results and Discussions

To assess the contribution of using ensemble of GIM-FS along with the cooperation of metaheuristics, we first compare the predictive ability in term of classification accuracy of the proposed approach with the single method: CPM-FS [20].

Fig. 4.a shows the average accuracy measure values of the two methods according to the subset size over the Colon dataset. From this figure, it can be clearly observed that ECPM-FS can achieve higher classification accuracy than the single CPM-FS regardless to the subset size. Furthermore, Fig. 4.b shows boxplot results through 40 runs on 9_tumors dataset.

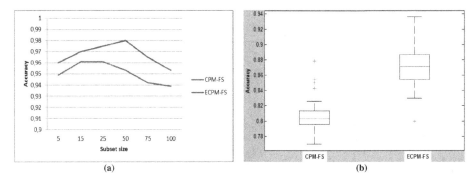

Fig. 4. Performance comparison: (a) comparison of ECPM-FS and CPM-FS through the subset size on Colon dataset, (b) Box plots of ECPM-FS and CPM-FS methods on 9-tumors dataset

Since the other studies mentioned in Table 3 were not endowed with sufficient information and configurations of their methods, a complete comparison cannot be done. Thus, we conduct a comparison based on the mentioned results in some recent studies in the literature. Table 3 summarizes the results obtained by the proposed methods and six other approaches [12, 26, 21, and 13] in terms of accuracy and the number of selected biomarkers.

Table 3. Comparison results based on classification accuracy and number of selected biomarkers (#)

	Our approaches		Other approaches from literature					
	CPM-FS [20]	**ECPM-FS**	IBPSO [12]	cuPSO [12]	CBBBOFS [26]	CIBBOFS [26]	PMSO [21]	EFS [13]
9_tumors	0.804 (25)	**0.8757 (25)**	0.783(1280)	0.85(149)	76.5(25)	0.7156 (25)	/	/
11_Tumors	0.85 (25)	0.874 (25)	0.931(2948)	0.936(535)	**0.8879 (25)**	0.8549 (25)	/	/
Prtate_Tumor	0.986 (25)	0.992 (25)	0.921(1294)	0.99(4)	**0.993 (25)**	0.9718 (25)	/	/
Colon	0.961 (25)	**0.975 (25)**	/	/	/	/	0.942(20)	0.87(20)
Leukemia	1(25)	**1(25)**	/	/	/	/	0.981(20)	0.98(71)
Ovarian	1(25)	**1(25)**	/	/	/	/	/	0.97(151)
Leukemia1	1(25)	**1(25)**	1(1034)	**1(5)**	/	/	/	/
Leukemia2	1(25)	**1(25)**	1(1292)	**1(5)**	1(25)	0.995 (25)	/	/
Brain_Tumor1	0.964 (25)	0.976 (25)	0.944 (754)	**0.977 (12)**	/	/	/	/
Brain_Tumor2	0.974(25)	**1(25)**	0.94(1197)	0.88 (161)	/	/	/	/
DLBCL	1(25)	**1(25)**	1(1042)	**1(3)**	/	/	/	/
SRBCT	1(25)	**1(25)**	1(431)	**1(5)**	/	/	/	/

In order to draw statistically meaningful conclusions, 30 independent runs have been performed for each dataset. The obtained mean values are recorded in Table 3. It can be seen from Table 3 that the ECPM-FS achieved significant improvement over the single CPM-FS and outperforms other approaches in almost all datasets.

Moreover, in order to analyze the robustness of signatures selected through ECPM-FS, we assessed the similarity between outputs of different independent runs of our method. The overall stability (S_{tot})is defined as the average over all pairwise similarity comparisons between the different resulting subsets of feature as follows [13]:

$$S_{tot} = \frac{2 \sum_{i=1}^{k} \sum_{j=i+1}^{k} S(f_i, f_j)}{k(k-1)}$$

Where f_i represents the outcome of the ECM-FS applied to subsample$_i$ ($1 \le i \le$ k=20), and S(f_i, f_j) represents a similarity measure between f_i and f_j. For feature subsets selection (as in our case), we use two well-known metrics: the Kuncheva index [27] and the Jaccard index [13], which are calculated as follows:

- Jaccard index= $\frac{|fi \cap fj|}{|fi \cup fj|}$

- Kuncheva index= $\frac{|fi \cap fj| - \frac{S}{N}}{S - \frac{S}{N}}$

where f_i and f_j are feature subsets of size S, obtained from a dataset of dimensionality N.

Accordingly, Fig. 5 represents the average robustness results of the ECPM-FS over eleven DNA microarray datasets, by means of two similarity measures.

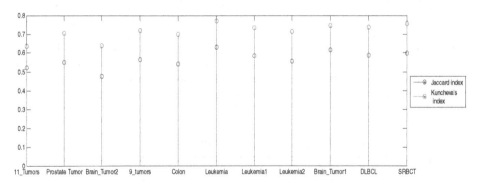

Fig. 5. Average similarities by means of Jaccard and Kuncheva indexes

From the above results, it can be clearly seen that the ECPM-FS competes with and even outperforms recent state-of-the-art methods in term of accuracy. Even though wrapper methods are not robust due to their stochastic behavior, especially in DNA-microarrays data which comprise redundant genes in high dimensional spaces. The Average similarities shown in Fig. 5 indicate that the proposed ECPM-FS is relatively robust over all datasets. The filter-based mechanisms in the initialization of populations and the repair process incorporated in our method minimize the degree of uncertainty in the search space procedure and avoid redundant genes. On the other side, the ensemble part allows to establish a consensus among several selections leading to a more robust and accurate selection. Therefore, the ECPM-FS allows primarily settling a tradeoff between the robustness and high performance of prediction.

Table 4. Top fifteen selected gene (gene index (Indx) and their frequency (Freq)) from the twelve datasets across 40 runs

Prostate Tumor	Indx	4823	8220	7652	6105	7451	7515	9949	10130	5815	9138	10125	7529	8765	120	181
	Freq	40	40	39	35	35	35	35	35	34	32	31	29	29	28	28
11_Tumors	Indx	542	581	7093	8037	9706	2768	4787	6139	7651	7735	7789	8168	11764	1808	2962
	Freq	40	40	40	40	40	38	38	38	38	38	38	38	38	36	36
9_tumors	Indx	80	1755	5032	5183	1361	1777	3372	4138	5147	15	1916	4604	3283	4996	1354
	Freq	40	40	40	40	39	39	38	38	38	37	37	37	36	36	35
Brain Tumor1	Indx	3586	5361	1183	4688	2610	2642	1497	1595	1965	5453	2532	5391	2478	244	505
	Freq	40	40	39	39	39	39	37	34	34	34	33	32	31	30	30
Brain Tumor2	Indx	276	423	687	915	1245	210	703	2313	5847	1696	10337	2846	4863	9568	9801
	Freq	40	40	40	40	40	39	39	39	39	38	38	37	37	37	37
Colon	Indx	1635	138	1060	1110	493	377	897	1263	1365	241	249	267	1549	1960	365
	Freq	40	39	39	39	38	37	37	37	37	36	36	36	35	35	34
DLBCL	Indx	1259	3257	3942	2164	226	409	1600	717	1670	773	856	3127	874	1122	5250
	Freq	40	40	40	39	34	33	31	30	28	27	27	27	26	26	26
Leukemia	Indx	4951	2354	4366	1834	4328	6855	1144	5501	1882	1902	2121	2642	6041	758	1685
	Freq	40	36	35	33	32	32	30	30	29	29	29	28	28	27	27
Leukemia1	Indx	1999	5142	618	1770	4009	1271	2350	3549	1611	3076	728	1426	4307	4688	3518
	Freq	40	39	38	37	37	35	34	34	33	33	32	32	32	32	31
Leukemia2	Indx	225	4992	7050	6746	7545	10355	832	4915	10174	1108	2079	5657	6118	4845	6720
	Freq	40	39	39	35	35	35	34	34	34	33	33	33	33	32	32
SRBCT	Indx	174	1389	1	1073	477	1613	368	1932	338	819	1315	107	108	1263	1700
	Freq	40	39	38	38	37	37	37	37	36	36	35	33	33	33	33
Ovarian	Indx	2313	1599	1679	1684	1688	2237	2240	181	1682	2238	2191	2193	1675	1689	1735
	Freq	40	39	39	39	39	39	39	38	38	38	37	37	36	36	36

Finally, we show the selected signatures using the ECPM-FS over 40 independent runs, as seen in Table 4. The genes belonging to the fifteen top selected genes were selected with high level frequency which confirms the ability of our method to select robust and significant genes. We also provide to biologists and clinician experts the biological interpretation of the selected signature from the SRBCT dataset, as shown in Table 5.

Table 5. Top fifteen selected genes from SRBCT dataset

Gene index	Gene ID	Hugo name	Gene description
174	4771	NF2	neurofibromin 2 (bilateral acoustic neuroma)
1389	2217	FCGRT	Fc fragment of IgG, receptor, transporter, alpha
1	1495	CTNNA1	catenin (cadherin-associated protein), alpha 1, 102kDa
1073	5045	FURIN	furin (paired basic amino acid cleaving enzyme)
477	1942	EFNA1	ephrin-A1
1613	8991	SELENBP1	selenium binding protein 1
368	7088	TLE1	transducin-like enhancer of split 2, homolog of Drosophila E(sp1)
1932	3159	HMGA1	high mobility group AT-hook 1

Table 5. (*Continued*)

338	*2737*	*GLI3*	*GLI family zinc finger 3*
819	4330	MN1	meningioma (disrupted in balanced translocation) 1
1315	6258	RXRG	retinoid X receptor, gamma
107	2619	GAS1	growth arrest-specific 1
108	7295	TXN	thioredoxin
1263	3316	HSPB2	heat shock 27kD protein 2
1700	2275	FHL3	ESTs, Moderately similar to skeletal muscle LIM-protein FHL3 [H.sapiens]

4   Conclusion and Future Work

This paper tackles the problem of selecting meaningful biomarkers from high-dimensional gene expression data in genomics. We propose a two-stage procedure based on an ensemble of cooperative parallel metaheuristics (ECPM-FS) for the discovery of a pre-specified number of useful markers. The best genes subsets are aggregated across the islands and are once again processed by a GIM-FS. Selected genes can then be used for further analysis, like cancer types classification. Experimental results from twelve DNA microarray datasets have shown that our approach performs better than other existing methods in the recent literature. Moreover, the proposed approach can be easily extended to any feature selection problem and it is suitable for large datasets.

The ongoing work concerns the collaboration with cancer biologists who can meaningfully interpret and test our results. As future work, we expect scaling up the proposed approach to cope with big data in genomics.

References

1. Bauer, D.C., et al.: Genomics and personalised whole-of-life healthcare. Trends in Molecular Medicine (2014)
2. Zhang, X., et al.: Integrative Omics Technologies in Cancer Biomarker Discovery. Omics Technologies in Cancer Biomarker Discovery 129 (2011)
3. Lundblad, R.L.: Development and Application of Biomarkers. CRC Press (2010)
4. Osl, M., et al.: Applied Data Mining: From Biomarker Discovery to Decision Support Systems. In: Trajanoski, Z. (ed.) Computational Medicine, pp. 173–184 (2012)
5. Fortino, V., et al.: A Robust and Accurate Method for Feature Selection and Prioritization from Multi-Class OMICs Data. PloS One 9 (9) (2014)
6. Somorjai, R.L., et al.: Class Prediction and Discovery Using Gene Microarray and Proteomics Mass Spectroscopy Data: Curses, Caveats, Cautions. Bioinformatics 19, 1484–1491 (2003)
7. Saeys, Y., Inza, I., Larrañaga, P.: A Review of Feature Selection Techniques in Bioinformatics. Bioinformatics 23, 2507–2517 (2007)
8. Wu, M.-Y., et al.: Biomarker Identification and Cancer Classification Based on Microarray Data Using Laplace Naive Bayes Model with Mean Shrinkage. IEEE/ACM Transactions on Computational Biology and Bioinformatics 9, 1649–1662 (2012)

9. Bolón-Canedo, V., et al.: A review of microarray datasets and applied feature selection methods. Information Sciences 282, 111–135 (2014)

10. Sudha George, G.V., et al.: Review on feature selection techniques and the impact of SVM for cancer classification using gene expression profile. International Journal of Computer Science & Engineering Survey 2, 16–27 (2011)

11. Cosmin, L., et al.: A Survey on Filter Techniques for Feature Selection in Gene Expression Microarray Analysis. IEEE/ACM Transactions on Computational Biology and Bioinformatics 9, 1106–1119 (2012)

12. Martinez, E., et al.: Compact cancer biomarkers discovery using a swarm intelligence feature selection algorithm. Computational Biology and Chemistry 34, 244–250 (2010)

13. Saeys, Y., Abeel, T., Van de Peer, Y.: Robust Feature Selection Using Ensemble Feature Selection Techniques. In: Daelemans, W., Goethals, B., Morik, K. (eds.) ECML PKDD 2008, Part II. LNCS (LNAI), vol. 5212, pp. 313–325. Springer, Heidelberg (2008)

14. Cadenas, J.M., et al.: Feature subset selection Filter–Wrapper based on low quality data. Expert Systems with Applications 40, 6241–6252 (2013)

15. Zhang, S., et al.: A new unsupervised feature ranking method for gene expression data based on consensus affinity. IEEE/ACM Transactions on Computational Biology and Bioinformatics (TCBB) 9(4), 1257–1263 (2012)

16. Boucheham, A., Batouche, M.: Robust biomarker discovery for cancer diagnosis based on meta-ensemble feature selection. In: Science and Information Conference (SAI), pp. 452–560. IEEE (2014)

17. Abeel, T., et al.: Robust biomarker identification for cancer diagnosis with ensemble feature selection methods. Bioinformatics 26, 392–398 (2010)

18. Ghorai, S., et al.: Cancer classification from gene expression data by NPPC ensemble. IEEE/ACM Transactions on Computational Biology and Bioinformatics (TCBB) 8(3), 659–671 (2011)

19. Izzo, D., Ruciński, M., Biscani, F.: The Generalized Island Model. In: Fernandez de Vega, F., Hidalgo Pérez, J.I., Lanchares, J. (eds.) Parallel Architectures & Bioinspired Algorithms. SCI, vol. 415, pp. 151–170. Springer, Heidelberg (2012)

20. Boucheham, A., Batouche, M.: Robust Hybrid wrapper/filter Biomarker Discovery based on Generalised Island Model from Gene Expression Data. International Journal of Computational Biology and Drug Design (in press)

21. García-Nieto, J., et al.: Parallel multi-swarm optimizer for gene selection in DNA microarrays. Appl. Intell. 37, 255–266 (2012)

22. Gutiérrez, A.L., et al.: Comparison of different PSO initialization techniques for high dimensional search space problems: A test with FSS and antenna arrays. Antennas and Propagation (EUCAP). In: Proceedings of the 5th European Conference on IEEE, pp. 965–969 (2011)

23. J.: R, Quinlan.: C4.5: programs for machine learning. Morgan Kaufmann Publishers 1 (1993)

24. Alba, E., et al.: Gene selection in cancer classification using PSO/SVM and GA/SVM hybrid algorithms. In: IEEE Congress on Evolutionary Computation, CEC 2007, pp. 284–290. IEEE (2007)

25. Huang, C.-L.: ACO-based hybrid classification system with feature subset selection and model parameters optimization. Neurocomputing 73, 438–448 (2009)

26. Yazdani, S., et al.: Feature subset selection using constrained binary/integer biogeography-based optimization. ISA Transactions 52, 383–390 (2013)

27. Kuncheva, L.I.: A stability index for feature selection. International Multi- Conference. In: Artificial Intelligence and Applications, pp. 390–395. ACTA Press Anaheim, CA (2007)

DEgenes Hunter - A Self-customised Gene Expression Analysis Workflow for Non-model Organisms

Isabel González Gayte[1], Rocío Bautista Moreno[2], and M. Gonzalo Claros[1,2]

[1] Departamento de Biología Molecular y Bioquímica, Universidad de Málaga,
29071 Málaga, Spain
[2] Plataforma Andaluza de Bioinformática, Centro de Supercomputación y
Bioinnovación, Universidad de Málaga,
29071 Málaga, Spain

Abstract. Data from high-throughput RNA sequencing require the development of more sophisticate bioinformatics tools to perform optimal gene expression analysis. Several R libraries are well considered for differential expression analyses but according to recent comparative studies, there is still an overall disagreement about which one is the most appropriate for each experiment. The applicable R libraries mainly depend on the presence or not of a reference genome and the number of replicates per condition. Here it is presented DEgenes Hunter, a RNA-seq analysis workflow for the detection of differentially expressed genes (DEGs) in organisms without genomic reference. The first advantage of DEgenes Hunter over other available solutions is that it is able to decide the most suitable algorithms to be employed according to the number of biological replicates provided in the sample. The different workflow branches allow its automatic self-customisation depending on the input data, when used by users without advanced statistical and programming skills. All applicable libraries served to obtain their respective DEGs and, as another advantage, genes marked as DEGs by all R packages employed are considered 'common DEGs', showing the lowest false discovery rate compared to the 'complete DEGs' group. A third advantage of DEgenes Hunter is that it comes with an integrated quality control module to discard or disregard low quality data before and after preprocessing. The 'common DEGs' are finally submitted to a functional gene set enrichment analysis (GSEA) and clustering. All results are provided as a PDF report.

Keywords: RNA-seq, R, pipeline, workflow, differential expression, bioinformatic tool, functional analysis.

1 Introduction

Nowadays, high-throughput technologies are well considered for genetic studies. For the analysis of gene expression profiles, data are obtained from RNA sequencing (RNA-seq) experiments. RNA-seq provides precise measurements of

F. Ortuño and I. Rojas (Eds.): IWBBIO 2015, Part II, LNCS 9044, pp. 313–321, 2015.

transcript levels at a wide dynamic range. The high amount of information generated by RNA-seq has to be studied by statistical approaches but, although there has been put much effort in this matter, there is still an overall disagreement about which statistical algorithm is the most appropriated to get the most reliable results [1].

The read counting variability is an issue, since it can lead to statistical misinterpretation of data [2]. Therefore, user should always be acquainted about the quality of the data. Furthermore, the result reliability is seriously affected by the experimental design: the number of replicates is crucial and has to be especially considered to study gene expression, even though some algorithms can work with few replicates. Hence, no algorithm is well suited for whatever experimental design.

Normalization is an essential step to study gene expression levels, but it is a fact that it can impact the downstream analysis. Several normalisation methods have been compared up to date, showing that the Trimmed Mean of M-values (TMM), among other normalisation approaches, shows a reasonable false-positive rate without any loss of power [3]. Therefore, all R-packages in DEgenes Hunter performs count data normalisation with TMM.

The gene expression bias is another aspect to consider since the proportion of up- and down-regulated genes among the differentially expressed genes (DEGs) can be highly variable. Notwithstanding, statistical inference packages often require the ratio of up- and down-regulated genes close to one for normalisation purposes. TMM normalisation works with the premise that most of the genes in a sample are not differentially expressed and, moreover, works with the assumption that DEGs up- and down-regulation ratio should be similar [3]. Not considering this premise could drive to misinterpretation of DEGs.

From the previous paragraphs, it can be deduced that data variability, normalisation, biases and experimental design are issues to be taken into account when performing accurate gene expression analyses. The workflow of DEgenes Hunter was intended to automate and simplify the RNA-seq analyses for skilled and non-skilled users. DEgenes Hunter detects the number of replicates provided in an experiment and creates the design matrix without any need of instructions. The software is also able to apply the most appropriate R libraries for the experimental design provided to study gene expression without user intervention. Since all R libraries employed are based on quite different rationales, common DEGs in all analyses seem to be highly reliable based on their low false discovery rate (FDR). The high reliability of the common DEGs group was confirmed by testing DEgenes Hunter with a real-world dataset from a non-model organism. To check DEgenes Hunter performance, the pipeline has been run with synthetic data, showing that it is able to overcome problems arising when disequilibrium between up- or down regulation is present. At the end, a pdf report with all results and additional information is provided.

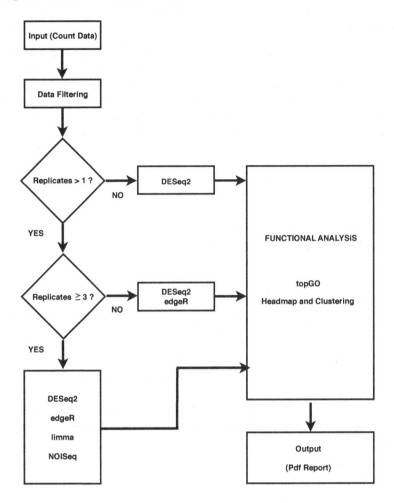

Fig. 1. DEgenes Hunter main workflow

2   Methods

2.1   Data Import

DEgenes Hunter requires as input data the matrix count. Raw counts of sequencing reads can be imported in a tab-delimited file in the form of a matrix of integer values, but a BAM file would be used in a near future. For the GSEA analysis, an annotation file as well as a file containing the mapping that associates gene ontology (GO) terms for each gene are required as tab-delimited files. Gene name or COG rows can also be present for more complete GSEA analyses.

2.2 Experimental Data

Triplicate real data of *Solea senegalensis* were obtained from SRX655595 and SRX655596 [4], where previous analyses indicated that 20% of genes were DEGs (Bautista et al, in preparation). Synthetic data were designed with a 20% of genes being DEGs with three unbalanced up- and down-regulation distributions (100/0 [all up-regulated], 50/50 [balanced] and 0/100 [all down-regulated]) to test the workflow robustness when data are extremely unbalanced. Real and synthetic data were used as source of one, two and three replicates per condition.

2.3 Normalisation and Data Filtering

Transcripts with less than two read mapped were discarded, as previously described [2]. Count data normalisation has been performed with TMM [5]. Normalised read counts are provided to the end user.

2.4 R Packages and DEG Analyses

R packages dedicated to DEG detection, based on different algorithms, are: *DESeq2* [6], *edgeR*, [7], *limma* [8] and *NOISeq* [9]. Packages included for graphic representations of common genes are *VennDiagram* [10] and *ggplot2* [11]. A data check is performed using *compcodeR* [12]. GSEA functional analyses, clustering and heatmaps were implemented using *topGO* [13] and *cluster* [14].

Any DEG should have an adjusted $P < 0.05$ and a $FDR < 0.01$; in the case of *NOISeq*, DEGs should have $q > 0.95$ ($q = 1 - P$). The minimum value of the logarithm of fold change (logFC) was set to 0.6. These parameters are customisable.

3 Results

3.1 Workflow Details

The workflow of DEgenes Hunter is shown on Fig. 1. After loading files, genes having less than two reads mapped are discarded. Then, the number of replicates determine the branch to be used: only *DESeq2* when there is no replicate, *edgeR* and *DESeq2* when there are two replicates, and *DESeq2*, *edgeR*, *limma* and *NOISeq* when three or more replicates are present. Since every package is based on different statistic principles, genes detected as differentially expressed by all R packages used in each case are considered as highly reliable, conforming the 'common DEGs' gene set, while all DEGs appeared in any analysis conforms the 'complete DEGs' group. The workflow has been provided with functions that allow the further analysis of 'common DEGs', for example with GSEA functional analyses (Fig. 2A), heatmaps (Fig 2B) and clustering (Fig. 2C).

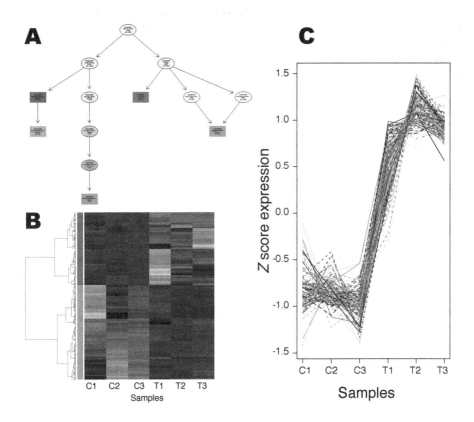

Fig. 2. Example analyses that can be performed with DEgenes Hunter on the 'common DEGs' group. A: A GSEA analysis performed with *topGO*, where rectangle colour represents the relative significance, ranging from dark red (most significant) to bright yellow (least significant). B: A typical heatmap that can also be used as a quality control to verify that control samples (C1, C2 and C3) and treatment samples (T1, T2 and T3) are grouped together. C: Expression clustering performed using *cluster* where the genes have similar expression levels among control samples, and a clearly higher value in treatment samples.

3.2 Performance Testing

Utility of 'common DEGs' group was confirmed comparing their FDR values. Figure 3 shows that the FDR for 'common DEGs' is considerably lower than for 'complete DEGs' and 'non-common DEGs' using separately any R package. Since there is no clear way to set the threshold for *qNOISeq* [15], it is very high in all cases.

Performance tests with synthetic data containing an unbalanced distribution of DEGs (Fig. 4) showed that the number of DEGs in the balanced (50/50) dataset was only slightly higher than in the unbalanced cases (ratios 100/0 and 0/100), indicating that the DEgenes Hunter workflow is robust in detecting DEGs.

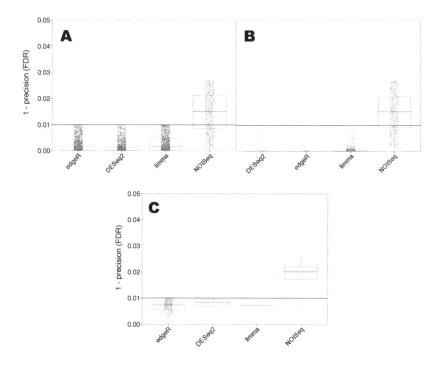

Fig. 3. Plot of FDR values of DEGs detected by packages (*DESeq2, edgeR, limma* and *NOISeq*) with real data in the most suitable conditions (three replicates) of 'complete DEGs' (A), 'common DEGs' (B) and 'non-common DEGs' (C) groups

3.3   PDF Report

To facilitate the analysis interpretation by the user, DEgenes Hunter creates a PDF report containing all details from the complete downstream analysis. This PDF contains DEG heatmaps for quality control and results regarding the 'common DEGs', genes marked as DEGs by only one of the R packages employed ('non-common DEGs') and the 'complete DEGs' groups. Result tables of all R-packages performed are included listing supplementary information about the DEGs. DEgenes Hunter also provides graphic plots for illustration of the obtained results. Normalisation tables are included for user's information. GSEA nodes plotted and expression clustering graphics (such as those in Fig 2) are provided in this PDF as well.

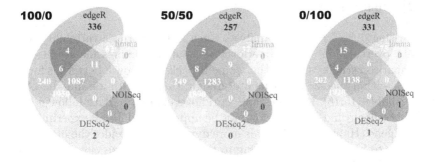

Fig. 4. Venn diagrams showing the numbers of DEGs found in synthetic data when different DEG ratios are used. 100/0 corresponds to all over-expressed/none repressed, 50/50 is the balanced ratio, and 100/0 corresponds to none over-expressed/all repressed.

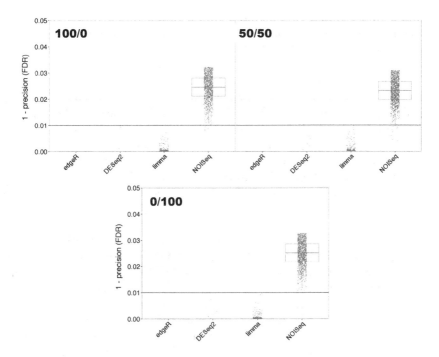

Fig. 5. Plots of FDR values of 'common DEGs' detected by packages (*DESeq2, edgeR, limma* and *NOISeq*) with synthetic data in different expression ratios. 100/0 corresponds to all over-expressed/none repressed, 50/50 is the balanced ratio, and 100/0 corresponds to none over-expressed/all repressed.

4   Discussion

Issues arising during RNA-seq analyses, such as the intrinsic variability of data, has to be overcome in order to perform optimal gene expression studies. Previous comparison studies showed that some R packages show a better performance when run with a specific experimental design, that is, with a minimum of replicates available [15,16]. Pipelines and workflows created to date follow a specific invariable scheme and can therefore not be appropriated for all experimental designs. DEgenes Hunter's workflow is more dynamic and decides the most suitable algorithms to be employed according to the automated detection of the number of biological replicates provided in the sample (Fig. 1).

For detecting the replicate numbers, no design matrix has to be supplied. The pipeline is able to create itself an appropriated design matrix. DEgenes Hunter is capable to work in a complete automatic manner without any need of advanced skills by the user. The software makes the most complicate decisions to get the most reliable DEGs in an experiment. Default values of global parameters are set considering manuals and previous comparison studies [1,16,15,2]. Nevertheless, since default values can be changed by the user, any analysis is customised firstly by the automatic selection of R algorithms according to the number of replicates provided, and secondly by the manual adjustment of global parameters.

R libraries used to calculate data dispersion in the different branches (Fig. 1) are based on quite different algorithms [2]. This is why 'common DEGs' are highly reliable, as was demonstrated by their FDR compared to the 'complete DEGs' group (Fig. 3 and Fig. 5). When analysing real data from *S. senegalensis*, the FDR values from the 'common DEGs' found were far under the 0.01 threshold (Fig. 3B). The same occurs with synthetic data, in spite of the most extreme over- and down-regulation ratios (Fig. 5). In comparison, genes marked as DEGs only by one of the R algorithms employed show much higher FDR values (Figs. 3 and 5). Therefore, instead of using just one of the most appropriate algorithm for each particular dataset [1], we demonstrate that the combination of the most suitable algorithms provides a smaller but more reliable collection of DEGs.

It is well known that the up- and down-regulated gene ratio can affect result reliability after data normalisation. TMM may show a poor performance when used with specific data types due to its assumption that most genes are not differentially expressed. Another premise of this normalisation strategy is that in the case that genes are differentially expressed, they are equally up- and down-regulated [3]. The more these statements departs from reality, the poorer results are obtained. Nevertheless, the present study shows that in the cases of the most extreme disequilibrium, the number of 'common DEGs' found (100/0 and 0/100 in Fig. 4) was only slightly lower than in the ideal case when the ratio was balanced (50/50 in Fig. 4). Moreover, these 'common DEGs' showed, such as in the case of our real data analysis, very low FDR values (Fig. 5). DEgenes Hunter is therefore able to detect a high amount of DEGs with a very low FDR value even under the worst starting conditions.

Finally, the PDF report generated by DEgenes Hunter gather expression results, tables, and graphics created by the different R packages (such as those of

Fig. 2), and offers an overview about the data quality: if input data were of poor quality it will be noticed, which may help to avoid mistakes of committing Type I errors. As a whole, the PDF serves to explain results to an unskilled scientist.

References

1. Gao, D., Kim, J., Kim, H., Phang, T.L., Selby, H., Tan, A.C., Tong, T.: A survey of statistical software for analysing rna-seq data. Human Genomics 5(1), 56 (2010)
2. Rapaport, F., Khanin, R., Liang, Y., Pirun, M., Krek, A., Zumbo, P., Mason, C.E., Socci, N.D., Betel, D.: Comprehensive evaluation of differential gene expression analysis methods for rna-seq data. Genome Biol. 14(9), R95 (2013)
3. Dillies, M.-A., Rau, A., Aubert, J., Hennequet-Antier, C., Jeanmougin, M., Servant, N., Keime, C., Marot, G., Castel, D., Estelle, J., et al.: A comprehensive evaluation of normalization methods for illumina high-throughput rna sequencing data analysis. Briefings in Bioinformatics 14(6), 671–683 (2013)
4. Benzekri, H., Armesto, P., Cousin, X., Rovira, M., Crespo, D., Merlo, M.A., Mazurais, D., Bautista, R., Guerrero-Fernández, D., Fernandez-Pozo, N., et al.: De novo assembly, characterization and functional annotation of senegalese sole (solea senegalensis) and common sole (solea solea) transcriptomes: integration in a database and design of a microarray. BMC Genomics 15(1), 952 (2014)
5. Robinson, M.D., Oshlack, A., et al.: A scaling normalization method for differential expression analysis of rna-seq data. Genome Biol. 11(3), R25 (2010)
6. Love, M., Anders, S., Huber, W.: Differential analysis of rna-seq data at the gene level using the deseq2 package (2013)
7. Robinson, M.D., McCarthy, D.J., Smyth, G.K.: edger: A bioconductor package for differential expression analysis of digital gene expression data. Bioinformatics 26(1), 139–140 (2010)
8. Law, C.W., Chen, Y., Shi, W., Smyth, G.K.: Voom: Precision weights unlock linear model analysis tools for rna-seq read counts. Preprint 2013 (2013)
9. Tarazona, S., García, F., Ferrer, A., Dopazo, J., Conesa, A.: Noiseq: A rna-seq differential expression method robust for sequencing depth biases. EMBnet Journal 17(B), 18–19 (2012)
10. Chen, H., Boutros, P.C.: Venndiagram: A package for the generation of highly-customizable venn and euler diagrams in r. BMC Bioinformatics 12(1), 35 (2011)
11. Wickham, H.: ggplot2: elegant graphics for data analysis. Springer (2009)
12. Soneson, C.: Compcoder-an r package for benchmarking differential expression methods for rna-seq data. Bioinformatics, btu324 (2014)
13. Alexa, A., Rahnenfuhrer, J.: topGO: enrichment analysis for gene ontology. R package version 2.8 (2010)
14. Maechler, M., Rousseeuw, P., Struyf, A., Hubert, M., Hornik, K.: cluster: Cluster Analysis Basics and Extensions. R package version 1.15.3 — For new features, see the 'Changelog' file (in the package source) (2014)
15. Soneson, C., Delorenzi, M.: A comparison of methods for differential expression analysis of rna-seq data. BMC Bioinformatics 14(1), 91 (2013)
16. Kvam, V.M., Liu, P., Si, Y.: A comparison of statistical methods for detecting differentially expressed genes from rna-seq data. American Journal of Botany 99(2), 248–256 (2012)

Bioinformatics Analyses to Separate Species Specific mRNAs from Unknown Sequences in *de novo* Assembled Transcriptomes

David Velasco, Pedro Seoane, and M. Gonzalo Claros

Departamento de Bioquímica y Biología Molecular,
Universidad de Málaga, Málaga MA 29071, Spain
{davvelsan,seoanezonjic,claros}@uma.es

Abstract. The use of RNA-Seq has transformed the way sequencing reads are analyzed, allowing for qualitative and quantitative studies of transcriptomes. These studies always include an important collection (usually > 40%) of unknown transcripts. In this study, we improve the capability of Full-LengtherNext, an algorithm developed in our laboratory to annotate, analyze and correct *de novo* transcriptomes, to detect of potentially coding sequences. Here we analyze five software implementations of coding sequence predictors and show that the use of high-quality sequences at the training stage, proper threshold selection during score interrogation and the algorithm adaptation to its input type have a profound effect on the accuracy of the prediction. TransDecoder, the best performing algorithm in our tests, was thus added to the Full-LenghterNext pipeline, significantly improving its coding prediction reliability. Moreover, these analyses served to make inferences about the quality of the sample and to extract the subset of species specific (perhaps novel) genes discovered in the transcriptome assembly. Indirectly, we also demonstrated that Full-LentherNext sequence classification is appropriate and worth taking into consideration.

1   Introduction

The advent of NGS entailed a paradigm change in the way sequencing technologies are applied to transcriptomics. Through RNA-Seq, we have higher exon mapping resolution and the ability to assemble a full transcriptome, even when lacking a reference genome to match against [19]. These advantages come at the cost of having to process very large amounts of sequencing data, requiring special considerations when performing a *de novo* assembly, and having to rely on custom heuristic algorithms and sophisticated stochastic models to analyze and asses the quality of the sample.

In sequencing projects, from data pre-processing to assembly, error factors accumulate at every step, translating into a net loss of sample resolution and annotating power of the analysis tools. Genomic, viral, vector and other ectopic nucleic acid material can dramatically enlarge the size of the final transcriptome assembly and produce a significant amount of contaminant transcripts

F. Ortuño and I. Rojas (Eds.): IWBBIO 2015, Part II, LNCS 9044, pp. 322–332, 2015.
© Springer International Publishing Switzerland 2015

during annotation [6]. Sequencing errors can seriously impede research oriented towards the discovery of single nucleotide polymorphisms (SNPs) [11] and low frequency variants [21]. There have been reports of obtaining wildly different results depending on the assembler used and the contiguity of the sample [15]. These problems are aggravated when assembling a *de novo* transcriptome from RNA-Seq data, necessitating specialized software for this particular purpose.

Even with the large amount of software available and the dedicated platforms in which they integrate constantly evolving and adapting to the new challenges, functional annotation of transcriptome sequence data remains a difficult task. Sequence annotation is usually performed on the basis of sequence homology analyses, adding a certain degree of error propagation to functional and structural studies [12,16]. While the availability of genome and transcriptome sequencing data for non-model organisms has increased at a very fast pace since the advent of NGS [5], they are still considerably under-annotated with regards to popular ones, leaving a significant amount of potentially protein coding sequences in the unknown limbo.

The pool of unknown transcripts remaining after functional annotation may be the result of sample contamination, errors during data processing, or novel or species specific genes. To solve this problem, various computational approaches have been implemented to find patterns unique to coding or non-coding regions [8,22,10,17,20,18]. Full-LengtherNext (FLN) is an in-development workflow solution designed for structural integrity annotation, chimera detection and quality control of assembled RNA-Seq data. It separates novel or specialized transcripts from potentially 'junk' sequences using a slightly modified version of the Test-Code statistic [8]. However, with sequencing technologies constantly evolving, this particular module required a review in the light of recent advancements published in the field.

2   Materials and Methods

2.1   Prediction Algorithms

In addition to the already implemented version of TestCode [8], four bioinformatics tools were selected for performance evaluation and sample analysis: two prokaryotic gene finders (GeneMarkHMM [2] and Glimmer [4]), and two mRNA coding region predictors (TransDecoder and ESTScan [13]). Software tools were launched accordingly to the developer instructions. ESTScan, Glimmer and GeneMark coding thresholds were dynamically adjusted to minimize false negative detection (below 0.5%). TransDecoder and the TestCode statistic cut-off scores were kept at their default values, per developer specification. Algorithm training, when required, was done with subsets of sequence samples not included in the experimental or testing groups.

2.2   Gold Standard Dataset

A Gold Standard (GS) was constructed with selected sequences of *Arabidopsis thaliana* (thale cress, 8, 265 seqs), *Drosophila melanogaster* (fruit fly, 8, 380 seqs),

Gallus gallus (chicken, 2, 174 seqs) and *Sus scrofa* (pig, 2, 575 seqs) from their NCBI-maintained UniGene transcriptome database. Uniprot sequences of each organism, verified to be complete, gap-less and having an N-terminal methionine, were clustered using the CD-HIT software [9] with a 40% identity threshold to eliminate redundancy, then used to generate a local BLAST database. Translated UniGene transcripts were matched against this database using BLASTx. From the resulting hits, those which had a 100% protein identify and at least 10 nucleotides up and downstream of each UTR were selected. For each organism, using only one UniGene per reference protein, the following experimental groups of sequences were designed:

- *Training*, containing 60% of the sequences selected for the particular organism. It served to train the predictive models, when required.
- *Full-Length*, consisting of the remainder 40% sequences. It served as a high-quality coding control.
- *Trimmed-Length*, a modification of the *Full-Length* group by removing the UTR regions plus 100 nucleotides on each end of every sequence. It served as a low-quality coding control to assess the response of algorithms on structurally incomplete coding sequences.
- *UTR*, consisting on *Full-Length* sequences devoid of the coding region. It served as a native, non-coding control.
- *Random*, including sequences between 250-5000 nucleotides long of equiprobable codon distributions. It served as a random, non-coding control.

2.3   Full-LengtherNext Transcriptome Analysis and Annotation

De novo assembled transcriptomes for *Olea europaea* (ReprOlive v1.0), *Pinus pinaster* [3] (SustainPineDB v3.0) and *Solea senegalensis* [1] (SoleaDB v4.0) available in our laboratory were analyzed with a development version of FLN platform (Seoane *et al*, in preparation). The assembled transcriptome was sequentially probed against user-provided, Swiss-Prot and TrEMBL databases for matching sequences. During annotation, any potential frame-shift errors, misalignments and chimeric constructs were evaluated and processed accordingly. Matching sequences were assigned an integrity status (e.g. *Complete, N-Terminal, Internal, C-Terminal*) and a quality (*Sure, Putative*) for such prediction. Any sequence without a match was classified as unknown and passed to a custom TestCode algorithm for coding prediction.

2.4   Preparation of *de novo* Transcriptome Samples

For each transcriptome, four experimental groups were constructed from their FLN annotation output:

- *Complete Sure*: Sequences classified as complete sure by FLN. These sequences have identified translation start and stop codons, usually retain extensive UTR regions and their CDS either exactly matches the reference in the database, or if they have a divergent region it is contained within the bounds of the reference. Only the longest sequence per matching accession number was considered (4,037 for olive, 8,158 for pine, 10,909 for sole). This group served as high-quality training set and positive control.

- *Complete Putative*: Sequences classified as complete putative by FLN. These sequences lack translation start or stop codons, have incomplete or missing UTR regions and may contain extensive divergent regions. Only the longest sequence per matching accession number was considered (1,578 for olive, 1,679 for pine, 2,617 for sole). This group served as low-quality training set and positive control.
- *Unknown*: Sequences without a match in any database (19,848 for olive, 110,501 for pine, 490,555 for sole). This group served as an experimental set of unknown sequences to query for the presence of potentially novel or highly divergent transcripts.
- *Random*, including sequences between 250-5000 nucleotides long of equiprobable codon distributions. It served as a random, non-coding control.

3   Results and Discussion

3.1   Software Comparison Using Known Coding Sequences

To asses the response of the prediction algorithms under ideal conditions (i.e. with a known output), we measured their performance using the Gold Standard dataset (see Section 2.2) as input. Their performance comparison, displayed in Figure 1, shows that ESTScan and TransDecoder were able predict coding sequences more accurately than any of the other tools across all organism samples. Our results indicate that training with high-quality coding sequences is enough to detect > 90% of the complete transcripts (Full-Length group) and > 70% of the incomplete (Trimmed-Length group), without any significant detection of false positives.

Interestingly, the Trimmed-Length panel in Figure 1 suggests that coding detection by the untrained TestCode algorithm, might have been less adversely affected than the other four trained algorithms by the structural changes performed on the Trimmed-Length group. This could be indicative of a dependency of these algorithms for patterns, such RBS motifs and translation start and stop sites, that are usually present only in complete, high-quality sequences. These specific structures, combined with the fortuitous presence of translation start and stop sites, could also explain the elevated rate of false positives shown in the UTR group in Figure 1.

A common technique to diagnose the performance and behavior of signal detection algorithms is to evaluate them in terms of sensitivity, specificity, precision and accuracy [7]. In our tests, the full-length sequences were used as true positive controls while the random sequences were used as the true negative controls. As shown in Table 1, ESTScan and TransDecoder denote a superior response, both showing a high rate of relevant information retrieval (sensitivity and specificity), exclusion rates (accuracy) and consistency of the results (precision). GeneMark

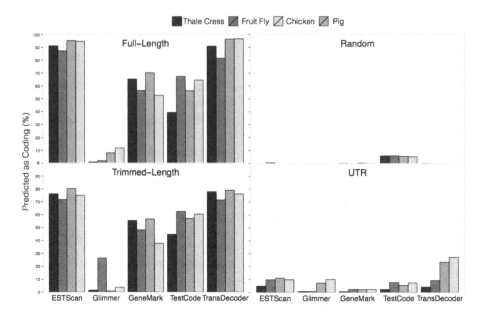

Fig. 1. Results of coding sequence prediction by the five predictor algorithms using the Gold Standard Dataset as a source of different types of known transcripts

very successfully excluded the true negative sequences from the sample (specificity), but fell short on the positive detection (sensitivity), which reflected poorly on its accuracy results. The TestCode results, albeit with a notably higher false positive detection ratio, showed a very similar performance to the GeneMark algorithm. In the case of Glimmer, due to its low coding detection rates, other than its sensitivity, the parametric values were not very informative (it is not surprising that overall low detection leads to low or no false negative detection).

3.2   Software Comparison Using *de novo* Transcriptomes

To analyze the performance of the prediction software in a more realistic scenario, we analyzed the *de novo* assembled transcriptomes of olive, pine and sole. These transcriptome samples differ from each other in the number of transcripts and the presence of contaminant sequences. Quantitatively, the sole transcriptome is significantly larger (622, 699 transcripts) than the ones of pine and olive (208, 105 and 73, 226, respectively). In addition, the sole transcriptome has a comparatively large amount of unknown sequences (490, 555) with respect to the other two (19, 848 and 110, 501 in olive and pine, respectively). Qualitatively, the group of unknown sequences in sole has been demonstrated to contain substantial genomic contamination (> 95%; [1]).

Our results corroborate that both ESTScan and TransDecoder can make accurate predictions in *de novo* transcriptome samples, provided that adequately

Table 1. Algorithm performance comparison for the five coding sequence predictors based on positive and negative detection rates. **Input**: Total number of input sequences. **TP**: True Positives. **FN**: False Negatives. **TN**: True Negatives. **FP**: False Positives. **Sensitivity**: measures the proportion of actual positives correctly identified, as $TP/(TP+FN)$. **Specificity**: measures the proportion of actual negatives correctly identified, as $TN/(TN+FP)$. **Precision**: evaluates the degree to which repeated measurements under unchanged conditions show the same results (i.e. reproducibility of the measurement), as $TP/(TP+FP)$. **Accuracy**: indicates proximity of measurement to the true value, as $(TP+TN)/(TP+FN+TN+FP)$. **Relative Error Quotient (REQ)**: assesses the prediction methods as $(FN \times W + FP)/TP1(1+W)$, where a low value represents a low proportion of errors and higher value represents a higher proportion [14].

Organism	Algorithm	Input	TP	FN	TN	FP	Sensitivity	Specificity	Precision	Accuracy	REQ
Thale Cress	ESTScan	5812	2648	258	2904	2	0.91122	0.99931	0.99925	0.95526	0.04909
	Glimmer	5812	27	2879	2906	0	0.00929	1.00000	1.00000	0.50465	53.31481
	GeneMark	5812	1900	1006	2898	8	0.65382	0.99725	0.99581	0.82553	0.26684
	TestCode	5812	1145	1761	2737	169	0.39401	0.94184	0.87139	0.66793	0.84279
	TransDecoder	5812	2640	266	2899	7	0.90847	0.99759	0.99736	0.95303	0.05170
Fruit Fly	ESTScan	6704	2927	425	3331	21	0.87321	0.99374	0.99288	0.93347	0.07619
	Glimmer	6704	68	3284	3352	0	0.02029	1.00000	1.00000	0.51014	24.14706
	GeneMark	6704	1891	1461	3349	3	0.56414	0.99911	0.99842	0.78162	0.38710
	TestCode	6704	2262	1090	3153	199	0.67482	0.94063	0.91914	0.80773	0.28492
	TransDecoder	6704	2734	618	3348	4	0.81563	0.99881	0.99854	0.90722	0.11375
Chicken	ESTScan	1740	828	42	869	1	0.95172	0.99885	0.99879	0.97529	0.02597
	Glimmer	1740	71	799	870	0	0.08161	1.00000	1.00000	0.54080	5.62676
	GeneMark	1740	611	259	867	3	0.70230	0.99655	0.99511	0.84943	0.21440
	TestCode	1740	490	380	822	48	0.56322	0.94483	0.91078	0.75402	0.43673
	TransDecoder	1740	839	31	870	0	0.96437	1.00000	1.00000	0.98218	0.01847
Pig	ESTScan	2060	975	55	1029	1	0.94660	0.99903	0.99898	0.97282	0.02872
	Glimmer	2060	123	907	1030	0	0.11942	1.00000	1.00000	0.55971	3.68699
	GeneMark	2060	545	485	1028	2	0.52913	0.99806	0.99634	0.76359	0.44679
	TestCode	2060	665	365	976	54	0.64563	0.94757	0.92490	0.79660	0.31504
	TransDecoder	2060	996	34	1030	0	0.96699	1.00000	1.00000	0.98350	0.01707
Summary	ESTScan	16316	7378	780	8133	25	0.90439	0.99694	0.99662	0.95066	0.05455
	Glimmer	16316	289	7869	8158	0	0.03543	1.00000	1.00000	0.51771	13.61419
	GeneMark	16316	4947	3211	8142	16	0.60640	0.99804	0.99678	0.80222	0.32616
	TestCode	16316	4562	3596	7688	470	0.55921	0.94239	0.90660	0.75080	0.44564
	TransDecoder	16316	7209	949	8147	11	0.88367	0.99865	0.99848	0.94116	0.06658

sized and optimal training sets are available to generate their predictive models (Figure 2, Left). However, ESTScan performance under suboptimal training conditions (Figure 2, Right) indicates its algorithms may be highly dependent on training set quality. Interestingly, ESTScan was able to regain its previous detection sensitivity in the larger sole sample, suggesting that the combination of training sample size and sequence quality is determinant of its accuracy in detection. In contrast, TransDecoder showed a higher resilience to training quality fluctuations and was able to make accurate predictions in all organism samples under both optimal and suboptimal training. These results imply that TransDecoder is the preferable choice of the two.

GeneMark average detection of complete sure sequences, under complete sure training conditions (Figure 2, Left), was significantly superior to its Full-Length results for the GS sequence tests, and almost comparable to the rates displayed by ESTScan and Transdecoder. However, complete putative coding detection was low and in the same range as the untrained TestCode algorithm. Under complete putative training conditions (Figure 2, Right), complete sure coding detection was greatly compromised, whereas low-quality sequence detection increased considerably. Together, these results might indicate an overly aggressive specialization of its learning algorithms towards the particular sequence type it was trained with, something undesirable when analyzing unknown sequences in *de novo* assembled transcriptomes, where sample heterogeneity is dominant.

Glimmer results for both high and low quality sequences, under optimal and suboptimal training conditions, barely reported any coding detection. This clearly shows the unsuitability of the software for transcript coding prediction. Nevertheless, it was interesting to see its consistently elevated detection for sequences in the unknown control (Unknown panels in Figure 2). While it is not possible to state with certainty, this increased response might suggest the presence of particular sequences types which Glimmer is more optimized to detect, such as sequences of genomic origin.

Finally, the TestCode results showed a remarkable consistency of this algorithm across all organism and control samples (Figure 2). This behaviour would support the existence of a period-3 pattern characteristic of coding sequences. Unfortunately, this detection method does not seem sufficient to correctly classify most sequences in a given sample and it incurs in severe false positive detection. While the trend is certainly there, the lack of a trained model impairs the algorithm detection capabilities in comparison to optimized, trained models, consigning this particular implementation of the TestCode to a more generalist role. However, it is our belief that the FLN implementation could be improved by dynamically recalculating the TestCode probability matrices accordingly to the input it receives. That is, by giving it a training step.

3.3   Validation of FLN Sure and Putative Quality Assessment

The fact that tool performance, particularly ESTscan, is sensitive to training with *Complete Sure* and *Complete Putative* sequences (Figure 2) indirectly suggests that a qualitative difference does indeed exist between the qualifiers 'sure' and 'putative' provided by FLN. GeneMark results in this regard were perhaps less clear about which group yielded a better performance, but its prediction outcomes under different training conditions support the existence of a qualitative difference among the two groups.

TransDecoder has an additional module that relays information about the integrity of the predicted sequence. In particular, it can determine which sequences can be considered structurally complete (i.e. when the predicted gene has in-frame start and stop codons). When comparing the amount sequences detected by TransDecoder as 'complete' with its total detection ratio in the FLN *Complete* groups (Figure 3), we observed that a majority of *Complete Sure*

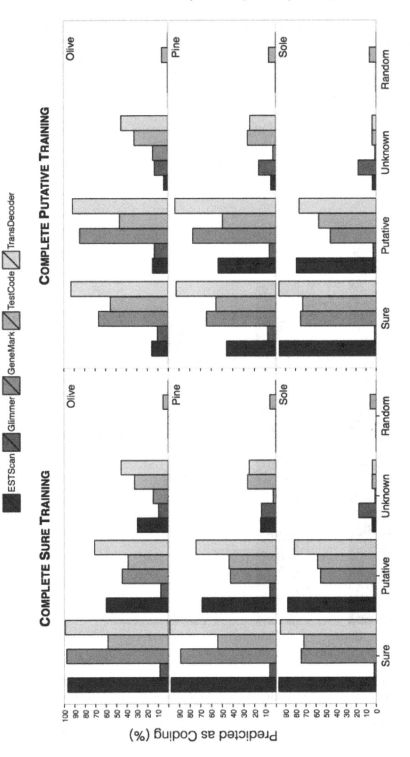

Fig. 2. Algorithm performance using *de novo* assembled transcriptome datasets, when trained with optimal (complete sure, left pannels) and suboptimal (complete putative, right pannels) sequences. TestCode results under Complete Putative Training are included only for consistency, since it does not require any training.

sequences were corroborated as being complete by TransDecoder, while in the *Complete Putative* group there was a significant reduction in all samples. Results that further support the quality assessment of the FLN algorithm.

3.4   Detection of Species-Specific Transcripts

The unknown or unannotated set of transcripts after a *de novo* assembly may be the consequence of genomic contamination, sequencing and assembly artifacts, or new and species-specific sequences. Because the reliability of TransDecoder in detecting coding sequences has been proved (Figure 1 and Figure 2), it follows that the number unknown of sequences predicted as coding by TransDecoder

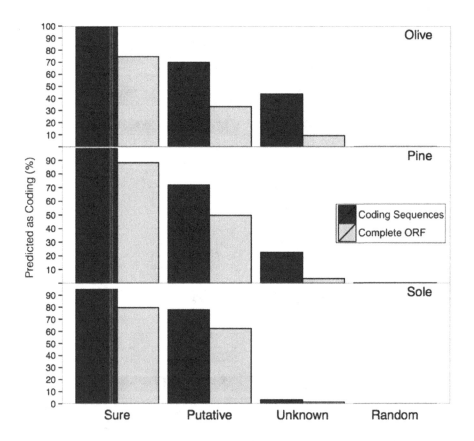

Fig. 3. Distribution of TransDecoder 'coding' and 'complete' qualifiers, when trained with *Complete Sure* transcripts, with respect to *Sure* and *Putative* qualifiers of FLN. 'Coding sequences' refers to the whole set of coding sequences predicted by TransDecoder. 'Complete ORF' refers to the subset of coding sequences that may contain a complete ORF.

$(8, 928$ in olive, $26, 675$ in pine and $19, 898$ in sole) could indeed be truly novel genes, or species-specific transcripts. This seemed corroborated in the case of the transcriptome of sole, where most of the unknown transcripts ($> 95\%$), later confirmed as of genomic origin [1], were ignored.

Furthermore, filtering TransDecoder results to retain only 'complete' sequences can help to further sieve the sample for the most likely coding transcripts. Complete transcript ratios in the Unknown group (Unknown in Figure 3) fall dramatically when compared to the total detection (olea $= 1, 836$, pine $= 3, 715$ and sole $= 6, 700$), suggesting that only a small fraction of these sequences are truly complete genes. Indeed, TransDecoder-detected complete sequences are tempting candidates for further analyses, and their relatively reduced numbers facilitate this task. Further steps to study these transcripts might include sequence clustering to eliminate redundancy, less stringent homology tests to capture loosely related sequences, and structural or functional characterization.

References

1. Benzekri, H., Armesto, P., Cousin, X., Rovira, M., Crespo, D., Merlo, M.A., Mazurais, D., Bautista, R., Guerrero-Fernández, D., Fernandez-Pozo, N., Ponce, M., Infante, C., Zambonino, J.L., Nidelet, S., Gut, M., Rebordinos, L., Planas, J.V., Bégout, M.L., Claros, M.G., Manchado, M.: De novo assembly, characterization and functional annotation of Senegalese sole (Solea senegalensis) and common sole (Solea solea) transcriptomes: integration in a database and design of a microarray. BMC Genomics 15, 952 (2014)
2. Besemer, J., Borodovsky, M.: Heuristic approach to deriving models for gene finding. Nucleic Acids Research 27(19), 3911–3920 (1999)
3. Canales, J., Bautista, R., Label, P., Gómez-Maldonado, J., Lesur, I., Fernández-Pozo, N., Rueda-López, M., Guerrero-Fernández, D., Castro-Rodríguez, V., Benzekri, H., Cañas, R.A., Guevara, M.A., Rodrigues, A., Seoane, P., Teyssier, C., Morel, A., Ehrenmann, F., Le Provost, G., Lalanne, C., Noirot, C., Klopp, C., Reymond, I., García-Gutiérrez, A., Trontin, J.F., Lelu-Walter, M.A., Miguel, C., Cervera, M.T., Cantón, F.R., Plomion, C., Harvengt, L., Avila, C., Gonzalo Claros, M., Cánovas, F.M.: De novo assembly of maritime pine transcriptome: implications for forest breeding and biotechnology. Plant Biotechnology Journal 12(3), 286–299 (2014)
4. Delcher, A.L., Bratke, K.A., Powers, E.C., Salzberg, S.L.: Identifying bacterial genes and endosymbiont DNA with Glimmer. Bioinformatics (Oxford, England) 23(6), 673–679 (2007)
5. Ellegren, H.: Genome sequencing and population genomics in non-model organisms. Trends in Ecology & Evolution 29(1), 51–63 (2014)
6. Falgueras, J., Lara, A.J., Fernández-Pozo, N., Cantón, F.R., Pérez-Trabado, G., Claros, M.G.: SeqTrim: a high-throughput pipeline for pre-processing any type of sequence read. BMC Bioinformatics 11, 38 (2010)
7. Fawcett, T.: An introduction to ROC analysis. Pattern Recognition Letters 27(8), 861–874 (2006)
8. Fickett, J.W.: Recognition of protein coding regions in DNA sequences. Nucleic Acids Research 10(17), 5303–5318 (1982)

9. Fu, L., Niu, B., Zhu, Z., Wu, S., Li, W.: CD-HIT: accelerated for clustering the next-generation sequencing data. Bioinformatics (Oxford, England) 28(23), 3150–3152 (2012)

10. Gao, J., Qi, Y., Cao, Y., Tung, W.W.: Protein coding sequence identification by simultaneously characterizing the periodic and random features of DNA sequences. Journal of Biomedicine & Biotechnology 2005(2), 139–146 (2005)

11. He, Z., Li, X., Ling, S., Fu, Y.X., Hungate, E., Shi, S., Wu, C.I.: Estimating DNA polymorphism from next generation sequencing data with high error rate by dual sequencing applications. BMC Genomics 14(1), 535 (2013)

12. Jones, C.E., Brown, A.L., Baumann, U.: Estimating the annotation error rate of curated GO database sequence annotations. BMC Bioinformatics 8, 170 (2007)

13. Lottaz, C., Iseli, C., Jongeneel, C.V., Bucher, P.: Modeling sequencing errors by combining Hidden Markov models. Bioinformatics 19(suppl. 2), ii103–ii112 (2003)

14. Martin, D.M.A., Berriman, M., Barton, G.J.: GOtcha: A new method for prediction of protein function assessed by the annotation of seven genomes. BMC Bioinformatics 5, 178 (2004)

15. Salzberg, S.L., Phillippy, A.M., Zimin, A., Puiu, D., Magoc, T., Koren, S., Treangen, T.J., Schatz, M.C., Delcher, A.L., Roberts, M., Marçais, G., Pop, M., Yorke, J.A.: GAGE: A critical evaluation of genome assemblies and assembly algorithms. Genome Research 22(3), 557–567 (2012)

16. Schnoes, A.M., Brown, S.D., Dodevski, I., Babbitt, P.C.: Annotation error in public databases: Misannotation of molecular function in enzyme superfamilies. PLoS Computational Biology 5(12), e1000605 (2009)

17. Stanke, M., Schöffmann, O., Morgenstern, B., Waack, S.: Gene prediction in eukaryotes with a generalized hidden Markov model that uses hints from external sources. BMC Bioinformatics 7, 62 (2006)

18. Wang, L., Park, H.J., Dasari, S., Wang, S., Kocher, J.P., Li, W.: CPAT: Coding-Potential Assessment Tool using an alignment-free logistic regression model. Nucleic Acids Research 41(6), e74 (2013)

19. Wang, Z., Gerstein, M., Snyder, M.: RNA-Seq: A revolutionary tool for transcriptomics. Nature Reviews Genetics 10(1), 57–63 (2009)

20. Yin, C., Yau, S.S.T.: Prediction of protein coding regions by the 3-base periodicity analysis of a DNA sequence. Journal of Theoretical Biology 247(4), 687–694 (2007)

21. Zagordi, O., Klein, R., Däumer, M., Beerenwinkel, N.: Error correction of next-generation sequencing data and reliable estimation of HIV quasispecies. Nucleic Acids Research 38(21), 7400–7409 (2010)

22. Zhang, M.Q.: Identification of protein coding regions in the human genome by quadratic discriminant analysis. Proceedings of the National Academy of Sciences of the United States of America 94, 565–568 (1997)

Evaluation of Combined Genome Assemblies:
A Case Study with Fungal Genomes

Mostafa M. Abbas*, Ponnuraman Balakrishnan, and Qutaibah M. Malluhi

KINDI Center for Computing Research, College of Engineering,
Qatar University, P.O. Box 2713, Doha, Qatar
mostafa_bioinfo@yahoo.com, baskrish1977@gmail.com,
qmalluhi@qu.edu.qa

Abstract. The rapid advances in genome sequencing leads to the generation of huge amount of data in a single sequencing experiment. Several genome assemblers with different objectives were developed to process these genomic data. Obviously, the output assemblies produced by these assemblers have different qualities due to their diverse nature. Recent research efforts concluded that combining the assemblies from different assemblers would enhance the quality of the output assembly. Based on this, our study combines the five best assemblies of three fungal genomes and evaluates the quality of the output assembly as compared to that produced by individual assemblers. The results conclude that the output assembly quality is influenced by the increase of the number of gaps in the input assemblies more than the increase in N50 size. Based on this conclusion, we propose a set of guidelines to get better output assemblies.

1 Introduction

Recent advances in Next-Generation Sequencing (NGS) have empowered high-throughput extraction of massive amount of genetic information with lower cost. These headways in genome sequencing technologies, for example, Illumina and Roche 454, lead to the generation of unprecedented amounts of biological data against its traditional counterpart of Sanger sequencing [1]. Furthermore, NGS technologies produce massive amounts of short DNA sequences called reads. Recently, numerous genome assembly tools [2, 3, 4, 5, 6, 7, 8, 9, 10] have been developed to handle NGS data. The genome assembly tools (i.e., assemblers) are dissent in their benefits and downsides in terms of quality of the output assemblies.

In recent efforts, many evaluation and comparative studies [11, 12, 13, 14, 15, 16] have been conducted to compare different assemblers. These efforts [12, 14, 16] concluded that the quality of assemblies can be further enhanced by combining the assemblies from different assemblers or the same assembler with different *k*-mer values. With this objective, several tools that combine assemblies have been developed and presented in the literature. Examples include GAM [17], minimus2 [18], MAIA [19], Reconciliator [20], Zoro [21], GAM-NGS [22], Metassembler [23], Mix [24],

---

* Corresponding author.

F. Ortuño and I. Rojas (Eds.): IWBBIO 2015, Part II, LNCS 9044, pp. 333–344, 2015.

GAA [25], GARM [26] and e-RGA [27]. The combining tools MAIA [19] and e-RGA [27] improved the assembly quality by using a closely related reference sequence. Reconciliator [20] and Zoro [21] targeted the reduction of errors in assemblies to improve their quality, The main objective of Mix [24] is to reduce contig fragmentation and maximize the cumulative contig length. The pipelines of GARM [26] and minimus2 [18] use NUCmer [28] to compute overlaps between contigs. GAM [17] and GAM-NGS [22] combine the assemblies from Sanger and NGS assemblers, respectively, and avoid the global alignment step. The GAA [25] is based on constructing an accordance graph for extracting the mapping information between the two input assemblies (one as a target and the other as a query). Metassembler [23] uses the mate-pair information and whole-genome alignments to combine the input assemblies to produce a better-quality output assembly. The output assembly merges the best locally superior assemblies on the genome.

The efficiency of each combining assembly tool is quantified by applying it on a set of assemblers against some datasets. The GAGE-B [14] evaluates the performance of two different combining tools minimus2 [18] and GAA [25] by applying them on a group of assemblies from different assemblers against a set of bacterial genomes. Similarly in [12], the combining tools GAM [17] and Metassembler [23] are evaluated by experimenting them on a group of assemblies from different assemblers against three eukaryotic genomes. In [14], the authors demonstrated that combining assemblies from the combining tools other than GAA do not produce improvements over input assemblies. The study presented in [12], demonstrate that the GAM tool produces poorly scoring resulting assembly for most metrics, whereas the Metassembler tool produces an assembly with marginal improvement in quality over the two source assemblies.

In [16], an evaluation study for five fungal draft genomes against seven NGS assemblers is presented. In this study, the assemblers are ranked based on their quality metrics of three groups. Furthermore, the work hints that the quality of assemblies may be enhanced further by the utilization of combining assembly tools.

In this work, we examine whether combining the assemblies leads to better assembly quality than that of individual assemblies. For that, we combine the five best assemblies at contigs level (as established in [16]) for three datasets using a Metassembler tool [23] and evaluate the performance of the combined assembly against individual assemblies based on some quality metrics.

2   Method

2.1   Data

For evaluating the Metassembler tool, we chose three fungal pathogens (Table 1) [29, 30, 31]. The three selected datasets have different sizes; two of them fall in the range of 90% of fungal genome sizes which is less than 60 Mb [32].

2.2   Assemblies

In [16], we have evaluated the performance of fungal genome assemblies from the following assemblers ABySS, IDBA-UD, Minia, SOAP, SPAdes, Sparse, and Velvet

Table 1. List of fungal genomes studied in the experiments

Species	Estimated Size (Mbp)	Estimated GC content %
Botryotinia fuckeliana (BcDW1) [29]	42.13	42
Eutypa lata (UCREL1) [30]	54.01	46.6
Puccinia striiformis f. sp. tritici (PST21) [31]	73.05	44.4

besides the draft genomes of these species that are submitted on the WGS project page in NCBI (http://www.ncbi.nlm.nih.gov/bioproject/webcite). Further, we have ranked the assemblies of these datasets based on different metrics. In this work, we have applied the Metassembler tool [23], using the default parameters, on the best five assemblies from each dataset in Table 1 at the contigs level. The assemblies are available at http://confluence.qu.edu.qa/display/download/bioinf.

2.3   Metrics

The quality metrics for evaluating draft genomes can be divided into three groups: goodness, problems and conservation [16]. In this work, we select three metrics:

- **Total length:** the total number of bases in contigs
- **N50:** The length of the smallest contig x, which makes the ratio of cumulative length of contigs from this length x to the longest contig in the assembly covers at least 50% of the bases of the assembly. An assembler with high N50 size value is obviously considered to be a high-quality assembler. The N50 size is a metric representing the goodness metrics group[16].
- **No of N's:** The total number of uncalled bases or gaps (N's) in the assembly bases. Mis-assemblies and gaps usually result from repeats as well as secondary structures, either in unrepresented GC-rich regions or in un-sequenced regions due to a low depth sequence coverage [33]. The higher the number of N's the lower the quality of the assembly. The No of N's is a metric representing the problems metrics group[16].

These metrics for the best five assemblies of each dataset are presented in Table 2. The assemblies in Table 2 are arranged in their descending order of the N50 metric. QUAST assessment tool [34] is used to measure these three metrics after removing any chaff contig, a single contig with a length less than 200 bp [13] from the output assembly. The high percentage of chaff contigs length leads to problems in further genomic analysis [13].

Table 2. The quality metrics of the best five assemblies from [16] for the studied datasets

Dataset	Assembly	Total length (Mbp)	N50 (Kbp)	No of N's (Kbp)
BcDw 1	ABySS	42.4	317.51	13.65
	SPAdes	42.35	175.15	0
	IDBA-UD	42.3	105.35	0
	df_1	42.07	95.64	0.25
	Velvet	42.13	57.74	0
UCREL 1	ABySS	53.76	119.67	116.72
	SPAdes	54.74	117.92	0
	IDBA-UD	55.07	79.3	0
	df_1	53.9	51.8	0.44
	Velvet	53.58	19.98	0
PST-21	df_1	73.05	3.96	0.23
	IDBA-UD	78.75	3.61	0
	ABySS	64.34	2.47	14.37
	Velvet	64.22	2.45	0
	Sparse	60.7	1.27	0

3   Results

In this study, we focus on evaluating the Metassembler tool [23] on fungal genomes. The Metassembler tool takes two assemblies as input(one being the target and the other being the query), and tries to enhance the output assembly. In our study, for a given dataset, we apply Metassembler tool on all combinations of the five assemblies such that each assembly plays one time as a target, whereas the other time as a query with the other assemblies. We refer to any combined assembly in the format target/query assembly. Tables 3, 4 and 5 show the application of Metassembler tool on the datasets BcDw 1, UCREL 1 and PST-21 respectively. In these tables, each cell represents the data of the combined assembly such that the column header represents the target assembly and the row header represents the query assembly. A bold item in Tables 3-5 refer to a metric of the combined assembly that is better than the corresponding metric for both input assemblies; target and query. The percentage of change in the output assembly with respect to the target assembly for a specific metric can be calculated by the equation $change = \frac{ouput-target}{target}$. For the case of N50, a positive change represents an improvement, while in the case of No. of N's, an improvement will be represented by a negative change.

Table 3. Combined assembly quality metrics for BcDW1

Metric	Assembly	ABySS	SPAdes	IDBA-UD	df_1	Velvet
Total length (Mbp)	ABySS		42.45	42.22	42.17	42.17
	SPAdes	42.34		42.24	42.14	42.17
	IDBA-UD	42.37	42.3		42.14	42.17
	df_1	42.35	42.27	42.24		42.17
	Velvet	42.37	42.28	42.23	42.15	
N50 (Kbp)	ABySS		287.74	252.44	216.41	204.65
	SPAdes	227.26		171.16	171.72	142.56
	IDBA-UD	213.09	**223.42**		**144.4**	**146.4**
	df_1	209.04	**210.6**	**145.39**		**142.67**
	Velvet	204.5	**210.78**	**136.33**	144.36	
No of N's (Kbp)	ABySS		0.2	1.04	**0.62**	1.93
	SPAdes	2.22		0	0.25	0
	IDBA-UD	2.24	0		0.25	0
	df_1	2.25	0.01	0.03		0.1
	Velvet	2.25	0	0	0.24	

The metrics collected from the combined assemblies of the BcDw1 dataset is given in Table.3. We can observe the following from Table 3:

- The difference between the total length of the largest and smallest assemblies is very low and never exceeds 0.4 Mbp.
- In all combinations, the N50 size of the combined assembly is better than the N50 size of the target assembly except when ABySS assembly is the target.
- If the target assembly has a better N50 (except for ABySS assembly) than the query, the N50 of the combined output assembly is better than the N50 of the target assembly with a percentage of improvement ranging from 20% to 50% (Figure 1).
- If the target assembly has a worse N50 than the query, then the N50 size of the combined assembly is better than the N50 of the target, but not better than the query with the following exceptions: the combined assembly of df_1/IDBA-UD, Velvet/IDBA-UD and Velvet/df_1 has N50 that is better than the target as well as query assemblies.
- For all combinations, the No. of N's of the combined assembly is better than (less than) or equal to the No. of N's of the target, except when ABySS or df_1 assemblies are used as the query.
- If ABySS assembly is used as a target or a query, the combined assembly notably enhances (decreases) the No. of N's of the ABySS at the cost of decreasing the N50 size of ABySS.
- The combined assembly of ABySS as a query will always have better N50 size and No. of N's than using it as a target.
- The SPAdes/ABySS combination produces the best combined assembly with the high N50 size of 287.74Kbp and a small No. of N's equal to 0.2 Kbp.

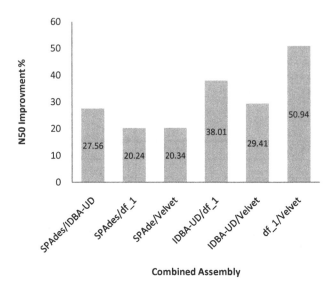

Fig. 1. The percentage of improvement in N50 with respect to the target assembly for BcDw 1, for the cases when the target assembly has better N50 than the query assembly

The metrics collected from the combined assemblies of UCREL 1 dataset is given in Table.4. We can observe the following from Table.4:

- The difference between the largest and smallest assembly is low and approximately equals to 2 Mbp.
- For all combinations, the N50 size of the combined assembly is better than the N50 size of the target assembly except in ABySS as a target. When ABySS is used as a target, the combined assemblies ABySS/SPAdes and ABySS/IDBA-UD have N50 size less than the N50 of both the target and query assemblies.
- If the target assembly has a better N50 (except for ABySS assembly), the N50 size of the combined assembly shows a small improvement of less than 1 % (Figure 2).
- The change in N50 and No. of N's of the combined assemblies mainly depends on the N50 and No. of N's of the target assembly. For example, when the target assembly is ABySS, SPAdes, IDBA-UD, df_1 and Velvet, the average of N50 sizes for different query assemblies are 56.77, 118.83, 79.72, 52.03 and 20.14; with a standard variation of 0.02, 0.05, 0.14, 0.12 and 0.05 respectively. Moreover, when the target assembly is ABySS, SPAdes, IDBA-UD, df_1 and Velvet, the average No. of N's is 7.41, 0, 0, 0.43 and 0 with standard variation 0.05, 0, 0, 0.01 and 0 respectively.
- If the target assembly has the worse N50, the N50 size of the combined assembly is better than the N50 of the target but not better than the query's.
- For all combinations, the No. of N's of the combined assembly is better than (less than) or equal to the No. of N's of the target.

- When ABySS is used as a target or query, the combined assembly notably enhances (decreases) the No. of N's of ABySS at cost of decreasing its N50 size.
- The SPAdes as a target produces a best combined assembly with the N50 size of 118Kbp and the smallest No. of N's equal to 0 Kbp.

Table 4. Combined assembly quality metrics for UCREL 1

Metric	Assembly	ABySS	SPAdes	IDBA-UD	df_1	Velvet
Total length (Mbp)	ABySS		54.08	54.36	53.55	53.01
	SPAdes	53.39		54.69	53.73	53.2
	IDBA-UD	53.41	54.37		53.82	53.3
	df_1	53.33	54.16	54.58		53.14
	Velvet	53.32	54.12	54.54	53.62	
N50 (Kbp)	ABySS		118.86	79.92	52.09	20.2
	SPAdes	56.76		79.62	52.08	20.11
	IDBA-UD	56.76	**118.76**		51.84	20.09
	df_1	56.76	**118.86**	**79.62**		20.15
	Velvet	56.81	**118.86**	**79.71**	**52.09**	
No of N's (Kbp)	ABySS		0	0	**0.42**	0
	SPAdes	7.43		0	0.43	0
	IDBA-UD	7.47	0		0.43	0
	df_1	7.39	0	0		0
	Velvet	7.36	0	0	0.42	

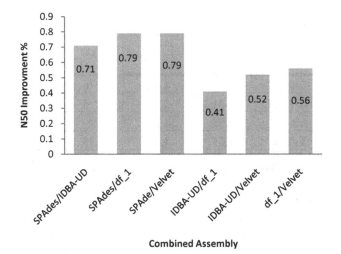

Fig. 2. The percentage of improvement in N50 with respect to the target assembly for UCREL 1, for the cases when the target assembly has better N50 than the query assembly

The metrics collected from the combined assemblies of PST-21 dataset is given in Table 5. We can observe the following from Table.5:

- The difference between the largest and smallest assemblies is very high and equal to 24.3 Mbp.
- For all combinations, the size of the combined assembly is less than the target as well as the query assembly.
- For a fixed target (query) assembly, the total length decreases, while the total length of the query (target) assembly decreases.
- For all combinations, the N50 size and No. of N's of the combined assembly are better than the N50 size and No. of N's of the target assembly at the cost of decreasing the total length of the target.
- If a target assembly has the better N50, then the N50 size of the combined assembly improves at cost of decreasing the total length (Figure 3).
- If the target assembly has worse N50, the N50 size of the combined assembly is better than the N50 of the target but not better than the query's, except for IDBA-UD/df_1 and Velvet/ABySS, where the combined assembly has N50 better than the target as well as query assemblies.
- For all combinations, the No. of N's of the combined assembly is better than or equal to the No. of N's of the target.

Table 5. Combined assembly quality metrics for PST-21

Metric	Assembly	df_1	IDBA-UD	ABySS	Velvet	Sparse
Total length (Mbp)	df_1		72.62	59.17	58.81	50.6
	IDBA-UD	68.52		60.12	59.96	51.14
	ABySS	65.38	69.52		57.48	50.37
	Velvet	65.4	69.82	56.74		48.32
	Sparse	63.5	66.96	56.27	55.16	
N50 (Kbp)	df_1		**4.02**	2.76	2.78	1.75
	IDBA-UD	**4.31**		2.73	2.72	1.73
	ABySS	**4.55**	**4.28**		**2.89**	1.78
	Velvet	**4.55**	**4.25**	**2.95**		1.87
	Sparse	**4.72**	**4.46**	**2.99**	**3.07**	
No of N's (Kbp)	df_1		0	5.7	0	0
	IDBA-UD	0.11		5.99	0	0
	ABySS	**0.09**	0		0	0
	Velvet	0.08	0	5.53		0
	Sparse	0.08	0	5.09	0	

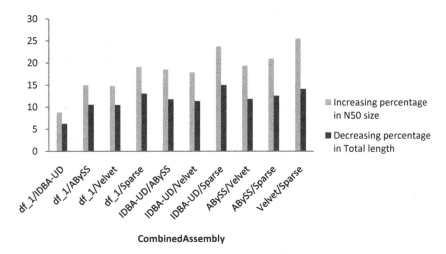

CombinedAssembly

Fig. 3. The percentage of changes in N50 size and Total length with respect to the target assembly for PST-21, for the cases when the target assembly has better N50 than the query assembly

4   Discussion and Conclusions

The observations that are presented in the previous section, support a set of conclusions about the steps to obtain the combined assembly with better performance using the Metassembler tool.

The influence of the N50 and No. of N's metrics in the input assembly on the combined assembly demonstrates that if the No. of N's is high in the input assembly, especially in the target assembly, the performance of the combined assembly will be low in terms of N50, which can be observed from ABySS assemblies. If the input assemblies have low (or zero) No. of N's, the combined assembly is better than the two input assemblies in terms of N50. The main reason behind this sensitivity in the combined assembly to the No. of N's is that the Metassembler tool depends on an alignment step which suffers from the existence of these N's (gaps).

In light of these observations, we can get a better combined assembly using the Metassembler tool (Although our experiments use Metassembler as a tool, it is expected that our results can be extended for any combined assembly tool that depends on an alignment step between the input assemblies) using the following guidelines for selecting the input assemblies:

1. Avoid the assemblies that have a large No. of N's.
2. If the assemblies contain different No. of N's, use as a target the assembly that has a lower No. of N's.
3. If the assemblies contain the same No. of N's, use as a target the assembly that has the higher N50 size.

The rapidly growing number of sequencing projects can take advantage of our results to get better quality assemblies by selecting more appropriate assemblies that can be used as the inputs for the Metassembler tool. The full results of this research work are freely available for download at http://confluence.qu.edu.qa/display/download/bioinf.

Acknowledgements. This publication was made possible by NPRP grant No. 4-1454-1-233 from the Qatar National Research Fund (a member of Qatar Foundation). The statements made herein are solely the responsibility of the authors. The authors are grateful to Prof. Michael Schatz and his student Alejandro Hernandez Wences for providing the Metassembler tool.

Authors' Contributions

Mostafa M Abbas and Qutaibah M Malluhi conceived and designed the study; Mostafa M Abbas ran the experiments; All authors analyzed the results; Mostafa M Abbas wrote the draft of the manuscript. All authors read and approved the final manuscript.

References

1. Mavromatis, K., Land, M.L., Brettin, T.S., Quest, D.J., Copeland, A., et al.: The Fast Changing Landscape of Sequencing Technologies and Their Impact on Microbial Genome Assemblies and Annotation. PLoS One 7(12), e48837 (2012)
2. Luo, R., Liu, B., Xie, Y., Li, Z., Huang, W., Yuan, J., He, G., Chen, Y., Pan, Q., Liu, Y., Tang, J., Wu, G., Zhang, H., Shi, Y., Liu, Y., Yu, C., Wang, B., Lu, Y., Han, C., Cheung, D.W., Yiu, S.M., Peng, S., Xiaoqian, Z., Liu, G., Liao, X., Li, Y., Yang, H., Wang, J., Lam, T.W., Wang, J.: SOAPdenovo2: an empirically improved memory-efficient short-read de novo assembler. GigaScience 1(1), 18 (2012)
3. Zerbino, D.R., Birney, E.: Velvet: Algorithms for de novo short read assembly using de Bruijn graphs. Genome Res. 18, 821–829 (2008)
4. Butler, J., MacCallum, L., Kleber, M., Shlyakhter, I.A., Belmonte, M.K., Lander, E.S., Nusbaum, C., Jaffe, D.B.: ALLPATHS: De novo assembly of whole-genome shotgun microreads. Genome Research 18, 810–820 (2008)
5. Simpson, J., Wong, K., Jackman, S., Schein, J., ABySS, A.: parallel assembler for short read sequence data. Genome, 1117–1123 (2009)
6. Chikhi, R., Rizk, G.: Space-efficient and exact de Bruijn graph representation based on a Bloom filter. Algorithms for Molecular Biology 8, 22 (2013)
7. Peng, Y., Leung, H.C., Yiu, S.M., Chin, F.Y.: IDBA-UD: A de novo assembler for single-cell and metagenomic sequencing data with highly uneven depth. Bioinformatics 28, 1420–1428 (2012)
8. Ye, C., Ma, Z.S., Cannon, C.H., Pop, M., Yu, D.W.: Exploiting sparseness in de novo genome assembly. BMC Bioinformatics 13(suppl. 6), S1 (2012)
9. Bankevich, A., Nurk, S., Antipov, D., Gurevich, A., Dvorkin, M., Kulikov, A., Lesin, V., Nikolenko, S., Pham, S., Prjibelski, A., Pyshkin, A., Sirotkin, A., Vyahhi, N., Tesler, G., Alekseyev, M., Pevzner, P.: SPAdes: A new genome assembler and its applications to single cell sequencing. Journal of Computational Biology 19(5), 455–477 (2012)
10. CLC bio, http://www.clcbio.com/

11. Earl, D., Bradnam, K., St John, J., Darling, A., Lin, D., Fass, J., Yu, H.O., Buffalo, V., Zerbino, D.R., Diekhans, M., Nguyen, N., Ariyaratne, P.N., Sung, W.K., Ning, Z., Haimel, M., Simpson, J.T., Fonseca, N.A., Docking, T.R., Ho, I.Y., Rokhsar, D.S., Chikhi, R., Lavenier, D., Chapuis, G., Naquin, D., Maillet, N., Schatz, M.C., Kelley, D.R., Phillippy, A.M., Koren, S., et al.: Assemblathon 1: A competitive assessment of de novo short read assembly methods. Genome Res. 21, 2224–2241 (2011)

12. Bradnam, K.R., Fass, J.N., Alexandrov, A., Baranay, P., Bechner, M., Birol, I., Boisvert, S., Chapman, J.A., Chapuis, G., Chikhi, R., Chitsaz, H., Chou, W.C., Corbeil, J., Del Fabbro, C., Docking, T.R., Durbin, R., Earl, D., Emrich, S., Fedotov, P., Fonseca, N.A., Ganapathy, G., Gibbs, R.A., Gnerre, S., Godzaridis, E., Goldstein, S., Haimel, M., Hall, G., Haussler, D., Hiatt, J.B., Ho, I.Y., et al.: Assemblathon 2: Evaluating de novo methods of genome assembly in three vertebrate species. Gigascience 2, 10 (2013)

13. Salzberg, S.L., Phillippy, A.M., Zimin, A., Puiu, D., Magoc, T., Koren, S., Treangen, T.J., Schatz, M.C., Delcher, A.L., Roberts, M.: GAGE: A critical evaluation of genome assemblies and assembly algorithms. Genome Res. 22(3), 557–567 (2012)

14. Magoc, T., Pabinger, S., Canzar, S., Liu, X., Su, Q., Puiu, D., Tallon, L.J., Salzberg, S.L.: GAGE-B: An evaluation of genome assemblers for bacterial organisms. Bioinformatics 29(14), 1718–1725 (2013)

15. Finotello, F., Lavezzo, E., Fontana, P., Peruzzo, D., Albiero, A., Barzon, L., Falda, M., Di Camillo, B., Toppo, S.: Comparative analysis of algorithms for whole-genome assembly of pyrosequencing data. Brief Bioinform 13(3), 269–280 (2011)

16. Abbas, M.M., Malluhi, Q.M., Balakrishnan, P.: Assessment of de novo assemblers for draft genomes: A case study with fungal genomes. BMC Genomics 15(suppl. 9), S10 (2014)

17. Casagrande, A., Del, F.C., Scalabrin, S., Policriti, A.G.: Genomic Assemblies Combiner: A Graph Based Method to Integrate Different Assemblies. Bioinformatics and Biomedicine (2009), 10.1109/BIBM.2009.28

18. Sommer, D., Delcher, A., Salzberg, A., Pop, M.: Minimus: A fast, lightweight genome assembler. BMC Bioinformatics 8, 64 (2007)

19. Nijkamp, J., Winterbach, W., van, D.B.M., Daran, J., Reinders, M., de Ridder, R.: Integrating genome assemblies with MAIA. Bioinformatics 26(18), i4339 (2010)

20. Zimin, A., Smith, D., Sutton, G., Yorke, J.: Assembly reconciliation. Bioinformatics 24, 42–45 (2008)

21. Argueso, J., Carazzolle, M., Mieczkowski, P., Duarte, F., Netto, O., Missawa, S., Galzerani, F., Costa, G., Vidal, R., Noronha, M., Dominska, M., Andrietta, M., Andrietta, S., Cunha, A., Gomes, L., Tavares, F., Alcarde, A., Dietrich, F., McCusker, J., Petes, T., Pereira, G.: Genome structure of a Saccharomyces cerevisiae strain widely used in bioethanol production. Genome Res. 19(12), 2258–2270 (2009)

22. Vicedomini, R., Vezzi, F., Scalabrin, S., Arvestad, L., Policriti, A.G.-N.: GAM-NGS: genomic assemblies combiner for next generation sequencing. BMC Bioinformatics 14(7), 1–18 (2013)

23. Metassembler,
    http://sourceforge.net/apps/mediawiki/metassembler/index.php
    ?title=Metassembler

24. Soueidan, H., Maurier, F., Groppi, A., Sirand-Pugnet, P., Tardy, F., Citti, C., Dupuy, V., Nikolski, M.: Finishing bacterial genome assemblies with Mix. BMC Bioinformatics 14(suppl. 15), S16 (2013)

25. Yao, G., Ye, L., Gao, H., Minx, P., Warren, W., Weinstock, G.: Graph accordance of next-generation sequence assemblies. Bioinformatics 28, 13–16 (2011)

26. Soto-Jimenez, L.M., Estrada, K., Sanchez-Flores, A.: GARM: Genome Assembly, Reconciliation and Combining Pipeline. Current Topics in Medicinal Chemistry 14(3), 418–424 (2014)
27. Vezzi, F., Cattonaro, F., Policriti, A.: e-RGA: enhanced reference guided assembly of complex genomes. EMBnet J. 17, 46–54 (2011)
28. Kurtz, A., Phillippy, A., Delcher, A., Smoot, M., Shumway, A., Antonescu, C., Salzberg, S.: Versatile and open software for comparing large genomes. Genome Biology 5(2), R12 (2004)
29. Blanco-Ulate, B., Allen, G., Powell, A.L., Cantu, D.: Draft genome sequence of Botrytis cinerea BcDW1, inoculum for noble rot of grape berries. Genome Announcements 1(3), e00252-13 (2013)
30. Blanco-Ulate, B., Rolshausen, P.E., Cantu, D.: Draft genome sequence of the grapevine dieback fungus Eutypa lata UCR-EL1. Genome Announcements 1(3), e00228-13 (2013)
31. Cantu, D., Segovia, V., Maclean, D., Bayles, R., Chen, X., Kamoun, S., Dubcovsky, J., Saunders, D.G., Uauy, C.: Genome analyses of the wheat yellow (stripe) rust pathogen Puccinia striiformis f. sp. tritici reveal polymorphic and haustorial expressed secreted proteins as candidate effectors. BMC Genomics 14, 270 (2013)
32. Gregory, T.R., Nicol, J.A., Tamm, H., Kullman, B., Kullman, K., Leitch, I.J., Murray, B.G., Kapraun, D.F., Greilhuber, J., Bennett, M.D.: Eukaryotic genome size database. Nucleic Acids Res. 35, D332-D338 (2007)
33. Tsai, I., Otto, T., Berriman, M.: Improving draft assemblies by iterative mapping and assembly of short reads to eliminate gaps. Genome Biol. 11(4), 41 (2010)
34. Gurevich, A., Saveliev, V., Vyahhi, N., Tesler, G.: QUAST: Quality assessment tool for genome assemblies. Bioinformatics 29, 1072–1075 (2013)

Using Multivariate Analysis and Bioinformatic Tools to Elucidate the Functions of a Cyanobacterial Global Regulator from RNA-Seq Data Obtained in Different Genetic and Environmental Backgrounds

José I. Labella[1,*], Francisco Rodríguez-Mateos[2], Javier Espinosa[1], and Asunción Contreras[1]

[1] Division of Genetics
[2] Department of Applied Matemathics,
University of Alicante, Apdo. 99, 03080, Alicante, Spain
ls.joseignacio@ua.es
http://dfgm.ua.es/genetica/investigacion/cyanobacterial_genetics/

Abstract. The cyanobacterial protein PipX, known as the coactivator of the nitrogen regulator NtcA, regulates multiple operons, many of them in an NtcA-independent manner. PipX has no DNA-binding activity and must interact with additional regulators to affect gene expression. Here we follow different bioinformatic approaches to analyze previous transcriptomic data from 10 different genetic or environmental backgrounds. Using standardized residuals from 8 mutant/control comparisons a set of 331 differentially expressed genes were obtained, which were clustered according to their patterns of expression. The clusters showed significant internal coherence, supporting inferences on PipX function and providing evidence of additional complexity for *in vivo* interactions. The Perl based program MultiGS, designed to help searches for regulatory motifs, was used to help finding NtcA motifs within the cluster containing nitrogen assimilation genes. The results pave the way to the identification of the yet unknown transcriptional regulator(s) involved in the NtcA-independent PipX regulons.

Keywords: PipX, NtcA, cluster analysis, motif elucidation.

1 Introduction

Operons controlled by the same regulatory proteins, that is, regulons, are expected to respond similarly to environmental conditions and mutations affecting the regulatory pathways involved. Transcriptomic data obtained by Next Generation Sequencing (NGS) from appropriate experimental strategies may provide valuable clues to characterize gene regulators in the context of complex regulons. In this context, the identification from transcriptomic data of groups of

---

* Corresponding author.

F. Ortuño and I. Rojas (Eds.): IWBBIO 2015, Part II, LNCS 9044, pp. 345–354, 2015.

genes sharing expression profiles can be a key step towards the identification of the transcriptional regulators involved. For a DNA-binding transcriptional factor, such as the cyanobacterial global regulator NtcA, the prediction is that the differentially regulated genes would be preceded by specific sites (NtcA binding motifs). However, the scenario is more complex if the transcriptional factor is a promiscuous co-regulator interacting with different DNA-binding proteins according to a variety of signals. In these cases mutations affecting the co-regulator would identify a modulon composed by a rather large and heterogeneous set of regulons, each one controlled by a different regulatory complex. This is the context for PipX, the PII interacting protein X, best characterized as the coactivator of the nitrogen regulator NtcA and suspected to bind to additional transcriptional regulators to control transcript levels (data not shown) [1,2].

2-oxoglutarate (2-OG), the signal of nitrogen deficiency, modulates the activity and/or binding properties of the signal transduction protein PII, the transcriptional activator NtcA, and PipX, the regulatory factor that can interact with either NtcA or PII [1,3]. The interaction between PipX and NtcA is known to be relevant under nitrogen limitation for activation of NtcA-dependent genes [1,2]. Under the physiological range of 2-OG levels, *pipX* mutations specifically impairing PipX-PII complexes favor formation of PipX-NtcA complexes. PipX residues Y32 and E4 are important for interactions with PII and NtcA proteins and Y32 for interactions with NtcA and these differences affect both NtcA-dependent and independent genes [3,4] . The participation of PipX in at least 3 NtcA-independent scenarios was proposed.

To extend and refine the previously identified PipX modulon and get additional insights into the functions of PipX, we present here an extended analysis based on the raw RNA-Seq data used in [5], incorporating CS3X as a mutant strain. We describe specific tools to refine motif searches and improve consensus binding motifs, in a close interaction between *in silico* analysis and biological data and interpretation. The biological implications of the new results are also discussed.

2   Methods

For details on the cyanobacterium *S. elongatus* strains (including mutant derivatives), growth conditions, RNA preparation and RNAseq analysis see [5].

Gene clustering was based on a mathematical algorithm dealing with comparisons (a total of 90) of the expression values (Reads per Kilobase) from the 10 dataset (5 strains in either nitrate or ammonia). A, B or C letters code were assigned depending on the comparisons values: > 1.5 times (A), < 0.5 times (B) or $[0, 5 - 1.5]$ times (C). The output was clustered in a dendrogram with bootstrapping ($n = 1000$). Genes were grouped into transcriptional units according to [6], except for genes belonging to the main ribosomal cluster (ORFs 2232-2221), that were considered as belonging to the same operon. Statistical analyses were performed using the R software [7] and the packages pvclust and Ape [8].

To identify NtcA binding sites, sequences of 75 nucleotides from the transcription start site (TSS) as defined by [5], or in its absence the initiation codon were searched with MEME [9] for palindromic motifs between 6 and 20 bp. Fragment selection was performed using our in-house Perl based program MultiGS (available upon request). In addition, we used a background consisting on a fourth-order Markov model of the entire genome. Expression values from clustered genes were represented in boxplots (A-I) from mutant/control comparisons.

3   Results and Discussion

3.1   Marker Allele Φ(C.S3-$pipX$) Affects Expression of Multiple Genes in *S. elongatus*

Strain CS3X, carrying the CS3 marker cassette in a *S. elongatus* wild type background, has been routinely used as the control or parental strain for CS3X^{E4A}, CS3X^{Y32A} and other strains carrying point mutations at *pipX* [4,5]. Since no detectable phenotype other than a slight reduction of PipX protein levels could be found with regards to the wild type, the CS3X strain was included in our previous transcriptome analysis with the double aim to minimize possible local effects of the CS3 marker on transcript levels and adding robustness to the analyses by providing additional data from two almost identical wild type strains. However, we noticed unexpected differences between the two control strains at particular transcripts of interest (data not shown).

To investigate the extent and relevance of the differences between CS3X and the *bona fide* wild type strain, we now treated CS3X as an additional mutant strain, for which we performed CS3X/WT comparisons to detect genes differentially regulated with a restrictive criterion (2.5 fold, see below). The differentially expressed genes can be observed in four regions of the scatterplot (Fig. 1, open circles), indicative of examples of up- and down-regulation in CS3X, with some genes been similarly altered in each of the tested conditions.

3.2   Insights into PipX Functions at Different Regulons Informed by Clustering of Gene Expression Patterns

The unexpected finding that the Φ(C.S3-$pipX$) allele significantly altered gene expression of given genes (data not shown), prompted us to investigate whether the corresponding regulatory defects were somehow related to PipX functions and thus there were coincidence between the transcripts abnormally regulated by the already characterized *pipX* null or point mutations and those abnormally regulated by the Φ(C.S3-$pipX$) allele. To investigate this issue, we obtained standardized residuals from linear regressions of (log-transformed) data from mutant versus control strains (CS3X^{E4A} or CS3X^{Y32A} vs CS3X, PipX-null vs WT, and CS3X vs WT) cultured with either ammonia or nitrate. Gene transcripts with residuals lower than 1.5 in any of the 8 mutant/control comparisons were considered as non-responsive. For subsequent analyses only those transcripts with

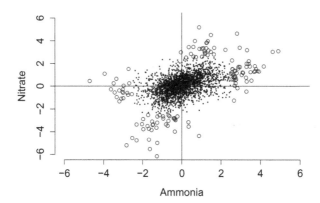

Fig. 1. Effect of Φ(C.S3-*pipX*) allele on nitrate and ammonia transcriptomes. The mutant/control log residuals in nitrate vs. ammonia are represented as a scatter plot. Genes below or above the cut-off (2.5 fold) are represented as dots and open circles, respectively.

residuals exceeding a threshold value of 2.5 for at least one out of the eight variables were considered to be differentially regulated. The resulting 331 differentially expressed genes were clustered into 9 main groups (designated A-I) according to their expression patterns (see methods). Clusters obtained in this way were subsequently analyzed for their internal coherence. The results are summarized in Fig. 2.

Operon Criterion. Although several factors, including the presence of antisense RNA account for exceptions, genes belonging to the same operons usually share a similar expression pattern. In agreement with this, 97 out of the 331 differentially regulated transcripts belonged to the 39 polycistronic transcription units identified and only 3 genes did not cluster with their operon partners. Thus, the majority of differentially regulated genes were found either as monocistronic units or clustering with their operon partners.

Expression Profile Criterion. Genes and operons similarly regulated by a particular type of PipX-containing regulatory complex would form regulons that are predicted to be similarly affected by different *pipX* alleles. In this context, we know that NtcA-activated transcripts are up-regulated in CS3X^{E4A} (nitrate and ammonia) and CS3X^{Y32A} (nitrate). These effects are only recognizable in Cluster A (see Fig. 3). Interestingly, the effect of alleles pipX-null and Φ(C.S3-*pipX*) are very similar, indicating that in the cells of CS3X growing in nitrate PipX levels are limiting for NtcA-PipX complexes and transcriptional activation of A genes. A similar PipX limitation can be invoked for Cluster G (Fig. 3), containing

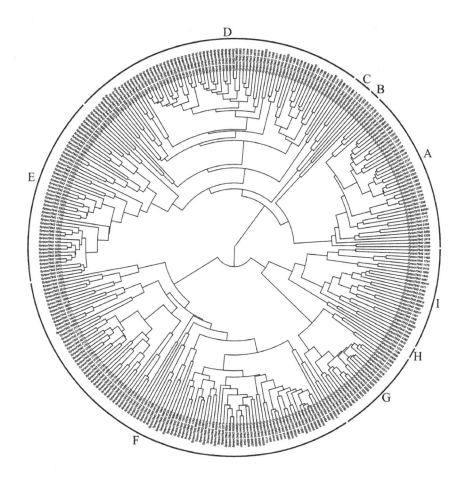

Fig. 2. Dendrogram showing the clusters A to I constructed according to expression profiles derived from *S. elongatus* strains grown in two nitrogen conditions

transcripts down-regulated in CS3XE4A and to a lesser extent in CS3X^{Y32A}. These findings and the impact of *pipX* inactivation on G genes are all consistent with a negative role for PipX and with NtcA repression. Remarkably, the impact of alleles *pipX*-null and *Φ*(C.S3-*pipX*) differed in the remaining groups in at least one of the Nitrogen regimes, indicating that NtcA is not involved in their regulation.

Functional Criterion. Genes involved in common biological processes are often part of regulons controlled by common regulatory factors. To explore this issue in clusters A-I, gene functions were assigned according to COG (cluster of orthologous genes) (data not shown). Genes related to nitrogen assimilation were very abundant (47%) in Cluster A, but virtually absent from the other clusters. Photosynthesis and/or Energy production genes were well represented in E (55%) and D (13%), Translation in F (24%) and G (74%), post-transcriptional modification, protein turnover, and chaperones in C. Finally, genes of unknown function were remarkably abundant in clusters B, H and I (75, 100 and 69%, respectively).

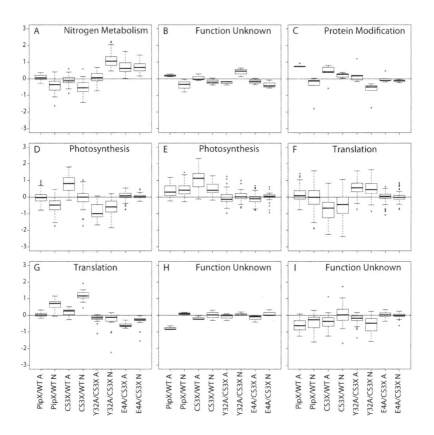

Fig. 3. Gene cluster boxplots of log10 mutant/control expression values with the most represented function indicated. WT, wild type; PipX, null-mutant; CS3X, Y32A and E4A, Φ(C.S3-$pipX$), Φ(C.S3-$pipX^{Y32A}$) and Φ(C.S3-$pipX^{E4A}$) strains, respectively.

Therefore, our clustering approach is supported by the biologically relevant criteria used here. The results emphasize the idea that different PipX regulatory complexes control different regulons. Although the molecular mechanism by which the Φ(C.S3-$pipX$) allele would affect expression of PipX regulons remain to be elucidated, treatment of CS3X as a mutant in the post-transcriptome analyses was highly informative. As a consequence, the 9 individual clusters in which we now subdivide the PipX modulon appear to be coherent. Further work is needed to test whether the regulatory differences between some of the pairs of closely related clusters are due to being the targets of different PipX-containing complexes or to the action of additional regulatory factors. Whatever the details, the results provide deeper insights into de PipX modulon and a good starting point for additional approaches to search for the elusive PipX complexes involved in NtcA-independent gene expression by *in silico* methods.

3.3   Development of the Perl-Based Program MultiGS and Searches for Regulatory Motifs

Our analyses strongly support the view of PipX as a global and promiscuous co-regulator with no DNA-binding activity involved in interactions with yet unknown proteins according to 2-OG and ATP/ADP levels, and probably additional signals. Therefore, the identification of unknown regulatory factors acting with PipX is a main challenge in cyanobacteria signal transduction. Since these PipX complexes would regulate the NtcA-independent genes identified by us, it is important to search for the hypothetical recognition sites of the corresponding DNA-binding regulator(s).

Incorrect annotation of operon TSSs and/or the use of different promoters in different environmental conditions and even genetic backgrounds add difficulty to motif searches. Therefore, it is important to be able to incorporate updates of TSS information, obtained with more informative experimental approaches, to the corresponding programs. In this context, we could often identify, in particular mutant backgrounds, relatively active promoters that were previously unnoticed and therefore missing in the reference databases. In addition, once the relevant TSS are taken in to account, decisions are also needed to determine the precise length and features of fragments to include, a labor-consuming task requiring automation. To increase the flexibility and efficiency of motif searches, we generated MultiGS (standing for Multi Genome Searcher) a Perl based program that reads a txt file with the annotated genome information. The input consists of a list of genes (gene-ID) with the coordinates of the sequence of interest relative to either the transcription start site or the first nucleotide of the start codon and the output is a FASTA file with the resulting DNA sequences. The genome searcher program can read an adapted GenBank txt genome annotation file, thus providing a simple and universal tool for preprocessing DNA motif searches.

To test MultiGS as a useful tool to help searching for DNA binding motifs, we focused in Cluster A because both the expression profile and functional criteria were consistent with NtcA activation. To this end, we considered the 30 operons belonging to Cluster A that were located in the main chromosome. MultiGS was used to extract the 75 nt long sequences upstream the TSS or the start codon and submitted to MEME [9] as FASTA format ouput. 22 out of 30 of the chromosomal transcriptional units were found to contain NtcA binding sites (see table 1) (p-val 10^{-17}). These NtcA boxes were mainly at -40 from the TSS, the preferred position for NtcA activation (Fig. 4). Therefore, MultiGS is a flexible tool for motif searches that would be useful for future searches within the NtcA-independent operons that would provide important breakthroughs.

Table 1. Group 4 ORFs. ID, gene description, and positions of the TSS and NtcA consensus sequence

Gene ID	Gene description	TSS Position	Box
Synpcc7942_0127	transcriptional regulator, Crp/Fnr family (NCBI); ntcA	-108	40.5 AGTAGGCAGTTGCTACA
Synpcc7942_0342	hypothetical protein (NCBI)	-45	40.5 AGTGGTGTTATGGACA
Synpcc7942_0365	response regulator receiver domain protein (CheY-like) (NCBI)	nd	-53.5 TCTAACAAAGATTACT
Synpcc7942_0442	ammonium transporter (NCBI); amt1	-102	40.5 TGTTACATCGATTACA
Synpcc7942_0840	conserved hypothetical protein (NCBI)	-44	40.5 TGTTACATCGATTACA
Synpcc7942_0841	putative flavoprotein involved in K+ transport (NCBI)	26	69.5 GGTCGGCGATTGATACG
Synpcc7942_0891	conserved hypothetical protein (NCBI)	-16	57.5 TCGTGTATCCAGTTACG
Synpcc7942_1032	conserved hypothetical protein (NCBI)	-26	40.5 CGTATCCCGAACTACA
Synpcc7942_1036	conserved hypothetical protein (NCBI)	-24	40.5 AGTAGCTACAGCTACG
Synpcc7942_1039	conserved hypothetical protein (NCBI)	-51	46.5 TGTAAGCAAGGCTACG
Synpcc7942_1240	Ferredoxin–nitrite reductase (NCBI)	-30	41.5 GGTAACAGAAACTACA
Synpcc7942_1477	conserved hypothetical protein (NCBI); transglutaminase like	nd	59.5 AGTGTCAGATGTTACG
Synpcc7942_1538	conserved hypothetical protein (NCBI)	-20	42.5 CGTTACCCTTGATACG
Synpcc7942_1636	pterin-4a-carbinolamine dehydratase (NCBI)	-25	40.5 CGTGCTTTTGCTACG
Synpcc7942_1713	probable hydrocarbon oxygenase MocD (NCBI)	-40	40.5 GGTAGCGGATCGCTACA
Synpcc7942_1764	conserved hypothetical protein (NCBI)	nd	11.5 CGTTGTCTCTGTTACC
Synpcc7942_1797	conserved hypothetical protein (NCBI)	-31	46.5 TGTGGTGGCGGTAACA
Synpcc7942_2107	ABC-type nitrate/sulfonate/bicarbonate transport systems periplasmic components-like (NCBI)	-16	42.5 TGTAACGACGGCTACA
Synpcc7942_2156	glutamine synthetase, type I (NCBI); glnA	-141	45.5 TGTATCAGGCTGTTACA
Synpcc7942_2279	ammonium transporter (NCBI); amtB	-36	42.5 AGTAGCAAAAGTTACG
Synpcc7942_2466	two component transcriptional regulator, winged helix family (NCBI); nrrA	-23	42.5 CGTAAAGGCGAATACA
Synpcc7942_2599	conserved hypothetical protein (NCBI)	nd	59.5 TGCAGCCAGTGCGACA

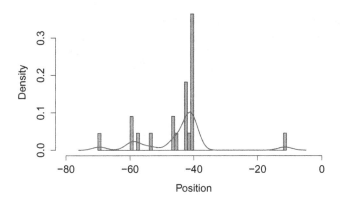

Fig. 4. Distribution of the NtcA motifs identified with MultiGS preprocessing and MEME in cluster A. Sites are positioned relative to the transcription start site.

Acknowledgments. This work was supported by the Spanish Ministry of Economy and Competitivity (Grant BFU2012-33364) and by the Valencia Regional Government (Grant ACOMP/2014/144 from Conselleria de Educación, Cultura y Deporte).

References

1. Espinosa, J., Forchhammer, K., Burillo, S., Contreras, A.: Interaction network in cyanobacterial nitrogen regulation: PipX, a protein that interacts in a 2-oxoglutarate dependent manner with PII and NtcA. Mol. Microbiol. 61(2), 457–469 (2006)
2. Espinosa, J., Forchhammer, K., Contreras, A.: Role of the Synechococcus PCC 7942 nitrogen regulator protein PipX in NtcA-controlled processes. Microbiology 153(3), 711–718 (2007)
3. Llacer, J.L., et al.: Structural basis for the regulation of NtcA-dependent transcription by proteins PipX and PII. Proc. Natl. Acad. Sci. USA 107(35), 15397–15402 (2010)
4. Laichoubi, K.B., Espinosa, J., Castells, M.A., Contreras, A.: Mutational Analysis of the Cyanobacterial Nitrogen Regulator PipX. PLoS One 7(4), e35845 (2012)
5. Espinosa, J., Rodrguez-Mateos, F., Salinas, P., Lanza, V.F., Dixon, R., de la Cruz, F., Contreras, A.: PipX, the coactivator of NtcA, is a global regulator in cyanobacteria. Proc. Natl. Acad. Sci. USA 111(23), E2423–E2430 (2014)
6. Billis, K., Billini, M., Tripp, H.J., Kyrpides, N.C., Mavromatis, K.: Comparative transcriptomics between Synechococcus PCC 7942 and Synechocystis PCC 6803 provide insights into mechanisms of stress acclimation. PLoS One 9(10), e109738 (2014)

7. Core Team, R.: R: A language and environment for statistical computing. R Foundation for Statistical Computing, Vienna (2014), http://www.R-project.org/
8. Maechler, M., Rousseeuw, P., Struyf, A., Hubert, M., Hornik, K.: Cluster Analysis Basics and Extensions. R package version 1.15.2 (2014), http://cran.stat.ucla.edu/web/packages/cluster/index.html
9. Bailey, T.L., et al.: MEME SUITE: Tools for motif discovery and searching. Nucleic Acids Res. 37(web server issue), W202–W208 (2009)

HvDBase: A Web Resource on *Hydra Vulgaris* Transcriptome

Daniela Evangelista[1,*,**], Kumar Parijat Tripathi[1,**], Valentina Scuotto[2], and Mario Rosario Guarracino[1]

[1] Laboratory for Genomics, Transcriptomics and Proteomics (LAB-GTP), High Performance Computing and Networking Institute (ICAR), National Research Council of Italy (CNR), Via Pietro Castellino, 111, Napoli, Italy
[2] Dipartimento di Scienze e Tecnologie, Università degli studi di Napoli Parthenope, Napoli, Italy
daniela.evangelista@na.icar.cnr.it

Abstract. The Whole Transcriptome Shotgun Sequencing (RNA-seq) uses the Next Generation Sequencing (NGS) capabilities to analyze RNA transcript counts and their quantification, with an extraordinary accuracy. In order to provide a comprehensive knowledge about the non-reference model organism *Hydra vulgaris* water polyp, we developed a comprehensive database in which the whole functional annotated transcriptome is integrated, and from which it is possible to have access and download all the information related to 15,522 transcripts. The study includes different functional annotations coming from 19 repositories, conveniently grouped by type, such as: i. Pathway: BBID, BioCarta, KEGG, Panther; ii. Domain: COG-Ontology, Interpro, PIR Superfamily, SMART, PFam; iii. Protein Interaction: BIND, MINT, UCSF-TFBS; iv. GO: GO-Term Cellular Components, GO-Term Biological Processes, GO-Term Molecular Functions; v. Miscellaneous: OMIM, EC Number, SP-PIR Keywords, UP-SEQ features. The easily-accessible nature of *HvDBase* makes this resource a valuable tool for rapidly retrieve knowledges at transcript level as well as useful to inspect differential expression of protein-coding genes in *Hydra vulgaris* transcriptome in the contex to the different experimental condition.
Web resource URL: http://www-labgtp.na.icar.cnr.it/HvDBase

Keywords: *Hydra vulgaris*, Annotations, Transcriptome, Database, PHP, MySQL.

1 Introduction

In 1744, *Hydra* regeneration was discovered by Abraham Trembley. From then onwards, this fresh water cnidarian polyp gained a status of potential model system for regeneration behaviour. *Hydra* is a medusozoan that diverged from

---

* Corresponding author.
** These authors contributed equally to this work.

F. Ortuño and I. Rojas (Eds.): IWBBIO 2015, Part II, LNCS 9044, pp. 355–362, 2015.
© Springer International Publishing Switzerland 2015

anthozoans at least 540 millions year ago. In the last two centuries, it immensely attracted the biologist as a model organism to unfold the biological mechanism involving nature of embryogenesis, neurogenesis, ecosystem-ecotoxicity, sex reversal, symbiosis, aging, feeding behavior, light regulation, multipotency of somatic stem cells, temperature-induced cell death, neuronal trans-differentiation etc [2]. In recent years, with the advent of genome and transcriptome sequencing methodologies [1], researchers generated the draft assembly of *Hydra* magnipappilata genome using shotgun approach. They observed *Hydra* genome is (A+T) -rich with 71 percent A+T, and also includes 57percent of transposable elements. The CA assembly (1.5gigabases (Gb)) has contig and scaffold N50 values of 12.8kilobases (kb) and 63.4kb, respectively. The RP assembly (1.0 Gb) has a contig N50 length of 9.7kb and a scaffold N50 length of 92.5kb. The CA assembly gives an estimated non-redundant genome size of 1.05Gb. The RP assembly gives an estimated non-redundant genome size of 0.9Gb. They also reported estimate that the *Hydra* genome contains 20,000 bona fide protein-coding genes (excluding transposable elements), based on expressed sequence tags (ESTs), homology and ab initio gene prediction [3]. To dissect the genetic cascades supporting the biological behaviour such as regeneration, multipotency of somatic stem cells, temperature-induced cell death, neuronal transdifferentiation shown by *Hydra* species, Yvan Wenger and Brigitte Galliot [4] carried out the transcriptomic analysis of *Hydra vulgaris*. They implemented a powerful strategy to combine Illumina and 454 reads and produced, with genome assistance, an extensive and accurate *Hydra* transcriptome. *Hydra vulgaris* is an organism of great ecological and biotechnological significance. In the absence of well annotated *Hydra vulgaris* transcriptome, interpreting extremely large transcriptomic data coming from RNA-Seq experiments in response to different experimental conditions, into biological knowledge is a problem, since biologist-friendly tools are lacking. There is only one database, Compagen, maintained by Bosch Laboratory at the University of Kiel [5], which is a comparative genomics platform for early branching metazoan animals. It stores various raw and processed sequence datasets along the evolutionary tree from sponges and cnidarians up to the tunicates and lower vertebrates. The Databases at Compagen[5] contains selected raw genomic and EST sequence datasets available from public domain (NCBI Trace archive, EST). In addition to the public datasets, Compagen [5] also provides already processed data like CAP3 assembled ESTs, Unigene collections of predicted peptides. This database does not provide any functional or gene annotation for *Hydra vulgaris* transcripts. In our lab, we develop a web application, to obtain the biological information for the whole transcriptome of *Hydra vulgaris*, with the help of an existing computational pipeline [6], the whole transcriptome of *Hydra vulgaris* is annotated and stored in *HvDBase*, which is a comprehensive web resource driven on a relational database. Here in this manuscript, we are presenting *HvDBase*, which contains functional and gene ontology annotations for around 35% of the total transcripts of *Hydra vulgaris*.

2   Material and Methods

2.1   *HvDBase* Annotation

HvDBase presently accommodates not only the gene ontological information, but also other available functional annotation about domains, metabolic pathways, protein-protein interactions, as well as relevant biological information from SwissProt and PIR protein databases with respect to each and every *Hydra vulgaris* transcript. With the help of *HvDBase*, end users can obtain a comprehensive biological information for the differentially expressed transcripts ids. Some biological information incorporated for each transcript within *HvDBase* regarding: i. Gene Ontology: controlled vocabularies (ontologies) that describe gene products in terms of their associated biological processes, cellular components and molecular functions in a species-independent manner; ii. Domain annotation: modular structure of the gene product, and evolutionary and molecular functional aspects of the transcripts, annotations for COG-Ontology, InterPro, PFAM and SMART domains are stored; iii. Metabolic Pathway annotation: several biological pathway information from KEGG, BBID, BioCarta, Panther; iv. Protein interaction: to study the interactions partner for the gene products, information from BIND, MINT, UCSF-TFBS databases; v. Miscellaneous: various biologically relevant information such as EC number, OMIM report, keywords and categories used in the UniProt-knowledgebase; sequence features such as regions or sites of interest in the expressed transcripts and gene products, for example post-translational modifications; binding sites, enzyme active sites, local secondary structure or other characteristics reported in the cited references with respect to *Hydra vulgaris* transcripts.

2.2   *HvDBase* Pipeline

In order to store *Hydra vulgaris* data and for facilitating the information recovery without browsing the jungle of on-line web repositories, we implemented in parallel the *HvDBase* database (DB) and a user-friendly web interface for the content visualization. It is organized in tables with a relational structure containing all the items handled with proper data type, for a better performance of the database with respect to speed and deployment. As reported in figure 1 - where the dotted box represents the background pipeline from which *Hydra vulgaris* transcriptomic annotation data were retrieved [6] - the box in the lower left corner shows the designed E/R diagram of the DB, as back-end of the system. This latter is closely related to the front-end side, which Graphical User Interface (GUI) of the Home Page is reported in Fig (1) along with the input and output of the web-pages. The results were organized in extremely easy-to-read tables, for helping external users to visualize and download the data in short span of time. The web interface allows the end-user to access to several sections.

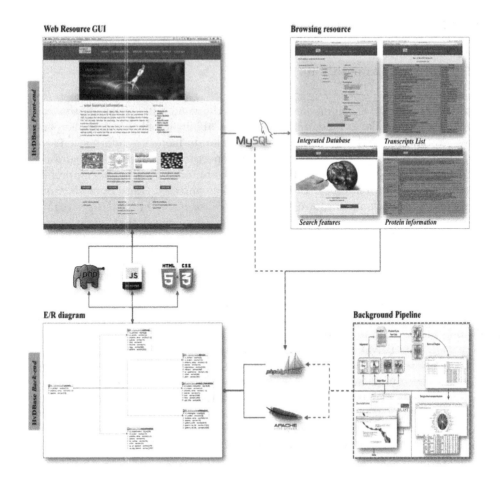

Fig. 1. *HvDBase* pipeline

HvDBase, indeed, consist of seven sections schematically shown in the block diagram (Fig. 2), as well as, in the Sitemap section of *HvDBase*. The core of the web portal is represented by the Transcriptome section. In particular, within this section, user can access to two distinct pages: Transcripts List or Database List. All the other sections were designed for all those users, who want to understand the concept of this web application: application content (see Resources sections); details related to the application development (see Pipeline section); teams miscellaneous information (see About us and Contact sections). Finally, the Search by Term section was implemented to swiftly recover, within the framework, all the information, which are beyond the knowledges of the end-user. All the information contained in such sections are processed in *HvDBase* through its easy and user-friendly web-interface, designed to rapidly retrieve data in a scientifically rigorous way.

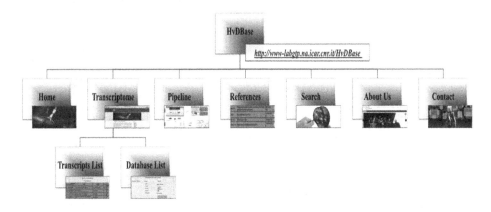

Fig. 2. *HvDBase* sitemap

2.3 Database Technology

The *HvDBase* resource grabs information from the database and inserts that information into the proper web page each time it is requested. If the information stored in the database changes, the web page connected to the database will also dynamically and automatically updates. It is a database-driven web site, more properly driven on a Relational Database Management System. *HvDBase* was developed using web server Apache/2.2.26; MySQL client version 5.3.28 - 10.04.1 (Ubuntu) and the free tool phpMyAdmin version 3.3.2 deb1 Ubuntu 0.2 to handle the administration of MySQL with InnoDB storage engine. The front-end is implemented using the scripting language PHP/5.2.6-3; the JavaScript technology for dynamic contents; the markup language HTML5 and style sheet CSS 3.0. The *HvDBase* code is validated according to the standard web of the international community W3C (World Wide Web Consortium: http://www.w3.org) and therefore, although optimized for Safari, it is easily accessible and clearly visible by all browsers and smartphones.

3 Results

3.1 Database Content

The *HvDBase* web resource allows to structure the information and to display it in sorted and filtered tables accompanied by thorough explanations. The data were collected from the literature and external database, then appropriately handled with ad hoc scripts. Overall, information currently contained in *HvDBase* are related to 19 different functional terms of 15,522 transcripts, that is around 35% of the total number of de novo genome based assembly of *Hydra* from RNA-seq transcriptomic data (Fig. 2). Currently, *Hydra* dataset contains 48.909 sequences, 45.269 of which are longer than 200 bp and have been deposited at the European Nucleotide Archive (ENA) [7] under the accession numbers HAAC01000001 - HAAC01045269 [4] .

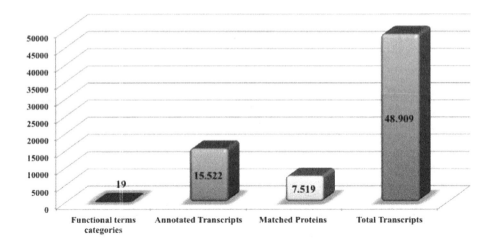

Fig. 3. *HvDBase* content at a glance

3.2   A Case Study

To display the performance of the web resource, in this section, we are presenting an example case study of the HAAC1000020 transcript and its functional annotation (Fig. 4) obtained through *HvDBase*. User, as already shown in the pipeline (Fig.1), by selecting the annotation categories of interest can retrieve the related information from the *Transcriptome page*. Here, we are showing the result page for the transcript query (HAAC1000020) functional annotation. For GO Ontology and metabolic pathways within *HvDBase* suggests the possible role of the HAAC1000020 in biological process (transportation), the molecular activity GO term suggest the importance of the transcript or its product in the formation of molecular machinery of transmembrane and voltage gated channels. This information is also supported by the Cellular content GO term as this transcript is associated with plasma membrane. The metabolic pathway annotation obtained from *HvDBase* for this transcript added further information to characterize this transcript in its role in signalling pathway. All these information in pieces combined to gather to suffeciently describe the biological activity, molecular mechanism and metabolic role of this unknown transcripts with in *Hydra vulgaris* . Similarly, all the unknown transcripts, generated through various transcriptomic experiment carried out on *Hydra vulgaris* can easily be characterized with the help of HvdBase web application in a very efficient and straight forward manner.

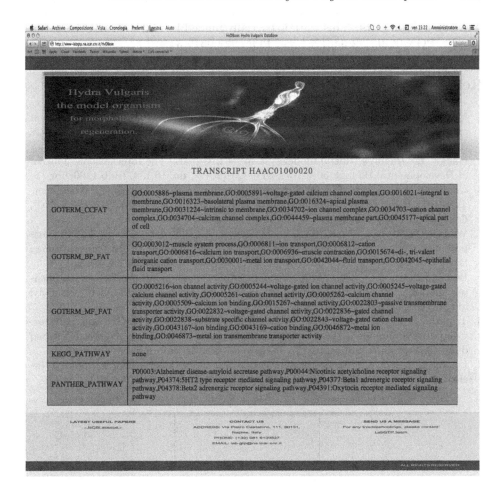

Fig. 4. Screenshot of the information related to the HAAC1000020 transcript

4   Conclusion

The creation of dedicated databases for non-reference model organisms is an important issue and always desirable. *HvDBase* is a pilot study useful to provide a comprehensive knowledge about the transcriptome of the non-reference model organism *Hydra vulgaris*. Through this web resource, researchers with low skills in bioinformatics are able to access and retrieve around 35% of the functional annotations of *Hydra vulgaris* transcriptome. *HvDBase* is a very easy-to-use web resource, freely available and without login requirements. As a modular and open-source platform, *HvDBase* can easily be extended and customized to future demands and developments. Currently, we are in the process of updating of *HvDBase* resource to make it much more informative, and, we are going to provide annotation for all other transcripts.

Acknowledgements. This work is partially funded by INTEROMICS flagship project, PON_02-00612-3461281 and PON_02-00619-3470457.

References

1. Mortazavi, A., Williams, B.A., Mccue, K., Schaeffer, L., Wold, B.: Mapping and quantifying mammalian transcriptomes by RNA-Seq. Nature Methods 5(7), 621–628 (2008), doi:10.1038/nmeth.1226
2. Galliot, B.: *Hydra*, a fruitful model system for 270 years. The International Journal of Developmental Biology 56(6-7-8), 411–423 (2012), doi:10.1387/ijdb.120086bg
3. Chapman, J.A., Kirkness, E.F., Simakov, O., Hampson, S.E., Mitros, T., Weinmaier, T., Steele, R.E.: The dynamic genome of *Hydra*. Nature 464(7288), 592–596 (2010), doi:10.1038/nature08830
4. Wenger, Y., Galliot, B.: RNAseq versus genome-predicted transcriptomes: A large population of novel transcripts identified in an Illumina-454 *Hydra* transcriptome. BMC Genomics 14(1), 204 (2013), doi:10.1186/1471-2164-14-204
5. Bosch, T.C.G., et al.: Compagen: A comparative genomic platform for early branching metazoa. Zoological Institute, University of Kiel, D-24118 Kiel, Germany, http://compagen.org
6. Tripathi, K.P., Evangelista, D., Cassandra, R., Guarracino, M.R.: Transcriptator: Computational pipeline to annotate transcripts and assembled reads from RNA-Seq data. (Accepted for publication in Lecture notes in Bioinformatics, Springer)
7. Leinonen, R., et al.: European Nucleotide Archieve (ENA): The European Nucleotide Archive. Nucleic Acids Res. 39(database issue), D28–D31 (2011)

Nucleotide Sequence Alignment
and Compression via Shortest Unique Substring*

Boran Adaş[1], Ersin Bayraktar[1], Simone Faro[2], Ibraheem Elsayed Moustafa[3],
and M. Oguzhan Külekci[3],**

[1] Department of Computer Enginering, İstanbul Technical University, Turkey
[2] Department of Mathematics and Computer Science, University of Catania, Italy
[3] Department of Biomedical Enginering, İstanbul Medipol University, Turkey
{adas,bayrakterer}@itu.edu.tr, faro@dmi.unict.it,
{iemoustafa,okulekci}@medipol.edu.tr

Abstract. Aligning short reads produced by high throughput sequencing equipments onto a reference genome is the fundamental step of sequence analysis. Since the sequencing machinery generates massive volumes of data, it is becoming more and more vital to keep those data compressed also. In this study we present the initial results of an on-going research project, which aims to *combine* the alignment and compression of short reads with a novel preprocessing technique based on shortest unique substring identifiers. We observe that clustering the short reads according to the set of unique identifiers they include provide us an opportunity to *combine* compression and alignment. Thus, we propose an alternative path in high-throughput sequence analysis pipeline, where instead of applying an immediate whole alignment, a preprocessing that clusters the reads according to the set of shortest unique substring identifiers extracted from the reference genome is to be performed first. We also present an analysis of the short unique substrings identifiers on the human reference genome and examine how labeling each short read with those identifiers helps in alignment and compression.

1 Introduction

Mapping short reads onto the reference genome is the fundamental initial step in the analysis of high-throughput sequencing data, where a large number of *alignment* software packages have been developed in the last decade [7]. In this paper we observe that clustering the short reads according to a set of unique identifiers of the reference genome they include provide an opportunity to improve both *alignment* and *compression* of short reads. To the best of our knowledge this is the first time this approach is used for the analysis of nucleotide sequences.

The general approach to achieve alignment fast in small memory footprint has appeared to be indexing the reference genome, and then seeking the occurrences of short reads one-by-one by using that index. It is not always possible to

---

* This work has been supported by the Scientific & Technological Research Council of Turkey (TÜBİTAK), BİDEB–2221 Fellowship Program, and also with the TÜBİTAK-ARDEB-1005 grant number 114E293.
** Corresponding author.

F. Ortuño and I. Rojas (Eds.): IWBBIO 2015, Part II, LNCS 9044, pp. 363–374, 2015.
© Springer International Publishing Switzerland 2015

exactly align each read since sequencing errors as well as differences between the sequenced individual and the reference are unavoidable. Thus, while mapping the reads, error-tolerant approximate matches should be considered. However, although there has been many efficient text indexing schemes for searching exact occurrences of the patterns, matching with symbol insertions, deletions, and mismatches is still an active research area.

Most of the aligners run with some parameters *limiting* the maximum number of mismatches/insertions/deletions allowed to occur while mapping a short read, and thus, especially large insertions or deletions are not easy to detect. With the ever increasing length of the short reads due to the technological advance of sequencing platforms, these limitations tend to become more severe. Underlining this fact, more recent aligners [10,1,12] as well as the new versions of the previous alignment packages [15,14] prefer to use k–mers of the short reads to roughly detect the mapping position on the reference genome, and then deploy a Smith-Waterman [18] style dynamic programming to achieve the task. In other words, instead of searching the whole read, the occurrences of k–mers extracted from the short read are scanned on the reference genome. When enough number of k–mers jointly points to a unique location, the Smith-Waterman algorithm is applied on the detected short region.

The point that is open for improvement in that approach is the optimization of the k value. The number of candidate regions increase with the short k values, and then it becomes difficult to decide on the correct region. Similarly, when k is set to a large value, sequencing errors or mutations are more likely to effect the performance, which is contrary to the basic idea behind the approach.

The ever increasing size of the data generated with high-throughput sequencing technologies requires to develop special methods to tackle with the problems of the huge genomic data sets [2]. In their *compressive genomics* definition, Loh *et al.* [16] stated *"algorithms that compute directly on compressed genomic data allow analyses to keep pace with data generation"*.

In that sense, compressing fastq files has been one of the most active research topics during the last few years [6], and many solutions have been proposed to represent those files as small as possible in size [5,8,4,11,3]. However, as stated in *compressive genomics* definition, the real challenge in fastq compression is more than the efficient archival of data, where we need support for operations to be achieved directly on compressed data such as efficient random access to any short-read as well as retrieving/extracting the reads mapped to a specific region of interest on the genome. The recent survey by Giancarlo *et al.* [9] lists the capabilities of the compressors in the genomic area in that sense.

The Idea and Our Contribution

A *unique substring* of the reference genome is a substring which is repeated only once in the whole sequence. In this paper we start by the observation that if a unique substring of the reference genome appears in a short read, then this short read can be mapped directly to the unique location of that substring identifier on the reference genome. This approach allows us to avoid to investigate any other k–mers since the detected substring is unique on the reference, and thus, points to its location unambiguously.

Moreover we observe that clustering the short reads according to a set of unique identifiers they include provide us an opportunity to *combine* compression and alignment. Thus, conforming to *compressive genomics* approach, we present an alternative path in high-throughput sequence analysis pipeline, where instead of applying an immediate whole alignment, a preprocessing that clusters the reads according to the set of shortest unique substrings identifiers extracted from the reference genome is performed first. At the end of this preprocessing operation each read is assigned to a substring identifier. That binding represents a rough alignment as we know the position of the unique substring on the reference, and therefore, the rough position of the read. Once each read is associated with its unique substring identifier, the user may use this information both for the alignment and compression, and even combining these two operations.

For the alignment, assume the user has a specific region of the interest on the genome, and wants to see the reads sequenced from this section. One simply selects the shortest unique substring identifiers of that region from the previously prepared dictionary, and retrieves the reads labelled with these substring identifiers. The labels of the reads tell the rough position of the read, and a Smith-Waterman type alignment may be called for full alignment information.

For the compression task, the user may create the buckets which represents regions on the genome. These buckets store the short reads which include the unique substrings identifiers of the selected region, and can be compressed efficiently due to their high redundancy originating from the fact that they all repeat the information from the same region.

A combined approach would be first to create the buckets and keep them compressed, and then, answer the alignment queries by extracting and generating the *full* alignment information of the reads from the related buckets.

Organization of the Paper

The paper is organized as follows. In Section 2 we briefly describe the process we used for identifying the set of the shortest unique substrings from the human genome and we analyze and describe in Section 3 the set extracted substrings. In Section 4 we describe our dictionary matching algorithm for mapping the set of short reads in their positions in the human genome. Finally we present our results in Section 5 and draw our conclusions in Section 6.

2   Shortest Unique Substring Identifiers of the Genome

Shortest Unique Substring (SUS) finding [17] has received significant attention very recently, and efficient methods have been developed to solve the problem [19,13]. Each position on a text has a corresponding SUS for sure, where there might be more than one SUS for some positions. Interested readers may refer to the regarding publications for the proofs and more detailed discussions. Formally we have the following definition.

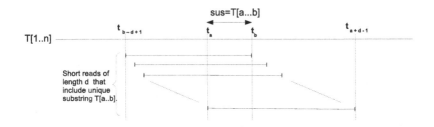

Fig. 1. Illustration of the short reads matching with the SUS identifier $T[a \ldots b]$ assuming a constant read length d

Definition 1 (Shortest Unique Substring). *Given a text $T[1, n]$ of length n, the shortest unique substring covering the specific location i, for any $1 \leq i \leq n$, is the shortest string of length ℓ, $T[a \ldots a+\ell-1]$, such that $1 \leq a \leq i \leq a+\ell-1 \leq n$ and $T[a \ldots a + \ell - 1] \neq T[b \ldots b + \ell - 1]$, for each $1 \leq b \leq n - \ell + 1$.*

The most obvious usage of SUS detection appears in displaying the results of a string search on a target text. Assume we are searching the occurrences of a keyword that appears more than once in the given text. Thus, while displaying the results, it is helpful to display a bit of the context including the detected position of the occurrence. In such a scenario, the length of the to-be-displayed context may be tuned according to the SUS of that position, which uniquely informs about the position of appearance.

In this study, we introduce a novel preprocessing based on the SUS signatures extracted from the reference genome that would help in sensitive read mapping and compression. With that purpose we extract the SUS identifiers from the reference genome, and build a SUS dictionary, where each substring is stored with the position of its occurrence on the reference. Notice that this is an operation that needs to be done on a target reference just once.

Fig.1 illustrates how SUS identifiers can be used in the alignment process. Assume that $T[a \ldots b]$ is such a SUS and d represents the short read length. The reads that include $T[a \ldots b]$ are shown in the figure. If we do not let any insertions or deletion during the mapping, the leftmost appropriate read including this SUS should map to $T[b-d+1 \ldots b]$, and similarly the rightmost one to $T[a \ldots a+d-1]$.

The good thing is that once we caught the SUS in the read, we have the flexibility to allow larger error thresholds, since we know exactly the address of the short read matching with the SUS. Thus, to let insertion and deletions, the region might be extended a bit further to the right and left, and then, the short reads may be aligned to that extended region via a cache-oblivious dynamic programming as performed in [10,12].

Careful readers will quickly realize that in this scenario the length of the SUS identifier should be less than or equal to read length d. In addition to that, we seek an exact match between the SUS and short reads. Surely, we know that the possibility of a mismatch becomes more significant as the length of the

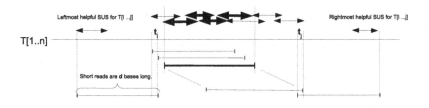

Fig. 2. A short read generally includes more than one SUS

SUS increases. Hence, long SUS identifiers are not supposed to help much, and we neglect in the SUS dictionary the ones that are longer than a predefined threshold λ during the operation. During our experiments in this study on human reference genome, we set that threshold to be $\lambda = 30$, which depends on the empirical experience that we can expect the sequencers today to be able to read that much of consecutive bases without any error.

Fortunately, a short read includes generally more than one SUS identifier as shown in Fig.2, where the sample read and the SUS candidates are marked bold. This becomes useful as we may still expect to have appropriate length SUS candidates, when we exclude the long SUS from the dictionary. Having more than one candidate helps in case of errors also, since an exact match of the short read at least with one of the SUS is enough to map it appropriately. For example in the Fig.2, the read can be located on the reference once one of the four possible SUS occur in it without an error. Below we give the formal definition of the SUS set of a region.

Definition 2 (Shortest Unique Substring Set of a Region). *Assume a region of interest $T[i\ldots j]$ on the reference genome $T[1\ldots n]$ is specified and the constant length of the short reads is d. The* SUS *set of the specified region is the list of the* SUS *strings from the* SUS *dictionary, whose beginning positions on the reference genome are between $i - d + 1$ and $j + d - 1$.*

With the concern of aligning all the reads corresponding to an arbitrary region of interest $T[i\ldots j]$, we seek the leftmost and rightmost SUS identifiers that are helpful to construct the region. With the term *helpful*, we mean there exists a chance that a short read including this SUS may cover at least one base from the target region. This is depicted in Fig.2 as when the selected leftmost (rightmost) SUS appears leftmost (rightmost) on a short read, that short read may cover the position t_i (t_j).

3   SUS Analysis of the Human Reference Genome

In this section we analyze the SUS identifiers we extracted from human reference genome GRCh38[1]. During our analysis we concatenated all chromosomes of the

---

[1] Available at
http://www.ncbi.nlm.nih.gov/projects/genome/assembly/grc/human/.

genome into a single string and changed everything other than A, C, G, T, N to N, and replaced consecutive repeating Ns with a single N letter. Considering the DNA sequencing technology, where short reads may originate from both the forward and reverse strands of the DNA, we appended the reverse complement of this string to its end, and thus, the resulting whole human genome is of length $5875280183 \approx 5.87$ billions bases.

For each position on this string, we have detected the corresponding SUS with the method of [13]. The operation took roughly 75 minutes on a machine with 256 GB memory and Intel Core 2 Quad processor running Linux Centos 6.2.

There may be more than one SUS (with the same length) for a position. We break the tie by choosing the leftmost one in such a case. Moreover one SUS may be shared by *consecutive* positions on the target string, and thus, we counted the number of distinct SUS in the SUS database of the whole genome. We found that $1924177251 \approx 1.92$ billion of the 5.87 billion items in the SUS database are unique when both the forward and reverse strands are taken into account.

Since long SUS identifiers are not useful in our strategy, we excluded the ones that are longer than the threshold value, which we set as 30 in our study (the longest SUS detected is nearly 1.2 million bases long). In addition, some of the SUS identifiers are either right or left extensions of neighbouring shorter ones. For instance assume a SUS is $T[a \ldots b]$, and while searching for the SUS covering position $b+1$, it might be the case that $T[a \ldots b+1]$ may be returned as the SUS of that position by the algorithm. We also get rid of such extension patterns, and create our final SUS dictionary composed of $963836205 \approx 1$ billion SUS identifiers.

A SUS has the potential to cover a position if it is in vicinity of d bases to that position. That is because, when that SUS appears at the very beginning or end of a short read, then that short read covers all λ positions to the right or left as shown in Figure 1. Certainly, it is much better for a position to have the chance of being covered by large number of distinct SUS.

For some positions on the human genome it might not be possible to detect a SUS identifier longer than the selected threshold 30 bases. If such positions does not have a neighbouring SUS in close vicinity, then the reads originating from this area has the danger of not being caught by any of the SUS identifiers from the dictionary. To measure this problem, we define below the theoretical SUS coverage of an individual position.

Definition 3 (Theoretical sus coverage of a position). *The leftmost short read possible to cover an inspected position i is $T[i-d+1 \ldots i]$, and the rightmost short read including position i is $T[i \ldots i+d-1]$. Notice that these short reads may be produced from both the forward and reverse strands by the sequencing equipment. Any SUS identifier $T[a \ldots b]$, such that $i-d+1 \leq a \leq b \leq i+d-1$, may appear in those short reads covering the position i. Therefore, we define the theoretical SUS coverage of position i as the total number of such SUS identifiers on both the forward and reverse strands.*

Figure 3 shows the theoretical SUS coverage of the human reference genome. The short reads that include the positions, which have 0 SUS coverage, have no chance of being identified by the proposed scheme. We call these reads *orphan*,

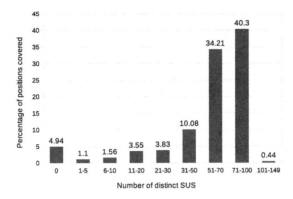

Fig. 3. The theoretical SUS coverage of the human reference genome

and observed that less than 5% of the genome remains orphan. Those orphan positions are non avoidable due to the repetitive nature of the genome, but they can be handled efficiently by the regular k–mer approaches.

4   SUS Dictionary Matching

In this section we describe the algorithm we used to match the SUS collected in the dictionary against the set of short reads. Before entering into details we observe that an important property of the SUS dictionary is that none of the items appear as a substring of another item. This property is formally stated by the following lemma.

Lemma 1. *Let* $S = \{s_1, s_2, \ldots s_m\}$ *be the* SUS *dictionary, where* s_i *is a unique substring of the reference genome* T. *There exists no* s_i *in* S, *which appears as a substring in any other* s_j, *with* $j \neq i$.

Proof. Assume s_i appears in s_j, where $i \neq j$. We know that s_i and s_j are unique on the reference genome by definition of SUS. We have also deleted from the set the right or left extensions of SUS identifiers while creating the dictionary, and thus, s_j cannot be a right or left extension of s_i. Hence, if s_i occurs in s_j, this means s_i is not unique, which contradicts the hypothesis.   □

Based on Lemma 1 we devised an algorithm for fast scanning of the short reads against the SUS dictionary. Specifically during the preprocessing phase it builds a data structure in order to index all the SUS in the dictionary. Then This index is used to speed up the searching process in the subsequent phase, where the short reads are searched, one by one, for any occurrence of the given SUS.

In our algorithm we make use of the *longest common prefix* of two sequences as define below.

Definition 4 (Longest Common Prefix). *Given two strings, x and y over the same alphabet, the longest common prefix array (*LCP*) between x and y, in symbol lcp(x, y), is the maximal length ℓ such that $x[1 \dots \ell] = y[1 \dots \ell]$, where $\ell \leq \max(|x|, |y|)$.*

For example, if $x = \text{ACATAC}$ and $y = \text{ACTTAGC}$ then $lcp(x, y) = 2$.

In the following we describe separately the preprocessing and the searching phase of our algorithm.

The Preprocessing Phase

Let S be the SUS dictionary and let R be the set of the short reads as described above. In this section we give a description of the data structure we use for matching the SUS against the short reads and briefly describe the preprocessing of the input data.

The set S of the SUS contains DNA sequences with a length between 12 and 30 bases. We observed on the human reference genome that the shortest SUS is of length 10 bases, where 10 or 11 bases long SUS identifiers are very few. Thus, we decided to consider 12 as the bottom threshold for SUS signatures and just extended the ones with 10 or 11 bases to reach length 12.

We indicate the minimum length of an SUS in S with the symbol $m = 12$. For each $s_i \in S$, let p_i be the prefix of length m of s_i, and let r_i be the suffix of s_i of length $|s_i| - m$. It is clear that $r_i = \varepsilon$ when $s_i = m$. In this context we can write $s_i = p_i \cdot r_i$ for each $s_i \in S$.

When preprocessing the set S we compute a fingerprint $f(s_i)$ for each $s_i \in S$. The fingerprint of an SUS s_i is computed by translating its prefix p_i in an integer number as $f(s_i) = \sum_{j=0}^{m-1} code(p_i[j]) \times 4^{m-1-j}$, where $code : \{\text{A}, \text{C}, \text{G}, \text{T}\} \rightarrow \{0, 1, 2, 3\}$ is a function which maps each character in an integer number. It is trivial to observe that the prefix of a SUS in S is uniquely described by a single fingerprint value. However there are SUS which share the same prefix, although they are different. Since the fingerprint value is computed on the prefix of length $m = 12$ of each SUS we have that $0 \leq f(s_i) < 2^{24}$ (where $2^{24} = 16.777.216$), for each $s_i \in S$.

During the preprocessing phase we construct an index table B of 2^{24} locations which is used to index all the sequences of length $m = 12$ over an alphabet of 4 elements. Then, for each s_i in S, we define a bucket, $b(s_i)$, containing useful information about the SUS and insert it in B according to its fingerprint. Thus each element $B[k]$ of the table is the set of buckets of all the SUS which share the same fingerprint k. More formally we have $B[k] = \{b(s_i) : s_i \in S \text{ and } f(s_i) = k\}$, for $0 \leq k < 2^{24}$. The set $B[k]$ is represented by a linked list where the buckets are lexicographically ordered according to the corresponding SUS. In this context we indicate with $prev(s_i)$ the SUS which precedes s_i in its linked list.

The bucket of each s_i in S is a triple $b(s_i) = \{i, lcp_i, r_i\}$, where

- i is the index of the SUS in the dictionary S. Such information is used to locate the SUS and its position in the reference genome.

- lcp_i is the longest common prefix between s_i and $prev(s_i)$.
- r_i is the suffix of s_i of length $|s_i| - m$.

The Searching Phase

During the searching phase we select each short read from the set R, one by one, and search it for the occurrence of any SUS in the dictionary S.

Let t be a short read in R and let n be the length of t. During the searching of t we open a substring w of length m over t, initially aligned with the left end of t so that $w = t[1 \ldots m]$. We call such a substring the *window* of t. Then the window is slided to the right character by character until it reaches the right end of t.

For each alignment of the window w at position i of t (so that $w = t[i \ldots i + m - 1]$), we check if any SUS in S has an occurrence beginning at position i of t. If no SUS occurs in t at position i the next iteration is started with a new alignment of the window at position $i + 1$.

For each iteration, say at position i, the algorithm computes the fingerprint k of the window $w = t[i \ldots i + m - 1]$. Then it easy to observe that only the SUS in the set $B[k]$ can occur at position i of t, since they share the same prefix as the window. Thus the algorithm checks the element of the set $B[k]$, one by one, until an occurrence is found or all possible candidates have been checked. The elements of the set $B[k]$ are checked by following a lexicographical order of the correspondent SUS.

Let $s_{i_1}, s_{i_2}, \ldots, s_{i_n}$ be the n SUS in the set $B[k]$, in lexicographical order. Since we already know that the first m characters of s_{i_1} are equal to $t[i \ldots i + m - 1]$, the algorithm scans the characters of the read t starting from position $i + m$ and comparing them with the corresponding characters in s_{i_1}, until the whole SUS is scanned or a mismatch is encountered. In the first case an occurrence is reported and the algorithm stops searching the read t. In the second case the algorithms discards s_{i_1} and continue comparing t with the next SUS s_{i_2}.

Suppose that the algorithm scanned j characters of s_{i_1}, starting from position $i + m$, before finding a mismatch. Thus we have $s_{i_1}[m + j - 1] = t[i + m + j - 1]$ and $s_{i_1}[m + j] \neq t[i + m + j]$.

We now recall that the value lcp_{i_2} is the maximal length of the shared prefix between s_{i_1} and s_{i_2}. Thus if $lcp_{i_2} < m + j$ we know that s_{i_2} cannot occur at position i of t. Moreover, for the same reason, none of the other SUS in the set $\{s_{i_2}, s_{i_3}, \ldots, s_{i_n}\}$ can occur at position i of t. Thus in this case the scanning is stopped and a new iteration is started with a new window.

In the other case, if $lcp_{i_2} \geq m + j$ the algorithm continues comparing t and s_{i_2} starting at position $i + m + j$ of t until the whole SUS is scanned or a mismatch is encountered.

When a new iteration on the new window $w' = t[i + 1 \ldots i + m]$ is started the algorithm can remember the length of the prefix which has been scanned in the previous iteration. Suppose j is the length of such a prefix, so that $s_{i_1}[m + j - 1] = t[i + m + j - 1]$ and $s_{i_1}[m + j] \neq t[i + m + j]$ and suppose $lcp_{i_2} < m + j$ so that a new iteration is started. Let k' be the new fingerprint value of the window w'.

By Lemma 1 we know that any SUS in $B[k']$, with a length less than $j - 1$, can occur at position $i + 1$. Thus the algorithm can discard from $B[k']$ all the SUS with a length less than $j - 1$.

This process stops when an occurrence of any SUS in S is found in t or when the starting position i of the window reaches the value $|t| - m$.

Observe that the computation of the fingerprint of a given window $w' = t[i + 1 \ldots i + m]$ can be computed in constant time from the fingerprint of the previous window $w = t[i \ldots i + m - 1]$ by the following relation

$$f(w') = (f(w) - code(t[i]) \times 4^{m-1}) + code(t[i + m])$$

Thus the computation of all windows along a short read of length d can be done in $\mathcal{O}(d)$ time. However each iteration of the searching process requires $\mathcal{O}(\lambda - m)$ time in the worst case. Thus the worst case time complexity for searching a short read of length d for any occurrence of the SUS in S is $\mathcal{O}((\lambda - m)d)$.

Despite its quadratic worst case time complexity it turns out from our experimental evaluation that the average number of text characters inspection during the search is linear.

5   Results

We have implemented the SUS pattern matching algorithm and applied on the short reads of the whole human genome $NA18507$ which was sequenced with Illumina HiSeq2500. The machine we have conducted this matching had $32GB$ of memory, LinuxMint 17 operating system. We only used a single CPU of the available four. It took ≈ 10 minutes to pass over the 4 million pair-end short reads to detect SUS identifiers, and the software used 16GB memory[2].

Table 1 summarizes what percent of the short reads could be identified with how many SUS signatures. 3.74% of the short reads include 1 to 5 distinct SUS signatures, and ≈ 50 % have at least 30 and at most 50 distinct SUS identifiers. Remember that maximum SUS length was set to 30 bases, and the read lengths in this experiment was 101 bases per short read. When one of the two pairs

Table 1. Percentages of the short reads including SUS identifiers on the first 4 million of the pair-end sequences of the NA18507

	% of short-reads identified by X SUS signatures					
unidentified	1–5	6–10	11–20	21–30	31–50	71–100
3.19	3.74	3.42	11.16	28.23	49.73	0.53

---

[2] It is noteworthy that although there is a lot to do for space usage reduction and execution time enhancement, we decided to apply those changes in final release of the software and did not pay much attention at this point to implement them in this proof–of–concept study.

in a pair-end tuple is identified with an SUS, we assume we can successfully align both pairs since we know that they are in a certain distance. Considering this fact, we have observed that, of the 4 million pair-end reads it is possible to identify ≈ 96 percent directly. This means those reads uniquely map to the area pointed by the SUS identifier they include. The remaining short reads in which no SUS could be located, it is necessary to run the regular k–mer approach to decide where they can map to. These are mostly the reads originating from highly repetitive areas of the genome or highly erroneous readings.

6   Conclusions and Future Works

We have introduced clustering of the short reads according to the SUS signatures extracted from the target species' reference genome. This clustering is supposed to help in two directions so as to improve the compression and alignment. Reordering the reads in the fastq file so that the ones having neighboring SUS signatures are kept close would keep the related items in the same bucket, and hence, better compression might be available. For the alignment, once an SUS is detected inside a short read, its position can be uniquely identified on the reference genome, and thus, more sensitive alignment might be possible with running the SW algorithm with a gretaer insertion–deletion flexibility. The proposed pipeline is shown in Figure 6. Within this study we have build the SUS dictionary for the human reference genome and developed an efficient SUS matching algorithm.

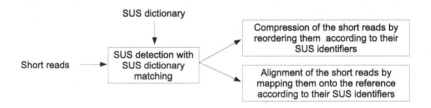

Fig. 4. Proposed sequence analysis pipeline

Next steps of the project will be building the actual alignment and compression blocks and benchmarking each against the current state–of–the–art solutions. Surely, decreasing the computational resource requirement at each step will be an important point, while it is not very much considered at this early proof-of–concept study.

References

1. Alkan, C., Kidd, J.M., Marques-Bonet, T., Aksay, G., Antonacci, F., Hormozdiari, F., Kitzman, J.O., Baker, C., Malig, M., Mutlu, O., et al.: Personalized copy number and segmental duplication maps using next-generation sequencing. Nature Genetics 41(10), 1061–1067 (2009)

2. Berger, B., Peng, J., Singh, M.: Computational solutions for omics data. Nature Reviews Genetics 14(5), 333–346 (2013)
3. Bonfield, J.K., Mahoney, M.V.: Compression of fastq and sam format sequencing data. PloS One 8(3), e59190 (2013)
4. Cox, A.J., Bauer, M.J., Jakobi, T., Rosone, G.: Large-scale compression of genomic sequence databases with the burrows–wheeler transform. Bioinformatics 28(11), 1415–1419 (2012)
5. Deorowicz, S., Grabowski, S.: Compression of dna sequence reads in fastq format. Bioinformatics 27(6), 860–862 (2011)
6. Deorowicz, S., Grabowski, S.: Data compression for sequencing data. Algorithms for Molecular Biology 8(1), 25 (2013)
7. Fonseca, N.A., Rung, J., Brazma, A., Marioni, J.C.: Tools for mapping high-throughput sequencing data. Bioinformatics 28(24), 3169–3177 (2012)
8. Hsi-Yang, F.M., Leinonen, R., Cochrane, G., Birney, E.: Efficient storage of high throughput dna sequencing data using reference-based compression. Genome Research 21(5), 734–740 (2011)
9. Giancarlo, R., Rombo, S.E., Utro, F.: Compressive biological sequence analysis and archival in the era of high-throughput sequencing technologies. Briefings in Bioinformatics, bbt088 (2013)
10. Hach, F., Hormozdiari, F., Alkan, C., Hormozdiari, F., Birol, I., Eichler, E.E., Sahinalp, S.C.: mrsfast: A cache-oblivious algorithm for short-read mapping. Nature Methods 7(8), 576–577 (2010)
11. Hach, F., Numanagić, I., Alkan, C., Sahinalp, S.C.: Scalce: Boosting sequence compression algorithms using locally consistent encoding. Bioinformatics 28(23), 3051–3057 (2012)
12. Hach, F., Sarrafi, I., Hormozdiari, F., Alkan, C., Eichler, E.E., Sahinalp, S.C.: mrsfast-ultra: a compact, snp-aware mapper for high performance sequencing applications. Nucleic Acids Research, gku370 (2014)
13. İleri, A.M., Külekci, M.O., Xu, B.: Shortest unique substring query revisited. In: Kulikov, A.S., Kuznetsov, S.O., Pevzner, P. (eds.) CPM 2014. LNCS, vol. 8486, pp. 172–181. Springer, Heidelberg (2014)
14. Langmead, B., Salzberg, S.L.: Fast gapped-read alignment with bowtie 2. Nature Methods 9(4), 357–359 (2012)
15. Li, H., Durbin, R.: Fast and accurate long-read alignment with burrows–wheeler transform. Bioinformatics 26(5), 589–595 (2010)
16. Loh, P.-R., Baym, M., Berger, B.: Compressive genomics. Nature Biotechnology 30(7), 627–630 (2012)
17. Pei, J., Wu, W.C.-H., Yeh, M.-Y.: On shortest unique substring queries. In: 2013 IEEE 29th International Conference on Data Engineering (ICDE), pp. 937–948. IEEE (2013)
18. Smith, T.F., Waterman, M.S.: Identification of common molecular subsequences. Journal of Molecular Biology 147(1), 195–197 (1981)
19. Tsuruta, K., Inenaga, S., Bannai, H., Takeda, M.: Shortest Unique Substrings Queries in Optimal Time. In: Geffert, V., Preneel, B., Rovan, B., Štuller, J., Tjoa, A.M. (eds.) SOFSEM 2014. LNCS, vol. 8327, pp. 503–513. Springer, Heidelberg (2014)

Modeling of the Urothelium with an Agent Based Approach

Angelo Torelli[1], Fabian Siegel[2], Philipp Erben[2], and Markus Gumbel[1,⋆]

[1] Institute for Medical Informatics, Mannheim University of Applied Sciences
{a.torelli,m.gumbel}@hs-mannheim.de
http://www.hs-mannheim.de
[2] Clinic of Urology, Medical Faculty Mannheim at the University of Heidelberg
{fabian.siegel,philipp.erben}@medma.uni-heidelberg.de
http://www.umm.de

Abstract. Novel models for cell differentiation and proliferation in the urothelium are presented. The models are simulated with the Glazier-Graner-Hogeweg technique using CompuCell3D. From a variety of tested models, the *contact model* is the best candidate to explain cell proliferation in the healthy urothelium. Based on this model, four variations were compared to highlight the key variations that best fit real urothelium. All simulations were quantified by a fitness function designed for the requirements of the urothelium. The findings suggest that adhesion and a nutrient dependent growth may play a crucial role in the maintenance of the urothelium. Aberrations in either adhesion or nutrient dependent growth led to the development of polyp-like formations. This work mimics the regeneration process and the steady state of the urothelium with a spatial and adhesion dependent approach for the first time.

Keywords: Agent-based, Glazier-Graner-Hogeweg, Monte Carlo, Simulation, Urothelium.

1 Introduction

Bladder cancer is the sixth most common cancer in men [1]. Although cancer can spread from neighboring organs to invade and grow in the bladder, the most common form of bladder cancer originates within the bladder's epithelial wall, the urothelium (see figure 7c). The urothelium is the tissue separating the intraluminal space, which in this case is filled with urine, and the interstitial fluid and is made of different layers, starting from a single layer of basal cells (in the sequel denoted as B and colored orange) and stem cells (S, blue) laying on top of the basal membrane (BM, red). These cell types are followed by three to five layers of intermediate cells (I, green). The urothelium ends at the bladder lumen with a single layer of specialized umbrella cells (U, gray), protecting its progenitors underneath [2]. This stratified epithelium of the bladder is anchored down on the lamina propria through the basal membrane, which is more a matrix than a

---

⋆ Corresponding author.

F. Ortuño and I. Rojas (Eds.): IWBBIO 2015, Part II, LNCS 9044, pp. 375–385, 2015.

membrane and is made up mostly of type-IV collagen, microfibrils and laminin [3]. The basal membrane also acts as a mechanical barrier, preventing malignant cells from invading the deeper tissues [4] and it is essential for angiogenesis [5].

Surprisingly, many characteristics of each bladder cell type are still unknown. For instance various hypotheses of cell lineages (The description of the history of the types of the cell progeny) of the urothelium have been postulated [6, 7, 8], but none has yet been proven. Only a few models *in silico* of the urothelium have been published, such as [9] in which the matrix metalloproteinase in the urothelium was analyzed by means of a cellular automaton.

In the course of the literature search, no previous work was found that simulates the urothelium in a healthy state for the purpose of better understanding of its architecture, lineage and function. Therefore, we extended the cellular automata approach. The aim for this work was: 1) to simulate the urothelium in both its normal healthy steady state, as well as 2) to test the stability of the tissue in a damaged or stressed state with the Glazier-Graner-Hogeweg (GGH) technique [10]. The GGH approch is a lattice-based Monte-Carlo simulation method that minimizes an energy function. Objects of the simulation are described and influenced by this energy function. We utilized the CompuCell3D framework [11], which is based on the GGH-model. Because of the lack of understanding of the function of each individual cell, various elementary models were set up parallel to each other to compare the biological facts with the hypotheses made in this work. The focus of our work is on tissue simulation and cell arrangements. Molecular aspects and physical forces, such as the expansion and contraction of the bladder, were not included yet.

This paper is organized in the following way: section *methods* explains the approach of developing models that in turn will compete against each other to be the model to best fit real anatomical tissue. This comparison will be based on a fitness value developed during this work and will also be explained in detail. Section *results* displays the benchmark of each of the models that will be discussed in the following sections *discussion* and *conclusion*.

2   Methods

The simulated models were run for an equivalent period of up to 2 years. All simulations start with two stem cells attached on the basal membrane. After about three to ten days, the urothelium should be regenerated and reaches a steady state [12]. 1440 Monte-Carlo-Steps (MCS) according to the GGH approach represents a day. The GGH-lattice has the dimensions $x \cdot y \cdot z$ equivalent to $150 \cdot 200 \cdot 1$ representing the total observed volume $V_{max} = 150 \cdot 200 \cdot 1 = 3 \cdot 10^4 \mu m^3$ for all simulations, with one voxel having the volume $1 \mu m$. The shape of a cell can be controlled in GGH-models by surface- and volume-constraint-parameters. These parameters were set in all models such that cells keep their target volume with little variation, but have a flexible surface. Thus, cells are deformable.

2.1   Biological Processes

Each of the developed representations of the urothelium consist of different components, which can be categorized into four biological processes: 1) birth, 2) death, 3) differentiation, and 4) sorting. These can then be combined – along with their parameter set – to create new, sophisticated models.

Birth- and Death Process.   The diffusion of nutrients in the urothelium has been included. The growth of the cell can either be infinite (abbreviated as IN) or nutrient dependent (NU). An infinite growth lets all cells proliferate according to its cell lineage (see Differentiation Process). In contrast, nutrient dependent growth lets cells only proliferate if a minimum of nutrients is available. Nutrients diffuse from the lamina propria through the basal membrane and are consumed by the adjacent cells. We have set the diffusion parameters such that nutrients reach into half of the urothelium. When cells grow and have reached their maximum (target) size they undergo mitosis. This implies a cell cycle period.

A cell can die, either through apoptosis, or mechanically through the process of voiding. Apoptosis is, in our simulations, set differently for each cell type to about 90, 30 and 10 days accordingly for the basal, intermediate and umbrella cells. Voiding of the bladder occurs every six hours. Here, apical cells that are in contact with the bladder lumen are randomly removed (washed out) with a probability of 2%. The stem cell does not undergo apoptosis, but can be washed away if it reaches the surface.

Differentiation Process.   Figures 1, 2 and 3 show possible cell division scenarios for each cell type. A cell can either divide, which is expressed by a plain line, or transform (differentiate), which is expressed by a dashed line. Division can occur either symmetrically (fig. 1a, 1b, 2a, 2b and 3a) or asymmetrically (fig. 1c and 2c). Tranformation happens either through contact (fig. 2e and 3c) or through time (fig. 2d). The special case of fusion has been included (fig. 3b), which is when more than one intermediate cell becomes an umbrella cell. This phenomenon is discussed in [13, 14]. All these approaches can form various cell lineages when they are combined. Different models were derived from these combinations and have been tested, but only four were compared (cf. section 3).

Sort Process.   As tissue formation is far from being random, we assume a sorting process. The tissue can sort itself either randomly (RA) or based on the differential adhesion hypothesis [15, 16]. In case of a random sort process, all cells have the same adhesion value of 1 independently of their cell type. When differential adhesion is applied, cells have a specific adhesion according to their cell types and contact to one another and other cell types. Table 1 shows the adhesion energies used. The energies were assigned according to values taken from the literature for two cell types (e. g. [16]) and adjusted for four cell types and its environment. The split axis is vertical and parallel to the basal membrane.

Fig. 1. Possible cell divisions for stem cells. S: Stem cell, B: Basal cell

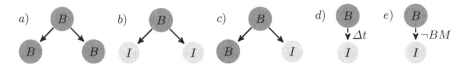

Fig. 2. Possible cell divisions for basal cells. B: Basal cell, I: Intermediate cell, BM: Basal membrane, Δt = cell cycle time.

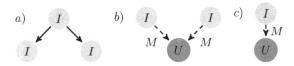

Fig. 3. Possible cell divisions for intermediate cells. I: Intermediate cell, U: Umbrella cell, M: Medium (urine)

Table 1. Differential adhesion energies used in all DAE-simulations. A small value represents stickiness, greater values less adhesion. M: Medium (urine), BM: basal membrane, S: stem cell, B: basal cell, I: intermediate cell, U: umbrella cell. Note that the matrix is symmetrical.

Types	M	BM	S	B	I	U
M	0	14	14	14	14	4
BM		0	1	3	12	12
S			6	4	8	14
B				5	8	12
I					6	4
U						2

2.2   Fitness Function

Finally, we have introduced a method for measuring the fitness of the model. This model is based on the arrangement and the volume of the simulated tissue compared with the observed biological data.

The fitness functions (see eq. 1) consist of a weighted fitness function for: 1) the arrangement of cells and 2) the volume of the cells. The fitness is measured at specific points in time t_i, typically starting from $t_1 = 20$ d (days) to the end of the simulation. All functions in this paper are dependent on time, such as $f(t_i)$, but for the sake of simplicity it will be referred to as f. Both functions are described in detail later in this section.

$$f = \frac{\bar{f}_a + f_v}{2} . \tag{1}$$

Thus, f enables us to quantify the results and to process the simulations in batch-mode, which serves as an input for parameter optimization. The fitness values were recorded throughout the entire simulation with a rate of a simulated half-day equaling 720 MCS, making one MCS equivalent to one minute.

Arrangement Fitness Function. The arrangement fitness function f_a surveys models to ensure that the cell strata are in the correct order. To ensure the correct order of the tissue, columnar samples are taken throughout the entire width of the tissue with steps of $20\mu m$. The columnar samples are vertical extractions of the tissue where fragments of little squares of two by two voxels are taken every $7\mu m$. Each of these voxels can be occupied by only one cell. To avoid duplicates, the most frequently occurring cell will be taken into account and inserted into a stack only once. This stack then undergoes further analysis done by a function applying Boolean terms. This analysis reaches an optimum of 1 if the urothelium reaches a state where the basal and stem cells layer is right above the basal membrane, followed by the various layers of intermediate cells and finally with one layer of umbrella cells before the lumen, and finally, the medium (urine), occupying the intraluminal space. In the worst case scenario, 0, the simulation does not create any cells at all. The arrangement fitness function is summarized in equation 2:

$$f_a = \begin{cases} \frac{1}{(1-L_B)+(L-L_I)+(1-L_U)+1} & \text{if } L_B + L + L_U > 0 \\ 0 & \text{otherwise .} \end{cases} \quad (2)$$

Where $(1 - L_B) + (L - L_I) + (1 - L_U)$ represents the number of cells that wandered away from their intended layer, described as followed:

- $L_B = 1$ if the first layer is made of cell type basal or stem, otherwise 0.
- $L_U = 1$ if the last layer is made of cell type umbrella, otherwise 0.
- L is the number of strata in between the first and last layer, while L_I is the subset of L counting the layers made of intermediate cells.

The function f_a is then calculated column by column on $n = 7$ different locations within the tissue. Out of these values, the average is taken as follows:

$$\bar{f}_a = \frac{1}{n} \sum_{i=1}^{n} f_a(i) . \quad (3)$$

Volume Fitness Functions. Since no information was found in the literature search regarding the volume of each of the urothelial cell types, values for the tissue height and for volume occupied by each cell type in percentage were calculated as an average from a variety of histological pictures of the urothelium that were found in these sources [17, 18, 19, 20, 21]. The average height of the urothelium is estimated to be $85\mu m$, which helps to determine the optimal volume a simulated urothelium should reach V_{opt} and is equal to

$150\mu m \cdot 85\mu m \cdot 1\mu m = 1.275 \cdot 10^4 \mu m^3$. Furthermore, umbrella cells were estimated to occupy 23%, intermediate cells occupy 67% and basal and stem cells together occupy 10% of the entire tissue volume respectively. With these values, it is possible to have an idea if the simulation is accurately representing the urothelium or not, by using S_B, S_I and S_U as the ideal volumes of each cell types. This is done by using a fitness function for each cell type f_B, f_I and f_U which are defined in equation 4.

$$f_B = \frac{1}{\frac{(S_B - I_B)^2}{a \cdot V_{max}} + 1}, \quad f_I = \frac{1}{\frac{(S_I - I_I)^2}{a \cdot V_{max}} + 1}, \quad f_U = \frac{1}{\frac{(S_U - I_U)^2}{a \cdot V_{max}} + 1}. \tag{4}$$

with

- $S_B = 10\% \cdot V_{opt}$, $S_I = 67\% \cdot V_{opt}$ and $S_U = 23\% \cdot V_{opt}$ are the ideal volumes of each cell types.
- I_B, I_I and I_U are the actual volumes of each cell type.
- $V_{max} = 3 \cdot 10^4 \mu m^3$ is the maximum volume all cell types can occupy.
- $a = 3\mu m^3$ is a factor that regulates the width of each volume fitness function.

The fitness functions f_B, f_I and f_U could be treated individually as shown in figure 4, but are merged together into one volume fitness function f_v that describes the entire volume of the tissue as shown in equation 5.

$$f_v = \begin{cases} (f_B + f_I + f_U)/3 & \text{if } 0 < I_B, \ I_I \text{ or } I_U < V_{max} \\ 0 & \text{otherwise} . \end{cases} \tag{5}$$

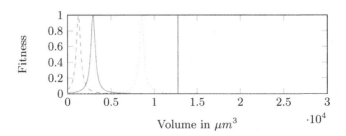

Fig. 4. Volume functions f_B (orange, dashed) with peak at 10% of V_{opt}, f_U (solid, gray) with peak at 23% of V_{opt}, and f_I (green, dotted) with peak at 67% of V_{opt} and V_{opt} as a vertical line (red). The x-axis extends from 0 to V_{max}.

3   Results

Among the various hypotheses we have tested, the so called *contact model* was very successful (cf. 5) based on the fitness function. Therefore, the *contact model* was further analyzed with four variations. We have also simulated many more models that can be derived from the differentiation process as shown in figures 1, 2 and 3. Their fitness was very poor and were consequently rejected for candidates for the healthy urothelium (data not shown).

The contact model states that stem cells living in the basal layer above the basal membrane are the main source of proliferation. From there, the different cell types are formed through transformation triggered by contact change, following a certain cell lineage hierarchy. It should be added that every model also includes the voiding and apoptosis process (described previously) and is therefore not marked as such separately. The variations of the contact model are the four combinations:

1. NU-RA: Nutrient-dependent growth and random sort process.
2. NU-DAE: Nutrient-dependent growth and differential adhesion sort process.
3. IN-RA: Infinite growth and random sort process.
4. IN-DAE: Infinite growth and differential adhesion sort process.

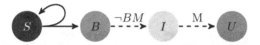

Fig. 5. Cell lineage of the contact model. S: Stem cell, B: Basal cell, BM: Basal membrane, I: Intermediate cell, M: Medium (urine), U: Umbrella cell

These are compared based on their fitness as shown in figure 6. Because the urothelium has a regeneration time between three and ten days, only fitness values after 20 days (or 28800 MCS) were taken into account. This ensures that all the simulations were in steady state. For each simulation run, the mean fitness of all points in time was calculated and these values were grouped by the model-variation and summarized as a box-and-whisker plot (see fig. 6).

Clearly, models with a random sort process, i. e. the same adhesion, do not form an urothelium-like tissue. We often observed simulations (IN-RA, NU-RA) where a stem cell had drifted to the surface of the urothelium and then washed out by the voiding process. With the disappearance of the cell source, the tissues disappear shortly after by an average of 60 days. However, the variance is remarkable. Figure 7a) shows an example of a simulation with no sort process, where only few basal cells are present and the protective umbrella cells engulf the urine and transport it inside the tissue jeopardizing the protective function of the urothelium.

On the other hand, contact models with a sort process based on differential adhesion show a significantly better behavior. On average over time, they reach a fitness of about 56% (IN-DAE) and 60% (NU-DAE). Nutrient-based growth leads to a striking smaller variance and extreme values for the overall fitness.

A representation of the NU-DAE-based urothelium *in silico* with a fitness of about 95% at 230 days is shown in figure 7b). The simulated tissue has a great similarity to the histological image of a healthy human urothelium, as depicted in figure 7c), in that it has the same number of layers and a corresponding order and size of cell types.

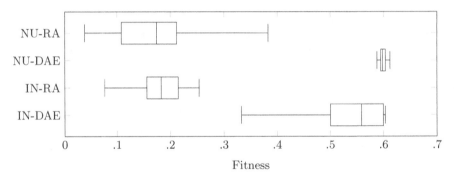

Fig. 6. Comparison of the four contact model variations in a box-and-whisker plot. Approximately 60% fitness were reached with differential-adhesion-energy-dependent (DAE) models. The highest results was accomplished by the model with an additional nutrient-dependent grow on top of the DAE process. Number of simulation runs per model: IN-RA: = 26, NU-DAE = 14, IN-DAE = 26, NU-RA = 37. The x-axis shows only part of the fitness which goes from 0 to 1.

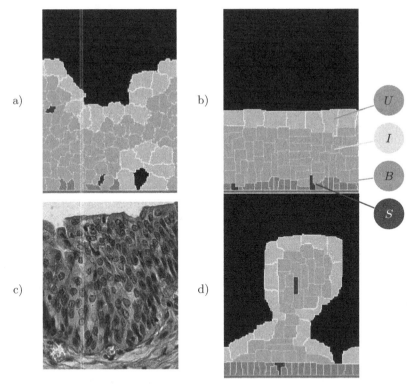

Fig. 7. a) Tissue formation without sort process (IN-RA). b) Simulation in steady state showing a healthy urothelium (NU-DAE). c) Histological image of the urothelium (Hematoxylin and eosin stain). d) Simulated formation of a polyp (IN-DAE).

Figure 7d) shows the deformation of the apical surface of an IN-DAE model. This malformation resembles a polyp, which is often encountered in the bladder lumen. These polyps are mostly benign, but they do have the potential of penetrating through the deeper tissues and of compromising the bladder and other organ function. These polyp-like structures are the reason for the bad fitness at a level as low as 33% observed in the IN-DAE models (cf. 6).

4   Discussion

The contact model was inspired by discoveries from Ho *et al.* [7] and Yamany *et al.* [8]. The suggested cell lineage of Ho *et al.* is especially similar to our approach. Both approaches allow progenitor cells to be responsible for the generation of intermediate cells that can further differentiate into superficial cells. This is supported by the data presented showing that the NU-DAE model reached about 60% of the fitness function.

It remains to be known how the contact model differs from or shares ideas with known hypotheses on cell proliferation for other tissues such as small intestine [22] or epidermis [23]. As the results show in figure 6, it was remarkable that the models IN-DAE and NU-DAE reached the highest fitness values, proving that differential adhesion energies are essential to maintain an organized and stable tissue throughout time. The difference between IN-DAE and NU-DAE can be noticed in the range of the fitness values. While in IN-DAE models the nutrient resources are never scarce the NU-DAE model has limited nutrients, permeating only to about half the ideal urothelium thickness, keeping the randomly escaping stem cells in check by starving them or eventually driving them back to the basal membrane. This rearrangement of the stem cell was due to the fact that without nutrients to divide there was no mitotic pressure to push the stem cell upwards leaving the stem cell enough time to reorganize accordingly to their differential adhesion values.

Urothelial carcinoma of the bladder can be divided into high-grade and low-grade tumors. High-grade tumors are very aggressive and have a risk of muscle invasion, they can present polypoid but can also present as a superficial carcinoma-in-situ. On the other hand low-grade tumors are usually papillary, confined to the mucosa and do not invade the detrusor muscle [14]. They may present as papillary tumors. The formation of polyps in the IN-DAE model demonstrate the potential benefit of this model for further biological questions in bladder cancer [24].

The cellular automaton approach from Kashdan and colleagues [9] showed that carcinogens are important for the development of bladder cancer. In contrast, this work shows the key components for maintaining a healthy urothelium *in silico*. It is also possible to integrate carcinogens fields in our models, as was done with the diffusion field of nutrients introduced in this paper.

It should also be mentioned that the contact model, as such, provides a sort mechanism. It prefers basal cells to be created close to the basal membrane and umbrella cells to be created at the surface. This is some sort of negative feedback mechanism for the differentiation process.

The fitness function we introduced allowed an easy evaluation and visualization of individual simulations. An average fitness of about 60% for a simulation run of NU-DAE models indicates that there could be better models. These simulations often have an excellent arrangement fitness of more than 90% but a poor volume fitness with less than 30% (data not shown). On the one hand, we believe that parameter optimization will improve the fitness. On the other hand, the narrow width of the volume fitness function's peak (compare figure 4) may cause a rapid decrease of the fitness when the volume of the cells shifts only a little away from the optimum. Clearly, a further validation of the *in silico* models is required. However, we found that data addressing turn over times and cell cycles are rare in the literature.

5   Conclusion

In this work we were able to successfully simulate the healthy urothelium in steady state with the GGH-method. A urothelium-dependent fitness function was defined, which enables the quantitative evaluation of the simulation's output. Models for the urothelium that use flexible cell shapes and non-lattice-based cell movements were applied for the first time. They showed that a sorting process and a limited proliferation mechanism appears to be essential. Changes in these parameters are possible mechanisms for the development of cancer.

Acknowledgment. We would like to thank Thomas Ihme for his CUDA server.

References

[1] Jemal, A., Bray, F., Center, M.M., Ferlay, J., Ward, E., Forman, D.: Global cancer statistics (March, April 2011)
[2] Wein, A.J., Kavoussi, L.R., Campbell, M.F., Walsh, P.C.: Campbell-Walsh urology, 10th edn. Elsevier Saunders, Philadelphia (2012)
[3] Paulsson, M.: Basement membrane proteins: structure, assembly, and cellular interactions. Critical Reviews in Biochemistry and Molecular Biology 27(1-2), 93–127 (1992)
[4] Liotta, L.A., Tryggvason, K., Garbisa, S., Hart, I., Foltz, C.M., Shafie, S.: Metastatic potential correlates with enzymatic degradation of basement membrane collagen. Nature 284(5751), 67–68 (1980)
[5] Kubota, Y., Kleinman, H.K., Martin, G.R., Lawley, T.J.: Role of laminin and basement membrane in the morphological differentiation of human endothelial cells into capillary-like structures. The Journal of Cell Biology 107(4), 1589–1598 (1988)
[6] Khandelwal, P., Abraham, S.N., Apodaca, G.: Cell biology and physiology of the uroepithelium. Am. J. Physiol. Renal. Physiol. (297), F1477–F1501 (2009)
[7] Ho, P.L., Kurtova, A., Chan, K.S.: Normal and neoplastic urothelial stem cells: Getting to the root of the problem. Nature Reviews Urology 9(10), 583–594 (2012)
[8] Yamany, T., van Batavia, J., Mendelsohn, C.: Formation and regeneration of the urothelium. Curr. Opin. Organ Transplant. 19(3), 323–330 (2014)

[9] Kashdan, E.: Hybrid discrete-continuous model of invasive bladder cancer. Mathematical Bioscience and Engineering 10(3), 729–742 (2013)

[10] Balter, A., Merks, R.M.H., Popławski, N.J., Swat, M., Glazier, J.A.: The glazier-graner-hogeweg model: Extensions, future directions, and opportunities for further study. Single-Cell-Based Models in Biology and Medicine, 151–167 (2007)

[11] Swat, M.H., Thomas, G.L., Belmonte, J.M., Shirinifard, A., Hmeljak, D., Glazier, J.A.: Multi-scale modeling of tissues using compucell3d. Computational Methods in Cell Biology, 325–366 (2012)

[12] Kreft, M.E., Sterle, M., Veranic, P., Jezernik, K.: Urothelial injuries and the early wound healing response: Tight junctions and urothelial cytodifferentiation. Histochemistry and Cell Biology 123(4-5), 529–539 (2005)

[13] Hicks, R.M.: The mammalian urinary bladder: An accommodating organ. Biol. Rev. Camb. Philos. Soc. 50(2), 215–246 (1975)

[14] MD, M.J.D.: Urothelial Tumors (ACS ATLAS OF CLINICAL ONCOLOGY) (American Cancer Society Atlas of Clinical Oncology). pmph usa (2004)

[15] Steinberg, M.: On the mechanisms of tissue reconstruction by dissociated cells. i. population kinetics, differential adhesiveness, and the ab- sence of directed migration. PNAS 48, 1577–1582 (1962)

[16] Glazier, G.: Simulation of the differential adhesion driven rearrangement of biological cells. Phys. Rev. E Stat. Phys. Plasmas Fluids Relat. Interdiscip. Topics 47(3), 2128–2154 (1993)

[17] Gasser, T.: Basiswissen Urologie, 5th edn. Springer, Heidelberg (2011)

[18] Rübben, H. (ed.): Uroonkologie. 6., aufl. 2013 edn. Springer, Berlin (2013)

[19] Hautmann, R., Huland, H.: Urologie: Mit 176 Tabellen, 4th edn. Springer, Heidelberg (2010)

[20] Hallscheidt, P., Haferkamp, A.: Urogenitale bildgebung. Springer, Berlin (2010)

[21] Schultz-Lampel, D., Goepel, M., Haferkamp, A.: Urodynamik, 3rd edn. Springer, Berlin (2012)

[22] Gumbel, M., Werner, O., Zajicek, G., Meinzer, H.P.: Simulation and 3-d visualization of the intestinal crypt. SCS Publications, pp. 310–312 (1998)

[23] Li, X., Upadhyay, A.K., Bullock, A.J., Dicolandrea, T., Xu, J., Binder, R.L., Robinson, M.K., Finlay, D.R., Mills, K.J., Bascom, C.C., Kelling, C.K., Isfort, R.J., Haycock, J.W., MacNeil, S., Smallwood, R.H.: Skin stem cell hypotheses and long term clone survival – explored using agent-based modelling. Sci. Rep. 3, 1904 (2013)

[24] Shah, J.B., McConkey, D.J., Dinney, C.P.N.: New strategies in muscle-invasive bladder cancer: On the road to personalized medicine. Clin. Cancer Res. 17(9), 2608–2612 (2011)

Systematic Comparison of Machine Learning Methods for Identification of miRNA Species as Disease Biomarkers

Chihiro Higuchi[1], Toshihiro Tanaka[1,2], and Yukinori Okada[2,*]

[1] Bioresource Research Center, Tokyo Medical and Dental University,
Tokyo, 113-8510, Japan
[2] Department of Human Genetics and Disease Diversity, Graduate School of Medical
and Dental Sciences, Tokyo Medical and Dental University, Tokyo, 113-8510, Japan
yokada.brc@tmd.ac.jp

Abstract. Micro RNA (miRNA) plays important roles in a variety of biological processes and can act as disease biomarkers. Thus, establishment of discovery methods to detect disease-related miRNAs is warranted. Human omics data including miRNA expression profiles have orders of magnitude with much more number of descriptors (p) than that of samples (n), which is so called "p >> n problem". Since traditional statistical methods mislead to localized solutions, application of machine learning (ML) methods that handle sparse selection of the variables are expected to solve this problem. Among many ML methods, least absolute shrinkage and selection operator (LASSO) and multivariate adaptive regression splines (MARS) give a few variables from the result of supervised learning with endpoints such as human disease statuses. Here, we performed systematic comparison of LASSO and MARS to discover biomarkers, using six miRNA expression data sets of human disease samples, which were obtained from NCBI Gene Expression Omnibus (GEO). We additionally conducted partial least square method discriminant analysis (PLS-DA), as a control traditional method to evaluate baseline performance of discriminant methods. We observed that LASSO and MARS showed relatively higher performance compared to that of PLS-DA, as the number of the samples increases. Also, some of the identified miRNA species by ML methods have already been reported as candidate disease biomarkers in the previous biological studies. These findings should contribute to the extension of our knowledge on ML method performances in empirical utilization of clinical data.

Keywords: Micro RNA, miRNA, machine learning, least absolute shrinkage and selection operator, LASSO, multivariate adaptive regression splines, MARS.

1. Introduction

MicroRNAs (miRNAs) are short endogenous noncoding RNAs (approximately 22 to 23 nucleotides) that play important roles in regulation of mRNA cleavage, stability,

---

* Corresponding author.

F. Ortuño and I. Rojas (Eds.): IWBBIO 2015, Part II, LNCS 9044, pp. 386–394, 2015.
© Springer International Publishing Switzerland 2015

and expression [1-3]. The miRNAs are abundant in multiple species, and currently, around 2,000 human miRNAs have been registered (miRBase release 22) [4]. Micro RNA (miRNA) plays important roles in a variety of biological processes of human diseases like cancers, autoimmune and infectious diseases [5-7]. Previous studies have also reported that miRNA can act as disease biomarkers to predict disease onset and progression, as well as promising resources of novel therapeutic drug targets [8,9]. Thus, establishment of discovery methods to systematically detect disease biomarker miRNAs has been warranted.

Human omics data including miRNA expression profiles have orders of magnitude with much more number of descriptors (p) than that of samples (n), which is so called "p >> n problem" [10,11]. Since traditional statistical methods mislead to localized solutions in the presence the of p >> n problem, application of machine learning methods that handle sparse selection of the variables are expected to solve this problem [12,13]. Among many machine learning methods, least absolute shrinkage and selection operator (LASSO) and multivariate adaptive regression splines (MARS) are popular. LASSO is an alternative regularized version of least square method and given sparse solution [12]. MARS performs non-linear regression with hinge function [13]. The solutions with LASSO are suggested separately for each disease. On the other hand, those with MARS are selected commonly for whole disease status. These data mining methods are expected to give a few variables from the result of supervised learning which can provide informative prediction ability of the endpoints such as human disease statuses. However, few studies have been reported on the actual applications of machine learning methods to miRNA data sets [14-16].

Aim of this study is to conduct empirical evaluations of the performances of the machine leaning methods to detect disease biomarker miRNAs. We obtained publicly available miRNA data sets of human clinical samples, and then applied LASSO and MARS to assess their empirical accuracy to predict disease status.

2.   Methods

2.1   Human miRNA Expression Data Sets

We obtained miRNA data sets obtained from human clinical samples from NCBI Gene Expression Omnibus (GEO) on November 30th, 2014. We selected miRNA data sets based on the following criteria; (i) data were obtained from multiple human samples with clinical status (diseases or healthy controls), (ii) each sample had one clinical status corresponding to one miRNA data set, (iii) the base studies were not longitudinal but single time-point observation, (iv) miRNA expression profiles were comprehensively obtained by microarrays, (v) details of the original studies were published as journal articles.

2.2   Machine Leaning Methods

We selected LASSO and MARS, the two major machine learning methods to provide sparse selection of the variables, to empirically evaluate prediction ability of disease statuses in the miRNA data sets. We also conducted partial least square method discriminant analysis (PLS-DA), as a control traditional method to evaluate baseline performance of discriminant methods. All calculation of LASSO, MARS and PLS-DA were performed using R 3.1.0 software environment for statistical computing [17]. We used R packages of glmnet 1.9-8 [18], earth 3.2-7 [19] and caret 6.0-37 [20], for LASSO, MARS, and PLS-DA, respectively.

We performed supervised learning of LASSO, MARS and PLS-DA, using the normalized miRNA expression profiles of the obtained data sets. For each of the data sets, we adopted clinical phenotypes of the samples as response variables, and expression profiles of the miRNAs as explanatory variables. We evaluated the performance of supervised learning by using leave-one-out cross-validation (LOOCV) [21]. Namely, for each iteration step, we divided the data set into one sample and the other samples. We conducted supervised learning using miRNA expression profiles of the other samples, and then, predicted the clinical status of the one sample. Average concordances of predicted clinical status of the subjects were considered as empirical prediction accuracy of the methods.

In the case of LASSO, which requires to penalty parameter to be able to discriminate disease status enough, we determined it with 10-fold cross validations using cv.glmnet function bundled in glmnet package. Furthermore, a value of ncomp, which is the number of components to include in the model and required with PLS-DA calculation, was determined by using LOOCV train function bundled in caret package. The biomarker candidates from supervised learning model were selected with coefficients for linear regression equation using coef.glmnet function bundled in glmnet package (LASSO), and with variable importance measure for hinge function suggested evimp function bundled in earth package (MARS).

3   Results

3.1   Selection of the Human miRNA Expression Data Sets

We obtained six miRNA expression data sets of human disease samples, considering a diversity of numbers of samples and descriptor such as numbers of miRNAs. The selected expression data were GSE29190 [22], GSE33857 [23], GSE34608 [24], GSE49012 [25], GSE50646 [26], GSE53992 [27]. We provide brief descriptions of the data sets as below, as well as the summary shown in Table 1.

1. GSE29190
   Transcriptome of peripheral blood mononuclear cells with active tuberculosis (TB; $n = 6$), latent TB ($n = 6$) and healthy controls ($n = 3$) [22]. MiRNA expressions were measured with Agilent-021827 Human miRNA Microarray (V3) (miRBase release 12.0 miRNA ID version).

2. GSE33857
   Peripheral blood of the patients with chronic hepatitis C (CHC; $n = 64$), chronic hepatitis B (CHB; $n = 4$), non-alcoholic steatohepatitis (NASH; $n = 7$), normal

liver subjects ($n = 12$), and Huh7 cell line ($n = 13$) [23]. MiRNA expressions were measured with Agilent-029297 Human miRNA Microarray v14 Rev.2 (miRNA ID version). We note that Huh7 were not mentioned in the original paper.

3. GSE34608

   Blood miRNAs from TB ($n = 8$) and sarcoidosis (SARC; $n = 8$) patients and healthy controls ($n = 8$) [24]. MiRNA expressions were measured with Agilent-019118 Human miRNA Microarray 2.0 G4470B (Feature Number version).

4. GSE49012

   The miRNA expression profiles of the liver tissues consisting of cirrhotic ($n = 22$) and normal ($n = 12$) [25]. MiRNA expressions were measured with Thermo Scientific Dharmacon microRNA human array.

5. GSE50646

   Peripheral blood of the patients affected with rheumatoid arthritis (RA), that consist of memory regulatory T cell (memory Treg; $n = 8$), memory T cell ($n = 8$), naïve T regulated cell (naïve Treg; $n = 8$), and naïve T cell ($n = 8$) [26]. MiRNA expressions were measured with gilent-031181 Unrestricted Human miRNA V16.0 Microarray (miRBase release 16.0 miRNA ID version).

6. GSE53992

   The miRNA expressions of the liver tissues consisting of normal ($n = 13$) and cholangiocarcinoma (CCA; $n = 33$) [27]. CCA samples consist of well differentiated carcinoma (WDC; $n = 12$), papillary carcinoma (PC; $n = 4$) and moderately differentiated carcinoma (MDC; $n = 17$). MiRNA expressions were measured with Agilent-031181 Unrestricted Human miRNA V16.0 Microarray 030840.

Table 1. Summary for the six miRNA expressions

GEO ID	Target disease	Tissue	Clinical Status	No. Samples	No. miRNAs
GSE29190	Tuberculosis (TB)	Lung	Normal, Active TB, Latent TB	15	834
GSE33857	Liver diseases	Liver	Normal, CHB, CHC, NASH, [Huh7]	100	877
GSE34608	Tuberculosis Sarcoidosis (SARC)	Lung	Control, SARC, TB,	34	738
GSE49012	Cirrhosis	Liver	Normal, Disease,	34	736
GSE50646	Rheumatoid arthritis (RA)	Blood	Memory Treg, Memory T cell, Naïve Treg, Naïve T cell	32	125
GSE53992	Cholangiocarcinoma (CCA)	Liver	Normal, WDC, PC, MDC	46	1368

3.2 Empirical Prediction Ability of the Machine Learning Methods

Overall discrimination results of the machine learning methods were indicated in Table 2. Calculated confusion matrices of the data sets were indicated in Table 3 (inaccurate predictions are colored with gray). "Accuracy" indicates the ratio of true positive against the entire number of samples. "Variable" indicates the number of miRNAs suggested by LASSO and MARS, which were selected as multiple values and indicated with ranges in the case of LASSO. Ncomp indicates the number of components required by PLS-DA, which was determined with cross validation. The miRNA expressions for each data sets were discriminated with 67% over of accuracy, except for GSE29190 (the TB study) which showed lower accuracy of 42% in average. We observed that LASSO and MARS showed relatively higher performance compared to that of PLS-DA, as the number of the samples increases. For example, MARS showed the highest performance in GSE53992 (the CCA study, $n = 46$), and LASSO showed the highest performance in GSE33857 (the liver disease study, $n = 100$). We note that the machine learning methods did not always demonstrate higher performances than the control method of PLS-DA. In several cases, the discrimination with MARS preferred than that of PLS-DA.

Table 2. Discrimination result with machine learning methods for six miRNA data sets

GEO ID	LASSO		MARS		PLS-DA		Category size	No. samples	No. miRNAs
	Accuracy	Variable	Accuracy	Variable	Accuracy	Ncomp			
GSE29190	0.267	1-3	0.467	4	0.600	2	3	15	834
GSE33857	0.860	4-11	0.860	20	0.840	19	5	100	877
GSE34608	0.875	3-8	0.917	4	0.958	5	3	24	719
GSE49012	1.000	6	0.971	7	1.000	5	2	34	738
GSE50646	0.906	1-6	0.938	6	0.969	8	4	32	125
GSE53992	0.674	2-12	0.739	6	0.674	8	4	46	1368

Table 3. Confusion matrix for six miRNA data sets

GEO ID : GSE29190		LASSO			MARS			PLS-DA		
Prediction		N	A	L	N	A	L	N	A	L
True	Normal (N)	0	1	2	2	1	0	0	1	2
	Active TB (A)	0	3	3	1	2	3	0	4	2
	Latent TB (L)	0	5	1	0	3	3	0	1	5

GED ID: GSE33857		LASSO					MARS					PLS-DA				
Prediction		N	B	C	A	H	N	B	C	A	H	N	B	C	A	H
True	Normal (N)	8	1	2	1	0	8	0	1	3	0	9	0	2	1	0
	CHB (B)	2	1	1	0	0	0	1	2	1	0	0	2	2	0	0
	CHC (C)	0	0	64	0	0	0	0	64	0	0	0	0	64	0	0
	NASH (A)	0	1	2	4	0	0	1	1	5	0	0	0	5	2	0
	Huh7 (H)	1	0	1	2	9	1	1	0	3	8	0	0	6	0	7

GEO ID : GSE34608		LASSO			MARS			PLS-DA		
Prediction		N	S	T	N	S	T	N	S	T
True	Normal (N)	8	0	0	8	0	0	8	0	0
	Sarcoidosis (S)	0	8	0	0	8	0	0	7	1
	Tuberculosis (T)	0	3	5	0	2	6	0	0	8

GEO ID : GSE49012		LASSO		MARS		PLS-DA	
Prediction		Normal	Disease	Normal	Disease	Normal	Disease
True	Normal	12	0	11	1	12	0
	Disease	0	22	0	22	0	22

GEO ID : GSE50646		LASSO				MARS				PLS-DA			
Prediction		A	B	C	D	A	B	C	D	A	B	C	D
True	Memory Treg (A)	8	0	0	0	8	0	0	0	8	0	0	0
	Memory T cell (B)	2	5	0	1	2	6	0	0	1	7	0	0
	Naive Treg (C)	0	0	8	0	0	0	8	0	0	0	8	0
	Naive T cell (D)	0	0	0	8	0	0	0	8	0	0	0	8

GEO ID : GSE53992		LASSO				MARS				PLS-DA			
Prediction		N	W	P	M	N	W	P	M	N	W	P	M
True	Normal (N)	7	4	0	2	10	2	0	1	8	4	0	1
	WDC (W)	5	6	0	1	5	7	0	0	4	7	0	1
	PC (P)	0	0	2	2	1	0	2	1	0	0	1	3
	MDC (M)	0	0	1	16	1	0	1	15	0	2	0	15

3.3 Selected Candidate Disease-Related miRNAs

The candidate miRNA biomarkers were selected by the machine learning methods were shown in Table 4. Overall, LASSO selected 92 miRNAs for the six studies, and MARS selected 47 miRNAs, of which 19 miRNAs are commonly selected. We found that some of the identified miRNA species commonly selected by the machine learning methods have already been reported as candidate disease biomarkers in the previous biological studies. For example, Jopping CL *et al.* reported that miR-122 act as modulation of hepatitis C for chronic hepatitis [28]. Nakasa *et al.* reported that and miR-146a has been reported for highly expressed synovial tissue of rheumatoid arthritis [29]. These findings should provide empirical validity of our selection results of the miRNA biomarkers, and contribute to the extension of our knowledge on the machine learning method performances in utilization of clinical data.

Table 4. miRNA biomarkers selected by LASSO and MARS

GEO ID	No. miRNAs selected by LASSO	No. miRNAs selected by MARS	Commonly selected miRNAs (No. miRNAs)
GSE29190	7	4	miR-18, miR-21, miR-520d (3)
GSE33857	25	20	miR-122, miR-149, miR-1539, miR-2116 (4)
GSE34608	16	4	miR-409, miR-451, miR-766, miR-942 (4)
GSE49012	6	7	miR-1234 (1)
GSE50646	17	6	miR-22, miR-31, miR-146a, miR-150 (4)
GSE53992	21	6	miR-548f, miR-769-5p, miR-4290 (3)

4   Discussion

In this study, we conducted one of the initial empirical evaluations of the machine learning method performances to predict disease status of the samples using human miRNA expression data sets. The machine learning methods of LASSO and MARS demonstrated relatively higher performance in concordance of the predicted and actual diseases status, compared to that obtained from the control method of PLS-DA, as the number of the samples in the data sets increases. Some of the identified miRNA species commonly selected by the two machine learning methods have already been reported as candidate disease biomarkers in the previous biological studies, which suggested empirical utility of the application of the machine learning methods to select biomarker miRNAs that can predict disease onset or prognosis.

LASSO is an excellent method of regression to discriminate binary category. The glmnet package is extended for multiclass discrimination to aggregate each regression equation for specific disease status or not so that it gives different solutions to every disease status. On the other hand, MARS discriminates whole disease status with hinge function, and has common variables with different disease specific coefficient. We guess that the partial match between the suggested miRNA was caused by the reason above.

Several studies reported identification of miRNA species as disease biomarkers using machine learning methods. Søkilde R *et al.* and Taguchi Y-h *et al.* demonstrated that LASSO is an effective method to find miRNAs as disease biomarkers [14,16]. In this study, we conducted comparative analysis using both LASSO and MARS. Additionally, we discuss to ensemble biomarker selection from miRNA expression data sets firstly.

Which should we use, LASSO or MARS, as the machine learning method in order to select biomarker candidates? While it would be difficult to conclude based on the current knowledge, one solution might be to use union sets of biomarker candidates suggested by LASSO and MARS. Considering the overlap of the selected candidates between the two methods, numbers of the biomarker candidates derived from union sets would still be small compared with the whole miRNAs. As we have previously mentioned, biomarker candidates suggested by LASSO could be more specifically selective for the binary target disease statuses than MARS. We note that only LASSO suggested miR-671-5P in the discrimination between NASH and others in GSE33857, while

MARS did not. This miRNA was reported as one of significant biomarker between NASH patients and non-NASH patients [30]. On the other hand, as an advantage of MARS, MARS suggests the selected smaller number of the disease biomarkers to discriminate total disease status than those suggested by LASSO as shown in Table 4.

In this article, we used LASSO and MARS to discriminate and select biomarker candidates form miRNA expressions. In recent years, many effective machine learning method are suggested. Ensemble methods such as random forests, Bayesian approaches and deep learning are expected as effective variable selection methods form omics expression data. We should investigate comprehensive analysis with the large number of the miRNA expression data sets using these additional machine learning methods, to further discover important roles of miRNAs in a variety of biological processes and disease biomarkers.

Acknowledgements. We thank Prof. Johji Inazawa, Mr. Kenji Yamane and Masahiro Kanai for their kind supports on the study. We also thank the two anonymous reviewers who provided thoughtful comments to our contributions. This study was supported by the Japan Science and Technology Agency (JST), Japan Society of the Promotion of Science (JSPS), Mochida Memorial Foundation for Medical and Pharmaceutical Research, Takeda Science Foundation, Gout Research Foundation, and The Tokyo Biochemical Research Foundation.

References

1. Ruvkun, G.: Molecular biology, Glimpses of a tiny RNA world. Science 294, 797–799 (2001)
2. Ambros, V., Bartel, B., Bartel, D.P., Burge, C.B., Carrington, J.C., et al.: A uniform system for microRNA annotation. RNA 9, 277–279 (2003)
3. Ebert, M.S., Sharp, P.: Roles for microRNAs in conferring robustness to biological processes. Cell 149, 215–424 (2012)
4. Kozomara, A., Griffiths-Jones, S.: miRBase: annotating high confidence microRNAs using deep sequencing data. Nucleic Acids Res. 42, D68–D73 (2014)
5. Medina, P.P., Nolde, M., Slack, F.: OncomiR addiction in an in vivo model of microRNA-21-induced pre-B-cell lymphoma. Nature 467, 86–90 (2010)
6. O'Connell, R.M., Kahn, D., Gibson, W.S., Round, J.L., Scholz, R.L., et al.: MicroRNA-155 promotes autoimmune inflammation by enhancing inflammatory T cell development. Immunity 33, 607–619 (2010)
7. Jangra, R.K., Yi, M., Lemon, S.: Regulation of hepatitis C virus translation and infectious virus production by the microRNA miR-122. J. Virol. 84, 6615–6625 (2010)
8. Kovalchuk, O., Filkowski, J., Meservy, J., Ilnytskyy, Y., Tryndyak, V.P., et al.: Involvement of microRNA-451 in resistance of the MCF-7 breast cancer cells to chemotherapeutic drug doxorubicin. Mol. Cancer Ther. 7, 2152–2159 (2008)
9. Guo, J.-X., Tao, Q.-S., Lou, P.-R., Chen, X., Chen, J., et al.: miR-181b as a potential molecular target for anticancer therapy of gastric neoplasms. Asian Pac. J. Cancer Prev. 13, 2263–2267 (2012)
10. Hastie, T., Tibshirani, R.: Efficient quadratic regularization for expression arrays. Biostatistics 5(3), 329–340 (2004)
11. Fan, C., Oh, D.S., Wessels, L., Weigelt, B., Nuyten, D.S., et al.: Concordance among gene-expression-based predictors for breast cancer. N. Engl. J. Med. 355, 560–569 (2006)

12. Tibshirani, R.: Regression shrinkage and selection via the lasso. J. Royal. Statist. Soc B 58, 267–268 (1996)
13. Friedman, J.: Multivariate adaptive regression splines. The Annals of Statistics 19, 1–67 (1991)
14. Søkilde, R., Vincent, M., Møller, A.K., Hansen, A., Høiby, P.E., et al.: Efficient identification of miRNAs for classification of tumor origin. J. Mol. Diagn. 16, 106–115 (2014)
15. Zhang, H., Yang, S., Guo, L., Zhao, Y., Shao, F., et al.: Comparisons of isomiR patterns and classification performance using the rank-based MANOVA and 10-fold cross-validation. Gene (2014)
16. Taguchi, Y.-H., Murakami, Y.: Universal disease biomarker: can a fixed set of blood microRNAs diagnose multiple diseases? BMC Res. Notes 7, 581 (2014)
17. R.A.: language and environment for statistical computing. R Foundation for Statistical Computing, Vienna, http://www.R-project.org/
18. Friedman, J.H., Hastie, T., Tibshirani, R.: Regularization Paths for Generalized Linear Models via Coordinate Descent. Journal of Statistical Software 33 (1), 1–22, http://www.jstatsoft.org/v33/i01/
19. Milborrow, S., Derived from mda:mars by Hastie, R., Tibshirani, R.: Uses Alan Miller's Fortran utilities with Thomas Lumley's leaps wrapper. earth: Multivariate Adaptive Regression Spline Models. R package version 3.2-7 (2014), http://CRAN.R-project.org/package=earth
20. Kuhn, M.: Contributions from Wing, J., Weston, S., Williams, A., Keefer, A., Engelhardt. A., et al.: caret: Classification and Regression Training. R package version 6.0-37 http://CRAN.R-project.org/package=caret.2014
21. Geisser, S.: Predictive Inference (1993) ISBN 0-412-03471-9
22. Wang, C., Yang, S., Sun, G., Tang, X., Lu, S., et al.: Comparative miRNA expression profiles in individuals with latent and active tuberculosis. PLoS One 6, e25832 (2011)
23. Murakami, Y., Toyoda, H., Tanahashi, T., Tanaka, J., Kumada, T., et al.: Comprehensive miRNA expression analysis in peripheral blood can diagnose liver disease. PLoS One 7, e48366 (2012)
24. Maertzdorf, J., Weiner III, J., Mollenkopf, H.J., TBornotTB Network and Bauer, T., et al.: Common patterns and disease-related signatures in tuberculosis and sarcoidosis. Proc. Natl. Acad. Sci. 109, 7853–7858 (2012)
25. Vuppalanchi, R., Liang, T., Goswami, C.P., Nalamasu, R., Li, L., et al.: Relationship between differential hepatic microRNA expression and decreased hepatic cytochrome P450 3A activity in cirrhosis. PLoS One 8, e74471 (2013)
26. Smigielska-Czepiel, K., van den Berg, A., Jellema, P., van der Lei, R.J., Bijzet, J., et al.: Comprehensive analysis of miRNA expression in T-cell subsets of rheumatoid arthritis patients reveals defined signatures of naive and memory Tregs. Genes Immun. 15, 115–125 (2014)
27. Plieskatt, J.L., Rinaldi, G., Feng, Y., Peng, J., Yonglitthipagon, P., et al.: Distinct miRNA signatures associate with subtypes of cholangiocarcinoma from infection with the tumourigenic liver fluke Opisthorchis viverrini. J. Hepatol. 61, 850–858 (2014)
28. Jopling, C.L., Yi, M., Lancaster, A.M., Lemon, S.M., Sarnow, P.: Modulation of hepatitis C virus RNA abundance by a liver-specific MicroRNA. Science 309, 1577–1581 (2005)
29. Nakasa, T., Miyaki, T., Okubo, S., Hashimoto, A., Nishida, M., et al.: Expression of micro RNA-146 in rheumatoid arthritis synovial tissue. Arthritis Rheum. 58, 1284–1292 (2008)
30. Estep, M., Armistead, D., Hossain, N., Elarainy, H., Goodman, Z., et al.: Differential expression of miRNAs in the visceral adipose tissue of patients with non-alcoholic fatty liver disease. Aliment. Pharmacol. Ther. 32(3), 487–497 (2010)

Numerical Investigation of Graph Spectra and Information Interpretability of Eigenvalues

Hector Zenil, Narsis A. Kiani, and Jesper Tegnér

Unit of Computational Medicine, Department of Medicine,
Centre for Molecular Medicine, Karolinska Institute
Stockholm, Sweden
{hector.zenil,narsis.kiani,jesper.tegner}@ki.se
http://www.compmed.se

Abstract. We undertake an extensive numerical investigation of the graph spectra of thousands regular graphs, a set of random Erdös-Rényi graphs, the two most popular types of complex networks and an evolving genetic network by using novel conceptual and experimental tools. Our objective in so doing is to contribute to an understanding of the meaning of the Eigenvalues of a graph relative to its topological and information-theoretic properties. We introduce a technique for identifying the most informative Eigenvalues of evolving networks by comparing graph spectra behavior to their algorithmic complexity. We suggest that extending techniques can be used to further investigate the behavior of evolving biological networks. In the extended version of this paper we apply these techniques to seven tissue specific regulatory networks as static example and network of a naïve pluripotent immune cell in the process of differentiating towards a Th17 cell as evolving example, finding the most and least informative Eigenvalues at every stage.

Keywords: Network science, graph spectra behavior, algorithmic probability, information content, algorithmic complexity, Eigenvalues meaning.

1 Background

The analysis of large networks raises in many of research fields, the ubiquity of large networks makes the analysis of the common properties of these networks important.In the most simplistic way can be seen or analyzed as a collection of vertices and edges but there are a very different way of representing the graph, using the eigenvalues and eigenvectors of matrices associated with the graph (Graph Spectra) rather than the vertices and edges themselves.In this study a graph or network G defined by pairs $(V(G), E(G))$,where $V(G)$ is a set of vertices (or nodes) and $E(G)$ represent edges(links). Let A be an $n \times n$ real matrix. An eigenvector of A is a vector such that $Ax = \lambda x$ for some real or complex number λ. λ is called the Eigenvalue of A belonging to Eigenvector v. The set of graph Eigenvalues of the adjacency matrix is called the spectrum of the graph. Spectral analysis is a widely used for a range of problems. In general,

F. Ortuño and I. Rojas (Eds.): IWBBIO 2015, Part II, LNCS 9044, pp. 395–405, 2015.

assigning meaning to Eigenvalues is very difficult. They are very context sensitive (i.e. relative to the graph type) and they are cryptic in the sense that they store many properties of a graph in a single number that does not lend itself to being easily used to reconstruct the properties it encodes. However, they are known to encode algebraic and topological information relating to a graph in various ways. In this paper we contribute toward the investigation of the interpretability of Eigenvalues, specifically with a general method to determine the type and the amount of information about a network that each Eigenvalue carries. We analyse growing networks ranging from complete graphs to complex random network and demonstrate the distinct behaviour of the eigenvalue spectra of different topology class. We will show the unique spectral properties of the major random graph models, *Erdös-Rényi* [6,7], *small-world* [15] and *scale free* [1].

2   Methodology

All graphs in this paper are undirected, so that the matrices are symmetrical and the Eigenvalues are real. They also have no loops, so the matrices have a zero diagonal and hence a zero trace, so that the Eigenvalues add up to zero. We are interested in investigating the behavior of $Spec(G)$ relative to the Kolmogorov complexity $K(G)$. Formally, the Kolmogorov complexity of a string s is $K(s) = \min\{|p| : U(p) = s\}$. That is, the length (in bits) of the shortest program p that when running on a universal Turing machine U outputs s upon halting. A universal Turing machine U is an abstraction of a general-purpose computer that can be programmed to reproduce any computable object, such as a string or a network (e.g. the elements of an adjacency matrix). By the *Invariance theorem* [10], K_U only depends on U up to a constant, so as is conventional, the U subscript can be dropped. Formally, $\exists \gamma$ such that $|K_U(s) - K_{U'}(s)| < \gamma$ where γ is a constant independent of U and U'. Due to its great power, K comes with a technical inconvenience (called *semi-computability*) and it has been proven that no effective algorithm exists which takes a string s as input and produces the exact integer $K(s)$ as output [8,3]. Despite the inconvenience K can be effectively approximated by using, for example, compression algorithms. Kolmogorov complexity can alternatively be understood in terms of uncompressibility. If an object, such as a biological network, is highly compressible, then K is small and the object is said to be non-random. However, if the object is uncompressible then it is considered algorithmically random.

Algorithmic Probability. There is another seminal concept in the theory of algorithmic information, namely the concept of *algorithmic probability* [14,9] and its related *Universal distribution*, also called Levin's *probability semi-measure* [9]. The algorithmic probability of a string s provides the probability that a valid random program p written in bits uniformly distributed produces the string s when run on a universal (prefix-free[1]) Turing machine U. In equation form this can be

---

[1] The group of valid programs forms a prefix-free set (no element is a prefix of any other, a property necessary to keep $0 < m(s) < 1$.) For details see [4,2].

rendered as $m(s) = \sum_{p:U(p)=s} 1/2^{|p|}$. That is, the sum over all the programs p for which U outputs s and halts. The algorithmic Coding Theorem [9] establishes the connection between $m(s)$ and $K(s)$ as $|-\log_2 m(s) - K(s)| < \mathcal{O}(1)$ (Eq. 1), where $\mathcal{O}(1)$ is an additive value independent of s. The Coding Theorem implies that [4,2] one can estimate the Kolmogorov complexity of a string from its frequency by rewriting Eq. (1) as $K_m(s) = -\log_2 m(s) + \mathcal{O}(1)$ (Eq. 2).

Kolmogorov Complexity of Unlabeled Graphs. As shown in [17], estimations of Kolmogorov complexity may be arrived at by means of the algorithmic Coding theorem, using a 2-dimensional lattice as tape for a 2-dimensional deterministic universal Turing machine. Hence $m(G)$ is the probability that a random computer program acting on a 2-dimensional grid prints out the adjacency matrix of G. Essentially it uses the fact that the more frequently an adjacency matrix is produced, the lower its Kolmogorov complexity and vice versa. We call this the *Block Decomposition Method* (BDM) as it requires the partition of the adjacency matrix of a graph into smaller matrices using which we can numerically calculate its algorithmic probability by running a large set of small 2-dimensional deterministic Turing machines, and thence, by applying the algorithmic Coding theorem, its Kolmogorov complexity. Then the overall complexity of the original adjacency matrix is the sum of the complexity of its parts, albeit with a logarithmic penalization for repetitions, given that n repetitions of the same object only adds $\log n$ to its overall complexity. Formally, the Kolmogorov complexity of a labeled graph G by means of BDM is defined as $K_{BDM}(G, d) = \sum_{(r_u, n_u) \in A(G)_{d \times d}} \log_2(n_u) + K_m(r_u)$, where $K_m(r_u)$ is the approximation of the Kolmogorov complexity of the subarrays r_u by using the algorithmic Coding theorem (Eq. (2)), and $A(G)_{d \times d}$ represents the set with elements (r_u, n_u) obtained when decomposing the adjacency matrix of G into non-overlapping squares of size d by d. In each (r_u, n_u) pair, r_u is one such square and n_u its multiplicity (number of occurrences). From now on $K_{BDM}(g, d = 4)$ will be denoted only by $K(G)$ but it should be taken as an approximation to $K(G)$ unless otherwise stated (e.g. when taking the theoretical true $K(G)$ value). More details of these measures and their application are given in [17]. The Kolmogorov complexity of a graph G is thus given by:

$$K'(G) = \min\{K(A(G_L))|G_L \in L(G)\}$$

where $L(G)$ is the group of all possible labelings of G and G_L a particular labeling. In fact $K(G)$ provides a choice for graph canonization, taking the adjacency matrix of G with lowest Kolmogorov complexity. Unfortunately, there is almost certainly no simple-to-calculate universal graph invariant, whether based on the graph spectrum or any other parameters of a graph. In [19], however, we proved that the calculation of the complexity of any labeled graph is a good approximation to its unlabeled version.

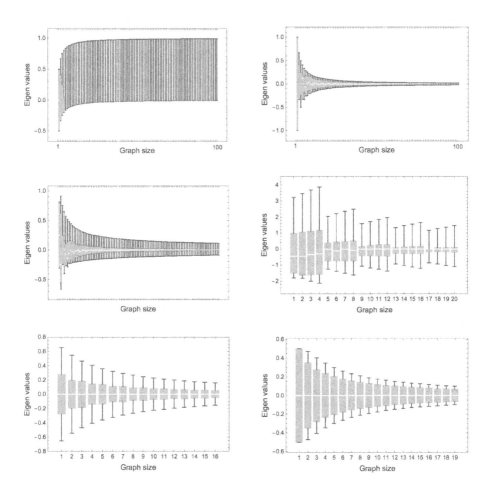

Fig. 1. Box distributions of Eigenvalues for growing regular graphs normalized by edge count. We call these "spectra signatures". Top left: Spectra signature of a growing complete graph showing that the Eigenvalues normalized by edge count do not carry any extra information than may be found in a simple Kolmogorov complexity graph. Top right: Spectra signature of a growing cycle graph showing a wider range of different Eigenvalues centered around $x = 0$. The next spectra signatures have an increasing number of different Eigenvalues but remain relatively simple given the regular structure of the graphs they represent. The diversity of Eigenvalues can be captured by classical Shannon entropy, but the non-trivial structure can only be captured by algorithmic complexity. Middle left: Spectra signature of a growing wheel graph. Middle right: Spectra signature of a growing fan graph. Bottom left: Spectra signature of a growing lattice graph. Bottom right: Spectra signature of a growing path graph. Obvious similarities between similar graphs can be recognized: cycles and wheels have similar patterns, grids and paths share some similarities too. However, star and fan graphs have spectra that show a greater degree of disparity than the spectra of the others.

3   Results

3.1   Most Informative Eigenvalues

It is clear that Eigenvalues carry different information and therefore can be of differential informative value. For example, take a complete graph of size n. To reconstruct it from its graph spectra it is enough to look at its largest Eigenvalue λ_1, simply because it indicates the size of the complete graph and therefore contains all the information about it– assuming that we know it is a complete graph. If we did not know it to be a complete graph then we would need to take into account the rest of the n Eigenvalues, but none of them on its own would suffice. That is only if a graph with $\lambda_1 \neq 0$ and $\lambda_i = -1$ with $i = 2$ to n uniquely determines a complete graph.

In Figs. 2, 3, 4 and 5, a sample of 4913 graphs distributed in 204 classes dividing (with possible repetition) the networks into bins of shared topological or algebraic properties, such as being a Moore, Haar, Cayley, tree or acyclic graph, display various (mostly significant) degrees of negative and positive correlation with one or more Eigenvalues. The number of graphs come from the graphs available in the *Mathematica* v.10 software built-in repository function GraphData[]. The most commonly found case was a negative correlation between largest Eigenvalues and graph information content. However, positive correlation and non-trivial differences between next largest and smallest Eigenvalues were found and their behavior is highly graph-topology dependent. This suggests

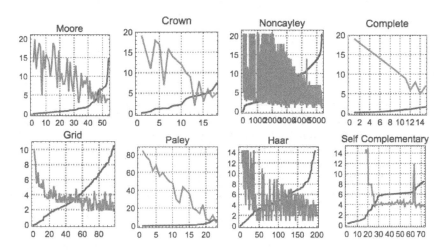

Fig. 2. Correlation plots of graph complexity vs largest Eigenvalues. On the X-axis are graphs (blue/darker curve) sorted by their algorithmic complexity (from lower to higher information content) normalized by graph edge count. On the Y-axis are the largest Eigenvalues for each graph (yellow/lighter curve). Both complexity and Eigenvalues are normalized by graph edge count as we are interested in structural information contained in both measures beyond information about the graph size.

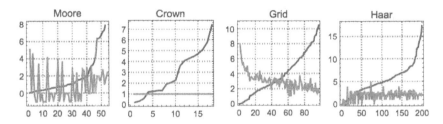

Fig. 3. Correlation plots of graph complexity vs second largest Eigenvalues. The second largest Eigenvalue displays a larger variety of correlations among graph classes, and depicted here is a case where it is found that the second value does not carry any information about Crown graphs. For Moore graphs the positive correlation is weak, and for Haar graphs it is null but noisy, unlike for Crowns. Specific statistics are given in Fig. 5 quantifying the correlations across all graph classes.

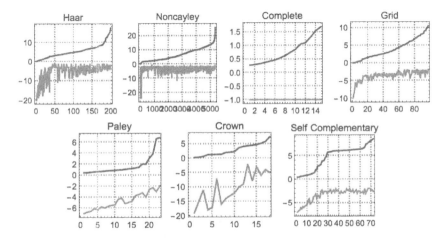

Fig. 4. Correlation plots of graph complexity vs smallest Eigenvalues. Smallest Eigenvalues tend to be positively correlated to graph information content. Depicted here is again a known example of a non-informative Eigenvalue for complete graphs, which is nonetheless informative in the sense that deleting the effect of size from its information content retrieves almost no information, hence all Eigenvalues and the complexity of the graph are basically flat (notice Y-axis scale). In another example, unlike the second largest Eigenvalue, it can be seen that the smallest Eigenvalue does carry information about Crown graphs.

that while the largest Eigenvalue encodes important structural information of the graph, all Eigenvalues may carry some information, with some being more or less informative than others. The complete graph is a trivial example of no correlation, where it is clear that the Eigenvalue is not providing any structural information about the graph other than its size, which is erased when normalized by edge count as it is in these plots, hence discounting by any edge count

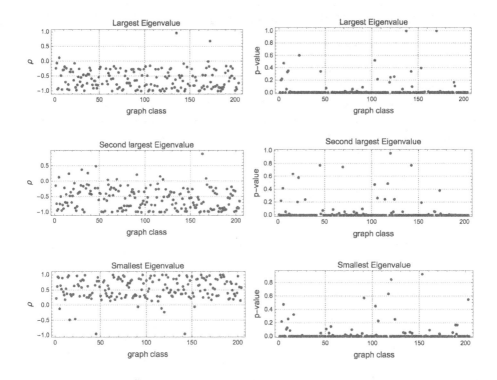

Fig. 5. Statistics (ρ) and p-value plots between graph complexity and largest, second largest and smallest Eigenvalues of 204 different graph classes including 4913 graphs. Clearly the graph class complexity correlates in different ways to different Eigenvalues but in most cases this correlation is strong and there is a clear tendency of the largest Eigenvalue to be negatively correlated to information content, then a quick transition at the second largest and finally a clear positive correlation with the smallest.

contribution. The degree and type of correlation can be found in Fig. 5, quantified by a typical Pearson correlation test.

If the Eigenvalue behavior of a graph G is flat, then its information-content is low or null, except perhaps because of the multiplicity of the value and the total number of occurrences of the same value, trivially indicating, for example, the size of the network, given that the number of Eigenvalues is equal to the number of vertices of G. This also means that Eigenvalues with flat behavior are less informative, a fact which enables clear discrimination between interesting and uninteresting Eigenvalues, beyond a simple consideration of numerical value (numerical values can be different and still not carry any information about a graph).

3.2   Graph Spectra Behavior of Evolving Networks

3.3   Spectra Signatures

We compared the *spectra signature* of an evolving graph to the Box plots of the Eigenvalues of the graph over time. Fig. 1, for example, shows the asymptotic behavior of each Eigenvalue for well known regular graphs and how the plots characterize them with various regular patterns, including cyclic behavior for a cycle graph. They also show how the accumulation of Eigenvalues is distributed differently for different graphs, with their rate of growth depending on the graph type. A complete graph G of size $n = |V(G)|$, for example, has graph spectra $(n-1)^1, (-1)^{n-1}$ with its values corresponding to the plot in Fig. 1(left). When the number of different Eigenvalues is small (i.e. their multiplicity is too high)

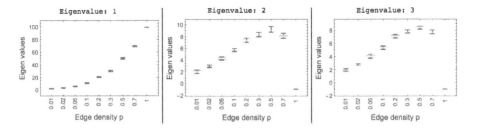

Fig. 6. Eigenvalues behavior. The largest Eigenvalue in a random E-R graph of size 100 vertices for edge density from 0 to 1 (*X*-axis) is the only one behaving differently from the rest. Some properties of the largest Eigenvalue are known, such as being an indicator of number of bifurcations, so the greater the edge count the greater its value. However, the next Eigenvalues all manifest a common behavior, reaching a maximum and describing a concave curve.

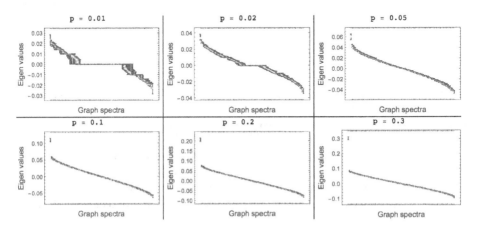

Fig. 7. Spectra signature of a random E-R graph of size 100 for edge density 0 to 1. Clearly for edge density 1, the random graph spectra are simply those of a complete graph.

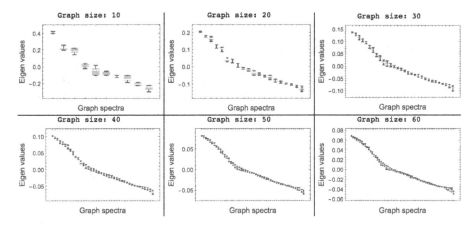

Fig. 8. Spectra signature of a Watts-Strogatz growing into a 100-node network with rewiring probability 0.05

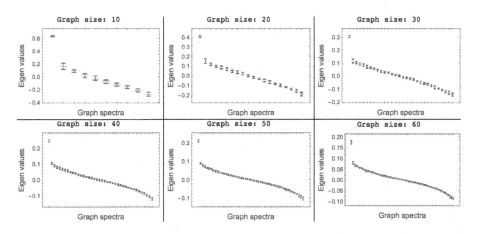

Fig. 9. Spectra signature of a growing Barabási-Albert network reaching a size of 100 nodes where a new vertex with 4 edges is added at each step

and they converge soon to a fixed normalized Eigenvalue, this is an indication that the Eigenvalue carries no information or is exhausted after a few evolving steps (i.e. no more information can be extracted, or the graph can be characterized after a few evolving steps) (see spectra signatures in Fig. 1). We undertook a novel numerical investigation of the Eigenvalues of growing graphs for different classes, shedding light on both known and possibly unexplored properties of Eigenvalues for some specific graph types. To this end we calculated what we defined as spectra signatures of random and complex networks prior to a deeper investigation concerning the information content of synthetic graphs and biological networks.

4   Conclusions

We have introduced a concept of spectra signatures based upon numerical calculations of growing networks with different group-theoretic and topological properties for the study of evolving network behavior. We have then moved toward the information content of these networks via estimating their Kolmogorov complexity by means of entropy, lossless compression and algorithmic probability (BDM).

We have introduced an analysis based on correlation comparisons of each Eigenvalue against the information content of a graph to reveal the most informative Eigenvalue for different graph classes. We found that the largest Eigenvalues are negatively correlated to graph complexity even after edge count normalization, while the smallest Eigenvalues are in general not correlated or positively correlated, with only a couple of cases of negative correlation. While most research has focused on a few of the largest Eigenvalues of a graph spectrum, we have shown that in actual fact the smallest Eigenvalues carry a high information content as often as the largest. The techniques introduced here can be extended to Laplacian matrices, but Laplacian matrices carry only redundant information about the degree of the vertices because the original graph can be reconstructed from the adjacency matrix alone. Thus the effect of $Spec(G)$ on the Laplacian or simple spectra of G with respect to $K(G)$ is negligible. For Kolmogorov complexity, we have $|K(A_L(G)) - K(A(G))| < c$, where $A_L(G)$ is the Laplacian matrix of G, $A(G)$ is the simple adjacency matrix of G and c is the algorithm implementing the Laplacian calculation $L = D(G) - A(G)$, where $D(G)$ is the diagonal degree matrix of G. We believe this is a novel approach to extracting meaning from and thus contributing to the solution of the problem of the interpretability of graph spectra, a fundamental step toward applications of graph spectra theory in network biology, especially in the context of evolving networks–given that some biological models are represented as Ordinary Differential Equations for which this approach, when applied to the Jacobian matrices of the ODEs, is thoroughly relevant. As introduced here, this approach promises to be able to reveal specifics about the behavior of a biological network over time through the study of Eigenvalues in relation to their information-content.

One future research direction is the investigation of behavioral differences in Eigenvalues of networks representing disease as compared to those of healthy networks, both as profiling techniques and as a tool for understanding the direction in which a healthy network may over time progress towards a disease state.

References

1. Farkasa, I., Derenyia, I., Jeongc, H., Nedac, Z., Oltvaie, Z.N., Ravaszc, E., Schubertf, A., Barabasi, A.L., Vicseka, T.: Networks in life: scaling properties and eigenvalue spectra. Physica A 314, 25–34 (2002)
2. Calude, C.S.: Information and Randomness: An Algorithmic Perspective, 2nd edn. EATCS Series. Springer (2010)

3. Chaitin, G.J.: On the length of programs for computing finite binary sequences. Journal of the ACM 13(4), 547–569 (1966)
4. Cover, T.M., Thomas, J.A.: Elements of Information Theory, 2nd edn. Wiley-Blackwell (2009)
5. Delahaye, J.-P., Zenil, H.: Numerical Evaluation of the Complexity of Short Strings: A Glance Into the Innermost Structure of Algorithmic Randomness. Applied Mathematics and Computation 219, 63–77 (2012)
6. Erdös, P., Rényi, A.: On Random Graphs I. Publ. Math. Debrecen 6, 290–297 (1959)
7. Gilbert, E.N.: Random graphs. Annals of Mathematical Statistics 30, 1141–1144
8. Kolmogorov, A.N.: Three approaches to the quantitative definition of information. Problems of Information and Transmission 1(1), 1–7 (1965)
9. Levin, L.A.: Laws of information conservation (non-growth) and aspects of the foundation of probability theory. Problems of Information Transmission 10(3), 206–210 (1974)
10. Li, M., Vitányi, P.: An Introduction to Kolmogorov Complexity and Its Applications, 3rd edn. Springer (2009)
11. Piperno, A.: Search Space Contraction in Canonical Labeling of Graphs (Preliminary Version), CoRR abs/0804.4881 (2008)
12. Skiena, S.: Implementing Discrete Mathematics: Combinatorics and Graph Theory with Mathematica, pp. 181–187. Addison-Wesley, Reading (1990)
13. Soler-Toscano, F., Zenil, H., Delahaye, J.-P., Gauvrit, N.: Calculating Kolmogorov Complexity from the Frequency Output Distributions of Small Turing Machines. PLoS One 9(5), e96223 (2014)
14. Solomonoff, R.J.: A formal theory of inductive inference: Parts 1 and 2. Information and Control 7,1–22, 224–254 (1964)
15. Watts, D.J., Strogatz, S.H.: Collective dynamics of 'small-world' networks. Nature 393, 440–442 (1998)
16. Zenil, H.: Network Motifs and Graphlets Wolfram Demonstrations Project (December 9, 2013), http://demonstrations.wolfram.com/NetworkMotifsAndGraphlets/
17. Zenil, H., Soler-Toscano, F., Dingle, K., Louis, A.: Graph Automorphisms and Topological Characterization of Complex Networks by Algorithmic Information Content, Physica A: Statistical Mechanics and its Applications 404, 341–358 (2014)
18. Zenil, H., Soler-Toscano, F., Delahaye, J.-P., Gauvrit, N.: Two-Dimensional Kolmogorov Complexity and Validation of the Coding Theorem Method by Compressibility (2013)
19. Zenil, H., Kiani, N.A., Tegnér, J.: Methods of Information Theory and Algorithmic Complexity for Network Biology (submitted to journal), arXiv:1401.3604 [q-bio.MN]
20. Zenil, H., Kiani, N.A., Tegnér, J.: Algorithmic complexity of motifs clusters superfamilies of networks. In: Proceedings of the IEEE International Conference on Bioinformatics and Biomedicine, Shanghai, China (2013)

Mixture Model Based Efficient Method for Magnetic Resonance Spectra Quantification

Franciszek Binczyk, Michal Marczyk, and Joanna Polanska

Data Mining Group, Institute of Automatic Control, Silesian University of Technology,
Akademicka 16, 44-100 Gliwice, Poland
{Franciszek.E.Binczyk,Michal.Marczyk,Joanna.Polanska}@polsl.pl

Abstract. Magnetic resonance spectroscopy (MRS) is a popular technique used in oncology to identify type of tumor and its progress with respect to its specific metabolism different than in normal tissue. In this study a complete pre-processing pipeline resulting in identification and quantification of chemical compounds has been proposed comprising novel method based on Gaussian mixture model (GMM). Model parameters were estimated with use of modified EM algorithm initialized by a new idea of spectrum segmentation. In order to make such solution ready to use in clinical applications an implementation based on GPU calculations is introduced. On simulated dataset we analyzed proposed methods by computational speed and data transfer time. On phantom data we compared our method to the two popular solutions: LC Model and Tarquin. It was observed that proposed algorithm outperforms both methods in sense of accuracy and precision of estimated concentration. The most efficient implementation was based on GPU with single precision calculations giving huge speed-up and satisfactory model accuracy comparing to CPU-based algorithm.

Keywords: Gaussian mixture model, GPU computing, Magnetic Resonance Spectroscopy, spectra pre-processing.

1 Introduction

Nuclear Magnetic Resonance (NMR) is a widely used technique in modern oncology. It allows for precise tumor diagnosis as well as its prognosis during therapy. According to the amount and type of carried information the NMR tests might be divided into: simple imaging (MRI), imaging of water molecules diffusion and perfusion, and most complex spectroscopy (MRS) [1]. MRS measurement gives a frequency spectrum that consists of peaks that represent different chemical compounds, which are products of cell metabolism. Peaks are located at different frequencies determined by chemical properties such as strong bound between nuclei (e.g. protium-protium) or electromagnetic neighborhood (electron clouds). The analysis of area covered by each peak leads to acquiring information about metabolite amount in analyzed tissue [1, 2]. By identification of metabolic trail it is possible to determine a type and possible spread/recurrence of the tumor, even while it is not visible in remaining NMR modalities. The main problem in spectroscopy is a limited size of the voxel. Since obtained

F. Ortuño and I. Rojas (Eds.): IWBBIO 2015, Part II, LNCS 9044, pp. 406–417, 2015.

signal is very complex, a single measurement point must be big enough to satisfy the compromise between the short time of medical examination and satisfactory information obtained. This imposes precise analysis of the signal to observe very small changes of cell metabolism to identify changes that occur only in a part of voxel volume (partial volume effect) [1].

There exists a number of ready to use software solutions, from which the most popular are: jMrui [3], Tarquin [4] and LC Model [5]. jMrui is based on singular value decomposition of free induction decay signal. The whole analysis is done at the time domain and as a result a set of damped harmonic exponents is obtained. The transformation to the frequency domain results in a single peak spectrum. Second method is based on a Hankel-Lanczos correction for singular value decomposition. Both mentioned algorithms do not require any prior knowledge. The third approach requires a database of metabolites measured at different experimental conditions (echo and relaxation time). The information about metabolite amount is done in a process of fitting analyzed signal peaks to the database.

In this paper a novel approach for modeling and quantification of MRS data is proposed. It was noticed that ideal NMR peaks in frequency spectrum should follow the Lorentz function, however due to the hardware approximation they are stored as Gaussians [1]. It is then possible to consider a spectrum as a Gaussian Mixture where each component stands for a single peak or group of close multiples that due to the low resolution are overlapped. Modeling algorithm is iterative and based on simple matrix operations, which allows to use fast graphics processor units (GPUs) or multi-core central processor units (CPUs) to increase computational speed. In this study we proposed parallel realizations of modified EM algorithm for estimating GMM parameters in spectra modeling. Using phantom data we compared results of signal quantification after using our pre-processing pipeline to two popular methods. We analyzed proposed implementations by computational speed and data transfer time on different platforms.

2   Methods

2.1   Data

To create virtual datasets with known size and number of peaks we used a virtual model creating 1D spectra [6]. Composition of compounds in artificial spectra was generated as described in [7]. We created 21 datasets with different number of peaks (from 100 to 300, 30000 points each) and 10 datasets with different length of the spectra (from 5000 to 50000 points, 200 peaks each).

The phantom dataset consists of 22 NMR spectra collected for a brain phantom at Centre of Oncology in Gliwice. The phantom was designed to imitate the chemical composition of healthy human brain. It consists of: 12.5 mMol of Naa, 10mMol of Creatine, 3 mMol of Choline, 7.5 mMol of Myo-inositol, 5mMol of Lactate and other compounds present in a human brain such as Glutamine, but not identified in daily clinical routine.

2.2   Data Pre-processing

Since the analysis of NMR spectra is based on estimation of area under each significant peak it is crucial to pre-process signal in a way that peak line shape would not be distorted (Fig. 1). The main distortion factors for NMR spectroscopy are: low signal resolution, eddy current, phase error and measurement noise. First element of the pre-processing pipeline was an improvement of signal resolution. It is done by addition of zero intensity signal at the end of time domain of free induction decay signal [1]. Since the zero value does not add any information in frequency domain it does not introduce any new peaks, but improves the line shape of existing signal. Length of added part must be a multiplicity of original signal size. Next steps are phase correction and eddy current correction. Eddy current distortion was removed with use of phase of reference water signal. The remaining phase error was corrected with use of modified self-tuned Automics algorithm [8]. Last step in MRS data pre-processing pipeline is noise filtering [9]. The main problem with majority of existing filters is that resulting signal is damped in comparison to the raw data. In MRS this is especially important while the amplitude of peak plays crucial role in quantification process. We use Savitzky-Golay filter [10] with two adjusted parameters: length of the moving window in which we perform signal approximation and order of approximation polynomial used.

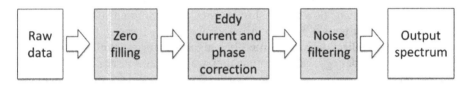

Fig. 1. Block diagram describing steps of proposed pre-processing pipeline

2.3   Gaussian Mixture Model

Quantitative information about metabolite amount in examined tissue is done on the basis of the real part of the signal called absorption spectrum, which is the best representation of the amount of molecules that absorbs emitted radio frequency pulse. Peak quantification may be divided to a problem of estimation of number of peaks and the precise estimation of their area under the signal. It is not trivial due to low signal resolution and existence of overlapping peaks. The possible solution is to represent signal with a mixture model where a single component represents signal peak. Model is constructed under the hypothesis that it is possible to find a number of components such that their sum would be a precise approximation of the original signal.

   NMR spectra contain information about frequency and their intensities (numbers of counts in a given frequency). We denote measurement points along the frequency axis by x_n. The numbers of counts corresponding to x_n are denoted by y_n, $n=1,2,...N$. N is the number of data points in the spectrum. We can define the mixture of Gaussian components as:

$$f(x) = \sum_{k=1}^{K} \alpha_k f_k(x, \mu_k, \sigma_k) \qquad (1)$$

where K stands for the number of Gaussian components, α_k, $k=1,2,...K$ are component weights, which sums to 1 and f_k is the probability density function of k-th Gaussian component:

$$f_k(x, \mu_k, \sigma_k) = \frac{1}{\sigma_k \sqrt{2\pi}} e^{\left(-\frac{(x-\mu_k)^2}{2\sigma_k^2}\right)} \qquad (2)$$

Procedure for fitting the Gaussian mixture model to the spectrum intensity y involves the following iterative expressions for parameters α_k, μ_k and σ_k:

$$p(k|n) = \frac{\alpha_k^i f_k(x_n, \mu_k^i, \sigma_k^i)}{\sum_{k=1}^{K} \alpha_k^i f_k(x_n, \mu_k^i, \sigma_k^i)} \qquad (3)$$

$$\alpha_k^{i+1} = \frac{\sum_{n=1}^{N} p(k|n) y_n}{\sum_{n=1}^{N} y_n} \qquad (4)$$

$$\mu_k^{i+1} = \frac{\sum_{n=1}^{N} p(k|n) y_n x_n}{\sum_{n=1}^{N} p(k|n) y_n} \qquad (5)$$

$$\sigma_k^{i+1} = \sqrt{\frac{\sum_{n=1}^{N} p(k|n) y_n (x_n - \mu_k^{i+1})^2}{\sum_{n=1}^{N} p(k|n) y_n}} \qquad (6)$$

In the above formulas i denotes successive iterations of algorithm and $p(k|n)$ is the conditional distribution of hidden variables, given data and parameters guess. Expressions (3)-(6) result from modifying original version of EM algorithm [11, 12].

To efficiently perform initial condition estimation for modeling of whole spectrum it was decided to divide signal into segments consisting of only limited number of major peaks. We find local maxima under assumption that peak intensity must be higher than a median of whole signal intensity. We define segment borders by finding minima between obtained peaks. For each segment a separate GMM decomposition is performed. After decomposition of each segment separately the whole spectrum is decomposed with use of components parameters' obtained in segmentation process.

Model is post-processed to ensure that signal peaks are represented by a single component. Simple merging routine is proposed that combines together two components if the area of their intersection is more than 90% of area of at least one. To quantify the area of the component in this step we take intensities that are closer than two standard deviations from the component mean.

2.4   Metabolite Quantification

For the analyzed phantom dataset a water signal spectrum was measured together with MRS thus it was possible to perform a very precise quantification in a process of scaling to water signal [5] according to the following formula:

$$Concentration = \frac{Peak\ area}{Water\ amount} * 55\ [mMol] \qquad (7)$$

2.5   EM Algorithm Implementation

The algorithm is implemented in Matlab R2013b programming environment as a m-file script named modEM_GPU with parallel computing capabilities. As an input user must provide vectors of measured values and their intensities. There is an option for providing initial conditions for the model parameters. If they are not set, initial parameters values for α and μ are drawn randomly from uniform distribution. Initial value of σ is set to cover distance between consecutive μ. Data are sent to GPU using gpuArray() Matlab function. Code is fully vectorized to increase the efficiency of platform used. Calculations are done on GPU mostly using arrayfun() and bsxfun() Matlab functions. Some basic arithmetic operations are implemented to increase speed. Instead of inner loop iterating over model components, we perform all operations using matrices. Algorithm stops when maximum number of iterations is reached or condition based on Euclidean distance between parameter estimates in consequent iterations is smaller than defined precision level. Data are sent back to host using gather() Matlab function. The outcome of the algorithm includes a set of estimated model parameters, values for log-likelihood function (logL) and Bayesian information criterion (BIC) [13]. Below we present the code for a main loop of modified EM algorithm run on GPU.

```
while count < max_iter && change > eps_change

    %variables for calculating precision
    old_alpha = alpha;
    old_sig2 = sig2;

    %temporary variables for big matrices operations
    x_tmp = repmat(x,1,KS);
    mu_tmp = repmat(mu,N,1);
    sig_tmp = repmat(sqrt(sig2),N,1);

    %conditional distribution of hidden variables (E-step)
    f = arrayfun(@norm_pdf,x_tmp,mu_tmp,sig_tmp)';
    scal = bsxfun(@times,(1./sig2).^(a_sig),...
    exp(-(dlt_sig./(2*sig2)))));
    f = bsxfun(@times,scal',f);
    px = max(alpha * f,5e-324);
    tmp = bsxfun(@times,alpha',bsxfun(@times,y,f));
    pk = bsxfun(@mrdivide,tmp,px);

    %new model parameters (M-step)
    denom = sum(pk,2)';
    alpha = max(denom/TIC,1e-10);
    mu = (pk*x)'./denom;
    tmp = arrayfun(@poww2,bsxfun(@minus,mu,x));
    sig2 = SW + SW2*sum(arrayfun(@timess,pk',tmp))./denom;
```

```
%check convergence
change = sum(abs(alpha-old_alpha))+
(sum((abs(sig2-old_sig2))./sig2))/KS;

    count = count + 1;
end
```

Full script can be downloaded from: http://cellab.polsl.pl/index.php/software/scripts-and-console-applications.

2.6   Hardware

To compare results of EM algorithm calculations we used two workstations with different CPUs and CUDA-enabled GPUs. Algorithm was tested on Intel i5 budget processor (1.7 GHz in normal work, 2.6 GHz in turbo mode) on laptop and a couple of two hi-end Intel Xeon processors (3.4 GHz in normal work, 3.6 GHz in turbo mode) on desktop server. We also used low-cost GeForce laptop graphic card GeForce 630M (96 CUDA cores, 800 MHz, 2 GB memory) and Tesla C2075 card (448 CUDA cores, 1.15 GHz, 6 GB memory) that is created specifically for high-performance GPU computing. Tesla is capable of running 1030 GFLOPs per second of single precision processing performance, while GeForce only 307 GFLOPs. Tesla card supports ECC and error correcting codes that can fix single-bit errors and report double-bit errors, but uses about 10% of memory size. Both devices support double precision calculations.

3   Results

3.1   Simulated Data

We tested efficiency of different implementations of EM algorithm for spectra with different length and number of peaks. We created simulation replicates by running algorithm on the same dataset with different initial conditions selected randomly. For each scenario we made 20 replicates on i5 and GeForce platforms and 50 replicates on Xeon and Tesla platforms. Since calculated confidence intervals were not distinguishable on the plots, we present only average values from the replicates. For each version of EM algorithm we ran exactly 200 iterations to provide the same experimental conditions for each version of algorithm. Using these settings there were no visual differences between obtained models for different versions of EM algorithm.

Computational time increases with a complexity of the spectrum defined by number of the peaks present in the signal. Increasing length of input data vectors also leads to increase of elapsed time. For datasets with 100 and 500 peaks we can get about 15x and 23x speed boost respectively using Tesla card comparing to budget CPU. For datasets with 5000 and 50000 points we can get about 10x and 17x speed boost respectively comparing the same configurations (Fig. 2).

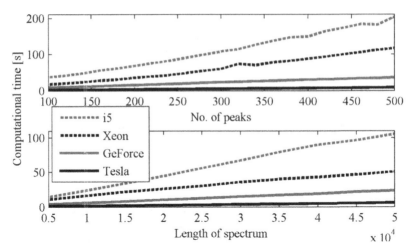

Fig. 2. Computational time for different realizations of modified EM algorithm. GPU usage results are reported for single precision calculations.

In Fig. 3 we compared computational time of two versions of algorithm with different definition of the most important variables type. On Tesla card applying single precision calculations gave about 1.3x speed increase comparing to the algorithm with double precision variables, while using GeForce card we got 1.5x speed boost. Increasing number of peaks in signal does not influent much on time advantage gain by using single precision calculations. Similar characteristic was observed in the scenario when length of spectrum was modified.

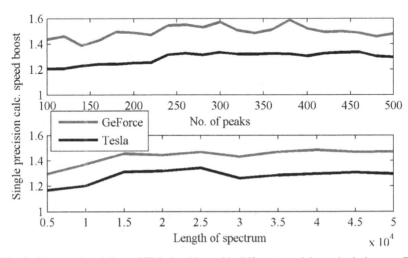

Fig. 3. Computational time of EM algorithm with different precision calculations on GPU

In many GPU computing applications data transfer between host and GPU may be a bottleneck in adjusting computational speed. We measured average time for data transfer using dataset with 200 peaks and 50000 points (Table 1). On GeForce platform we made 100 simulation replicates and on Tesla platform 200 replicates. Generally, using built-in Matlab functions it is much faster to write data to GPU than read it. In analysis of 1D spectra by our algorithm data transfer time is considerably slower than computational time of EM iterations performed only on GPU.

Table 1. Average time of data transfer between host computer and GPU compared to computational time on GPU in seconds. Percentage values are presented in brackets.

Platform	GeForce 630M	Tesla C2075
Host to GPU	0.0018 (0.008 %)	0.001 (0.016 %)
GPU comp.	23.67 (99.488 %)	6.04 (98.951 %)
GPU to Host	0.12 (0.504 %)	0.063 (1.032 %)

3.2   Phantom Dataset

All 22 spectra from analyzed set were pre-processed with use of proposed analysis pipeline. The length of additional zero values signal used to enhance signal resolution was set to 6146 points. The addition of wider signal did not improve peak line shape but only increased size of the data that could influence the time of GMM decomposition. Phase correction and eddy current adjustment did not require any external parameters adjustment. Savitzky-Golay filter parameters were experimentally adjusted to provide minimal damping of the peaks across spectrum giving moving window of length equals 25 points in which we approximate signal using polynomial of 5th degree. The exemplary results for one spectrum of applied pre-processing pipeline are presented at Fig. 4.

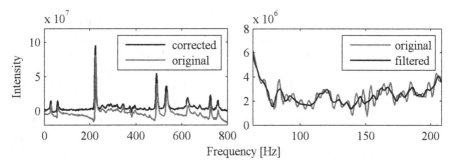

Fig. 4. Results of applied pre-processing. Left side: phase and eddy current correction, right side: noise filtering.

Pre-processed signal was decomposed with use of GMM by modified EM algorithm, initialized with results of spectrum segmentation. The number of segments was equal to 13 in average (varies from 11 to 16 in concurrent spectra) for all spectra, found with use of defined threshold. Exemplary division of spectrum into segments is presented in the upper panel of Fig. 5. The total number of components found in the segmentation process were equal to 61 in average (varies from 55 to 65 in concurrent spectra). Estimated components parameters' were then used as an input values for whole spectrum decomposition. In the lower panel of Fig. 5 we present an example of GMM-based decomposition.

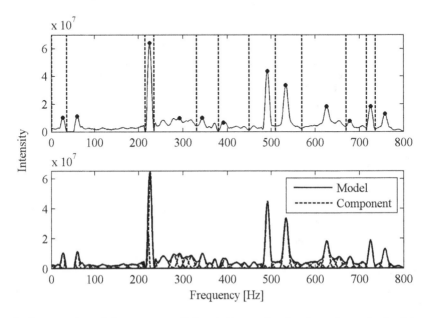

Fig. 5. Segmentation of the spectrum. Points indicates the signal maximum in the interval, dotted line – borders of the interval.

Signal decomposed into GMM was then quantified with use of described approach. Obtained amounts of metabolites were compared to the real values taken from phantom dataset description (NAA – 12.5 mMol, Creatine – 10 mMol, Choline – 3 mMol, Lactate – 5 mMol and Inositol – 7.5 mMol). The accuracy and precision of measured concentrations for three different methods are considered as mean value and variance and their 95% confidence intervals (Table 2).

New implementation of GMM-based method drastically speed-ups both parts of the algorithm comparing to the original one. Due to the small number of estimated segments we can get only 4 times faster segmentation algorithm by using parallel CPU calculations. Comparing time of full model building after 5000 iterations of EM algorithm we can get about 15 times faster method using GPU computing (dominance grows with increase of number of iterations).

Table 2. Accuracy and precision of quantification results measured by mean and variance and its 95 % confidence intervals of metabolite amount calculated using different methods on phantom dataset

Metabo-lite	Quantification method								
	LC model			Tarquin			GMM based		
	Mean	95% CI		Mean	95% CI		Mean	95% CI	
NAA	12.54	12.36	12.72	12.56	12.14	12.97	**12.49**	12.44	12.55
Creatine	**10.00**	9.89	10.11	9.99	9.78	10.19	9.97	9.92	10.02
Choline	**2.99**	2.96	3.03	2.91	2.63	3.19	**2.99**	2.97	3.01
Lactate	**5.02**	4.93	5.11	4.95	4.72	5.18	**5.02**	4.97	5.07
Inositol	7.59	7.48	7.70	7.56	7.35	7.77	**7.45**	7.41	7.50
	Var	95% CI		Var	95% CI		Var	95% CI	
NAA	0.42	0.32	0.59	0.94	0.72	1.34	**0.13**	0.10	0.19
Creatine	0.25	0.19	0.36	0.46	0.36	0.66	**0.12**	0.09	0.17
Choline	0.08	0.06	0.11	0.63	0.49	0.91	**0.05**	0.04	0.07
Lactate	0.21	0.16	0.30	0.53	0.41	0.75	**0.12**	0.09	0.17
Inositol	0.25	0.19	0.35	0.47	0.37	0.68	**0.10**	0.08	0.15

4 Discussion

Gaussian components fitted to MRS spectrum correctly define locations and intensities corresponding to compounds present in the analyzed sample. Each Gaussian component covers a certain range in the measurement scale so it can better capture structural properties of the spectrum than just a height of the peak. By use of a mixture model we can properly estimate area under the peak for overlapping peaks. GPU implementation of the EM algorithm significantly increases potential of current biological experiments and the range of application of the proposed method. The main positive aspect of GPU-based approach is the possibility to increase model precision without the rapid increase of computational time and to quickly analyze big datasets. It is very important since the development of new stronger in vivo scanners results in obtaining much complex signals.

In the analysis of five mainly used metabolites present in phantom dataset GMM based method outperforms LC Model and Tarquin algorithms. We can achieve better accuracy and precision of estimating metabolite concentration through detailed signal pre-processing and proper modeling. In our method applying of zero filling step provides peak line shapes more similar to Gaussian distribution which cause better representation of the signal peak using Gaussian mixture model. The biggest impact on the difference in pre-processing stage between results after using three methods was observed in phase correction part that influences peak line shape the most.

In MRS data obtained intensity values are continuous, so creating one dimensional data by treating spectrum as a histogram and using original EM algorithm brings a risk of accuracy decrease. Even if we discretize intensity values, we create a data vector of enormous size. Also in our problem we need much more model components than in common applications of GMM. Running original EM algorithm on created one dimensional MS data results in about 300 times worse computational time with comparison to GPU implementation of modified algorithm. Original version of algorithm, which was previously implemented in CUDA environment [14], is not recommended for modeling biological spectra. We intend to write in the near future our modified EM algorithm in CUDA C environment, but it needs thorough planning of memory management while performing GPU computing, which was not needed in Matlab implementation.

5   Conclusions

Proposed GMM based method together with pre-processing pipeline gives very good accuracy and precision of quantification of metabolite amounts. Comparing GMM based method to two popular tools: Tarquin and LC Model, we conclude that proposed methodology achieved comparable accuracy, but much better precision, which was achieved by complex and adapted pre-processing pipeline. GPU computing approach enables efficient and intuitive implementation of modified EM algorithm. Vectorization and parallelization of the code drastically speed-ups algorithm in comparison to the standard implementation. The most efficient implementation was based on GPU with single precision calculations giving satisfactory model accuracy.

Acknowledgments. This work was financially supported by internal grant of Silesian University of Technology for young researchers BKM/524/RAU1/2014/18 (FB), BKM/524/RAU1/2014/32 (MM) and BK/265/RAU1/2014/10 (JP). We are very thankful to prof. Maria Sokol, dr Lukasz Boguszewicz and Agnieszka Skorupa for access to the phantom data. All the calculations were carried out using GeCONiI infrastructure funded by project number POIG.02.03.01-24-099/13.

References

1. de Graaf, R.A.: Vivo NMR Spectroscopy: Principles and Techniques. Wiley (2013)
2. Jacobsen, N.E.: NMR Spectroscopy Explained: Simplified Theory, Applications and Examples for Organic Chemistry and Structural Biology. Wiley (2007)
3. Lupu, M., Todor, D.: A singular value decomposition based algorithm for multicomponent exponential fitting of NMR relaxation signals. Chemometrics and Intelligent Laboratory Systems 29, 11–17 (1995)
4. Wilson, M., Reynolds, G., Kauppinen, R.A., Arvanitis, T.N., Peet, A.C.: A constrained least-squares approach to the automated quantitation of in vivo (1)H magnetic resonance spectroscopy data. Official Journal of the Society of Magnetic Resonance in Medicine / Society of Magnetic Resonance in Medicine 65, 1–12 (2011)
5. Provencher, S.W.: Estimation of metabolite concentrations from localized in vivo proton NMR spectra. Magnetic Resonance in Medicine: Official Journal of the Society of Magnetic Resonance in Medicine / Society of Magnetic Resonance in Medicine 30, 672–679 (1993)

6. Coombes, K.R., Koomen, J.M., Baggerly, K.A., Morris, J.S., Kobayashi, R.: Understanding the characteristics of mass spectrometry data through the use of simulation. Cancer Informatics 1, 41–52 (2005)
7. Marczyk, M., Polanska, J., Polanski, A.: Comparison of Algorithms for Profile-Based Alignment of Low Resolution MALDI-ToF Spectra. In: Gruca, D.A., Czachórski, T., Kozielski, S. (eds.) Man-Machine Interactions 3, vol. 242, pp. 193–201. Springer International Publishing (2014)
8. Binczyk, F., Tarnawski, R., Polanska, J.: Improvement in the accuracy of Nuclear Magnetic Resonance spectrum analysis by automatic tuning of phase correction algorithms. In: International Work-Conference Bioinformatics and Biomedical Engineering, pp. 778–788 (2014)
9. Müller, N., Jerschow, A.: Nuclear spin noise imaging. Proceedings of the National Academy of Sciences 103, 6790–6792 (2006)
10. Savitzky, A., Golay, M.J.E.: Smoothing + Differentiation of Data by Simplified Least Squares Procedures. Anal. Chem. 36, 1627 (1964)
11. McLachlan, G.J., Peel, D.: Finite mixture models. Wiley, New York (2000)
12. Polanski, A., Kimmel, M.: Bioinformatics. Springer, London (2007)
13. Schwarz, G.: Estimating the dimension of a model. The Annals of Statistics 6, 461–464 (1978)
14. Machlica, L., Vanek, J., Zajic, Z.: Fast Estimation of Gaussian Mixture Model Parameters on GPU Using CUDA. In: 12th International Conference on Parallel and Distributed Computing, Applications and Technologies, pp. 167–172 (2011)

Noise and Baseline Filtration in Mass Spectrometry

Jan Urban and Dalibor Štys

Institute of Complex Systems, South Bohemian Research Center of Aquaculture
and Biodiversity of Hydrocenoses, Faculty of Fisheries and Protection of Waters,
University of South Bohemia in České Budějovice, Zámek 136, Nové Hrady, 37333
urbanj@frov.jcu.cz

Abstract. Mass spectrometry (MS) produce terabytes of measurements daily
around the world. Systemic (instrumental and chemical) and random noise complicate the dataset. Correct interpretation of mass spectrometry (MS) is aected by
presented noise across all kinds of MS techniques. The noise addition may produce fake peaks or hide small intensities in the measurements. Thus, MS data are
crowded and have a uneven baseline. In tandem with chromatography, the systemic noise causes extraneous peaks or rising baseline during gradient elution.
The interpretation of MS is not trivial mainly because of the vast amount of
noise especially in complex samples. It is necessary to consider approaches for
noise subtraction. Common algorithms based on thresholding or wavelet transformation are not resistant to the losses of information from their principle. Thresholding methods, even in the adaptive form, still discard parts under threshold
level(s) from the whole measurement. The wavelet transformations directly
change the information content and are sensitive to the window length. Therefore, some information could not be used for for further analyzing process, including peak detection. Omitting the presence of baseline (also called background, systemic noise or mobile phase) impedes objective analysis. Behavior of
the baseline content is not constant in time axis. The results may be measured by
increase of data mining output, both qualitatively and quantitatively.

Keywords: Mass spectrometry, Mass spectrum, Mass spectra, Chromatography, Baseline filtration, Noise filtration, Denoising.

1 Introduction

The comprehensive comparison of complex mixtures of similar compounds by
LC-MS has been a major issue in the 1980s and 1990s (1,2,3) and became again highly interesting with extensions to -omics approaches from genomics to proteomics and
metabolomics. Thus, procedures for peak detection and segmantation are required for
any further data analysis. The major problem in peaks detection is complexity of the
signal and different noise sources (4).

Contemporary paradigms of real systems assume that any natural or artificial
process under study fulfills the general set of nature laws. Those laws are a~priori
stochastic (probabilistic) descriptions, where a deterministic case is just a special case
of stochasticity (with the probabilities equal to one). The stochastic behavior is given

F. Ortuño and I. Rojas (Eds.): IWBBIO 2015, Part II, LNCS 9044, pp. 418–425, 2015.
© Springer International Publishing Switzerland 2015

by our inability to measure (observe) exact values of all system attributes with infinite accuracy. All individual objects of interest are ordered to the proper subtraction of general laws, usually just by parameterization. It is necessary to mention that any thought construction above the object behavior never works with the real object itself, only with the abstract object. A comprehensive abstract model is necessary for consequential processing and analysis. The reason of mathematically described data model is to encapsulate behavior hypothesis into appropriate mathematical space.

However, even the measured data were obtained by measurement device which was designed according to some model of physical (chemical, biological, mathematical) process of the measure and they are always quantized in the value domain. It is done by analog-digital converters on the input of (control, storage and processing) computers and at many other instances which reflects primarily our inability to measure with infinitesimal accuracy and precision. Discretization and quantization are sampling processes of mapping of continuous values by a finite set(s). Sampling is the reduction of a continuous signal to a discrete signal. In practice, the continuous signal is sampled using an analog-to-digital converter (ADC), a non-ideal device with various physical limitations. The sampling is just the measuring of the signal with some (usually predefined) sampling frequency. Quantization is the process of approximating (mapping) a continuous range of values by a finite set of discrete values (discrete number of levels). Output of the sampling and quantization is a sequence of piecewise constant values or rectangular pulses.

A discrete signal is a series consisting of a sequence of quantities. A digital signal is a discrete signal for which also the amplitude is discrete and that takes on only a discrete set of value. Discrete-valued signals are always an approximation to the original continuous-valued signal. Continuity does not really exist - it is not possible to observe or measure continuously. Therefore, all possible datasets are already discrete models according to the theory of systems. The domains of definition of each system attributes are considered as finite according to our finite knowledge of the real world and finite amounts of each measurable range.

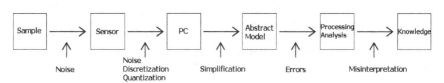

Fig. 1. Illustration of information flux during usual experiment analysis

Liquid chromatography in tandem with mass spectrometry (LC-MS) produce terabytes of measurements daily around the world. Systemic (instrumental and chemical) and random noise complicate the dataset. Correct interpretation of mass spectrometry (MS) is affected by presented noise across all kinds of MS techniques. The noise addition may produce fake peaks or hide small intensities in the measurements. This systemic noise causes extraneous peaks or rising baseline during gradient elution. The interpretation of LC-MS is not trivial mainly because of the vast amount of noise especially in complex samples. It is necessary to consider approaches for denoising and baseline subtraction. Common algorithms based on thresholding or wavelet transformation are not resistant to the losses of information from their principle. Thresholding methods, even in the adaptive form, still discard parts under threshold level(s) from the whole measurement. The wavelet transformations directly change the information content and are sensitive to the window length. Therefore, some information could not be used for further analyzing process, including peak detection.

The filtration is necessary processing step to emphasize features which are relevant for other steps, especially segmentation of the measurement into individual eluted compounds. Noise additions are produced not only by random errors (random noise) but also by influence of baseline from the Liquid chromatography. Sum of the noise and the signal may produce false interpretation or hide the signal under reasonable level. Therefore, baseline in LC-MS negatively affects the measurement analysis and represents the systematic noise in nonlinear level on the time axis. Generally, any filtration which uses hard fixed threshold values are problematic. Its results are often inconsistent between runs, instrumentation and methods because the values from nearest threshold neighborhood may be easily misclassified.

Chemical noise (e.g. sodium adducts) results from mobile phase impurities. It is more difficult to remove then random noise, because they have a pattern similar to the signal. Chemical noise can reduce mass accuracy by shifting peaks centroids. Noise produced by random errors is caused by minor variation of the distribution surface. Systemic errors become more noticeable as they create borders effects, that are systematically over or underestimated.

2   Approach

Noise is not considered as the error in value (increase or decrease of the signal), but as the presence of signal in the individual values. Therefore, random noise is random in the signal position and the intensity value means amount of repetition of its occurrence on that position. During the measurement of every single value is the value obtained as integration of signal over short time period of the measurement itself. Therefore, there is usually hundreds or thousands of small measurements of individual values. In other words, there is enough repetitions of measurements on the signal position for estimation of the probability density function (PDF). This is the main idea, the real signal is just a disturbance in the noise PDF.

Because presented noises have to correspond to some probabilistic distributions, it is possible to approximate the distributions and identify the parameters of the distribution. Using this parameters helps to describe measurement more accurately. This information could be used in addition for filtration or for further analysis process. The object of interest is such description of the signal which reduce influence of presented noises. Precise contributions of noises are unknown because the stochastic behavior. However, the characteristics can be estimated by probabilistic analysis. Afterwards, with the knowledge of noise characteristics and measurement data output can be also estimated the pure signal mapping.

Basically, estimations of exact values of noise contribution may not be accurate even if the characteristics of noises could be estimated well. Consequently, quantitative error and two kind of qualitative errors could be made. In quantitative case, this is the common error of the estimation solutions, the estimated value of intensity differs to real but a priori unknown value. The qualitative errors of estimation are the same as in another detection tasks or in hypothesis testing. They are known as false reject (false negative in some literature) and false alarm (false positive, false accept). False reject happens where the signal is present but not recognized. On the other hand, false alarm means positive value of signal when the signal is not present. Generally, the quantity is not precise because of random noise and none processing of the data causes no false reject but remains all of false alarms. To reduce quantitative errors, it is advisable to repeat the measurement of the same sample many times. Therefore, it is not the key problem of analysis of single one measurement. But the estimation of exact values will produce the qualitative errors. It can decrease the false alarms but increase the false rejects. The most of filtration methods is designed to decrease the false alarms. The optimal rate between false alarms and false rejects is nontrivial question.

From this point of view, error ratio based directly on given task is more suitable. The error ratio produced by filtration algorithms could be tuned via some parameters, but the relation between them is generally not evident. Especially when there are several steps in the filtration which can be tuned independently. Instead of value estimation of real signal intensity, which is error-full, I propose to evaluate probability factor that the measured data output is the real signal value.

Understanding to the measurement is more straightforward according to the estimated probability, because this information will be available for all values and positions. Therefore, there is only one parameter which characterizes quality of the measurement data output during interpretation itself. No other parameters like SNR or intensity levels in blank measurement need to be evaluated and tuned. In praxis, there are basically two principles how to deal with this probability information. The first one, a fixed threshold value can be tuned for any step of further output analysis. When the higher threshold is set, the total number of credible data points decreases as well as number of false alarms but possibility of false rejects occurrence increases. When the lower threshold is used the opposite situation happens, naturally.

The second principle is to use whole probabilistic information in further output analysis [5]. This case is more advisable because no part of measurement data output and no probabilistic information are discarded. Of course, all analysis steps have to support processing of uncertain data characterized by probability values. Unfortunately, the most of common analysis algorithms assume exact data only although no real data are accurate and noise-free. This approach is focused on proper characterisation of presented noise. Noise contribution produced by many sources is characterised separately in position of 2D signals. Information about the both of noise characterisations were integrated into probability factor.

The intensity y produced by the spectrometer is usually shown as two 2D graphs, $y1(t)$ and $y2(m)$, where t is time and m is the mass to charge ratio. We take a more general approach, looking at peeks in 3D, $y(t,m)$. At each point (t,m), where t is the

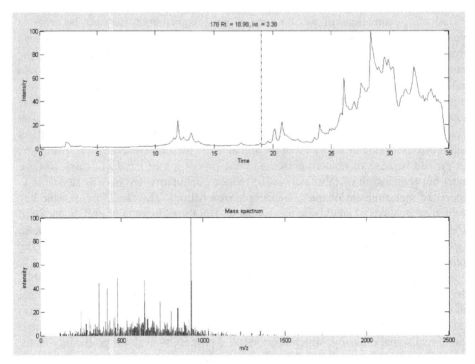

Fig. 2. Typical graphical representation of LC-MS measurement. Top plot is Total Ion Current chromatogram (TIC) with selected retention time (dash-dotted line). Bottom plot is Mass Spectrum (MS) in selected retention time. Individual discrete bars are plotted with different colour to distinguish them graphically (optionally).

retention time and m is mass m/z, the output intensity of the spectrometer is composed of several parts:

$y(t,m) = s(t,m) + q(t,m) + r(t,m)$
$ps(t,m) = $ probablility that $y(t,m)$ is useful signal, $s(t,m)$
$pq(t,m) = $ probablility that $y(t,m)$ is not random noise, $r(t,m)=0$,

where $y(t,m)$ is the useful signal, $s(t,m)$ is the useful signal, $q(t,m)$ is the systematic noise, and $r(t,m)$ is the random noise.

Typical measurement output data from HPLC/MC is set of points in three dimensional space which is defined by axes: retention time, molecular mass and intensity. Analytes elute in every retention time point from HPLC column, obviously because of delay proportional to some chemical property, and enter the MS ionization chamber. Intensity for each detectable mass is measured inside the MS and this value represented amount of ionized molecules of individual mass in exact retention time point. This value may be only the natural number or zero. The retention time and molecular mass are the attributes of our system. Let mark those attributes by sign a_k, k=0,1. Whole set could be described as $A = \{a_k | k = 0, 1\}$, where a_k are names of corresponding attribute. Each attribute a_k is acquiring a value. In abstract systems the value of k–th attribute a_k is represented by k–th variable v_k from V_k, k = 0, 1, where V_k is domain of definition of k–th variable, it is set of all values that variable v_k can reach.

Attribute A_0 represent reference attribute, supposed to be retention time. For its variable v_0 can be used common sign t from T, T = t_0, t_1, t_2, ..., t_e, where e is natural number. On the set T could (but do not have to) be defined a difference which represents the time period from time point t_i to time point t_j . Value of the 2–nd attribute a_1 is represented by variable v_1 and means molecular mass. For its variable v_1 can be used common sign m from M, M=m_0,m_1,m_2, ...,m_n, where n is natural number. Now, we obtained two sets which describe the values of two axis: retention time and molecular mass [6]. Every individual measurement run generates intensity values for all possible pairs retention time t and molecular mass m. Therefore, this generation process can be symbolically described as mapping:

$$y : T \times M \rightarrow [t \text{ from } T, m \text{ from } M \mid y(t,m) \text{ from } I, I = 0, 1, ..., imax,$$

where I is set of natural numbers with zero and the value of mapping y(t,m) means intensity of molecular mass m in retention time t. Exact value of imax is limited by saturation level of MS detector. Our abstract system is then defined in domain by ordered pair (T,M) [7].

Every measurements data output has its errors. Two basic principles of error occurrence will be discussed. The first one, called systematic noise, is in LC/MS produced by presence of mobile phase that carry analytes through the HPLC column into MS. The effect of the mobile phase is dependent on measurement device, its setup and mobile phase composition. During the analysis of the measurement, it is necessary to keep in mind that in measurement data output the systematic noise influence is always present. The other one, called random noise, includes all unwanted sources of transient disturbances and it is always present too. Both of them affect the signal transparency and can be also described according to the theory of systems. Systematic noise can be described as mapping q : T $\times$M $\rightarrow$ [t from T, m from M $\mid$ q(t,m) from I, I = 0, 1, ..., imax, and random noise can be described formally in the same way marked by sign r(t,m). The conditions for noise mappings are equal to conditions for mapping of signal generating process. If we were able to define the generating process as mapping as well as both noises then we can also define mapping of pure analytes contribution s(t,m). Consequently, the relations between all mappings will be y(t,m) = s(t,m) + q(t,m) + r(t,m). Our object of interest is description of s(t,m). Thus we have to reduce influence of presented noises, which can produce false peak or hide analytes signal under reasonable level. Precise contributions of both noises are unknown due to their stochastic character but their characteristics can be estimated by probabilistic analysis. Afterwards, with the knowledge of noise characteristics and measurement data output can be also estimated the pure analytes mapping:

$$s'(t,m) = y(t,m) - q'(t,m) - r'(t,m).$$

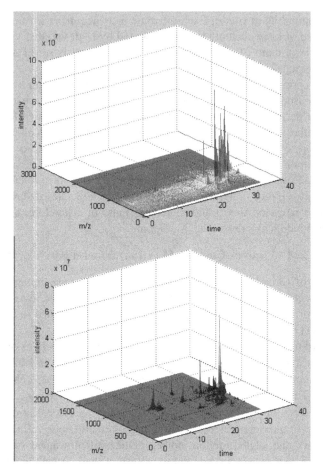

Fig. 3. Top 3D plot is raw LC-MS measurement, bottom 3D plot is the same measurement filtered via probabilistic approach, showing only signals with probability >75 %

Instead of value estimation of analyte signal intensity s'(t,m) which contains errors, we evaluate probability factor p(t,m) that the measurement data output y(t,m) is analyte intensity s'(t,m):

$$p(t,m) = p\ [y(t,m) = s(t,m) \mid \lambda q, \lambda r],$$

where λq and λr are estimated characteristic of mapping q(t,m) and mapping r(t,m) respectively. It should be noted here that up to 98% of measured information is noise. The probability factor p(t,m) means probability that analyte with molecular mass m in retention time t has intensity y(t,m). Probability p(t,m) is multiplication of two independent probabilities. The first one is probability pr(t,m) that measurement data output y(t,m) is not produced by random noise r(t,m):

$$pr(t,m) = p\ [y(t,m) = s(t,m) + q(t,m) \mid \lambda r]$$

The second probability is probability pq(t,m) that measurement data output y(t,m) is not produced by systematic noise q(t,m):

$$pq(t,m) = p [y(t,m) = s(t,m) + r(t,m) \mid \lambda q] ,$$

and the final probability p(t,m) is p(t,m) = pr(t,m) * pq(t,m). With probability factor p(t,m) which can be evaluated precisely with well noise characteristics, the error ratio can be tuned directly for any task. Subsequent filtration and/or analyzing steps can utilise this probability in their outputs via probability theory formulas.

Acknowledgement. The study was financially supported by the Ministry of Education, Youth and Sports of the Czech Republic - projects 'CENAKVA' (No. CZ.1.05/2.1.00/01.0024), 'CENAKVA II' (No. LO1205 under the NPU I program); and by the Postdok JU (CZ. 1.07/2.3.00/30.0006).

References

[1] Hearn, M.T.W. (ed.): HPLC of Proteins, Peptides and Polynucleotides. Contemporary Topics and Applications. Wiley, New York (1991)
[2] Snyder, L.R., Dolan, J.W.: High-Performance Gradient Elution: The Practical Application of the Linear-Solvent-Strength Model. Wiley, New York (2006)
[3] Mant, C.T., Hodges, R.S.: HPLC of Biological Macromolecules, pp. 433–511. Marcel Dekker, New York (2002)
[4] Du, P., Kibbe, W.A., Lin, S.M.: Imroved peak detection in mass spectrum by incorporating continuous wavelet transform-based pattern matching. Bioinformatics 22(17), 2059–2065 (2006)
[5] Urban, J., Vaněk, J., Soukup, J., Štys, D.: Expertomica metabolite profiling: Getting more information from LC-MS using the stochastic systems approach. Bioinformatics 25(20), 2764–2767 (2009)
[6] Urban, J., AfsethNK, Š.D.: Mass assignment, centroiding, and resolution – fundamental definitions and confusions in mass spectrometry. TrAC Trends in Analytical Chemistry 53, 126–136 (2014)
[7] Žampa, P.: The principle and the law of causality in a new approach to system theory. In: Cybernetics and Systems 2004, pp. 3–8. Austrian Society for Cybernetics Studies, Vienna (2004), ISBN 3-85206-169-5

BioWes – From Design of Experiment, through Protocol to Repository, Control, Standardization and Back-Tracking

Antonín Bárta, Petr Císař, Dmytro Soloviov, Pavel Souček, Dalibor Štys,
Štěpán Papáček, Aliaksandr Pautsina, Renata Rychtáriková, and Jan Urban

Institute of Complex Systems, South Bohemian Research Center of Aquaculture
and Biodiversity of Hydrocenoses, Faculty of Fisheries and Protection of Waters, University of
South Bohemia in České Budějovice, Zámek 136, Nové Hrady, 373 33, Czech Republic
abarta@frov.jcu.cz

Abstract. The amount of data produced by current experiments in systems
biology is enormous. Some database and software support for biology experiments exist but usually deal just with some part of data and metadata management. Primary data are only occasionally analyzed in-depth and shared. Studies
showed that cost of data sharing is cheaper than experimental work. Experimental costs are several orders higher than data. Therefore, being up to date is a
worldwide problem for all users of biology applications. In practice, the problem extends to general fields such as knowledge mining, experiment quality and
repeatability, and the philosophical epistemology of biological problems.

The BioWes project is inspired by several similar projects that try to solve a
substantial contemporary problem of sharing big amount of experimental data.
There are several projects that offer the solution for data sharing (for different
types of data). The problem is that the amount data produced by experimentalists
is constantly increasing and the speed of internet will always be a step behind.
The effective and easier way how to share experimental data between researchers
is to share metadata. Metadata means the overall knowledge about the experiment
that consist of complex information of experimental procedure and knowledge
that can be extracted from data automatically or manually by post-processing. The
data itself is meaningless without any additional knowledge concerning the experiment. There is no project that can offer the whole concept of experimental data
sharing and data processing based on the sharing of knowledge.

Keywords: Database, Repository, Metadata, Data management, Protocols,
Experiment setup, Design of experiment.

1   Introduction

The project BioWes is inspired by several similar projects that try to solve a
substantial contemporary problem of sharing big amount of experimental data. There
are several projects that offer the solution for data sharing (for different types of data).

F. Ortuño and I. Rojas (Eds.): IWBBIO 2015, Part II, LNCS 9044, pp. 426–430, 2015.

The problem is that the amount of data produced by experimentalist is constantly increasing and the speed of internet will always be a step behind [1]. The effective and easier way how to share experimental data between researchers is to share the metadata. Metadata means the overall knowledge about the experiment that consist of complex information of experimental procedure and knowledge that can be extracted from data automatically or manually by post-processing. Data itself is meaningless without any additional knowledge concerning the experiment. There is no project that can offer the whole concept of experimental data sharing and data processing based on the sharing of knowledge (see Fig.1).

The main reason of sharing metadata data is to save money and time necessary for experimentation and to compare the results between different experimenters. Data sharing and especially metadata sharing can be understood as the advertisement of the experiments of a particular experimenter. Experimental data sharing and comparison can help to improve experimental procedures and defining of standards in this area [1, 2].

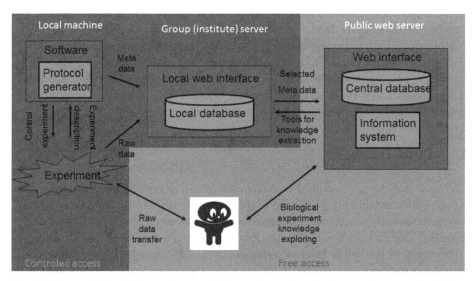

Fig. 1. Scheme of the sharing and usage of data and metadata database on the example of Bio-Wes system

2   Protocol Generator

Scheme of the usage of BioWes system are shown on Fig. 1. Software interface, which is in the direct contact with a user (experimenter) is called Protocol Generator. It is a standalone application that should ensure the repeatability and correctness of the biological experiments. The tool is designed to lead the experimenter through the particular type of experiment as a supervisor and to help him. Protocol generator has two purposes: the first one is to check that the procedure of the experiment has been done precisely and the second one is to produce all important settings that are part of the experiment in the form of report on the experiment. The method to ensure precise realization of experiment is to check if all the necessary parameters and steps of the experiment have been set and done. The list of necessary parameters and steps for the particular type of experiment comes up from the analysis of biological experiments from different research institutes.

The description of the experiment can be created by the user for specific experiment. Graphical user interface Protocol designer has been implemented for this purpose. The protocol template can be created by any BioWes user who can define all the important conditions of the experiment. The user can use 10 basic components for definition of the protocol. The protocol of experiment can be shared among the people who realize the experiment instead of students to ensure the repeatability of the experiment.

The template can be later modified for new experiment to speed up the process.
Main advantage of the electronic protocol is that there is a direct link between the protocol and experimental data. Both are stored in the central database and can be used for obtaining future data.

Protocol generator supports external plugins for mining information about setting of devices from external files. The plugins can read the information about some parameters of experiment form files produced by measurement device (magnification of microscope) and fill it into the protocol. Plugins are using open interface and therefore it can be created by the users for specific devices.

3   Local Database

The main purpose of local data management tool is to organize and store the raw experimental data directly on the site of the institution (experimenter). The local data management tool provide the functionality of data storing, searching, filtration and reporting. The tool is connected to the Protocol generator to support the reporting of experiments on the higher level of metadata. Local data management is realized as a specialized database that will be optimized to the type of experimental data produced by particular institution. The database with uniform interface is modified according to the needs of the particular experimenter to reach the aims of different experiments. The local management tool provide the communication module (interface) to global data management tool (web based data sharing). The global data management tool is used for metadata sharing between different institutions and the public. The process of communication between local and global data management tools will be under full control of the institution. Therefore only metadata can be shared with the rest of scientific community or the public. The advantage of this approach is the direct control of „what I am sharing with the others". [3]

4   Sharing and Management

The central data storage is realized as combination of local data storage (located at the institution) for raw data and one central data storage selected metadata. The data structures in central database will be defined generally to cover all the different metadata types and to upgrade the structures in the future. [4]

The central data storage serve as the first option for searching the experimental data through metadata and allow the user to find the proper experiments and results. All the metadata are available using the XLM data structures exchange.The BioWes experimental data and metadata management allows:– fast information search – the usage of data structures indexing and advanced algorithms of query execution plans minimize the time response of data storage – standardized system control – the commercial database ensure the safe and secured operation of the database.

The central database will be in cooperation with the information system. The information system provide the parameterization of the central storage, user accounts control and policy. Data from central database are presented using the web presentations to the different kind of the scientific and general audience controlled by the access rights.

The interface is used as an interpreter between central data storage, local data storage and visualization framework.

The user friendliness of the central database is supported by the visualization framework. The visualization framework will be implemented as several software modules and interface for extensions of information system. The visualization framework allow simple and intelligible visualization and comparison of the metadata and results of searches. It is based on the mix of existing modules and third party modules. The third party modules could be plugged into the central data storage for the user of the system. The modules are focused on raw data processing, data mining and aggregation of the metadata. Standard interface of the central data storage will allow the user to upload the extracted information back into the central storage, describe it and integrate it into current structures.

One part of the web solution will be the offer of tools for raw data post-processing. These tools are highly specialized for experimental data processing. They can be used by anyone to produce metadata from raw experimental data and share metadata using BioWes web solution. The list of the tools can be extended by any third party tool for biological data processing. Several tools we are working on directly under the project:

– Cell time lapse image processing and representation
– LC-MS measurements filtration and analysis
– Software for behavior analysis of aquatic organisms

Acknowledgement. The study was financially supported by TACR project TA01010214 BioWes, by the Ministry of Education, Youth and Sports of the Czech Republic - projects 'CENAKVA' (No. CZ.1.05/2.1.00/01.0024), 'CENAKVA II' (No. LO1205 under the NPU I program); and by the Postdok JU (CZ. 1.07/2.3.00/30.0006).

References

[1] Haug, K., et al.: MetaboLights—an open-access general-purpose repository for metabolomics studies and associated meta-data. NAR, gks1004 (2012)

[2] Freire, J., Bonnet, P., Shasha, D.: Computational reproducibility: State-of-the-art, challenges, and database research opportunities. In: Proceedings of the 2012 ACM SIGMOD International Conference on Management of Data. ACM (2012)

[3] Mayer-Schönberger, V., Cukier, K.: Big data: A revolution that will transform how we live, work, and think. HMH (2013)

[4] Borgman, C.L.: The conundrum of sharing research data. JASIST, 1059–1078 (2012)

Measurement in Biological Systems
from the Self-organisation Point of View

Dalibor Štys, Jan Urban, Renata Rychtáriková, Anna Zhyrova, and Petr Císař

Institute of Complex Systems, Faculty of Fisheries and Protection of Waters
University of South Bohemia, Zámek 136, 373 33 Nové Hrady, Czech Republic
stys@jcu.cz
http://www.frov.jcu.cz/cs/ustav-komplexnich-systemu-uks

Abstract. Measurement in biological systems became a subject of concern as a consequence of numerous reports on limited reproducibility of experimental results. To reveal origins of this inconsistency, we have examined general features of biological systems as dynamical systems far from not only their chemical equilibrium, but, in most cases, also of their Lyapunov stable states. Thus, in biological experiments, we do not observe states, but distinct trajectories followed by the examined organism. If one of the possible sequences is selected, a minute sub-section of the whole problem is obtained – sometimes in a seemingly highly reproducible manner. But the state of the organism is known only if a complete set of possible trajectories is known. And this is often practically impossible. Therefore, we propose a different framework for reporting and analysis of biological experiments, reflecting the view of non-linear mathematics. This view should be used to avoid overoptimistic results, which have to be consequently retracted or largely complemented. An increase of specification of experimental procedures is the way for better understanding of the scope of paths, which the biological system may be evolving. And it is hidden in the evolution of experimental protocols. Our system bioWes is a tool for objectivization of this knowledge.

Keywords: Measurement, statistics, self-organisation, biological systems.

1 Introduction

Measurement in biological systems became a subject of concern as a consequence of numerous reports on limited reproducibility of experimental results [1,2]. By detailed examination, it was often found that many of the results are not exactly fake, but represent a selection from actually obtained results. In the same time, articles are accompanied by statistical analysis, which seemingly confirms the normal, Gaussian, distribution of results.

To reveal this inconsistency, we have discussed the main features of biological systems from the mathematical point of view. Biological systems are dynamical systems maintained far from not only their equilibrium, but in most cases also of recurrent, Lyapunov stable states [3]. In other words, in biological experiments, we do not observe states, but distinct trajectories followed by the examined

F. Ortuño and I. Rojas (Eds.): IWBBIO 2015, Part II, LNCS 9044, pp. 431–443, 2015.
© Springer International Publishing Switzerland 2015

organism. These trajectories are characterized by a sequence of distinct spatial structures of the organism, i.e., a sequence of cell states.

There are two points of view from which this problem must be approached: (1) the properties of the experimental system itself and (2) technical potential of the measurement.

2   Technical Limits of the Information Content of the Measurement

A new system theory was introduced by Pavel Žampa [5]. The main differences of the Žampa´s systems theory from other system theories are (1) inclusions of the input and output into the system description, (2) a definition of the system attribute as a distinct concept from the system variable, and (3) an introduction of the system time as a time of measurement. From that, the concept of the complete immediate cause as the list of values of all system attributes at all time instants, preceding the examined system time necessary for its description, naturally arises. Here we summarise selected parts of Žampa´s system theory needed for discussion in this article.

The adequate model of the time, which we call a real time, is a variable t,

$$t \in T, \tag{1}$$

whose definition set is a non-empty set T of all time events

$$t_k, \text{ where } k \in \{0, 1, 2, ..., F\}. \tag{2}$$

If there exist a relation $t_i < t_j$ for each two time instants $t_i, t_j \in T$, we say that the time instant t_i precedes the time instant t_j.

We shall denominate the studied system attributes (abstract variables) – such as a coordinate of position, coordinate of speed, position of a switch, verity of a statement – by symbols a_i, where $i = 1, 2, ..., n$. The set of all abstract variables will be denominated by symbol A, and thus it holds:

$$A = \{a_i \mid i \in I\}, \tag{3}$$

where I is an appropriate nonempty index set. Adequate model of the i-th attribute $a_i, i \in I$ is an abstract variable of the i-th attribute v_i

$$v_i \in V_i, \text{ for } i \in I, \tag{4}$$

whose definition set is a nonempty set V_i, where $i \in I$, with elements, which we shall call values of the i-th attribute. The introduction of the variable $t \in T$ and the set of variables $v_i \in V_i$ for $i \in I$, formalised the notion of time and other attributes of the system. We may now introduce a system variable v defined by relations

$$v = (v_1, v_2, ..., v_n), \tag{5}$$

$$v \in V, \tag{6}$$

and

$$V = V_1 \times V_2 \times ... \times V_n \tag{7}$$

Thus, the system variable is an ordered set of n variables of system attributes. Then, the ordered set

$$(T, V) \tag{8}$$

is the basis of the mathematisation of the problem of the definition of the system trajectory as a mapping. The trajectory of an abstract system corresponds to the mapping z

$$z : T \times I \to \bigcup_{i \in I} V_i \text{ such that } z(t, i) \in V_i, i \in I \tag{9}$$

If we denote the set of all system trajectories as Ω, we may write

$$\Omega = \{z \mid z : T \times I \to \bigcup_{i \in I} V_i \text{ such that } z(t, i) \in V_i, i \in I\}. \tag{10}$$

The system event is marked B and defined as a sub-set of the set Ω of all system trajectories:

$$B \subset \Omega. \tag{11}$$

We usually demarcate the system event as a set of trajectories z having a certain property $V(z)$, which we describe as

$$B = \{z \mid V(z)\}. \tag{12}$$

Then, we may define an abstract deterministic system $\mathscr{D}$ as

$$\mathscr{D} = (T, V, z). \tag{13}$$

For technical as well as internal mechanism reasons, neither of the measurable systems is truly deterministic. By introduction of probabilistic instead of deterministic mapping $P(B)$, which is defined on the potency $\mathcal{B}$ of the set Ω of all trajectories $B \in \mathcal{B}$ we may define stochastic (abstract) system $\mathscr{S}$ as

$$\mathscr{S} = (T, V, P). \tag{14}$$

Finally, we shall formalise the problem of causality in a measured system (Fig. 1). The system is measured at system times, nevertheless, it is also evolving between them. In the same time, for the definition of the future evolution of the system, it may not be sufficient to consider one time instant, no matter how good our system model is.

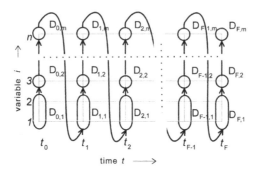

Fig. 1. The concept of causal relations and the importance of system model [6]. There is depicted a set of variables measured at a time instant (represented by circles an ovals) and causal relation in the behaviour between measuring times. To determine the set of measured values at time t_l we must consider not only limited set of measured values at times $t_j < t_l$ but also appropriate causality in intermediate times, not defined by the set of system time T. A set of variables at the time of measurement $D_{k,l}$ is then defined by two indexes k, l, where the first expresses the causality within the measurement time interval, while the other is the index of the system time from the set I. Then, unity $C_{k,l} \subset \bigcup_{(i,j)<(k,l)} D_{i,j}$ is *the complete immediate cause*, the set of all variables necessary for prediction of the set $D_{k,l}$. From that, among other conclusions, follows that at least for technical reasons may not be understood without the knowledge of an appropriate system model.

We generally assume that the system trajectory is defined as mapping z with a definition set D,

$$D = T \times I, \tag{15}$$

which is defined in parts by its internal mechanism. We thus define each segment $z \mid D_{k,l}$ of the trajectory z exactly once and that in the dependency on the segment $z \mid C_{k,l}$, where for $C_{k,l}$ holds

$$C_{k,l} \subset \bigcup_{(i,j)<(k,l)} D_{i,j}. \tag{16}$$

Thus, the cause $C_{k,l}$ determines the consequence $D_{k,l}$ and is understood as a *complete immediate cause* of the consequence $D_{k,l}$.

Thus, we need an appropriate model of the system mainly for two reasons: to determine (1) system behaviour between the time instants and (2) the time extent of the complete immediate cause.

3   Phenomenological Variables and Measurement in Chemistry

The problem of Žampa´s system theory is the definition of a truly appropriate system model. In most cases the models are rather limited, yet, in mechanics or electronics, these limitations may be often overcome. In this discussion we illustrate the problem of measurement in chemistry, which has clear consequences for measurement in biological systems.

In each physico-chemical textbook is introduced the idea of the chemical potential μ_i and the activity a_i which are the real measures of the contribution of each molecule to total Gibbs energy G of the examined system. It is not from the first sight controversial to write the total Gibbs energy as

$$G = \sum_{i=0}^{n} \mu_i, \tag{17}$$

where index i determines the individual chemical component of n components present in the mixture. We should, however, expand μ_i as

$$\mu_i = \mu_{0,i} + \nu_i RT \ln(a_i) = \mu_{0,i} + \nu_i RT \ln(c_i * \gamma_i) = \mu_{0,i} + \\ \nu_i RT \ln(c_i) + \nu_i RT \ln(f(c_1, c_2...c_n, T, p, V...)), \tag{18}$$

where $\mu_{0,i}$ is the standard chemical potential of the i-th component of concentration c_i, $\gamma_i = f(c_1, c_2...c_n, T, p, V...)$ is its activity coefficient, ν_i is the respective stoichiometric coefficient, and R is the universal gas constant. The activity coefficient is in principle a function of concentrations c_i of all components in the mixture and all other relevant state variables such as temperature T, pressure p, volume V, etc. The difference between the ideal course, where the $\gamma_i = 1$, and a real situation may be demonstrated on such simple examples as distillation of spirits and the existence of the azeotropic mixture.

From that comes the following moral for the construction of the model of chemical system: we cannot construct our system using concentrations of components as orthogonal variables if we want to obtain a multidimensional plane G vs. $\ln(x_i)$. In fact, such a construction is almost never practically possible and the problem is solved by plotting experimental results, giving a complicated surface far from the ideal one.

The experimentally determined shape of the real state function gives us indices which features we should seek, namely in terms of molecular interactions and their influences on phase behaviour. In the terminology introduced in Chapter 2 we must consider that the definition of system attributes A and the definition of the appropriate variables V is inseparable from the system model, i.e., from the mapping z or the set of probability densities P.

In context of our improper and idealised model, our simply measurable concentrations are *phenomenological variables*. However, in the context of a proper model, describing all molecular interaction and following, e.g., phase changes in macroscopic behaviour, concentrations will be *internal orthogonal variables* of the system. But we completely lose the simple definition of chemical potential μ_i through logarithm concentration. The relation between chemical potential and concentration may be regained only through the logic of statistical mechanics, specifically through definition of the grand canonical ensemble, and becomes rather impractical. The proper and quite unsatisfying conclusion is that even in chemistry we have the choice between a rather simplified model of low predictive value and the **phenomenological model** arising from interpolation between experimental variables in the multidimensional space of variables.

4   Properties of the Model of a Biological Experimental System

Biological systems are permanently out of equilibrium state and, moreover, non-homogeneous. In Chapter 3 was shown that even in the real world of equilibrium chemistry, our idealised models bring us only a very limited insight into the mechanism controlling the system. Nevertheless, they are good communication tools and have been a good start for following, more specific models. In order to get a similar common language, we need adequate discussion tools for biological measurements.

The most important problem is the long coexistence of several different phases in close distances and its sudden change upon a signal like that for cell division. Biological patterns have been attributed to the periodically repeating solution of the reaction-diffusion equation since the pioneering work of Turing [7,8]. Similarly, the method of cellular automata has been in parallel developed, i.e. [9]. In both cases, the final states have been only discussed. These states are not homogeneous in the standard chemical sense, but satisfy the conditions of Poincaré recurrence [10]. Such systems are structured and ergodic, which means that, over sufficiently long periods of time, all accessible microstates are achieved by the system. See, i.e., Birkhoff [11] for the exact formulation, but one must be aware of the fact that there is a vivid discussion of this problem in physics.

Fig. 2. Model of the reaction diffusion system using cellular automata [12]. The calculation results in an interchanging, dynamic image, which resembles waves, spirals, and related patterns in the Belousov-Zhabotinsky reaction, *Dicyclostelium* colony growth, and other excitable media. The parameters of the models are various rules, threshold interval, etc. The dynamics is sensitive namely to density and type of initial states.

There is an obvious physico-chemical problem with these simulations: they ignore many obvious facts, namely coexistence of multiple phases in the living organism. Also, biological systems are **non-ergodic**. For example, a living cell never visits all possible physical and chemical structures in a time between cell divisions. Quite the contrary – the cell division seems to be well controlled by various mechanism of its timing (i.e., [13]). These are two main reasons for insufficiency of contemporary models to provide qualitative basis for methodology of measurement in biological systems.

One of the most visible analogies to the behaviour of biological systems may be found in the domain of cellular automata [12,15,16]. The path through the zone of attraction to various attractor basins may be relatively easily searched through correct mathematical simulation of discrete states. Some of these findings have been tracted under the name of discrete dynamic networks by Stuart Kaufmann [17] (Fig. 3). Perhaps a bit unjustly, in the light of incompleteness of the theory enabling inclusion of the phase transition [14], the discrete dynamic networks have been criticized for giving a very little insight into the physico-chemical mechanisms generating living cells. A cellular automaton including three qualitatively different processes – the Dewndeys hodgepodge model of the Belousov-Zhabotinsky reaction [18] (Fig. 4) – is an example. The hodgepodge machine has been much less thoroughly analysed than the Turing pattern, since it is more considered as a "mathematical recreation" than a serious model.

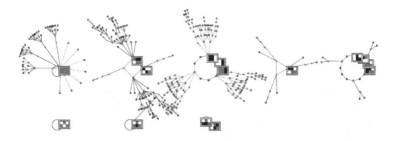

Fig. 3. Illustration of basins of attraction in a simple discrete system [12]. The basin of attraction is a periodically behaving, recurrent, dynamical system, in which structures periodically change or remain stable. In this example, we obtain just 8 non-equivalent basins of attraction. From various initial configurations through variant but defined paths, the system arrives to one of the basins of attraction. In certain cases (in a fixed point), the image remain still. In other cases, there is a periodic interchange between configurations. As seen, paths may be quite long, merge, and diverge.

Since 1990, the recognition of patterns has been extensively studied with the development of machine vision. Each of the recognition methods is based on a certain assumption about the algorithmic principles of generation of the observed image. In the least demanding case of some machine learning approaches, there is an assumption of image statistics. We believe that in this realm we shall seek inspirations for the definition of the proper model of the observed process.

We have recently contributed to the discussion on general identifiers of observed structures by introduction of a point information gain, point information gain entropy, and point information gain entropy density [4,20]. This approach is based on most general assumption of origins of observed structures both through

Fig. 4. Sketch of the basin of attraction of hodgepodge machine (calculated as implemented in the Netlogo software [19]). One of the peculiarities of the hodgepodge model is that for its proper functionality it requires only to select an interval of allowed state values (threshold range), or, better expressed, a proper relation between the number of allowed states and the stepwise "chemical reaction" component of the reaction-diffusion simulation. Therefore, it is a good example of a semi-discrete behaviour.

self-organising processes in the observed object and its projection into the dataset by the measuring device.

5   Measurement in Biological Systems

From the reasoning above we may begin to analyse conditions of the design of a proper measurement of biological systems. The main questions are:
(a) What might be the system attributes – the set A – and what might be variables – the set V – representing them?
(b) What is the proper model of the system?

We may start with the problem of discreteness. Usage of discrete entities such as agents or pixels has provided many good analogies to observations in biological and social systems. Unfortunately, the nature is not exactly discrete but only partly discrete, i.e., each animal is composed of – distinct – organs, organs of – distinct – cells etc. And these organs, cells etc. are again existent in discrete states. The immediate objection to the previous statements is that organs or cells are not as distinct as elementary pixels in the computer simulation and their states can not be described with natural numbers. However, to some extent, the biological experience does that and the biological literature is full of precise statements on cell, organ, organism, etc. states.

Similarly, as we are unable to measure at the infinite number of the time instants, we usually measure only a sub-set of possible variables. If only one of the **system trajectories** or, even worse, one of the cell states is selected and reported, a minute sub-section of the whole problem is obviously obtained. This is neither a new nor surprising problem – in Feymann's worlds "you should always decide to publish it whichever way it comes out" [21]. But in biological systems the problem is more serious, as it is complicated by the strong non-linearity which is selecting the basin of attraction by initial conditions. Thus, the scope of results is limited, sometimes in a highly reproducible manner. The problem of irreproducibility of biological results arises from "not publishing whichever comes out" further amplified by the fact that a very constrained sub-set of outcomes is obtained at given time, with given strain and set of chemicals. But that is only due to a very specific set-up when many conditions are not recorded.

In Chapter 3 we have discussed the problem of chemical activity vs. concentration with the conclusion that with the usage of an adequate model, concentration may be a good orthogonal coordinate of the system model. Just, the surface describing the multidimensional state equation $G = f(c_1, c_2, ...c_n, p, V, T, ...)$ might be quite complicated. In the case of systems highly sensitive to initial conditions we might expect an occurrence of highly divergent trajectories originating from very well determined initial conditions. And these trajectories themselves might be confined to rather small region of the phase space. In this case, we may define true phenomenological variables as a set of variables leading to the same state trajectory. In other words, the set of trajectories determining the system event B is not arbitrary, but determined by system internal behaviour – which may be also understood as the best possible model. We propose to name such a distinct system event as *a system phenomenon Φ*, where

$$\Phi \subset \Omega, \tag{19}$$

and propose that signs of the system phenomena should be defined and examined. As we explained before, there is a good, theoretically substantiated, reason that even a quite small subset of the measured system variables will give a good stochastic model. Based on this good discrimination, we may define a decomposition F of the set Ω into disjunct subsets Φ:

$$\Phi \subset F. \tag{20}$$

The stochastic system $\mathscr{S} = (T, V, P)$ may be replaced by the phenomenological stochastic system $\mathscr{F} = (T, V, P_\phi)$, where P_ϕ is the probability of transition between individual phenomena at given combination (T, V).

In case of knowledge of an appropriate non-linear model, it is enough to know the resulting trajectory at the given set of conditions. Or, in other words, we can examine the position of the border between the two zones of attraction experimentally, instead of common "constructive" examination of conditions, i.e., variable values, at which the system works in a desired way.

The state of the organism is known only if a complete phenomenological model $\mathscr{F}$ is known. This is in most cases practically impossible. It may be said with certain exaggeration that the only statistically relevant biological experiment is a record of the distribution of stock exchange indexes. Thus, we propose a different framework for reporting and analysis of biological experiments. The most precise possible description of each experimental step is a must. Under any circumstances we must anticipate that the biological object may follow a wildly different trajectory due to subtle differences in the set-up, which were not reported. The behaviour of biological systems has to be studied in the framework of mathematics of non-linear systems outside the Lyapunov stability. And this is so far almost unstudied problem, namely since it is not understood as such.

Instead of showing a biological example – which would be difficult to explain in a sketchy way – we demonstrate our idea using a much simpler example of chemical self-organisation, the Belousov-Zhabotinsky reaction. In Fig. 5 we show the performance of this well known experiment using chemicals obtained from two different providers. Although the ferroin indicator was in both cases declared to be of p.a. quality, we have obtained two wildly different self-organising structures. To give an illustrative explanation let us consider following: if the presence of the contamination in the chemical is guaranteed to be less than 0.1% , there is still 10^{19} of undetected molecules in each mole of the chemical. In case of sensitivity to initial conditions, the difference in the behaviour is not surprising.

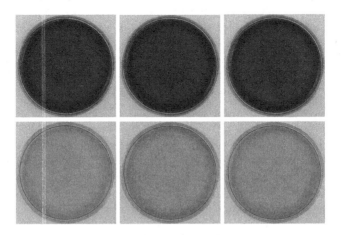

Fig. 5. The course of the Belousov-Zhabotinsky reaction as a consequence of different purity of a reactant. The experiments were performed as described in the commercially available kit [22] with the difference in a supplier of ferroin. *Upper row* – supplier Penta (the time instants indexes i were 25, 50, and 100), *lower row* – supplier Fluka ($i = 5$, 15, and 30). Distance between time instants was 10 s.

6   Conclusions

There are two main domains in any experiment – the design of the experiment and analysis of its results. In this article, we show that the usage of tools for design of experiment, which have been well established in physical and many chemical experiments is no guarantee for the proper experiment design in the biological (and many chemical) experiments.

Proper discussion of this discrepancy should be made in the frame of the theory of dynamic systems (Chapter 2). The cybernetic systems analysis anticipates that each of the experiments may be performed from the beginning, in another words that the instant t_0 of the experimental time is equal to $t = 0$ of the time of the system dynamics – e.g., on/off change of the state of an electrical switch. Or, more exactly, we assume that all time instants included in the complete immediate cause $C_{k,l}$ determining the behaviour of the system at time of the measurement t_k lie within the time extent of the set T. Also, in a standard experiment, we consider that we are able to determine values of sufficient number of variables from the set V, which allow us to build good model of the experiment.

In equilibrium physical chemistry, we rather assume the occurrence of the system on the manifold of the state equation. This also assumes that for actual values of state variables any history of the system is irrelevant, i.e., $C_{k,l} = D_{k,l}$.

In biology, we should instead a-priori assume that the complete immediate cause $C_{k,l}$ may not only contain all time instants covered by the measurement, but that it may even include some events which occurred at time instants before the experiment has started.

The contemporary unspoken common assumption is that biological system may by described as a special chemical system. Since as early as 17th century, some thinkers have been assuming that biological systems may be modelled as (in present terminology) cybernetic machines. The possibility that the biological system may be understood as an equilibrium chemical system is clearly incorrect. But also the fact that a biological system may be, to some extent, modelled as a cybernetic system comes from the biased interpretation of its non-linearity and semi-discreteness. It is for that reason that we observe only few basins of attraction and a few paths through the state space which lead to them. Finally, the path, within which the phenomenological stochastic system $\mathscr{F} = (T, V, P_\phi)$ is evolving in time, may include a much smaller part of the whole phase space than that in which a mechanical system with the Gaussian probability density function is evolving. This may be mistaken with a good reproducibility of the biological experiment when it is repeated with the same set of chemicals and within a short time interval of repetition. It leads on one hand to relative sloppiness in the definition of experimental conditions and on the other hand to bad surprises, when the experiment needs to be reproduced or transferred to production line [1,2].

The proper conclusion is that a biological experiment will never be complete, simply because we can never reverse the time and we shall not know the true starting point. One of the possibilities how to proceed in a proper analysis of the biological experiment is to seek conditions at which the system begins to follow another trajectory. It is similar to the qualitative analysis of the system of non-linear differential equations when nullclines are sought [23]. However, we must be aware of the fact that our system is in-part discrete, which means that we rather seek trajectories to the basin of attraction in a cellular automaton as shown by Wuensche [12]. This factor of semi-discretion leads to the positive role of noise in biology [24], which we may briefly describe as a constant faltering of the system, which may more frequently occupy trajectory acquiring a broader part of the phase space.

The biological system is also periodically internally re-started and re-synchronised. Let us mention the control of the bacterial cell cycle by MinD/MinE system [13] as an illustrative example. We have recently shown [25] that the method of shaking influences strongly the outcome of the self-organisation in the Belousov-Zhabotinsky reaction.

An obvious solution to the problem of measurement in biological systems is to record as many experimental outcomes as possible and publish them. This does not satisfy the human desire of understanding the system, i.e., making a model for the given observation. Yet, we may gradually come close to the suitable phenomenological description of the relatively narrow distribution of the distinct possible outcomes. This possibility comes from strong non-linearity and results in tendency to classify biological phenomena qualitatively, i.e., giving it a name such as "stress behaviour", "resting state", etc. These are the phenomena Φ discussed in Chapter 5. The persistent problem is how to define them properly. In our opinion, many jewels are hidden in **experimental protocols and their evolution**, whose analysis may lead to the proper classification and construction of a really suitable model. Experimental protocols often evolve from a simple set-up of chemical type into elaborate knowledge including provider of chemicals and many tricks, often unspoken. But only such analysis eventually leads to a successful biotechnological procedure or a relatively reproducible experiment.

Our knowledge-based data repository bioWes [26] provides solution to the problem. Its key component is the protocol generator which records the evolution of protocols. To each individual protocol is attached the respective dataset. We believe that the bioWes approach may lead to true understanding of biological systems as well as to, e.g., acceleration of development and increase of reliability of biotechnological drugs.

Acknowledgement. This work was partly supported by the Ministry of Education, Youth and Sports of the Czech Republic – projects CENAKVA (No.CZ.1.05/ 2.1.00/01.0024) and CENAKVA II (No. LO1205 under the NPU I program), by Postdok JU CZ.1.07/2.3.00/30.0006, and GAJU Grant (134/2013/Z 2014 FUUP). Authors thank to Petr Jizba, Jaroslav Hlinka, Harald Martens, Štěpán Papáček and Tomáš Náhlík for important discussions.

References

1. Prinz, F., Schlange, T., Asadullah, K.: Believe It or Not: How Much Can We Rely on Published Data on Potential Drug Targets? Nat. Rev. Drug Discov. 10, 712 (2011)
2. Begley, C.G., Ellis, L.M.: Drug Development: Raise Standards for Preclinical Cancer Research. Nature 483, 531–533 (2012)
3. Lyapunov, A.M.: The General Problem of the Stability of Motion. Kharkov Mathematical Society (1892) (in Russian)
4. Štys, D., Náhlík, T., Urban, J., Vaněk, T., Císař, P.: The Cell Monolayer Trajectory from the System State Point of View. Mol. BioSyst. 7, 2824–2833 (2011)
5. Žampa, P., Arnošt, R.: Alternative Approach to Continuous Time Stochastic Systems Defnition. In: Proc. of the 4th WSEAS Conference, Wisconsin, USA (2004)
6. Žampa, P.: Handouts for the lectures. University of West Bohemia

7. Turing, A.M.: The Chemical Basis of Morphogenesis. Philos. T. Roy. Soc. 237(641), 37–72 (1952)
8. Cross, M.C., Hohenberg, P.C.: Pattern Formation Outside of Equilibrium. Rev. Mod. Phys. 65, 851–1112 (1993)
9. Greenberg, J.M., Hastings, S.P.: Spatial Patterns for Discrete Models of Diffusion in Excitable Media. SIAM J. Appl. Math. 34, 515–523 (1978)
10. Poincare, H.: Sur le problème des trois corps et les équations de la dynamique. Acta Math. Stockh. 13, 17 (1890)
11. Birkhoff, G.D.: Proof of the Ergodic Theorem. Proc. Natl. Acad. Sci. USA 17(12), 656–660 (1931)
12. Wuensche, A.: Exploring Discrete Dynamics. Luniver Press (2011)
13. Loose, M., Fischer-Friedrich, E., Ries, J., Kruse, K., Schwille, P.: Spatial Regulators for Bacterial Cell Division Self-Organize into Surface Waves in Vitro. Science 320(5877), 789–792 (2008)
14. Gross, D.H.E.: A New Thermodynamics from Nuclei to Stars. Entropy 6, 158–179 (2004)
15. Shalizi, C.R., Shalizi, K.L., Crutchfield, J.P.: An Algorithm for Pattern Discovery in Time Series, arXiv preprint cs/0210025 (2002)
16. Crutchfield, J.P.: Between Order and Chaos. Nature Phys. 8, 17–24 (2012)
17. Kauffman, S.A.: The Origins of Order, Self-Organization and Selection in Evolution. Oxford University Press (1993)
18. Dewdney, A.K.: The Hodgepodge Machine Makes Waves. Scientific American 225, 104 (1988)
19. Wilensky, U.: NetLogo B-Z Reaction model, Center for Connected Learning and Computer-Based Modeling, Northwestern Institute on Complex Systems, Northwestern University, Evanston, IL (2003), http://ccl.northwestern.edu/netlogo/models/B-ZReaction
20. Štys, D., Korbel, J., Rychtáriková, R., Soloviov, D., Císař, P., Urban, J.: Point Information Gain, Point Information Gain Entropy and Point Information Gain Entropy Density as Measures of Semantic and Syntactic Information of Multidimensional Discrete Phenomena, http://arxiv.org/pdf/1501.02891v1.pdf
21. Feynman, R.: Surely you´re joking, Mr. Feynman. W. W. Norton & Company (1985)
22. Belousov-Zhabotinski Reaction Do-it-Yourself Kit (2010), http://drjackcohen.com/BZ01.html
23. Klipp, E., Liebermeister, W., Wierling, C., Kowald, A., Lehrach, H., Herwig, R.: Systems Biology: A Textbook. WileyVCH Verlag GmbH, Weinheim (2009)
24. Tsimring, L.S.: Noise in Biology. Rep. Progr. Phys. 77, 26601 (2014)
25. Zhyrova, A., Rychtáriková, R., Náhlík, T., Štys, D.: The Path of Aging: Self-Organisation in the Nature and the 15 Properties. Proc. Purplsoc (in press)
26. http://www.biowes.org

FRAP & FLIP: Two Sides of the Same Coin?

Štěpán Papáček[1], Jiří Jablonský[1], Ctirad Matonoha[2], Radek Kaňa[3],
and Stefan Kindermann[4]

[1] Institute of Complex Systems, South Bohemian Research Center of Aquaculture
and Biodiversity of Hydrocenoses, Faculty of Fisheries and Protection of Waters,
University of South Bohemia in České Budějovice, Zámek 136, 373 33 Nové Hrady,
Czech Republic
spapacek@frov.jcu.cz, jiri.jablonsky@gmail.com
[2] Institute of Computer Science, Academy of Sciences of the Czech Republic,
Pod Vodarenskou vezi 2, 182 07 Prague 8, Czech Republic
matonoha@cs.cas.cz
[3] Institute of Microbiology, Academy of Sciences of the Czech Republic,
Centre Algatech, 379 81 Trebon, Czech Republic
kana@alga.cz
[4] Industrial Mathematics Institute, Johannes Kepler University of Linz,
Altenbergerstr. 69, 4040 Linz, Austria
kindermann@indmath.uni-linz.ac.at

Abstract. The aim of this study is to point out the similarity and differences of data processing based on either FRAP (Fluorescence Recovery After Photobleaching) or FLIP (Fluorescence Loss In Photobleaching) experimental techniques. The core idea, closely related to the sensitivity analysis, is based on discerning between relevant and irrelevant data. Presented mathematical model allows to visualize the mutual relation between the FRAP and FLIP methods. The whole concept resides in the processing of full spatio-temporal data instead of the space averaged time series (FRAP recovery curves). The method theoretically confirms the empirical knowledge, that the mobility of fluorescent molecules can be determined with both FRAP and FLIP methods (using the full data approach). Our analysis, based on the idealized theoretical case study, supports the conclusions of our recent experiments and thus it validates the reliability of our new approach. The presented finding are expected to be reflected into experimental protocol setup as well as we will continue working on the further enhancing the method of parameter identification.

Keywords: FRAP, FLIP, sensitivity analysis, parameter identification.

1   Introduction

Imagine three interconnected activities: (i) design of a measurement technique and equipment, (ii) setting of an experimental protocol and the proper experimental data acquisition, (iii) data processing (according to some mathematical model of a presumably known underlying process(es)). The close collaboration

F. Ortuño and I. Rojas (Eds.): IWBBIO 2015, Part II, LNCS 9044, pp. 444–455, 2015.
© Springer International Publishing Switzerland 2015

between the researchers dedicated to each one of the above mentioned activities is the key factor of the final success of an interdisciplinary research project.

However, it seems, there is a gap between the design and fabrication of a sophisticated equipment for the data acquisition in biology and biomedicine and the quality of both the experimental design and further data processing, i.e., between the aforementioned points (i, ii and iii). In many cases, a large amount of data is (routinely) generated without a clear purpose. Afterwards, these data are used only qualitatively in order to illustrate some hypothetical (a priori chosen) process under consideration. In contrast, we are prone not only to repeat the experiments but also to modify the experimental protocol or even to combine several experimental techniques. We aim to establish a link between experimental (mainly empirical or data driven) results and a theoretical study (based on a simplified mathematical model) of the studied processes. Moreover, we want to quantify our results and to analyze simultaneously both the data and the experimental protocol. Our ultimate goal is to enhance or even optimize, in some sense, the whole research process. For this purpose, the BioWes system [1] has been developed.

In this paper, we focus on the photobleaching techniques and the mobility of photosynthetic proteins. We are elaborating a reliable software CA-FRAP[1] for the processing of spatio-temporal images acquired by the so-called FRAP (Fluorescence Recovery After Photobleaching) and FLIP (Fluorescence Loss in Photobleaching) methods, see e.g., [3,4,5], using confocal laser scanning microscopy (CLSM). CLSM is an advanced technique allowing to obtain high-resolution optical images with deep selectivity, rejecting the information coming from the out-of-focus planes. However, the small energy level emitted by the fluorophore and the amplification performed by the photon detector introduces a non-negligible measurement noise making the subsequent parameter identification problem highly unstable due to the high sensitivity of result on the initial data, i.e., ill-posed in Hadamard's sense [6,7]. The analysis of the ill-posedness of the parameter identification of reaction-diffusion models based on spatio-temporal FRAP images was treated in papers [8,9], as far as we know for the first time in FRAP-related literature. We note that the lemma about the data selection and the theorem exhibiting larger sensitivities (and hence smaller confidence intervals) of the parameter identification problem with full spatio-temporal data compared to that with integrated data (FRAP recovery curve) is provided in Appendix.

2   Preliminaries

2.1   FRAP & FLIP Images Acquisition and Their Biological Interpretation

In order to study the mobility of photosynthetic proteins in microalgae and cyanobacteria, the photobleaching techniques, namely FRAP and FLIP, are used

---

[1] Software CA-FRAP uses the so-called UFO system [2], for more details mail to: matonoha@cs.cas.cz.

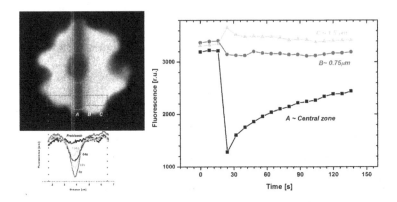

Fig. 1. Analysis of the photobleaching images revealing the phycobilisome dynamics in the red alga *P. cruentum*. Upper left: Representative image taken 8 s after bleaching. Three rectangular regions of interest (ROI) with different distances from the central zone are labeled as follows. ROI-A: the central zone, ROI-B 0.75 μm, ROI-C 1.5 μm. Lower left: The recovery dynamics in various one-dimensional bleach profiles at different times after bleaching (as labeled) is clearly seen. The curves were constructed based on two-dimensional images by averaging the data along the axes parallel to bleach stripe. On the right: three FRAP recovery curves, i.e., time dependence of space-averaged values of fluorescence in three zones (A, B, C).

by microbiologists. Both techniques are based on the measuring the change in fluorescence intensity in a region of interest (ROI) in response to a high-intensity laser pulse provided by CLSM. While the ROI in FRAP is inside the bleached area, the ROI in FLIP is somewhere outside the bleached area. We suppose that the laser pulse (bleach) causes an irreversible loss in fluorescence in the bleached area without any damage to intracellular structures. After the bleach, we observe either the recovery or loss in fluorescence in ROI corresponding to FRAP or FLIP, respectively. Based on the spatio-temporal FRAP data, the effective diffusion coefficient reflecting the mobility of photosynthetic proteins of different microbial species is estimated using either a closed form model [3,10,11] or simulation based model [5,8,9,12,13]. While the former kind of models needs some conditions to be assumed, the latter is more general although computationally more expensive.

We aim to reveal the phycobilisomes mobility and the regulation of microalgae light-harvesting system in general. Therefore, we analyze the phycobilisome dynamics in several red alga strains [14]. The upper left part in Fig. 1 shows the first post-bleach image taken 8 s after application of high laser intensity across the vertical axis reducing the phycobilisome fluorescence in *P. cruentum* to about 40% of the initial value due to the destruction of a portion of the phycobilin pigments.

The lower left part in Fig. 1 represents a one-dimensional projection of fluorescence changes along the axes parallel to bleach stripe in a selected area of the thylakoid membrane. There is a central zone (labeled as A) flanked by two rectangular areas with distances 0.75 μm (B), and 1.5 μm (C) from the center, respectively. While the phycobilisome fluorescence in the central zone A slowly recovered with time after bleaching, fluorescence intensity out of the bleached area (especially in zone C – 1.5 μm) decreased with time after bleaching. This loss of fluorescence out of the central zone represents an additional proof of the diffusive exchange of non fluorescent phycobilisomes from the central (bleached) zone with fluorescent phycobilisomes from the outer nonbleached zones. At a distance of about 0.75 μm from the central zone, the movement of fluorescent phycobilisomes into and out of the 0.75 μm zone was equilibrated, resulting in a constant (space-averaged) fluorescence during the whole time-lapse experiment, i.e., there is low or zero sensitivity. This latter result points out the importance of an adequate choice of ROI and corresponds to the existence of a data space, where the signal sensitivity is low, see Fig. 2 for the illustration of this concept, and Section 3 for the analytical justification.

2.2   FRAP & FLIP Data Structure and (Pre)Processing

The data structure related to the FRAP & FLIP photobleaching techniques consists of a time series of rectangular matrices (2D fluorescence profiles), where each entry quantifies the fluorescence intensity u at a particular spatial point in a finite 2D domain (e.g., by a number between 0 and 255):

$$u(x_l, t_j)_{l=1}^{N_x}, \quad j = 0, ..., N_t,$$

where l is the spatial index uniquely identifying the pixel position where the signal u is measured and j is the time index (the initial condition corresponds to $j = 0$). Usually, the measured points are uniformly distributed both in time and space, i.e., on an equidistant 2D mesh. In the following we adopt the simplified notation consisting in using only one index i for all data in the space-time domain, i.e., $u(x_i, t_i)_{i=1}^{N_{\text{Data}}} \in \mathbb{R}^{N_{\text{Data}}}$.

Almost without exception, the experimental biologists use the so-called FRAP recovery curve, i.e., the space averaged fluorescence signal, instead of the spatio-temporal data $u(x_i, t_i)$, see e.g., Fig. 1 (right). The subsequent calculation of the effective diffusion coefficient is then simpler. On the other hand, the sensitivity analysis and error analysis of respective methods is often misunderstood, cf. [13]. This is the reason why we present a rigorous study of the data processing step in Section 3. We also compare the *full data approach* with the *integrated data approach*, cf. Theorem 1 in Appendix.

2.3   Reaction-Diffusion System as Initial Boundary Value Problem

Consider the general reaction-diffusion equation for a previously not fixed number of interacting components whose concentration profiles are $(u_i(x,t))_{i=1}^n$, e.g., all mobile with different time-dependent diffusion coefficients $D_i{}^2$

$$\frac{\partial}{\partial t} u(x,t) = D\Delta u(x,t) - Ku(x,t) , \tag{1}$$

where $u(x,t) = [u_1(x,t), \ldots, u_n(x,t)]^T$ is a vector of concentration profiles, D is a diagonal matrix and K is a (singular) matrix of reaction rates. Boundary conditions could be, e.g.,

$$u(x,t) = 0, \quad \text{or} \quad \frac{\partial}{\partial n} u(x,t) = 0 \quad \text{on } \partial\Omega \times [0,T]. \tag{2}$$

For two components we get

$$\frac{\partial}{\partial t} u_1 = D_1(t)\Delta u_1 - k_{\text{on}} u_1 + k_{\text{off}} u_2$$

$$\frac{\partial}{\partial t} u_2 = D_2(t)\Delta u_2 - k_{\text{off}} u_2 + k_{\text{on}} u_1$$

with initial conditions $u_i(x,0) = u_{i0}\phi(x)$, $i \in \{1,2\}$, where $\phi(x)$ is some given initial shape. The matrix of reaction rates is now $K = \begin{bmatrix} k_{\text{on}} & -k_{\text{off}} \\ -k_{\text{on}} & k_{\text{off}} \end{bmatrix}$, k_{on} and k_{off} are the association and dissociation rates, respectively, cf. [15].

The model parameter estimation problem is further formulated as an ordinary least squares problem (4) in the following section. The detailed analysis is presented elsewhere [8,16]. Here, we need the uncertainty assessment which is based on the evaluation of the Fisher information matrix.

3   Model Parameter Estimation and Sensitivity Analysis

3.1   Parameter Estimation Based on Spatio-temporal Data

We aim to present a parameter estimation problem with spatio-temporal experimental observation in a comprehensive mathematical framework allowing simultaneously to determine both the parameter value p (generally $p \in \mathbb{R}^q$, $q \in \mathbb{N}$) and the corresponding confidence interval proportional to the output noise and a quantity related to the sensitivity, see (7). The data are represented by a (measured) signal on a Cartesian product of the space-points $(x_i)_{i=1}^n$ and time-points $(t_j)_{j=1}^m$; let $N_{\text{Data}} := m \times n$ be the total number of spatio-temporal data points.

---

[2] The first attempt to identify both model type and model parameters in inverting FRAP data, i.e., to get along more or less justified assumptions about model type or parameters, is presented in [15].

We define the operator $S : \mathbb{R}^q \to \mathbb{R}^{N_{\text{Data}}}$ that maps parameter values $p_1, \ldots, p_q$ to the solution of the underlying initial-boundary value problem, e.g. (1-2), evaluated at points (x_i, t_j): $S(p) = \{u(x_i, t_j, p) \in \mathbb{R}, \quad 1 \leq i \leq n, \quad 1 \leq j \leq m\}$.

Some commonly used FRAP methods do not employ all the N_{Data} measured values at points $\{(x_i, t_j), \; i = 1, \ldots, n, \; j = 1, \ldots, m\}$. They either employ some of the values or perform some preprocessing, e.g. space averaging, see [11,16]. Hence, we further define the observation operator $G : \mathbb{R}^{N_{\text{Data}}} \to \mathbb{R}^{N_{\text{data}}}$ that evaluates the set of values $S(p)$ on a certain subset of the full data space ($N_{\text{data}} \leq N_{\text{Data}}$): $G(S(p)) = (z(x_l, t_l, p))_{l=1}^{N_{\text{data}}}$.

We now define the forward map $F : p \to z(x_l, t_l, p)_{l=1}^{N_{\text{data}}}$. Here, $F = G \circ S$ represents the parameter-to-output map, defined as the composition of the PDE solution operator S and the observation operator G.[3] Our regression model is now

$$F(p) = \text{data}, \tag{3}$$

where the data are modeled as contaminated with additive Gaussian noise

$$\text{data} = F(p_T) + e = (z(x_l, t_l, p_T))_{l=1}^{N_{\text{data}}} + (e_l)_{l=1}^{N_{\text{data}}}.$$

Here $p_T \in \mathbb{R}^q$ denotes the true values and $e \in \mathbb{R}^{N_{\text{data}}}$ is a data error vector which we assume to be normally distributed with variance σ^2, i.e., $e_i = \mathcal{N}(0, \sigma^2)$, $i = 1, \ldots, N_{\text{data}}$.

Given some data, the aim of the parameter estimation problem is to find p_T, such that (3) is satisfied in some appropriate sense. Since (3) usually consists of an overdetermined system (there are more data points than unknowns), it cannot be expected that (3) holds with equality, but instead an appropriate notion of a solution is that of a least-squares solution $\hat{p}$ (with $\| \, . \, \|$ denoting the Euclidean norm on $\mathbb{R}^{N_{\text{data}}}$):

$$\| \, \text{data} - F(\hat{p}) \, \|^2 = \min_p \| \, \text{data} - F(p) \, \|^2. \tag{4}$$

3.2 Sensitivity Analysis and Confidence Intervals

For the sensitivity analysis we require the Fréchet-derivative $F'[p_1, \ldots, p_q] \in \mathbb{R}^{N_{\text{data}} \times q}$ of the forward map F, that is

$$F'[p_1, \ldots, p_q] = \left(\frac{\partial}{\partial p_1} F(p_1, \ldots, p_q) \; \cdots \; \frac{\partial}{\partial p_q} F(p_1, \ldots, p) \right)$$

$$= \begin{pmatrix} \frac{\partial}{\partial p_1} z(x_1, t_1, p) & \cdots & \frac{\partial}{\partial p_q} z(x_1, t_1, p) \\ \cdots & \cdots & \cdots \\ \cdots & \cdots & \cdots \\ \frac{\partial}{\partial p_1} z(x_{N_{\text{data}}}, t_{N_{\text{data}}}, p) & \cdots & \frac{\partial}{\partial p_q} z(x_{N_{\text{data}}}, t_{N_{\text{data}}}, p) \end{pmatrix}.$$

---

[3] For the one-point Mullineaux method [3], only the point with the spatial coordinate $x = 0$ is measured, i.e., $G_M : z(t_j, p) := z(0, t_j, p) = u(0, t_j, p)$, $j = 1, \ldots, N_{\text{data}} = m$. For the second method, we reduce the data space taking the so-called relevant data only [16], i.e., $G_{\text{PDE}} : z(x_l, t_l, p) = u(x_i, t_j, p)$, $i = 1, \ldots, n^* \leq n$, $j = 1, \ldots, m^* \leq m$, $l = 1, \ldots, N_{\text{data}} = m^* \times n^*$.

A corresponding quantity is the Fisher information matrix (FIM)

$$M[p_1, \ldots, p_q] = F'[p_1, \ldots, p_q]^T F'[p_1, \ldots, p_q] \in \mathbb{R}^{q \times q}. \tag{5}$$

Based on the book of Bates and Watts [17], we can estimate confidence intervals. Suppose we have computed $\hat{p}$ as a least-squares solution in the sense of (4). Let us define the residual as

$$res^2(\hat{p}) = \|F(\hat{p}) - \text{data}\|^2 = \sum_{i=1}^{N_{\text{data}}} [\text{data}_i - z(x_i, t_i, \hat{p})]^2. \tag{6}$$

Then according to [17], it is possible to quantify the error between the computed parameters $\hat{p}$ and the true parameters p_T.

Having only one single scalar parameter p as unknown (e.g., for one diffusion coefficient of one component system), the Fisher information matrix M collapses into the scalar quantity $\sum_{i=1}^{N_{\text{data}}} \left[\frac{\partial}{\partial p} z(x_i, t_i, p) \mid_{p=\hat{p}} \right]^2$, and the $1 - \alpha$ confidence interval for full observations is described as follows

$$(\hat{p} - p_T)^2 \sum_{i=1}^{N_{\text{data}}} \left[\frac{\partial}{\partial p} z(x_i, t_i, p) \mid_{p=\hat{p}} \right]^2 \leq \frac{res^2(\hat{p})}{N_{\text{data}} - 1} f_{1, N_{\text{data}}-1}(\alpha), \tag{7}$$

where $f_{1, N_{\text{data}}-1}(\alpha)$ corresponds to the upper α quantile of the Fisher distribution with 1 and $N_{\text{data}} - 1$ degrees of freedom. In (7), several simplifications are possible. Note that according to our noise model, the residual term $\frac{res^2(\hat{p})}{N_{\text{data}}-1}$ is an estimator of the error variance [17] such that the approximation $\frac{res^2(\hat{p})}{N_{\text{data}}-1} \sim \sigma^2$ holds if N_{data} is large. Moreover, we remind the reader that the Fisher distribution with 1 and $N_{\text{data}} - 1$ degrees of freedom converges to the χ^2-distribution as $N_{\text{data}} \to \infty$. Hence, the term $f_{1, N_{\text{data}}-1}(\alpha)$ can approximately be viewed as independent of N_{data} as well and of a moderate size.

3.3   Theoretical Results on Data Space Selection and DOE

In FRAP community, there are many rather empirical recommendations related to the design of experiment, e.g., how to set the bleach shape and size, the monitored region location and size, the time span of the measurement, cf. [5,11]. However, based on the sensitivity analysis (by maximizing the sensitivity of the measured output on the estimated parameter) we would have a tool for the optimal DOE (design of experiments), i.e., for an optimal choice of some design factors. There is also another issue with great practical relevance (mainly when the cost of observations and experimental runs can not be neglected) residing in an adequate choice of the observation region. This issue is closely related with the amount of data to process. The key concept relies on the idea to select the data space where the sensitivity is "sufficiently high".

For all the results following, we always consider only the constant diffusivity p as unknown, and focus on the one-dimensional Fick diffusion of one component only. Hence, (7) is the central estimate which we use in the sequel.

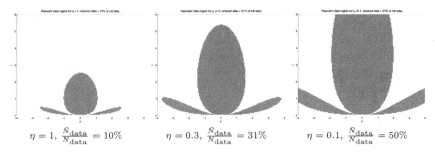

$\eta = 1,\ \dfrac{\bar{N}_{\text{data}}}{N_{\text{data}}} = 10\%$ $\eta = 0.3,\ \dfrac{\bar{N}_{\text{data}}}{N_{\text{data}}} = 31\%$ $\eta = 0.1,\ \dfrac{\bar{N}_{\text{data}}}{N_{\text{data}}} = 50\%$

Fig. 2. Data space selection according to the sensitivities. Gray regions correspond to regions where (for a previously chosen threshold) the relevant data are taken. White regions represent irrelevant data points, i.e., the region where the sensitivity is lower than a certain value. The factor $1 + \eta$, cf. (11), and the ratio $\frac{\bar{N}_{\text{data}}}{N_{\text{data}}}$ indicate the enlargement of the confidence interval of the estimated parameter, and the fraction of (relevant) data points taken over all data points, respectively.

We mention that most of the analysis can be extended to the case of more unknown parameters and to the two-dimensional case (i.e., diffusion on a surface) as well. However, the case of an unknown non-constant diffusivity is out of the scope of this paper.

The confidence interval estimate (7) gives us useful information on the quality of a least-squares estimate. The sensitivity (or FIM) is the central ingredient in this estimate, being the main factor controlling the error $|\hat{p} - p_T|$. As a next step, starting from (7), we might change the data observation and consider the question if we can use less data without increasing the error too much.

Now let us assume that we have a different kind of data (i.e., less data points, also called *reduced data* in the following)

$$u(\bar{x}_i, \bar{t}_i)_{i=1}^{\bar{N}_{\text{data}}} \in \mathbb{R}^{\bar{N}_{\text{data}}},$$

where

$$\bar{N}_{\text{data}} < N_{\text{data}} \quad \text{and} \quad \{(\bar{x}_i, \bar{t}_i)\} \subset \{(x_i, t_i)\}$$

and we compute a least-squares estimate $\bar{p}_c$ using these data. Without loss of generality let us impose the following ordering of the data $\{(\bar{x}_i, \bar{t}_i)_{i=1}^{\bar{N}_{\text{data}}}\} = \{(x_i, t_i)_{i=1}^{\bar{N}_{\text{data}}}\}$, i.e., that the new data are just the first $\bar{N}_{\text{data}}$ values of the original data.

The corresponding confidence interval estimate (7) still holds with a (usually larger) confidence interval

$$(\bar{p}_c - p_T)^2 \sum_{i=1}^{\bar{N}_{\text{data}}} (\frac{\partial}{\partial p} u(x_i, t_i))^2 \leq \frac{res^2(\bar{p}_c)}{\bar{N}_{\text{data}} - 1} f_{1, \bar{N}_{\text{data}} - 1}(\alpha). \tag{8}$$

Note that for the residual we now have to take the Euclidean norm in $\mathbb{R}^{\bar{N}_{\text{data}}}$ in the corresponding expression (6), i.e., a sum of the form $\sum_{i=1}^{\bar{N}_{\text{data}}}$.

In [16] we developed a novel approach aiming to reduce the amount of data without shortening the confidence interval. The main result is presented in Appendix, cf. (11). We see almost the same type of estimates as in the full data case (7), but the upper bound is enlarged by the factor $(1 + \eta)$, where $\eta > 0$ is chosen according to (10). The whole idea and its utility is demonstrated in the following example.

Example 1. In this example we visualize our previous theoretical results, see Fig. 2. To do so, we compute a least-squares estimate $\hat{p}$ using different kind of data. Assume the Fick diffusion equation for 1D domain, i.e., $\frac{\partial}{\partial t} u(x, t) = p\Delta u(x, t)$, $x \in \mathbb{R}$ and $u(x, 0) = u_0(x) = u_{0,0} e^{-\frac{2x^2}{r_0^2}}$. Here $u_{0,0} \geq 0$ is the maximum depth at time t_0 for $x = 0$, and $r_0 > 0$ is the half-width of the bleach at normalized height (depth) e^{-2}, i.e. $\frac{u_0(r_0)}{u_{0,0}} = e^{-2}$, cf. [3]. In a first experiment we consider a rectangular spatio-temporal data grid with space interval $x_i \in [-4, 4]$ and time interval $t_i \in [0, 6]$. The grid size in both space and time direction was set to $\Delta x = \Delta t = 0.05$. We simulated data by assuming an exact diffusion coefficient $p_T = 2$ with bleach radius $r_0 = 0.1$ and computed the data for the 1D case. These data were perturbed by normally distributed additive noise (white Gaussian noise) with standard deviation $\sigma = 0.05$. Based on these data we computed a least-squares estimate $\hat{p}$ of the diffusion coefficient using a Gauss-Newton procedure. We refer to this setup as the full data case.

Next, we computed a similar estimate using less data. For this, we considered the data point selection criterion (10) for various values of $\eta = 1, 0.3, 0.1$. The data regions were selected by first ordering all data points according to their sensitivities $(\frac{\partial}{\partial p} u(x_1, t_1))^2 \leq (\frac{\partial}{\partial p} u(x_2, t_2))^2 \leq \dots$ Then we selected the reduced data from the region of points (x_i, t_i), $i = 1, \dots, \bar{N}_{\text{data}}$ using this ordering and the smallest $\bar{N}_{\text{data}}$ such that (10) holds for the choice of η. In a sense, we take the most sensitive data points as relevant data. In Fig. 2, the gray region corresponds to that region where the data points were selected and the complementary region, the white one, represents irrelevant data that can be neglected without much loss of quality. Note that for the case $\eta = 0.1$ (third column), we have virtually the same standard error as for the full data case even though we have only used 50% of the original data.

This *lotus flower* like plot corresponds quite well with the experimental findings, cf. the subsection 2.1. For instance, the relevant data points for the single bleach FRAP method are distributed in a narrow central region. Conversely, the FLIP relevant data region is outside the bleached area and due to the loss of sensitivity for the increasing time since the bleaching the necessity of repeated bleaching is evident.

4   Concluding Remarks

In this paper we emphasize the benefits bringing the interconnection of three activities in biology and biomedicine: (i) analysis and choice of a specific measurement technique, e.g., FRAP & FLIP, (ii) design of experiment, i.e., setting of an experimental protocol, (iii) development of a reliable mathematical model, e.g., reaction-diffusion model. This holistic approach can be very rewarding.[4]

The key concept of the presented approach is the sensitivity analysis. It allows to solve both the problem of parameter estimation in the reaction-diffusion systems (in terms of mean values and confidence intervals) and the problem of optimal data space selection (taking only the relevant data). Moreover, with a reliable model, the experimental factors or variables (e.g., the bleach size) can be optimized as well.

In a toy example we can see that the empirical recommendations concerning FRAP & FLIP are justified, i.e., we see the necessity of repeated bleaching in FLIP. As a by-product, we present the theorem claiming that the data set represented by the FRAP recovery curves (the integrated data approach) leads to a larger confidence interval compared to the spatio-temporal data. This statement should be promoted in the FRAP community, because it goes against their common knowledge and professional experience.

For the photosynthetic proteins mobility studies, the relevant data correspond to both FRAP and FLIP region of interest (see gray regions in Fig. 2). Nevertheless, in contrast to the single bleach FRAP method, the FLIP relevant data region is smaller due to the loss of sensitivity for the increasing time from the bleaching. We strongly suggest to use FLIP data as a control data set. A possible discrepancy between the estimated parameters points out the existence of so far unconsidered process in FRAP data. We are working on identification of this process as well as on improvement of experimental techniques. Once proven the concept, it will be implemented into the special software processing the photobleaching data.

Acknowledgement. This work was supported by the Ministry of Education, Youth and Sports of the Czech Republic – projects CENAKVA (No. CZ.1.05/ 2.1.00/01.0024) and CENAKVA II (No. LO1205 under the NPU I program), by TA ČR TA01010214, by Postdok JU (CZ.1.07/2.3.00/30.0006), and by the long-term strategic development financing of the Institute of Computer Science (RVO: 67985807). Work of R.K. has been supported by an institutional project ALGATECH plus (MSMT, LO1416) and by project GACR (no. P501-12-0304).

References

1. http://www.biowes.org/
2. Lukšan, L., Tůma, M., Matonoha, C., Vlček, J., Ramešová, N., Šiška, M., Hartman, J.: UFO 2014 - Interactive system for universal functional optimization. Technical Report V-1191, Institute of Computer Science, Academy of Sciences of the Czech Republic, Prague (2011), http://www.cs.cas.cz/luksan/ufo.html

---

[4] The communication between the members of mathematical and biological community is complicated, but there exist some successful stories. The condition *sine qua non* is the Richard P. Feynman first principle: "...you must not fool yourself..." [18].

3. Mullineaux, C.W., Tobin, M.J., Jones, G.R.: Mobility of photosynthetic complexes in thylakoid membranes. Nature 390, 421–424 (1997)

4. Mueller, F., Mazza, D., Stasevich, T.J., McNally, J.G.: FRAP and kinetic modeling in the analysis of nuclear protein dynamics: What do we really know? Current Opinion in Cell Biology 22, 1–9 (2010)

5. Sbalzarini, I.F.: Analysis, Modeling and Simulation of Diffusion Processes in Cell Biology. VDM Verlag Dr. Muller (2009)

6. Hadamard, J.: Lectures on the Cauchy Problem in Linear Partial Differential Equations. Yale University Press, New Haven (1923)

7. Engl, H., Hanke, M., Neubauer, A.: Regularization of Ill-posed Problems. Kluwer, Dortrecht (1996)

8. Papáček, Š., Kaňa, R., Matonoha, C.: Estimation of diffusivity of phycobilisomes on thylakoid membrane based on spatio-temporal FRAP images. Mathematical and Computer Modelling 57, 1907–1912 (2013)

9. Kaňa, R., Matonoha, C., Papáček, Š., Soukup, J.: On estimation of diffusion coefficient based on spatio-temporal FRAP images: An inverse ill-posed problem. In: Chleboun, J., Segeth, K., Šístek, J., Vejchodský, T. (eds.) Programs and Algorithms of Numerical Mathematics 16, pp. 100–111 (2013)

10. Axelrod, D., Koppel, D.E., Schlessinger, J., Elson, E., Webb, W.W.: Mobility measurement by analysis of fluorescence photobleaching recovery kinetics. Biophys. J. 1069, 1055–1069 (1976)

11. Ellenberg, J., Siggia, E.D., Moreira, J.E., Smith, C.L., Presley, J.F., Worman, H.J., Lippincott-Schwartz, J.: Nuclear membrane dynamics and reassembly in living cells: targeting of an inner nuclear membrane protein in interphase and mitosis. The Journal of Cell Biology 138, 1193–1206 (1997)

12. Irrechukwu, O.N., Levenston, M.E.: Improved Estimation of Solute Diffusivity Through Numerical Analysis of FRAP Experiments. Cellular and Molecular Bioengineering 2, 104–117 (2009)

13. Papáček, Š., Jablonský, J., Matonoha, C.: On two methods for the parameter estimation problem with spatio-temporal FRAP data. In: Chleboun, J., Segeth, K., Šístek, J., Vejchodský, T. (eds.) Programs and Algorithms of Numerical Mathematics 17, pp. 100–111 (2015)

14. Kaňa, R., Kotabová, E., Lukeš, M., Papáček, Š., Matonoha, C., Liu, L.N., Prášil, O., Mullineaux, C.W.: Phycobilisome Mobility and Its Role in the Regulation of Light Harvesting in Red Algae. Plant Physiology 165(4), 1618–1631 (2014)

15. Mai, J., Trump, S., Ali, R., Schiltz, R.L., Hager, G., Hanke, T., Lehmann, I., Attinger, S.: Are assumptions about the model type necessary in reaction-diffusion modeling? A FRAP application. Biophys. J. 100(5), 1178–1188 (2011)

16. Kindermann, S., Papáček, Š.: On data space selection and data processing for parameter identification in a reaction-diffusion model based on FRAP experiments. Submitted to Abstract and Applied Analysis (2015)

17. Bates, D.M., Watts, D.G.: Nonlinear regression analysis: Its applications. John Wiley & Sons, New York (1988)

18. Feynman, R.P.: Surely You're Joking, Mr. Feynman!: Adventures of a Curious Character. W.W.Norton & Company (1997)

Appendix

Full Spatio-Temporal Data vs. FRAP Curve

A common procedure in the evaluation of photobleaching experiments is the *integrated data approach* leading to the space averaged FRAP curve, i.e., the signal $v(t_j)_{j=1}^{N_t}$, cf. (9), instead of the spatio-temporal data $u(x_l, t_j)$, is used for the further parameter identification. The comparison of both approaches in view of sensitivity and confidence intervals is presented as the following Theorem 1.

Theorem 1. Let us consider that the full data $u(x_l, t_j)$, $l = 1, \ldots, N_x$, $j = 1, \ldots, N_t$ are given on a rectangular spatio-temporal grid, where N_x is the number of points in the monitored region (either 2D or 1D finite domain) and N_t is the number of time points. The (space-)integrated data

$$v(t_j) = \frac{1}{N_x} \sum_{l=1}^{N_x} u(x_l, t_j), \quad j = 1, \ldots, N_t \qquad (9)$$

are given on a temporal grid. Assume that the data error is normally distributed and that the number of data points is large enough such that $\frac{res^2(D_c)}{N_{\text{data}}-1}$ is almost equal to the data error variance in either case and that the Fisher-quantile satisfy $\frac{f_{1, N_{\text{data}}-1}(\alpha)}{f_{1, N_t-1}(\alpha)} \sim 1$. Then the full data approach has equal or smaller confidence interval bounds for the parameter estimates of $\hat{p}_T$.

The explanation resides in fact that although the data error variance is smaller for the integrated data case, the sensitivity is smaller as well, and this more than compensates the benefit of a smaller variance, see [16] for more details.

Relevant vs. Irrelevant Data

Lemma 1. Let $\bar{N}_{\text{data}}$ be the number of reduced (relevant) data points and suppose that η is chosen according to

$$\eta \geq \frac{\sum_{i=\bar{N}_{\text{data}}+1}^{N_{\text{data}}} (\frac{\partial}{\partial p} u(x_i, t_i))^2}{\sum_{i=1}^{\bar{N}_{\text{data}}} (\frac{\partial}{\partial p} u(x_i, t_i))^2}. \qquad (10)$$

Then using reduced data, we find a confidence interval of the form

$$(\bar{p}_c - p_T)^2 \sum_{i=1}^{N_{\text{data}}} (\frac{\partial}{\partial p} u(x_i, t_i))^2 \qquad (11)$$

$$\leq \left[\frac{f_{1, \bar{N}_{\text{data}}-1}(\alpha)}{f_{1, N_{\text{data}}-1}(\alpha)} \right] (1+\eta) \left(\frac{res^2(\bar{p}_c, R_c)}{\bar{N}_{\text{data}} - 1} f_{1, N_{\text{data}}-1}(\alpha) \right).$$

The proof is given in [16]. The relevance of this observation is that we have almost the same type of estimates as in the full data case, cf. (7), but we might ignore some (irrelevant) data points. More precisely, if (10) holds with small η, we call the new data $u(\bar{x}_i, \bar{t}_i)_{i=1}^{\bar{N}_{\text{data}}}$ the *relevant data* and the complement $\{u(x_i, t_i)_{i=1}^{N_{\text{data}}}\} \setminus \{u(\bar{x}_i, \bar{t}_i)_{i=1}^{\bar{N}_{\text{data}}}\}$ the *irrelevant data*.

MicroRNA Target Prediction Based Upon Metastable RNA Secondary Structures

Ouala Abdelhadi Ep Souki[1], Luke Day[1], Andreas A. Albrecht[2],
and Kathleen Steinhöfel[1]

[1] King's College London, Department of Informatics
[2] Middlesex University London, School of Science and Technology
{ouala.abdelhadi_ep_souki,luke.day,kathleen.steinhofel}@kcl.ac.uk,
a.albrecht@mdx.ac.uk

Abstract. In this work, we present `RNAStrucTar`, a miRNA target prediction tool that analyses putative mRNA binding sites within 3'UTR secondary structures representing metastable conformations. The first stage consists of generating conformations that can be classified as deep local minima. The second stage incorporates duplex structure prediction through sequence alignment and energy computation. Target site accessibility related to different sets of metastable conformations is also taken into account. An overall interaction score computed from multiple binding sites is returned. The approach is discussed in the context of single nucleotide polymorphisms (SNPs). We selected 20 instances of type [mRNA;SNP;miRNA] reported in recent literature where methods such as PCR and/or luciferase reporter assays are utilised. If the two main scores returned by `RNAStrucTar` are combined, 16 instances are correctly classified according to experimental findings from the literature, with two false classifications and two indifferent outcomes. When additionally combined with `STarMir` results (14 correct, but partly on different instances), then at least one of both methods supports the experimental findings on 18 instances, with one indifferent outcome and one prediction in favour of the experimentally established weaker binding.

Keywords: microRNA, target prediction, RNA secondary structures, metastable conformations, single nucleotide polymorphisms.

1   Introduction

MicroRNAs (miRNAs) are short non-coding RNAs known to possess important post-transcriptional regulatory roles. They bind to messenger-RNAs (mRNAs) that contain specific complementary target sub-sequences and this way blocking the mRNA translation into proteins [3]. Hundreds of targeted genes associated with cancer, cardiovascular disease, viral infections and other diseases have been experimentally verified [20,36]. However, the number of discovered miRNAs is increasing [22], and each miRNA is thought to regulate a few hundred targeted genes in mammals [37]. Therefore, the problem of finding genes that are regulated by miRNAs is of a great importance to a better understanding of biological processes. Identifying experimentally miRNA targets is a laborious process

F. Ortuño and I. Rojas (Eds.): IWBBIO 2015, Part II, LNCS 9044, pp. 456–467, 2015.
© Springer International Publishing Switzerland 2015

with time-consuming and expensive experiments, and therefore computational methods for miRNA target prediction are applied for narrowing down potential candidates for experimental validation. The first computational target prediction tools were proposed in 2003 [13,23], and over the past decade a variety of computational methods has been developed for identifying putative targets of miRNAs.

Computational methods perform the predictions based upon features extracted from experimental data. Early target prediction algorithms primarily focus on sequential features, such as PicTar [2], TargetScan [23], and miRanda [13]. Most methods require the target sequence to have a near perfect complementarity to a region in the miRNA sequence, which is named seed and defined by the first 2–8 nucleotides, starting at the 5' end. The prediction accuracy can be improved by taking into account evolutionary conservation of binding sites in both sequences [33]. Thermodynamic stability of miRNA–mRNA duplex structures is also one of the most frequently used features, where the overall change in Gibbs free energy is employed as an indicator of how strongly bound the sequence pair is.

Advanced methods take into account the secondary structure of mRNAs. One of the key concepts underlying such tools is the accessibility of potential binding sites. Some tools avoid intra-molecular base pairing by omitting the computation of folded structures within the monomers and therefore rely almost exclusively on the free energy of the duplex formation. The assumption that the mRNA is in linear form certainly reduces the computational complexity. However, recent studies suggest that this assumption describes only part of the binding process and that prediction tools can be improved by incorporating the folded structure of the mRNA into the prediction algorithm [21,34]. In reality, either the binding site must not be involved in any base pairing with other parts of the same mRNA, or there should be an energetic penalty for freeing base pairing interactions within the mRNA in order to make the target site accessible for the binding. This energy cost has to be considered as a part of the total hybridization energy [19,25]. More detailed reviews of features used in existing miRNA target prediction tools can be found in [1,4,31,33].

The standard assumption in miRNA target predictions is that the functional state of the mRNA is the minimum free energy (MFE) structure. However, recent literature argues in favour of the existence of multiple active RNA conformations instead of a unique MFE conformation as the single biologically active state [14]. Long et al. [25] overcame the limitation of MFE structure prediction by averaging over 1,000 structures sampled from a statistically representative sample from the Boltzmann-weighted ensemble of RNA secondary structures by using Sfold [12]. In [21], the authors compute the accessibility in relation to the probability that the target region is unpaired in thermodynamic equilibrium and, additionally, based on the ensemble of all possible structures in thermodynamic equilibrium, where RNAfold [18] is utilised. Marín and J. Vaníček [28] argue that considering only the MFE structure neglects the possibility that the miRNA binds to a 3'UTR structure with a slightly higher energy than the MFE structure, but with better accessibility. The authors compute pair probabilities from the canonical ensemble of secondary structures generated by RNAplfold [5] and obtain comparable or better results obtained for MFE structures only.

Within the past few years, analysing concentration levels of miRNAs and their putative targets has become a major topic in miRNA research. In [35], the authors provide experimental evidence that the typical number of gene copies present in a single cell lies between 5–20. with most genes having less than a 100 copies. Ragan *et al.* [32] published results on miRNA target predictions that utilize information about miRNA and mRNA concentration levels. For the miRNA target prediction tool TargetScan, Garcia *et al.* [9] demonstrate how predictions may improve if target abundance is accounted for in binding scores.

Our work [10] assumes the existence of multiple active RNA structures different from the MFE. We studied the problem of miRNA bindings to metastable secondary structures in the context of Single Nucleotide Polymorphisms (SNPs) and mRNA concentration levels. Our analysis showed that the number of metastable structures and features of miRNA bindings to metastable conformations could provide additional information supporting the differences in expression levels of mRNAs and their corresponding SNP variants. As a consequence, miRNA target predictions using metastable conformations in a pre-processing step of wet lab experiments may improve the confidence about expected miRNA-mRNA bindings. RNAStrucTar, the method described in the present paper, is a new miRNA target prediction tool that analyses putative mRNA binding sites - in contrast to [10] - by a specific energy evaluation of duplex structures based upon secondary structures representing metastable conformations.

2   Methods

Our work assumes the existence of multiple active RNA conformations instead of a unique MFE conformation as the single biologically active state. The second basic feature of our approach relates to the presence of multiple copies of each individual mRNA. There are two main stages in RNAStrucTar: The first stage is the generation of metastable conformations, and the second stage comprises of miRNA target prediction based upon an energy assessment that incorporates target accessibility related to an input set of secondary structures. A flowchart describing the particular steps is given in Figure 1.

2.1   Metastable Secondary Structures

Metastable secondary structures are generated by using standard tools provided by the Vienna RNA server [16]. The RNAsubopt tool by Wuchty *et al.* [38] generates all suboptimal foldings of a sequence in a partial energy landscape defined by an energy range ΔE above the MFE structure. A method to elucidate the basin structure of landscapes by means of tree-structures representing local minima and their connecting saddle points is provided by the Barriers program [15]. The input to Barriers is a list of conformations sorted by energy values.

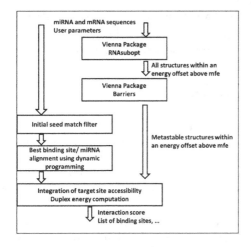

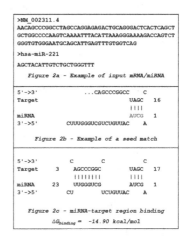

Fig. 1. RNAStrucTar flowchart   **Fig. 2.** RNAStrucTar example

The RNAsubopt tool together with the Barriers program allows the user to identify the set of metastable conformations MS within an energy range ΔE above the MFE conformation. We denote the number of local minima by ms = |MS|.

2.2 Identification of Putative Nucleation Sites

RNAStrucTar first scans the mRNA sequence in search for putative nucleation sites, considering complementariness with the seed region of the miRNA. This seed match step is commonly used and considered as a speed-up factor that accelerates the algorithm while it differs from tool to tool in terms of perfectness of matches. A flexible miRNA seed window - nucleotides 2 to 8, counting from the 5' end of miRNA - is used to scan the mRNA sequence for potential target sites. All results shown in this work were obtained by using 3-mers or 4-mers complementarity to miRNA positions 2–5 in the seed match step; see Figure 2. However, the algorithm allows looking for shorter or longer matches and also matches with varying starting positions.

2.3 [Binding Region, miRNA]-Duplex Structure Prediction

After the seed regions are identified, the upstream flanking region of the seed region is extracted for the next step. Among common features of prediction programs are dynamic programming and the alignment of the miRNA seed region to the target mRNA. We propose a dynamic programming approach for finding minimum energy alignments between the full length of the miRNA and the target sequence for each putative binding site. For this purpose, a modified version of RNAduplex [26] is adapted to compute the optimum duplex structure for each putative binding site. For each such site j, the binding pattern and its free energy $\Delta G^j_{\text{binding}}$ are computed according to the seed alignment from the previous step; see Figure 2. At the end of this step, weak binding sites are filtered out by applying an energy threshold ϑ with the default setting -10kcal/mol. This way we obtain k binding sites that satisfy the condition $\Delta G^j_{\text{binding}} \leq \vartheta$, $j = 1, ..., k$.

2.4   Integration of Target Site Accessibility

Similar to [21] and [25], we adopted the simplifying assumption that the binding of a miRNA to a longer target mRNA should cause a local structural alteration at the target site, but but has no long-range effects on the overall target secondary structure. This leads to a breakage of intramolecular bonds within the target region. For each input secondary structure $F_i \in MS$, $i = 1, ..., ms$, the energy contribution $\Delta G^j_{\text{open},i}$ of the deleted bindings is computed for each site j by using RNAeval [16] and according to standard data of the Nearest Neighbour Model [7,29]. Given F_i, we denote by $F_{\text{open},i}$ the associated secondary structure where all base pair bindings within j are removed. We then define

$$\Delta G^j_{\text{open},i} = \text{RNAeval}(F_i) - \text{RNAeval}(F_{\text{open},i}). \tag{1}$$

2.5   miRNA-Target Score Derived from a Single Binding Site

At this stage, we estimate the free energy of the miRNA:mRNA duplex structure by using RNAeval. For each putative binding site j and for each input secondary structure F_i, we generate an artificial RNA sequence that consists of the original mRNA (3'UTR) sequence, a linker sequence XXXX, and the miRNA sequence. The corresponding folding is denoted by $F^j_{\text{concat},i}$. The score $S(\text{miRNA},3'\text{UTR},F_i,j) = S_{i,j}$ is then defined by

$$S_{i,j} = \text{RNAeval}(F^j_{\text{concat},i}) - \text{RNAeval}(F_i). \tag{2}$$

We integrate the scores of multiple conformations F_i for the $h_j \leq ms$ negative values of $S_{i,j} < 0$ by setting

$$S_j = -\log \sum_{s=1}^{h_j} e^{-S_{i_s,j}}, \ j = 1, ..., k; \tag{3}$$

see Figure 3. In Figure 3 we assume for simplicity $h_j = ms$ for all $j \leq k$. The setting according to Eqn. 3 is inspired by the PITA score [21].

2.6   MicroRNA Target Prediction Scores

Some existing target prediction methods check the presence of multiple target sites and take the number of target sites into account for a final score.

$$\begin{pmatrix} S_{1,1} & \cdots & S_{1,k} \\ \vdots & \vdots & \vdots \\ S_{ms,1} & \cdots & S_{ms,k} \end{pmatrix}$$

$$\begin{pmatrix} S_{1,1} & \cdots & S_{1,k} \\ \vdots & \vdots & \vdots \\ S_{ms,1} & \cdots & S_{ms,k} \\ \hline S_1 & \cdots & S_k \end{pmatrix}$$

$$\left(\begin{array}{ccc|c} S_{1,1} & \cdots & S_{1,k} & \\ \vdots & \vdots & \vdots & \\ S_{ms,1} & \cdots & S_{ms,k} & \\ \hline S_1 & \cdots & S_k & S_{\mathrm{tot}} \end{array}\right)$$

Fig. 3. Integration of multiple binding sites and conformation into a single score

We explored different ways to account for the occurrence of multiple binding sites and we ended up with a scoring function where the emphasis is on combining strong duplex conformations with a user-defined target region. To integrate multiple sites with S_j-scores for a given miRNA and a fixed 3'UTR into an overall miRNA:target interaction score, we define

$$S_{\mathrm{tot}} = -\log \sum_{j=1}^{k} e^{-S_j}. \tag{4}$$

We note that by definition $S_j < 0$ (assuming $h_j \geq 1$) for all $j \leq k$, see Eqn. 3. We recall that each $S_j \leq 0$ represents information about $h_j \leq$ ms bindings to metastable conformations F_i, $i = 1, ..., h_j$, which justifies the notation S_{tot} as total score.

Additionally, other alternative scoring functions were also analysed: For each conformation F_i, the linear sum S_i of $k_i \leq k$ values of $S_{i,j} < 0$ is computed, and for $h \leq$ ms values of $S_i < 0$, the average value is denoted by S_{sum}. Thus, we define

$$S_i = \sum_{t=1}^{k_i} S_{i,j_t}; \tag{5}$$

$$S_{\mathrm{sum}} = \frac{\sum_{s=1}^{h} S_{i_s}}{h}. \tag{6}$$

In addition to S_{tot} and S_{sum}, RNAStrucTar allows the user to calculate a score S_{u} derived from a binding site u that contains a user defined position (usually, where the SNP is located) within the input RNA sequence.

$$S_{\mathrm{u}} = \frac{\sum_{s=1}^{h_u} S_{i_s,u}}{h_u}. \tag{7}$$

Again, similar to the PITA score [21], we define

$$S_{\mathrm{P}} = \frac{\sum_{s=1}^{h} -\log \sum_{t=1}^{k_i} e^{-S_{i_s,j_t}}}{h}. \tag{8}$$

We emphasise that based upon Eqns. 3–8 the values of S_j, S_{tot}, S_i, S_{sum}, S_{u}, and S_{P} are either negative or not defined (e.g., if $h_j = 0$ for some j or $k_i = 0$ for some i).

2.7   Metastable Conformations Sets

In our analysis [10], along with the energy offset ΔE above the MFE conformation, we tried to restrict metastable states to deep local minima. The parameter D indicates the depth of a local minimum or - in other terms - the escape height from a local minimum, which is taken in barrier trees as the distance to the nearest saddle point. We found that out of the three different parameters we introduced, the average depth and the average opening energy of metastable conformations may provide supporting information for a stronger separation between miRNA bindings to the two alleles defined by a given SNP. Here, we aim at individual miRNA–mRNA binding predictions over samples of metastable conformations defined by these parameters. Therefore, we order the metastable conformations with respect to:

(a) The depth $D(F_i)$ in descending order (deepest first).
(b) The absolute value of the opening energy $\Delta G^u_{\text{open},i}$ of the user defined target region, ranked in ascending order.

We obtain the following two sets, where N is a user defined parameter:

(a) **Set A**: the N deepest metastable conformations among MS.
(b) **Set B**: the N most easily accessible conformations in the user defined target region among the deepest metastable conformations.

3   Results

3.1   Test Dataset

The tools `RNAsubopt` and `Barriers` generate a huge amount of secondary structures, even for a small offset ΔE above the MFE value. Consequently, a large scale test or a genome wide prediction analysis is not possible at this stage. In order to test our approach, we use the same data acquisition method as in [10], i.e., we use miRNA–mRNA pairs from published experimental work where SNPs are linked to specific diseases. SNPs can be located in miRNA binding regions, and consequently they could affect gene expression. `RNAStrucTar` can be used to evaluate how SNPs affect miRNA regulation by using as input the wild type and the SNP variant. Our aim is to determine the ability of `RNAStrucTar` to provide supportive information for a stronger discrimination between miRNA bindings to the two alleles defined by a given SNP (also denoted as RS sequences). The selection of test sequences was governed by the need of having miRNA–mRNA interactions with a high level of experimental validation, for example, by being based upon PCR and/or luciferase reporter assays. We analysed 20 instances of [mRNA/3'UTR;RS;miRNA] interactions, where 14 instances were used in [10] and defined in Section 2 there. The 6 remaining instances were sourced from [8,11,17,24,30,39]. The sequence IDs were retrieved from the NCBI database and the NCBI Single Nucleotide Polymorphism Database (dbSNP) of nucleotide sequence variation. We also utilised `mirdSNP` [6] and `mirTarbase` [20] for retrieving information related to wild type and variant alleles, ensuring this way a maximum consistency between the publication and the different databases. The sequence length refers to data directly obtained from the NCBI database together with transcript information provided by the ENSEMBL database, and the length ranges between 124 nt and 1167 nt. Some of the publications were sourced from the Human microRNA Disease Database (HMDD) [27]. All results shown in this work were obtained using 3-mers or 4-mers complementarity to the miRNA positions 2–5 in the seed match step. The setting of ΔE depends on the length of the 3'UTR and was selected in such a way that a sufficiently large number of metastable conformations is available. Experimental findings suggest that the typical number of gene copies lies between 5–20. Therefore, tests were carried out with N=10 and N=20. For each case, the SNP position was used as the user defined position in order to obtain the score S_u.

3.2   Energy Scores

We note that the publications of experimental work where the test data are taken from differentiate for each allele pair between *weaker* bindings (expression levels) and *stronger* bindings for the miRNA under consideration. Consequently, we calculate energy scores for the weaker and stronger allele, respectively, where it depends on the particular instance which one of the wild type or RS sequence produces the stronger or weaker interaction. Thus, for a given input [mRNA/3'UTR;RS;miRNA], `RNAStrucTar` returns the scores S_{tot} from Eqn. 4, S_{sum} from Eqn. 6, S_P from Eqn. 8, and S_u from Eqn. 7 (binding site u contains SNP position) for two alleles, and subsequently the differences are calculated:

$$\Delta S_{tot} = S_{tot}^{stronger} - S_{tot}^{weaker}; \tag{9}$$

$$\Delta S_{sum} = S_{sum}^{stronger} - S_{sum}^{weaker}; \tag{10}$$

$$\Delta S_P = S_P^{stronger} - S_P^{weaker}; \tag{11}$$

$$\Delta S_u = S_u^{stronger} - S_u^{weaker}. \tag{12}$$

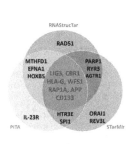

Fig. 4. Comparison of predictions between `RNAStrucTar` and two existing methods

Negative values of ΔS are expected for a target prediction to be classified as correct. Here, we focus on Case A, although Case B is briefly discussed. The results obtained for the twenty instances and Case A by using 3-mers complementarity regarding the seed match and with setting N = 10 are summarized in Table 1. Overall, the score S_{sum} differentiates better than the other scores and it differentiates particularly well between the two alleles on 14 instances, while the S_{sum} scores are indifferent (-1 kcal/mol $< \Delta S_{sum} \leq 0$ kcal/mol) in four cases (SPI1, IL23R, REV3L, ORAI1). For the two other cases (HTR3E and FGF20), S_{sum} is in favour of the weaker allele (W-allele). If Case

B is taken into account for N = 10, S_{sum} returns a strong and correct prediction for REV3L (Case A is also in favour of the S-allele, but above -1kcal/mol).

The score S_u gives positive predictions for 12 instances and eight indifferent predictions. However, if S_{sum} and S_u are taken together for Case A, the predictions are in favour of the correct S-allele by at least one of the scores on 16 instances and four indifferent scores.

The two other scores S_{tot} and S_P are in favour of the weaker allele on three instances, and the number of instances where the scores are indifferent is 9 for S_{tot} and 10 for S_P, although a negative value of ΔS is returned on 14 instances for S_{tot} and 13 instances for S_P, but not always below -1kcal/mol. We conclude that these two scores are less sensitive to binding patterns when compared to S_{sum}.

Table 1. Summary of miRNA binding prediction by RNAStrucTar. A'+' ('-') indicates that the score supports the allele with the stronger (weaker) miRNA binding, with ΔS threshold -1kcal/mol for '+'. A '0' means $-1\text{kcal/mol} < \Delta S \leq 0\text{kcal/mol}$.

	LIG3	CBR1	HTR3E	SPI1	HLA-G	MTHFD1	PARP1	WFS1	EFNA1	IL-23R
L(3'UTR) nt	124	284	302	369	386	393	769	779	843	851
W-allele	A	G	A	T	C	A	C	A	A	A
S-allele	C	A	G	C	G	G	T	G	G	C
miRNA	221	574	510	569	148a	197	145	668	200c	let-7e
SNP pos	83	133	76	330	233	120	607	253	154	309
ΔS_{tot}	+	+	+	0	+	0	0	0	-	0
ΔS_{sum}	+	+	-	0	+	+	+	+	+	0
ΔS_P	+	+	-	0	+	0	0	0	-	0
ΔS_u	+	+	+	0	+	+	+	+	+	0
	RYR3	AGTR1	FGF20	HOXB5	RAD51	REV3L	ORAI1	RAP1A	APP	CD133
L(3'UTR) nt	880	888	903	952	978	985	1034	1078	1120	1167
W-allele	G	C	T	G	A	C	T	C	C	A
S-allele	A	A	C	A	G	T	C	A	T	C
miRNA	367	155	433	7	197	25	519a	196a	147	135b
SNP pos	839	86	182	141	718	460	86	366	171	667
ΔS_{tot}	0	0	-	+	+	0	-	+	+	0
ΔS_{sum}	+	+	-	+	+	0	0	+	+	+
ΔS_P	0	0	-	+	+	0	0	+	+	0
ΔS_u	0	0	0	0	+	+	0	+	0	+

3.3   Comparison to other Computational Methods

We compare our predictions to those produced by PITA [21] and STarMir [25]. For the twenty instances we consider, the PITA tools returns predictions in favour of the S-allele on 13 instances, with six indifferent scores and one prediction in favour of the W-allele (the seven instances are: PARP1, RYR3, AGTR1, FGF20,

RAD51, REV3L, ORAI1). Thus, HTR3E, SPI1, and IL23R are correct by `PITA`, but not by `RNAStrucTar` (S_{sum} only); PARP1, RYR3, AGTR1, and RAD51 are correct by `RNAStrucTar`, but not by `PITA`; both tools fail on FGF20, REV3L, and ORAI1.

The equivalent of S_{sum} for `STarMir` predictions returns score differences in favour of the S-allele on 14 instances, with no indifferent outcomes, but six false predictions on MTHFD1L, EFNA1, IL23R, FGF20, HOXB5, and RAD51. Thus, HTR3E, SPI1, REV3L, and ORAI1 are correct by `STarMir`, but not by `RNAStrucTar`; MTHFD1L, EFNA1, HOXB5, and RAD51 are correct by `RNAStrucTar`, but not by `STarMir`; both tools fail on IL23R and FGF20.

In Figure 4 we combine the results of the three methods. If we classify an instance as positively predicted if at least two of the methods return a prediction in favour of the S-allele, then 16 correct predictions are made. If only one positive return by a single method is required, then 19 correct predictions are produced by the three methods (only FGF20 is rejected).

4   Conclusion

We present in this paper RNAStrucTar, a miRNA target prediction tool which incorporates target site accessibility related to metastable secondary structures close to the MFE conformation. We tested our method on 20 miRNA:mRNA interaction pairs that have been experimentally evaluated in the literature. We found that a combination of the two main scores returned by `RNAStrucTar` supports the experimental findings on 16 instances, with four indifferent outcomes and no false classifications. If `STarMir` results (14 correct, but partly on different instances) are taken into account, then experimental findings are supported on 18 instances. Thus, we think that `RNAStrucTar` may provide additional, useful information for miRNA target predictions. A user friendly version of `RNAStrucTar` with an appropriate interface is under development.

References

1. Alexiou, P., Maragkakis, M., Papadopoulos, G.L., Reczko, M., Hatzigeorgiou, A.G.: Lost in translation: An assessment and perspective for computational microRNA target identification. Bioinformatics 25(23), 3049–3055 (2009)
2. Krek, A., Grün, D., Poy, M.N., Wolf, R., Rosenberg, L., Epstein, E.J., MacMenamin, P., da Piedade, I., Gunsalus, K.C., Stoffel, M., Rajewsky, N.: Combinatorial microRNA target predictions. Nature Genetics 37, 495–500 (2005)
3. Bartel, D.P.: MicroRNAs: Genomics, biogenesis, mechanism, and function. Cell 116(2), 281–297 (2004)
4. Bartel, D.P.: Micrornas: Target recognition and regulatory functions. Cell 136(2), 215–233 (2009)
5. Bernhart, S.H., Hofacker, I.L., Stadler, P.F.: Local RNA base pairing probabilities in large sequences. Bioinformatics 22(5), 614–615 (2006)
6. Bruno, A., Li, L., Kalabus, J., Pan, Y., Yu, A., Hu, Z.: Mirdsnp: A database of disease-associated SNPs and microRNA target sites on 3'UTRs of human genes. BMC Genomics 13(1), 44 (2012)

7. Chen, J.L., Dishler, A.L., Kennedy, S.D., Yildirim, I., Liu, B., Turner, D.H., Serra, M.J.: Testing the nearest neighbor model for canonical RNA base pairs: Revision of GU parameters. Biochemistry 51(16), 3508–3522 (2012)

8. Cheng, M., Yang, L., Yang, R., Yang, X., Deng, J., Yu, B., Huang, D., Zhang, S., Wang, H., Qiu, F., Zhou, Y., Lu, J.: A microRNA-135a/b binding polymorphism in CD133 confers decreased risk and favorable prognosis of lung cancer in Chinese by reducing CD133 expression. Carcinogenesis 34(10), 2292–2299 (2013)

9. Garcia, D.M., Baek, D., Shin, C., Bell, G.W., Grimson, A., Bartel, D.P.: Weak seed-pairing stability and high target-site abundance decrease the proficiency of lsy-6 and other microRNAs. Nat. Struct. Mol. Biol. 18, 1139–1146 (2011)

10. Day, L., Abdelhadi Ep Souki, O., Albrecht, A.A., Steinhfel, K.: Accessibility of microRNA binding sites in metastable RNA secondary structures in the presence of SNPs. Bioinformatics 30(3), 343–352 (2014)

11. Delay, C., Calon, F., Mathews, P., Hebert, S.: Alzheimer-specific variants in the 3'UTR of amyloid precursor protein affect microrna function. Molecular Neurodegeneration 6(1), 70 (2011)

12. Ding, Y., Chan, C.Y., Lawrence, C.E.: Sfold web server for statistical folding and rational design of nucleic acids. Nucleic Acids Research 32, W135–W141 (2004)

13. Enright, A., John, B., Gaul, U., Tuschl, T., Sander, C., Marks, D.: MicroRNA targets in drosophila. Genome Biology 5(1), R1 (2003)

14. Johnson, E., Srivastava, R.: Volatility in mRNA secondary structure as a design principle for antisense. Nucleic Acids Research 41, e43 (2012)

15. Flamm, C., Hofacker, I.L., Stadler, P.F., Wolfinger, M.T.: Barrier trees of degenerate landscapes. Zeitschrift für Physikalische Chemie 216, 155–173 (2002)

16. Gruber, A.R., Lorenz, R., Bernhart, S.H., Neuböck, R., Hofacker, I.L.: The Vienna RNA websuite. Nucleic Acids Research 36(suppl. 2), W70–W74 (2008)

17. Hikami, K., Kawasaki, A., Ito, I., Koga, M., Ito, S., Hayashi, T., Matsumoto, I., Tsutsumi, A., Kusaoi, M., Takasaki, Y., Hashimoto, H., Arinami, T., Sumida, T., Tsuchiya, N.: Association of a functional polymorphism in the 3'-untranslated region of SPI1 with systemic lupus erythematosus. Arthritis & Rheumatism 63(3), 755–763 (2011)

18. Hofacker, I., Fontana, W., Stadler, P., Bonhoeffer, L., Tacker, M., Schuster, P.: Fast folding and comparison of RNA secondary structures. Monatshefte für Chemie/Chemical Monthly 125(2), 167–188 (1994)

19. Hofacker, I.L.: How microRNAs choose their targets. Nature Genetics (10), 1191–1192 (2007)

20. Hsu, S.-D., Tseng, Y.-T., Shrestha, S., Lin, Y.-L., Khaleel, A., Chou, C.-H., Chu, C.-F., Huang, H.-Y., Lin, C.-M., Ho, S.-Y., Jian, T.-Y., Lin, F.-M., Chang, T.-H., Weng, S.-L., Liao, K.-W., Liao, I.-E., Liu, C.-C., Huang, H.-D.: MiRTarbase update 2014: an information resource for experimentally validated miRNA-target interactions. Nucleic Acids Research 42(D1), D78–D85 (2014)

21. Kertesz, M., Iovino, N., Unnerstall, U., Gaul, U., Segal, E.: The role of site accessibility in microRNA target recognition. Nature Genetics 39(10), 1278–1284 (2007)

22. Kozomara, A., Griffiths-Jones, S.: miRBase: Annotating high confidence microRNAs using deep sequencing data. Nucleic Acids Research 42(D1), D68–D73 (2014)

23. Lewis, B.P., Shih, I.H., Jones-Rhoades, M.W., Bartel, D.P., Burge, C.B.: Prediction of mammalian microRNA targets. Cell 115(7), 787–798 (2003)

24. Li, Y., Nie, Y., Cao, J., Tu, S., Lin, Y., Du, Y., Li, Y.: G–A variant in miR-200c binding site of EFNA1 alters susceptibility to gastric cancer. Molecular Carcinogenesis 53(3), 219–229 (2014)

25. Long, D., Lee, R., Williams, P., Chan, C.Y., Ambros, V., Ding, Y.: Potent effect of target structure on microRNA function. Nat. Struct. Mol. Biol. 14(4), 287–294 (2007)
26. Lorenz, R., Bernhart, S., Honer zu Siederdissen, C., Tafer, H., Flamm, C., Stadler, P., Hofacker, I.: Vienna RNA package 2.0. Algorithms for Molecular Biology 6(1), 26 (2011)
27. Lu, M., Zhang, Q., Deng, M., Miao, J., Guo, Y., Gao, W., Cui, Q.: An analysis of human microRNA and disease associations. PLoS One 3(10), e3420 (2008)
28. Marín, R.M., Vaníček, J.: Optimal use of conservation and accessibility filters in microRNA target prediction. PLoS One 7(2), e32208 (2012)
29. Mathews, D.H., Disney, M.D., Childs, J.L., Schroeder, S.J., Zuker, M., Turner, D.H.: Incorporating chemical modification constraints into a dynamic programming algorithm for prediction of rna secondary structure. PNAS USA 101(19), 7287–7292 (2004)
30. Minguzzi, S., Selcuklu, S.D., Spillane, C., Parle-McDermott, A.: An NTD-associated polymorphism in the 3'UTR of MTHFD1L can affect disease risk by altering mirna binding. Human Mutation 35(1), 96–104 (2014)
31. Peterson, S.M., Thompson, J.A., Ufkin, M.L., Sathyanarayana, P., Liaw, L., Congdon, C.B.: Common features of microRNA target prediction tools. Frontiers in Genetics 5, 23 (2014)
32. Ragan, C., Zuker, M., Ragan, M.A.: Quantitative prediction of miRNA-mRNa interaction based on equilibrium concentrations. PLoS Comput. Biol. 7(2), e1001090 (2011)
33. Rajewsky, N.: MicroRNA target predictions in animals. Nature Genetics 38(suppl. (6s)), S8–S13 (2006)
34. Robins, H., Li, Y., Padgett, R.W.: Incorporating structure to predict microRNA targets. PNAS USA 102(11), 4006–4009 (2005)
35. Subkhankulova, T., Gilchrist, M., Livesey, F.: Modelling and measuring single cell RNA expression levels find considerable transcriptional differences among phenotypically identical cells. BMC Genomics 9(1), 268 (2008)
36. Vergoulis, T., Vlachos, I.S., Alexiou, P., Georgakilas, G., Maragkakis, M., Reczko, M., Gerangelos, S., Koziris, N., Dalamagas, T., Hatzigeorgiou, A.G.: TarBase 6.0: capturing the exponential growth of miRNA targets with experimental support. Nucleic Acids Research 40, D222–D229 (2011)
37. Wienholds, E., Plasterk, R.H.: MicroRNA function in animal development. FEBS Letters 579(26), 5911–5922 (2005)
38. Wuchty, S., Fontana, W., Hofacker, I.L., Schuster, P.: Complete suboptimal folding of RNA and the stability of secondary structures. Biopolymers 49(2), 145–165 (1999)
39. Zhang, S., Chen, H., Zhao, X., Cao, J., Tong, J., Lu, J., Wu, W., Shen, H., Wei, Q., Lu, D.: REV3L 3'UTR 460 T>C polymorphism in microRNA target sites contributes to lung cancer susceptibility. Oncogene 32, 242–250 (2013)

Inference of Circadian Regulatory Pathways Based on Delay Differential Equations

Catherine F. Higham and Dirk Husmeier

School of Mathematics and Statistics, College of Science and Engineering,
University of Glasgow, Glasgow G12 8QQ, Scotland, UK
{Catherine.Higham,Dirk.Husmeier}@glasgow.ac.uk

Abstract. Inference of circadian regulatory network models is highly challenging due to the number of biological species and non-linear interactions. In addition, statistical methods that require the numerical integration of the data model are computationally expensive.

Using state-of-the-art adaptive gradient matching methods which model the data with Gaussian processes, we address these issues through two novel steps. First, we exploit the fact that, when considering gradients, the interacting biological species can be decoupled into sub-models which contain fewer parameters and are individually quicker to run. Second, we substantially reduce the complexity of the network by introducing time delays to simplify the modelling of the intermediate protein dynamics.

A Metropolis-Hastings scheme is used to draw samples from the posterior distribution in a Bayesian framework. Using a recent delay differential equation model describing circadian regulation affecting physiology in the mouse liver, we investigate the extent to which deviance information criterion can distinguish between under-specified, correct and over-specified models.

Keywords: Bayesian Inference, Gaussian Processes, Adaptive Gradient Matching, Circadian Regulation, Delay Differential Equations.

1 Introduction

1.1 Biological Background/Motivation

The circadian clock is a molecular mechanism, involving interlocked, transcriptional feedback loops, that synchronises biological processes with the day/night cycle and is found in many organisms, see [19]. Mathematical models are being developed to describe the dynamics of the clock transcriptional network and its downstream regulation, for Arabidopsis, see [11], [12], and for the mouse liver and adrenal gland, see [8], [9]. The field is now sufficiently mature for us to consider validating, comparing and extending these models in the presence of experimental data.

F. Ortuño and I. Rojas (Eds.): IWBBIO 2015, Part II, LNCS 9044, pp. 468–478, 2015.
© Springer International Publishing Switzerland 2015

1.2   Network Inference

Statistical pathway inference techniques aim to make inference in a network where the vertices are molecular components such as genes or gene products and the edges represent regulatory interactions between these components. Statistical models for exploring large spaces are typically linear for reasons of speed but come at the cost of over-simplifying the non-linear features of the network. When the network is known, differential equations (DEs) are widely used to model biochemical dynamics and capture a wealth of detail about the the network. Fitting approaches which directly solve the DEs, see [17], are currently infeasible for large systems especially when model comparison is required. Hence, recent work, to select between network models focuses on an intermediate approach, incorporating prior knowledge about the structure of biochemical DE models into an inference framework, see [10].

Differential equation models are used extensively in science and engineering. A common requirement is to estimate model parameters by fitting them to observed data collected over time. This involves repeatedly finding a solution to the DEs which, because these systems are typically non-linear, involves numerical approximation. Numerical integration is computationally expensive, in all but the simplest cases, and hence, there is much interest in methods that avoid this step. One alternative approach focuses on gradient matching with Gaussian Processes (GPs). This is currently a very active area (e.g. [2], [4], [18], [7]).

Gradient matching uses an alternative model of the data, an interpolant, and matches the derivative of this interpolant with the DE outputs, thus avoiding explicit numerical integration, see [13]. GPs are a natural choice for the alternative data model, and in particular they admit exact derivative expressions. A GP may be fitted to the data and the DE parameters found by matching the GP derivative, for which there is an analytical expression, to the DE derivative. However the accuracy of the original method proposed in [2], was limited by the lack of regularisation from the DE parameter inference to the GP inference. In work by [4], all parameters are consistently inferred in the context of the whole model, rather than in a piecewise heuristic manner. This introduces, in effect, a coupling mechanism between the Gradient Process and the DEs which enhances the learning of the parameters associated with the DEs.

Here, we contribute to this field by developing novel methodological advances and illustrating them on a core clock model for mouse liver and adrenal gland developed by [9]. This model comprises five clock genes and is based on expression and experimentally verified circadian cis-regulatory sites, see figure 1. The intermediate protein dynamics are modelled using time delays, vastly simplifying the network complexity. The expression of each clock gene is described by a delay differential equation with a production term that depends on the concentrations of core clock regulatory components and a decay term. The adaptive gradient matching (AGM) statistical model developed in [4] is our framework for Bayesian inference. We introduce modularisation by exploiting the fact that when gradient matching, the system reduces to five equations, one for each species, which are no longer coupled (as the right-hand side does not require any of the other left-hand side values).

A Metropolis-Hastings scheme is devised to sample from the posterior probability densities for the model parameters (including the parameters for the DEs
and the hyper-parameters for the GPs).

We generate data from the full model (M_0), see [9], using the parameters
described by the authors. The aim of this work is to investigate whether deviation information criterion (DIC) which considers both the measure of fit and the
measure of complexity based on the posterior samples, can distinguish between
under-specified, correct and over-specified models for a range of alternative hypothesis.

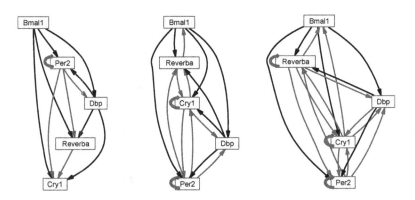

Fig. 1. Regulatory network Model M_0 (*middle*) containing five core clock genes (*boxes*).
Direction of activation (*bold black line*) or inhibition (*regular red line*) is indicated by
the (*arrowhead*). Regulatory network Model M_r with reduced number of edges (*left*).
Regulatory network Model M_a with added number of edges (*right*).

2   Methods

We describe here our approach to Bayesian model fitting and selection for systems of differential equations (ordinary and time delay) using adaptive gradient
matching with a Gaussian process.

2.1   Reaction Graph

Consider a regulatory network of genes which may activate or inhibit transcription directly or indirectly. This network can be considered as a reaction graph
where the nodes are the molecular components such as genes or gene products and the edges are the regulatory interactions between these components.
Changes over time in the gene and gene product states can be modelled using
differential equations. To simplify the modelling of the intermediate protein dynamics we introduce time-delayed variables. This reduces the network to gene
transcripts and is convenient for parameter estimation and model selection since

this data is more readily available than protein data. Proteins are sometimes treated as missing data and modelled using latent variables; employing time-delayed variables is an alternative approach [8].

Consider a set of T arbitrary time points $t_1 < \ldots < t_T$ and K gene transcription states. We define $\mathbf{x}_k \equiv \mathbf{x}_k(t) \equiv [x_k(t_1), x_k(t_2), \ldots, x_k(t_T)]^\top$ as the transcription state sequence for the kth state.

2.2   The Dynamical Model

Let K denote the number of genes in the network and i, where $i = 1 \ldots N_k$, the associated set of N_k regulatory genes. The time series for the kth gene transcription state is represented by a set of K differential equations of the form

$$\dot{\mathbf{x}}_k(t) \equiv \frac{d\mathbf{x}_k(t)}{dt} = \mathbf{f}_k(\mathbf{x}_k(t), \mathbf{x}_i(t), \boldsymbol{\theta}).$$

Here $\boldsymbol{\theta} = \{\boldsymbol{\theta}_1 \ldots \boldsymbol{\theta}_K\}$ is our general notation for the parameters of the differential equations. Specifically, we have

$$\mathbf{f}_k(\mathbf{x}_k(t), \mathbf{x}_i(t), \boldsymbol{\theta}_k) = \prod_{i=1 \ldots N_k} \left(\frac{1 + av_i \mathbf{x}_i(t_{\tau_i})/a_i}{1 + \mathbf{x}_i(t_{\tau_i})/a_i} \right)^{p_i} - d_k \mathbf{x}_k(t). \qquad (1)$$

In this notation, p_i is the number of clock-controlled elements and $(1/a_i)^{p_i}$ represents the basal production rate of species i. When species i is an activator, av_i scales the activation and when species i is a repressor, $av_i = 0$. We use t_{τ_i} to indicate the value $t - \tau_i$, thus accounting for the delay including translation, post-translational modifications, complex formation and nuclear translocation. The parameter d_k is the degradation rate of species k.

2.3   The Observation Model

Let $\mathbf{y}_k(t) = \mathbf{x}_k(t) + \epsilon(t)$ be noisy observations of this process, where $\epsilon(t)$ is assumed to have a zero-mean Gaussian distribution with variance σ_k^2 for each of the model states. Assuming independence over the observation times, we have an observation model

$$p(\mathbf{y}_k|\mathbf{x}_k) = \prod_t p(\mathbf{y}_k(t)|\mathbf{x}_k(t), \sigma_k^2) = \prod_t \mathcal{N}(\mathbf{x}_k(t), \sigma_k^2 \mathbf{I}). \qquad (2)$$

Here $\mathcal{N}(\mathbf{x}_k(t), \sigma_k^2 \mathbf{I})$ denotes the probability density function for a Gaussian random variable with mean $\mathbf{x}_k(t)$ and variance $\sigma_k^2 \mathbf{I}$.

2.4   Adaptive Gradient Matching

Instead of obtaining $\mathbf{x}_k$ by solving the dynamical system, recent gradient matching approaches in [2] and [4], put a GP prior on $\mathbf{x}_k$ and match the GP derivatives with the derivatives arising from the differential equations. This approach leads naturally to a decoupling of $\mathbf{x}_k$ and a reduction in complexity for parameter estimation.

2.5   Gaussian Process Modelling

A Gaussian Process (GP) is a stochastic process governing the properties of a function. A GP is defined by a mean and correlation function called a kernel. In this work we use the most commonly used kernel, the "squared-exponential" kernel. With a GP prior on $\mathbf{x}_k$, $p(\mathbf{x}_k|\phi_k) = N(\mathbf{x}_k|\mu_k, \mathbf{C}_{\phi_k})$ where μ_k is the data mean, $\mathbf{C}_{\phi_k}$ is the correlation function and ϕ_k are the hyper-parameters of the GP. The derivative of a GP is also a GP and the conditional distribution for the state derivatives is $\mathcal{N}(\mathbf{m}_k, \mathbf{A}_k)$ where $\mathbf{m}_k = {}'\mathbf{C}_{\phi_k}\mathbf{C}_{\phi_k}(\mathbf{x}_k - \mu_k)$ and $\mathbf{A}_k = \mathbf{C}''_{\phi_k} - {}'\mathbf{C}_{\phi_k}\mathbf{C}^{-1}_{\phi_k}\mathbf{C}'_{\phi_k}$, respectively, see [15]. Here, the matrix $\mathbf{C}''_{\phi_k}$ denotes the auto-covariance for each state derivative, and the matrices $\mathbf{C}'_{\phi_k}$ and ${}'\mathbf{C}_{\phi_k}$ denote the cross-variances between the kth state and its derivative. For further details concerning the derivation and analytical form of these expressions, see [4].

2.6   Statistical Model

Bayes' theorem links what we would like to know, the posterior probability distributions for the unknown parameters, to the likelihood of seeing the observations given the model and its parameters and the prior information about the parameters. Following [4] we propose the following adaptive gradient matching (AGM) model over states $\mathbf{x}_k$, their derivatives $\dot{\mathbf{x}}_k$, observations $\mathbf{y}_k$ and parameters associated with the DE, $\boldsymbol{\theta}_k$, with the observational noise, σ_k^2, and with the GP, ϕ_k,

$$p(\mathbf{y}_k, \mathbf{x}_k, \boldsymbol{\theta}_k, \phi_k, \gamma_k^2, \sigma_k^2) = p(\mathbf{y}_k|\mathbf{x}_k, \sigma_k^2)p(\mathbf{x}_k|\boldsymbol{\theta}_k, \phi_k, \gamma_k^2)p(\boldsymbol{\theta}_k, \phi_k, \gamma_k^2, \sigma_k^2). \quad (3)$$

The term $p(\mathbf{x}_k|\boldsymbol{\theta}_k, \phi_k, \gamma_k^2)$ combines the DE gradient with the GP gradient in a compatability function and arises from a products of experts approach described in [4]

$$p(\mathbf{x}_k|\boldsymbol{\theta}_k, \phi_k, \gamma_k^2) \propto \frac{\exp[-1/2(\mathbf{f}_k - \mathbf{m}_k)^\top (\mathbf{A}_k + \gamma_k^2\mathbf{I})^{-1}(\mathbf{f}_k - \mathbf{m}_k)]}{(2\pi)^{n/2}|\mathbf{A}_k + \gamma_k^2\mathbf{I}|^{1/2}}. \quad (4)$$

The function $\mathbf{f}_k$ is defined in equation (1), $\mathbf{A}_k$ and $\mathbf{m}_k$ are defined in section 2.5, n is the number of time points (also equal to the number of rows in $\mathbf{A}_k$), γ_k is the slack parameter controlling the coupling of GP and DE, and $\mathbf{I}$ is the identity matrix.

2.7   Sampling

We use a Metropolis-Hastings scheme to draw samples from the posterior distribution. Denoting $q_1(\boldsymbol{\Theta}_k)$ and $q_2(x_k)$, where $\boldsymbol{\Theta}_k = \{\boldsymbol{\theta}_k, \phi_k, \gamma_k^2, \sigma_k^2\}$, as the proposal distributions for the parameters, $\boldsymbol{\Theta}_k$, and the states, $\mathbf{x}_k$, the proposal moves are accepted or rejected according to the standard Metropolis-Hastings criteria

$$P_{accept} = \min\left\{1, \frac{\pi(\mathbf{y}_k, \mathbf{x}'_k, \boldsymbol{\Theta}'_k)}{\pi(\mathbf{y}_k, \mathbf{x}_k, \boldsymbol{\Theta}_k)}\right\} \quad (5)$$

where $\pi = \frac{p(\mathbf{y}_k, \mathbf{x}_k, \theta_k, \phi_k, \gamma_k^2, \sigma_k^2)}{q_1(\mathbf{\Theta}_k) q_2(\mathbf{x}_k)}$ and the numerator is defined in equation (3).

The parameters $\mathbf{\Theta}_k$ are proposed simultaneously from a multivariate Gaussian using the efficient adaptive MCMC algorithm described by [5], adapted for our non standard posterior distribution. The priors for all parameters are informed gamma priors. Here, $\mathbf{x}_k$ and ϕ_k are initialised using a GP regression fit with maximum likelihood to the data $\mathbf{y}_k$, see [14]. The proposal function for $\mathbf{x}_k$ is $\mathcal{N}(\boldsymbol{\mu}_{\mathbf{x}|\mathbf{y}}, \boldsymbol{\Sigma}_{\mathbf{x}|\mathbf{y}})$, dropping the subscript k for convenience, and where $\boldsymbol{\mu}_{\mathbf{x}|\mathbf{y}} = (\mathbf{C}_\phi^{-1} + (\sigma^2 I)^{-1})^{-1} (\sigma^2 I)^{-1} \mathbf{y}$, $\boldsymbol{\Sigma}_{\mathbf{x}|\mathbf{y}} = (\mathbf{C}_\phi^{-1} + (\sigma^2 I)^{-1})^{-1}$.

2.8 Model Selection Using Deviance Information Criterion

For competing parametric statistical models, the deviance information criterion (DIC), as described in [16], considers both the measure of fit and the measure of complexity. The deviance (associated with the observed likelihood $p(\mathbf{y}|\theta)$ is defined by

$$D(\theta) = -2 \log p(\mathbf{y}|\theta) + 2 \log h(\mathbf{y}),$$

where $h(\mathbf{y})$ depends only on the data. The model complexity or effective dimension, p_D, is defined as

$$p_D = \bar{D}(\theta) - D(\bar{\theta}),$$

and

$$DIC = \bar{D}(\theta) + p_D,$$
$$= 2\bar{D}(\theta) - D(\bar{\theta})$$

where $\bar{D}(\theta)$ is the expected value of $D(\theta)$ and $\bar{\theta}$ is the expected value of θ. For model comparison, we set $h(\mathbf{y}) = 1$ for all models so that $D(\theta) = -2 \log p(\mathbf{y}|\theta)$. For $D(\theta)$ available in closed form, $\bar{D}(\theta)$ can be approximated from the MCMC run by taking the sample mean of the simulated values of $D(\theta)$.

In our case, as $\mathbf{y}$ is conditioned on θ and $\mathbf{X} = \mathbf{x}_1, \ldots, \mathbf{x}_k$, we use the complete DIC suggested in [3]

$$DIC(\mathbf{y}, \mathbf{X}) = -4\mathrm{E}\left[\log p(\mathbf{y}, \mathbf{X}|\theta) | \mathbf{y}, \mathbf{X}\right] + 2 \log p(\mathbf{y}, \mathbf{X}|\mathrm{E}[\theta|\mathbf{y}, \mathbf{X}]).$$

The intuitive idea is that models with a smaller DIC score are preferred to models with a larger DIC score. Models are penalised by the value of $\bar{D}$ but also (in common with other information criteria) by the effective number of parameters p_D. Since $\bar{D}$ will decrease as the number of parameters in a model increases, the p_D term compensates for this effect by favouring models with a smaller number of parameters.

An advantage of DIC, over other criteria such as Bayes factors, is that DIC is easily calculated from the samples generated by a Markov chain Monte Carlo simulation.

3   Results

3.1   Mouse Liver Model

The model selection framework is applied to the five component clock model for mouse liver and adrenal gland developed in [9]. Data is generated from this delay differential equation model using numerical integration over an interval of 24 hours with the published parameter set. Clean data were then sampled in 2 hour intervals and corrupted with additive Gaussian noise which corresponds to a signal-to-noise ratio of 10, see figure 2. In this experiment, we fit a GP with a periodic kernel to each of the five time series in order to provide an initial estimate for the GP hyper-parameters, $\phi_k = \{l_k, sf_k\}$. Here l_k is the length scale parameter and sf_k is the vertical scale parameter. The GP fitting, illustrated in figure 2, also provides an initial estimate for σ_k^2.

The AGM framework combined with the Metropolis-Hastings scheme, outlined in section 2.7, is used to obtain posterior samples for the parameters of the statistical model comprising DE parameters $\theta_k = \{ap_i, av_i, d_k, \tau_i\}$, see equation (1) for more details, GP parameters, ϕ_k, the noise parameter σ_k^2 and the slack parameter, γ_k^2, associated with the AGM framework, see equation (4). In total, for five network species (Bmal1, Rev-erba, Per2, Cry1 and Dbp), see figure 2, there are 54 parameters to be learnt. Note that p_i, the number of clock controlled elements in the x_k regulatory region, are taken as given and not inferred in these experiments. Details of the priors used for each parameter type are given in section 2.7. Two MCMC chains were run for 2×10^6 iterations and convergence was monitored using the potential scale reduction factor (PSRF) discussed in [1].

3.2   Model Selection Experiment

To illustrate these new tools in the context of model selection, we propose two alternative models M_r (one edge per species is *removed*) and M_a (one edge per species is *added* up to a maximum of five edges) to the true model, M_0, and then apply the inference framework to each model using the dataset described above. The specific details as to which edges were removed, added or changed are described in table 1. The DIC score for each model was estimated using 10,000 posterior samples taken from the end of the MCMC chains. Each entry in the table required approximately 2 hours of CPU time on a HPC cluster. Models with a smaller DIC are preferred to models with a larger DIC. For comparison between models, the scores for the alternative models are adjusted by subtracting the value arising from the fitting of the true model M_0. A positive adjusted score indicates that the true model is preferred to the alternative model and a negative adjusted score indicates that the alternative model is preferred to the true model. For this experiment, the DIC differences for M_r were positive for all species (Reverba (adj DIC=0.1), Per2 (adj DIC=0.9), Cry1 (adj DIC=8.6) and Dbp (adj DIC=21.4)) indicating that the true model is preferred to the alternative model, M_r for all species, strongly for Cry1 and Dbp, and weakly for Reverba and Per2. Bmal1 was not included in this experiment as it only has one interaction in the true model.

Over all 5 species, the total adjusted DIC score is 31 suggesting that the true model is preferred to an alternative model with edges missing as outlined for M_r, see table 1. For M_a, the DIC differences are negative for Bmal1 but positive for Reverba, Per2, Cry1 and Dbp (adjusted DICs -1.8, 0.6, 18.7 and 25.4 respectively), see table 1. The total adjusted DIC score is 44.7 suggesting that the true model is preferred to the specified model M_a.

3.3   Parameter Estimation

Comparison of the posterior densities for the model parameters between the species, see figure 3, suggests that uncertainty increases with the number of parameters. Generally recovery of the true value (comparison possible when using synthetic data) is good with the distributions lying over the true value. The occasional parameter is very different. In the case of parameter cp for Reverba, this is explained by over-fitting to the noisy data. The method allows for further investigation of individual differences or deviations from the true value.

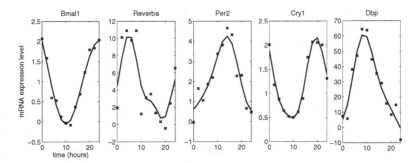

Fig. 2. Synthetic expression level time series data *crosses* generated for five species over 24 hours and the initial GP fit *line*

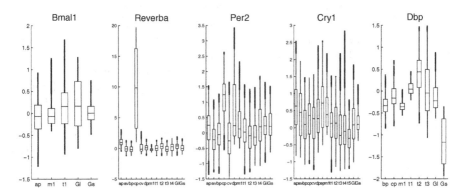

Fig. 3. Posterior probability densities for model parameters shown as box plot distributions. The values have been shifted so that "0" signifies the true parameter.

Table 1. Adjusted DIC scores for the true model M_0 and alternative models M_r and M_a. The production term (listed below) summarises for each species, the direction of the edges acting on that species. For example act1*rep3*rep4 is short-hand for the the activation of Bmal1(1), and the repression of Per2(3) and Cry1(4) on Dbp, M_0.

Species	M_0	M_r	M_a
Bmal1	0	n/a	-1.8
Reverba	0	0.1	1.8
Per2	0	0.9	0.6
Cry1	0	8.6	18.7
Dbp	0	21.4	25.4
Total	0	31.0	44.7

Production terms	M_0	M_r	M_a
Bmal1 (1)	rep2	n/a	rep1*rep4
Reverba (2)	act1*rep3*act5*rep4	act1*rep3*act5	act1*rep3*act5*rep4*rep2
Per2 (3)	act1*rep3*act5*rep4	act1*rep3*act5	act1*rep3*act5*rep4*rep2
Cry1 (4)	act1*rep3*act5*rep2*rep4	act1*rep3*act5*rep2	act1*rep3*rep5*rep2*rep4
Dbp (5)	act1*rep3*rep4	act1*rep3	act1*rep3*rep4*rep2

4   Conclusion

We present here new tools for model selection in gene regulatory networks, with an emphasis on improving computational efficiency for large-scale simulations. Model selection for complex networks is a demanding topic that is challenging state-of-the-art methodology. There is an on-going requirement from experimentalists to revise models as more data becomes available and to choose between alternative hypothesis. For example linking clock mechanics to down-stream activities such as metabolism.

For large networks, methods requiring numerical integration are infeasible. Here we use a state-of-the-art gradient matching approach which substantially reduces the computational expense [2]. The relative cost of AGM to numerical integration of the DE, in the examples considered here is conservatively ten times faster. This improvement makes model selection between several models a realistic objective. Here we looked in detail at one true model and two alternative models. Future work could accommodate many more alternatives.

In terms of model selection, we show that it is possible to distinguish between the true model and a under- or over- specified model where the number of edges have been reduced or added.

Future work will compare the DIC measure with other statistics for model selection and investigate the sensitivity of model selection to factors such as the prior information.

Acknowledgements. This work was funded by the EU under the FP7 scheme as part of the TiMeT project. TiMet - Linking the Clock to Metabolism is a Collaborative Project (Grant Agreement 245143) funded by the European Commission FP7, in response to call FP7-KBBE-2009-3.

References

1. Brooks, S.P., Gelman, A.: General methods for monitoring convergence of iterative simulations. Journal of Computational and Graphical Statistics 7, 434 (1998)
2. Calderhead, B., Girolami, M., Lawrence, N.D.: Accelerating Bayesian inference over nonlinear differential equations with Gaussian processes. Advances in Neural Information Processing Systems (NIPS) 21, 217–224 (2009)
3. Celeux, G., Forbes, F., Robert, C.P., Titterington, M.: Deviance information criteria for missing data models. Bayesian Analysis 1, 651–674 (2006)
4. Dondelinger, F., Husmeier, D., Rogers, S., Filippone, M.: ODE parameter inference using adaptive gradient matching with Gaussian processes. In: Proceedings of the Sixteenth International Conference on Artificial Intelligence and Statistics, pp. 216–228 (2013)
5. Haario, H., Laine, M., Mira, A., Saksman, E.: DRAM: Efficient adaptive MCMC. Statistics and Computing 16, 339–354 (2006)
6. Higham, C.F., Husmeier, D.: A Bayesian approach for parameter estimation in the extended clock gene circuit of Arabidopsis thaliana. BMC Bioinformatics 14, S3 (2013)
7. Holsclaw, T., Sansó, B., Lee, H.K.H., Heitmann, K., Habib, S., Higdon, D., Alam, U.: Gaussian process modeling of derivative curves. Technometrics 55, 57–67 (2012)
8. Korenčič, A., Bordyugov, G., Košir, R., Rozman, D., Goličnik, M., Herzel, H.: The interplay of cis-regulatory elements rules circadian rhythms in mouse liver. PLoS One 7, e46835 (2012)
9. Korenčič, A., Košir, R., Bordyugov, G., Lehmann, R., Rozman, D., Herzel, H.:: Timing of circadian genes in mammalian tissues. Scientific Reports 4 (2014)
10. Oates, C.J., Dondelinger, F., Bayani, N., Korola, J., Gray, J.W., Mukherjee, S.: Causal network inference using biochemical kinetics. Bioinformatics 30, i468–i474 (2014)
11. Pokhilko, A., Fernández, A.P., Edwards, K.D., Southern, M.M., Halliday, K.J., Millar, A.J.: The clock gene circuit in Arabidopsis includes a repressilator with additional feedback loops. Molecular Systems Biology 8, 574 (2012)
12. Pokhilko, A., Mas, P., Millar, A.J.: Modelling the widespread effects of TOC1 signalling on the plant circadian clock and its outputs. BMC Systems Biology 7, 23 (2013)
13. Ramsay, J.O., Hooker, G., Campbell, D., Cao, J.: Parameter estimation for differential equations: A generalized smoothing approach. Journal of the Royal Statistical Society: Series B (Statistical Methodology) 69, 741–796 (2007)
14. Rasmussen, C.E., Nickish, H.: Gaussian processes for machine learning (GPML) toolbox. Journal of Machine Learning Research 11, 3011–3015 (2010)
15. Solak, E., Murray-Smith, R., Leithead, W., Rasmussen, C., Leith, D.: Derivative observations in gaussian process models of dynamic systems. In: Advances in Neural Information Processing Systems (NIPS), pp. 1033–1040 (2003)
16. Spiegelhalter, D.J., Best, N.G., Carlin, B.P., Van Der Linde, A.: Bayesian measures of model complexity and fit. Journal of the Royal Statistical Society: Series B (Statistical Methodology) 64, 583–639 (2002)

17. Vyshemirsky, V., Girolami, M.A.: Bayesian ranking of biochemical system models. Bioinformatics 24, 833–839 (2008)
18. Wang, Y., Barber, D.: Gaussian processes for Bayesian estimation in ordinary differential equations. In: Journal of Machine Learning Research - Workshop and Conference Proceedings (ICML), vol. 32, pp. 1485–1493 (2014)
19. Zhang, E.E., Kay, S.A.: Clocks not winding down: unravelling circadian networks. Nature Reviews. Molecular Cell Biology 11, 764–776 (2010)

Simplifying Tele-rehabilitation Devices for Their Practical Use in Non-clinical Environments*

Daniel Cesarini, Davide Calvaresi, Mauro Marinoni, Paquale Buonocunto, and Giorgio Buttazzo

Scuola Superiore Sant'Anna, Pisa, Italy
{d.cesarini,d.calvaresi,m.marinoni,p.buonocunto,g.buttazzo}@sssup.it
http://retis.sssup.it

Abstract. The lack of success of tele-monitoring systems in non-clinical environments is mainly due to the difficulty experienced by common users to deal with them. In particular, for achieving a correct operation, the user is required to take care of a number of annoying details, such as wearing them correctly, putting them in operation, using them in a proper way, and transferring the acquired data to the medical center. In spite of the many technological advances concerning miniaturization, energy consumption reduction, and the availability of mobile devices, many things are still missing to make these technologies simple enough to be really usable by a broad population, and in particular by elderly people. To bridge this gap between users and devices, a smart software layer could automatically manage configuration, calibration, and data transfer without requiring the intervention of a formal caregiver. This paper describes the key features that should be implemented to simplify the needed initial calibration phase of sensing systems and to support the patient with a multimodal feedback throughout the execution of the exercises. A simple mobile application is also presented as a demonstrator of the advantages of the proposed solution.

Keywords: eHealth, patient-centric, sensors, monitoring, bio-feedback.

1 Introduction

The average age of the population is increasing and so is the need for rehabilitation and motor therapy sessions. This increment of prospective patients together with the problem of the decreasing availability of public money for the healthcare sector, is likely to turn into a degradation of the quality of care to the whole population. A tendency of the last years is to enhance the usage of self-care procedures, such as motor therapy sessions and rehabilitation exercises performed outside of "formal" healthcare structures. To monitor the correct execution of the intended exercises, and provide valuable self-correction information, scientists are proposing a broad range of technologies, wearable and not, that promise

---

* This work has been partially suported by Telecom Italia under the grant agreement POR CRO FSE 20072013, SISTAG, no. 16.

F. Ortuño and I. Rojas (Eds.): IWBBIO 2015, Part II, LNCS 9044, pp. 479–490, 2015.

to be usable in such contexts. Moreover, the cross- contamination among heterogeneous areas, like psychology, medicine, arts and technology is giving birth to
new approaches and methodologies, which can be combined to build systems that
are more effective in stimulating people's regarding the rehabilitation activities.

In particular e-Health and tele-rehabilitation, represent clear efforts aimed at
offloading hospitals and clinics from time consuming and costly services. In fact,
several operations related to rehabilitation can be carried out independently by
the patient at home or followed by an *informal caregiver*. By informal caregiver
we mean a relative or someone paid to assist the patient, who has no specific
knowledge on rehabilitation. However, a series of issues arise if patients have to
carry out rehabilitation exercises outside hospitals, as no specialized personnel
is observing and supporting them directly how to:

- how to *monitor* the exercise execution?
- how to provide a valuable *feedback* to the patient?
- how to *inform the patient / informal caregiver* about the execution performance?

Considering specific tele-rehabilitation sessions, focusing on motor tasks, we
propose the adoption of a set of wearable sensors coupled with an application
running on a smartphone to directly help the patient or the informal caregiver.
In this work, we do not focus on the communication between the system and the
physician therapist but rather on the patient side infrastructure.

The rest of the paper is organized as follows: Section 2 describes the components of a wearable tele-rehabilitation system; Section 3 the system requirements;
Section 4 presents the proposed system; finally, Section 5 concludes the paper
and discusses some future perspectives.

2   Wearable Tele-Rehabilitation System Components

In this paper we refer to tele-rehabilitation systems for patients that have to recover from injuries or rehabilitate after some kind of orthopedic surgery. These
systems are generally composed of three main components as schematically
shown in Figure 1: (i) a set of wearable sensors, (ii) a system to process sensor
data in real-time, integrate the data, present information to the patient, and
communicate with the remote physician, and (iii) an interface for the physician
to overview the rehabilitation work of the patients.

The wearable sensor nodes, whose number depends on the number of joints
that need to be monitored, send data to a central mobile unit that performs sensory integration reconstructing the posture of the monitored limbs and possibly
recognizes the performed actions. The central unit can be a mobile system that
runs an application under the Android operating system. The mobile application processes the data and provides indications and feedback to the user. This
work focuses on the system in charge of monitoring the patient activity, and on
providing a guide and feedback for the execution of rehabilitation sessions, while
a discussion of other components of the system can be found in [5].

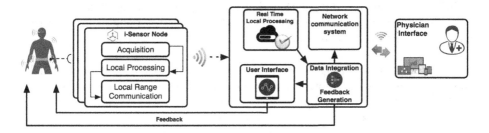

Fig. 1. The tele-rehabilitation reference model

A wide range of wearable devices has been developed for rehabilitation purposes in the last years [12,14]. The introduction of the MEMS (Micro-Electro-Mechanical Systems) enabled the development of motion sensors, like accelerometers, gyroscopes, and magnetometers. These sensors can be combined to create an Inertial Measurement Unit (IMU), that allows the acquisition of the body dynamics without external hardware to instrument the environment [13]. However, these devices suffer from non-negligible measurement errors, that give rise to a drift in the reconstructed signals.

Another possibility is to use vision-based systems [4]. For instance the Vicon [22] uses data captured from a set of cameras to compute 3D positions with a high precision. However they are quite expensive and require to instrument the scene and have a line of sight. On the other hand, low-cost solutions like the Microsoft Kinect [7], are more inaccurate. Another solution is represented by the use of exoskeletons, which are rigid structures mounted on the body where the interconnections are monitored with potentiometers or encoders to monitor the patients joint. These systems provide high precision, but are expensive and intrusive.

The following section briefly analyzes the requirements for the patient-side technological components of the tele-rehabilitation system.

3 System Requirements

When dealing with people that are not specialized in handling technology, devices, interfaces, and procedures have to be designed in such a way that they are unobtrusive, easy to use, and robust. In the next sections we briefly overview the requirements of the patient-side components of the system: the monitoring subsystem, the user interface, and the system for handling exceptions.

3.1 Monitoring Subsystem

Monitoring movements and actions is the primary task that needs to be executed in order to acquire information about how patients are executing motor tasks. This acquisition can be performed using a broad range of devices and technologies. However, data gathered through any kind of equipment needs to be sufficiently precise to enable analysis and to understand whether movements have been performed correctly.

Continuous Monitoring. Several health problems can only be detected by sporadic events that cannot be predicted in advance. For this reason some patients are required to be monitored continuously for 24 or 48 hours. The monitoring is a process composed of several phases and processes: gathering data through sensors, cross-checking the information, data analysis, data storage, and results notice to caregiver [3]. Such processes have to be continuous to provide relevant information about all actions performed by the patient and his/hers psycho-physical status.

Fixation. The setup procedure before the exercise must be simple and able to handle the issues related to the correct placement of sensor nodes. This is crucial to produce meaningful and accurate data for the analysis. The *fixation* of the sensors on the body should be easy to handle even for people with significant loss of functions [10]. The sensors need to be attached to each limb that needs to be monitored, and if interested in monitoring a single joint two sensors are needed, one for each segment starting from the joint. The user interface should guide the patient in such a delicate phase.

Accuracy and Precision. When monitoring limb positions and movements, there are crucial aspects to guarantee valid and useful measurements. For example, in knee tele-rehabilitation applications, the flexion/extension angle must be typically monitored with a precision of 1 degree [9].

Self Calibration and Auto-orientation. If the users operate in a non-controlled environment, it becomes impossible to exactly know in advance whether sensors are worn on the intended limbs. This problem can be solved by assessing the orientation of the sensors with respect to the limbs. Another issue is related to the nature of the used inertial sensors, accelerometers and gyroscopes, which suffer from errors in motion estimation because of measurement noise and fluctuation of offsets, thus requiring a specific *calibration* phase, that should be implemented in an automatic and easy way.

Real-Time Acquisition and Processing. Data acquisition and processing have to be managed by proper real-time kernels, since sensory data must be analyzed as they are produced to enable a prompt feedback generation for the user. A time synchronization is also needed among multiple nodes to know the precise time at which each sample is acquired. In fact, small time differences between samples affect the error in the sensor data integration phase.

Wearability and Comfort. Since the patient must carry several sensor nodes for long periods and during physical activities, the comfort becomes an important design consideration to provide the highest degree of convenience. This requires the development of unobtrusive devices with a small form factor. However, the node size is limited by the battery dimension because the node must have enough autonomy needed to perform the whole rehabilitation exercise without recharge. Also the weight is important because it can modify how the patient performs the exercise thus distorting the information.

Wireless Communication. Wireless communication technology is essential in these types of systems to avoid the encumbrance of wires and leave the user free to move during the execution of exercises.

Local Storage on the Node. is needed to save data in the case of temporary loss of connection in the wireless communication channel.

Energy. Energy consumption is a crucial problem in wireless devices, especially when they are required to be used for several hours (or days) for a continuous monitoring activity. Several solutions can be adopted to reduce energy consumption, as lowering the acquisition rates or turning off specific devices when they are not used. Also, a wireless recharge capability of the nodes is highly desired to simplify the task to non expert users.

Cost. The costs of the monitoring system becomes crucial in the cases in which this technology is adopted in a large scale. The current technology allows reducing the cost by an order of magnitude with respect to some commercial devices available today on the market.

3.2   Interaction Between System and User

The interaction between the system and the user needs to be accurately designed and implemented. In particular the feedback generation and the user interface are crucial, and will be analyzed next.

Feedback Generation for the User. Feedback can be broadly classified into different categories [19], depending on the *modality* and the *point in time* at which is provided. Concerning the modality, it can be *unimodal* (e.g., visual, auditory, haptic) or *multimodal* (a different combination of unimodal feedback). Concerning the time dimension, feedback can be *anticipatory*, if generated to help the user in predicting the correct time at which an action has to be performed (as in the case of a metronome); *contemporary*, if generated to provide a real-time representation or evaluation of the performance of the currently executed action; *posterior*, if generated after the task execution to provide a final evaluation of the exercise performance.

Visual feedback can be provided in the form of a virtual mirror [16], on which a 3D avatar represents a virtual image of the patient performing the same actions reconstructed by sensors; the avatar can also be coupled with a "ghost" avatar, acting as a guideline for the motion that needs to be performed. Rodger et al. [15] use a *auditive* feedback to produce walking sounds to help alleviating gait disturbances in Parkinson's disease. Sound was also used in other contexts to provide a feedback for different motor activities, e.g., to guide swimmers in increasing the degree of symmetry [6] or enhance a rehabilitation system [20]. *Haptic* feedback, in particular in the form of vibrotactile feedback, was used to help users in executing different locomotor performance tasks [18]. *Multimodal* feedback, namely the combination of the aforementioned three methods is thoroughly discussed in [19]. Motor-task learning and motor rehabilitation require different kinds of feedback: in learning, emphasis needs to be put on how the movement has to be performed, whilst in rehabilitation the feedback should provide information about erroneous performance, thus preventing wrong movements and motivating the process of rehabilitation [23].

Providing a feedback during the execution of exercises is important not only to guide the motion and provide immediate information about possible errors, but also to motivate the patient in a given direction of motion.

Easy Interface. Guiding the patient or the informal caregiver is essential if no formal caregiver is present during the rehabilitation session. In this case, to attain clinically relevant improvements [8] the system must be designed to provide a continuous guide for the patient through a graphical user interface. Before each exercise detailed illustrative instructions can help the patient in understanding how it has to be executed, while during the execution, visual and auditive cues can notify the user about possible errors, reached "check points", and progress state of the exercise. Taking into account that during the execution of the exercises sensor nodes are associated with mobile devices (tablets or smartphones), any type of information can be conveniently displayed on the mobile devices' screen. In fact, translating raw data from sensors to visual or auditory information allow to have "messages" that are immediately understood by the user. Once the exercise's execution is complete, the system should provide a short summary on the user performance, to allow him/her to understand the progress made in the rehabilitation process.

3.3 Handling of Errors and Unforeseen Conditions

To simplify the interaction with non expert users, the system should be able to detect at least two anomalous conditions that can derive from an incorrect usage of the system or from the malfunctioning of the system's devices:

- **Sensor misplacement detection and support**. The system should automatically detect whether the user is wearing the sensors correctly and, in the case or placement or orientation errors, provide indications on how to solve the problem.
- **Automatic detection of malfunctioning of devices**. The system should be able to detect a set of problems related to the wearable devices, such as low energy in the battery, values out of range, missing data, connection errors, etc. Once the problem is detected, the system should support the user to solve it, suggesting for instance to recharge the battery, or contact the help service for technical assistance.

4   Proposed System

The monitoring system described in this paper has been designed taking into account the main concepts and requirements described above. The system is composed of a set of wireless sensor nodes, a mobile application, which incorporates a graphic interface, an auditive feedback, auto-calibration functions for the sensors, and a communication module to send data to a remote server. These components are described below.

4.1   Sensor Nodes

The sensor nodes employed in the system use an inertial measurement unit (IMU) incorporating three accelerometers, three gyroscopes, and three magnetometers [2]. These signals are directly integrated onboard to provide an accurate estimation of the sensor orientation. The device has a very low-power consumption guaranteeing a continuous acquisition operation for at least 3 days at full sampling rate. It also provides a good balance between lifetime, dimensions ($4 * 3 * 0.8$ cm) and weight (30 g). The node and its internal circuitry are shown in Figures 2(a) and 2(b).

(a) Closed node   (b) Node's interior

Fig. 2. Sensor node

The nodes can be easily mounted on limbs using elastic bands, taking into consideration that a good node attachment technique increases the overall accuracy of measurements. Furthermore, the sensor nodes are characterized by an overall easy setup and handling, as they are equipped with a wireless recharging circuitry, an easy to calibrate IMU, and a Bluetooth 4.0 radio, that enables them to seamlessly work with modern Android devices. Figure 3 shows the block diagram of the node internal architecture, composed of an ARM-Cortex M0 processing unit, a Bluetooth 4.0 transceiver, an SD card slot for local storage, a 9-axis

IMU, a power manager with battery charge regulator, a Li-Po battery, and two chargers: one USB and one wireless.

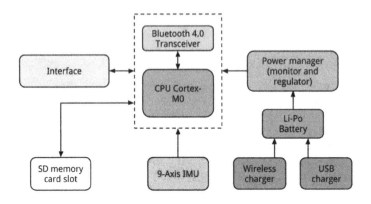

Fig. 3. The sensor node's block diagram.**TBD**

4.2   The Application Running on the Mobile Device

The mobile app running on the Android-based device incorporates a visual interface, an auditive feedback generation module, an auto-calibration unit, and a sensor data integration module. Data arrive to the application through the Bluetooth connection that is automatically established when the application is activated.

Auto-calibration Unit. The auto-calibration unit includes a set of functions, some of which have already been developed, while others are still under development. In particular, "fixed errors" are corrected performing a static calibration procedure of the nodes before being handed by patients, while "random errors", related to electrical noise or thermal drift are managed following the procedure illustrated by Gietzalt et al. [11], and should be further investigated to enable a sensor node to automatically compute the offset and adjust his calibration algorithm. Calibration techniques are paramount for the accuracy of measured data, especially in dynamic conditions. Designing hardware and software auto-calibration techniques is still an open research area. Once the data from each sensor is accurately processed and filtered, it can be integrated into a kinematic model of the body and used to reconstruct the parameters that are relevant for the rehabilitation process.

Processing. The processing activities running on the mobile device are in charge of integrating the various sensory data coming from the sensors, to reconstruct the posture of the monitored limbs, analyzing the reconstructed signals for detecting critical conditions, and evaluating the performance of the actions to generate an instructive feedback for the user. Such an evaluation is done by comparing the actual trajectory with a reference trajectory acquired in the presence of a physician (or physiotherapist), using techniques as Dynamic Time Warping [1] to derive the error signal for the feedback generation module.

The Visual Interface. The visual interface is developed to guide the patients in the execution of the exercises and is depicted in Figure 4. It includes a 3D Avatar (acting as a virtual mirror) and a simple 2D display for a more quantitative signal representation. The 3D Avatar replicates the movements executed by the patient or a set of prerecorded motions to act as a guideline for the what needs to be done. The 2D representation indicates the current angular value of a joint and a target position.

The Auditive Feedback. is embedded into the mobile application and reproduced either over loudspeakers or user worn earphones. Audio is produced by a module based on the PureData library for Android (libpd). The sound provides information about the difference between the target position and the current patient position, underlining and reinforcing the visual information form the simple 2D representation described before.

Communication Module. The communication module exploits the capabilities of the smart-phone to transfer all data about exercise execution to the physician or the official caregiver, according to a more general framework presented in [5].

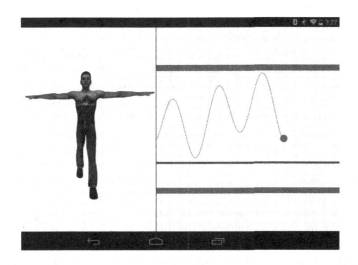

Fig. 4. The mobile application interface

4.3   The Complete System Usage

The proposed rehabilitation cycle that exploits the sensor nodes and the mobile phone is schematically represented in Figure 5. The user is required to wear the sensors, the system setups itself and then the exercise can begin, either carried out independently or with the help of a caregiver that follows instructions provided by the application. At the end of each exercise a quick report, acting as a post-execution, final effect feedback, is generated and provided to the patient and/or the informal caregiver.

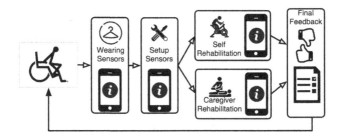

Fig. 5. The proposed rehabilitation cycle

5   Conclusions

This paper presented a mobile wearable monitoring system that can be effectively used during tele-rehabilitation sessions to monitor and evaluate the performance of motor exercises. The system has been designed and implemented according to a set of requirements aimed at simplifying the procedures and helping patients that are not familiar with technology.

Some features include a specific help to select the exercise, measure its execution performance, detect and notify sensor misplacements, and provide a valuable multimodal feedback that, at present, consists of visual and auditive stimuli.

In the future, to further reduce the impact of technology, we aim at incorporating energy-harvesting devices into the sensor nodes to prolong their lifetime and possibly avoid explicit battery recharge cycles. Other issues considered for a future research concern the following problems: (i) reducing the number of wearable sensors with respect to the number of interesting limb segments, by integrating a kinematic model of the body in the posture reconstruction algorithm; (ii) automatic detection of the limb where each sensor is mounted on, following the approach described in [21]; (iii) exploiting kinematic constraints of the human limbs to automatically estimate joints' axes and positions from inertial measurement data, following the approach [17] by T. Seel et al.

We wish to conclude the paper remembering that psychological aspects are of primary importance in a rehabilitation process. In fact, patients are often affected by post event depression and need to be motivated and encouraged to adhere to long term medical therapies. Rehabilitation exercises are usually considered boring for patients because of their repetitive nature. An interesting solution could be to present an exercise as a form of game, incorporating both pedagogical and entertainment elements, with increasingly difficult levels to make the patient feel the challenge and increase its involvement in the rehabilitation process.

References

1. Blackburn, J., Ribeiro, E.: Human motion recognition using isomap and dynamic time warping. In: Elgammal, A., Rosenhahn, B., Klette, R. (eds.) Human Motion 2007. LNCS, vol. 4814, pp. 285–298. Springer, Heidelberg (2007)
2. Buonocunto, P., Marinoni, M.: Tracking limbs motion using a wireless network of inertial measurement units. In: Proceedings of the 9th IEEE International Symposium on Industrial Embedded Systems (SIES 2014), Pisa, Italy, June 18-20 (2014)
3. Calvaresi, D., Claudi, A., Dragoni, A.F., Yu, E., Accattoli, D., Sernani, P.: A goal-oriented requirements engineering approach for the ambient assisted living domain. In: Proceedings of the 7th International Conference on PErvasive Technologies Related to Assistive Environments, p. 20. ACM (2014)
4. Cardinaux, F., Bhowmik, D., Abhayaratne, C., Hawley, M.S.: Video based technology for ambient assisted living: A review of the literature. Journal of Ambient Intelligence and Smart Environments 3(3), 253–269 (2011)
5. Cesarini, D., Buonocunto, P., Marinoni, M., Buttazzo, G.: A telerehabilitation framework for lower-limb functional recovery. In: International Conference on Body Area Networks, BodyNets2014, London, UK, IEEE Computer Society (2014)
6. Cesarini, D., Hermann, T., Ungerechts, B.: A real-time auditory biofeedback system for sports swimming. In: Proc. of the Int'l Conf. on Auditory Display 2014, Sonification for Sports and Performance Workshop, NY, US (June 22, 2014)
7. Chang, Y.-J., Chen, S.-F., Huang, J.-D.: A kinect-based system for physical rehabilitation: A pilot study for young adults with motor disabilities. Research in Developmental Disabilities 32(6), 2566–2570 (2011)
8. de Blok, B.M., de Greef, M.H., ten Hacken, N.H., Sprenger, S.R., Postema, K., Wempe, J.B.: The effects of a lifestyle physical activity counseling program with feedback of a pedometer during pulmonary rehabilitation in patients with copd: A pilot study. Patient Education and Counseling 61(1), 48–55 (2006)
9. Dejnabadi, H., Jolles, B., Aminian, K.: A new approach to accurate measurement of uniaxial joint angles based on a combination of accelerometers and gyroscopes. IEEE Transactions on Biomedical Engineering 52(8), 1478–1484 (2005)
10. Fong, D.T.-P., Chan, Y.-Y.: The use of wearable inertial motion sensors in human lower limb biomechanics studies: A systematic review. Sensors 10(12), 11556–11565 (2010)
11. Gietzelt, M., Wolf, K.-H., Marschollek, M., Haux, R.: Automatic self-calibration of body worn triaxial-accelerometers for application in healthcare. In: Second International Conference on Pervasive Computing Technologies for Healthcare, PervasiveHealth 2008, pp. 177–180. IEEE (2008)
12. Hadjidj, A., Souil, M., Bouabdallah, A., Challal, Y., Owen, H.: Wireless sensor networks for rehabilitation applications: Challenges and opportunities. Journal of Network and Computer Applications 36(1), 1–15 (2013)
13. Kim, J., Yang, S., Gerla, M.: Stroketrack: wireless inertial motion tracking of human arms for stroke telerehabilitation. In: Proceedings of the First ACM Workshop on Mobile Systems, Applications, and Services for Healthcare, p. 4. ACM (2011)

14. Pawar, P., Jones, V., Van Beijnum, B.-J.F., Hermens, H.: A framework for the comparison of mobile patient monitoring systems. Journal of Biomedical Informatics 45(3), 544–556 (2012)
15. Rodger, M., Young, W., Craig, C.: Synthesis of walking sounds for alleviating gait disturbances in parkinson's disease. IEEE Transactions on Neural Systems and Rehabilitation Engineering 22(3), 543–548 (2014)
16. Roosink, M., Robitaille, N., McFadyen, B.J., Hébert, L.J., Jackson, P.L., Bouyer, L.J., Mercier, C.: Real-time modulation of visual feedback on human full-body movements in a virtual mirror: development and proof-of-concept. Journal of NeuroEngineering and Rehabilitation 12(1), 2 (2015)
17. Seel, T., Schauer, T., Raisch, J.: Joint axis and position estimation from inertial measurement data by exploiting kinematic constraints. In: 2012 IEEE Int'l Conf. on Control Applications (October 2012)
18. Sienko, K.H., Balkwill, M.D., Oddsson, L.I., Wall, C.: The effect of vibrotactile feedback on postural sway during locomotor activities. Journal of Neuroengineering and Rehabilitation 10(1), 93 (2013)
19. Sigrist, R., Rauter, G., Riener, R., Wolf, P.: Augmented visual, auditory, haptic, and multimodal feedback in motor learning: A review. Psychonomic Bulletin & Review 20(1), 21–53 (2013)
20. Torres, A.V., Kluckner, V., Franinovic, K.: Development of a sonification method to enhance gait rehabilitation. In: 4th Interactive Sonification Workshop Proceedings of ISon (2013)
21. Weenk, D., Van Beijnum, B.-J.F., Baten, C.T., Hermens, H.J., Veltink, P.H.: Automatic identification of inertial sensor placement on human body segments during walking. Journal of Neuroengineering and Rehabilitation 10(1), 31 (2013)
22. Windolf, M., Götzen, N., Morlock, M.: Systematic accuracy and precision analysis of video motion capturing systems—exemplified on the Vicon-460 system. Journal of Biomechanics 41(12), 2776–2780 (2008)
23. Zijlstra, A., Mancini, M., Chiari, L., Zijlstra, W.: Biofeedback for training balance and mobility tasks in older populations: A systematic review. J. Neuroeng. Rehabil. 7(1), 58 (2010)

Non-intrusive Patient Monitoring for Supporting General Practitioners in Following Diseases Evolution

Davide Calvaresi[1], Daniel Cesarini[1], Mauro Marinoni[1], Paquale Buonocunto[1],
Stefania Bandinelli[2], and Giorgio Buttazzo[1]

[1] Scuola Superiore Sant'Anna,Pisa, Italy
{d.calvaresi,d.cesarini,m.marinoni,p.buonocunto,g.buttazzo}@sssup.it
http://retis.sssup.it
[2] Geriatric Unit, Azienda Sanitaria Firenze (ASF), Florence, Italy

Abstract. Current patient follow-up practices held by General Practitioners (GPs) are often unstructured. Due to the high number of patients and time limitations, data collection and trend analysis is often performed only for a small number of critical patients. An increasing demand is coming from the physician community for having a set of supporting tools for reducing the time needed to process patient data and speed-up the diagnosis process. Furthermore, the possibility of monitoring patient activities at home would provide less biased and more significant data. Unfortunately, however, current solutions are not able to collect reliable data without the intervention of formal caregivers. This paper proposes an improved version of some medically-backed techniques in an unobtrusive platform to monitor patients at home. Data are automatically collected and analyzed to provide GPs with the current status of the monitored patients and their health trend, contributing in a more precise and reliable decision making.

Keywords: eHealth, patient, tele-monitoring, General Practitioners.

1 Introduction

Demographic changes in terms of aging of population and evolution of habits led to an observable increase of the incidence of chronic diseases. In particular, elderly population is developing new needs and demanding an increasing support aimed at detecting and handling diseases as soon as possible. In fact, if not correctly and promptly detected and addressed, several diseases can become chronic, requiring personalized, specialized and costly treatments.

Currently, General Practitioners (GPs) pay much effort in gaining highly precise, single values of patients physiological state, such as blood pressure, blood values, or diagnostics exams. With respect to analyzing single values, more insights can be gained from the evolution of those values. However a too sparse sampling, over time, can hinder the significance of a study on the evolution of

F. Ortuño and I. Rojas (Eds.): IWBBIO 2015, Part II, LNCS 9044, pp. 491–501, 2015.
© Springer International Publishing Switzerland 2015

values. In fact, some studies [13] exploit a set of measurements over time to develop trend analysis, intercept the occurrence of changes in health status.

A particular condition characterized by a high risk of developing into a more dramatic and difficult one is the so called *frailty*, defined as "a distinctive health state related to the ageing process in which multiple body systems gradually lose their in-built reserves and enhances the risk of adverse outcomes in the elderly." [1] Several situations of frailty [11], requiring support by specialized personnel, can be identified among elderly people [2]. The state and the evolution of selected subjects can be described by different indexes, obtained from various motion and vital signals. Based on these indexes, people requiring healthcare interventions can be classified into a set of categories. The main categories are [9]: Chronic conditions (47%), Acute illness (25%), Trauma/injury or poisoning (8%), Dental (7%), Routine preventative health care (6%), Pregnancy/birth (4%), others (3%). The "chronic segment" is huge not only in terms of number of affected patients, but also in terms of costs for the health system. For a multitude of reasons, in most countries, the existing health structures cannot host all chronic patients. Nevertheless, these patients need a continuous support and monitoring in order to stabilize their condition and detect critical events in time. In addition, even though comorbidities should to be considered for these patients, they are often neglected by the existing practices, hiding possible situations of risk.

The role of the GPs is crucial, as most of non-critical patients live at home. Usually, to update anamneses, GPs use to evaluate patients in their own ambulatories. Unfortunately, mainly for monetary reasons, each GP is in charge of providing basic care for a high number of patients, with the consequence that the time reserved for each of them is often too low to acquire a complete personal state, both in physical and cognitive terms. The difficulty to deliver care given by GPs to patients over a long time period is exacerbated in rural and isolated environments, where ambulatories are not easily reachable and are only used in cases of acute illness, and not for prevention. On one hand, we have to consider that most patients can live independently even in the presence of little physical and health problems. On the other hand, difficulties related to the distance that patients need to cover to reach the ambulatory in terms of time, costs, mobility, etc. Even when these difficulties can be overcome, there is still a high chance that the data collected by the GPs are related only to a subset of events and patient states, similarly to the well known *white coat syndrome* [18], as some may not be manifest or reported by the patients, when they are visited.

Evidence exists about a correlation between motor activities and physical and mental state [21]. Thus, also in the case of elderly, the analysis of motor behavior can provide valuable information for GPs about possible alterations of health state. In fact, there are evidences from large-scale pilot studies [21] and [10] about the benefits of a continuous data collection regarding motion, during patients' daily living, to support the classical sparse and occasional medical examinations. Therefore, continuous monitoring performed at home could be essential not only

---

[1] http://www.bgs.org.uk/index.php/frailty-explained

to follow the evolution of overall patient state, but also to catch sporadic events that could reveal the occurrence of new comorbidities.

The public health system can be described as a chain of loosely coupled (sometimes disjoint) entities. New forms of professional associations of practitioners, such as the Italian "Associazione Funzionale Territoriale" (AFT), focus on providing an integrated care, in which some forms of specializations are coupled with General Practitioners duties. This new fragment of health system can benefit from a technological contribution. Information and Communication Technology (ICT) can play an important role to link together such entities and maintain a coherent and updated history of each patient, as well as to enable new monitoring and therapy procedures. Indeed, telehealth has already been announced several times as a solution to the above mentioned problems. However, telehealth requires the development of a new approach to simplify GPs' duties.

In this work we explore the possibility of integrating technology in the practices of GPs, to restructure some of the processes. According to [17], motor performance analysis can provide reliable information, and it can be used to develop health factors and indexes. Until now, such an analysis is carried out manually by health professionists [21]. There is, however, the possibility to execute some physical *tests* [2] automatically, and still achieve meaningful and accurate results. Exploiting the increasing availability of wearable motion sensors and vision based systems, we propose the development of an unobtrusive system for monitoring human activities in real-life scenarios. The aggregation of the acquired data can provide valuable information on the state evolution of patients, leading to enrich the number of parameters relevant for the diagnoses performed by physicians. In particular, we propose two specific procedures for gathering patient motor activities in two different contexts.

The next paragraphs are organized as follows: Section 2 presents the proposed methodology; Section 3 describes the motor tests considered in this work; Section 4 illustrates two possible scenarios in which the tests can be executed; finally Section 5 concludes the paper and provides some future steps.

2   Reference Methodology

The typical current practice followed by GPs in their ambulatory is to visit patients to gather both specific and general information by executing classical physical examinations and data transcriptions. Generally, the visit is followed by offline activities that have to be performed in a second phase to avoid interference with the visit. This process is schematically illustrated in Figure 1.

Possible offline operations are: *information organization, trends and alert analysis, etc.* However, such a phase split requires much more effort and time, with respect to a single phase procedure.

---

[2] The term *test* is used to denote a procedure performed on a patient to record a set of parameters aimed at assessing his/her activity.

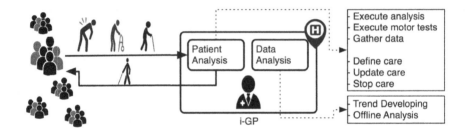

Fig. 1. Set of current GPs' practices

The proposed approach is to adopt a reference methodology considered as the current state of the art [21,10] and exploit the available technology to automatize the various required steps, including the execution of motor tests.

The reference methodology considered in this paper is schematically illustrated in Figure 2.

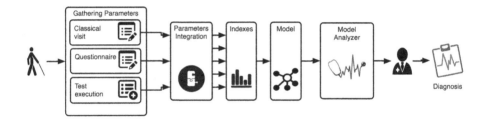

Fig. 2. Block diagram of the reference methodology

According to this approach, classical data obtained by physical examinations are integrated with patient's general details (acquired through proper questionnaires) and other parameters deriving from the execution of specific motor tests. Indexes indicating patient's general status are then derived by a set of algorithms and inserted into a model that can be analyzed to support the physician in diagnosis and therapy definition.

The next section presents two motor tests that can be included in the reference methodology described above.

3   Considered Motor Tests

Motor tests, in particular of lower extremities, can provide valuable information about particular conditions and patient state [20]. For instance, the analysis of the locomotion activity can produce the following indexes [21,10]: *Global fitness, Stride Regularity in medio-lateral direction, Forward stepping smoothness, Stepping symmetry,* and *Forward stepping regularity.*

Our goal is to define a set of procedures for deriving these indexes exploiting pervasive and wearable technologies.

The following sections describe two specific tests considered as a state of practice in the medical environment: the gait speed and the sit-to-stand test.

3.1   Gait Speed

The first test we present is also known as *"Gait speed calculation"*. In such a practice the patient walks over a known distance and time is recorded. The patient has to walk at usual pace, starting from a standing static condition and the average speed is calculated dividing the covered distance by the execution time. To standardize the procedure, distances are expressed in meters and time in seconds. The covered distance for this test is generally between 4 and 10 meters. A model to be used for this test is proposed by Guralnkik et al [12] for 4-m gait speed. The execution of the gait speed test is schematically illustrated in Figure 3.

The aforementioned test is generally integrated by additional patients' variables, like sex, age, ethnicity, height, weight, body mass index (BMI), smoking and alcohol habits, use of mobility aids, etc. The **prediction's** accuracy of such a complex model does not significantly differ from the results obtained with a more simplified model with takes into account only patient's *sex* and *age* coupled with his *motor performance* related to the executed test [21] [17].

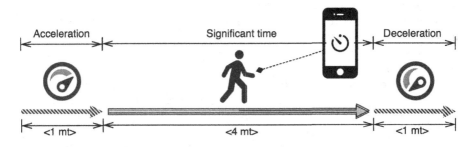

Fig. 3. Gait Speed test

3.2   Sit-to-Stand

Another interesting and standardized procedure is the *"Sit-to-Stand"* (SiSt) test [3]: the patient is seated on an armless adjustable chair, whose seat must be placed at knee height; the patient must fold the arms across his/her chest, while his/hers feet have to be positioned at approximately 10 cm of distance from each other, with the shanks positioned at an approximately 10 degrees angle relative to the vertical axis. The complete movement/cycle is composed of sitting, flexion, extension, standing, flexion, extension, sitting phases, as shown in Figure 4. This simple test can be utilized differently: considering the total time needed to perform the test [12], averaging the measures over the five repetitions obtaining more significant data [8] or as numbers of times the patient stands in 30 seconds to assess endurance and leg strength.

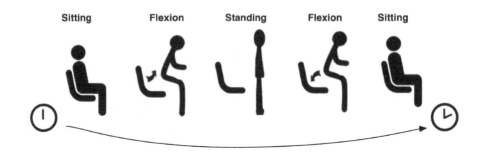

Fig. 4. Sit-to-Stand test

3.3 Technology for Index Extraction

To derive the physical status indexes described above in a way that is both cost effective and easy to use, while providing high-quality data two technology solutions can be adopted: patients could use a single smartphone, exploiting its on-board kinematic sensors, carrying it in their pocket; or, they could wear simple independent wireless motion sensors, as done in [5,19,6], connected to a smartphone through a wireless link.

For example, in the case of *gait speed test*, a waist mounted accelerometer, combined with other similar sensors [23,15] can provide traveled distance and speed, using dead reckoning techniques [22], and medio-lateral and frontal symmetry, exploiting auto-correlation of the signal along the medio-lateral direction [1]. Similarly, for the *sitting-standing test*, we can obtain another series of characterizing parameters, such as the time from sitting to standing, by segmenting the signal provided by a waist mounted accelerometer, detecting the overall sequence of positions of the waist while performing the complete test. Another fruitful choice is to use visual sensors [16], such as the Microsoft Kinect or other cameras endorsed with a depth sensor, taking advantage of image processing and analysis. With respect to kinematic sensing, visual sensing has less precision and resolution, but more information can be derived about body postures, as evidenced by several existing studies [4,14].

The next section describes two scenario in which this technologies can be used.

4 Scenarios

Exploiting the large variety of sensing and communication devices offered by modern technology in the *gait speed* and *sit to stand* practices, GPs will be enabled to quickly perform more precise tests on the patients. In addition, by processing the acquired data through proper algorithms, it is possible to provide physicians with a valuable support for a faster and more reliable diagnosis. In the following, we propose two possible scenarios where the tests described above can be effectively adopted. In fact, *the reduction of the time of treatments* is a key milestone, as pointed out in [7]. They involve different actors for the execution of this tests.

The first scenario is illustrated in Figure 5 and considers a single patient wearing a personal sensing system in his/her own environment.

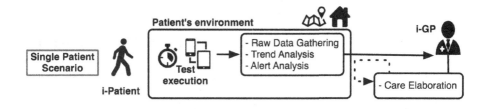

Fig. 5. Proposed Single patient scenario node in healthcare chain

The second scenario, illustrated in Figure 6, considers the possibility of using more complex and costly equipment that can be shared by more users in specialized healthcare centers managed by volunteer associations with help of semi-formal caregivers.

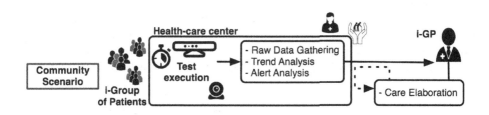

Fig. 6. Proposed Community scenario node in healthcare chain

4.1 Single Patient Scenario

In the single patient scenario, tests are performed using an inertial wearable sensor. The required operations to perform the test are depicted in Figure 7.

For the first setup, the user starts downloading the app suggested by the physician, Task (a). Then, once the user wants to execute the test he/she moves

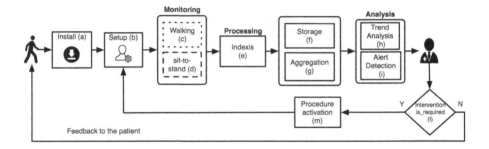

Fig. 7. Single patient scenario's procedure

to Task (b), wearing the sensor. Then, the user is guided by the app in performing Task (c), walking, in the case of the gait test, or Task (d), SiSt, in the case of the sit-to-stand test. In both cases, sensory data are sent to the smartphone, where a set of algorithms process them to compute the performance indexes, Task (e). Such indexes are then stored, Task (f), and aggregated in time, Task(g), for performing trend analysis, Task (h), and alert detection, Task (i).

In the case of a detection of critical or anomalous situations, the related parameters are notified to the GP, who can evaluate the situation, Task (i), and decide the appropriate procedure to be activated, notifying the user.

4.2   Associations Scenario

The second scenario we propose consists in delegating the data gathering phase to associations or charity organizations that have among their personnel at least some semi-formal caregivers. The practice for data extraction could be the same, using simple sensors and smartphones, as already described for the first scenario, or improved using more complex equipment. When using more expensive sensors, procedures can involve multiple users. Moreover, provided that huge amounts of data will be collected for the executed tests, statistical analysis can be performed

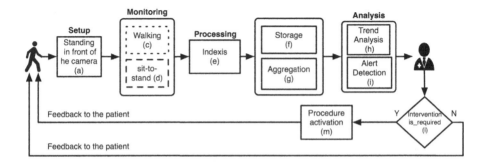

Fig. 8. Community scenario's procedure

to derive useful results for further research studies. The sequence of involved tasks is reported in Figure 8. The setup phase, Task (a), requires the patient just to stand in front of a camera in order to be recognized. After that, the normal test execution is similar to the process described in the first scenario. Given the simplicity of the setup, the time required for each patient (recognition and tests execution) is relatively short, (in the order of seconds).

5 Conclusions

This paper presented a set of technological solutions for achieving faster, simpler as well as more reliable and reproducible motor testing procedures. Two established test procedures, namely the "Gait Speed" and "Sit-to-Stand" tests, have been considered to compute specific patients health indexes derived by healthcare experts. Taking into account some critical factors we proposed to simplify and innovate the presented tests execution and management. Indeed we investigated how to relax some technological constraints to achieved the same results. According to current resources in terms of actors and technology we identify some requirements and deployment scenarios. Based on the several studies in the literature, we believe that the proposed unobtrusive and simplified data collection, the automatic production of trend analysis for diagnosis support especially for chronically suffering patients, will bring considerable benefits to the public health system. We envision that the time required to perform the entire process of data collection, analysis, diagnosis and care delivering will be reduced. In the future, we plan to implement the proposed approach and develop new techniques and tests that will be integrated into the presented working-chain.

The development of a diffuse adoption of sensing and monitoring technology, carried out by the deployment of the equipment needed in the described scenarios, involving a huge number of patients and participants, could represent a valuable opportunity to test new algorithms against real data, thus attracting even more researchers to invest effort and knowledge in the development of new and smarter solutions.

References

1. Avvenuti, M., Casella, A., Cesarini, D.: Using gait symmetry to virtually align a triaxial accelerometer during running and walking. Electronics Letters 49(2), 120–121 (2013)
2. Bavazzano, A., Badiani, E., Bandinelli, S., Benvenuti, F., Berni, G., Biagini, C.A., Briganti, N., Cai, M., Canavacci, L., Carriero, G., Cavallini, M.C., Francesconi, P., Gabbani, L., Galantini, P., Giraldi, M., Lombardi, B., Magnolfi, S., Michelotti, F., Nastruzzi, A., Rossi, C., Tomei, A., Tonelli, L.: Frailty in elderly people. Technical report, Regional Health Council, Tuscany Region, Italy (2015)
3. Beauchet, O., Dubost, V., Herrmann, F., Rabilloud, M., Gonthier, R., Kressig, R.W.: Relationship between dual-task related gait changes and intrinsic risk factors for falls among transitional frail older adults. Aging Clinical and Experimental Research 17(4), 270–275 (2005)

4. Benetazzo, F., Iarlori, S., Ferracuti, F., Giantomassi, A., Ortenzi, D., Freddi, A., Monteriu, A., Capecci, M., Ceravolo, M.G., Innocenzi, S., Longo, S.: Low cost rgb-d vision based system for on-line performance evaluation of motor disabilities rehabilitation at home. In: Proceedings of the 5th Forum Italiano on Ambient Assisted Living ForItAAL. IEEE (2014)

5. Buonocunto, P., Marinoni, M.: Tracking limbs motion using a wireless network of inertial measurement units. In: 2014 9th IEEE International Symposium on Industrial Embedded Systems (SIES), pp. 66–76. IEEE (2014)

6. Burns, A., Greene, B.R., McGrath, M.J., O'Shea, T.J., Kuris, B., Ayer, S.M., Stroiescu, F., Cionca, V.: Shimmer–a wireless sensor platform for noninvasive biomedical research. IEEE Sensors Journal 10(9), 1527–1534 (2010)

7. Calvaresi, D., Claudi, A., Dragoni, A.F., Yu, E., Accattoli, D., Sernani, P.: A goal-oriented requirements engineering approach for the ambient assisted living domain. In: Proceedings of the 7th International Conference on PErvasive Technologies Related to Assistive Environments, p. 20. ACM (2014)

8. Cao, H., Plichta, M.M., Schäfer, A., Haddad, L., Grimm, O., Schneider, M., Esslinger, C., Kirsch, P., Meyer-Lindenberg, A., Tost, H.: Test–retest reliability of fmri-based graph theoretical properties during working memory, emotion processing, and resting state. Neuroimage 84, 888–900 (2014)

9. Conway, P., Goodrich, K., Machlin, S., Sasse, B., Cohen, J.: Patient-centered care categorization of us health care expenditures. Health Services Research 46(2), 479–490 (2011)

10. Deshpande, N., Metter, E.J., Bandinelli, S., Guralnik, J., Ferrucci, L.: Gait speed under varied challenges and cognitive decline in older persons: A prospective study. Age and Ageing, afp093 (2009)

11. Fried, L.P., Tangen, C.M., Walston, J., Newman, A.B., Hirsch, C., Gottdiener, J., Seeman, T., Tracy, R., Kop, W.J., Burke, G., et al.: Frailty in older adults evidence for a phenotype. The Journals of Gerontology Series A: Biological Sciences and Medical Sciences 56(3), M146–M157 (2001)

12. Guralnik, J.M., Ferrucci, L., Pieper, C.F., Leveille, S.G., Markides, K.S., Ostir, G.V., Studenski, S., Berkman, L.F., Wallace, R.B.: Lower extremity function and subsequent disability consistency across studies, predictive models, and value of gait speed alone compared with the short physical performance battery. The Journals of Gerontology Series A: Biological Sciences and Medical Sciences 55(4), M221–M231 (2000)

13. Hope, C., Lewis, C., Perry, I., Gamble, A.: Computed trend analysis in automated patient monitoring systems. British Journal of Anaesthesia 45(5), 440–449 (1973)

14. Iarlori, S., Ferracuti, F., Giantomassi, A., Longhi, S.: RGB-D video monitoring system to assess the dementia disease state based on recurrent neural networks with parametric bias action recognition and DAFS index evaluation. In: Miesenberger, K., Fels, D., Archambault, D., Peňáz, P., Zagler, W. (eds.) ICCHP 2014, Part II. LNCS, vol. 8548, pp. 156–163. Springer, Heidelberg (2014)

15. Mannini, A., Sabatini, A.M.: Walking speed estimation using foot-mounted inertial sensors: Comparing machine learning and strap-down integration methods. Medical Engineering & Physics 36(10), 1312–1321 (2014)

16. Obdrzalek, S., Kurillo, G., Ofli, F., Bajcsy, R., Seto, E., Jimison, H., Pavel, M.: Accuracy and robustness of kinect pose estimation in the context of coaching of elderly population. In: 2012 Annual International Conference of the IEEE Engineering in Medicine and Biology Society (EMBC), pp. 1188–1193. IEEE (2012)

17. Peters, D.M., Fritz, S.L., Krotish, D.E.: Assessing the reliability and validity of a shorter walk test compared with the 10-meter walk test for measurements of gait speed in healthy, older adults. Journal of Geriatric Physical Therapy 36(1), 24–30 (2013)
18. Pickering, T.G., Eguchi, K., Kario, K.: Masked hypertension: A review. Hypertension Research 30(6), 479 (2007)
19. Roetenberg, D., Luinge, H., Slycke, P.: Xsens mvn: full 6dof human motion tracking using miniature inertial sensors. Xsens Motion Technologies BV, Tech. Rep. (2009)
20. Stenholm, S., Guralnik, J.M., Bandinelli, S., Ferrucci, L.: The prognostic value of repeated measures of lower extremity performance: Should we measure more than once? The Journals of Gerontology Series A: Biological Sciences and Medical Sciences, glt175 (2013)
21. Studenski, S., Perera, S., Patel, K., Rosano, C., Faulkner, K., Inzitari, M., Brach, J., Chandler, J., Cawthon, P., Connor, E.B., et al.: Gait speed and survival in older adults. JAMA 305(1), 50–58 (2011)
22. Tian, Z., Zhang, Y., Zhou, M., Liu, Y.: Pedestrian dead reckoning for marg navigation using a smartphone. EURASIP Journal on Advances in Signal Processing 2014(1), 1–9 (2014)
23. Trojaniello, D., Cereatti, A., Pelosin, E., Avanzino, L., Mirelman, A., Hausdorff, J.M., Della Croce, U.: Estimation of step-by-step spatio-temporal parameters of normal and impaired gait using shank-mounted magneto-inertial sensors: application to elderly, hemiparetic, parkinsonian and choreic gait. Journal of Neuroengineering and Rehabilitation 11(1), 152 (2014)

Interactive Business Models for Telerehabilitation after Total Knee Replacement:
Preliminary Results from Tuscany

Francesco Fusco and Giuseppe Turchetti

Institute of Management-Management and Innovation (MAIN) Scuola Superiore Sant'Anna,
Pisa, Italy
{f.fusco,g.turchetti}@sssup.it

Abstract. To date Total Knee Replacement (TKR) is one of the most performed procedures in Italy; likewise, rehabilitation after TKR accounts for 182 million of euro each year. The deployment of ICT was able to increase the efficiency in several areas, but in healthcare sector still fails to be widely adopted. According to management literature, business modelling is crucial for a product success and the stakeholder engagement is valuable as well. In this direction, we designed 4 telerehabilitation business/governance models through brainstorming session and developed them interviewing a large sample of the stakeholders involved into the telehealth arena. Whereas the decision makers highlighted the need of gradual changes in healthcare, the preliminary results showed the interest in exploring innovative governance pathways able to directly involve the patients in the healing process and reduce waiting lists over the regional healthcare service. Future research aims to capture the others stakeholders perspectives.

Keywords: Business model, telemedicine, canvas, stakeholder, total knee replacement, telerehabilitation, rehabilitation.

1 Introduction

Total Knee Replacement (TKR) has an important role in healthcare expenditure, as it is the 21st most performed surgical procedure in Italy in 2013 [1,2]. In this sense, "Istituto Superiore di Sanità" observed the TKR performed in Italy has more than doubled from 2001 (26'694) and 2011 (63'125). "Agenzia Nazionale per i servizi Sanitari" (Age.na.s.) forecasts this trend to further increase in the next years, with a reduction in the average age of the target population for knee prostheses surgery [3]. Therefore, the increment in TKR procedures is going to have an effect on the knee rehabilitation as well. Although rehabilitation showed to improve the patients' recovery after surgery [4], the expected increment in TKR procedures could lead to serious concerns with respect to the socio-economic sustainability of knee rehabilitation. According to Piscitelli et al [5], rehabilitation for TKR currently accounts for almost 182 million of Euros per year. Since years Information and Communication Technologies (ICT) caused a revolution in user's everyday life. Although mobile

F. Ortuño and I. Rojas (Eds.): IWBBIO 2015, Part II, LNCS 9044, pp. 502–511, 2015.

communication and internet diffusion have widely spread in several fields, with an improvement in terms of GDP and productivity growth [6], they have not reached the same diffusion and impact in healthcare. In this sense, the drivers responsible for the success of telemedicine programs are still uncertain and further research is required. According to Osterwald, business model for innovative services/products should support the value creation for the whole society (i.e. patients, caregivers, medical personnel and decision makers) rather than exclusively for the firm that provides them [7]. Although the aims of the stakeholders could partially vary, none of them can be neglected. Therefore, we aim to detect the best business model able to optimize the value creation for most of the telerehabilitation stakeholders. In the current work, we report the preliminary results coming from the public healthcare decision makers interviews that we performed in the "Area Vasta Nord-Ovest" (ESTAV) in Tuscany (i.e. Livorno, Pisa, Lucca, Viareggio and Massa-Carrara health care districts).

2   Methods

The whole project was based on the design science research methodology (DSRM) [8], which is a five steps approach aimed to define the objectives, the design, the development, the demonstration and the evaluation of a new solution. The context and the objectives have been described in the previous section; the current work addresses the following two phases: design and development. Design was dealt with brainstorming sessions with researchers of the Institute of Management and a pool of experts composed by employees of a primary telecommunication company, a physiatrist and a TKR patient. The development phase is still ongoing as we are currently running face-to-face interviews to decision makers, physiotherapists, patients and caregivers. The present work is focusing on decision makers. During the interviews, they were asked to provide their believes related to the Osterwald's canvas sections [7] (key activities, key partners, key resources, customer relationship, customer segments, value proposition, channels, cost structure) for each business/governance model proposed. At the really end of the interviews they were asked to rank the models according to their preference. The current work will provide an insight on those sections of the Osterwald's canvas which are mostly of interest to the decision makers' (i.e. key activities, key resources, customer/patient segments).

2.1   Key Activities

This section of the interviews proposed three different scenarios, composed by not dividable couple of activities (one acquisition of the device activity and an activity to manage telerehabilitation sessions), ranging from full control of the telerehabilitation service (i.e. acquire the device and completely follow-up the patients) to delegate the telerehabilitation tasks to the Telco (i.e. payment of the service based on the pay-as-you-go scheme and leave the complete control of the telesessions and data to the Telco providing the service). While the intermediate option provided a partial control on the service (i.e. rental of the devices , the Telco manages the telesessions, but reports the data to the healthcare units). The interviewed persons were asked to state their preference for the 3 sets of activities. Additionally, they were asked to provide any further activity they would include and to match the two preferred activities listed in the previous questions.

2.2   Customer/Patient Segments

Decision makers were asked, basing on their experience, to provide the best profile for patients to candidate for telerehabilitation service for each model. The items enclosed in this section were: gender, maximum age, education, cohabiting status and working status.

2.3   Key Resources

To capture the importance of single resources items for the success of telerehabilitation service, we administered them the Visual Analogue Scale (VAS) for the following resources flows: local health authority (ASL) expenditure, regional healthcare reimbursement, national authority payment/incentives, incentives and patients' payment. The VAS value ranged from 0 to 10 – where 0 stands for "not important resources flow" while 10 represents "extremely important resource flow".

3   Results

3.1   Designing Phase: Models for Telerehabilitation

The brainstorming sessions during the designing phase resulted into 4 different governance/business models, which were structured increasing the level of innovation from the first model up to the fourth one. From the second model up to the fourth, we designed our models within a pay-for-performance incentives frame. This choice is due to the evident easiness in collecting data through telemedicine and the opportunity to have an impartial tool to report performance data. The administered models are further described in following subsections.

Model I. The first model (Figure 1) is conservative as most of the process flows follow the usual healthcare pathway in place in Italy. To summarize, the Italian National Healthcare Service (Ita-NHS) is organized into regional healthcare services (Regional-NHS). The regional-NHS budget is based on regional taxations and "intramoenia" activities (i.e. private care procedures performed into public healthcare units - part of the revenues of the physician is shared with the healthcare unit)[9,10] and it is divided among the primary care units (i.e. ASL) according to specific goals. The regional-NHS covers almost the whole cost, with a small fixed fee to be paid by patients (i.e. "ticket").The ASL provides half of the number of the rehabilitation service in usual care and the other half at patients' houses through telerehabilitation (mixed UC/TR).

Model I

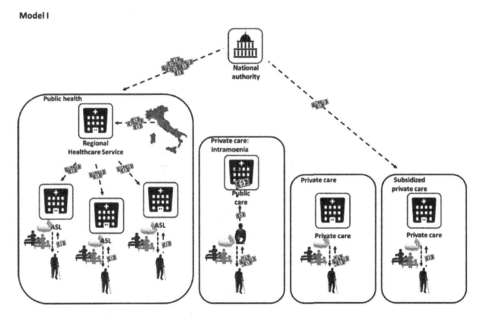

Fig. 1. Telerehabilitation business/governance models: Model I

If the regional budget is not sufficient to cover the whole healthcare service expenditure, the National Authorities (i.e. Ministry of Health and Ministry of Economy and Finance) cover the necessary amount of resources to deliver healthcare in such regions. Likewise, national authorities can subsidize private care units, who deploy the same mixed UC/TR service, to preserve the accessibility to the minimum level of healthcare service. In addition, both "intramoenia" service and private care would deliver the UC/TR service. The devices (i.e. a couple of sensors and tablet) are provided by the Telco supporting the service to each ASL.

Model II. The second model (Figure 2) supposes a partnership between the Telco and the ASL. The first will provide the service and the latter will perform the healthcare sessions according to mixed UC/TR scheme. The data obtained during the telesessions are used to observe patients' adherence to the prescribed treatment. The adherence rate is adjusted on socio-demographic characteristics of the patients, in order to avoid inequity between different kinds of patients (i.e. younger patients are more likely to understand and use the service rather than elderly ones. In addition, patients with co-morbidities could have a lower adherence unrelated to their willingness). The telesessions data will be used to scale-down the patients co-payment according to their adherence rate. In other terms, the higher is the patient's adherence, the lower is the patient's payment. The telesessions data are forwarded to the national authority, which will cover the part of the ticket not paid by the patients with good and normal adherence. From societal perspective the incentives would be justified due to the ability of the service to ensure a high quality care service (because of the ensured adherence to treatment) and the reduction of productivity loss (patients would perform the telesession after the working day).

Model II

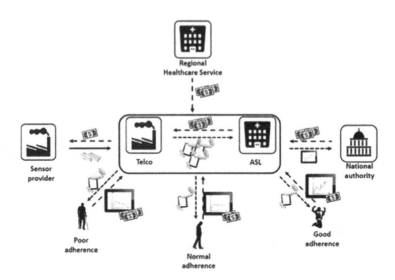

Fig. 2. Telerehabilitation business/governance models: Model II

Model III. The third model (Figure 3) is based again on a partnership between the Telco and the ASL. However, in this case the medical personnel is encouraged to push the innovation into their unit thanks to incentives according to their performance. The patients will follow the mixed UC/TR treatment, and their data are used to assess who among the medical personnel achieved the best adherence in his/her group of patients. Also in this case the scheme pay-for-performance is adopted to cover medical personnel's incentives and those expenditure belonging to an improvement of the performance in the healing process supported by telesession data.

Model IV. The fourth model (Figure 4) supposes that the healthcare units could have a different rate of access to their services resulting into long queues for a treatment in some geographical areas and a temporary unemployment of medical personnel in other areas. Therefore, we proposed a partial digitalization of physiotherapists, able to perform telesession or teleconsultation for those healthcare units with a lack of physiotherapists or for those areas with difficult access (e.g. rural areas). In the case of telesession the virtual physiotherapist will remotely monitor the patients. While in the teleconsultation case, the physiotherapist will provide support to other medical personnel to perform the rehabilitation session. The early stage of the model adoption supposes semi-virtual physiotherapists due to the usual care activities in the healthcare units. However, we believe this model could result into a new occupation category like full time virtual-physiotherapist. The healthcare units receiving the remote procedures will pay a fee to those providing remotely the medical personnel, who shares part of the revenue to the forming healthcare unit. Finally, the national authority will provide incentives due to the improvement in efficiency of treatment (i.e. reduction of waiting list and transportation of patients into a different healthcare unit) to the pool of experts and to the healthcare units they belong to.

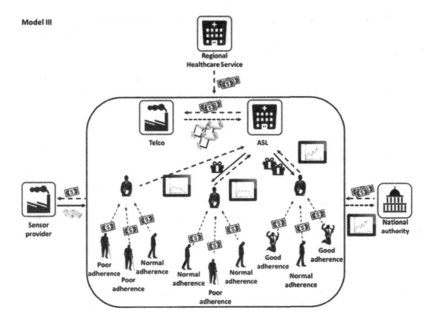

Fig. 3. Telerehabilitation business/governance models: Model III

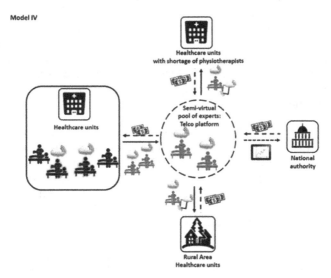

Fig. 4. Telerehabilitation business/governance models: Model IV

3.2 Development Phase: Insights from Decision Makers in Tuscany

The development phase is still ongoing, in this section we report the major remarks from interviews with decision makers of the public rehabilitation units in ESTAV. Currently we have interviewed 4 out of 6 of the directors of rehabilitation units (Livorno, Pisa, Lucca and Massa-Carrara), the average age is 56.6 years and the 80% of them is male. All of the interviewed persons have at least a personal computer and use it for the following reasons: working purposes (within the healthcare unit and to contact their patients) and in their spare time.

Key Activities. All the stakeholders interviewed agreed toward the intermediate set of activities. The first and the third models have registered the same preferences towards the intermediate option (3 choose it as first option and 1 as second option); likewise, the second and fourth registered the same trend. However, both of them showed a higher willingness to delegate the telerehabilitation session to Telco, especially in the fourth model. However, in the first three models when asked to match the preferred activities across the proposed scenarios, they preferred the rental option (only one preferred the pay-as-you-go option). On the other hand, the 50% of stakeholders choose the intermediate option to manage the telesessions, while the other 50% choose the full control option. The fourth model led decision makers to prefer the pay-as-you-go option (3 out of four), while only one still preferred the rental. Finally the intermediate option to manage telesession is the most preferred also in this case (50% of the decision makers).

Customer/Patient Segments. The target population for a telerehabilitation service would not vary across the different business/governance models. There is a large consensus on: gender, education level, cohabiting status and working status. Briefly, the best profile for a telerehabilitation service user is male or female, cohabiting, with a secondary school degree (i.e. high-school) and works as freelance professional or at least as employee. On the other hand, the responses about the maximum age range from 65 years (50%) to 70 years (50%). Although most of the features did not vary across the different models, one decision maker raised the concern about gender and age for the second and third model. He/She stated that the performance schemes would be a problem for older patients; additionally, he/she supposed it would work better for male patients rather than females.

Key Resources. When asked to describe the importance of some resources items, the responses varied accordingly with the proposed models. Although in the most conservative scenario (Model I), the regional-NHS reimbursement has a pivotal role, it becomes decreasingly important according to the innovativeness of the model. On the other hand, the importance of national authority payment/incentive increases from model I to model III. Nevertheless, in the fourth model there is not a defined trend, the ASL expenditure was the most important item, while the importance of the other items level out.

Model Comparison. At the end of the interview, all the decision makers were asked to express their preference list for the showed models. The most preferred one was the second model; although none of the remaining models prevailed on the others, there was a slightly preference for the first one as second choice.

4   Discussion

Since years business modelling has played a pivotal role in determining firm success; in this sense Chelsbourgh stated the importance of business models up to be responsible, if not properly designed, for failure even for good products [11]. Considering the lack of value co-creation to be the most important issue in the new services development [12] and the co-creation pathway to be determined by the firm's willingness to collaborate with the other stakeholders to achieve a shared benefit [11]; our work aimed to provide 4 business/governance models and refine them according to the stakeholders needs and insights through face-to-face interviews. Here we propose some remarks form the preliminary results outlined in this paper. When asked to describe whom of the TKR population would benefit from telerehabilitation, decision makers agreed on patients' characteristics. In addition, changing the models does not seem to have an effect on the maximum age of the target population and its working status. In other terms, decision makers seemed to appreciate the advantages of telerehabilitation on patients' management independently from which model would be implemented. The only concern about age and sex was raised in model II from a decision maker, which supposed telerehabilitation would yield better results in younger male patients. However, the stakeholders agreed about the working and the cohabiting status, since telerehabilitation service could reduce productivity lost especially for patients and caregivers working as freelance professional. The control of follow-up telesession was a crucial point in the key resources section. None of the interviewed was interested in completely acquire the devices in any of the proposed models. However, they would prefer to control the telesessions; nevertheless, the trade-off between buying or renting the devices, sacrificed the full control on telesession in favour of the rental scheme. The only model able to modify this trend was the fourth, where it was preferred to completely delegate the telerehabilitation session to the Telco. The key resources section registered a high variability, especially for the third-party payer reimbursements. Although the regional-NHS decreases of importance from model I to IV, the impact of the national authority incentives increases over the 4 models. The incentives for telerehabilitation seems to be more likely to be accepted if aimed to reduce the patients' co-payment (model II), rather than encourage physiatrists and physiotherapists to push the innovation over the healthcare units. This could be related to the difficulties in defining the incentive size and nature (i.e. positive incentive for higher performance, penalties for low performances), leading to the reluctance to risk in this kind of incentives schemes [13]. Therefore, if there is a willingness to innovate, involving more the patients into the healing process (model II), there is still some reluctance to completely overturn the healthcare system. Whereas the interviewed persons choose the second model (as first choice in the 50% and 25% as second choice of the interviewed), they highlighted the need of a gradual introduction of innovative services in healthcare through usual governance pathway (model I). The paper showed the preliminary results of a broader work. Although the decision makers enrolment is still ongoing, here we reported the insights of the majority of them. Nevertheless, we could draw some not definitive conclusion about telerehabilitation in ESTAV. The top-management of public healthcare units positively responded to hypothetical scenarios with ICTs able to reduce expenditures and waiting lists. There is a growing trust towards third-party service provider (i.e. Telerehabilitation experts deployed by a Telco), proved by the willingness to partially delegate the service (i.e. the intermediate dominated the other options in the key activities section). Likewise, the acceptance of innovation in public health is relying on national authority legislations and incentives.

The patients' age does not seem to be a constraint from decision makers perspective, who were confident in telerehabilitation usability for the elderly. Further conclusion will be drawn in the future steps as in this sense the other stakeholders perspective is too important to be neglected; in addition, economic evaluations based on the different models are claimed to prove the telerehabilitation models sustainability. The future research will focus on gathering believes of the medical personnel, patients and caregivers. In addition, a decision model [14] will simulate the socio-economic burden of these models.

Acknowledgments. Authors thank Miss Maral Mahdad for the enlightening conversations, the support in the designing and questionnaire development phases. We also thank the healthcare units' directors from Livorno, Lucca, Massa-Carrara and Pisa districts for participating in the interviews. Finally, we thanks Telecom Italia spa, who funds the project.

References

1. Ministero della Salute. Tavole Rapporto SDO (2012),
   http://www.salute.gov.it/portale/documentazione/p6_2_8_3_1.j
   sp?lingua=italiano&id=16
2. Torre, M., Luzi, I., Romanini, E., Zanoli, G., Tranquilli Leali, P., Masciocchi, M., Leone, L.: The Italian Arthroplasty Registry (RIAP): State of the art. G. Ital. di Ortop. e Traumatol. 39, 90–95 (2013)
3. Cerbo, M., Fella, D., Jefferson, T., Migliore, A., Paone, S.: P.M., L, V.: Agenas - Le protesi per la sostituzione primaria totale del ginocchio in Italia,
   http://www.salute.gov.it/imgs/C_17_pagineAree_1202_listaFile
   _itemName_4_file.pdf
4. Bohannon, R.W., Cooper, J.: Total knee arthroplasty: Evaluation of an acute care rehabilitation program. Am. Acad. Phys. Med. Rehabil. 74, 1091–1094 (1993)
5. Piscitelli, P., Iolascon, G., Di Tanna, G., Bizzi, E., Chitano, G., Argentiero, A., Neglia, C., Giolli, L., Distante, A., Gimigliano, R., Brandi, M.L., Migliore, A.: Socioeconomic burden of total joint arthroplasty for symptomatic hip and knee osteoarthritis in the Italian population: A 5-year analysis based on hospitalization records. Arthritis Care Res (Hoboken) 64, 1320–1327 (2012)
6. Gruber, H., Koutroumpis, P.: Mobile telecommunications and the impact on economic development. In: Economic Policy Fifty-Second Panel Meeting, pp. 22–23 (2011)
7. Osterwalder, A., Pigneur, Y., Smith, A., Movement, T.: Business Model Generation: A Handbook for Visionaries, Game Changers, and Challengers (2010)
8. Peffers, K., Tuunanen, T., Rothenberger, M.A., Chatterjee, S.: A design science research methodology for information systems research. J. Manag. Inf. Syst. 24, 45–77 (2007)
9. Regione Toscana. Delibera Giunta Regionale 595 del 30.5.05,
   http://www.uisp.it/firenze/files/areaperlagrandeta/LEGHEAREE
   ATTIVITA/AreaAnziani/Documenti/DeliberaGRn595.pdf
10. Regione Toscana. Accordo quadro regione Toscana – trasporti sanitari,
   http://www.misericordiaonline.org/servizi/contratti/page/AQR
   Toscana.pdf

11. Chesbrough, H.W.: Open Business Models: How to Thrive in the New Innovation Landscape (2006)
12. Lin, F.R., Hsieh, P.S.: Analyzing the sustainability of a newly developed service: An activity theory perspective. Technovation 34, 113–125 (2014)
13. Kahneman, D., Knetsch, J.L., Thaler, R., Kahneman, B.D.: Fairness as a Constraint on Profit Seeking: Entitlements in the Market. Am. Econ. Rev. 76, 728–741 (1986)
14. Fusco, F., Turchetti, G.: A Cost-Effectiveness Analysis for Total Knee Arthroplasty Telerehabilitation: Proof of Concept of A Decision Model. Value Heal. 17, A380 (2014)

Ontological Personal Healthcare
Using Medical Standards

Yeong-Tae Song, Neekoo Torkian, and Fatimah Al-Dossary

Dept. of Computer and Information Sciences
Towson University, MD USA
ysong@towson.edu, ntorki1@students.towson.edu

Abstract. According to The Center for Managing Chronic Diseases [1], 7 out of 10 deaths among Americans each year are from chronic diseases. Chronic diseases generally do not come with a cure, which require long-term supervision, monitoring, and various treatments that necessitate coordination among medical professionals. Information and Communication Technology (ICT) may help improve such process and be able to facilitate diverse clinical decision-making processes. In an attempt to achieve improved and evidence based care, we propose to make the connection between patients and clinicians using relevant standards. In the approach, patients collect relevant medical data periodically using personalized mobile healthcare system (PMHS) and share the data with clinical institutions to be compliant with the meaningful use of medical records. The collected medical data and measurement results are analyzed and the analysis result may be used to come up with personalized recommendations for preventing healthy people to become chronic disease patient. The ontology focuses on organizing measurement results semantically. For semantic use of the collected medical records, we have used Protégé-OWL for building ontology and used SPARQL query to retrieve meaningful information from the patient-provided medical data.

In this paper, we propose an ontology-based framework called personalized disease tracking framework (PDTF) for monitoring and managing chronic diseases such as diabetes.

Keywords: Personal medical record, ontology, chronic disease, HL7 CDA, interoperability, diabetes, personalized disease tracking framework.

1 Introduction

Typically chronic diseases have long-lasting conditions that may need long term monitoring and control. To stay healthy, identifying vital signs of such conditions is essential. Once identified, monitoring of the conditions is crucial to control such diseases before actually becoming the patient of such diseases or the symptoms of it getting worse to reach some type of complications. With the advent of potable medical devices (PMDs) such as *Wrist Blood Pressure Monitor or portable body fat analyzer,* it is possible to capture and record personal medical data. Such systems can help users who wish to monitor and control their symptoms. The collected medical data may be

F. Ortuño and I. Rojas (Eds.): IWBBIO 2015, Part II, LNCS 9044, pp. 512–526, 2015.
© Springer International Publishing Switzerland 2015

shared with clinicians for treatment when needed. For that the collected data need to be in some interoperable format so it can be possible to share with any clinicians who can provide treatment. To be meaningful, the collected data from home based healthcare must be utilized to detect or monitor vital signs of chronic diseases. The collected data intended to be a part of the interoperable electronic health records (EHRs) [9] and the compliance with meaningful use (MU) [17]. In this paper, we showed how to utilize and analyze the collected data. We used diabetes as an example.

Diabetes mellitus (DM) is a chronic disease that occurs when the pancreas does not produce enough insulin. Based on The European Chronic Diseases Alliance (ECDA) [2], 25.8 million people in the United States have diabetes. Based on Centers for Disease Control (CDC) [3] a person with diabetes has about twice the risk of dying in contrast to a person of similar age without diabetes. As described by CDC [3], diabetes is one of the chronic diseases that may lead to complications such as heart disease, blurred vision, kidney failure, etc. Such complications can be prevented by properly monitoring their medical data and symptoms. We have used portable medical devices (PMDs) and mobile medical data collection software such as personal mobile healthcare system (PMHS), shown in the figure 2, for that purpose. With PMHS, all measurements can be stored in one place and be able to produce interoperable medical record for sharing. The produced medical record may be used to monitor vital signs for chronic disease such as diabetes.

Since physicians' availabilities are limited, patients need to be in charge of their personal health monitoring and recording system so they have access to their own records and be able to understand what   clinicians say. The PDTF helps them monitor their health status with or without physician involvement. Integration of medical data into sharable format requires the compliance to the standards such as Health Level 7 [18], ICD-9, or SNOMED. The PDTF is proposed to bring an evidence-based framework for monitoring and help prevent diabetes. For that, the definition of the relations among the measurement attributes such as blood pressure, blood glucose, etc. needs to be done so that can be used to analyze vital signs of the diseases in order to prevent them from happening or worsening. For monitoring, patients need to input their measurements periodically using PMDs and PMHS for analysis.

For the analysis, we described how ontology can be used to analyze the relations among measurement attributes. For this approach, we used PMDs for the measurements and input their readings into PMHS that can store them in clinical document architecture (CDA) format. We also discussed how ontology is used to analyze the meaning of the changes between measurements (e.g., changes in blood glucose in 2 weeks) taken from different dates. For that, the measurement information in each CDA document is used to populate web ontology language (OWL) file for analysis. SPARQL query is used to retrieve the meanings from the measurements. These can be used to determine how likely a diabetic person has the risk of getting other diseases. Based on the measurements, we give personalized recommendation to the person so it may help them reduce the risk of getting other symptoms by changing their lifestyle such as diet and exercise.

The rest of the paper is organized as follows. Section 2 discussed the background. Section 3 discussed our personal mobile healthcare application. Section 4 discussed our proposed personal disease tracking framework. And finally we conclude our study in the section 5.

2   Background

As discussed in various articles [4,6,7,8,10,12,13,14], semantic technology can be used to handle semantic rich medical data. It can be used to integrate medical data and visualize so we can have better understanding of the factors that affect patients' symptoms. Ontology is one of the technologies that can be used to define relations among the constituent factors in medical domain. It has been used for knowledge representation and semantic integration in many areas. In the medical domain, V. Bicer, G. et al. [5] introduced semantic standard for their clinical data specification for interoperability in healthcare domain. The use of ontologies in the medical field simplifies communication and solves interoperability issue while finding relationships constituent attributes for the domain. Ontology can also define concepts so data integration is more comprehensible in ontological scenarios.

There have been a number of ontological healthcare applications such as home based data management by N. Lasierra et al., [6], ontology-based healthcare context information model by J. Kim and K. Chung [7], and ontology-based personalization of health-care knowledge by D. Riaño et al.[8]. These applications allow patients at home to collect data and participate in personal care.

I. Berges et al. [9] discussed about transferring medical records through an ontology based system. Remote care of patients with chronic diseases has been reported in some publications. F. Latfi et al. [10] proposed an ontology-based system that monitors patients suffering from a degenerative disease. V. S. Fook et al. [11] presented an ontology-based system to monitor patients from dementia. T. Sampalli et al. [14] proposed an ontology for patient profile of chronic diseases with detailed methodology for the development. For personalized care ontology, some solutions have been proposed, which combine some rules with ontology models. They tried to achieve data integration through ontology. F. Paganelli and D. Giuli [12] proposed a context management middleware in tele-monitoring scenario to support Health Information Exchange (HIE). EJ Ko et al. [13] proposed an ontology-based solution with context-aware framework. Context-aware and personalized-care are the two significant challenges addressed in this work. J. Kim and K.-Y. Chung [7] proposed a way to utilize user collected data for the analysis with context-aware inference engine that is referencing some ontology model. D. Riaño et al [8] proposed the Knowledge-Based HomeCare eServices for an Aging Europe (4KCARE) project, which includes the use of ontologies for representing diseases and symptoms. N. Lasierra et al. [6] proposed a home-based monitoring system, which is an application that enables medical data to be exchanged between patient and healthcare professional who are located in a medical center. At the home site, patients take their measurements to record their health condition. On the healthcare site, the server device is used to manage information that has been taken from the home sites and make necessary analysis and feedbacks.

In this paper, we proposed a personalized disease tracking framework (PDTF) using ontology. Individuals take measurements from PMDs to monitor their current health condition. Measured data will be stored in PMHS application for convenient access by the individuals. The PMHS can produce collected medical record in a CDA [18] document for sharing and further analysis. Since CDA document is based on XML language, it may be parsed with the help of JDOM [15] for easy update of an OWL file. Depending on the measurements, vital sign(s) may be identified for the purpose of notification and recommendation. Vital signs may help the individuals to determine whether there is a risk of getting another disease(s) as accompanying complication. The proposed PDTF may be used to monitor individual's health condition with or without the help from physicians. Medical standards such as CDA document combined with ontology can help produce interoperable medical records and reduce potential medical errors.

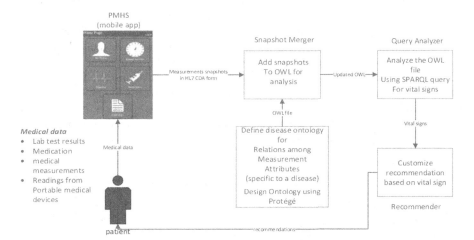

Fig. 1. Personalize Disease Tracking Framework

3   Personal Mobile Healthcare System (PMHS)

The PMHS is an Android based application that can collect personal medical data using smartphones or tablets. Most of data should come from PMDs for accuracy and reliability. At this point, the interfaces between the PMHS and PMDs are not automated, which means the data should be manually inputted to the app. There are number of categories that PMHS can deal with: blood glucose, blood pressure, lab test result, pulse, chronic disease to monitor, and kinds of medication. When a set of measurements is done, the PMHS produces measurements result in interoperable HL7 CDA format as shown in the figure 2. The contents of the CDA document can be determined by the users - partial record by specifying dates and kinds of measurements or entire measurements. Every time a set of measurement is done, the PMHS creates a CDA file called a snapshot to be merged with OWL file which contains ontology for the target chronic disease. For the extraction of the data from the CDA files can be done through the Document Object Model for Java (dom4) [15].

4   Personalized Disease Tracking Framework (PDTF)

The proposed framework, as shown in the figure 1, is designed to be used with any chronic disease monitoring. The ontology design can be done for a specific chronic disease or general enough to be used with a group of chronic diseases. In this study, we focused on diabetes as an example. The PDTF consists of 3 modules – Snapshot Merger, Query Analyzer and Recommender - and 1 mobile application - PMHS. The PMHS, as shown in the figure 2, is currently available on Android platform only. It takes input from the users, typically readings from PMDs. The list of PMDs we have considered is shown in the table 6. It accumulates personal medical record locally and export part or all of the data in CDA document to be used at Snapshot Merger and Query Analyzer. Since it is in HL7 CDA [18] format, it can be shared with other clinical institutions or physicians. However, the primary purpose of the PDTF is to provide self-monitoring capability. It can be designed to focus on one or multiple chronic diseases depending on the number of diseases the user wishes to monitor.The ontology in the Query Analyzer is designed for diabetes so the relations among measurement attributes are tuned for that. The outcome of the analyzer is the physical condition – categorized as patterns as shown in the table 5, if exist, based on the records in the snapshots submitted by the users. If the outcome is normal, there will be no recommendation other than "continue to do what you have been doing". If there exist

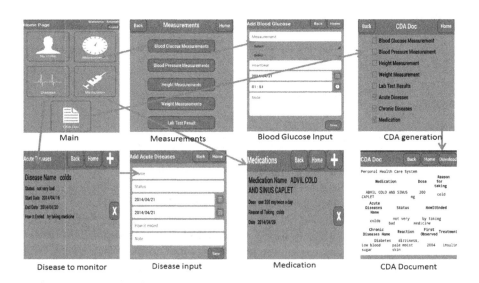

Fig. 2. Personal Mobile Healthcare System (PMHS)

any pattern that needs user's attention, the PDTF would give corresponding recommendation. When that happens, the user needs to take another set of measurements and submit to the Snapshot merger for the analysis after a predetermined period of time, say a couple of weeks. The differences between previous snapshot and current snapshot would be used to determine vital signs that may be categorized into patterns in the table 5. The process continues until no more vital signs (patterns) exist or the user needs to consult with physicians when it reaches predetermined boundary values.

4.1   Vital Sign Analysis

The collected medical data stored in the snapshot are used to update the OWL file for analysis. Once the patient input his/her measurements, the PMHS can produce a snapshot in CDA document. Each CDA document contains a set of measurements with time stamps. Vitals sign analysis is done through the analysis of the relations among the measurements and the changes in values between snapshots. By considering interval between snapshots and the range of differences, appropriate recommendations may be provided to the patients. We have included the factors that have major impact on chronic diseases in the snapshots in CDA format. For our experiment, we have used the following factors: blood glucose level, blood pressure, pulse, body temperature, respiratory rate, and body mass index (BMI). An example of CDA document is shown in the figure 3. Like many other chronic diseases, diabetes may cause various complications if not managed properly as indicated in the table 5.

For efficient monitoring, we categorized sets of abnormal values in the snapshots as problem patterns. For example, if a patient has high BMI and high blood glucose level, it can be categorized as pattern B as shown in the table 5 and corresponding recommendation may be given, which may include reducing simple carbs food such as pasta and white rice and increase exercise with the aim of reducing weight for a certain period of time, say a couple of weeks. After that, the patient needs take another set of measurements or snapshot for the comparison in the query analyzer. It keeps track of any changes to measure the effectiveness of the recommendation. A sample recommendation is shown in the figure 10.

The snapshots are used to update the OWL file in the Snapshot Merger module. The updated OWL file will have the history of measurements with time intervals between measurements. For the sake of medical record sharing, the OWL file may be imported to a SPARQL engine such as the Virtuoso [16]. In the Virtuoso server, it could be converted to a RDF triple store for sharable SPARQL queries with user interfaces. In our Query Analyzer, a set of SPARQL queries is applied to the updated OWL file to find vital signs or patterns from the measurements. When some vital sign is identified, the Recommender prepares personalized recommendations by referencing basic personal medical data such as blood pressure, pulse, body temperature, body mass index (BMI), and blood glucose level. The pattern analysis by SPARQL query is shown in the figure 8. The last column in the figure shows the patterns that the user falls under.

```
<ClinicalDocument xmlns="urn:hl7-org:v3" xmlns:msxsl="urn:schemas-microsoft-
com:xslt" "/>
<templateId root="2.16.840.1.113883.10.20.1"/>
<id root="66ac79c1-6a18-486c-98ec-f389967b7356"/>
<code code="34133-9" displayName="Summarization of episode note" "/>
<title>HealthVault Continuity of Care Document</title>
<id root="1.3.6.1.4.1.311.68.1"/>
<td>Blood<br/>glucose</td>
<td>70</td><td/></tr>
<td>Blood Pressure</td><td>Systolic: 115mm[Hg],Diastolic: 72mm[Hg],Pulse:
72/min</td>
<td/></tr><tr><td>2015/01/13</td>
<td> Height </td><td> 71 inches </td><td/></tr><tr><td>2015/01/13</td>
<td>Weight </td><td> 195 lb </td><td/></tr></tbody></table></text><entry
typeCode="DRIV">
<organizer classCode="CLUSTER" moodCode="EVN">
<id root="d06f8e7a-cba2-4a29-a3a6-0f91cb690add"/>
<code code="434912009" codeSystem="2.16.840.1.113883.6.96"
displayName="Blood glucose"/>
```

Fig. 3. A Sample CDA document by PMHS

Figure 4 through 7 shows 5 snapshots of the patient Jess and the differences between them after the patient followed the recommendations. We compared the snapshots one by one to see if the patient's physical condition has changed after following the recommendation.

Table 1. Snapshot 1 and 2 Comparison

Snapshot Number	Body Temperature	Systolic B.P	Diastolic B.P	A1C	FPG	OGTT	Pulse	BMI	Pattern
S1	98 F	95	145	5.9%	100	140	85	27	A
S2	99 F	86	127	6.1	102	142	88	27	A

Patient	BloodGlucoseTestNumber	Measurem...	BMI	A1C	F..	OG...	GlucoseLevel	Pattern		Pati...	BloodPressureTestNumber	Me...	BMI	Systoli...	DiastolicBloodPre...	BloodPressureLevel	Pattern
Jess	BloodGlucoseTest1	"10.16.2C"27" "5.9%"^"1("14CHighRiskInDiabetes						A		Jess	BloodPressureTest1	"10."27"95 mn"145 mm Hg"^High/Hypertension					A
Patient	BloodGlucoseTestNumber	Measurem...	BMI	A1C	F.	OG...	GlucoseLevel	Pattern		Pati...	BloodPressureTestNumber	Me...	BMI	Systoli...	DiastolicBloodPre...	BloodPressureLevel	Pattern
Jess	BloodGlucoseTest2	"10.20.2C"27" "6.1 %"^"1("142HighRiskInDiabetes						A		Jess	BloodPressureTest2	"10."27"86 mn"127 mm Hg"^PreHigh/PreHypertension					.

Fig. 4. S1&S2 comparison

As shown in the table 1, snapshot1 (s1) has high BP, BG, and BMI. S1 has the pattern A according to the table 5. Possible recommendation for such patient could be reducing fatty foods, regular physical activities (1 hour walking), avoiding smoking, and sodium. After following the recommendation for a fixed amount of time, the numbers are changed to s2. Since the blood pressure level is reduced, the recommendation helped reducing the blood pressure. But still the blood glucose and BMI are high. Figure 4 shows the result from SPARQL query. As shown, blood pressure test 2 doesn't show any pattern which means that it no longer shows a particular pattern.

We continue the measurement until all others don't show any abnormal values. Table 2 and Figure 5 show the comparison between s2 and s3.

Table 2. Snapshot 2 and 3 Comparison

Snapshot Number	Body Temperature	Systolic B.P	Diastolic B.P	A1C	FPG	OGTT	Pulse	BMI	Pattern
S2	99 F	86	127	6.1	102	142	88	27	A
S3	97.9 F	84	124	5.9%	97	140	65	24.2	B

Patient	BloodGlucoseTestNumber	Measurem.	BMI	A1C	F..	OG..	GlucoseLevel	Pattern	Pati..	BloodPressureTestNumber	Me..	BMI	Systoli.	DiastolicBloodPre.	BloodPressureLevel	Pattern
Jess	BloodGlucoseTest2	"10.20.2C"27"	6.1 %"^"1("142HighRiskInDiabetes	A	Jess	BloodPressureTest2	"10."27"86 mn"127 mm Hg"^PreHigh/PreHypertension	.
Jess	BloodGlucoseTest3	"10.25.2C"24.:"5.9 %"^"9("14CPreDiabetesRisk	B	Jess	BloodPressureTest3	"10."24"84 mn"124 mm Hg"^PreHigh/PreHypertension	.

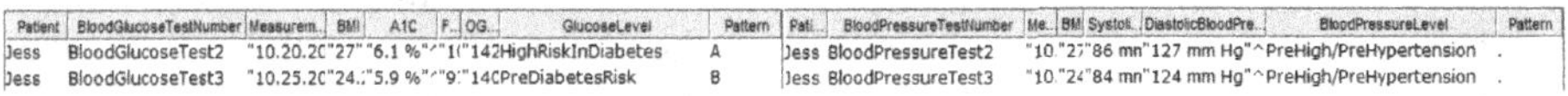

Fig. 5. S2&S3 comparison

Based on table 2 and figure 5, s3 doesn't show high blood pressure but blood glucose and BMI are high which means the risk of getting pre-diabetes. This belongs to pattern B. The corresponding recommendation may be given to the patient until no more pattern is showing in the comparison table. Table 3 shows the comparison between s3 and s4.

Table 3. Snapshot 3 and 4 Comparison

Snapshot Number	Body Temperature	Systolic B.P	Diastolic B.P	A1C	FPG	OGTT	Pulse	BMI	Pattern
S3	97.9 F	84	124	5.9%	97	140	65	24.2	B
S4	96 F	82	123	5.7%	98	139	68	24.5	-

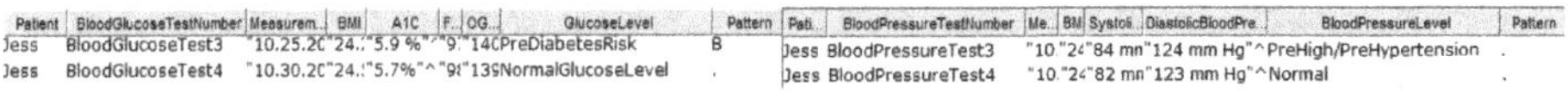

Patient	BloodGlucoseTestNumber	Measurem.	BMI	A1C	F..	OG..	GlucoseLevel	Pattern	Pati..	BloodPressureTestNumber	Me..	BMI	Systoli.	DiastolicBloodPre.	BloodPressureLevel	Pattern
Jess	BloodGlucoseTest3	"10.25.2C"24.:"5.9 %"^"9("14CPreDiabetesRisk	B	Jess	BloodPressureTest3	"10."24"84 mn"124 mm Hg"^PreHigh/PreHypertension	.
Jess	BloodGlucoseTest4	"10.30.2C"24.:"5.7%"^"9("135NormalGlucoseLevel	.	Jess	BloodPressureTest4	"10."24"82 mn"123 mm Hg"^Normal	.

Fig. 6. S3&S4 comparison

As shown in table 3 and figure 6, there is a difference in numbers that shows the blood glucose and BMI as well as blood pressure in normal range. The changes in s4 may indicate that the provided recommendation was effective. The recommendation for now is to keep staying on the diet and exercise. Another comparison is shown in table 4 and figure 7.

Table 4. Snapshot 4 and 5 Comparison

Snapshot Number	Body Temperature	Systolic B.P	Diastolic B.P	A1C	FPG	OGTT	Pulse	BMI	Pattern
S4	96 F	82	123	5.7%	97	139	68	24.5	-
S5	99 F	80	120	5.5%	93	135	70	24	-

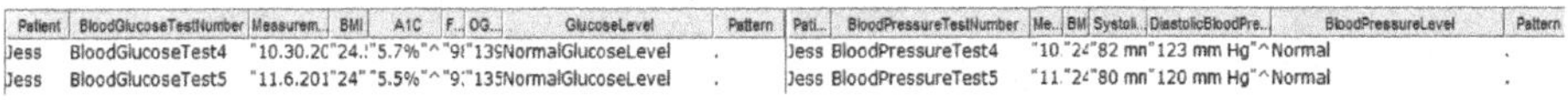

Patient	BloodGlucoseTestNumber	Measurem.	BMI	A1C	F..	OG..	GlucoseLevel	Pattern	Pati..	BloodPressureTestNumber	Me..	BMI	Systoli.	DiastolicBloodPre.	BloodPressureLevel	Pattern
Jess	BloodGlucoseTest4	"10.30.2C"24.:"5.7%"^"9("135NormalGlucoseLevel	.	Jess	BloodPressureTest4	"10."24"82 mn"123 mm Hg"^Normal	.
Jess	BloodGlucoseTest5	"11.6.201"24"5.5%"^"9("135NormalGlucoseLevel	.	Jess	BloodPressureTest5	"11."24"80 mn"120 mm Hg"^Normal	.

Fig. 7. S4&S5 comparison

The final result is shown in s5. All factors are showing in normal range and there is no pattern belongs to Jess. The recommendation still remains the same so as to sustain the normal range.

The PDTF helps users to monitor their health status so they can continue to live healthy without getting any help from physicians. Monitoring vital signs may help prevent any other diseases from occurring as complication. As one of the main purposes of ontology is to show the relations between the elements that constitute a thing (a domain), we have designed an OWL file for tracing vital signs that may be related to each constituent elements. The query results also show the effect of the recommendations given in the previous analysis through the changes between them. Since the identified patterns are used to determine the kind of recommendation to the patient, it can be considered as evidence-based preventative care for the people with diabetic or high risk of becoming diabetes.

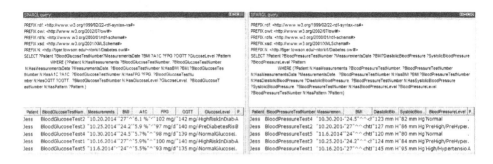

Fig. 8. Pattern analysis results using SPARQL query

4.2   Categories of Patterns

The vital signs in the snapshots can be categorized into one of the following 4 patterns as shown in the table 5. If everything is normal, then the snapshot doesn't belong to any of the following categories. The patterns in the table are based on WebMD [19].

Table 5. Patterns of physical condition and potential complications

Pattern	Signs	Leading to Other diseases and symptoms
A	High BG(Diabetes risk), high BP(Hypertension risk), High BMI(Obesity)	Ischemic Heart Disease (IHD), Pre-Diabetes, Diabetes Risk, Obesity, Kidney Failure, Heart Disease, Vision Problem, Frequent Urination, Excessive Hunger, Excessive Thirst, Hypertension
B	High BG, High BMI	Pre-Diabetes, Diabetes Risk, Obesity, Kidney Failure, Heart Disease, Vision Problem, Frequent Urination, Excessive Hunger, Excessive Thirst
C	High BP, High BG	Pre-Diabetes, Diabetes Risk, Obesity, Kidney Failure, Heart Disease, Vision Problem, Frequent Urination, Excessive Hunger, Excessive Thirst, Hypertension
D	High BG, Low Temp	Pre-Diabetes, Diabetes, Hypothermia, Shivering

Pattern A

Based on Health Grade [20] high BG, BP, and BMI may lead to other diseases such as Ischemic Heart Disease (IHD), which is also known as Coronary Artery Disease. It is a condition that affects the blood flow into the heart. The risk factors in the development of IHD are smoking, diabetes mellitus and hypertension.

Recommendations

According to [20], there are some risk factors that should be considered in order to improve the pattern A condition. For example, fatty diet (rich saturated fat foods), smoking, sodium diet, stress should be avoided. In addition, more physical activities, and maintaining a healthy body weight are considered important. Such recommendation could help lower the risk of IHD while patients monitor and manage the other factors such as blood glucose, physical activity, blood pressure, smoking and other tobacco use, and the amount of cholesterol and fat intake.

Pattern B

Based on WebMD, high BG, and BMI may lead to other diseases such as diabetes. Complications of diabetes may include kidney failure, non-healing ulcers of the foot, generalized weakness, deterioration in vision, frequent infections. Diabetes also is characterized by excessive thirst (Polydipsia), frequent urination (Polyuria), and excessive hunger (Polyphagia). Smoking can aggravate this problem even further. They are also prone to develop hypertension heart disease.

Recommendations

The monitoring and control of pre-diabetes or diabetes through the combination of diet and exercise are important to keep it from development. The prevention of the disease may happen if adequate exercise and proper diet are provided. As described in WebMD, over weight increases the possibility of getting type 2 diabetes. Some recommendations for that are: at least 30 minutes exercise a day, 5 days a week regular exercise, and low-fat, low carb, high-fiber diet [21]. Regular measurement as a part of monitoring and control is equally important to keep the pattern B under control.

Pattern C

Based on [19], high BG, and BP may be the cause some diseases such as hypertension. Hypertension is a major cause of kidney failure, affecting eyesight, and cause damage to the blood vessels in the retina. It is also a risk factor for other diabetes related complications.

Recommendation

Since it is diabetes related hypertension, control of blood glucose should be the major concern in improving pattern C. Control of blood pressure, reducing stress and anxiety, avoiding fatty foods and salty foods, smoking, alcohol, and drug are considered helpful factors for the improvement of the pattern. Regulated diet and exercise may help in delaying the progress of the disease.

Pattern D

As mentioned in [19], abnormally low body temperature (hypothermia) can be serious. Hypothermia is a frequent sign in patients with diabetes. Hypothermia may occur from shock, alcohol or drug use, or from diseases such as diabetes. High blood glucose level is also a sign of diabetes.

Recommendation
Constant monitoring of blood glucose level and avoiding alcohol and caffeine consumption may help improve the pattern D condition. Prevention is always the best care when it comes to chronic disease. As represented by CDC [3] by controlling vital signs such as A1C, the risk of eye, kidney, and heart diseases can be reduced by 40%. Blood pressure can reduce the risk of heart disease and stroke by 33%–50%. Improving control of cholesterol can reduce cardiovascular complications by 20%–50%.

We have shown the usage of relations in the ontology in determining patterns in the figure 8.

Structure of the Ontology
To improve the usefulness of the ontology, we established relations between parameters by storing and linking data. The ontology in this paper is developed by storing, linking concepts and related data for diabetes and the relations between vital signs and diabetes leading to other complications. Finding these relations may help person to prevent further development to other diseases. The ontology in the figure 9 contains information from five main domains including: patient, questionnaire, measurements, recommendation, and pattern. Person domain presents general information of patients such as age, gender telephone number, measurement dates, and address. Questionnaires is a set of questions that patient has answered. Measurement data include blood glucose level, blood pressure, pulse, body temperature, and BMI. Recommendation domain contains recommendations based on person's physical condition and the resulting pattern from the measurement.

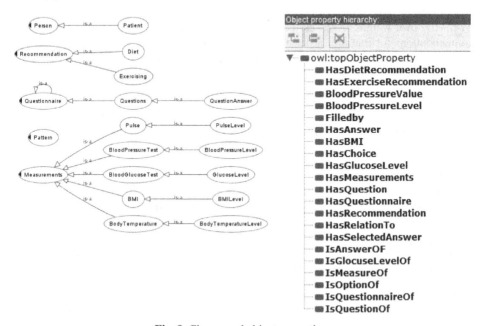

Fig. 9. Classes and object properties

Classes

In Protégé environment, classes specify the concept of a domain. In PDTF, we used OWL language which is associated with a set of individuals. Figure 9 shows the ontology classes and object properties. We have classified that into 5 main classes and 12 sub-classes. These sub-classes are individuals of each class and are related to the class.

Object Properties

Object properties are ways to relate Objects. This relation (also called predicates) is established between two objects (also called individuals). It has a domain and a range which links individuals from the domain to individuals from the range. Object Properties generally use the syntax object1 *ObjectProperty* object2.

Table 6. Portable Medical Devices (PMDs)

Measurements	PMDs
Blood Glucose	OneTouch Ultra Mini Blood Glucose Meter
Blood Pressure	Omron 7 Series Wrist Blood Pressure Monitor
BMI	Omron Portable Body Fat Analyzer
Pulse	IRONMAN Pulse Calculator
Body Temperature	Reli-On - 60-Second Basal Thermometer

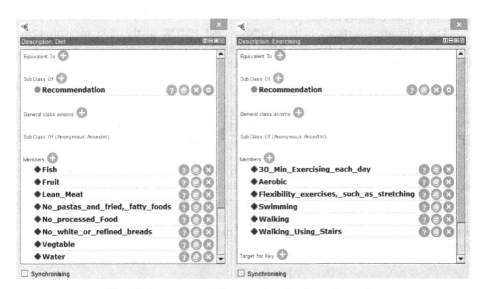

Fig. 10. A recommendation template for diet and exercise

5   Conclusion

In this paper, we proposed an ontology-based personalized disease tracking framework (PDTF) that is designed to help users to monitor and trace their potential chronic disease or symptoms so it can help prevent them from occurring or getting complications from the chronic disease they are already suffering. We have developed personalized mobile healthcare system (PMHS), an Android based mobile app, for collecting and recording personal medical records. It can produce measurements result in Health Level 7 CDA form for interoperability. Snapshot merger module was used to update OWL file for analysis. SPARQL query analyzer module that utilizes the OWL tile was used to detect vital signs from the snapshots so it can offer corresponding recommendations in the recommender module. PMDs and PMHS were used to collect medical data as snapshots. Ontology was used to define the relations among the attributes in a chronic disease. In this paper, we used diabetes as an example but it can be applicable to any other diseases by defining the relations that are specific to the target disease. The SPARQL query was designed to analyze the snapshots from users and provided customized recommendations. To support evidenced-based healthcare, the effect of the recommendation was analyzed by comparing snapshot values. The designed ontology was used to identify the patterns that each snapshot falls under and relates those to corresponding recommendations. Periodically measured clinical data from patients can be used to improve the quality of patient care since it can be shared with clinicians.

Future Work

The future work consists of four parts:

- The software interfaces between PMDs and PMHS so the data measurement process can be automated. Bluetooth may be used to connect smartphones and PMDs. The PMD with communication capability can transmit measured data from PMD to smartphones or tablets.

- The integration with an electronic health record system so the CDA documents can be a part of personal medical record. With personal medical record system such as PMHS alone, it is not sufficient to implement electronic health record system. The CDA document from patient may be used to complement electronic medical record system.

- The integration with coding system such as ICD-9-CM, LOINC and SNOMED so the medical records can be unambiguous. Without standardized codes, the observation, problem description, and medical history may not be accurate.

- The integration with a smart e-learning system so the users can learn about their potential diseases. The education of patients may improve the outcome of the monitoring or any treatment.

References

1. What is Chronic Disease? | Center for Managing Chronic Disease - Putting People at the Center of Solutions, `http://cmcd.sph.umich.edu/what-is-chronic-disease.html`
2. International Diabetes Federation. The European Chronic Diseases Alliance - ECDA, `http://www.idf.org/regions/EUR/ecda`
3. CDC - Chronic Disease - Diabetes - At A Glance, `http://www.cdc.gov/chronicdisease/resources/publications/AAG/ddt.htm`
4. Dataversity (August 2012), `http://www.dataversity.net/precision-medicine-is-semantic-medicine/`
5. Bicer, V., Laleci, G., Dogac, A., Kabak, Y.: Artemis message exchange framework. ACM SIGMOD Record 34(3), 71 (2005)
6. Lasierra, N., Alesanco, A., Garcia, J., O'Sullivan, D.: Data management in home scenarios using an autonomic ontology-based approach. In: 2012 IEEE International Conference on Pervasive Computing and Communications Workshops (2012)
7. Kim, J., Chung, K.: Ontology-based healthcare context information model to implement ubiquitous environment. Multimedia Tools and Applications 71(2), 873–888 (2011)
8. Riaño, D., Real, F., López-Vallverdú, J.A., Campana, F., Ercolani, S., Mecocci, P., Annicchiarico, R., Caltagirone, C.: An ontology-based personalization of health-care knowledge to support clinical decisions for chronically ill patients. Journal of Biomedical Informatics 45(3), 429–446 (2012)
9. Berges, I., Bermudez, J., Illarramendi, A.: Toward Semantic Interoperability of Electronic Health Records. IEEE Transactions on Information Technology in Biomedicine 16(3), 424–431 (2012)
10. Latfi, F., Lefebvre, B., Descheneaux, C.: Ontology-Based Management of the Telehealth Smart Home, Dedicated to Elderly in Loss of Cognitive Autonomy. In: Proceedings of the OWLED 2007 Workshop on OWL: Experiences and Directions, Innsbruck, Austria (June 2007)
11. Siang Fook, V., Siew Choo Tay, M., Jayachandran, J.B., Zhang, D.: An Ontology-based Context Model in Monitoring and Handling Agitation Behaviour for Persons with Dementia. In: Fourth Annual IEEE International Conference on Pervasive Computing and Communications Workshops, PERCOMW 2006 (2006)
12. Paganelli, F., Giuli, D.: An Ontology-Based System for Context-Aware and Configurable Services to Support Home-Based Continuous Care. IEEE Transactions on Information Technology in Biomedicine 15(2), 324–333 (2011)
13. Ko, E., Lee, H., Lee, J.: Ontology-Based Context Modeling and Reasoning for U-HealthCare. IEICE Transactions on Information and Systems 90(8), 1262–1270 (2007)
14. Sampalli, T., Shepherd, M., Duffy, J.: A Patient Profile Ontology in the Heterogeneous Domain of Complex and Chronic Health Conditions. In: 2011 44th Hawaii International Conference on System Sciences (2011)
15. JDOM, `http://www.jdom.org`
16. Virtuoso Open-Source Wiki: Virtuoso Open-Source Edition, `http://virtuoso.openlinksw.com`
17. Meaningful Use Definition, `http://www.cms.gov/`
18. Health Level Seven (HL7) CDA release 2

19. WebMD - Better information. Better health, `http://www.webmd.com/`
20. Ischemic Heart Disease - Symptoms, Causes, Treatments, `http://www.healthgrades.com/conditions/ischemic-heart-disease`
21. Type 2 Diabetes: Causes, Tests, Treatments, `http://www.webmd.com/diabetes/guide/diabetes_symptoms_types`

Applications of High Performance Computing in Bioinformatics, Computational Biology and Computational Chemistry

Horacio Péréz-Sánchez[1], Afshin Fassihi[1], José M. Cecilia[1], Hesham H. Ali[2], and Mario Cannataro[3]

[1] Bioinformatics and High Performance Computing Research Group (BIO-HPC)
Computer Science Department
Universidad Católica San Antonio de Murcia (UCAM), Guadalupe E30107, Spain
{hperez,afassihi,jmcecilia}@ucam.edu
[2] Department of Computer Science
College of Information Science and Technology
University of Nebraska at Omaha, USA
hali@unomaha.edu
[3] Bioinformatics Laboratory
Department of Medical and Surgical Sciences
University Magna Graecia of Catanzaro, Italy
cannataro@unicz.it

Abstract. In the last 10 years, we are witnessing one of the major revolutions in parallel systems. The consolidation of heterogeneous systems at different levels -from desktop computers to large-scale systems such as supercomputers, clusters or grids, through all kinds of low-power devices- is providing a computational power unimaginable just few years ago, trying to follow the wake of Moore's law. This landscape in the high performance computing arena opens up great opportunities in the simulation of relevant biological systems and for applications in Bioinformatics, Computational Biology and Computational Chemistry. This introductory article shows the last tendencies of this active research field and our perspectives for the forthcoming years.

Keywords: HPC, Parallel Computing, Heterogeneous Computing, Bioinformatics, Computational Biology, Computational Chemistry.

1 Introduction

The integration of the latest breakthroughs in biochemistry and biotechnology from one side and high performance computing and computational modelling from the other, enables remarkable advances in the fields of healthcare, drug discovery, genome research, systems biology and so on. By integrating all these developments together, scientists are creating new exciting personal therapeutic strategies for living longer and having healthier lifestyles that were unfeasible not that long ago.

F. Ortuño and I. Rojas (Eds.): IWBBIO 2015, Part II, LNCS 9044, pp. 527–541, 2015.
© Springer International Publishing Switzerland 2015

Those efforts have created new research fields such as *Bioinformatics and Computational Biology*, defined most broadly as informatics in the domains of biology and biomedical research. Bioinformatics spans to many different research areas, such as life sciences, where there are many examples of scientific applications for discovering biological and medical unknown factors that could greatly benefit from increased computational resources.

Indeed, as computing resources available on current systems are limited, this limitation becomes a serious constraint, then hindering it from successfully taking the next step forward. For instance, applications such programs of Molecular Dynamics (MD) [1], employed to analyse the dynamical properties of macromolecules such as folding and allosteric regulations, or software used for solving atom-to-atom interactions for drug discovery, such as AutoDock [2] and FlexScreen [3], could clearly benefit from enhanced computing capabilities and also from some novel algorithms approaches like those inspired by nature such as Genetic algorithms [4] or Ant Colony Optimization techniques [5].

There are other applications, which are actually working well, but their execution is too slow for providing feedback in real-time to the users, and thus limiting the effectiveness and comfort of such applications. In this group, we may cite biomedical image processing applications, such as X-ray computed tomography or mammography for breast cancer detection. The low-performance of these applications can drastically affect the patient's health. For instance, imagine a patient who is waiting for mammography results. She is actually waiting for a breast cancer diagnostic, thus the acceleration of the diagnosis time becomes paramount for the patient's health.

High Performance Computing technologies are at the forefront of those revolutions, making it possible to carry out and accelerate radical biological and medical breakthroughs that would directly translate into real benefits for the society and the environment. In this regard, Graphics Processing Units (GPUs) are providing a unique opportunity to tremendously increase the effective computational capability of the commodity PCs, allowing desktop supercomputing at very low prices. Moreover, large clusters are adopting the use of these relatively inexpensive and powerful devices as a way of accelerating parts of the applications they are running. Since June 2011, when the fastest supercomputer was the Tianhe-1A, placed in the National Supercomputing Centre at Tianjin (China)[1], including up to 7.168 NVIDIA® Tesla™ M2050 general purpose GPUs, several supercomputers have followed this trend.

However, the inclusion of those accelerators in the system has a great impact on the power consumption of the system, as a high-end GPU may increase the power consumption of a cluster node up to 30% which is actually a big issue. This is a critical concern especially for very large data centres, where the cost dedicated to supply power to such computers represents an important fraction of the Total Cost of Ownership (TCO) [6]. The research community is also aware of this and it is making efforts in striving to develop reduced-power installations.

---

[1] http://www.top500.org/

Thus, the GREEN500 list[2] shows the 500 most power efficient computers in the world. In this way, we can see a clear shift from the traditional metric FLOPS (FLoating point Operations Per Second) to FLOPS per watt.

Virtualization techniques may provide significant energy savings, as they enable a larger resource usage by sharing a given hardware among several users, thus reducing the required amount of instances of that particular device. As a result, virtualization is being increasingly adopted in data centres. In particular, cloud computing is an inherently energy-efficient virtualization technique [7], in which services run remotely in a ubiquitous computing cloud that provides scalable and virtualized resources. Thus peak loads can be moved to other parts of the cloud and the aggregation of a cloud's resources can provide higher hardware utilization [8]. Public cloud providers offer their services in a *pay as you go* fashion, and provide an alternative to physical infrastructures. However, this alternative only becomes real for a specific amount of data and target execution time.

The rest of the paper is organized as follows. We briefly introduce some HPC architectures we think are at the forefront of this revolution in Section 2. In Section 3 we present some relevant Bioinformatics applications that are using HPC alternatives to deal with their computational issues before we summarize our findings and conclude with suggestions for future work.

2 HPC Architectures

Traditionally, high performance computing has been employed for addressing bioinformatics problems that would otherwise be impossible to solve. A first example was the preliminary assembly of the human genome, a huge effort in which the Human Genome Project was challenged by Celera Genomics with a different approach, consisting in a bioinformatics post analysis of whole sets of shotgun sequencing runs instead of the long-standing vector cloning technique, arriving very close to an unpredictable victory [9].

In this Section, we briefly summarize the high performance systems that has been commonly used in Bioinformatics. Among them, we highlight throughput-oriented architectures such as GPUs, large-scale heterogeneous clusters and clouds or distributed computing systems, with a discussion about how the latest breakthroughs in these architectures are being used in this field.

2.1 GPU Computing

Driven by the demand of the game industry, graphics processing units (GPUs) have completed a steady transition from mainframes to workstations to PC cards, where they emerge nowadays like a solid and compelling alternative to traditional computing platforms. GPUs deliver extremely high floating point performance and massively parallelism at a very low cost, thus promoting a new

---

[2] http://www.green500.org/

concept of the high performance computing (HPC) market; i.e. *heterogeneous computing* where processors with different characteristics work together to enhance the application performance taking care of the power budget. This fact has attracted many researchers and encouraged the use of GPUs in a broader range of applications, particularly in the field of Bioinformatics, where developers are required to leverage this new landscape of computation with new programming models which ease the developers task of writing programs to run efficiently on such platforms altogether [10].

The most popular microprocessor companies such as NVIDIA, ATI/AMD or Intel, have developed hardware products aimed specifically at the heterogeneous or massively parallel computing market: Tesla products are from NVIDIA, Firestream is AMDs product line and Intel Xeon Phi comes from Intel. They have also released software components, which provide simpler access to this computing power. CUDA (Compute Unified Device Architecture) is NVIDIAs solution as a simple block-based API for programming; AMDs alternative was called Stream Computing and Intel relies on X86-based programming. More recently (in 2008), the OpenCL[3] emerged as an attempt to unify all of those models with a superset of features, being the best broadly supported multi-platform data-parallel programming interface for heterogeneous computing, including GPUs, accelerators and similar devices.

Although these efforts in developing programming models have made great contributions to leverage the capabilities of these platforms, developers have to deal with a massively parallel and high throughput-oriented architecture[11], which is quite different than traditional computing architectures. Moreover, GPUs are being connected with CPUs through PCI Express bus to build heterogeneous parallel computers, presenting multiple independent memory spaces, a wide spectrum of high speed processing functions, and communication latency between them. These issues drastically increase scaling to a GPU-cluster, bringing additional sources of latency. Therefore, programmability on these platforms is still a challenge, and thus many research efforts have provided abstraction layers avoiding to deal with the hardware particularities of these accelerators and also extracting transparently high level of performance, providing portability across operating systems, host CPUs and accelerators. For example, libraries interfaces for programming with popular programming languages like OMPSs for OpenMP [4] or OpenACC[5] API, which describes a collection of compiler directives to specify loops and regions of code in standard programming language such as C, C++ or Fortran.

3 Applications

This section describes some bioinformatics applications from the High Performance computing point of view.

---

[3] http://www.khronos.org/opencl/

[4] http://openmp.org/wp/

[5] http://www.openacc-standard.org/

3.1   Virtual Screening

In this Section, we summarize the main technical contributions for the parallelization of Virtual Screening (VS) methods on GPUs available on the bibliography. Concretely, we pay special attention to the parallelization of docking methods on GPUs.

In terms of implementations, the trend seems to be reusing available libraries when possible and implement the achievements into existing simulation packages for VS. Among the most-used strategies are either implementing the most time-consuming parts of previously designed codes for serial computers, or redesigning the whole code from scratch. When porting VS methods to GPUs, we should realize that not all methods are equally amenable for optimization. Programmers should check carefully how the code works and whether it is suited for the target architecture. Irrespective of CUDA, most authors maintain that the application will be more accessible in the future thanks to new and promising programming paradigms which are still in the experimental stage or are not yet broadly used. Among them, we may highlight OpenCL or DirectCompute.

Dock6.2   In the work of Yang et al. [12] a GPU accelerated amber score in Dock6.2 is presented. They report up to 6.5x speedup factor with respect to 3,000 cycles during MD simulation compared to a dual core CPU. [6].

The lack of the single-precision floating point operations in the targeted GPU (NVIDIA GeForce 9800GT) produces small precision losses compared to the CPU, which the authors assume as acceptable. They highlight the thread management utilizing multiple blocks and single transferring of the molecule grids as the main factor that dominates the performance improvements on GPU.

They use another optimization technique, such as dealing with the latency attributed to thread synchronization, divergence hidden and shared memory through tiling, that authors state may double the speedup of the simulation. We miss a deeper analysis on the device memory bandwidth utilization. It is not clear whether the pattern accesses to device memory in the different versions of the designs presented here are coalesced or not, which may drastically affect the overall performance.

They finally conclude that the speedup of Amber scoring is limited by the Amdahl's law for two main reasons: (1) the rest of the Amber scoring takes a higher percentage of the run time than the portion parallelized on the GPU, and (2) partitioning the work among SMs will eventually decrease the individual job size to a point where the overhead of initializing an SP dominates the application execution time. However, we do not see any clear evaluation that supports these conclusions.

Autodock.   In the paper of Kannan et al. [13] the migration to NVIDIA GPUs of part of the molecular docking application *Autodock* is presented. Concretely,

---

[6] http://dock.compbio.ucsf.edu/DOCK_6/

they only focus on the Genetic Algorithm (GA) which is used to find the optimal docking conformation of a ligand with respect to a protein. They use single-precision floating point operation arguing that, "GA depends on relative goodness among individual energies and single precision may not affect the accuracy of GA path significantly". All the data relative to the GA state is maintained on the GPU memory, avoiding data movement through the PCI Express bus.

The GA algorithms need random numbers for the selection process. They decide to generate the random numbers on the CPU instead of doing it on the GPU. The explanation of that is two-fold according to the authors: (1) it enables one-to-one comparisons of CPU and GPU results, and (2) it reduces the design, coding and validation effort of generating random numbers on GPU.

A very nice decision is what the authors call *CGPU Memory Manager* that enables alignment for individual memory request, support for pinned memory and join memory transfer to do all of them in just one transfer. Regarding the fitness function of the GA, authors decide to evaluate all the individuals in a population regardless of modifications. This avoids warp divergences although it makes some redundant work.

Three different parallel design alternatives are discussed in this regard. Two of them only differ in the way they calculate the fitness function, assigning the calculation of the fitness of an individual either to a GPU thread or GPU block. A good comparison between them is provided. The last one includes an extra management of the memory to avoid atomic operations which drastically penalizes the performance.

All of these implementations are rewarded with up to 50x in the fitness calculation, but they do not mention anything about global speedup of the Autodock program.

Genetic Algorithms Based Docking. In literature is also available [14] an enhanced version of the PLANTS [15] approach for protein-ligand docking using GPUs. They report speedup factors of up to 50x in their GPU implementation compared to an optimized CPU based implementation for the evaluation of interaction potentials in the context of rigid protein. The GPU implementation was carried out using OpenGL to access the GPU's pipeline and Nvidia's Cg language for implementing the shaders programs (i.e. Cg kernels to compute on the GPU). Using this way of programming GPUs, the developing effort is too high, and also some peculiarities of the GPU architecture may be limited. For instance, the authors say that some of the spatial data structures used in the CPU implementation can not directly be mapped to the GPU programming model because of missing support for shared memory operations [14].

The speedup factors observed, especially for small ligands, are limited by several factors. First, only the generation of the ligand-protein conformation and the scoring function evaluation are carried out on the GPU, whereas the optimization algorithm is run on the CPU. This algorithmic decomposition implies time-consuming data transfers through PCI Express bus. The optimization algorithm used in PLANTS is the Ant Colony Optimization (ACO) algorithm [16].

Concretely, authors propose a parallel scheme for this algorithm on a CPU cluster, which use multiple ant colonies in parallel, exchanging information occasionally between them [17]. Developing the ACO algorithm on the GPU as it has been shown in [18] can drastically reduce the communications overhead between CPU and GPU.

3.2   Parallel Processing of Microarray Data

High throughput experimental platforms, such as mass spectrometry, microarray, and next generation sequencing, are producing an overwhelming amount of data yielding to the so called omics world. Among the many omics disciplines, genomics, proteomics, and interactomics and are gaining an increasing interest in the scientific community for the study of diseaases at the molecular level.

The increasing availability of omics data poses new challenges for efficient data storage and integration as well as data preprocessing and analysis: managing omics data requires both space for data storing as well as procedures for data preprocessing and analysis. Main challenges are: (i) the efficient storage, retrieval and integration of experimental data; (ii) their efficient and high-throughput preprocessing and analysis; (iii) the building of reproducible "in silico" experiments; (iv) the annotation of omics data with pre-existing knowledge (e.g. extracted from ontologies like Gene Ontology, or from specialized databases); (v) the integration of omics and clinical data.

Well-known high performance computing techniques, such as Parallel and Grid Computing, and emerging computational models such as Graphics Processing and Cloud Computing, are more and more used in bioinformatics and life sciences [19], with the main aim to face the complexity of bioinformatics algorithms and to allow the efficient analysis of huge data.

Affymetrix Power Tools. Affymetrix Power Tools[7] are a set of command line programs provided by Affymetrix that implement different algorithms for preprocessing Affymetrix arrays. In particular **apt-probeset-summarize** implements summarization and normalization methods (e.g. RMA, RMA-SKETCH and PLIER) for gene expression arrays, while **apt-dmet-genotype** supports probe-set summarization of binary CEL files and the management of resulting preprocessed files (.CHP) related to genotyping arrays, i.e. microarray detecting single nucleotide polymorphisms (SNPs) and copy number viariations (CNVs). The system developed in [20] uses a master/slave approach, where the master node computes partitions of the input dataset (i.e. sets a list of probesets intervals) and calls in parallel several slaves each one wrapping and executing the apt-probeset-summarize program, that is applied to the proper partition of data. Such system showed a nearly linear speedup up to 20 slaves.

---

[7] www.affymetrix.com

DMET Console. DMET Console[8] is a graphical software provided by Affymetrix that supports probe-set summarization of a complete dataset of binary .CEL files, the management of resulting preprocessed files (.CHP), and the building of a tabular dataset containing the genotype call for all the probesets and all the samples of an experiment [21,22].

DMET-Analyzer. DMET-Analyzer [23] is a platform-independent software built in Java that supports the automatic statistics test of the association between SNPs and sample conditions. It has a simple graphical user interface that allows users to upload DMET files produced by DMET Console and produces as output a list of candidate SNPs. DMET-Analyzer supports the visualization of the SNPs detected on the entire dataset as a heatmap to give an immediate visual feedback to the user. It implements a Hardy-Weinberg equilibrium calculator that can be used for testing the genetic model. Finally, it annotates significant SNPs with information provided by Affymetrix libraries and with links to the dbSNP database (for basic information about SNPs) and to the PharmaGKB pharmacogenomics knowledge base, giving various information (e.g. pathways) related to pharmacogemomics.

coreSNP. coreSNP is a parallel software tool [24] for the parallel preprocessing and statistical analysis of DMET data that extends the DMET-Analyzer sequential algorithm presented in [23]. It allows to statistically test, in a parallel and automatic way and for each probe, the significance of the presence of SNPs in two classes of samples using the well known Fisher test. It automatizes the workflow of analysis of SNPs data avoiding the use of multiple tools and optimizes the execution of statistical tests on large datasets. Efficiency in executing statistical tests is obtained by parallelizing the core algorithm of DMET-Analyzer on multicore architectures and by defining efficient ad hoc data structures. The scalable multi-threaded implementation of coreSNP allows to handle the huge volumes of experimental pharmacogenomics data in a very efficient way, while its easy to use graphical user interface and its ability to annotate significant SNPs, allow biologists to interpret the results easily.

Cloud BioLinux. Cloud BioLinux[9] is a publicly accessible Virtual Machine that provides high-performance bioinformatics computing on different Cloud platforms [25]. It provides both pre-configured command line applications and graphical-enabled applications. Users may access Cloud BioLinux applications on the Amazon EC2 cloud, or they can upload Cloud BioLinux images to private Eucalyptus clouds. Moreover, users can run Cloud BioLinux on a local desktop

---

[8] The Affymetrix DMET (drug metabolism enzymes and transporters) Plus Premier Pack is a novel microarray platform for gene profiling, designed specifically to detect in human samples the presence/absence of SNPs on 225 genes that are related with drug absorption, distribution, metabolism and excretion (ADME)

[9] `cloudbiolinux.org`

computer using VirtualBox, a virtualization software available for main operating systems.

3.3 Big Data Analytics and Network Models in Biomedical Informatics

The increasing availability of big biological and medical data continues to present major challenges as well as great opportunities in biomedical research. Such challenges and opportunities are only expected to continue to grow. New technologies, such as next-generation sequencing, and new health-related policies, such as the ones associated with electronic medical records; promise to produce even more data in the near future. Such explosion of biomedical data requires an associated increase in the scale and sophistication of the automated systems and intelligent tools needed to enable the researchers to take full advantage of available data. This comes during a time when the notion of big data analytics is emerging as a significant research area in computing and information technology to address the problem of mining useful knowledge from raw data in various application domains, with big data defined loosely as data that is high in volume, velocity, variety, and veracity. One of the main research questions in biomedical sciences has been how to extract useful knowledge from such massive raw data. The development of innovative tools to integrate, analyze and mine such data sources is a key step towards achieving large impact levels. In addition, advanced tools for visualizing biomedical data at various processing stages are critical in maximizing the value of its utilization. As a result, attention has been shifting recently from a pure focus on data generation technologies to a more balanced approach significant emphasize on data analysis and data visualization tools. Such tools are critical in taking full advantage of the public and private data currently available to all biomedical researchers. In the realm of biomedical informatics and systems biology, there is a growing need for innovative multidisciplinary research that can handle the continuously growing big data while retaining the capability to analyze and predict anomalies in the system, be the system on the cellular, organismal, or social level. The use of advanced network modeling and analysis has shown to be particularly effective in integrating different types of data elements in the biomedical domain and explore the interrelationships among such elements. In addition, due to the large size of most biological and medical data, the need to utilize high performance computing systems is also emerging as an essential component of how to efficiently analyze raw data in biomedical research. Hence, the proper integration of carefully selected/developed algorithms along with efficient utilization of high performance computing systems form the key ingredients in the process of reaching new discoveries from biological data.

Network Models in Biomedical Informatics. Since the explosive influx of biological data obtained from high-throughput medical instruments, the ability to leverage the currently available data to extract useful knowledge has become one of the most challenging and exciting aspects in biomedical research. The analysis of such data is particularly complex not only due to its massive size but also due to its heterogeneity and inherent noise associated with several data gathering steps. The utilization of biological networks to model and integrate large-scale heterogeneous biomedical data continues to grow, especially with the systems biology approach taking center stage in many bioinformatics applications. Networks have been used to model various types of relations in Bioinformatics including gene interactions, protein-protein interactions, gene-disease relationship, and correlations among gene expressions. Networks constructed from high-throughput biological data are rising in popularity and importance in systems biology for their ability to effectively model relationships between entities. In addition, since networks and essentially are graphs, the ability to take full advantage of existing graph theoretic properties and algorithms makes network modeling a reliable approach for examining large complex systems. More importantly, critical elements and structures obtained from networks that model biological data have been shown to correspond to critical biological elements and cellular functions associated with the domain from which the data is obtained. Network structure has been tied to cellular function since the discovery of the link between high degree nodes and essential proteins in the interactome of yeast [26]. Initial studies performed on protein-protein interaction networks indicated that these networks follow the rather useful power-law degree distribution, meaning that most nodes in the network are poorly connected with a few nodes are very well connected; such nodes are known informally as hubs [27] [26]. Hubs have been found in the yeast protein-protein interaction network (also known as an interactome) to correspond to essential genes and have been found to be critical for maintenance of structure in other biological networks as well, such as the metabolome and the correlation network [26].

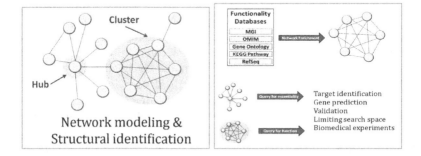

Fig. 1.

Other standard network/graph parameters such as clustering coefficient can point toward the modularity of the network4, and previous studies to identify modules in clustered networks indicate that when found, tend to correspond to genes or gene products working together toward some discrete function, such as a protein complex in an interactome or as a regulatory cohort [27]. Many algorithms currently exist that are able to find clusters within networks that employ clustering via random seed selection and growing, spectral clustering, or clustering coefficient. It is worth nothing that while gene clusters tend to correspond to biological functions, the actual structures they form in the network can be mined based solely on network structure, often without the help of biological annotation data. Thus, the link between network structure and function can be exploited to identify known and unknown network elements.

Gene Expressions and Correlation Networks. Gene expression can be modeled as a network where nodes represent gene probes and edges represent relationships between expression arrays per each gene probe. Many studies have examined these so-called correlation or co-expression networks, particularly how they are built and validated in silico. In 2005, Horvath et al. proposed a method for creating correlation networks using a weighted gene expression schema, identifying a soft-thresholding approach that resulted in a scale-free network distribution [28]. Extensive research performed by Dempsey and Ali also found links between structure and function in networks built from correlation [29] [26]. In other words, there is a growing body of research that indicates the correlation network is a valid method for identifying co-expression and potentially coregulation from high-throughput gene expression data. Correlation networks can be effectively used to model complex biological data in which nodes represent genes or gene products and edges connecting the nodes represent degrees of correlation associated with their expression levels. Although loaded with biologically-relevant signals, correlation networks do contain noise plus they are too large for simple data mining tools. In this project, we implement different types of filters to reduce the network size and sort out signals from noise. The use of gene correlation networks has emerged to assist in the discovery of previously unknown genetic relationships and the identification of significant biological functions. Such networks provide a useful mechanism to model experimental results obtained from expression data and capture a snapshot of the expression as well as the temporal changes in various experiments. In addition, gene Ontology is often integrated with biological networks within the analysis process as a source of domain knowledge. The application of correlation networks to characterize biological data has been and remains a unique method for determining changes in biological relationships over time. The correlation network is an all vs. all graphical model that examines the degree of relation over pairs of data points, such as genes in microarray expression data. Certain characteristics of the network, such as density, hubs, cliques, and pathways, can be used to filter noise and to identify causative sub-networks of biologically relevant genes or complexes. More recently correlation networks are being used to model biological relationships over time, for example in the progression of disease, the effect of pharmaceuticals on systems of the body, and in the process of aging [30].

Biological Networks and High Performance Computing. Although extremely effective, modeling big biomedical data using networks can be computationally extensive. To start with the number of nodes in such networks could be very large, however, the main computational intensity results from the large size of the solution space. For a network of size n, the number of possible relationships is of order $O(n^2)$, while the search space for the clusters or dense subgraphs is bounded only by the $O(2^n)$ exponential function. While heuristics are naturally used to conduct the clustering analysis, dealing with large networks remain expensive computationally, hence the need for high performance computing. The figure below show the impact of using up to 1000 processors to create and analyze a correlation network use to model brain tissues in aging mice. Due to the high degree of parallelism associate with most graph operation, significant speedup can be obtain by increasing the number of processors; in this case from 2 weeks to few minutes.

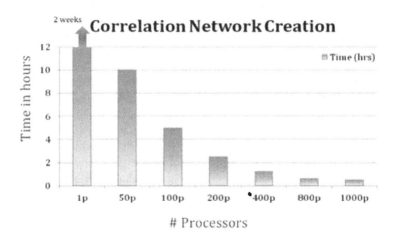

Fig. 2.

The need for high performance computing approaches to effectively solve critical problems in Bioinformatics present an exciting set of opportunities for computational researchers to get involved in Bioinformatics research. The next generation of Bioinformatics tools will be expected to take advantage of parallel computing techniques, the availability of computational resources via the clouds, and many ideas developed in the high performance computing domain to address big data analytics challenges in biomedical research.

4   Conclusions and Outlook

Applications with a real impact on the society, such as those in the field of Bioinformatics and Computational Biology, can take advantage from improvement in high performance computing to overcome computational limitations they have by definition. Those applications are developed to give the opportunity to create new exciting personal therapeutic strategies for living longer and having healthier lifestyles that were unfeasible not that long ago.

This work summarizes the main trends in HPC applied to Bioinformatics. We show several successful stories and application fields, in which relevant biological problems have been solved (or are being targeted) thanks to the computational power available in current processors. We have also pointed out the main drawbacks in the HPC arena, which may limit this good alliance between Bioinformatics and HPC systems. Among them we may highlight power consumption, high learning-curve in emerging programming models to leverage their computational power and the total cost of ownership. We think that the investigations on improvement Bioinformatics application's performance on HPC systems will be also of high technological interest as those applications have novel computational patterns that can lead the next generation of heterogeneous computing systems.

Acknowledgements. This work was supported by the Fundación Séneca la Región de Murcia under Project 18946/JLI/13. This work has been funded by the Nils Coordinated Mobility under grant 012-ABEL-CM-2014A, in part financed by the European Regional Development Fund (ERDF). This work was partially supported by the computing facilities of Extremadura Research Centre for Advanced Technologies (CETA−CIEMAT), funded by the European Regional Development Fund (ERDF). CETA−CIEMAT belongs to CIEMAT and the Government of Spain. The authors also thankfully acknowledge the computer resources and the technical support provided by the Plataforma Andaluza de Bioinformática of the University of Málaga. We also thank NVIDIA for hardware donation under CUDA Teaching Center 2014. This work has been partially supported by the projects PRIN 2010-2011 2010NFEB9L_003 and PON04a2_D DICET-INMOTO-ORCHESTRA funded by MIUR.

References

1. Klepeis, J.L., Lindorff-Larsen, K., Dror, R.O., Shaw, D.E.: Long-timescale molecular dynamics simulations of protein structure and function. Current Opinion in Structural Biology 19(2), 120–127 (2009)
2. Morris, G.M., Goodsell, D.S., Huey, R., Olson, A.J.: Distributed automated docking of flexible ligands to proteins: Parallel applications of AutoDock 2.4. Journal of Computer-Aided Molecular Design 10(4), 293–304 (1996)
3. Fischer, B., Merlitz, H., Wenzel, W.: Increasing Diversity in In-Silico Screening with Target Flexibility. In: Proceedings of Computational Life Sciences, First International Symposium, CompLife. CompLife 2005, pp. 186–197. Springer (2005)
4. Halperin, I., Ma, B., Wolfson, H., Nussinov, R.: Principles of docking: An overview of search algorithms and a guide to scoring functions. Proteins 47(4), 409–443 (2002)

5. Korb, O., Stützle, T., Exner, T.E.: Accelerating Molecular Docking Calculations Using Graphics Processing Units. Journal of Chemical Information and Modeling 51, 865–876 (2011)
6. Fan, X., Weber, W.-D., Barroso, L.A.: Power provisioning for a Warehouse-Sized Computer. In: Proceedings of the 34th annual International Symposium on Computer Architecture, ISCA 07, pp. 13–23. ACM (2007)
7. Hewitt, C.: ORGs for Scalable, Robust, Privacy-Friendly Client Cloud Computing. IEEE Internet Computing 12(5), 96–99 (2008)
8. Berl, A., Gelenbe, E., Di Girolamo, M., Giuliani, G., De Meer, H., Dang, M.Q., Pentikousis, K.: Energy-efficient cloud computing. The Computer Journal 53(7), 1045–1051 (2010)
9. Venter, J.C., Adams, M.D., Myers, E.W., Li, P.W., Mural, R.J., Sutton, G.G., Smith, H.O., Yandell, M., Evans, C.A., Holt, R.A.: The Sequence of the Human Genome. Science 291(5507), 1304–1351 (2001), http://dx.doi.org/10.1126/science.1058040
10. Garland, M., Le Grand, S., Nickolls, J., Anderson, J., Hardwick, J., Morton, S., Phillips, E., Zhang, Y., Volkov, V.: Parallel Computing Experiences with CUDA. IEEE Micro 28, 13–27 (2008)
11. Garland, M., Kirk, D.B.: Understanding throughput-oriented Architectures. Communications of the ACM 53, 58–66 (2010)
12. Yang, H., Zhou, Q., Li, B., Wang, Y., Luan, Z., Qian, D., Li, H.: GPU Acceleration of Dock6's Amber Scoring Computation. Advances in Computational Biology 680, 497–511 (2010)
13. Kannan, S., Ganji, R.: Porting Autodock to CUDA. In: 2010 IEEE Congress on Evolutionary Computation (CEC), pp. 1–8 (2010)
14. Korb, O., Stützle, T., Exner, T.E.: Accelerating molecular docking calculations using graphics processing units. Journal of Chemical Information and Modeling 51(4), 865–876 (2011)
15. Korb, O., Stützle, T., Exner, T.E.: PLANTS: Application of Ant Colony Optimization to Structure-Based Drug Design. In: Dorigo, M., Gambardella, L.M., Birattari, M., Martinoli, A., Poli, R., Stützle, T. (eds.) ANTS 2006. LNCS, vol. 4150, pp. 247–258. Springer, Heidelberg (2006)
16. Dorigo, M.: Optimization, learning and natural algorithms. Ph.D. dissertation, Politecnico di Milano, Italy (1992)
17. Manfrin, M., Birattari, M., Stützle, T., Dorigo, M.: Parallel ant colony optimization for the traveling salesman problem. In: Dorigo, M., Gambardella, L.M., Birattari, M., Martinoli, A., Poli, R., Stützle, T. (eds.) ANTS 2006. LNCS, vol. 4150, pp. 224–234. Springer, Heidelberg (2006)
18. Cecilia, J.M., García, J.M., Ujaldón, M., Nisbet, A., Amos, M.: Parallelization strategies for ant colony optimisation on gpus. In: NIDISC 2011: 14th International Workshop on Nature Inspired Distributed Computing. Proc. 25th International Parallel and Distributed Processing Symposium (IPDPS 2011), Anchorage, Alaska, USA (May 2011)
19. Cannataro, M.: Computational Grid Technologies for Life Sciences, Biomedicine and Healthcare. IGI Global Press, Hershey (2009)
20. Guzzi, P.H., Cannataro, M.: Parallel pre-processing of affymetrix microarray data. In: Guarracino, M.R., Vivien, F., Träff, J.L., Cannataro, M., Danelutto, M., Hast, A., Perla, F., Knüpfer, A., Di Martino, B., Alexander, M. (eds.) Euro-Par Workshop 2010. LNCS, vol. 6586, pp. 225–232. Springer, Heidelberg (2011)

21. Sissung, T.M., English, B.C., Venzon, D., Figg, W.D., Deeken, J.F.: Clinical pharmacology and pharmacogenetics in a genomics era: the DMET platform. Pharmacogenomics 11(1), 89–103 (2010), http://dx.doi.org/10.2217/pgs.09.154

22. Burmester, J.K., Sedova, M., Shapero, M.H., Mansfield, E.: Dmet microarray technology for pharmacogenomics-based personalized medicine. Microarray Methods for Drug Discovery, Methods in Molecular Biology 632, 99–124 (2010)

23. Guzzi, P.H., Agapito, G., Di Martino, M.T., Arbitrio, M., Tagliaferrri, P., Tassone, P., Cannataro, M.: DMET-analyzer: Automatic analysis of affymetrix DMET data. BMC Bioinformatics 13, 258 (2012),
http://dx.doi.org/10.1186/1471-2105-13-258

24. Guzzi, P.H., Agapito, G., Cannataro, M.: coresnp: Parallel processing of microarray data. IEEE Transaction on Computers 63(12), 2961–2974 (2014)

25. Krampis, K., Booth, T., Chapman, B., Tiwari, B., Bicak, M., et al.: Cloud biolinux: Pre-configured and on-demand bioinformatics computing for the genomics community. BMC Bioinformatics 13(42), 3448–3449 (2012)

26. Dempsey, K., Ali, H.: On the robustness of the biological correlation network model. In: International Conference on Bioinformatics Models, Methods and Algorithms (BIOINFORMATICS 2014), pp. 186–195 (2014)

27. Dempsey, K., Currall, B., Hallworth, R., Ali, H.H.: A new approach for sequence analysis: Illustrating an expanded bioinformatics view through exploring properties of the prestin protein. In: Handbook of Research on Computational and Systems Biology, pp. 202–223 (2011)

28. Zhang, B., Horvath, S.: A general framework for weighted gene co-expression network analysis. Statistical Applications in Genetics and Molecular Biology 4(1) (January 2005), http://dx.doi.org/10.2202/1544-6115.1128

29. Dempsey, K., Currall, B., Hallworth, R., Ali, H.: An intelligent data-centric approach toward identification of conserved motifs in protein sequences. In: Proceedings of the First ACM International Conference on Bioinformatics and Computational Biology. BCB 2010, pp. 398–401. ACM, New York (2010),
http://doi.acm.org/10.1145/1854776.1854839

30. Dempsey, K., Bonasera, S., Bastola, D., Ali, H.: A novel correlation networks approach for the identification of gene targets. In: 2011 44th Hawaii International Conference on System Sciences (HICSS), pp. 1–8 (January 20011)

Computing Biological Model Parameters
by Parallel Statistical Model Checking*

Toni Mancini, Enrico Tronci, Ivano Salvo, Federico Mari, Annalisa Massini,
and Igor Melatti

Computer Science Department, Sapienza University of Rome, Italy

Abstract. Biological models typically depend on many *parameters*.
Assigning suitable values to such parameters enables *model individualisation*. In our clinical setting, this means finding a model for a given patient.
Parameter values cannot be assigned arbitrarily, since *inter-dependency*
constraints among them are not modelled and ignoring such constraints
leads to biologically meaningless model behaviours. Classical parameter
identification or estimation techniques are typically not applicable due to
scarcity of clinical measurements and the huge size of parameter space.
Recently, we have proposed a statistical algorithm that finds (almost) all
biologically meaningful parameter values. Unfortunately, such algorithm
is computationally extremely intensive, taking up to months of sequential computation. In this paper we propose a parallel algorithm designed
as to be effectively executed on an arbitrary large cluster of multi-core
heterogenous machines.

1 Introduction

Systems biology models aim at providing quantitative information about time
evolution of biological species. One of the main goals of systems biology in a
health-care context is to *individualise* models in order to compute patient-specific
predictions (see, e.g., [24]) for the time evolution of species (e.g., hormones).

Depending on the system at hand, many modelling approaches are currently
investigated. For example, see [22,21] for an overview on discrete as well as
continuous modelling approaches, and [49] for a survey on stochastic modelling
approaches. In biological networks modelled with a system of *Ordinary Differential Equations (ODEs)* depending on a set of parameters (as in, e.g., [35,50,39])
model individualisation can be done by assigning suitable values to the model
parameters. Such biological models depend on many (easily hundreds of) parameters, whose values cannot be chosen arbitrarily because of *inter-dependency*
constraints among them (see, e.g., [26]) that, usually, are not explicitly known
and thus are not modelled. If model parameter values are chosen ignoring such
constraints, then the resulting model behaviour is biologically meaningless.

Model identification (see, e.g., [27]) techniques are typically used to compute
values for model parameters so that a suitable error function measuring mismatch
between model predictions and experimental data is minimised (*parameter estimation*). If such a value exists and is unique, the model is said *identifiable*. In a

---

* This work has been partially supported by the EC FP7 project PAEON (Model
Driven Computation of Treatments for Infertility Related Endocrinological Diseases,
600773).

F. Ortuño and I. Rojas (Eds.): IWBBIO 2015, Part II, LNCS 9044, pp. 542–554, 2015.

clinical setting, for each patient, only a small number of measurements is available, since they can be costly, invasive and time-consuming. Therefore, although in principle model identification techniques could be used to compute *patient-specific* model parameters, in practice, because of the large amount of measurements needed (see, e.g., [9]), they are typically used to compute a *default parameter value* that averages among the behaviours of many patients (as, e.g. in [39]). *Parameter estimation* approaches cannot be used either, since with such a few data they would not take into due consideration inter-dependencies among model parameters [26].

Motivations. To overcome scarcity of measurements, we proposed a two-phase approach [47]. First, an *off-line* phase that accounts for parameter inter-dependencies [26] greatly narrows down the search space to a set S of parameter values yielding *biologically meaningful* model behaviours. Second, an *on-line* phase computes a patient-specific model by selecting in S those parameter values that minimise mismatch with respect to patient measurements. This enables fast patient-specific predictions for the time evolution of each species of interest.

In general, to decide if time evolution of species concentration is biologically meaningful takes a domain expert. To build a general purpose tool that can automatically search through millions of model parameter values, in [47] we proposed a criterion which regards as *Biologically Admissible (BA)* those parameter values entailing time evolutions with a second order statistics *close enough* to that of the model default parameter values.

The computation of the set S of BA values for the model parameters requires to explore the set of all possible values for the parameter vector, that is typically huge. Since an exhaustive exploration would be unfeasible, in [47] a Statistical Model Checking (SMC) based approach is proposed. Nonetheless, the exploration of the parameter space may take *months* of sequential computation.

Main Contributions. Our main contribution is a SMC parallel algorithm and its distributed multi-core implementation to compute the set of all BA parameter values. We propose a master-slave architecture where a single master process (*Orchestrator*) implements the SMC algorithm and delegates to a high number of slaves (*BA Verifiers*) the numerical integration of the ODE system defining the model. As a consequence, our parallel algorithm will benefit from the availability of many heterogeneous computational units (e.g., a data-centre in the cloud).

The SMC algorithm proposed in [47] would require too much synchronisation in a parallel context. Here, we define a new random sampling process at the basis of our SMC algorithm, which enables massive parallelisation of BA Verifiers.

We developed our distributed multi-core tool in the C language using Message Passing Interface (MPI) [42]. We evaluate effectiveness of our approach by using it on the *GynCycle* model in [39], an ODE model which predicts blood concentration of several species during female menstrual cycle. We show that our implementation achieves high efficiency even when using dozens of computational cores (e.g., efficiency is 74% when using 80 cores).

Related Work. The input to our algorithm consists of a system model along with the *default value* for its parameters. The *GynCycle* model considered in our case study has been presented in [39] and the default (inter-patient) values for its parameters have been computed in [11] using model identification (often referred to as *parameter identification* in our setting) techniques [27].

In recent years, parallel and distributed computing has received attention in order to cope with the complexity of biological systems. See [5] for a survey of parallel methods to solve ODEs, parallel model checking, and parallel simulations in biological applications.

Statistical Model Checking mainly addresses system verification of stochastic systems with respect to probabilistic temporal properties or continuous stochastic properties (see, e.g., [41]). Several parallel and distributed approaches to SMC have been introduced (see, e.g., [3,40]), some of them motivated by the complexity of biological models [4]. Here, we focus on deterministic biological systems modelled with ODEs, and we apply SMC techniques (along the lines of [18]) to infer statistical completeness of our set of BA parameters.

Parallel approaches close to ours are those in [7,44,8], where the problem of computing all (discretised) model parameter values meeting given LTL properties has been investigated. We extend such works in two directions. First, the above mentioned papers focus on piecewise affine ODE systems, whereas we can handle any (possibly non-linear) ODE system. Second, they aim at computing a maximal set of parameters satisfying a given LTL property. Thus, when the model changes, a new LTL property has to be provided by domain experts. Our approach infers such a system property by the default value for the model parameters, thus decreasing the amount of input needed from domain experts.

A key feature of parameter identification approaches is their ability to give information about parameter *identifiability* (see, e.g., [9] and citations thereof). Gradient-based methods, as, e.g., the classical one in [25], provide a local optimum solution to the parameter estimation problem. Global methods, such as [28], provide a global optimum solution whereas heuristics approaches as evolutionary algorithms (see, e.g., [6,45]), provide near-global optimal solutions. All such approaches do not provide information about parameter identifiability. When observations are scarce, parameters usually become non-identifiable. Studying the correlation among system parameters can reduce the number of data needed for identifiability, see for example [37,26]. Our goal here is to support model individualisation from clinical measurements. This means that we need to compute model parameters from a few (say, 3) observations about a small subset (4 in our case study) of the species occurring in the model (33 in our case). Because of scarcity of measurements, neither model identification approaches nor parameter estimation approaches can be used in our setting.

Model checking based parameter estimation approaches have been investigated for example in [20,12,38]. Such approaches differ from ours, since they do not address the problem of automatically restricting the search space. Model checking techniques have been widely used in systems biology, to verify time behaviours. Examples are in [23,19,14,16,35]. Such approaches focus on verifying a given property for the model trajectories, whereas our main problem here is to compute *all* biologically plausible values for the model parameters.

We note that computing the set of *all* model parameter values that satisfy a given property is closely related to that of computing *all* control strategies satisfying a given property. In a discrete time setting this problem has been addressed, for piecewise affine systems and safety properties, in [33,2,1,34].

2 Background

Unless otherwise stated, all forthcoming definitions are based on [47,43]. Throughout the paper, we denote with $[n]$ the set $\{1, 2, \ldots, n\}$ of the first n natural numbers and with $\mathbb{R}^+, \mathbb{R}^{\geq 0}$ and $\mathbb{R}$ the sets of, respectively, positive, non-negative and all real numbers. We also denote with $(\mathbb{R}^{\geq 0} \times \mathbb{R}^{\geq 0})^*$ the set of pairs $(a, b) \in \mathbb{R}^{\geq 0} \times \mathbb{R}^{\geq 0}$ such that $a \geq b$.

2.1 Parametric Dynamical Systems

We model biological systems using dynamical systems. Usually, a dynamical system comes equipped with a function space that models both *controllable* (e.g., treatments) and *uncontrollable* inputs (*disturbances*). Here, we do not address treatments or disturbances and accordingly we omit inputs from Def. 1.

Definition 1 (Parametric Dynamical System). *A* Parametric Dynamical System *(or, simply, a Dynamical System)* S *is a tuple* $(\mathcal{X}, \mathcal{Y}, \Lambda, \varphi, \psi)$, *where:*

- $\mathcal{X} = X_1 \times \ldots \times X_n$ *is a non-empty set of* states *(state space of* S*);*
- $\mathcal{Y} = Y_1 \times \ldots \times Y_p$ *is a non-empty set of* outputs *(output value space);*
- Λ *is a non-empty set of* parameters *(parameter value space);*
- $\psi : \mathbb{R}^{\geq 0} \times \mathcal{X} \to \mathcal{Y}$ *is the* observation function *of* S*;*
- $\varphi : (\mathbb{R}^{\geq 0} \times \mathbb{R}^{\geq 0})^* \times \mathcal{X} \times \Lambda \to \mathcal{X}$ *is the* transition map *of* S*. Intuitively,* $\varphi(t_2, t_1, x, \lambda)$ *is the state reached by the system (with parameter values* λ*) at time* t_2 *starting from the state* $x \in \mathcal{X}$ *at time* t_1*.*

Remark 1. To simplify notation, unless otherwise stated, we assume that the set of parameters Λ has the form $\mathcal{X} \times \Gamma$ (where Γ is a non-empty set). Therefore, a parameter $\lambda = (x_0, \gamma) \in \Lambda$ embodies information about the initial state x_0 of a system trajectory. Such a system trajectory is a function of time $x(\lambda)(t)$, which, for each $t \in \mathbb{R}^{\geq 0}$, evaluates to $\varphi(t, 0, x_0, \gamma)$. In the following, abusing notation, we write $x(\lambda, t)$ instead of $x(\lambda)(t)$. Analogously, we write $x_i(\lambda, t)$ $[y_i(\lambda, t)]$ for the time evolution $x_i(\lambda)(t)$ $[y_i(\lambda)(t)]$ of the ith state [output] component with parameters γ starting in x_0 from time 0.

Example 1. Dynamical systems whose dynamics is described by a system of Ordinary Differential Equations (ODEs) depending on parameters are currently of great interest as a mathematical model for biological networks (see, e.g., [15,39]). In this paper, we will use as a case study the *GynCycle* model presented in [39]. It is a ODE model for the feedback mechanisms between Gonadotropin-Releasing Hormone (GnRH), Follicle-Stimulating Hormone (FSH), Luteinizing Hormone (LH), development of follicles and corpus luteum, and the production of Estradiol (E2), Progesterone (P4), Inhibin A (IhA), and Inhibin B (IhB) during the female menstrual cycle. The model aims at predicting blood concentrations of LH, FSH, E2, and P4 during different stages of the menstrual cycle. The model is intended as a tool to help in preparing and monitoring clinical trials with new drugs that affect GnRH receptors (*quantitative and systems pharmacology*).

In our *black-box* approach, the system transition map models our call to a solver (namely, *Limex* [13]) computing a solution to the ODEs defining our dynamical system. This is along the lines of simulation based system level formal verification as in [29,31,30,32,46,10].

2.2 Biological Admissibility

In general, given a value λ for the (vector of) model parameters, it takes a domain expert to decide if a time evolution $x(\lambda, t)$ is *biologically meaningful*. Indeed, many parameter values lead to time evolutions for the model species that are not compatible with the laws of biology. Our goal is to build a general purpose tool that automatically filters out biologically meaningless parameter values. Following [47], we provide a formal criterion for biological admissibility, by asking that the time evolution of $x(\lambda, t)$ is *similar enough* to that of $x(\lambda_0, t)$, that is the one entailed by the model default parameter value λ_0. To this end, we introduce three measures of how similar two trajectories are.

Given a function f from $\mathbb{R}$ to $\mathbb{R}$ and $\alpha, \tau \in \mathbb{R}$, we denote with $f^{\alpha,\tau}$ the function defined by $f^{\alpha,\tau}(t) = f(\alpha(t + \tau))$ for all t. Here, α and τ are used to model, respectively, a stretch and a shift of f. Given two functions f and g from $\mathbb{R}$ to $\mathbb{R}$, the *cross-correlation* (see, e.g., [48]) $\langle f, g \rangle(\xi)$ between f and g is a function of ξ (where $\xi \in \mathbb{R}$ is the *time lag*) defined as: $\langle f, g \rangle(\xi) = \int_{-\infty}^{+\infty} f(t)g(t + \xi)dt$. We consider the *normalised zero-lag cross-correlation* function $\rho_{f,g}$, defined as $\rho_{f,g} = \frac{\langle f,g \rangle(0)}{\|f\|\|g\|}$, where, for any f, $\|f\|$ is the L^2 norm of f, i.e., $\sqrt{\langle f, f \rangle(0)}$. The higher $\rho_{f,g}$ the more *similar* f and g (e.g., f and g have the same peaks). In particular, $\rho_{f,g}$ is 1 if f is equal to g up to an amplification factor.

Let $\mathcal{S}$ be dynamical system with n state variables and a default parameter value λ_0. Given a parameter value λ and a finite horizon $h \in \mathbb{R}^{\geq 0}$, let $x_i(\lambda_0, t)$ and $x_i(\lambda, t)$ be the time evolutions of species x_i (for each $i \in [n]$) under parameters λ_0 and λ respectively. Being time evolutions, both $x_i(\lambda_0, t)$ and $x_i(\lambda, t)$ are defined for $0 \leq t \leq h$. Anyway, to easily match the above general definition of cross-correlation, we define such functions on the whole set of real numbers, as being 0 for any $t < 0$ or $t > h$. In order to model biological admissibility, we define the following three functions (i ranges over $[n]$, $\alpha, \tau \in \mathbb{R}$):

$$\rho_{\lambda_0,\lambda,i}(\alpha, \tau) = \rho_{x_i(\lambda_0), x_i^{\alpha,\tau}(\lambda)} \qquad \mu_{\lambda_0,\lambda,i}(\alpha, \tau) = \left| \frac{\int_0^h (x_i(\lambda_0, t) - x_i^{\alpha,\tau}(\lambda, t))dt}{\int_0^h x_i(\lambda_0, t)dt} \right|$$

$$\chi_{\lambda_0,\lambda,i}(\alpha) = \left| (\|x_i(\lambda_0)\|^2 - \|x_i^{\alpha,\tau}(\lambda)\|^2) \right| / \|x_i(\lambda_0)\|^2$$

The *normalised zero-lag cross-correlation* $\rho_{\lambda_0,\lambda,i}(\alpha, \tau)$ measures the similarity of the trajectories $x_i(\lambda_0, t)$ and $x_i(\lambda, t)$ as for qualitative aspects (for example, if they have the same peaks), when $x_i(\lambda, t)$ is subject to stretch α and time-shift τ. The *normalised average differences* $\mu_{\lambda_0,\lambda,i}(\alpha, \tau)$ and the *normalised squared norm differences* $\chi_{\lambda_0,\lambda,i}(\alpha, \tau)$ are two measures of the average distance between $x_i(\lambda_0, t)$ and $x_i(\lambda, t)$, when $x_i(\lambda, t)$ is subject to stretch α and time-shift τ.

In Def. 2, we use these functions to formalise the notion of Biologically Admissible (BA) parameter λ with respect to a default parameter λ_0. Intuitively, λ is BA if the three measures above are all above or below certain thresholds.

Definition 2 (Biologically Admissible parameter). *Let λ_0, $\lambda \in \mathcal{X} \times \Lambda$ be two parameters. Let $\mathbb{A} \subseteq \mathbb{R}^+$, $\mathbb{B} \subseteq \mathbb{R}$ be two sets of real numbers such that $1 \in \mathbb{A}$ and $0 \in \mathbb{B}$. Given a tuple $\Theta = (\theta_1, \theta_2, \theta_3)$ of positive real numbers, we say that λ is Θ-biologically admissible with respect to λ_0, notation $adm_{\mathbb{A},\mathbb{B}}(\lambda_0, \lambda, \Theta)$, if there exist $\alpha \in \mathbb{A}$ and $\tau \in \mathbb{B}$ such that, for all $i \in [n]$: $(\rho_{\lambda_0,\lambda,i}(\alpha, \tau) \geq \theta_1) \wedge (\mu_{\lambda_0,\lambda,i}(\alpha, \tau) \leq \theta_2) \wedge (\chi_{\lambda_0,\lambda,i}(\alpha, \tau) \leq \theta_3)$.* □

3 Computation of Admissible Parameters

Our goal is to compute the set S of (with high confidence) all Biologically Admissible (BA) parameter values with respect to a default parameter value λ_0 validated by the model designer as biologically meaningful.

Since small differences in values are meaningless from a biological point of view, we consider a (grid-shaped) *discretised parameter space* $\hat{\Lambda}$ that is a finite subset of the set of possible parameter values Λ. An exhaustive search on $\hat{\Lambda}$ would be unfeasible, due to the large number of parameters to identify (75 in our case study) that makes $\hat{\Lambda}$ huge (10^{75} elements if we consider 10 possible values for each parameter). To overcome such an obstruction, we follow an approach inspired by Statistical Model Checking (SMC) [18,17]. Statistical Hypothesis Testing is used in [47] to compute, with high statistical confidence, the set S of all BA values with respect to a default value λ_0 for the model parameters.

Given arbitrary values in $(0, 1)$ for ε (probability threshold) and δ (confidence threshold), the SMC algorithm in [47] computes the set S of BA parameters by randomly sampling the discretised parameter space $\hat{\Lambda}$ and adding to S those parameter values $\lambda \in \hat{\Lambda}$ which are shown (by simulation) to be BA. The algorithm terminates when set S remains unchanged after $N = \lceil \ln \delta / \ln(1 - \varepsilon) \rceil$ attempts. At this point, following [18], in [47] it is proved that, with statistical confidence $1 - \delta$, the probability that the sampling process will extract a BA parameter vector value not already in S is less than ε.

Unfortunately, the SMC algorithm proposed in [47] *cannot* be extended to work in a parallel context efficiently, as too much synchronisation would be required. Here, we define a new random sampling process at the basis of our SMC approach, which enables massive parallelisation of the computation of S. Our new algorithm has been explicitly designed as to be easily deployed on a cluster of heterogeneous multi-core machines connected by a network.

3.1 Algorithm Outline

An overall high-level view of our algorithm deployed on multiple machines connected by a network is shown in Fig. 1. The parallel algorithm that we present here consists of one *Orchestrator* and many *BA Verifiers*.

Orchestrator. The orchestrator initialises the set S of BA parameter values to the singleton set $\{\lambda_0\}$. Then, at each iteration, it randomly chooses N parameter values $\lambda_1, \ldots, \lambda_N \in \hat{\Lambda}$ independently (where $N = \lceil \ln \delta / \ln(1 - \varepsilon) \rceil$) and delegates the verification of each of them to an idle BA Verifier. After having collected all the N answers, the Orchestrator adds to the set S those parameter values

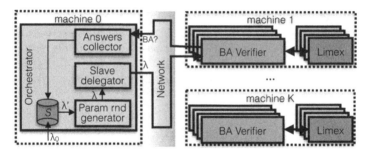

Fig. 1. Parallel algorithm Architecture

returned as BA. If S changes (i.e., at least one of the N randomly generated parameters is BA and not already in S), a new round of this process starts, otherwise the set S computed so far is returned as the final result.

The sampling space $\hat{\Lambda}$ is given by the set of discretised values for the model parameters. Our sampling strategy (Sect. 3.2) guarantees that any parameter value $\lambda \in \hat{\Lambda}$ can be extracted with non-zero probability, as required by [18,47]. To speed up our procedure, we give a higher probability to parameter values that differ from some parameters already in S for a small number of components.

Note that, at each iteration, the Orchestrator adds up to N parameters to S. Thus, increasing the number of parallel BA Verifiers helps a faster growth of the set of BA parameter values S.

BA Verifiers. Each BA Verifier repeatedly takes a parameter λ as input from the Orchestrator, checks if it is Biologically Admissible, and sends back the answer (consisting of the result and the parameter value) to the Orchestrator. To check whether parameter λ is admissible, the BA Verifier in charge runs its *own* instance of the *Limex* solver to compute the time evolutions of all species under parameter λ and checks whether the normalised zero-lag cross-correlation, the normalised average differences, and the normalised squared norm differences for all species are above or below the given thresholds $\Theta = (\theta_1, \theta_2, \theta_3)$, as prescribed by Def. 2.

We observe that the computation distributed to the BA Verifiers is the heaviest part, since it entails to numerically solve the system of differential equations (for a given discretisation of the time output period $[0, h]$ into a finite set T of time-points) and to compute the functions defined in Sect. 2.2 by numerical integration. In order to speed up their computation, BA Verifiers invoke the *Limex* solver just once for each parameter value λ: given the requested finite output time set T and the sets $\mathbb{A}$ and $\mathbb{B}$ for the allowed stretch and time-shift factors, they simulate the system $\mathcal{S}$ computing the trajectory $(t, x(\lambda, t))$ for all time points in a set $T_{\mathbb{A}, \mathbb{B}}$ defined as $T \cup \{t' \mid t' = \alpha(t + \tau), t \in T, \alpha \in \mathbb{A}, \tau \in \mathbb{B}\}$. The set $T_{\mathbb{A}, \mathbb{B}}$ contains all time instants in which species values are to be known in order to evaluate whether parameter λ satisfies Def. 2.

3.2 Parameter Probability Space

The probability distribution over the discretised parameter space $\hat{\Lambda}$ used by the Orchestrator to generate new parameter values to examine is parametric to the set S of BA parameter values found so far. To speed up the finding of new BA parameter values (with respect to, e.g., uniform sampling), parameter values that are close to those in S are most likely to be chosen.

Given a set S, we extract the N values $\lambda_1, \ldots, \lambda_N$ to examine at each iteration of the Orchestrator independently as follows. For all $i \in [1, N]$: 1) We randomly choose $\lambda_i' \in S$ considering a uniform probability distribution over S. 2) We randomly choose the maximum number h_i of components in which λ_i will differ from λ_i'. In this case, the set $[n]$ is considered distributed as a power-law of the form $\mathbf{Pr}[h] = a h_i^{-b}$, with $b > 1$ and a being a normalisation constant. This implies that, with high probability, λ_i will differ from λ_i' in a small number of components. 3) We randomly choose a subset H_i of h_i different components in $[n]$, assuming a uniform distribution over the set of subsets of cardinality h_i. 4) Finally, the parameter value λ_i is such that for all $j \in H_i$ $\lambda_{i,j}$ is choosen in $\hat{\Lambda}_j$ uniformly at random and $\lambda_{i,j} = \lambda_{i,j}'$ for all $j \in [n] \setminus H_i$.

This sampling technique defines a probability space $(\hat{\Lambda}, \mathcal{P}(\hat{\Lambda}), \mathbf{Pr}^S)$ parametric with respect to a set $S \subseteq \hat{\Lambda}$. By multiplying the (conditional) probabilities of steps 1)–4) above, we have: $\mathbf{Pr}^S[\lambda] = \frac{1}{|S|} \sum_{\lambda' \in S} a \, |d(\lambda, \lambda')|^{-b} \binom{n}{|d(\lambda, \lambda')|}^{-1} \prod_{i \in d(\lambda, \lambda')} \frac{1}{|\hat{\Lambda}_i|}$, where $d(\lambda, \lambda')$ is the set of the components on which λ and λ' differ. Note that $\mathbf{Pr}^S[\lambda]$ is non-zero for all λ.

3.3   Algorithm Correctness

The guarantee that, upon termination, with high statistical confidence, all BA parameter values are in S depends only on the fact that the sampling process consecutively fails N times to find a BA parameter value outside S, and not on how the set S has been populated in the previous iterations of the algorithm.

Stemming from the above considerations, we show the following theorem, stating the correctness of our parallel algorithm.

Theorem 1. *Given a dynamical system $\mathcal{S}$ as in Def. 1, a finite subset $\hat{\Lambda}$ of Λ, a value $\lambda_0 \in \hat{\Lambda}$, a tuple Θ of biological admissibility thresholds, two real numbers ε and δ in $(0, 1)$, and two finite sets of real numbers $\mathbb{A}$ and $\mathbb{B}$ (with $1 \in \mathbb{A}$ and $0 \in \mathbb{B}$), our parallel algorithm is such that:*

1. it terminates;

2. upon termination, it computes a set $S \subseteq \hat{\Lambda}$ of Θ-Biologically Admissible parameter values;

3. with confidence $1 - \delta$: $\mathbf{Pr}^S[\{\lambda \in \hat{\Lambda} \setminus S \mid \mathrm{adm}_{\mathbb{A}, \mathbb{B}}(\lambda_0, \lambda, \Theta)\}] < \varepsilon$. $\square$

4   Experimental Results

The computational effectiveness of our distributed multi-core implementation has been evaluated on the *GynCycle* model [39]. Such a model has 114 parameters, 75 of which are patient-specific (at least for our purposes), and consists of 41 differential equations defining the time evolution of 33 species.

We implemented our tool in the C programming language using Message Passing Interface (MPI) [42] to enable the communication between the Orchestrator and BA Verifiers spread on multiple machines connected by a network.

4.1   Experimental Setting

Experiments have been carried out on a cluster of 7 Linux heterogeneous machines: 1 machine equipped with 2 × Intel(R) Xeon(R), 2.83 GHz and 8GB of RAM (category A), 2 machines equipped with 2 × Intel(R) Xeon(R), 2.66 GHz and 8GB of RAM (cat. B), and 4 machines equipped with 2 × Intel(R) Xeon(R), 2.27 GHz and 16GB of RAM (cat. C). We used a maximum number of 81 CPU cores (7 out of the 8 available cores for machines of categories A and B and 15 out of the 16 available cores for machines of cat. C). The single Orchestrator process was always run on a core of the machine in cat. A.

We set both ε and δ to 10^{-3}. The stretch factor α (see Def. 2 in Sect. 2.2) ranges in the set $\mathbb{A} = \{0.90, 0.95, 1.00, 1.05, 1.10\}$, while the set $\mathbb{B}$ for the shift factor τ (see Def. 2 in Sect. 2.2) consists of all values from -3 to 3 days multiple of 6 hours. The discretisation $\hat{\Lambda}$ of Λ has been obtained by uniformly discretising the range of each parameter into 5 values. We set Limex to compute time evolutions for all species over $h = 90$ days, returning values with a time step of 15 minutes. Integrals for cross-correlation and norms have been computed numerically with a time step of 15 minutes.

In [47], suitable values for the biological admissibility thresholds $\theta_1, \theta_2, \theta_3$ have been considered, in order to largely cover the set of model meaningful biological behaviours. Here we are interested in evaluating the *speedup* and the *efficiency* of our distributed multi-core algorithm. To this end, in order to execute multiple experiments in reasonable time, we set the biological admissibility thresholds $\theta_1, \theta_2, \theta_3$ to, respectively, 0.99, 0.01, 0.01. Such values are way overly restrictive from a biological point of view, and allow us to compute only a *tiny fraction* (only 8 parameter values) of the set of the BA parameters shown in [47] (which consists of several thousands of Biologically Admissible (BA) parameter values). Anyway, the overall number of random parameter values generated and examined in our case (27620) is sufficiently large to let us correctly evaluate the computational performance of our algorithm.

4.2   Experimental Results

Table 1 shows the overall computation time (column *"time"*) when varying the number of BA Verifiers (col. *"# proc."*) used in parallel by our algorithm. Each BA Verifier runs on a *different* core of a machine in our cluster. To make the different values comparable (given the stochastic nature of our algorithm and the heterogeneity of our cluster machines), we started all runs using the *same* random seed and used the *same* proportion of machines of each category in all runs (col. *"# cores"*). To neutralise biases due to the heterogeneity of our cluster machines, we determined the computation time of our algorithm when using a *single* BA Verifier (sequential time) by carrying out three runs allocating the (single) BA Verifier on a core of a machine of each category. Such computation times are listed in Table 2. From such data we have computed the completion time in the first line of Table 1 by averaging the three sequential execution times, using the proportion of the number of cores for each machine category as weights.

Column *"speedup"* in Table 1 shows the speedup achieved by our algorithm. For each number v of parallel BA Verifiers, the speedup is the ratio t_v/t_1, where t_v and t_1 are the computation times shown in Table 1 when using, respectively,

Table 1. Computation times

#proc.	# cores A B C	time (h:m:s)	speedup	eff.
1	– – –	*238:16:55*	1×	100%
26	2 4 20	9:16:57	25.67×	98.73%
52	4 9 39	5:16:25	45.18×	86.88%
80	6 14 60	4:1:12	59.27×	74.09%

Table 2. Sequential time

machine cat. for the sequential alg.	time (h:m:s)
A	194:47:45
B	206:19:15
C	250:5:18

v and 1 BA Verifiers. Column "*eff.*" shows the efficiency of our algorithm and is computed, as typically done in the evaluation of parallel algorithms, by dividing the speedup by the number of the parallel BA Verifiers used.

From Table 1 we can see that our distributed multi-core implementation scales well with the number of used parallel BA Verifier instances. The observed lack of efficiency, mostly due to network delays, is typical in a cluster setting. We note that high-performance parallel simulation typically has efficiency values in the range 40%-80% (e.g., see [36]). Accordingly, an efficiency of 74% (last row of Table 1) is to be considered state-of-the-art.

5 Conclusions

We presented a parallel algorithm which efficiently computes the set of Biologically Admissible (BA) parameters for an ODE-based biological model. In our approach, this is a crucial step to enable fast computation of patient-specific predictions from clinical trials. The main ingredient of our parallel algorithm is a novel random sampling process which allows the parallel execution of an arbitrarily high number of processes to check their biological admissibility (which is the most computationally demanding part). Such processes are independent and communicate only with an orchestrator. Our results show that our distributed multi-core implementation scales well with the number of available cores.

References

1. Alimguzhin, V., Mari, F., Melatti, I., Salvo, I., Tronci, E.: A map-reduce parallel approach to automatic synthesis of control software. In: Bartocci, E., Ramakrishnan, C.R. (eds.) SPIN 2013. LNCS, vol. 7976, pp. 43–60. Springer, Heidelberg (2013)
2. Alimguzhin, V., Mari, F., Melatti, I., Salvo, I., Tronci, E.: On-the-fly control software synthesis. In: Bartocci, E., Ramakrishnan, C.R. (eds.) SPIN 2013. LNCS, vol. 7976, pp. 61–80. Springer, Heidelberg (2013)
3. AlTurki, M., Meseguer, J.: PVESTA: A parallel statistical model checking and quantitative analysis tool. In: Corradini, A., Klin, B., Cîrstea, C. (eds.) CALCO 2011. LNCS, vol. 6859, pp. 386–392. Springer, Heidelberg (2011)
4. Ballarini, P., Forlin, M., Mazza, T., Prandi, D.: Efficient parallel statistical model Checking of Biochemical Networks. In: Proc. of PDMC, EPCTS 2014, pp. 47–61 (2009)
5. Ballarini, P., Guido, R., Mazza, T., Prandi, D.: Taming the complexity of biological pathways through parallel computing. Briefings in Bioinformatics 10(3), 278–288 (2009)

6. Balsa-Canto, E., Peifer, M., Banga, J.R., Timmer, J., Fleck, C.: Hybrid optimization method with general switching strategy for parameter estimation. BMC Systems Biology 2, 26 (2008)
7. Barnat, J., Brim, L., Černá, I., Dražan, S., Šafránek, D.: Parallel model checking large-scale genetic regulatory networks with DiVinE. ENTCS 194(3), 35–50 (2008)
8. Barnat, J., Brim, L., Šafránek, D., Vejnár, M.: Parameter scanning by parallel model checking with applications in systems biology. In: Proc. of HiBi/PDMC, pp. 95–104. IEEE (2010)
9. Chis, O.-T., Banga, J.R., Balsa-Canto, E.: Structural identifiability of systems biology models: A critical comparison of methods. PLoS ONE, 6(11) (2011)
10. Della Penna, G., Intrigila, B., Tronci, E., Venturini Zilli, M.: Synchronized regular expressions. Acta Inf. 39(1), 31–70 (2003)
11. Dierkes, T., Röblitz, S., Wade, M., Deuflhard, P.: Parameter identification in large kinetic networks with BioPARKIN. CoRR, abs (2013)
12. Donaldson, R., Gilbert, D.: A model checking approach to the parameter estimation of biochemical pathways. In: Heiner, M., Uhrmacher, A.M. (eds.) CMSB 2008. LNCS (LNBI), vol. 5307, pp. 269–287. Springer, Heidelberg (2008)
13. Ehrig, R., Nowak, U., Oeverdieck, L., Deuflhard, P.: Advanced extrapolation methods for large scale differential algebraic problems. In: High Performance Scient. and Eng. Comp. LNCSE (1999)
14. Gong, H., Zuliani, P., Komuravelli, A., Faeder, J.R., Clarke, E.M.: Analysis and verification of the hmgb1 signaling pathway. BMC Bioinform. 11(S-7), S10 (2010)
15. Gong, H., Zuliani, P., Komuravelli, A., Faeder, J.R., Clarke, E.M.: Computational modeling and verification of signaling pathways in cancer. In: Horimoto, K., Nakatsui, M., Popov, N. (eds.) ANB 2010. LNCS, vol. 6479, pp. 117–135. Springer, Heidelberg (2012)
16. Gong, H., Zuliani, P., Wang, Q., Clarke, E.M.: Formal analysis for logical models of pancreatic cancer. In: Proc. of 50th CDC, pp. 4855–4860. IEEE (2011)
17. Grosu, R., Smolka, S.A.: Quantitative model checking. In: Preliminary Proc. of ISoLA, pp. 165–174 (2004)
18. Grosu, R., Smolka, S.A.: Monte carlo model checking. In: Halbwachs, N., Zuck, L.D. (eds.) TACAS 2005. LNCS, vol. 3440, pp. 271–286. Springer, Heidelberg (2005)
19. Heath, J., Kwiatkowska, M.Z., Norman, G., Parker, D., Tymchyshyn, O.: Probabilistic model checking of complex biological pathways. Theor. Comput. Sci. 391(3), 239–257 (2008)
20. Hussain, F., Dutta, R.G., Jha, S.K., Langmead, C.J., Jha, S.: Parameter discovery for stochastic biological models against temporal behavioral specifications using an sprt based metric for simulated annealing. In: Proc. of 2nd ICCABS, pp. 1–6. IEEE (2012)
21. Ingalls, B., Iglesias, P.: Control Theory and Systems Biology. MIT Press (2009)
22. De Jong, H.: Modeling and simulation of genetic regulatory systems: A literature review. Journal of Computational Biology 9, 67–103 (2002)
23. Kwiatkowska, M., Norman, G., Parker, D.: Using probabilistic model checking in systems biology. ACM SIGMETRICS Perf. Eval. Rev. 35(4), 14–21 (2008)
24. Langmead, C.J.: Generalized queries and bayesian statistical model checking in dynamic bayesian networks: Application to personalized medicine. In: Proc. of CSB, pp. 201–212 (2009)
25. Levenberg, K.: A method for the solution of certain non-linear problems in least squares. The Quarterly of Applied Math 2, 164–168 (1944)
26. Li, P., Vu, Q.D.: Identification of parameter correlations for parameter estimation in dynamic biological models. BMC Systems Biology 7(1), 91 (2013)

27. Ljung, L.: System Identification (2Nd Ed.): Theory for the User. Prentice Hall PTR, Upper Saddle River (1999)

28. Stahl, S., Brusco, M.: Branch-and-Bound Applications in Combinatorial Data Analysis. Statistics and Computing. Springer (2005)

29. Mancini, T., Mari, F., Massini, A., Melatti, I., Merli, F., Tronci, E.: System level formal verification via model checking driven simulation. In: Sharygina, N., Veith, H. (eds.) CAV 2013. LNCS, vol. 8044, pp. 296–312. Springer, Heidelberg (2013)

30. Mancini, T., Mari, F., Massini, A., Melatti, I., Tronci, E.: Anytime system level verification via random exhaustive hardware in the loop simulation. In: Proc. of DSD, pp. 236–245 (2014)

31. Mancini, T., Mari, F., Massini, A., Melatti, I., Tronci, E.: System level formal verification via distributed multi-core hardware in the loop simulation. In: Proc. of PDP (2014)

32. Mancini, T., Mari, F., Massini, A., Melatti, I., Tronci, E.: SyLVaaS: System level formal verification as a service. In: Proc. of PDP. IEEE (2015)

33. Mari, F., Melatti, I., Salvo, I., Tronci, E.: Synthesis of quantized feedback control software for discrete time linear hybrid systems. In: Touili, T., Cook, B., Jackson, P. (eds.) CAV 2010. LNCS, vol. 6174, pp. 180–195. Springer, Heidelberg (2010)

34. Mari, F., Melatti, I., Salvo, I., Tronci, E.: Model based synthesis of control software from system level formal specifications. ACM TOSEM 23(1), 1–42 (2014)

35. Miskov-Zivanov, N., Zuliani, P., Clarke, E.M., Faeder, J.R.: Studies of biological networks with statistical model checking: Application to immune system cells. In: Proc. of BCB, pp. 728–729. ACM (2007)

36. Phillips, J.C., Sun, Y., Jain, N., Bohm, E.J., Kalé, L.V.: Mapping to irregular torus topologies and other techniques for petascale biomolecular simulation. In: Proc. of SC14, pp. 81–91. IEEE (2014)

37. Raue, A., Kreutz, C., Maiwald, T., Bachmann, J., Schilling, M., Klingmüller, U., Timmer, J.: Structural and practical identifiability analysis of partially observed dynamical models by exploiting the profile likelihood. Bioinformatics 25(15), 1923–1929 (2009)

38. Rizk, A., Batt, G., Fages, F., Soliman, S.: On a continuous degree of satisfaction of temporal logic formulae with applications to systems biology. In: Heiner, M., Uhrmacher, A.M. (eds.) CMSB 2008. LNCS (LNBI), vol. 5307, pp. 251–268. Springer, Heidelberg (2008)

39. Röblitz, S., Stötzel, C., Deuflhard, P., Jones, H.M., Azulay, D.-O., van der Graaf, P., Martin, S.W.: A mathematical model of the human menstrual cycle for the administration of GnRH analogues. Journ. of Theor. Biology 321, 8–27 (2013)

40. Sebastio, S., Vandin, A.: Multivesta: statistical model checking for discrete event simulators. In: Proc. of ValueTools, pp. 310–315 (2013)

41. Sen, K., Viswanathan, M., Agha, G.: On statistical model checking of stochastic systems. In: Etessami, K., Rajamani, S.K. (eds.) CAV 2005. LNCS, vol. 3576, pp. 266–280. Springer, Heidelberg (2005)

42. Snir, M., Otto, S., Huss-Lederman, S., Walker, D., Dongarra, J.: MPI-The Complete Reference, Vol. 1: The MPI Core, 2nd edn. MIT Press (1998)

43. Sontag, E.D.: Mathematical Control Theory: Deterministic Finite Dimensional Systems (2nd Edition). Springer, New York (1998)

44. Streck, A., Krejci, A., Brim, L., Barnat, J., Safranek, D., Vejnar, M., Vejpustek, T.: On parameter synthesis by parallel model checking. IEEE/ACM Trans. on Comput. Biology and Bioinf. 9(3), 693–705 (2012)

45. Sun, J., Garibaldi, J.M., Hodgman, C.: Parameter estimation using metaheuristics in systems biology: A comprehensive review. IEEE/ACM Trans. Comput. Biology Bioinform. 9(1), 185–202 (2012)

46. Tronci, E., Mancini, T., Mari, F., Melatti, I., Salvo, I., Prodanovic, M., Gruber, J.K., Hayes, B., Elmegaard, L.: Demand-aware price policy synthesis and verification services for smart grids. In: SmartGridComm, pp. 236–245. IEEE (2014)
47. Tronci, E., Mancini, T., Salvo, I., Sinisi, S., Mari, F., Melatti, I., Massini, A., Davì, F., Dierkes, T., Ehrig, R., Röblitz, S., Leeners, B., Krüger, T.H.C., Egli, M., Ille, F.: Patient-specific models from inter-patient biological models and clinical records. In: Proc. of FMCAD, pp. 207–214 (2014)
48. Vaseghi, S.V.: Advanced Digital Signal Processing and Noise Reductio. John Wiley & Sons (2006)
49. Wilkinson, D.J.: Stochastic Modelling for Systems Biology. Chapman & Hall (2006)
50. Zuliani, P., Platzer, A., Clarke, E.M.: Bayesian statistical model checking with application to Stateflow/Simulink verification. Formal Methods in System Design 43(2), 338–367 (2013)

Mobile Access to On-line Analytic Bioinformatics Tools

Sergio Díaz Del Pino, Tor Johan Mikael Karlsson, Juan Falgueras Cano,
and Oswaldo Trelles

[1] Computer Architecture Department, Málaga University, Louis Pasteur 35,
29071 Málaga, Spain
[2] Integromics S.L, Parque Tecnológico de Ciencias de la Salud, Avenida de la Innovación,
nº 1, 18100 Armilla, Granada, Spain
[3] Computer Sciences Department, Malaga University, Louis Pasteur 35, 29071 Malaga, Spain
`{sergiodiazdp,tjkarlsson,juanfc,ortrelles}@uma.es`

Abstract. The use of mobile devices grow continuously in many aspects of our everyday lives. It is essential that bioinformatics and biomedicine adapt to this trend because these platforms can provide universal access to computational resources independently of their location. We report the implementation of a light-weight client which allows researchers to launch complex analysis experiments and monitor the progress using their mobile devices. Starting from the analysis of existing functionality in current bioinformatics clients for web-services, we have selected, designed and implemented a complete set of functions for accessing computational resources in bioinformatics and biomedicine through mobile devices.

Keywords: Bioinformatics, mobile-based client, Web Services, Cloud Computing.

1   Introduction

It is a commonplace statement that bioinformatics is mostly a web-based domain. The major players in the field; i.e. EBI (EMBL European Bioinformatics Institute, 2013), NCBI (National Center for Biotechnology Information, 2013), INB (The Spanish Institute for Bioinformatics, 2013), etc. provide web-based services to access databases and data analysis applications located in their servers through some form of web-based interface.

On the other hand, the rapid proliferation of mobile devices, including smartphones and tablets, makes the development of new scientific applications for these platforms an urgent issue. Since most mobile devices are endowed with some sort of web-browser it would be reasonable to expect that resources and clients from bioinformatics would be easily adapted to mobile devices. However experience dictates that, from the bioinformatics user perspective, the transition is anything but satisfactory. Moving or adapting web-based applications to mobile devices requires a detailed study on the capabilities of these new devices to define the best way to cover the required functionality.

In this document we describe our experience in client development for mobile devices in bioinformatics. In particular, web-based bioinformatics platforms (e.g. MOWServ 2 (Mateos, 2010) was used as the starting point to move into the mobile's environment

F. Ortuño and I. Rojas (Eds.): IWBBIO 2015, Part II, LNCS 9044, pp. 555–565, 2015.
© Springer International Publishing Switzerland 2015

(see Section 2). In Section 3, we will describe the developed prototype and finally we will discuss and conclude in Section 4. The prototype is being evaluated in the context of the Mr.SymBioMath EU-project and will become part of the services offered by the Spanish National Institute for Bioinformatics.

2   Background

Although there are several bioinformatics applications already ported to mobile devices, such as: Biocatalogue (http://bit.ly/15m2Rtp), SimAlign (http://bit.ly/188GlGo), Oh BLAST it!... (http://bit.ly/1yYS32q) (full list in supplementary material). However, although such applications are certainly useful, the amount of effort needed to develop specific applications for all bioinformatics software would be unreasonable. This paper reports a general approach in which is only needed to register a new application describing its metadata information to make it available for the general user.

2.1   Considerations for Mobile Interfaces

Many studies address the appropriate transition between WIMP (Windows, Icons, Menus, and Pointer) interfaces and mobile (or touch user) interfaces (GUI), and the details, such as the substitution of the legacy hovering effects with distinct types of animation (Cheung *et al.*, 2012). The interaction style and the metaphor to use must be to redesign.

In particular, biomedical informatics applications require specific considerations regarding the distraction issue in order to be suitable in complex settings (Deegan, Robin, 2013); and they also require particularly suited graphic design to ease handling and avoid cluttering, especially in genetic related displaying (Pfeifer, 2012; Plumlee, 2006). In this environment, achieving trustworthy and usable interactive devices requires the application of Human Computer Interaction (HCI) research techniques, and the awareness of HCI issues throughout the lifecycle, from design through procurement, training and use (Acharya, 2010).

Development of mobile applications is an emerging area and follows few standards, if any. Due to the success of the iOS platform, Android and others, smartphones account for half of all mobile phones in 2013, http://bit.ly/1DTeK9e). It is possible to characterize the development of mobile applications in the following programming models:

- Native applications are developed entirely in the device's native language. This kind of applications is highly coupled to a specific platform, but has access to all the internals of the device (GPS, accelerometer, contacts, calendar, etc.).
- Web-based applications are platform independent and only require a web browser. An important drawback is that they do not have access to some parts of the device.
- Hybrid applications are a merge of the two previous models. There provide an application container developed in the native language, which is in charge of

loading the HTML5 application and displaying it to the user. This model shares some of the advantages of native applications such as access to device's APIs and App Store distribution and also some advantages of HTML5 applications (cross-platform).

2.2 Repositories of Bioinformatics Services: Browsing, Discovering and Invocation

In the very dynamic world of bioinformatics and biomedicine, the number of services is not only high but growing continuously as new tools and data types appear. The large number of such available resources suggests the need for some type of intelligent software organization to facilitate the integrated exploitation of tools (i.e. to avoid the construction of specific interfaces for the services or help the user to discover appropriate tools).

Repositories or catalogues of services and datatypes appear as the option of choice to organise this information. Repositories store the meta-data of the services (parameters, data types, documentation, etc.) in a centralized way. Currently, there are a number of these meta-data repositories in the bioinformatics field, with BioCatalogue (Bhagat et al., 2010) being one of the most representative.

Several tools for browsing, discovering and service invocation from metadata repositories are available (i.e. jORCA (V. Martín-Requena et al, 2010), MOWServ, Taverna (T. Oinn, et al. 2004). In general, these applications offer solutions for typical laptops and desktops computers which are not suitable for mobile devices whose screen is much smaller. For example, searching for a resource can be an issue, especially on large ones lists. Navigating through the entire tool/data type list searching for a specific item can be a tedious and inefficient task. In addition, data entry will be done by touching the screen and not by clicking with a mouse.

2.3 Additional Functionality

The previously mentioned functionality is the most common one in a bioinformatics platform, but additional functionality is also needed:

- File up- and down-load: the system requires a component which allows the users to up- and down-load their files into/from the system (decoupling the data transfer from the invocation call). The component should be able to work for multiple files at the same time; cancel/resume on-going processes and to provide progress information.
- User accounting: the system requires also a component to manage user authentication
- Multi-repository manager: this component should allow the user to select a repository among a list of available repositories. New catalogues can be added by modifying configuration files.

3   System and Methods

3.1   The Proposed Architecture

In the previous section, we have specified the traditional functionality of a Web based application in the bioinformatics and biomedical application domain. This section outlines the development of a mobile application to satisfy such requirements under the current capabilities of these devices.

This work extends our previous developments for desktop and laptops computers, in particular MOWServ 2, by replacing, adapting or re-designing components in MOWServ 2 (the client side) by specific mobile modules sharing the same server side (see Figure 1). The client side is composed of different modules, developed in Javascript, whose main functionality is:

- Browsing the service catalogue.
- Discovering services.
- Invoke services
- File management

Regarding the functionality, our starting point was an initial analysis of what MOWServ2 provided. Due to the successful performance of this software, the decision was to maintain the original server-side architecture. The main advantage is that we were able to use the same server side and thereby reduced the time required for developing the new mobile application.

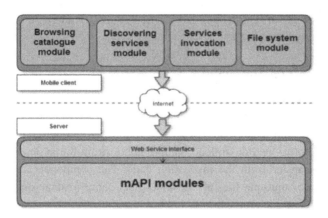

Fig. 1. System Architecture. The application, divided in modules, use a web service to connect with mAPI.

In short, our design consists of a bottom part where MAPI and its web-service act as the main communicators with the repositories and services themselves, and a client-side which process the information and offers the user experience through different modules.

MAPI (Karlsson, 2013) -a framework for the development of Web Services- allow us to use service inputs and outputs in a different format than the one accepted by the native service in a transparent way. The use of MAPI for the invocation also permits the execution of any kind of service that has a corresponding worker (Service invocation module).

3.2   Implementation

After considering the three possible programming models explained in Section 2.1 and benchmarking the available programming options, we determined that web-based development is the best option for our application. Web-based developments are the best choice for those applications that require portability. The state of web-based development tools using HTML5 and JavaScript have matured and offer functional and well-designed frameworks while at the same time is compatible for all major platforms (iOS, Android, Windows Phone).

For this reasons this kind of development is cheaper and, perhaps, faster but results are more limited when it comes to native characteristics of the platform. Native development environments and the amount of available libraries, on the other hand, allow really fast coding and testing but such developments are limited to the specific platform.

3.3   User Interface

The transition between an application thought for a WIMP (Windows, Icons, Menus and Pointer) desktop environment to a touch screen on mobile devices implies losing some of the abilities in the former model. There are also strong limitations such as screen sizes but especially the human-machine interaction should be re-evaluated to improve user experience.

All the previous considerations must be taken into account, due to the importance of the user interface in the relationship between users and electronic devices. In fact, to do some kind of work you normally have to use more than one application. Therefore, we evaluated the best option to present the information from MOWServ 2 in a mobile device.

3.3.1   Browsing the Catalogue
A "TableView" has been used to represent the tree as a list of folders and services. This provides the user a quick overview of the main categories of the entire tree. Then the user is able to navigate through the tree by selecting the categories to follow a path to a specific service. Once a category is selected, a new panel appears from right to left containing the children categories and services.

3.3.2   Discovering Services
A text-box component has been used to discover services filtering the tree depending on user input, showing only the matching services and categories Since the Magallanes tool also demands a string to be used as the search criteria, the idea is the same (results not shown) but invoking a web-service to complete the discovering.

3.3.3   Invocation

The invocation of a Web Service with the mobile application is done in the same way as for MOWServ 2 (i.e. through the MAPI service). Naturally, before making the service call, the user needs to fill the parameters data out (strings and numeric values), paths to data files (including special characters such as backslashes, dots, etc.), or select one entry from a list of files stored remotely. Secondary parameters will be filled with default values as described in the Web Service catalogue. Once the user has filled all the required parameters, it is possible to invoke the service. Following a similar criteria as for the input file, the final result is stored in a remote endpoint. Our philosophy is to provide mechanisms to download results from external storage systems.

3.3.4   File System

For the file system, the solution we have devised is to decouple the up- and down-load of data from the Web Services call. Typically, data for the service calls will not be stored on the mobile devices due to the large data sizes involved in current bioinformatics analysis. In a first step, users must upload the data files to the data storage endpoints using services such as Globus Online. The result is a reference that is used later in the call to the service (i.e. call-by-reference). By following this approach, the user does not need to send the entire input data-files to the service during the call, but sends it before the execution so that it is stored in the server before performing the call.

4   Results

Life Cycle. The typical life-cycle for browsing, discovering, and invoking services in the cloud environment from external clients is illustrated in Figure 2. The process consists –in general- of several steps; The starting point (1) is to register the service(s) into the repository using the Flipper (Torreño, 2014) application. Once the service is registered, its metadata information becomes available for the Client. In order to (2) discover the appropriated service and once users select a given service (3) they provide interfaces to complete the service parameters by accept the user parameter for tuning the application behaviour, based on the metadata retrieved from the repository.

Data consumed and results produced by services are stored in the data storage and can be used instantly as data entry for other services.

Diving into the mORCA application

In order to demonstrate the potential of providing access through mobile devices (i.e. developing such applications), we present two different exercises: The first one is focused on service discovery and the second on enacting a typical service (blast). In fact, these two exercises together conform a simple workflow. In the first service, we will retrieve a sequence that will be used as input for the Blast service.

The mobile client is available at http://bit.ly/1yfr8NC

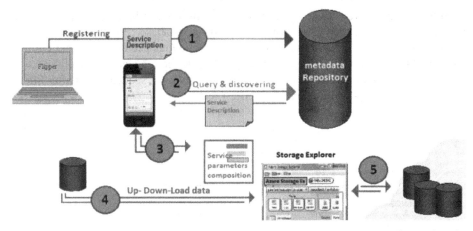

Fig. 2. Life-cycle for (1) service metadata registering; (2) service or datatype discovering,(3) service parameters composition; (4) up- and down-load data into/from the data storage (5) and to communicate to enact the Web Services. More details in the main text.

The first example involves the discovering and retrieval of a given service: 'GetAminoAcid' service. This service can retrieve a biological aminoacid sequence from a database. Initially, it is necessary to (a) log into the system for security reasons. (b) Select the repository where we are going to work (i.e. Bitlab repository). (c) Locate the service by browsing the catalogue (Step 2 in figure 2). After the service is selected, a fast query to the web-service is done and the (d) interface is generated in the client (Step 3 in fig 2). Once the parameters are filled correctly the service can be (e) invoked. Finally, the results of the invocation are showed in the screen and stored in the server for future uses.

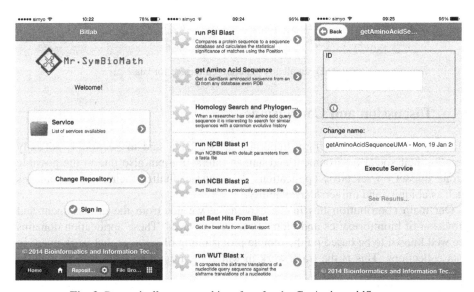

Fig. 3. Dynamically generated interface for the GetAminoacidSequence

In the second exercise, we will execute the well-known Blast application to compare the retrieved sequence against the SwissProt database. For this example, the sequence is already available in the server by our previous exercise but other options are available, i.e. uploading the data or copy and paste the sequence manually.

We use the 'cloud' icon to select our previously obtained sequence and after filling the rest of parameters, we can launch the Blast service.

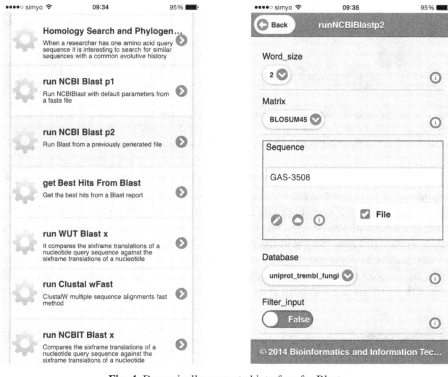

Fig. 4. Dynamically generated interface for Blast

5   Discussion and Conclusions

Mobile devices are increasingly important, as we have said in Section 1, not only because their social environment and human interaction but also due to the possible strong impact on scientific applications. We must capitalize their main advantages such as ubiquity and universal availability.

Our main contribution lies in the experience we got from the development and provision of bioinformatics and biomedical web-services. These application domains are well known to be based on the remote provisioning of services that work over big data collections. This is the framework in which the migration from web to mobiles has been addressed.

Some of the abilities of the interaction language are lost in the translation process, but other possibilities emerge in the new environment. These possibilities do not exist nor were possible / necessary in the legacy software. One of the most notorious is the hovering effect -impossible in the touching interaction- that gives the user an informative feedback when their mouse pointer is over some interactive item. On the other side, effects such as "pinching" (thumb and finger approaching), rotating, two-finger moving, etc. are difficult on WIMP interfaces. However, they are not impossible; we have the example of Apple's Magic Trackpad which can provide this kind of effects in a desktop environment.

Screen sizes are also much more limited and constrain the amount of information that can be displayed, not only because of size but also because the user attention is not only on the device. Another consideration is related to the precision the input devices (fingers) have compared to the exactness of the mouse pointer.

Although some Android-based and Windows Phone-based mobiles have file managers, they are not enough for big data files upload (even if the user is under a Wi-Fi connection). For this reason we decided to separate the file transfer from the service invocation.

We have made an initial evaluation based on real users whom we have asked to test the application. The users were not trained in advance. They did not know any details neither of the interface nor the steps to follow. There were only given a two paragraph description of the goal and the file with the data needed. The first exercise was fairly simple and only intended to be a first step and to provide familiarity with the interface. The second one required the users to change parameters before launching a "Clustal" multi-sequence comparison and implied a two step approach using the corresponding tools from the application

After the completion of the experiments, the user filled out a questionnaire form with 21 points, answering questions about experience with mobile devices, knowledge of the field. In particular, the question about "tasks in which you felt lost" was particularly relevant and helpful for developers.

In summary, this paper reports a prototype mobile application able to browse various service repositories, generate user interfaces dynamically based on service metadata and execute/monitor service progress. The development of the application was preceded by an exhaustive evaluation of an existing web-service client for traditional laptops / desktop computers. The user interface was re-designed to fit the usability aspects of mobile devices.

Acknowledgements. This work has been partially financed by the National Institute for Bioinformatics (www.inab.org) (INB-GN5) PT13-0010012, RIRAAF: Research network of allergies and drugs adverse reactions (RD12/0013/006) and Mr.SymBioMath (IAPP-code 324554) funded by the EU and the JDA-P10-TIC6108. The authors would like to thank the Bitlab research team for invaluable help and contribution with pieces of software.

References

Mateos, J.M., Martinez, A., Trelles, O.: MOWSERV 2: Friendly and extensible web platform for bioinformatics tools integration. In: Proceedings of X Spanish Bioinformatics Symposium (international), pp. 253–256 (2010) ISBN-978-84-614-4481-6

Martín-Requena, V., Rios, J., García, M., Ramírez, S., Trelles, O.: jORCA: Easily integrating bioinformatics Web Services. Bioinformatics 26(4), 553–559 (2010)

Wilkinson, M.D., Links, M.: BioMOBY: An open source biological web services proposal. Briefings in Bioinformatics 3(4), 331–341 (2002)

Karlsson, J., Trelles, O.: MAPI: A software framework for distributed biomedical applications. J. Biomedical Semantics 4, 4 (2013)

Ríos, J., et al.: Magallanes: A web services discovery and automatic workflow composition tool. BMC Bioinformatics 10, 334 (2009)

Martínez, A., Gordon, P., Sensen, C., Trelles, O.: Towards closing the gap between user data and standardized input. In: Network Tools and Applications in Biology (NETTAB 2009) (2009)

Plumlee, M.D., Ware, C.: Zooming versus multiple window interfaces: Cognitive costs of visual comparisons. ACM Trans. Comput.-Hum. Interact. 13(2), 179–209 (2006)

Vardoulakis, L.P., Karlson, A., Morris, D., Smith, G., Gatewood, J., Tan, D.: Using mobile phones to present medical information to hospital patients. In: Proceedings of the SIGCHI Conference on Human Factors in Computing Systems, CHI 2012, pp. 1411–1420. ACM, New York (2012)

Deegan, R.: Managing distractions in complex settings. In: Proceedings of the 15th International Conference on Human-computer Interaction with Mobile Devices and Services, MobileHCI 2013, pp. 147–150. ACM, New York (2013)

Iacovides, I., Cox, A.L., Blandford, A.: Supporting learning within the workplace: Device training in healthcare. In: Proceedings of the 31st European Conference on Cognitive Ergonomics, ECCE 2013, pp. 30:1–30:4. ACM, New York (2013)

Acharya, C., Thimbleby, H., Oladimeji, P.: Human computer interaction and medical devices. In: Proceedings of the 24th BCS Interaction Specialist Group Conference, BCS 2010, pp. 168–176. British Computer Society, Swinton (2010)

Cheung, V., Heydekorn, J., Scott, S., Raimund, D.: Revisiting hovering: Interaction guides for interactive surfaces. In: Proceedings of the 2012 ACM International Conference on Interactive Tabletops and Surfaces, ITS 2012, pp. 355–358. ACM, New York (2012)

Tirado, O.T., Trelles, O.: Easily registering bioinformatics services metadata. In: ECCB 2014: The 13th European Conference on Computational Biology. Methods and technologies for computational biology, France (2014)

Internet Trends (2013), http://www.kpcb.com/insights/2013-internet-trends (October 7, 2013)

The 'Mobile Only' Internet Generation (2010), http://www.slideshare.net/OnDevice/the-mobile-only-internet-generation (October 7, 2013]

Pew's Internet Mobile | Pew Research Center's Internet & American Life Project, EMBL European Bioinformatics Institute (2013), http://www.pewinternet.org/Commentary/2012/February/Pew-Internet-Mobile.aspx, http://www.ebi.ac.uk/ (October 7, 2013)

National Center for Biotechnology Information (2013), http://www.ncbi.nlm.nih.gov/ (October 7, 2013)

The Spanish institute of bioinformatics, Native, HTML5 or Hybrid, Understanding Your Mobile Application Development Options (2013), `http://www.inab.org/`, `http://s3.amazonaws.com/dfc-wiki/en/images/c/c2/Native_html5_hybrid.png` (October 7, 2013)

Globus Online | Reliable File Transfer. No IT required, Smartphones account for half of all mobile phones, dominate new phones purchase in the US (2013), `https://www.globusonline.org/`, `http://www.nielsen.com/us/en/newswire/2012/smartphones-account-for-half-of-all-mobile-phones-dominate-new-phone-purchases-in-the-us.html` (October 11, 2013)

Oinn, T., et al.: Taverna: A tool for the composition and enactment of bioinformatics workflows. Bioinformatics 20(17), 3045–3054 (2004)

isDNA: A Tool for Real-Time Visualization of Plasmid DNA Monte-Carlo Simulations in 3D

Adriano N. Raposo and Abel J.P. Gomes

Instituto de Telecomunicações, Universidade da Beira Interior,
Av. Marquês D'Avila e Bolama, 6200-001 Covilhã, Portugal
anraposo@ubi.pt, agomes@di.ubi.pt
https://www.it.ubi.pt/medialab

Abstract. Computational simulation of plasmid DNA (pDNA) molecules, owning a closed-circular shape, has been a subject of study for many years. Monte-Carlo methods are the most popular family of methods that have been used in pDNA simulations. However, though there are many software tools for assembling and visualizing DNA molecules, none of them allows the user to visualize the course of the simulation in 3D. As far as we know, we present here the first software (called isDNA) allowing the user to visualize 3D MC simulations of pDNA in real-time. This is sustained on an adaptive DNA assembly algorithm that uses Gaussian molecular surfaces of the nucleotides as building blocks, and an efficient deformation algorithm for pDNA's MC simulations.

Keywords: pDNA, Monte-Carlo, simulation, visualization, software.

1 Introduction

As known, DNA plays an important role in life sciences research. DNA is made of four subsidiary molecules called *nucleotides*: *adenine* (A); *cytosine* (C); *guanine* (G); and *thymine* (T). A DNA molecule is usually represented by its base-pairs sequence because there is a unique correspondence between the nucleobases in the two strands, in addition to the fact that the atomic structure of the nucleotides is widely known. Thus, a DNA molecule can be assembled using either its atoms or, alternatively, its nucleotides. The advantage in using nucleotides as building blocks of DNA is that we have a speedup of 34×, since a nucleotide has about 34 atoms in general, what is relevant for visualisation purposes [1].

In turn, pDNA molecules own a closed-circular shape, being widely used in several fields including biotechnology, pharmaceutical sciences, and medical research. In a simple way, we can say that pDNA molecules are DNA molecules whose double helix's extremeties are tied up one to another. pDNA molecules are very dynamic and flexible, and the term "closed-circular" does not mean this type of molecules exhibits a perfect circular conformation. Instead, pDNA may acquire a large variety of conformations including *supercoiled*, or even *knotted* conformations. On the other hand, the production of pDNA in laboratory is done in two major steps: *fermentation* and *purification*. In the fermentation step, the

F. Ortuño and I. Rojas (Eds.): IWBBIO 2015, Part II, LNCS 9044, pp. 566–577, 2015.

pDNA that we want to produce is replicated by a bacterium. The purification step serves the purpose of separating the pDNA of interest in relation to the DNA of the bacteria and other contaminants. This is where pDNA computer simulations can play an important role.

In this sense, computational methods based on laboratory data have been developed to simulate possible conformations for pDNA molecules under certain thermodynamic conditions. The MC method, due to its reliability, is probably the most popular simulation tool for pDNA. The principle behind the MC method is to make the molecule achieve an elastic energy equilibrium state in as few iterations as possible without compromising the effectiveness and reliability of the method. The deformation method traditionally used in MC simulations, known as *crankshaft move*, has a very low acceptance ratio of trials, i.e., many trials are rejected. Even worse, it is the fact that the crankshaft move presents a very unnatural behavior, as it features very sudden motions along large portions of the molecule. To solve these problems, Raposo and Gomes developed a more efficient deformation algorithm for pDNA's MC simulations [2].

In this paper, we combine assembling and visualization algorithms together with the Raposo-Gomes move into a new software tool, called isDNA, which is seemingly the first to allow real-time 3D visualization of pDNA's MC simulations. For that purpose, this new software implements an adaptive assembly algorithm for DNA [1], and an efficient deformation algorithm of pDNA in MC simulations [2]. The pDNA deformation algorithm adopted by isDNA also contributes to the smoothness of the 3D animation with small and controlled changes in the molecule between iterations.

isDNA is implemented as a C++ package, which already includes a simple GLUI/OpenGL graphical user interface (GUI) capable of loading GBK (GenBank) files, allowing also 3D interaction with the user. The main canvas of the GUI shows in real-time the animated course of the pDNA simulation in 3D. Besides, pDNA conformations resulting from MC simulations can be saved into text files. These conformation files can later been used to re-assemble the pDNA molecules. The isDNA source code is publicly available at: https://github.com/ISDNA.

2 Related Work

Let us now briefly review DNA assembly methods, as well as traditional deformation algorithms used in Monte Carlo simulations of pDNA.

2.1 DNA Assembly Methods: Predictive versus Adaptive

The *predictive methods* for DNA assembly include two different approaches: the Cambridge meeting method [3]; and the wedge angle method [4]. These approaches try to predict possible trajectories of DNA molecules just from their base-pair sequences. This prediction is based on previously obtained geometric values between consecutive DNA nucleotides. This DNA assembling paradigm is specifically suitable for scenarios where we want to know the likely conformation (or topology) of DNA base-pair sequences. It is important to clarify that neither the Cambridge methods [3] nor the wedge angle methods [4] are adequate for pDNA simulation purposes, in largely because they assume that the trajectory of the DNA axis in the process of dinucleotide stacking is determined by the sequence of nucleotides. Examples of Cambridge-based software packages are: FREEHELIX [5]; 3DNA [6]; w3DNA [7]; Curves [8]; and Curves+ [9]. In respect to wedge angle software packages, we have the following: CURVATURE [10]; DNACURVE (http://www.lfd.uci.edu/~gohlke/dnacurve); NAB [11]; ADN Viewer [12]. However, because they use predictive DNA assembly methods, none of these tools allows 3D visualization of Monte Carlo simulations of pDNA.

In turn, the *adaptive methods* include those due to Raposo and Gomes [1], and to Hornus et al. [13]. These methods are able to adapt specific base-pair sequences to arbitrary conformations, i.e., these methods allow the base-pair sequence to be assembled along an arbitrary DNA trajectory. In this sense, we can say that adaptive DNA assembling methods are particularly suitable for simulation scenarios of DNA, in particular pDNA. This is so because a simulation process produces a sequence of DNA conformations, so that we need to proceed to the DNA assembling for each conformation.

The software presented in this paper belongs to the family of adaptive assembly algorithms; more specifically, it implements the stacking algorithm previously proposed by Raposo and Gomes [1]. In the literature, we find only a few other codes that implement the DNA assembling for arbitrary conformations, namely: NAB [11]; and GraphiteLifeExplorer [13]. Currently, NAB is bundled as part of AmberTools v.13 [14]. Interestingly, GraphiteLifeExplorer enables the user to model DNA molecules of arbitrary length by modeling its helical axis as a quadratic or cubic Bézier curve in space. Finally, it is also important to clarify that, even presenting adaptive capabilities, neither NAB nor GraphiteLifeExplorer allow real-time 3D visualization of pDNA's MC simulations. The real-time capability is achieved by isDNA mainly because the DNA assembly algorithm adopted [1] uses pre-triangulated molecular surfaces of the nucleotides as building blocks (see Figure 1, instead of their atoms, dramatically reducing the number of graphical objects to be assembled and rendered. Taking the pUC19 molecule as an example, adopting an atomistic representation implies rendering approximately 180,000 atoms. In turn, using a nucleotide-based representation for the same molecule, we only have to render about 5,000 building blocks.

2.2   Deformation Methods for pDNA Simulations

In pDNA simulations, it is usual to simplify a DNA molecule through a linear skeleton (i.e., polygonal line consisting of with equal sized segments connecting consecutive nucleotides) that represents its topological conformation. Then, new pDNA conformations are randomly generated by applying deformations to molecule's skeleton. These trial conformations are then subject to evaluation for acceptance/rejection by the simulation method. However, the deformation method used to generate the new trials has been, in essence, the same since it was first introduced in the context of lattice polymer chains. This move was later

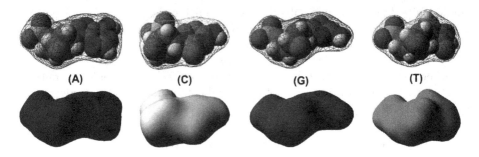

Fig. 1. DNA building blocks: (A) adenine, (C) cytosine, (G) guanine, and (T) thymine. (Image courtesy of Raposo and Gomes [2]).

adapted for pDNA simulation using MC methods. This deformation method is known as the *standard* crankshaft move, with its origins dating back to the early 1960s [15][16].

Later, Klenin et al. [17][18] proposed a biased crankshaft move that starts by randomly choosing two vertices v_m and v_n of the skeleton, rotating then randomly all the vertices —and, consequently, all connecting segments— between v_m and v_n by the angle θ around the axis defined by the line connecting v_m and v_n (Figure 2 (left)). The value of θ is uniformly distributed over a certain interval, and must be continuously adjusted during the simulation to guarantee that about half the steps are accepted [17].

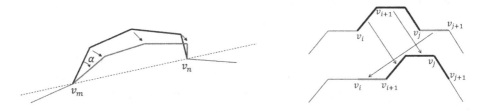

Fig. 2. Crankshaft move (left) and reptation move (right). (Image courtesy of Raposo and Gomes [2]).

In the literature, we can find an alternative deformation method that was thought to increase the acceptance ratio of the simulation trials, which is known as *reptation motion* [18] (Figure 2 (right)). In a simple way, this deformation move is just a sub-chain translation. First, two vertices v_i and v_j are randomly chosen. Then, the sub-chain between v_i and v_j is translated by one segment length along the chain contour. The segment placed immediately after v_j is also translated to fill the gap between v_i and v_{i+1}.

More recently, Raposo and Gomes [2] presented a more efficient unbiased move for pDNA, whose skeleton is a closed polyline. This move, implemented in the isDNA software package, not only preserves the size of each segment and its connectivity, but also is very effective in maximizing the acceptance ratio of the trials and stabilizing the molecule, thereby allowing steady, gradual temperature changes during the simulation. Our method also generates natural and realistic animations that can be used in real-time simulation and visualization.

Other types of motion can be adopted if Metropolis' microscopic reversibility is satisfied, i.e., if the probability of each trial conformation is the same as that one of the reverse movement [19].

3   Real-Time Adaptive DNA Assembly

For 3D visualization of pDNA simulations in real-time, isDNA implements the DNA assembly algorithm introduced by Raposo and Gomes [1]. This DNA assembly algorithm uses four three-dimensional building blocks representing DNA nucleotides (Figure 1), namely, adenine (A), cytosine (C), thymine (T), and guanine (G). Each building block is a pre-triangulated isosurface generated by a triangulation algorithm for molecules [20].

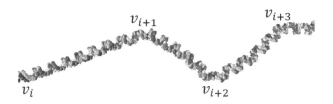

Fig. 3. Piece of assembled DNA. (Image courtesy of Raposo and Gomes [2]).

The DNA axis is approximated by a polyline whose segments have a length of $H = 3.3$ Å (the axial distance between two consecutive nucleotides) [21]. Then, the assembly procedure for nucleotides can be thought of as the operation of wrapping helicoidal DNA backbones around cylinders along the DNA skeleton segments as follows (Figure 3):

1. Given a nucleobase n_i, two geometric instances of nucleotides must be generated, the first for the building block b_i and the second for the mate building block B_i.
2. The base pair $b_i B_i$ is aligned and positioned at the origin laying on the plane $z = 0$.
3. The base pair $b_i B_i$ is aligned with the plane perpendicular to segment i. This alignment is also done about the origin o of the coordinate system.
4. The base pair $b_i B_i$ is translated to the plane perpendicular to segment i.
5. Finally, b_i and B_i are displaced to their correct positions in relation to the midpoint of the corresponding segment i of the DNA axis.

For complete details on the adaptive DNA assembly algorithm implemented in isDNA, the reader is referred to [1].

4   Plasmid DNA Monte-Carlo Simulation Method

The MC simulation method, originally introduced by Metropolis et al. [19], generates pDNA conformations combining energy calculations, random conformational changes, and statistics. This simulation method is considered the standard in pDNA simulations.

4.1   Monte-Carlo Method

MC simulation methods are iterative methods that try to reach a thermodynamic equilibrium state of pDNA molecules based on elastic energy calculations and statistics. So, in each iteration, these methods perform random deformations on the pDNA skeleton until one of the conformation trials is accepted. The acceptance of a new conformation is based on the principle of elastic energy minimization, but the conformation trial can be also accepted if it has a certain probability of occurrence. In isDNA, we used the same MC simulation method and parameters as those used in [22]. For more details on the Monte-Carlo methods, the reader is referred to [2].

4.2   Deformation Algorithm

The pDNA deformation algorithm that was implemented in isDNA has been recently presented as a more efficient way of generating pDNA conformation trials for pDNA's MC simulation procedures [2]. When compared to the traditional deformation methods used in MC simulations, this new deformation method has a higher acceptance ratio of trials, and generates smoother and more controlled deformations, what enhances the real-time animation of the simulation course.

The deformation algorithm implemented in isDNA uses a linear skeleton (i.e., a polyline) with equal sized segments (corresponding to approximately to 30 base pairs of the double helix [23]), henceforth called the DNA skeleton. The pDNA skeleton can assume any closed unknotted conformation, being the completely relaxed circular conformation the simplest of those conformations. Then, the first step of the algorithm is to determine the number of segments of the DNA skeleton, ensuring around 30 base pairs per segment.

Let $\mathbf{P}_k$ a three-dimensional closed polyline representing the DNA skeleton. Now, we need to find a new polyline $\mathbf{P}_{k+1}$ by deforming $\mathbf{P}_k$, but keeping the same number s of equal sized segments and connectivity. From the set of vertices $\{\mathbf{v}_i\}$, $i = 0, ..., s-1$ of the polyline, we choose a random vertex $\mathbf{v}_m$, $0 \leq m \leq s-1$ as the current *mobile vertex*, i.e., the vertex that most moves in the current trial conformation. It happens that any move of $\mathbf{v}_m$ implies moving its closest neighbors $\mathbf{v}_{m-1}$ and $\mathbf{v}_{m+1}$, here called *semi-mobile vertices*. The remaining neighbors $\mathbf{v}_{m-2}$ and $\mathbf{v}_{m+2}$ remain fixed, i.e., they do not move in a deformation step. Therefore, in each deformation step, only three vertices will be displaced: $\mathbf{v}_m$, $\mathbf{v}_{m-1}$ and $\mathbf{v}_{m+1}$.

But, $\mathbf{v}_m$ cannot be freely moved around, unless within the sphere $\mathbf{N}_m$ centered at $\mathbf{v}_m$ itself; the radius of its sphere is $r = 2\,\Delta$, where $\Delta = 3.3$ Å stands for the distance between two consecutive base pairs. In true, the new position of $\mathbf{v}_m$ is obtained randomly in the intersection of the three spheres, $\mathbf{N}_m$, $\mathbf{S}_{m-2}$ and $\mathbf{S}_{m+2}$, with the latter two spheres with radius $2l$ centered on the fixed vertices $\mathbf{v}_{m-2}$ and $\mathbf{v}_{m+2}$, respectively, where l is the length of each DNA skeleton segment. Note that the small radius r of sphere $\mathbf{N}_m$ guarantees a transition from $\mathbf{P}_k$ to $\mathbf{P}_{k+1}$ without noticeable jumps.

In order to calculate the new position of $\mathbf{v}_m$, we first need to convert the Cartesian coordinates (x, y, z) to spherical coordinates (d, θ, ϕ) relative to $\mathbf{v}_{m-2}$, where d is the distance between $\mathbf{v}_{m-2}$ and $\mathbf{v}_m$. Then, one randomly generates a new position for $\mathbf{v}_m$ as $(d + \Delta d, \theta + \Delta\theta, \phi + \Delta\phi)$, where $\Delta d \in [-r, r]$ and $\Delta\theta, \Delta\phi \in [-\pi, \pi]$. The new position of $\mathbf{v}_{m+1}$ is calculated in a similar manner.

So, the algorithm also converts the Cartesian coordinates of $\mathbf{v}_{m-1}$ to spherical coordinates (l, α, β) relative to $\mathbf{v}_{m-2}$, where l is the radius of the three spheres $\mathbf{s}_m$, $\mathbf{s}_{m-1}$, and $\mathbf{s}_{m-2}$ centered on $\mathbf{v}_m$, $\mathbf{v}_{m-1}$, and $\mathbf{v}_{m-2}$, respectively. Moving $\mathbf{v}_{m-1}$ to a new position must be done without changing its distance l to $\mathbf{v}_{m-2}$ and $\mathbf{v}_m$. That is, the new $\mathbf{v}_{m-1}$ must lie on the circumference on the intersection of the two surfaces bounding $\mathbf{s}_m$ and $\mathbf{s}_{m-2}$. If $\Delta d = 0$, the new position of $\mathbf{v}_{m-1}$ relative to $\mathbf{v}_{m-2}$ is given by $(l, \alpha + \Delta\theta, \beta + \Delta\phi)$; otherwise, the new location of $\mathbf{v}_{m-1}$ is $(l, \alpha + \Delta\theta + \Delta\psi, \beta + \Delta\phi)$, where $\Delta\psi$ is the angle of the angular motion of $\mathbf{v}_{m-1}$ on $\mathbf{s}_{m-2}$ resulting from the translational displacement Δd of $\mathbf{v}_m$ along the line defined by $\mathbf{v}_m$ and $\mathbf{v}_{m-2}$. We compute $\Delta\psi$ by rearranging the equation that describes the reciprocal motion of the piston with respect to the crank angle (cf. [24], p.44).

Being aware that knots can occur when random deformations are applied to pDNA conformations, we must check for the existence of knots and reject the deformation if we find one or more knots. For knot detection isDNA uses the method of Harris and Harvey [25].

As an example of isDNA output, Figure 4 shows a snapshot of a 3D animated MC simulation performed by isDNA using the pUC19 molecule. The MC simulation conditions were the same as the ones presented in [2] when the temperature of the experiments decreases. This snapshot was taken about 1 minute after the beginning of the simulation, with a circular conformation used as the initial conformation of the simulation. For more details on the MC deformation algorithm implemented in isDNA, the reader is referred to [2].

Fig. 4. Snapshot of a 3D animated MC simulation of the pUC19 molecule performed by isDNA

5   The Software

isDNA aims at helping users to perform real-time MC-based pDNA simulation and 3D visualization. However, we can see isDNA as a bundle with two different components: the isDNA GUI (Graphical User Interface) and the isDNA API (Aplication Programming Interface). The idea is to satisfy the needs of two different categories of end users: (a) those needing a simulation tool with embedded 3D visualization capabilities; and (b) those wanting to integrate the isDNA API into third-party 3D visualization tools.

In terms of system requirements, both isDNA GUI and API can be used in general-purpose personal computers without high-performance requirements (i.e., CPU multi-threading and GPU computing) or graphics acceleration. The results presented in this paper were obtained using a laptop equipped with an Intel Core i5-2430M CPU, 2.4GHz clock, 4GB of RAM, and an Nvidia GEFORCE GT 520MX graphics card with 1GB of memory, running Microsoft Windows 7.

5.1   The GUI

The isDNA GUI is very simple and spartan, because it only provides the essential functionalities for 3D visualization and interaction with the pDNA molecules. More specifically, the user interface includes: a 3D canvas where the molecules are rendered; one button that allows the user to rotate de molecules; two buttons to translate the user point of view (one of them for zooming purposes); a select box that allows the user to choose the pDNA molecule; two check buttons that hide and show the molecule components, one for the skeleton and the other for the nucleotides; and, finally, a button to close the application. It it worthy to mention that the user interface was implemented using the GLUI User Interface Library, a GLUT-based C++ user interface library that provides controls such as buttons, checkboxes and radio buttons to OpenGL applications. This means that those who want to recompile isDNA will need the GLUI library that is available at `http://glui.sourceforge.net`.

The aim of this user interface is just to provide the users with a *ready-to-use* solution for pDNA simulation and real-time visualization in 3D. In the future we intend to enhance isDNA user interface with much more functionalities that, at this moment, are just possible for users that integrate isDNA API into their own, or into third-party, visualization tools. Among those new functionalities, we count having a way of saving and loading pDNA conformations and GBK files directly from the user interface. The isDNA API is presented in more detail in the following section.

5.2   The API

The API briefly described in this section is the core of the isDNA software package. This section is particularly useful for those users who want to know more details about how the algorithms are implemented, and also for those users who want to include the isDNA functionalities into their own software tools. However, at this point, it is important to remember that this is a C++ API, and thus it is only usable in a C++/OpenGL development context. As previously mentioned, the isDNA API source code is available at `https://github.com/ISDNA`.

The two essential classes in the isDNA API are `Dna` and `BuildingBlock`. The `Dna` class is the core class of the API. In this class, we use constants to define DNA geometric properties and MC simulation parameters. The most important methods implemented in the `Dna` class are the following:

- `Dna(char*,char*)` - The constructor of the class. The first parameter is the path to the GBK containing the DNA base pairs sequence. The second parameter is the path to the file that contains the conformation to be loaded.
- `void draw(void)` - Draws the molecule.
- `void drawSkeleton(void)` - Draws the skeleton of molecule.
- `void drawAxis(void)` - Draws the axis of the molecule.
- `void circularConformation(void)` - Builds a circular conformation.
- `void buildAxis(void)` - Builds the axis of the molecule.
- `double randomMoveNew(void)` - Implements the deformation method used by the MC simulation.
- `double twist(void)` - Calculates and returns the *twist* value of the molecule.
- `double writhe(void)` - Calculates and returns the *writhe* value of the molecule.
- `long double bendingEnergy(void)` - Calculates and returns the *bending energy* value of the molecule used in MC simulations.
- `long double torsionalEnergy(void)` - Calculates and returns the *torsional energy* value of the molecule used in MC simulations.
- `int countKnots(void)` - Counts and returns the number of knots in the molecule.
- `void saveConformation(char* file)` - Saves the current conformation of the molecule to the file specified in the parameter.
- `int loadConformation(char* file)` - Load a conformation for the molecule from the file specified in the parameter.
- `void translate(double x, double y, double z)` - Translates the molecule.
- `void rotate(double angle, Point p, Point q)` - Rotates de molecule.

The `BuildingBlock` class is very important because the building blocks used to assemble the molecules are instances of it. The methods implemented in this class are the following:

- `BuildingBlock(char)` - The constructor of the class. The parameter is the nucleotide letter (A, C, G or T).
- `char getNucleotide(void)` - Returns the nucleotide of the building block.
- `GLfloat* getVertex(void)` - Returns the array of vertices of the building block surface mesh.
- `GLfloat* getNormals(void)` - Returns the array of normal vectors of the building block mesh triangles.

- `GLint getVertexCount(void)` - Returns the number of vertices in the building block mesh.
- `GLfloat* getColor(void)` - Returns the color of the building block according to its nucleotide.
- `void draw(void)` - Draws the building block.

isDNA API also implements other generic classes for 3D geometric purposes like, for example, `Vector`, `Point` and `Tuple`, which are not presented here for sake of brevity. Note that isDNA API does not include molecular surface triangulation functionalities. Instead, isDNA API includes four pre-triangulated mesh templates, one for each type of nucleotide building block.

5.3   Features and Future Work

This section presents a summary of the key features of isDNA, as well as an example of a simulation performed with it. At this point, it is also worthy to say that all the simulations presented in [2] were performed using isDNA, which was also used to generate the pictures published therein. Summing up, the most relevant features of isDNA are:

- Real-time visualization of pDNA MC simulations in 3D;
- Loading of GBK (GenBank) files containing the base-pairs sequences of pDNA molecules;
- The user can choose between starting the simulations from a circular conformation or, alternatively, from a previously saved conformation generated, for example, by a former isDNA simulation;
- The user can save conformations into a text file at any point of the pDNA MC simulations;
- The user is allowed to adjust the MC simulation parameters according to their own needs;
- The isDNA API can be fully integrated as a library in the source code of any C++/OpenGL software;
- The isDNA GUI allows the user to choose between a 3D representation and a simplified skeleton representation of a given pDNA molecule;
- The isDNA GUI allows the user to interact with the pDNA molecules in 3D.

In the future, we intend to improve the GUI with more user-friendly functionalities like menus and buttons to save or load pDNA conformations and GKB files, which at this point are only accessible to the API users. We also intend to include adequate functionalities to export the graphical results to image files with illustration quality. Another improvement can be the implementation of a 3D curve smoothing algorithm (like the one due to Kummerle and Pomplun [22]), in order to eliminate sharp kinks that may occur in the simulated conformations.

6   Conclusions

This paper presents isDNA, a novel software tool specifically developed for real-time 3D visualization of pDNA molecules subject to Monte Carlo simulations. As far as the authors know, this is the first software that has these funcionalities, i.e., the existing tools for DNA simulation do not include realistic 3D representation of the molecules and the real-time visualization of the simulation course. These objectives were achieved with the implementation of an adaptive algorithm for DNA assembly that uses instances of the molecular surfaces of the nucleotides as building blocks [1], together with the adoption of an efficient deformation algorithm for pDNA MC simulations [2]. isDNA includes a GUI with basic 3D interaction functionalities that provides a ready-to-use pDNA simulation and visualization tool. Nevertheless, the isDNA API can be integrated into third-party C++/OpenGL sofwtare tools.

References

1. Raposo, A.N., Gomes, A.J.: 3D molecular assembling of B-DNA sequences using nucleotides as building blocks. Graphical Models 74(4), 244–254 (2012)
2. Raposo, A.N., Gomes, A.J.P.: Efficient deformation algorithm for plasmid DNA simulations. BMC Bioinformatics 15(1), 301 (2014)
3. Dickerson, R.E.: Definitions and nomenclature of nucleic acid structure components. Nucleic Acids Research 17(5), 1797–1803 (1989)
4. Bolshoy, A., McNamara, P., Harrington, R.E., Trifonov, E.N.: Curved dna without a-A: experimental estimation of all 16 dna wedge angles. Proceedings of the National Academy of Sciences 88(6), 2312–2316 (1991)
5. Dickerson, R.E.: DNA bending: The prevalence of kinkiness and the virtues of normality. Nucleic Acids Research 26(8), 1906–1926 (1998)
6. Lu, X.J., Olson, W.K.: 3DNA: A software package for the analysis, rebuilding and visualization of three-dimensional nucleic acid structures. Nucleic Acids Research 31(17), 5108–5121 (2003)
7. Zheng, G., Lu, X.J., Olson, W.K.: Web 3DNA–a web server for the analysis, reconstruction, and visualization of three-dimensional nucleic-acid structures. Nucleic Acids Research 37(web server issue), W240–W246 (2009)
8. Lavery, R., Sklenar, H.: The definition of generalized helicoidal parameters and of axis curvature for irregular nucleic acids. Journal of Biomolecular Structure and Dynamics 6(1), 63–91 (1988)
9. Lavery, R., Moakher, M., Maddocks, J.H., Petkeviciute, D., Zakrzewska, K.: Conformational analysis of nucleic acids revisited: Curves+. Nucleic Acids Research 37(17), 5917–5929 (2009)
10. Shpigelman, E.S., Trifonov, E.N., Bolshoy, A.: CURVATURE: Software for the analysis of curved DNA. Computer Applications in the Biosciences 9(4), 435–440 (1993)
11. Macke, T.J., Case, D.A.: 25. In: Modeling Unusual Nucleic Acid Structure, pp. 379–393. American Chemical Society (1998)
12. Herisson, J., Ferey, N., Gros, P.E., Gherbi, R.: ADN-Viewer: A 3D approach for bioinformatic analyses of large DNA sequences. Cellular and Molecular Biology 52(6), 24–31 (2006)

13. Hornus, S., Levy, B., Lariviere, D., Fourmentin, E.: Easy DNA modeling and more with GraphiteLifeExplorer. PLoS One 8(1), e53609 (2013)
14. Salomon-Ferrer, R., Case, D.A., Walker, R.C.: An overview of the AMBER biomolecular simulation package. Wiley Interdisciplinary Reviews: Computational Molecular Science 3(2), 198–210 (2013)
15. Verdier, P., Stockmayer, W.: Monte Carlo calculations on the dynamics of polymers in dilute solution. The Journal of Chemical Physics 36(1), 227–235 (1962)
16. Hilhorst, H., Deutch, J.: Analysis of Monte Carlo results on the kinetics of lattice polymer chains with excluded volume. The Journal of Chemical Physics 63(12), 5153–5161 (1975)
17. Klenin, K., Vologodskii, A., Anshelevich, V., Dykhne, A., Frank-Kamenetskii, M.: Computer simulation of DNA supercoiling. Journal of Molecular Biology 63(3), 413–419 (1991)
18. Vologodskii, A.V., Levene, S.D., Klenin, K.V., Frank-Kamenetskii, M., Cozzarelli, N.R.: Conformational and thermodynamic properties of supercoiled DNA. Journal of Molecular Biology 227(4), 1224–1243 (1992)
19. Metropolis, N., Rosenbluth, A.W., Rosenbluth, M.N., Teller, A.H., Teller, E.: Equation of State Calculations by Fast Computing Machines. The Journal of Chemical Physics 21(6), 1087–1092 (1953)
20. Raposo, A.N., Queiroz, J.A., Gomes, A.J.P.: Triangulation of molecular surfaces using an isosurface continuation algorithm. In: Proceedings of the 2009 International Conference on Computational Science and its Applications, pp. 145–153. IEEE Computer Society (2009)
21. Bates, A., Maxwell, A.: DNA Topology, 2nd edn. Oxford University Press (2005)
22. Kummerle, E.A., Pomplun, E.: A computer-generated supercoiled model of the pUC19 plasmid. European Biophysics Journal 34(1), 13–18 (2005)
23. Vologodskii, A.: Monte Carlo simulation of DNA topological properties. In: Monastyrsky, M. (ed.) Topology in Molecular Biology. Biological and Medical Physics, Biomedical Engineering, pp. 23–41. Springer, Heidelberg (2007)
24. Heywood, J.: Internal Combustion Engine Fundamentals. McGraw-Hill, Inc., New York (1988)
25. Harris, B.A., Harvey, S.C.: Program for analyzing knots represented by polygonal paths. Journal of Computational Chemistry 20(8), 813–818 (1999)

Transport Properties of RNA Nanotubes Using Molecular Dynamics Simulation

Shyam R. Badu, Roderik Melnik, and Sanjay Prabhakar

MS2Discovery Interdisciplinary Research Institute, M NeT Laboratory,
Wilfrid Laurier University, Waterloo, ON, N2L 3C5 Canada
{sbadu,rmelnik}@wlu.ca

Abstract. We present novel molecular dynamics studies of transport properties of RNA nanotubes. Specifically, we determine the velocity trajectories for the phosphorous atom at the phosphate backbone of the RNA nanotube, the oxygen atom at sugar ring, and the ^{23}Na$^+$and ^{35}Cl$^-$ions in physiological solutions. At the constant temperature simulation it has been found that the fluctuation of the velocities is small and consistent with simulation time. We have also presented the velocity autocorrelation function for the phosphorous atom in RNA nanotubes that provides better insight into the diffusion direction of the system in physiological solution. We compare our results calculated computationally with the available experimental results.

1 Introduction

The RNA nanoclusters have a wide range of current and potential applications in a variety of fields, and in particular in nanomedicine. As a result, it is becoming increasingly important to study the properties of these systems in solutions. Particularly, in studies of transport phenomena, properties, and characteristics, including diffusion coefficients and velocity autocorrelation functions for these systems, are the subject of interest. Several experimental studies have been performed to analyze the diffusion coefficients of the biomolecular systems [1,2,11,12]. Furthermore, substantial efforts were devoted to computational studies where molecular dynamics simulations and other methodologies were applied to DNA polymers (e.g. [8,14]). In some such cases the average diffusion coefficient for the solvent as a function of the distance from the solute was reported. By using the molecular dynamics simulation the diffusion coefficient for the single strand RNA can also be calculated [15].

The self-diffusivity of the system of molecules under such studies in molecular dynamics simulations is defined by the random motion of the molecules in the media in which the change of the mass flux is zero. The self-diffusion coefficient can be calculated in two different ways. One is from the root mean square deviation and the other one is from the velocity autocorrelation function. More specifically, once the mean square deviation of the system is defined as

$$MSD(t_1,t_2) = \frac{1}{N} \sum_{i=1}^{N} \|x_i(t_2) - x_i(t_1)\|^2, \tag{1}$$

F. Ortuño and I. Rojas (Eds.): IWBBIO 2015, Part II, LNCS 9044, pp. 578–583, 2015.

Fig. 1. (Color online.)The VMD generated structure of the RNA nanotube with 5 rings in a physiological solution (water molecules are not shown)

then the self-diffusion coefficient can be expressed as

$$D_s = \lim_{t \to 0} \frac{1}{6Nt} \sum_{i=1}^{N} \|x_i(t_2) - x_i(t_1)\|^2. \tag{2}$$

The autocorrelation function of the system can be expressed as

$$VACF(t) = \frac{<v(t)v(0)>}{(v(0))^2}. \tag{3}$$

In the study of the dynamic properties of the system consisting of RNA nanoclusters we use the structures modeled in the earlier studies [4,9,16]. However, unlike these earlier papers our focus here is on the transport properties such as velocity trajectories, autocorrelation functions and the diffusion coefficient of the nanocluster in physiological solutions The building blocks for these RNA nanoclusters are based on the RNAIi/RNAIIi complexes which are taken from the protein data bank with the pdb code (2bj2.pdb) [6]. A typical example of the RNA nanocluster in the physiological solution is presented in Figure 1.

This contribution is organized as follows. In section 2, we describe computational details used in our analysis of RNA nanoscale systems and highlight the main features of this analysis. The results are presented and discussed in Section 3, while concluding remarks are found in Section 4.

2 Computational Details

In order to perform all atom molecular dynamics simulations on the RNA nanoclusters we used CHARMM27 force field [10] implemented by NAMD package [7] The potential of the system used during the molecular dynamics simulation using CHARMM force filed can be expressed as follows:

$$V_{total} = \sum_{bond} K_b(r - r_0)^2 + \sum_{angle} K_\theta(\theta - \theta_0)^2 + \sum_{dihedral} K_\phi(1 + cos(n\phi - \gamma))$$

$$+ \sum_{Hbond} (\frac{C_{ij}}{r_{ij}^{12}} - \frac{D_{ij}}{r_{ij}^{10}}) + \sum_{Vanderwaals} (\frac{A_{ij}}{r_{ij}^{12}} - \frac{B_{ij}}{r_{ij}^{10}}) + \sum \frac{q_{ij}}{\varepsilon r_{ij}}, \qquad (4)$$

where the first term corresponds to bonds, second corresponding to angle parameters, the third term corresponds to the potential energy and interactions arised from the dihedral angles in the molecular system, the fourth term defines the interaction coming from the hydrogen bonds which includes the base pairing as well as the hydrogen bonding between the RNA and the water molecules. Finally the last term in the potential expression represents the long distance interactions known as the (van der Waals' interactions). As it was done for the nanoclusters in [4,9] the modeling of RNA nanotube including pre and post processing of the input-output files have been performed by using the visualization software software VMD and gnuplot. The main features of our analysis here are to calculate the velocity trajectory along the path of molecular dynamics simulation and then to calculate the autocorrelation function which can later be used for calculation of the diffusion properties of the RNA nanocluster in physiological solutions. We note that the RNA-nanotube was solvated by the water in a water box. The size of the box is taken in such a way that the distance wall of the water box is at a distance larger than the cut off radius used in the MD simulation. In order to make the system neutral we have added 594, 924, 1254 and 1584 $^{23}Na^+$ for two ring, three ring, four ring and five ring nanotubes, respectively. Furthermore, in order to make the solution equivalent to the physiological solution we have added extra $^{23}Na^+$ and $^{35}Cl^-$ ions. This system was simulated at constant temperature and pressure using NAMD software. The temperature in the system is controlled by the Langevin method [5] with damping $\eta = 5$ ps^{-1}.

3 Results and Discussion

Recently, we succeeded in describing RNA nanoclusters of variable sizes by using the molecular dynamics techniques [3,4]. Specifically, we used the RNA building blocks and self assembled them to construct the RNA nanotube using the VMD and the protocols available in the software NAMD. Now, for these nanoclusters we have calculated the trajectories for the velocities, focusing on atoms that may influence substantially dynamical properties of these RNA nanoclusters. Hence, this contribution represents a new steps in the study of transport properties including diffusion phenomena in the RNA nanocluster that can be analyzed via molecular dynamics simulations. In particular the trajectories for the phosphorous atom in the phosphate backbone, the sodium ion in physiological solution, the chloride ion and the oxygen atom at sugar ring of the

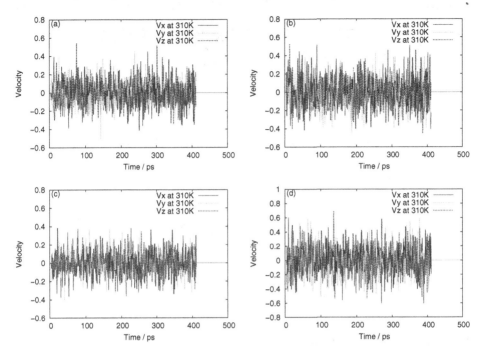

Fig. 2. (color online)The trajectories of the velocity for (a) Phosphorus at the phosphate backbone (b) Sodium ion in a physiological solution (c) Chloride ion in a physiological solution and (d) The oxygen atom at a sugar ring

five ring RNA nanotube are presented in Figure 2. For each of the atoms the velocities have been tracked consistently along all directions i.e x, y, and z. From the plots it is clear that the variation in the velocities during the molecular dynamics simulations is small and consistent in all directions, given constant temperature. This consistency of the velocity components during the MD simulation is due to the consistency of the temperature, along with some variations due to damping. The nature of the fluctuation of the velocities during molecular dynamics simulations is similar to the fluctuations observed for the temperature reported in earlier studies [3,4]. This feature has been observed because the classical velocities are proportional to the temperature of the system. In Figure 3, we have also presented the results for the velocity autocorrelation function in three different directions for the phosphorus atom in the RNA nanotube using molecular dynamics velocity trajectories in physiological solutions. From the plots in Figure 3, we conclude that the variations of the velocity autocorrelation function(VACF) in x direction is significantly different than those in the y and z directions. These velocity autocorrelation functions are primary characteristics for the estimates of diffusion coefficients of the molecular systems under considerations.

Up till now, very limited experimental studies have been done for the diffusion properties of the RNA nanoclusters. However, it is worthwhile noting that some experimental studies on the conformational diffusion coefficient for a typical biopolymer has recently been performed by using the experimental technique force spectroscopy [13]. This shed further light on the inter chain motion in complex biological polymers.

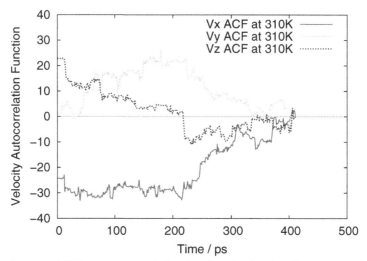

Fig. 3. (color online) Velocity autocorrelation function for the phosphorous atom in the RNA nanotube using molecular dynamics simulation

4   Conclusions and Outlook

In this contribution, we presented new results on dynamic properties of RNA nanotubes, in particular the velocity trajectories and velocity autocorrelation functions for P, O, $^{23}Na^+$, and $^{35}Cl^-$ atoms during molecular dynamics simulations of RNA nanotubes in physiological solutions. Such systems are of particular interest in nanomedical applications [4,9]. In typical NVT runs of constant temperature molecular dynamics simulations for these RNA nanotubes, we found that the velocity is fluctuating in all directions rather uniformly. We have presented the VACFs, focusing on the phosphorous atom at the phosphate backbone of the RNA nanotubes. At the same time, it would be very interesting to calculate the VACFs for other atoms in order to better understand the trend of their velocity variations in different directions. Using these autocorrelation functions deduced from velocity trajectories, a detailed study of diffusion characteristics of such systems represents an important avenue of future work that should provide additional insight into transport phenomena in RNA nanotubes and their applications in biomedicine as well as other fields. Finally, we note that, despite efficient coarse-graining procedures, our current studies have been limited by severe computational challenges and were naturally limited in time scales smaller than realistically required for many biological processes. Longer simulations would provide a better understanding about the transport properties of the RNA nanoclusters. Undoubtedly, these studies should encourage experimentalists to do such kind of measurements that would provide the diffusion parameters on the RNA nanoclusters.

Acknowledgements. Authors are grateful to the NSERC and CRC Program for their support and Shared Hierarchical Academic Research Computing Network (SHARC-NET: www.sharcnet.ca) for providing the computational facilities. Finally, we would like to thank Dr. P. J. Douglas Roberts for helping with technical SHARCNET computational aspects.

References

1. Ando, T., Skolnick, J.: Crowding and hydrodynamic interactions likely dominate in vivo macromolecular motion. PNAS 107(43), 18457–18462 (2010)
2. Arrio-Dupont, M., Foucault, G., Vacher, M., Devaux, P.F., Cribier, S.: Translational diffusion of globular proteins in the cytoplasm of cultured muscle cells. Biophysical Journal 78(2), 901–907 (2000)
3. Badu, S.R., Melnik, R., Paliy, M., Prabhakar, S., Sebetci, A., Shapiro, B.A.: High performance computing studies of RNA nanotubes. In: Proceedings of IWBBIO-2014 (International Work-Conference on Bioinformatics and Biomedical Engineering), pp. 601–607 (2014)
4. Badu, S.R., Melnik, R., Paliy, M., Prabhakar, S., Sebetci, A., Shapiro, B.A.: Modeling of RNA nanotubes using molecular dynamics simulation. Eur. Biophys. J. 43(10-11), 555–564 (2014)
5. Feller, S.E., Zhang, Y., Pastor, R.W., Brooks, B.R.: Constant pressure molecular dynamics simulation: The langevin piston method. The Journal of Chemical Physics 103(11), 4613–4621 (1995)
6. Lee, A.J., Crothers, D.M.: The solution structure of an RNA looploop complex: The ColE1 inverted loop sequence. Structure 6(8), 993–1007 (1998)
7. MacKerell, Bashford, D., Bellott, Dunbrack, Evanseck, J.D., Field, M.J., Fischer, S., Gao, J., Guo, H., Ha, S., Joseph-McCarthy, D., Kuchnir, L., Kuczera, K., Lau, F.T.K., Mattos, C., Michnick, Ngo, T., Nguyen, D.T., Prodhom, B., Reiher, W.E., Roux, B., Schlenkrich, M., Smith, J.C., Stote, R., Straub, J., Watanabe, M., Wirkiewicz-Kuczera, J., Yin, D., Karplus, M.: All-atom empirical potential for molecular modeling and dynamics studies of proteins. J. Phys. Chem. B 102(18), 3586–3616 (1998)
8. Makarov, V.A., Feig, M., Andrews, B.K., Pettitt, B.M.: Diffusion of solvent around biomolecular solutes: A molecular dynamics simulation study. Biophysical Journal 75(1), 150–158 (1998)
9. Paliy, M., Melnik, R., Shapiro, B.A.: Molecular dynamics study of the RNA ring nanostructure: A phenomenon of self-stabilization. Phys. Biol. 6(4), 046003 (2009)
10. Phillips, J.C., Braun, R., Wang, W., Gumbart, J., Tajkhorshid, E., Villa, E., Chipot, C., Skeel, R.D., Kal, L., Schulten, K.: Scalable molecular dynamics with NAMD. J. Comput. Chem. 26(16), 1781–1802 (2005)
11. Verkman, A.S.: Solute and macromolecule diffusion in cellular aqueous compartments. Trends in Biochemical Sciences 27(1), 27–33 (2002)
12. Wojcieszyn, J.W., Schlegel, R.A., Wu, E.S., Jacobson, K.A.: Diffusion of injected macromolecules within the cytoplasm of living cells. PNAS 78(7), 4407–4410 (1981)
13. Woodside, M., Lambert, J., Beach, K.: Determining intrachain diffusion coefficients for biopolymer dynamics from single-molecule force spectroscopy measurements. Biophysical Journal 107(7), 1647–1653 (2014)
14. Yang, X., Melnik, R.: Effect of internal viscosity on brownian dynamics of dna molecules in shear flow. Computational Biology and Chemistry 31, 110–114 (2007)
15. Yeh, I.C., Hummer, G.: Diffusion and electrophoretic mobility of single-stranded {RNA} from molecular dynamics simulations. Biophysical Journal 86(2), 681–689 (2004)
16. Yingling, Y.G., Shapiro, B.A.: Computational design of an RNA hexagonal nanoring and an RNA nanotube. Nano Lett. 7(8), 2328–2334 (2007)

Molecular Dynamics Simulations of Ligand Recognition upon Binding Antithrombin: A MM/GBSA Approach

Xiaohua Zhang[1], Horacio Péréz-Sánchez[2,*], and Felice C. Lightstone[1,*]

[1] Biosciences and Biotechnology Division, Physical and Life Sciences Directorate,
Lawrence Livermore National Laboratory (LLNL), Livermore, CA, USA
{zhang30,lightstone1}@llnl.gov
[2] Bioinformatics and High Performance Computing Research Group, Department of
Computer Science, Universidad Católica San Antonio de Murcia (UCAM), Spain
hperez@ucam.edu

Abstract. A high-throughput virtual screening pipeline has been extended from single energetically minimized structure Molecular Mechanics/Generalized Born Surface Area (MM/GBSA) rescoring to ensemble-average MM/GBSA rescoring. For validation, the binding affinities of a series of antithrombin ligands have been calculated by using the two MM/GBSA rescoring methods. The correlation coefficient (R^2) of calculated and experimental binding free energies has been improved from 0.36 (for single-structure MM/GBSA rescoring) to 0.69 (for ensemble-average one). Decomposition of the calculated binding free energy reveals the electrostatic interactions in both solute and solvent play an important role in determining the binding free energy. The increasing negative charge of the compounds provides a more favorable electrostatic energy change but creates a higher penalty for the solvation free energy. Such a penalty is compensated by the electrostatic energy change, which results in a better binding affinity. The best binder has the highest ligand efficiency.

Keywords: MM/GBSA, Molecular dynamics, Binding Affinity, Antithrombin, Heparin.

1 Introduction

High-throughput virtual screening is an important tool for computer-aided drug discovery. We have developed a high throughput virtual screening pipeline for *in-silico* screening of a virtual compound database using high performance computing (HPC) [1]. The previous pipeline consists of four modules: receptor/target preparation, ligand preparation, VinaLC docking calculation [2], and single-structure MM/GBSA rescoring. All modules are parallelized to exploit typical cluster-type supercomputers. The MM/GBSA method is selected for rescoring because it is the fastest force-field based method that computes the free energy of binding, as compared to the other computational free energy methods, such as free energy perturbation (FEP) or thermodynamic

---

[*] Corresponding author.

F. Ortuño and I. Rojas (Eds.): IWBBIO 2015, Part II, LNCS 9044, pp. 584–593, 2015.
© Springer International Publishing Switzerland 2015

integration (TI) methods [3]. The MM/GBSA method has been widely exploited in free energy calculations [4, 5]. One of the notable features of this pipeline is an automated receptor preparation scheme with unsupervised binding site identification, which enables automatically running the whole pipeline with little or no human intervention. Perez-Sanchez and co-workers have developed a similar approach to improve drug discovery using massively parallel GPU hardware instead of supercomputers [6]. Their GPU-based program, BINDSURF [7], takes advantage of massively parallel and high arithmetic intensity of GPUs to speed-up the calculation in low cost desktop machine.

In this study, we have extended our pipeline from single-structure to ensemble-average MM/GSBA rescoring. To validate the new approach, we have gathered a panel of antithrombin ligands (Figure 1), including heparin and non-polysaccharide scaffold compounds. For the purpose of comparison, both single-structure and ensemble-average MM/GSBA rescoring are employed in the binding affinity calculations of antithrombin ligands. We must point out that estimation/calculations of entropy term are tricky. In most scenarios, the entropy term is neglected in the calculation for relative free binding energies. Quite a few researchers dispute the benefits of including the entropy term, which can be a major source of error due to the drawback of the entropy calculation method [8, 9], despite others who advocate its usage [10]. We choose to neglect the entropy term in our calculations.

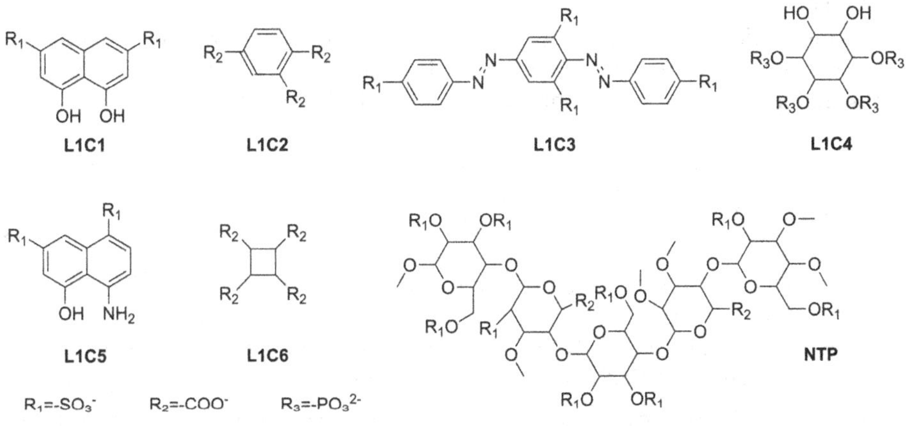

Fig. 1. Compounds targeting antithrombin. Compound NTP is a synthetic pentasaccharide compound from the crystal structure (PDB ID: 1AZX).

Antithrombin is a glycoprotein that plays a crucial role in the regulation of blood coagulation by inactivating several enzymes of the coagulation system and, thus, is an important drug target for the anticoagulant treatment. Antithrombin has two major isoforms, α and β, in the blood circulation [11]. α-Antithrombin is the dominant form of antithrombin and consists of 432 amino acids with 4 glycosylation sites, where an oligosaccharide occupyies each glycosylation site [12]. Heparin is the first compound that was identified and used as an anticoagulant and antithrombotic agent. It is a sulfated polysaccharide containing a specific pentasaccharide fragment (Figure 1, NTP) that binds and activates the antithrombin [13]. This binding localized the function of antithrombin to inhibition of serine proteases in the coagulation cascade in the bloodstream, which allows coagulant activity in damaged tissue outside the vascular system [12].

Due to increasing interests in clinical applications, computational studies have been carried out to investigate the structure and behavior of antithrombin. Verli and co-workers performed molecular dynamics simulations to study the induced-fit mechanism of the antithrombin-heparin interaction and effects of glycosylation on heparin binding [14, 15]. Several detailed conformational changes associated with heparin binding to antithrombin were revealed. They also confirmed an intermediate state between the native and activated forms of antithrombin. Because of the weak surface complementarity and the high charge density of the sulfated sugar chain, the docking of heparin to its protein partners presents a challenging task for computational docking. Wade and Bitomsky developed a protocol that can predict the heparin binding site correctly [16]. Navarro-Fernandez and colleagues screened a large database *in silico* by using FlexScreen docking program [17] and identified a new, non-polysaccharide scaffold able to interact with the heparin binding domain of antithrombin [18]. They predicted D-myo-inositol 3,4,5,6-tetrakisphosphate (Figure1, L1C4) to strongly interact with antithrombin, which was confirmed by experimental binding affinity study.

Here, we carried out molecular dynamics (MD) simulations of ligand recognition upon binding antithrombin and calculate the ligand binding affinity, using the ensemble-average MM/GBSA rescoring method, as a complementary study to previous docking works [16, 18]. The advantage of long-time MD simulations is that more configurational space can be explored and dynamical properties of systems can be revealed.

2   Method

The initial structure for the antithrombin complex with Compound NTP is obtained from the PDB bank (PDB ID: 1AZX). The initial structures for the antithrombin complex with the other 6 ligands (Figure 1) are obtained from the FlexScreen docking program [17]. The MM/GBSA calculations are applied to these initial structures by using our in-house developed pipeline [1, 2] and Amber molecular simulation package [19]. The Amber forcefield f99SB [19] is employed in the calculation for the antithrombin receptor. Ligands use the Amber GAFF forcefield [20] as determined by the Antechamber program [21] in the Amber package. Partial charges of ligands are calculated using the AM1-BCC method [22]. The fourth module of the pipeline is employed for the single-structure MM/GBSA calculation, where the receptor-ligand complexes are energetically minimized by the MM/GBSA method implemented in the Sander program of the Amber package [23]. The atomic radii developed by Onufriev and coworkers (Amber input parameter igb=5) are chosen for all GB calculations [24]. For the ensemble-average MM/GBSA rescoring, energetically minimized structures from single-structure MM/GBSA rescoring are served as initial structures. The systems are heated from 0 K to room temperature, 300 K. The MD simulations with a time step of 2 fs for the integration of the equations of motion are carried out at room temperature. The systems are equilibrated at room temperature for 500 ps. Each MD trajectory is followed to 100 ns after equilibrium. Binding affinities of antithrombin and its 7 ligands are calculated by post-processing the ensembles of structures extracted from MD trajectories using MM/GBSA calculations. In the MM/GBSA calculation, the binding free energy between a receptor and a ligand is calculated using the following equations:

$$\Delta G_{bind} = G_{complex} - G_{receptor} - G_{ligand} \tag{1}$$
$$\Delta G_{bind} = \Delta H - T\Delta S \approx \Delta E_{gas} + \Delta G_{sol} - T\Delta S \tag{2}$$
$$\Delta E_{gas} = \Delta E_{int} + \Delta E_{ELE} + \Delta E_{VDW} \tag{3}$$
$$\Delta G_{sol} = \Delta G_{GB} + \Delta G_{Surf} \tag{4}$$

The binding free energy (ΔG_{bind}) is decomposed into different energy terms. Because the structures of complex, receptor, and ligand are extracted from the same trajectory, the internal energy change (ΔE_{int}) is canceled. Thus, the gas-phase interaction energy (ΔE_{gas}) between the receptor and the ligand is the sum of electrostatic (ΔE_{ELE}) and van der Waals (ΔE_{VDW}) interaction energies. The solvation free energy (ΔG_{sol}) is divided into the polar and non-polar energy terms. The polar solvation energy (ΔG_{GB}) is calculated by using the GB model. The non-polar contribution is calculated based on the solvent-accessible surface area (ΔG_{Surf}). A value of 80 is used for the solvent dielectric constant, and the solute dielectric constant is set to 1. The calculated binding free energy (ΔG_{bind}) is the sum of the gas-phase interaction energy and solvation free energy because we neglect the entropy term. The experimental binding free energy is estimated from the experimental dissociation constant (Kd) by the equation:

$$\Delta G_{Exp} = RT \cdot \ln(Kd) \tag{5}$$

where R is the gas constant, and T is the temperature.

3   Results and Discussion

The calculated binding free energies of seven antithrombin ligands using the ensemble-average MM/GBSA rescoring are shown in Table 1 together with their corresponding experimental values. Each calculated binding free energy is averaged from snapshots extracted from 100 ns MD trajectories. Except for Compound L1C1, all the antithrombin ligands have experimental binding free energies. As determined experimentally, Compound L1C4 is the best binder with a Kd value of 0.088 uM [18]. As predicted by the MM/GBSA method, Compound L1C4 has the most negative binding free energy (-308.01 kcal/mol), which is in agreement with the experimental results. The second best binder as predicted by the MM/GBSA calculation is Compound NTP with a calculated binding free energy of -279.57 kcal/mol, confirming the experimental ranking relative to Compound L1C4. Compound L1C2 is predicted to have the worst binding free energy of the six ligands, which is also in agreement with its experimental ranking value. In summary, the MM/GBSA calculations rank the binding affinities of all six antithrombin ligands in same exact order as that of experimental binding free energy rankings.

Table 1. Calculated and experimental binding free energies (kcal/mol) of antithrombin ligands

Cmpd	ΔE_{ELE}	ΔE_{VDW}	ΔE_{aa}	ΔG_{Surf}	ΔG_{GB}	ΔG_{GB-ELE}	ΔG_{Sol}	ΔG_{Bind}	Kd(uM)	ΔG_{Exp}
L1C1	-552.67	-23.68	-576.35	-2.75	480.12	-72.55	477.37	-98.97	-	-
L1C2	-442.99	0.47	-442.52	-1.20	417.60	-25.39	416.41	-26.11	13700	-2.54
L1C3	-836.77	-39.96	-876.73	-4.06	781.94	-54.83	777.88	-98.85	10.02	-6.81
L1C4	-1599.09	33.02	-1566.07	-2.98	1261.05	-338.04	1258.07	-308.01	0.088	-9.62
L1C5	-613.30	-19.00	-632.31	-2.57	525.82	-87.48	523.25	-109.06	0.69	-8.40
L1C6	-818.73	8.21	-810.52	-1.54	752.94	-65.79	751.41	-59.11	17.52	-6.48
NTP	-2598.87	-60.89	-2659.76	-7.58	2387.77	-211.09	2380.20	-279.57	0.104	-9.52

The calculated binding free energies of six antithrombin ligands using the ensemble-average MM/GBSA rescoring have been plotted against the free energies derived from experimental dissociation constants. The correlation coefficient (R2) is 0.69, which indicates good correlation between the calculated and experimental values (Figure 2). By comparison, the correlation coefficient calculated by single-structure MM/GBSA is only 0.36, and Compound NTP is predicted to be the best binder, instead of Compound L1C4. Thus, using the ensemble-average MM/GBSA rescoring method significantly improves the accuracy of the prediction over the single-structure MM/GBSA rescoring.

As shown in Figure 1, all antithrombin ligands contain negatively charged groups, suggesting electrostatic interactions should be a key factor in the binding affinity. Compound NTP has a total charge of -11, and Compound L1C4 has a total charge of -8. By decomposing the binding free energy, Compound NTP and L1C4 have the largest electrostatic energy changes upon binding in both gas phase (ΔE_{ELE}) and GB solvent (ΔG_{GB-ELE}). The energy change upon binding in gas phase is equivalent to the energy change upon binding for the solute. Thus, in other words, Compound NTP and L1C4 have the largest electrostatic energy changes

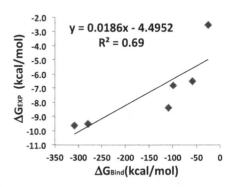

Fig. 2. The scatter plot of calculated MM/GBSA binding free energy versus experimental binding affinity estimated from dissociation constant

upon binding in solute and solvent. In contrast, Compound L1C2 has the smallest electrostatic energy changes in solute and solvent. Although Compound L1C4 has the least favorable van der Waals energy change upon binding, the electrostatic energy change compensates significantly. For all ligands, the van der Waals energy changes (ΔE_{VDW}) upon binding are less than the electrostatic energy changes (ΔE_{ELE}) by 1-2 orders of magnitude. The contribution of the van der Waals energy change has been overpowered by the electrostatic energy change. Non-polar contribution of solvation free energy of Compound NTP and L1C3 are more negative than that of the other compounds because the sizes of Compound NTP and L1C3 are larger than the other compounds. Nevertheless, non-polar contributions for all compounds are small. The non-polar contribution is overwhelmed by the polar contribution of solvation free energy. Thus, the two major factors to determine the binding affinity are the electrostatic energy change and solvation free energy change. The larger the total charge of the compound, the larger the penalty cost is for solvation free energy. However, high penalty for large total charge of compound has been paid by the large favorable electrostatic energy changes. Although the electrostatic energy change of Compound L1C4 is less than that of Compound NTP, Compound L1C4 needs less compensation for the solvation free energy. Thus, Compound L1C4 is a better binder than Compound NTP.

Hydrogen bonding analysis determines the numbers of hydrogen bonds to antithrombin that are persistent at >20% of the time. Compound NTP has 40 hydrogen bonds to antithrombin, while L1C4 has 25 hydrogen bonds. For Compounds L1C1, L1C2, L1C3, L1C5, and L1C6, that number of hydrogen bonds are 5, 10, 12, 12, and 12, respectively. Taking the molecular weight into account and using a similar approach as Reynolds' ligand efficiency method [25], Compound L1C4 has the highest ligand efficiency.

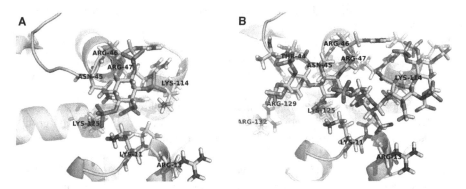

Fig. 3. Initial structures of Compounds L1C4 (A) and NTP (B) complexed with antithrombin

Compound L1C4 forms double hydrogen bonds with Arg47 (Figure 3A). One hydrogen bond (O6-HH21-NH2) has 94.81% persistence, and the other one (O6-HE-NE) has 89.55%. The average hydrogen bond distances between the heavy atoms are 2.74 Å and 2.70 Å, respectively. Compound L1C4 has strong hydrogen bonds with Arg47, and one of the four phosphate groups from Compound L1C4 is locked to the Arg47. According to the hydrogen bonding analysis, Compound L1C4 is also hydrogen bonded to Arg46, Arg13, Lys114, Lys11, Lys125, and Asn45, which are key residues to the binding process. We find that the binding of Compound L1C4 to antithrombin is non-specific. Except for the phosphate group locked to Arg47, the other three phosphate groups of Compound L1C4 can rotate so that key residues can form hydrogen bonds to different oxygen atoms of phosphate at different times during the MD trajectory. Notably, Arg13 starts far away from Compound L1C4 in the initial conformation. After 8 ns of MD simulation, Arg13 begins to make hydrogen bonds with the phosphate group of Compound L1C4, which suggests that long-time MD simulations are essential to obtaining accurate binding affinities. As shown in Figure 3B, Compound NTP makes hydrogen bonds to antithrombin mainly via its negatively charged sulfate groups. Compound NTP forms high persistent hydrogen bonds with Arg13, Arg129, Arg47, and Asn45 (70~88%) and forms medium persistent hydrogen bonds with Arg132, Lys125, and Thr44 (43~66%). Only relatively weak hydrogen bonds are observed with Arg46, LYS114 and LYS11.

Navarro-Fernandez and colleagues have also predicted the interacting residues by using FlexScreen scoring function [17]. For compound L1C4, they identified Lys11, Asn45, Arg46, Arg47, Lys114, and Lys125 as key residues, but Arg13 is missing from their list. For the Compound L1C4 docking calculation, they used the receptor structure from the X-ray crystal structure of antithrombin complexed with Compound NTP. Thus, the initial receptor structure for Compound L1C4 docking is biased. Docking calculations usually hold the receptor protein rigid. A few docking programs are able to have set side-chains of key residues in receptor as flexible. However, most docking programs cannot sample the larger configuration space for the whole ligand-receptor complex. From our MD simulations, we find that Arg13 is quite flexible and can be adjusted to accommodate both large (e.g. Compound NTP) and small (e.g. Compound L1C4) compounds, which shows the advantages of using MD simulations over simple docking calculations.

Judging from the hydrogen bond analysis on Compound L1C4 and NTP, Arg47, Arg13, and Asn45 play crucial roles in the antithrombin binding process. Antithrombin provides multiple sulfate/phosphate binding sites, consisting of mostly positively charged residues (arginine, lysine) and neutral charged residues that can provide rich hydrogen bond donors/acceptors (asparagine). All four phosphate groups of Compound L1C4 form hydrogen bonds with antithrombin while not all the sulfate groups of Compound NTP can form hydrogen bonds with antithrombin. As pointed out above, introducing positively charged group in the ligand will result in a penalty for solvation free energy. If adding a positively charged group cannot form favorable interactions (e.g. hydrogen bonding), ligand efficiency will be reduced, explaining why Compound L1C4 has higher ligand efficiency than Compound NTP.

Comparing the results from single-structure and ensemble-average MM/GBSA rescoring, the latter yields more accurate results. The ensemble-average MM/GBSA rescoring ranks the binding affinities of antithrombin ligands in the order that agrees with the experimental results. The advantage of ensemble-average MM/GBSA rescoring is that the binding affinity is averaged from an ensemble of structures extracted from long-time MD simulations. Long-time MD simulations can explore more configuration space and find energetically favorable configurations, which could offset the bias of initial structures. This can be verified in the MD trajectory of Compound

L1C4. Arg13 is observed to form hydrogen bonds with the phosphate group of Compound L1C4 after 8 ns of MD simulation.

The ensemble-average MM/GBSA rescoring method is relatively accurate compared to single-structure MM/GBSA. However, the method to obtain an ensemble of structures, long-time MD simulations, is computationally intensive. Since the drug-like virtual libraries often contain millions of compounds, high-throughput virtual screening could be very costly, if using a more accurate but expensive method at the very beginning of the screening. To bridge the gap, our virtual screening pipeline uses a down-select scheme to screen large virtual compound libraries. A standard procedure to run the pipeline is to down-select compounds after they pass each screening method as implemented in the pipeline. The first screening method in the pipeline is VinaLC docking, which can dock one million compounds in 1.4 h on about 15K CPUs [2]. Top ranked poses of down-selected compounds after docking are rescored using a single-structure MM/GBSA rescoring method. Finally, the most expensive ensemble-average MM/GBSA rescoring method in the pipeline can be applied to an amenable number of compounds down-selected after single-structure MM/GBSA rescoring, providing the accuracy needed for a fewer number of compounds.

4   Conclusion

In this article, we introduce a new addition, ensemble-average MM/GBSA rescoring, to our virtual screening pipeline. As a proof of concept, we calculated the binding affinities of seven antithrombin ligands by employing the previous single-structure MM/GBSA rescoring method and newly developed ensemble-average MM/GBSA rescoring method. The correlation coefficient of calculated and experimental binding affinities is improved from 0.36 to 0.69 when using ensemble-average MM/GBSA rescoring. The rank order of calculated binding free energies using ensemble-average MM/GBSA rescoring exactly matches the experimentally derived free energies. We demonstrate that long-time MD trajectory can explore more configuration space and find energetically favorable configurations so that it can offset the bias of initial structures and improve the accuracy of binding affinity prediction. The electrostatic interactions in both solute and solvent contribute favorably to the binding free energy. Adding more negatively charged groups to the ligand provides more favorable electrostatic energy change. However, it creates a higher penalty for the solvation free energy simultaneously. The penalty can be compensated for by forming more hydrogen bonds as more negatively charged groups are added into the ligand. The negatively charge groups added to the ligand must actively interact with receptor by forming hydrogen bonds to achieve high ligand efficiency. Compound L1C4 has higher ligand efficiency because it uses all its phosphate groups to form hydrogen bonds with antithrombin while Compound NTP does not.

Acknowledgements. The authors thank Livermore Computing for the computer time and Laboratory Directed Research and Development for funding (12-SI-004). We also thank Livermore Computing Grand Challenge for extensive computing resources.

This work was performed under the auspices of the United States Department of Energy by the Lawrence Livermore National Laboratory under Contract DE-AC52-07NA27344. Release Number LLNL-JRNL-666361. This work was partially supported by the Fundación Séneca la Región de Murcia under Project 18946/JLI/13. This work has been funded by the Nils Coordinated Mobility under grant 012-ABEL-CM-2014A, in part financed by the European Regional Development Fund (ERDF).

References

1. Zhang, X., Wong, S.E., Lightstone, F.C.: Toward Fully Automated High Performance Computing Drug Discovery: A Massively Parallel Virtual Screening Pipeline for Docking and Molecular Mechanics/Generalized Born Surface Area Rescoring to Improve Enrichment. J. Chem. Inf. Model. 54, 324–337 (2014)
2. Zhang, X., Wong, S.E., Lightstone, F.C.: Message passing interface and multithreading hybrid for parallel molecular docking of large databases on petascale high performance computing machines. J. Comput. Chem. 34, 915–927 (2013)
3. Beveridge, D.L., DiCapua, F.M.: Free Energy Via Molecular Simulation: Applications to Chemical and Biomolecular Systems. Annu. Rev. Biophys. Biophys. Chem. 18, 431–492 (1989)
4. Kollman, P.A., Massova, I., Reyes, C., Kuhn, B., Huo, S., Chong, L., Lee, M., Lee, T., Duan, Y., Wang, W., Donini, O., Cieplak, P., Srinivasan, J., Case, D.A., Cheatham, T.E.: Calculating Structures and Free Energies of Complex Molecules: Combining Molecular Mechanics and Continuum Models. Acc. Chem. Res. 33, 889–897 (2000)
5. Hou, T.J., Wang, J.M., Li, Y.Y., Wang, W.: Assessing the Performance of the MM/PBSA and MM/GBSA Methods. 1. The Accuracy of Binding Free Energy Calculations Based on Molecular Dynamics Simulations. J. Chem. Inf. Model. 51, 69–82 (2011)
6. Pérez-Sánchez, H., Cano, G., García-Rodríguez, J.: Improving drug discovery using hybrid softcomputing methods. Applied Soft Computing 20, 119–126 (2014)
7. Sanchez-Linares, I., Perez-Sanchez, H., Cecilia, J., Garcia, J.: High-Throughput parallel blind Virtual Screening using BINDSURF. BMC Bioinformatics 13, S13 (2012)
8. Rastelli, G., Rio, A.D., Degliesposti, G., Sgobba, M.: Fast and accurate predictions of binding free energies using MM-PBSA and MM-GBSA. J. Comput. Chem. 31, 797–810 (2010)
9. Greenidge, P.A., Kramer, C., Mozziconacci, J.C., Wolf, R.M.: MM/GBSA Binding Energy Prediction on the PDBbind Data Set: Successes, Failures, and Directions for Further Improvement. J. Chem. Inf. Model. 53, 201–209 (2013)
10. Lafont, V., Armstrong, A.A., Ohtaka, H., Kiso, Y., Mario Amzel, L., Freire, E.: Compensating Enthalpic and Entropic Changes Hinder Binding Affinity Optimization. Chemical Biology & Drug Design 69, 413–422 (2007)
11. Turko, I.V., Fan, B., Gettins, P.G.W.: Carbohydrate isoforms of antithrombin variant N135Q with different heparin affinities. FEBS Lett. 335, 9–12
12. Jin, L., Abrahams, J.P., Skinner, R., Petitou, M., Pike, R.N., Carrell, R.W.: The anticoagulant activation of antithrombin by heparin. Proc. Natl. Acad. Sci. U. S. A. 94, 14683–14688 (1997)
13. Thunberg, L., Bäckström, G., Lindahl, U.: Further characterization of the antithrombin-binding sequence in heparin. Carbohydr. Res. 100, 393–410 (1982)

14. Verli, H., Guimarães, J.A.: Insights into the induced fit mechanism in antithrombin–heparin interaction using molecular dynamics simulations. J. Mol. Graph. Model. 24, 203–212 (2005)
15. Pol-Fachin, L., Franco Becker, C., Almeida Guimarães, J., Verli, H.: Effects of glycosylation on heparin binding and antithrombin activation by heparin. Proteins: Structure, Function, and Bioinformatics 79, 2735–2745 (2011)
16. Bitomsky, W., Wade, R.C.: Docking of Glycosaminoglycans to Heparin-Binding Proteins: Validation for aFGF, bFGF, and Antithrombin and Application to IL-8. J. Am. Chem. Soc. 121, 3004–3013 (1999)
17. Merlitz, H., Wenzel, W.: Comparison of stochastic optimization methods for receptor–ligand docking. Chem. Phys. Lett. 362, 271–277 (2002)
18. Navarro-Fernández, J., Pérez-Sánchez, H., Martínez-Martínez, I., Meliciani, I., Guerrero, J.A., Vicente, V., Corral, J., Wenzel, W.: In Silico Discovery of a Compound with Nanomolar Affinity to Antithrombin Causing Partial Activation and Increased Heparin Affinity. J. Med. Chem. 55, 6403–6412 (2012)
19. Ponder, J.W., Case, D.A.: Force fields for protein simulations. Protein Simulations 66, 27–85 (2003)
20. Wang, J.M., Wolf, R.M., Caldwell, J.W., Kollman, P.A., Case, D.A.: Development and testing of a general amber force field. J. Comput. Chem. 25, 1157–1174 (2004)
21. Wang, J., Wang, W., Kollman, P.A., Case, D.A.: Automatic atom type and bond type perception in molecular mechanical calculations. J. Mol. Graph. Model. 25, 247–260 (2006)
22. Jakalian, A., Bush, B.L., Jack, D.B., Bayly, C.I.: Fast, efficient generation of high-quality atomic Charges. AM1-BCC model: I. Method. J. Comput. Chem. 21, 132–146 (2000)
23. Case, D.A., Cheatham III, T.E., Darden, T., Gohlke, H., Luo, R., Merz Jr., K.M., Onufriev, A., Simmerling, C., Wang, B., Woods, R.J.: The Amber biomolecular simulation programs. J. Comput. Chem. 26, 1668–1688 (2005)
24. Onufriev, A., Bashford, D., Case, D.A.: Exploring protein native states and large-scale conformational changes with a modified generalized born model. Proteins-Struct. Funct. Bioinf. 55, 383–394 (2004)
25. Reynolds, C.H., Tounge, B.A., Bembenek, S.D.: Ligand binding efficiency: Trends, physical basis, and implications. J. Med. Chem. 51, 2432–2438 (2008)

Predicting Atherosclerotic Plaque Location in an Iliac Bifurcation Using a Hybrid CFD/Biomechanical Approach

Mona Alimohammadi[1,*], Cesar Pichardo-Almarza[1,2], Giulia Di Tomaso[1], Stavroula Balabani[1], Obiekezie Agu[3], and Vanessa Diaz-Zuccarini[1]

[1]UCL Mechanical Engineering, Multiscale Cardiovascular Engineering Group (MUSE), Torrington Place, WC1E 7JE. London, UK
[2]Xenologiq Ltd, Canterbury, CT2 7FG, UK
[3]University College Hospital, Vascular Unit, NW1 2BU, London, UK
mona.alimohammadi.10@ucl.ac.uk

Abstract. Experimental evidence indicates that haemodynamic stimuli influence some properties of the arterial endothelium, such as cell geometry and permeability, leading to possible accumulation of blood-borne macromolecules and initiation of atherosclerosis. Patient-specific computational models are able to capture complex haemodynamic characteristics to explore and analyse the development of these diseases *in silico*. Patient-specific models are particularly beneficial in the case of aortic dissection (AD), a condition in which the aortic wall is split in two, creating a true and a false lumen. In this condition, the proportion of blood through the main vessel and the main aortic branches is substantially modified and malperfusion (lack of blood supply) of the downstream vessels is often observed. Furthermore, AD alters the haemodynamics downstream of the lesion, potentially leading to the formation of atherosclerotic plaques at the iliac bifurcation. In order to correctly approximate the haemodynamic changes and analyse the role they play in the development of atherosclerosis formations in AD patients, a combined multiscale methodology is required. In this study, both, blood flow through an iliac bifurcation of a patient suffering from type-B aortic dissection and endothelium behavior are analysed, in order to investigate atherosclerosis formation.

1   Introduction

Cardiovascular diseases are the leading cause of death in developed countries [1]. Thoracic aortic diseases occur between 16 to 10 per 100000 per year for both men and women in Europe [2], among which, aortic dissection (AD), and its associated vascular diseases and further complications, are relatively poorly understood. One of these complication is calcification of the aortic wall, which carries with it significant risks and increased mortality rates [3, 4]. AD is initiated by the formation of a cleavage in the intimal layer of the aortic vessel wall, allowing blood to enter the vessel wall,

---

* Corresponding author.

F. Ortuño and I. Rojas (Eds.): IWBBIO 2015, Part II, LNCS 9044, pp. 594–606, 2015.

forming a pathway between the layers and creating a so called false lumen (FL), as opposed to the true lumen (TL). AD is highly prevalent in patients with hereditary connective tissue disorders such as Ehler-Danlos and Marfan syndrome [1, 5, 6], which would potentially affect the aortic wall distensibility, stiffness and tissue fragility [7, 8]. It has been reported that 31% of patients suffering from AD have a history of plaque formation [9] and atherosclerosis has been shown to be a major postoperative factor in the mortality rate for dissections involving the descending part of the aorta (Stanford type-B) [10] (as opposed to the ascending aorta: Stanford type-A). Calcification is present in both the TL and FL, as well as downstream of the dissected region [11, 12]. The dissection causes a pathological haemodynamic environment, prone to calcification of the vessel wall and atherosclerotic plaque formation downstream of the dissected thoracic aorta at the iliac bifurcation.

The process of atherosclerotic plaque formation is complex, multifactorial and systemic. Experimental evidence indicates that haemodynamic stimuli influence some properties of the arterial endothelium, such as cell geometry and permeability, leading to possible accumulation of blood-borne macromolecules (e.g. Low Density Lipoproteins – LDL) and initiation of atherosclerosis. In the early stages of the disease, an accumulation of lipid-laden macrophages (foam cells) is observed in the subendothelium. As the disease develops, smooth muscles cells and fibrous tissue accumulate. Formation of lesions is promoted by plasma proteins carrying elevated levels of cholesterol and triglycerides. Clinical manifestations of atherosclerosis, including coronary artery disease, cerebrovascular disease, and peripheral arterial disease, occur in 2 of 3 men and 1 in 2 women over the age of 40 [13].

The management of these combined, complex conditions proves to be challenging for clinicians. Patient-specific computational models are able to capture complex haemodynamic characteristics to explore and analyse the development of these diseases *in silico*. Patient-specific models are particularly beneficial in the case of AD, as the proportion of blood through the aorta and the aortic branches is substantially modified and malperfusion (lack of blood supply) of the downstream vessels is often observed [5, 14]. AD also alters the haemodynamics downstream of the lesion, potentially leading to the formation of atherosclerotic plaques at the iliac bifurcation. To correctly approximate the haemodynamic changes and analyse the role they play in the development of atherosclerosis formation in AD patients, a combined multiscale methodology is required. Several studies have been carried out to analyse haemodynamic parameters such as pressure, flow and wall shear stress (WSS) through the iliac arteries [15, 16]. Kim et al. [17] used a lagrangian method for the boundary conditions (BCs) for example, while others have used electrical analogues to represent the characteristics of the downstream and upstream vasculature in dissected aortae.

In this study, simulation of the blood flow through an iliac bifurcation of a patient suffering from type-B aortic dissection is used to estimate locations of atherogenesis using a mathematical model of atherosclerosis which includes the modelling of LDL transport from the lumen into the arterial wall (through the endothelium) using a three-pore modelling approach [18] and based in previous work [19, 20].

In order to capture detailed haemodynamic values that will affect the formation of plaque in individual patients, realistic fluid simulations are required and this requires appropriate model definition. In this work, the pressure and flow rate at each boundary change during the cardiac cycle, a feature that is captured by the simulation by using dynamic boundary conditions. Coupling the three-dimensional (3D) domain with Windkessel models is an approach that can account for the interdependent time varying flow and pressure values along the patient's cardiac cycle [21, 22]. The challenge in applying this approach is the estimation of the Windkessel parameters (resistance or compliance) and there is no agreed methodology on how to select these [23].

In this paper, a numerical simulation to investigate the possibility of plaque formation in an individual patient is produced. The blood flow through a patient-specific iliac bifurcation is simulated by coupling the 3D domain to the three element Windkessel models at each of the (3D) domain's outlets. The Windkessel parameters are tuned using the invasive pressure measurements acquired on the same patient prior to the simulation. These results are then coupled to an endothelial permeability model, as explained in Sections 2 and 3 in order to predict locations of atherosclerosis, using a 'virtual follow-up' approach.

2   Methods

2.1   Geometry

An iliac bifurcation from a 54 year-old female patient suffering from type-B dissection and atherosclerosis was segmented from MSCT images (Ethics Reference number 13/EM/0143), obtained at University College Hospital (UCH). The 3D geometry was reconstructed from approximately 200 CT slices using ScanIP (Simpleware Ltd. UK). In order to segment the region of interest (the fluid domain), multiple thresholds and flood fills were applied to select the iliac arteries (IAs). Dilatation was used for ensure all pixels within the domain were selected. In order to miminise pixelisation artefacts, the final mask was smoothed using recursive Gaussian and median filters. Existing atherosclerotic formations observed in the images (identified by particularly bright regions at the vessel walls) were virtually removed to obtain a patent lumen. This geometry was then used to predict locations of atherosclerosis in the patient, using a 'virtual follow-up' approach. The final geometry can be seen in Figure 1a, which shows the flow inlet at the distal abdominal aorta (DA) and four outlets of left and right, internal and external IAs (R_{ext}, R_{int}, L_{ext}, L_{int}). Figure 1b shows the atherosclerotic plaque visible in the lumen (upper panel) and segmented using ScanIP. The blue and yellow mask (lower panel), representing the lumen and plaque were added together to create a single lumen (virtually removing the plaque regions). All boundaries were cropped to provide flat surfaces.

2.2   Computational Fluid Dynamics

Blood was considered to be an incompressible fluid with a density of 1056 kg m^{-3}. To model the non-Newtonian properties of the blood flow, the Carreau-Yasuda model was used with parameters defined by Gijsen et al. [24]. The flow was considered to be laminar. The pressure wave at the DA from a previous study on the same patient [22]

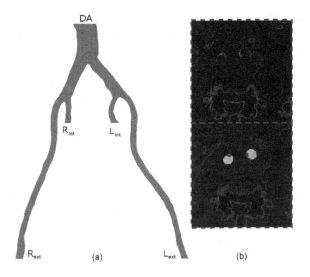

Fig. 1. (a) Geometry of the iliac bifurcation. (b) CT scan slices – upper panel: raw, lower panel: showing lumina (yellow) and atherosclerotic plaques (blue).

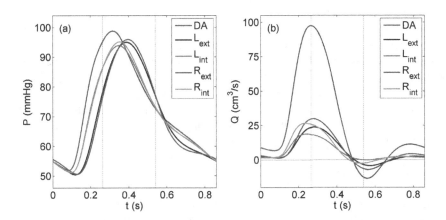

Fig. 2. (a) Pressure waves at each of the 3D domain outlets. (b) Flow waves at each of the 3D domain outlets. Vertical grey lines show time points considered in following figures.

(Fig. 2a) was used as inlet boundary condition. A three-element Windkessel model was coupled to each of the outlets. The zero-dimensional Windkessel model uses a circuit analogy, comparing pressure and flow to voltage and current, respectively. Therefore, solving the equation for a circuit of a parallel resistor (R_2) and capacitor (C) in series with a resistor (R_1) yields:

$$P = (R_1 + R_2)Q - \frac{R_2 C dp}{dt} + \frac{R_1 R_2 dQ}{dt} \qquad (1)$$

The derivative terms are calculated using the backward Euler method. The vessel wall was assumed to be rigid with no slip condition and a time step of 5 *ms* was used for the simulation.

In order to define the parameters of the Windkessel models, a tuning methodology was used to match invasive pressure measurements, which were available 3 *cm* upstream and 20 *cm* downstream, of the iliac bifurcation. Detailed analysis of this methodology was reported in a previous study[22].

2.3   Atherogenesis

A mathematical model describing the processes related to the early stages of atherosclerosis was implemented. Here, low density lipoprotein (LDL) is transported from the artery lumen into the arterial wall, taking into account the effects of local wall shear stress (WSS) on the endothelial cell layer and its pathways of volume and solute flux (see [19] and [20] for more details). The endothelium is described with a three-pore modelling approach, taking into consideration the contributions of the vesicular pathway, normal junctions, and leaky junctions. The fraction of leaky junctions is calculated as a function of the local WSS based on previous models and published experimental data [18, 19] and is used in conjunction with pore theory to determine the transport properties of this pathway.

It is assumed that the endothelium is influenced by the mechanical stimuli exerted by the blood flow. Experimental findings show that in areas of altered hemodynamics, endothelial cells do not have a typical cobblestone shape, but a more circular shape instead and exhibit increased permeability. The model uses a relationship between endothelial permeability and local WSS based on the Endothelial Cell Shape Index (ECSI) taken after experimental findings [25]. ECSI is related to the cellular shape and takes values from zero to one, i.e. a circle has an ECSI of one whilst a straight line has an ECSI of zero. The relationship between WSS and ECSI was modelled as shown in equation (2):

$$ECSI = 0.380e^{-0.3537\,WSS} + 0.225e^{-0.3537\,WSS} \tag{2}$$

To model the LDL transport through the endothelium, a modified version of the Kedem Ketchalsky's equations for membrane transport was used [19, 20]:

$$J_v = L_p(\Delta p_{end} - \sigma_d \Delta \Pi) \tag{3}$$

$$J_s = P_i\big(c_{lum} - c_{w,end}\big)\frac{Pe}{e^{Pe} - 1} + J_v(1 - \sigma)\bar{c} \tag{4}$$

$$Pe = \frac{J_v(1-\sigma)}{P_i} \tag{5}$$

where J_v is the volumetric flux through the endothelium, L_p is the hydraulic conductivity, Δp_{end} is the pressure difference through the endothelium, σ_d is the osmotic reflection coefficient and $\Delta\Pi$ is the osmotic pressure. The solute flux, J_s, can be divided into a convective component and a diffusive component (Equation 5). P_i is the diffusive permeability, Pe the modified Peclet number, $c_{w,end}$ the LDL concentration at the sub-endothelial layer and σ the solvent drag coefficient.

As previously mentioned, three main penetration pathways were considered: leaky junctions, normal junctions and vesicular pathways [19]. The model considers that the bulk volume flux through the endothelial membrane is given by:

$$J_v = J_{v,lj} + J_{v,nj} \tag{6}$$

where $J_{v,lj}$ is the flux through leaky junctions and $J_{v,nj}$ is the flux through normal junctions. Following the 'three-pore theory' [18], solute flux only occurs through endothelial leaky cell junctions and vesicles:

$$J_s = J_{s,lj} + J_{s,v} \tag{7}$$

where the solute flux through the vesicular pathway ($J_{s,v}$) is calculated as 10% of the solute flux through the leaky junction pathway ($J_{s,lj}$) [18].

Leaky cells have high permeability to LDL, which can be linked to the magnitude of WSS acting on the endothelium. Experimental findings show that in areas of low WSS and high ECSI, the number of mitotic cells (MC) is increased, leading to [18]:

$$MC = 0.003739 e^{14.75 \cdot ECSI} \tag{8}$$

Within the endothelium, it has been shown that the quantity of leaky mitotic cells is approximately 80.5% and since these represent approximately 45.3% of the total number of leaky cells (LC) in that area [26], the number of LC was calculated as:

$$LC = 0.307 + 0.805 MC \tag{9}$$

The ratio of endothelium (ϕ) covered by LCs is calculated as:

$$\phi = \frac{LC \; \pi R_{cell}^2}{unit \; area} \tag{10}$$

where R_{cell} is the radius. Using these values, the transport properties of the endothelium can be determined. The total hydraulic conductivity of the endothelial leaky junctions ($L_{p,lj}$) is defined as:

$$L_{p,lj} = \phi \cdot L_{p,slj} \tag{11}$$

where $L_{p,slj}$ is the hydraulic conductivity of a single leaky junction, w and l_{lj} are the half-width (20 nm) and the length (2 μm) of the LJ [27, 28].

$$L_{p,sj} = \frac{w^2}{3\mu_p l_{lj}} \tag{12}$$

3   Results

Fig. 2 shows the pressure and flow waves at each of the domain boundaries for one cardiac cycle. Due to the dynamic boundary conditions, the pressure waves are out of phase, with the internal IAs lagging behind the DA, and the external IAs delayed further. Correspondingly, the flow rates are out of phase with one another, with the external IAs lagging behind the internal IAs. Note that the flow maxima occur prior to the pressure maxima. The maximum flow rate entering the domain at the DA is approximately in phase with the external IAs.

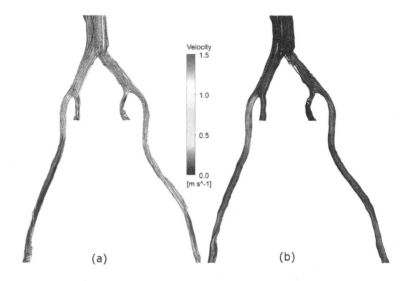

Fig. 3. Streamlines at (a) Peak systolic flow and (b) maximum negative flow

Fig. 3 shows the streamlines at two time points in the cardiac cycle; peak systolic flow and the point of minimum flow, as indicated by the dashed vertical lines in Fig. 2. At peak systole (Fig. 3a), the velocity is elevated after the first bifurcation and further increased after the second bifurcation through both L_{int} and R_{int} branches. A small region of disturbed flow can be seen at the curvature of the L_{int} branch (Fig. 3a). The velocity through the R_{ext} branch is marginally higher than the L_{ext} branch, and overall fairly uniform streamlines can be observed throughout the domain at peak systole. At the negative flow time-point (where the negative flow can be observed in Fig. 2b), the velocity is low throughout the domain, but is relatively uniform, in contrast to streamlines in the dissected thoracic aorta at a similar time point [22].

Fig. 4 shows pressure contours along the bifurcation at different time points. At peak systole (Fig. 4a), the pressure drop across the domain is ~ 25 mmHg. The pressure drops continuously throughout the domain. Small elevations in pressure at the wall can be observed at the stagnation points in each of the internal/external bifurcations. At the point of maximum negative flow (Fig. 4b), the pressure throughout the domain is fairly uniform.

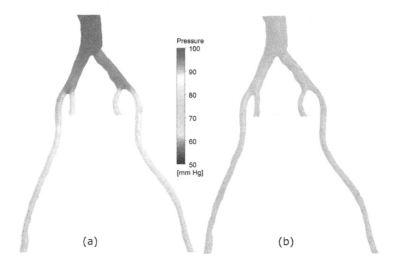

Fig. 4. Pressure contours at (a) Peak systolic flow and (b) maximum negative flow

Wall shear stress (WSS) is an important factor to investigate. It is considered to be a key haemodynamic parameter in both AD initiation and plaque formation [25, 29, 30]. A number of WSS indices are commonly used in the analysis of transient physiological flow simulations, in order to describe the WSS characteristics as a single spatial distribution.

The time-averaged wall shear stress (TAWSS) is defined by the following expression:

$$TAWSS = \frac{1}{T} \int_0^T |\tau(t)| \, dt \tag{12}$$

Where $|\tau(t)|$ is the magnitude of WSS vector at time t. This equation is applied at each location on the vessel wall to give the TAWSS distribution. An alternative index that provides insight into the nature of the oscillatory forces acting on the endothelium is the oscillatory shear index (OSI) [31]:

$$OSI = \frac{1}{2}\left(1 - \frac{\left|\frac{1}{T}\int_0^T \tau(t)dt\right|}{TAWSS}\right) \tag{13}$$

The TAWSS is shown in Fig. 5 in both left-posterior and right-anterior views. A small region of high TAWSS can be observed at the iliac bifurcation in Fig. 5a. Otherwise, the TAWSS is relatively low in the DA and left and right iliac arteries prior to the secondary bifurcations. Around the external/internal bifurcations, there are scattered regions of very high TAWSS, which persist along the external iliac arteries.

Fig. 6 shows the OSI in both views. In the descending aorta, the OSI is relatively high, at around 0.3, and the same is true of the right IA. In the left IA, there is a region of very low OSI. The OSI in the L_{int} artery is also low, and elsewhere the OSI is scattered.

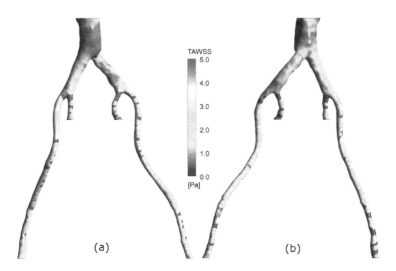

Fig. 5. TAWSS (a) Left posterior view (b) Right anterior view

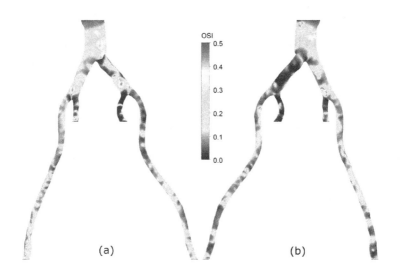

Fig. 6. OSI (a) Left posterior view (b) Right anterior view

Figures 7 and 8 show the endothelial cell shape index (ECSI) and the hydraulic conductivity ($L_{p,lj}$, units $mPa^{-1}s^{-1}$) of the leaky junctions in the arterial wall. ECSI values (Fig. 7) are between 0.1 and 0.6, which is consistent with the assumptions for modelling leaky junctions proposed by Olgac et al. [18] and Di Tomaso et al. [19]

$L_{p,lj}$ values (Fig. 8) are between 3.9×10^{-12} and 8.3×10^{-10} $mPa^{-1}s^{-1}$) with values larger than 1.19e-11 $mPa^{-1}s^{-1}$ (as reported by Tedgui and Lever [26] for normal junctions) in regions with lower WSS values.

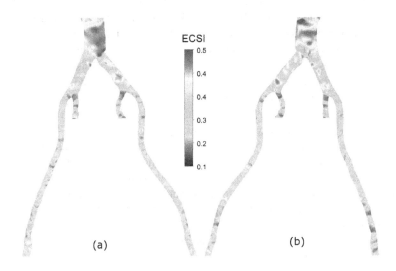

Fig. 7. Endothelial cell shape index (ECSI) for (a) Left posterior view (b) Right anterior view

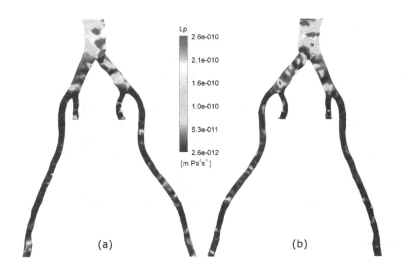

Fig. 8. Hydraulic conductivity of leaky junctions in the endothelium ($L_{p,lj}$). (a) Left posterior view (b) Right anterior view

4   Discussion

The acquired pulse pressure, as well as the maxima and minima of the pressure waves at each outlet, were close to those measured invasively, giving confidence that the boundary conditions were appropriately prescribed. The simulation results were processed to analyse the WSS, and other haemodynamic parameters were derived.

As expected, regions with low WSS show higher values for ECSI and L_p (having a similar pattern as OSI). Higher values of ECSI imply a higher number of endothelial cells producing leaky junctions, which in turn, would increase the permeability of the endothelium in those regions. L_p is inversely related to the resistance to mass flow provided by the endothelium, so higher values of L_p are associated with higher values of endothelial permeability. Thus regions where L_p is higher are expected to be more prone to develop atherosclerosis. Normal junctions in the endothelium will decrease in regions where the LDL flux through leaky junctions increases. This is normally most pronounced around the flow separation and reattachment points, where WSS is low. The increase in LDL flux through leaky junctions is the result of the increased number of leaky junctions themselves, combined with a decreased flow resistance through the entire leaky junction pathway. Because normal junctions and leaky junctions are parallel pathways, the decreased flow resistance of the leaky junction pathway leads to a decrease in the LDL flux through the normal junction pathway.

The areas that the model identified as prone to atherosclerosis formation were compared to the plaque distribution observed in the patient, prior to the virtual removal. The technique was able to predict the key locations of plaque formation, providing validation for the approach.

5   Conclusions

Given the difficulties in preventive diagnosis and clinical management of the disease, patient-specific computational models of atherosclerosis can offer much needed help by allowing the study of plaque formation *in silico* for individual patients. Understanding the effects that different haemodynamic modelling approaches and haemodynamic parameters have on the accuracy of detection of the atherosclerotic areas can lead to an improvement of *in silico* modelling techniques for atherosclerosis. These scenarios are currently being tested in the vascular services unit at UCH.

Acknowledgments. This work was supported by the EPSRC grant "Personalised Medicine Through Learning in the Model Space" (grant number EP/L000296/1).

References

1. Erbel, R., Alfonso, F., Boileau, C., Dirsch, O., Eber, B., Haverich, A., Rakowski, H., Struyven, J., Radegran, K., Sechtem, U., Taylor, J., Zollikofer, C., Klein, W.W., Mulder, B., Providencia, L.A.: Task Force on Aortic Dissection, European Society of Cardiology: Diagnosis and management of aortic dissection. Eur. Heart J. 22(18), 1642–1681 (2001)
2. Bastien, M., Dagenais, F., Dumont, É., Vadeboncoeur, N., Dion, B., Royer, M., Gaudet-Savard, T., Poirier, P.: Assessment of management of cardiovascular risk factors in patients with thoracic aortic disease. Blood Pressure Monitoring 17, 235–242 (2012)
3. Itani, Y., Watanabe, S., Masuda, Y.: Aortic calcification detected in a mass chest screening program using a mobile helical computed tomography unit. Relationship to risk factors and coronary artery disease. Circ. J. 68, 538–541 (2004)

4. Iribarren, C.: Patients with vascular calcifications are at increased risk of cardiovascular events: implications for risk factor management and further research. J. Intern. Med. 261, 235–237 (2007)

5. Tsai, T.T., Trimarchi, S., Nienaber, C.A.: Acute aortic dissection: perspectives from the International Registry of Acute Aortic Dissection (IRAD). Eur. J. Vasc. Endovasc. Surg. 37, 149–159 (2009)

6. Dietz, H.C., Cutting, G.R., Pyeritz, R.E., Maslen, C.L., Sakai, L.Y., Corson, G.M., Puffenberger, E.G., Hamosh, A., Nanthakumar, E.J., Curristin, S.M.: Marfan syndrome caused by a recurrent de novo missense mutation in the fibrillin gene. Nature 352, 337–339 (1991)

7. Pyeritz, R.E.: The Marfan Syndrome 51, 481–510 (2000)

8. Braverman, A.C.: Aortic dissection: Prompt diagnosis and emergency treatment are critical. Cleveland Clinic Journal of Medicine 78, 685–696 (2011)

9. Coady, M.A., Rizzo, J.A., Elefteriades, J.A.: Pathologic variants of thoracic aortic dissections. Penetrating atherosclerotic ulcers and intramural hematomas. Cardiol. Clin. 17, 637–657 (1999)

10. Tsai, T.T., Fattori, R., Trimarchi, S., Isselbacher, E., Myrmel, T., Evangelista, A., Hutchison, S., Sechtem, U., Cooper, J.V., Smith, D.E., Pape, L., Froehlich, J., Raghupathy, A., Januzzi, J.L., Eagle, K.A., Nienaber, C.A.: International Registry of Acute Aortic Dissection: Long-term survival in patients presenting with type B acute aortic dissection: insights from the International Registry of Acute Aortic Dissection. Circulation 114, 2226–2231 (2006)

11. LePage, M.A., Quint, L.E., Sonnad, S.S.: Aortic dissection: CT features that distinguish true lumen from false lumen. Am. J. Roentgenology 177, 207–211 (2001)

12. Willoteaux, S., Lions, C., Gaxotte, V., Negaiwi, Z., Beregi, J.P.: Imaging of aortic dissection by helical computed tomography (CT). European Radiology 14, 1999–2008 (2004)

13. Robinson, J.G., Fox, K.M., Bullano, M.F., Grandy, S.: The SHIELD Study Group: Atherosclerosis profile and incidence of cardiovascular events: A population-based survey. BMC Cardiovasc. Disord. 9, 46 (2009)

14. Svensson, L.G., Kouchoukos, N.T., Miller, D.C., Bavaria, J.E., Coselli, J.S., Curi, M.A., Eggebrecht, H., Elefteriades, J.A., Erbel, R., Gleason, T.G., Lytle, B.W., Mitchell, R.S., Nienaber, C.A., Roselli, E.E., Safi, H.J., Shemin, R.J., Sicard, G.A., Sundt III, T.M., Szeto, W.Y., III Wheatley, G.H.: Expert Consensus Document on the Treatment of Descending Thoracic Aortic Disease Using Endovascular Stent-Grafts. Ann. Thorac. Surg. 85, S1–S41 (2008)

15. O'Rourke, M.J., McCullough, J.P.: An investigation of the flow field within patient-specific models of an abdominal aortic aneurysm under steady inflow conditions. Proceedings of the Institution of Mechanical Engineers, Part H: Journal of Engineering in Medicine 224, 971–988 (2010)

16. Alishahi, M., Alishahi, M.M., Emdad, H.: Numerical simulation of blood flow in a flexible stenosed abdominal real aorta. Scientia Iranica 18, 1297–1305 (2011)

17. Kim, H.J., Figueroa, C.A., Hughes, T.J.R., Jansen, K.E., Taylor, C.A.: Augmented Lagrangian method for constraining the shape of velocity profiles at outlet boundaries for three-dimensional finite element simulations of blood flow. Comp. Meth. Appl. Mech. Eng. 198, 3551–3566 (2009)

18. Olgac, U., Kurtcuoglu, V., Poulikakos, D.: Computational modeling of coupled blood-wall mass transport of LDL: effects of local wall shear stress. American Journal of Physiology-Heart and Circulatory Physiology 294, H909–H919 (2008)

19. Di Tomaso, G., Díaz-Zuccarini, V., Pichardo-Almarza, C.: A multiscale model of atherosclerotic plaque formation at its early stage. IEEE Trans. Biomed. Eng. 58, 3460–3463 (2011)

20. Diaz-Zuccarini, V., Di Tomaso, G., Agu, O., Pichardo-Almarza, C.: Towards personalised management of atherosclerosis via computational models in vascular clinics: technology based on patient-specific simulation approach. Healthcare Technology Letters, pp. 1–6 (2014)

21. Brown, A.G., Shi, Y., Marzo, A., Staicu, C., Valverde, I., Beerbaum, P., Lawford, P.V., Hose, D.R.: Accuracy vs. computational time: translating aortic simulations to the clinic. J. Biomech. 45, 516–523 (2012)

22. Alimohammadi, M., Agu, O., Balabani, S., Díaz-Zuccarini, V.: Development of a patient-specific simulation tool to analyse aortic dissections: Assessment of mixed patient-specific flow and pressure boundary conditions. Med. Eng. Phys. 36, 275–284 (2014)

23. Shi, Y., Lawford, P., Hose, R.: Review of Zero-D and 1-D Models of Blood Flow in the Cardiovascular System. BioMed. Eng. OnLine 10, 33 (2011)

24. Gijsen, F., Van de Vosse, F.N., Janssen, J.D.: The influence of the non-Newtonian properties of blood on the flow in large arteries: Steady flow in a carotid bifurcation model. J. Biomech. 32, 601–608 (1999)

25. Levesque, M.J., Liepsch, D., Moravec, S., Nerem, R.M.: Correlation of endothelial cell shape and wall shear stress in a stenosed dog aorta. Arteriosclerosis 6, 220–229 (1986)

26. Tedgui, A., Lever, M.J.: Filtration through damaged and undamaged rabbit thoracic aorta. Am. J. Phys. 247, H784–H791 (1984)

27. Bird, R.B., Stewart, W.E., Lightfoot, E.N.: Transport Phenomena. John Wiley and Sons (2007)

28. Sun, N., Wood, N.B., Hughes, A.D., Thom, S.A.M., Xu, X.Y.: Influence of pulsatile flow on LDL transport in the arterial wall. Ann. Biomed. Eng. 35, 1782–1790 (2007)

29. Gao, F., Guo, Z., Sakamoto, M., Matsuzawa, T.: Fluid-structure Interaction within a Layered Aortic Arch Model. J. Biol. Phys. 32, 435–454 (2006)

30. Gerdes, A., Joubert-Hübner, E., Esders, K., Sievers, H.H.: Hydrodynamics of aortic arch vessels during perfusion through the right subclavian artery. Ann. Thorac. Surg. 69, 1425–1430 (2000)

31. Ku, D.N., Giddens, D.P., Zarins, C.K., Glagov, S.: Pulsatile flow and atherosclerosis in the human carotid bifurcation. Positive correlation between plaque location and low oscillating shear stress. Arteriosclerosis 5, 293–302 (1985)

Identification of Biologically Significant Elements Using Correlation Networks in High Performance Computing Environments

Kathryn Dempsey Cooper, Sachin Pawaskar, and Hesham H. Ali

College of Information Science and Technology,
University of Nebraska at Omaha, NE 68182 USA
{kdempsey,spawaskar,hali}@unomaha.edu

Abstract. Network modeling of high throughput biological data has emerged as a popular tool for analysis in the past decade. Among the many types of networks available, the correlation network model is typically used to represent gene expression data generated via microarray or RNAseq, and many of the structures found within the correlation network have been found to correspond to biological function. The recently described gateway node is a gene that is found structurally to be co-regulated with distinct groups of genes at different conditions or treatments; the resulting structure is typically two clusters connected by one or a few nodes within a multi-state network. As network size and dimensionality grows, however, the methods proposed to identify these gateway nodes require parallelization to remain efficient and computationally feasible. In this research we present our method for identifying gateway nodes in three datasets using a high performance computing environment: quiescence in *Saccharomyces cerevisiae*, brain aging in *Mus Musculus*, and the effects of creatine on aging in *Mus musculus*. We find that our parallel method improves runtime and performs equally as well as sequential approach.

Keywords: high performance computing, parallel algorithms, correlation networks, gateway nodes.

1 Introduction

As the popularity of network modeling for big biological data grows, the need for algorithms and methods that can analyze these data grows with it. Network modeling in biological data came of age in 2001 with the finding of small world property in complex system [12]; protein-protein interaction networks were one of the models analyzed. Then came the structure-function correspondence: in PPI's, hub nodes are speculated to be linked with essential genes or proteins [3, 11, 12]; nodes in a clique tend to correspond to proteins in complex [3,7,10,16] , and the disassortativity of hubs could suggest that hub proteins are ancestral in nature [17]. The correlation network, where genes are represented as nodes, finds some measure of correlation between gene expression patterns to determine a relationship [13]. For example, linear relationships can be captured by the Pearson Correlation coefficient; networks built using this

F. Ortuño and I. Rojas (Eds.): IWBBIO 2015, Part II, LNCS 9044, pp. 607–619, 2015.

measure have been found to tend toward assortativity [17], to have a lower hub lethality rate [5], and to contain clusters whose manipulation suggests that the expression system is robust to minor changes [6,7].

The goal of the identification of gateway nodes is to identify the key genes in mechanistic changes between states. Gene expression experiments, particularly where sample size is large, provide an ideal experimental setup where comparison of states (treated, untreated or different time points) can occur while other key parameters are held consistent (tissue type, organism type and strain, etc). As such, in this research, we identify three datasets and the gateway nodes between the states found within them. Then, we take this gateway node analysis approach and parallelize it.

The recent integration of high performance computing approaches and bioinformatics or biomedical informatics methods approaches have allowed for massive strides in systems biology, or the identification of the mechanistic dynamics of a system as a whole. Previous work in this area, for example, has improved sequence assembly via Energy Aware parallelization, which minimizes energy and computational resources while improving runtime [21]. This marriage of computing and biological expertise is critical in the advancement of technologies designed to diagnose and prevent diseases, and as such, continued research in this area is critical.

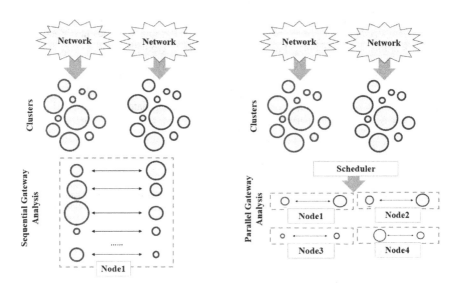

Fig. 1. The sequential versus naively parallel gateway nodes analysis. On the left, we have two networks, which after clustering, need to have a sequential pairwise comparison of clusters. In the parallel approach, a scheduler (the master node) takes the number of jobs and distributes them evenly among nodes (worker nodes).

2   Experimental Suite

In this research, three datasets are presented to highlight the computational and biological power of the parallel gateway analysis. Known datasets were drawn from NCBI's Gene

Expression Omnibus [9]. The first uses a model organism, *Saccharomyces cerevisiae*; this dataset is chosen for the vast array of knowledge available about the yeast organism, which allows for a more confident biological assessment of the gateway node functionality without actually performing any experiments *in vivo*. The second, GSE5078, is one of the datasets used in the original gateway node analysis; this dataset is used largely to determine if the same gateway nodes are identified sequentially versus in parallel. The final dataset is chosen for its large network size (more specifically, larger number of clusters) to highlight the scalability of the parallel method.

- *GSE5078*: Generated by Verbitsky *et al.* 2004 [14]; this dataset includes expression data from Bl/6 mice hippocampus separated into two groups: Young (YNG), at 2 months, and Middle-Aged (MID) at 15 months. Both sets have 9 samples.
- *GSE8542*: Generated by Aragon *et al.* 2008 [18]; this dataset includes expression data from BY4742 yeast separated into two groups: quiescent (QUI) or non-quiescent (NON).

RandomClique: Six sets of "clusters" made by random generation of 100 cliques between the sizes of 5 and 100 nodes. The clusters were grouped into six sets, R1, R2, R3, R4, R5, and R6, all consisting of 100 cliques each. Comparisons of the faux networks were performed in the following matchups: R1 vs. R2 (R1-R2), R3 vs. R4 (R3-R4), and R5 vs. R6 (R5-R6).

2.1 Network Creation and Manipulation

Networks were created by pairwise calculation of the Pearson Correlation coefficient (as described in [8]) with a correlation (ρ) threshold of 0.85 to 1.00; hypothesis testing was performed using the Student's t-test and values with p-value > 0.0005 were thrown out. The resulting network uses gene probes as nodes and correlated expression patterns as edges. As a creation quality check, the networks were checked for duplicate and self-edges; none were found.

Clustering of the networks was performed with AllegroMCODE v.1.0 [16]. Nodes with degree less than 15 were not used in cluster finding, and a scoring cutoff of 0.2 (the default) was used. Clusters with a minimum K-core of 10 were found using a maximum search depth of 10.

2.2 Gateway Node Identification

Per each dataset, gateway nodes are calculated as described in Dempsey 2014 and briefly here: Networks are first clustered to identify the dense groups within the network, and then the clusters are compared to determine if any nodes are shared between them. If nodes are shared, the number of edges between them and the clusters they connect are determined to calculate a gatewayness score. This gatewayness score is calculated as:

$$gatewayness_{nodeA} = \frac{\deg ree_{nodeA}}{\deg ree_{all\,gatewaynodes}}$$ (Equation 1)

In this equation the gateway node A being studied is defined as any node shared between two clusters of a different state and the total degree of all gateway nodes is the sum of the degree of any node shared between two clusters of a different state. If node A is the only gateway node between two clusters 1 and 2 and has a degree of 50, the gatewayness score will be 50/50 = 1.00, or 100%. If there are two gateway nodes A and B, where the degree of A is 45 and the degree of B is 55, the gatewayness of A will be 45/(45+55) = 0.45 or 45%, and the gatewayness of B will be 55/(45+55) = 0.55 or 55%. Thus, gatewayness is a measure of the responsibility of a node's connectivity between two clusters of a different state.

One way to reduce the runtime of the gateway nodes analysis in large networks is by only allowing clusters of a certain density to be analyzed; for example, if a network has 100 clusters, a density filter can be imposed (say, where the edge density of the cluster is used to remove clusters); in previous studies, using a cluster density filter of 65% can remove up to 60% of the clusters analyzed. However, it is most beneficial to compare all possible clusters instead of imposing further restriction (and thus possibly removing more biological information), which our parallel algorithm approach allows for.

Fig. 2. Parallel implementation process flow diagram

2.3   High Performance Computing Environment

The gateway node analysis is an easily parallelizable problem – the algorithm takes a pair of clusters, compares the nodes between them, and when nodes overlap between clusters, calculates the edge intersection between the two clusters. The runtime for this analysis increases in linear time increases when the size, density, or number of clusters increases. However, the problem can be scheduled to different processors by simply determining how many comparisons need to be made and then delegating them to respective worker nodes from one master.

As shown in Figure 1, the sequential approach and parallel approach differ only in the determination of gateway nodes. First, networks are created or downloaded (the networks are assumed to originate from the same set of probes – genes, gene products, proteins, etc., or such that nodes can be paired together according to some mapping function). Next, networks are clustered – using any type of clustering function desired – and the resulting clusters are forwarded to the gateway analysis. In this study, we use a specific clustering approach known for its identification of small, dense clusters (MCODE), but any type of clustering can be made. Since this approach borrows from previous studies, the same clustering method used in Dempsey *et al.* 2013 was

```
Int main(int argc, char **argv)
{
    int rank;
    MPI_Init(argc, argv);
    MPI_Comm_rank(MPI_COMM_WORLD, &rank);
    if (rank == 0) {
        Init();
        exec_master(); // Builds tasks & sends to worker
    } else {
        exec_worker();
    }
    MPI_Finalize();
}
```

```
static void exec_worker(void) {
    char rundate[16], runtime[16], cmd[256];
    Work work;
    MPI_Status status;
    while (true) { // Receive a message from the master
        MPI_Recv(&work,1,Worktype,MASTER,MPI_ANY_TAG,MPI_COMM_WORLD,&status);
        if (status.MPI_TAG == EXITTAG) { // Check tag of the received message.
            return;
        }
        pBlProg->BuildCommandString(&work, cmd);
        ExecuteTask(cmd);

        // Send the result back to the master task
        strcpy(work.sNodeName, sProcessorName);
        work.iNode = RANK;
        strcpy(work.sRunDate, rundate);
        strcpy(work.sRunTime, runtime);
        work.iET = sw.ElapsedTime();
        MPI_Send(&work,1,Worktype,MASTER,WORKTAG,MPI_COMM_WORLD);
    }
}
```

Fig. 3. Pseudo-code of parallel implementation

used for comparison. Finally, the gateway analysis approach uses anywhere from 1-64 nodes to identify gateway nodes using code written in Perl.

2.4   Parallel Implementation

The input dataset, consists of cluster files as mentioned above which are stored in their respective directories. Let us say that Organism1 cluster files are in Dir1 and contains **m** cluster files, and Organism2 cluster files are in Dir2 and contain **n** cluster files. The scheduling engines master process reads creates tasks for gateway analysis by comparing these files against each other. It takes two clusters as input and outputs any gateway nodes and their scores; a wrapper is used sequentially to run the script and deliver all possible combinations of clusters. The Big O of our parallel approach is O (m*n). The master thread sends each task with the 2 files as input to worker processors running gateway analysis algorithm. The master thread manages the execution order of the gateway analysis step. Figure 2 below shows the process flow of our parallel implementation and the pseudo Code of this implementation is shown Figure 3. The code was implemented on the Tusker Cluster described below as well. Tusker is a 40 TF cluster consisting of 106 Dell R815 nodes using AMD 6272 2.1GHz processors, connected via Mellanox QDR Infiniband and backed by approximately 350 TB of Terascala Lustre-based parallel filesystem. All experiments were run on this cluster.

Table 1. Top ten gateway nodes for mouse and yeast networks

Network	ID	Dataset	Nodes	Edges	Density	Clusters	Clustering Runtime
Young	YNG	GSE5078	12368	72967	0.095%	35	31.434 seconds
Middle-Aged	MID	GSE5078	12340	79176	0.104%	36	20.298 seconds
Non-quiescent	NON	GSE8542	1541	2515	0.212%	11	1.793 seconds
Quiescent	QUI	GSE8542	2543	5363	0.166%	62	2.671 seconds

3   Results

The results of our naively parallel gateway node analysis study are below. Table 1 describes the network sizes, edge density, number of clusters, clustering parameters, and clustering runtime. While the numbers of nodes and edges differ greatly due to difference in genome sizes, the density of the networks are relatively similar, and all networks are sparse. Using the same parameters to identify clusters in each network reveals a similar number of clusters in the mouse network (35 in the YNG and 36 in the MID) compared to the yeast network which has a more varied number (11 in the NON and 62 in the QUI). Clustering runtime appears to have no relationship with density, but rather seems to be linked to overall network size via edge count.

Table 2. Top ten gateway nodes for mouse and yeast networks

MOUSE - 0% Density		MOUSE - 65% Density		YEAST – 0% Density	
Gene ID	Gatewayness Score:	Gene ID	Gatewayness Score:	Gene ID	Gatewayness Score:
Map3k2	100.00%	Sla	100.00%	MCM21	33.33%
Pira1	100.00%	Matn3	100.00%	CPR5	33.33%
Ace	100.00%	Dio1	100.00%	TIM11	33.33%
Cts7	100.00%	Fbp1	100.00%	YGR164W	33.33%
Six3	100.00%	Ceacam12	100.00%	CBP4	33.33%
Immp1l	100.00%	Ptprb	100.00%	RPL1B	33.33%
Ythdf2	100.00%	Plin4	100.00%	GTR2	25.00%
Krt25	100.00%	Cldn1	100.00%	HGH1	25.00%
Tsks	100.00%	Akr1c21	100.00%	CRH1	25.00%
Vil1	100.00%	Ltc4s	100.00%	CLC1	25.00%

3.1   Model Organism – *S. cerevisiae* Gateway Nodes

There were 97 gateway nodes identified in the sequential and all parallel runs of the yeast network dataset; there were no gateway nodes with a score of 1.00. The gateway nodes identified in each respective run did not change with processor number. The density threshold used for yeast was 0%, meaning that any clusters that overlapped with one another were considered. While the yeast networks are relatively small, in larger networks, this all to all comparison with no density filter is desired. A density filter is typically used to reduce the amount of clusters to compare to improve runtime of gateway identification, but via naïve parallelization of the approach, all clusters can be compared. Further, this can be used to determine the distribution of gateway nodes and their relative functional impact according to cluster density, if such a relationship exists.

Gene list analysis of the gateway nodes [15] was performed using PantherDB's tool (version 8.1) [19, 20]. Gateway nodes were functionally classified according to Molecular Function, Biological Process, and Pathway. The results of these classifications are shown in Figures 4 and 5. The classifications of genes in terms of Molecular Function (MF) and Biological Process (BP) are largely standard with the majority of genes involved in metabolic processes and catalytic activity (the profile of BP and MF classification in the mouse dataset is very similar – see Figure 4). However, in the pathway classification set, telling evidence of gateway biological impact emerges. The EGF receptor signaling pathway has been implicated as an upstream regulator in astroglial cells in the transition from quiescence to reactivity [1]. The PDGF signaling pathway plays a similar role; stimulation of cell growth and proliferation; quiescence stems out of the metabolism by activation of certain elements [2]. Glycolysis, the third most pathway identified via the gateway node classification, plays a major role in the shift from non-quiescence to quiescence. Glucose levels available in media can be used to stimulate the shift from non-quiescence to quiescence; [2] suggests that this is due to the inherent changes caused in glucose metabolism when glucose is lacking or present in media.

3.2   Known Dataset – GSE5078 Gateway Nodes

There were 172 gateway nodes identified in the sequential and all parallel runs of the mouse network dataset at 0% density threshold in mice; for the 65% density threshold, 25 gateway nodes were identified. In parallel and sequential runs for both parameterizations, all gateway nodes matched. Functional classification of gateway nodes at 0% density threshold and 65% density thresholds are shown in Figures 5 and 6. The gateway nodes identified at 65% match up with those identified in Dempsey *et al*. The gateway nodes identified within this dataset have been found to be related to aging. One example of this is Klotho and Ins2 (not listed in the top 10 gateway nodes, shown in Table 2), which are involved in the insulin signaling pathway, which has long been known to be involved in biological aging.

Of the top twenty gateway nodes identified in the mouse datasets for 0% and 65% densities, nine (45%) are protein binding molecules (*Map3k2, Ace, Six3, Kr25, Vil1,Sla, Fbp1,Ptprb,* and *Cldn1*), meaning that their gene products they bind with other proteins; nearly all of the gateway genes identified are pleiotropic, or having a number of roles in the cell. This follows with the concepts proposed in Dempsey *et al*., that gateway nodes are tied to the mechanistic changes in expression that occur to restore homeostasis in changing environments within the cell.

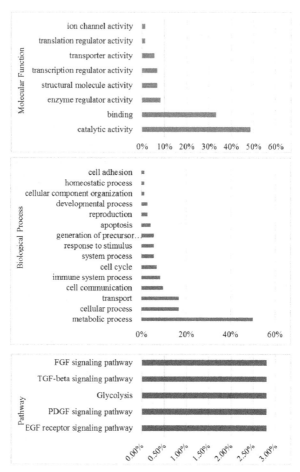

Fig. 4. The functional classifications of the yeast gateway nodes at 0% density. Blue – Biological process, Red – Molecular Function, and Green – Pathway. The axis is the percentage of genes in the gateway node list with that annotation compared to the background (mouse genome).

3.3   Scalability

The smaller model network analyses (mouse and yeast) both ran in minimal time sequentially – 144 seconds for yeast, 305 seconds for mouse at 0%, and 309 seconds at 65%. While this time requirement hardly calls for parallelization, extending the gateway node analysis into larger and more dimensional studies will require analysis of much larger networks and datasets at many more states. Systems biology approaches nearly guarantee that the data available will continue. Regardless, parallelization of the gateway node analysis in these models shows good scalability, as shown in Figure 7.

The random networks are designed to represent the scalability of these larger networks, and on this larger view, the scalability of this naively parallel approach

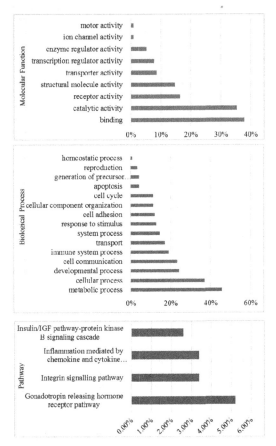

Fig. 5. Left: The functional classifications of the mouse gateway nodes at 0% density. Blue – Biological process, Red – Molecular Function, and Green – Pathway. The axis is the percentage of genes in the gateway node list with that annotation compared to the background (mouse genome).

does not disappoint. For the R1-R2 analysis, the runtime takes 68 minutes using 1 processor, and 1 minute and 25 seconds using 64 processors, a speedup of 48.6. The runtime and speedup for the random runs are shown in Figures 8 and 9. The naively parallel approach described reduces runtime, particularly as networks get larger.

4 Discussion

In recent years, modeling of high throughput biological data via network or graph theoretic modeling has emerged as a popular tool for analysis. The correlation network model, used to represent gene expression data, is one of many different types of models that rely on correlation of expression patterns to form internal graph structures. One

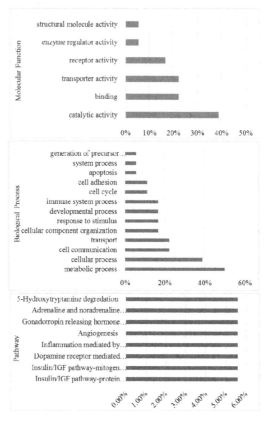

Fig. 6. Left: The functional classifications of the mouse gateway nodes at 65% density. Blue – Biological process, Red – Molecular Function, and Green – Pathway. The axis is the percentage of genes in the gateway node list with that annotation compared to the background (mouse genome).

of these structures, the gateway node, has been found to represent co-regulation with distinct groups of genes at different conditions or treatments. The structure that results typically represents 1-10% of the original network, making them a desirable target for deciphering the mechanistic changes between states or environments. As network size and dimensionality grows, however, the methods proposed to identify these gateway nodes require parallelization to remain efficient and computationally feasible. In this research we have presented our method for identifying gateway nodes in three datasets using a high performance computing environment: quiescence in *Saccharomyces cerevisiae,* brain aging in *Mus Musculus,* and the effects of creatine on aging in *Mus musculus.* The results show that our parallel method improves runtime and performs equally as well as sequential approach, meaning that as network dimensionality and size increases, we will have the tools required to analyze the entire system.

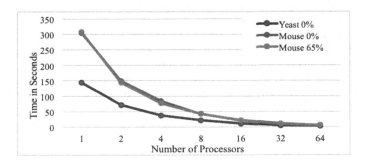

Fig. 7. Scalability for the Yeast and Mouse networks. The x-axis represents the number of processors used and the y-axis represents the time in seconds.

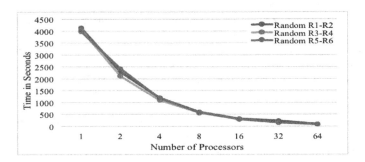

Fig. 8. Scalability for the Random networks. The x-axis represents the number of processors used and the y-axis represents the time in seconds.

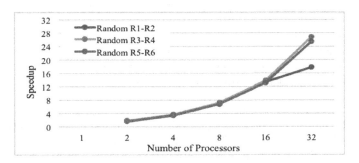

Fig. 9. Speedup for the Random Networks. The x-axis represents the number of processors

References

1. Smith, T.F., Waterman, M.S.: Identification of Common Molecular Subsequences. J. Mol. Biol. 147, 195–197 (1981)
2. May, P., Ehrlich, H.-C., Steinke, T.: ZIB Structure Prediction Pipeline: Composing a Complex Biological Workflow through Web Services. In: Nagel, W.E., Walter, W.V., Lehner, W. (eds.) Euro-Par 2006. LNCS, vol. 4128, pp. 1148–1158. Springer, Heidelberg (2006)

618   K.D. Cooper, S. Pawaskar, and H.H. Ali

3. Foster, I., Kesselman, C.: The Grid: Blueprint for a New Computing Infrastructure. Morgan Kaufmann, San Francisco (1999)
4. Czajkowski, K., Fitzgerald, S., Foster, I., Kesselman, C.: Grid Information Services for Distributed Resource Sharing. In: 10th IEEE International Symposium on High Performance Distributed Computing, pp. 181–184. IEEE Press, New York (2001)
5. Foster, I., Kesselman, C., Nick, J., Tuecke, S.: The Physiology of the Grid: an Open Grid Services Architecture for Distributed Systems Integration. Technical report, Global Grid Forum (2002)
6. National Center for Biotechnology Information, http://www.ncbi.nlm.nih.gov
7. Liu, Chen, Johns, Neufeld: Epidermal growth factor receptor activation: an upstream signal for transition of quiescent astrocytes into reactive astrocytes after neural injury. J. Neurosci. 26(28), 7532–7540 (2006)
8. Laporte, D., Lebaudy, A., Sahin, A., Pinson, B., Ceschin, J., Daignan-Fornier, B., Sagot, I.: Metabolic status rather than cell cycle signals control quiescence entry and exit. J. Cell Biol. 192(6), 949–957 (2011), doi:10.1083/jcb.201009028
9. Barabasi, A.L., Oltvai, Z.N.: Network biology: Understanding the cell's functional organization. Nature Reviews. Genetics 5(2), 101–113 (2004)
10. Bult, C.J., Eppig, J.T., Kadin, J.A., Richardson, J.E., Blake, J.A., and the members of the Mouse Genome Database Group.: The Mouse Genome Database (MGD): mouse biology and model systems. Nucleic Acids Res. 36(database issue), D724–D728 (2008)
11. Dempsey, K., Ali, H.: On the discovery of Cellular subsystems in correlation networks using centrality measures. Current Bioinformatics 7(4) (2014)
12. Duraisamy, K., Dempsey, K., Ali, H.: S. Bhowmick.: A noise reducing sampling approach for uncovering critical properties in large scale biological networks. In: High Performance Computing and Simulation 2011 International Conference (HPCS), Istanbul, Turkey, July 4-8 (2011)
13. Dong, J., Horvath, S.: Understanding network concepts in modules. BMC Systems Biology 1, 24 (2007)
14. Ewens, W.J., Grant, G.R.: Statistical methods in bioinformatics, 2nd edn. Springer, New York (2005)
15. Edgar, R., Domrachev, M., Lash, A.E.: Gene Expression Omnibus: NCBI gene expression and hybridization array data repository. Nuc. Acid Res. 30(1), 207–210 (2002)
16. Enright, A.J., Van Dongen, S., Ouzounis, C.A.: An efficient algorithm for large-scale detection of protein families. Nucleic Acids Research 30(7), 1575–1584 (2002)
17. Hao, D., Li, C.: The dichotomy in degree correlation of biological networks. PloS One 6, e28322 (2011), doi: 10.1371/journal.pone.0028322
18. Jeong, H., Mason, S.P., Barabasi, A.L., Oltvai, Z.N.: Lethality and centrality in protein networks. Nature 411(6833), 41–42 (2001)
19. Opgen-Rhein, R., Strimmer, K.: From correlation to causation networks: A simple approximate learning algorithm and its application to high-dimensional plant gene expression data. BMC Systems Biology 1, 37 (2007)
20. Verbitsky, M., Yonan, A.L., Malleret, G., Kandel, E.R., Gilliam, T.C., Pavlidis, P.: Altered hippocampal transcript profile accompanies an age-related spatial memory deficit in mice. Learning & Memory (Cold Spring Harbor, N.Y.) 11(3), 253–260 (2004)
21. Subramanian, A., Tamayo, P., Mootha, V., Mukherjee, S., Ebert, B., Gilette, M., Paulovich, A., Pomeroy, S., Golub, T., Lander, E., Mesirov, J.P.: Gene set enrichment analysis: A knowledge-based approach for interpreting genome-wise expression profiles. Proc. Natl. Acad. Sci. 102(43), 15545–15550 (2005)

22. Yoon, J.S., Jung, W.H.: A GPU-accelerated bioinformatics application for large-scale protein interaction networks. APBC poster presentation (2011)
23. Newman, M.: Assortative Mixing in Networks. Phys. Rev. Lett. 89(20), 208701 (2002)
24. Aragon, A.D., Werner-Washburne, M.: Characterization of differentiated quiescent and non-quiescent cells in yeast stationary-phase cultures. Mol. Biol. Cell 19(3), 1271–1280 (2008)
25. Miu, H., Muruganujan, A., Thomas, P.: PANTHER in 2013: Modeling the evolution of gene function, and other gene attrbutes, in the context of phylogenetic trees. Nucl. Acids Res. 41(database issue), D377–D386 (2012)
26. Thomas, P., Kejariwal, A., Guo, N., Mi, H., Campbell, M.J., Muruganujan, A., Lazareva-Ulitsky, B.: Applications for protein sequence-function evolution data: mRNA/protein expression analysis and coding SNP tools. Nuc. Acids Res. 34(suppl. 2), W645-W650
27. Pawaskar, S., Warnke, J., Ali, H.: An energy-aware bioinformatics application for assembling short-reaads. In: High Performance Computing Systems, HPCS 2013, pp. 154–160. IEEE (2012)

Enhancing the Parallelization of Non-bonded Interactions Kernel for Virtual Screening on GPUs

Baldomero Imbernón, Antonio Llanes, Jorge Peña-García, José L. Abellán,
Horacio Pérez-Sánchez, and José M. Cecilia

Bioinformatics and High Performance Computing Research Group (BIO-HPC),
Computer Science Department,
Universidad Católica San Antonio de Murcia (UCAM), Spain
{bimbernon,allanes,jpena,jlabellan,hperez,jmcecilia}@ucam.edu

Abstract. Virtual Screening (VS) methods can considerably aid clinical research, predicting how ligands interact with drug targets. Most VS methods suppose a unique binding site for the target, usually derived from the interpretation of the protein crystal structure. But it has been demonstrated that in many cases, diverse ligands interact with unrelated parts of the target and many VS methods do not take into account this relevant fact.However, this fact increases the computationally complexity exponentially. In this work we enhance the parallelization of non-bonded interactions kernel for VS methods on Nvidia GPU architectures. We show several parallelization strategies that lead to a speed up factor of 15x compared to previous GPU implementations.

Keywords: Virtual Screening, GPUs, HPC.

1 Introduction

The discovery of new drugs can enormously benefit from the use of Virtual Screening (VS) methods [4]. The different approaches used in VS methods differ mainly by the way they model the interacting molecules but all of them have in common that they screen databases of chemical compounds containing up to millions of ligands [3]. Larger databases increase the chances of generating hits or leads, but the computational time needed for the calculations increases not only with the size of the database but also with the accuracy of the chosen VS method. Fast docking methods with atomic resolution require a few minutes per ligand [15], while more accurate molecular dynamics-based approaches still require hundreds or thousands of hours per ligand [14]. Therefore, the limitations of VS predictions are directly related to a lack of computational resources, a major bottleneck that prevents the application of detailed, high-accuracy models to VS.

In most of the VS methods the biological system is represented in terms of interacting particles. For the calculation of the interaction energies, classical potentials are commonly used, separated into bonded and non-bonded terms. The

F. Ortuño and I. Rojas (Eds.): IWBBIO 2015, Part II, LNCS 9044, pp. 620–626, 2015.
© Springer International Publishing Switzerland 2015

latter describe interactions between all the elements of the system. The relevant non-bonded potentials used in VS calculations are the Coulomb and the Lennard-Jones potentials, since these describe very accurately the most important short and long range interactions between protein and ligand atoms [5].

In VS methods the most intensive computations are spent in the calculation of non-bonded kernels. For example, in Molecular Dynamics it takes up to 80% of the total execution time [7]. Thus this part can be considered as a bottleneck, and it has been shown that its parallelization and optimization [10] permits VS methods to deal with more complex systems, simulate longer time scales or screen larger databases.

High Performance Computing (HPC) solutions like GPUs [6,11] have demonstrated they can increase considerably the performance of the different VS methods, as well as, the quality and quantity of the conclusions we can get from screening [12]. In this work, we enhance the simulation of the Lennard-Jones potentials kernels of virtual screening on Graphics Processing Units. Our starting point is the work described in [12]. Then, we provide two different new alternatives for this kernel. The former evenly distributes the atoms into warps to avoid warp divergences and increase the Kepler's SMx occupancy. The latter is a new design that focuses on a region of interest to calculate the potential instead of calculating the interactions with the whole protein. Although this latter approach compromises the quality of the results, the computationally time for the simulation is decreased drastically.

The rest of the paper is structured as follows: First, we introduce the sequential baselines of the targeted kernel before our CUDA parallel designs are introduced to the reader. Then, a preliminary experimental results are shown to finish with some conclusions and directions for future work.

2 The Lennard-Jones Potential on the Graphics Processing Units

This Section briefly summarizes the Lennard-Jones potential calculation kernel, beginning from the sequential code, and analyzing different CUDA alternative designs. We first briefly review the main characteristics of CUDA [9], for the benefit of readers who are unfamiliar with the programming model. CUDA is based on a hierarchy of abstraction layers; the *thread* is the basic execution unit; threads are grouped into *blocks*, each of which runs on a single multiprocessor, where they can share data on a small but extremely fast memory. A *grid* is composed of blocks, which are equally distributed and scheduled among all multiprocessors. The parallel sections of an application are executed as *kernels* in a SIMD (Single Instruction Multiple Data) fashion, that is, with all threads running the same code. A kernel is therefore executed by a grid of thread blocks, where threads run simultaneously grouped in batches called *warps*, which are the scheduling units.

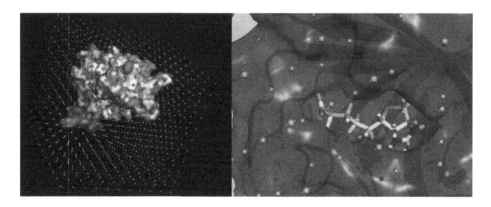

Fig. 1. (A) Representation of the grid for the protein streptavidin. Length of the side of the cube (L) is 50 Å, spacing between grid points d is 5 Å, and the total number of grid points is equal to 11^3. (B) Biotin in the binding pocket of streptavidin.

2.1   The Roadmap for the Optimization

In our previous work, we pointed out the main bottleneck of Virtual Screening methods are related to the computation of full non-bonded interactions Kernels and how GPUs can yield speedups of up to 260 times [2]. Nevertheless, those kernels need to perform up to N^2 interaction calculations (being N = total number of particles in the system) and even using GPUs, the required computation time grows polynomially with N, which definitely limits the successful simulation of large systems.

Taking a look at different algorithmic alternatives, we decided to evaluate grid kernels as an alternative to enhance the computational complexity of our simulations [8]. In [13], we empirically demonstrated that up to 200x speedup factor could be obtained by those grid kernels, even though they were executed in sequential architectures. An additional 30x speedup factor was provided by applying GPUs to these grid kernels.

We briefly summarizes the calculation of non-bonded interactions using grids (see Algorithm 1). The protein is placed inside a cube of minimal volume $Vol = L^3$ that encloses it. A three dimensional grid is created dividing the cube into $(N-1)^3$ smaller cubes of identical volume, each one of side length $d = L/N$, so that the total number of grid points is N^3. A graphical depiction of the grid for streptavidin can be seen in Figure 1 (A) and in more detail for the ligand biotin on its binding pocket in Figure 1(B).

Once the protein grid is loaded into memory, the calculation of the Lennard-Jones potential for the protein-ligand system is performed as follows; for each ligand atom i with charge q_i at point P_i we calculate which are the eight closest protein grid point neighbours. Next, an interpolation procedure is applied to estimate the value of the Lennard-Jones potential due to all protein atoms at P_i. The same procedure is applied to all ligand atoms summing them up.

Algorithm 1. Sequential pseudocode using grids for the calculation of the Lennard-Jones potentials

1. **for** $i = 1$ to $N\_simulations$ **do**
2.   **for** $j = 1$ to $nlig$ **do**
3. $index = positionToGridCoordinates(VDWGRidInfo, j)$
4.   **for** $k = 0$ to $numNeighbours(VDWGRid[index])$ **do**
5. $vdwTerm+ = vdwEnergy(j, VDWGRid[index][k])$
6.   **end for**
7.   energy[i * nlig + j] =vdwTerm;
8.   **end for**
9. **end for**

Algorithm 2. GPU pseudocode for the calculation of the Lennard-Jones potentials

1. **for all** nBlocks **do**
2. $rlig = rotate(clig[myAtom], myQuaternion)$
3. $ilig = shift(myShift, rlig)$
4. $index = positionToGridCoordinates(VDWGridInfo, ilig)$
5.   **for** $k = 0$ to $numNeighbours(VDWGRid[index])$ **do**
6. $vdwTerm+ = vdwEnergy(ilig, VDWGRid[index][k])$
7.   **end for**
8. $energy\_shared[myAtom] = vdwTerm$
9. $totalEnergy = parallelReduction(energy\_shared)$
10.   **if** $threadId == (numThreads\%nlig)$ **then**
11. $energy[mySimulation]+ = totalEnergy$
12.   **end if**
13. **end for**

2.2   Our Departure GPU Design

The Algorithm 2 shows the pseudocode of the Lennard-Jones potentials, where $VDWGridInfo$ and $VDWGrid$ are the grid description and the grid data, both stored in the GPU global memory. Each thread calculates the energy of a single atom. Each thread applies the rotation and displacement corresponding to the simulation over the ligand model in order to obtain the current atom position (lines 2-3). Then, it calculates the grid position, calculates the lennard-Jones potential using the neighbors stored in the $VDWGrid$ and stores the result in shared memory (lines 4-9) . The parameters needed by the $VDWEnergy$ procedure is previously stored in the GPU constant memory. Finally, threads of the same simulation sum up their results by a parallel reduction (line 9) and one of these threads accumulates the final result in global memory (lines 11-12).

2.3   Improving Memory Usage and Throughput

This section summarizes two alternative GPU designs. The former reduces the memory usage for our simulation. One of the main problem in docking simulation is the memory usage whenever large proteins are simulated to interact with ligands. Even more so for our docking perspective as it scans the full protein surface using several different ligand conformation.

In our previous design, the full protein is stored in GPU device memory as all the protein's atoms actually interact with every atom of the ligand. However, this interaction can be limited to those atoms that are closer to the ligand as they have more influence in whole kernel calculation. In this way, the neighborhood to a particular ligand's conformation is stored at a given time. The number of atoms in the neighborhood is a degree of freedom which may affect to both: performance and accuracy.

The latter design is based on the computation distribution among all threads that execute a given kernel. In this design, a conformation is identified to a warp not like our previous design which identified a conformation per block. In this way, we can take advantage of the new shuffle instructions that allows inter-thread communication among threads within the same warp.

3 Experimental Results

This section briefly shows the experimental results obtained with our implementations. First of all we review our experimental set up.

In this work, we target a Kepler-based architecture Tesla K40c. Kepler is the last generation of Nvidia GPU architecture [1]. Compared to previous designs (Fermi), it extends the number of cores within a multiprocessor from 32 to 192 and the scheduling units from 2 to up to 8 warps at a time. In addition the L2 cache doubles its size. Moreover, it also introduces new capabilities for irregular computation like dynamic parallelism and Hyper Q. Despite the resource additions (resulting in a far higher transistor count per unit area), Kepler GPUs are three times more power-efficient than previous generations. This is mainly achieved keeping the frequency below 1 GHz and using a manufacturing process of 28 nm. It has up to 2880 cores running at 0.88 GHz, giving a raw processing power up to 5068 GFLOPS. The memory speed is 3,0 GHz with a 384-bits memory bus width that provides a bandwidth of 288 GB/sec. The memory size is 12 GB of GDDR5 with ECC capabilities.

On the software side, The CUDA programming model is used to program the Nvidia Tesla K40c. More precisely, CUDA toolkit 6.5 leverages the Nvidia architecture. Moreover, we carried out VS calculations for the direct prediction of binding poses using four different ligands, taken form the Protein Data Bank (PDB), that conveniently represent chemical diversity of large compound databases.

The Figure 3 shows the execution times for our GPU implementations. Our departure point it was able to deal with the whole molecule but it developed to leverage previous generation of Nvidia GPU architectures. Depending of the protein and ligand size our both alternatives behave differently. The former approach is the less accurate solution although it also consume 80% less memory than the others. The latter is however the fastest implementation and it fully reproduce the non-bonded interactions.

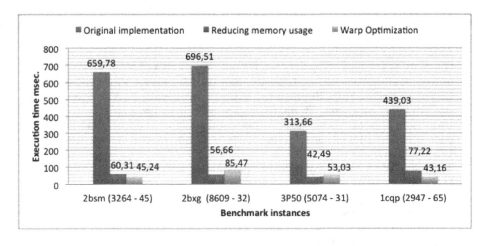

4   Conclusions and Outlook

In this work we have presented different GPU implementations to enhance the most expensive part of Virtual Screening methods, i.e. the interactions between all the elements of the system, and particularly, the Lennard-Jones potential. Our experimental results reveal that the optimization on new generation of GPUs can lead to an important speed-up factor compared to previous implementations. Moreover, the realistic simulation protein-ligand interactions can exceed the memory capacity of the GPU and thus clever ideas need to be implemented to place all the information on reduced GPU memories.

In the next steps we want to include these implementations on our docking program called BINDSURF to deal with the simulation of larger systems. Lastly, we are also working on improved scoring functions to include efficiently metals and aromatic interactions, and a GPU based method, already developed in our group for implicit solvation models.

Acknowledgements. This work has been funded by grants from the Fundación Séneca of the Región of Murcia (18946/JLI/13) and by the Nils Coordinated Mobility under grant 012-ABEL-CM-2014A, in part financed by the European Regional Development Fund (ERDF). We also thank Nvidia for the hardware donation under CUDA Teaching Program. Moreover, this work was also partially supported by the computing facilities of Extremadura Research Center for Advanced Technologies (CETACIEMAT), funded by the European Regional Development Fund (ERDF). CETACIEMAT belongs to CIEMAT and the Government of Spain. The authors also thankfully acknowledge the computer resources and the technical support provided by the Plataforma Andaluza de Bioinformática of the University of Málaga.

References

1. AMD. Nvidia corporation. the kepler architecture (2013)
2. Guerrero, G.D., Pérez-Sánchez, H., Wenzel, W., Cecilia, J.M., García, J.M.: Effective Parallelization of Non-bonded Interactions Kernel for Virtual Screening on GPUs. In: Rocha, M.P., Rodríguez, J.M.C., Fdez-Riverola, F., Valencia, A. (eds.) PACBB 2011. AISC, vol. 93, pp. 63–69. Springer, Heidelberg (2011)
3. Irwin, J.J., Shoichet, B.K.: ZINC–a free database of commercially available compounds for virtual screening. Journal of Chemical Information and Modeling 45(1), 177–182 (2005)
4. Jorgensen, W.L.: The Many Roles of Computation in Drug Discovery. Science 303, 1813–1818 (2004)
5. Jorgensen, W.L., Chandrasekhar, J., Madura, J.D., Impey, R.W., Klein, M.L.: Comparison of simple potential functions for simulating liquid water. The Journal of Chemical Physics 79(2), 926–935 (1983)
6. Kadau, K., Germann, T.C., Lomdahl, P.S.: Molecular Dynamics Comes of Age: 320 Billion Atom Simulation on BlueGene/L. International Journal of Modern Physics C 17(12), 1755–1761 (2006)
7. Kuntz, S.K., Murphy, R.C., Niemier, M.T., Izaguirre, J., Kogge, P.M.: Petaflop computing for protein folding. In: In Proceedings of the Tenth SIAM Conference on Parallel Processing for Scientific Computing, pp. 12–14
8. Meng, E.C., Shoichet, B.K., Kuntz, I.D.: Automated docking with grid-based energy evaluation. Journal of Computational Chemistry 13(4), 505–524 (1992)
9. NVIDIA Corporation. NVIDIA CUDA C Programming Guide 6.5 (2014)
10. Pérez-Sánchez, H., Wenzel, W.: Optimization methods for virtual screening on novel computational architectures. Curr. Comput. Aided Drug. Des. 7(1), 44–52 (2011)
11. Prakhov, N.D., Chernorudskiy, A.L., Gaiin, M.R.: VSDocker: A tool for parallel high-throughput virtual screening using AutoDock on Windows-based computer clusters. Bioinformatics 26(10), 1374–1375 (2010)
12. Sánchez-Linares, I., Pérez-Sánchez, H., Cecilia, J.M., García, J.M.: High-throughput parallel blind virtual screening using bindsurf. BMC Bioinformatics 13(suppl. 14), S13 (2012)
13. Sánchez-Linares, I., Pérez Sánchez, H.E., García, J.M.: Accelerating grid kernels for virtual screening on graphics processing units. In: PARCO, pp. 413–420 (2011)
14. Wang, J., Deng, Y., Roux, B.: Absolute Binding Free Energy Calculations Using Molecular Dynamics Simulations with Restraining Potentials. Biophys. J. 91(8), 2798–2814 (2006)
15. Zhou, Z., Felts, A.K., Friesner, R.A., Levy, R.M.: Comparative performance of several flexible docking programs and scoring functions: enrichment studies for a diverse set of pharmaceutically relevant targets. Journal of Chemical Information and Modeling 47(4), 1599–1608 (2007)

Prediction of Functional Types of Ligands for G Protein-Coupled Receptors with Dynamically Discriminable States Embedded in Low Dimension

Yu-Hsuan Chen and Jung-Hsin Lin

School of Pharmacy, National Taiwan University, Taipei, 100, Taiwan
Research Center for Applied Sciences, Academia Sinica, Taipei, 115, Taiwan
Institute of Biomedical Sciences, Academia Sinica, Taipei, 115, Taiwan
jlin@ntu.edu.tw, jhlin@gate.sinica.edu.tw

Abstract. In principle, the differential dynamics of a protein perturbed by various ligands should be able to reflect ligands' different functions. However, in the field of G protein-coupled receptor (GPCR), the phenomenon of conformational heterogeneity, i.e., the sharing of conformations traversed by differently liganded receptors, poses a challenge for delineating ligand's action on perturbing protein dynamics. In a previous work, we have conduct multiple molecular dynamics (MD) simulations of the agonists- and antagonists-bound human A_{2A} adenosine receptor ($A_{2A}AR$) starting from an intermediate state conformation to maximize the sensitivity of ligand-perturbed dynamics. Conformational heterogeneity can be visualized directly by the Markov state model (MSM) analysis, which is a two-stage procedure first by performing clustering based on conformational similarity to form microstates and then kinetic lumping based on state inter-convertibility to aggregate microstates into macrostates. To delineate the geometric properties of these macrostates, we embedded them onto the low dimensional space constructed with a non-linear dimensionality reduction scheme. While the crystal structures of the G-protein coupled receptor in different states (fully active, intermediate, inactive) can be projected onto divisible regions in the first two dimensions of the isomap embedding, conformations from three "purer" states (agonist-enriched, apo-enriched, antagonist-enriched) cannot be very clearly separated with this two-dimensional embedding. Dimensionality higher than two may still be needed to specify dynamically discriminable states even with nonlinear dimensionality reduction techniques.

1   Introduction

The time evolution of a biomolecular system in explicit solvent typically involves degrees of freedom of several tens of thousands to even millions. However, the dynamics of biomolecules are usually highly cooperative and it has been shown possible to define collective variables of much smaller numbers to depict the major events and features of the biomolecular dynamics systems.[1]

Non-linear dimensionality reduction (NLDR) belongs to one important class of general approaches to detect the dimensionality of the intrinsic manifold and to provide the projection scheme to do the embedding to the low dimension. Several classical approaches, e.g.,

F. Ortuño and I. Rojas (Eds.): IWBBIO 2015, Part II, LNCS 9044, pp. 627–634, 2015.
© Springer International Publishing Switzerland 2015

isomap,[2] diffusion map,[3] multiscale singular value decomposition (MSVD),[4,5] etc, have become popular in of pattern recognition, stellar spectra, and other fields in the past decade.

There have been several enhanced sampling methods with various biasing schemes to accelerate the convergence of sampling in the constructing of free energy landscape, e.g., metadynamics,[6] adaptively biasing force (ABF) method,[7] accelerated molecular dynamics (AMD),[8] etc. However, these methods are either only applicable to relatively small systems if moderate computing resources are available, or only provide limited improved sampling efficiency. For example, we have not seen simulations of protein folding with the AMD method.

To mitigate the problem of efficient searching in high dimensions, we will first adopt techniques in non-linear dimensionality reduction (NLDR) to embed the high dimensional phase points onto the lower-dimensional manifolds. Several classical approaches, e.g., isomap, diffusion map, multiscale singular value decomposition (MSVD), etc, have become popular in of pattern recognition, stellar spectra, and other fields in the past decade. However, the applications of such techniques in the molecular simulations are still quite limited. Besides, sometimes these popular NLDR techniques do not generate consistent results. Figure 1 shows two examples of NLDR with isomap. From Figure 1(b) &(e), it can be seen that isomap successfully identify the intrinsic dimensionality of both swiss roll and sphere to be 2. However, the low dimension embedding with isomap on the sphere data reveals serious distortions.

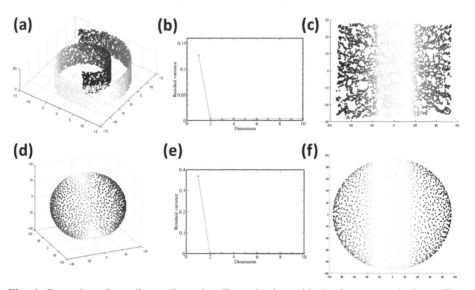

Fig. 1. Examples of non-linear dimensionality reduction with the Isomap method. (a) The three-dimensional plot of the Swiss Roll data. (b) Residual variance versus embedding dimensionality of the Swiss Roll data. (c) Embedding of the three dimensional Swiss Roll data onto the low dimensional manifold. The coloring sequences and the local metrics are conserved from the mapping of (a) to (c). (d) The three-dimensional plot of a hemisphere. (e) Residual variance versus embedding dimensionality of the hemisphere data. (c) Embedding of the three dimensional hemisphere data onto the low dimensional manifold. The coloring sequences are conserved from the mapping of (d) to (f), but the local metrics are seriously distorted.

Table 1. Comparison of multiscale singular value decomposition and isomap on various data sets.   The data set can be retrieved at http://www.math.duke.edu/~mauro/code.html#MSVD.

Source Data	MSVD	Isomap
9-d sphere with little noise	9	10
9-d sphere with large noise	9	10
9-d sphere with large noise, single point	10	10
Sphere and segment	3	3
Spiral and plane	3	3
47-dimensional sphere, 8000 points, large noise	5	6
Meyerstaircase	1	3
6-D Cube with low sampling rate	6	6
10-D cube with high noise	10	10
9-D sphere with high noise	9	10
Two lines and a plane	3	5 (3 when K=33)
Isomap Faces	3	3
CBCLFaces1	5 ~ 7	6
CBCLFaces2	4 ~ 7	6
HandVideo (shrink, due to memory problem)	3 ~ 7	2 or 6
FaceVideo	4 ~ 7	3
ScienceNews	15	15
Patches from image of Lena	4	4

Table 1 shows the comparison of isomap and a more recent approach, dubbed "multiscale singular value decomposion", on 18 very different datasets. It was shown that isomap sometimes would predict the wrong intrinsic dimensionality (mild mistakes shown in blue, severe mistakes shown in red). Isomap employs the kNN algorithm to approximate the local geodesics, and therefore it requires some input parameters, namely, k, or the distance cutoff, ε. The results of Table 1 indicate that MSVD can help isomap to decide the suitable parameter. However, the MSVD scheme does not provide (yet!) a readily accessible approach for low dimension embedding. Therefore, one possible approach could be to combine MSVD and isomap to do the low dimension embedding.

2   Adenosine Receptor

Adenosine receptor is a member of class A GPCRs. In humans, there are four major subtypes of adenosine receptor named A1, A2A, A2B, and A3 adenosine receptors. Each of them possesses different functions. A1 and A3 receptors are coupled to the Gi protein, and therefore the activation of A1 and A3 receptors results in the inhibition of adenylyl cyclase (AC) and the reduced formation of cyclic adenosine monophosphate (cAMP). On the other hand, A2A and A2B receptors are coupled to the Gs protein. Activation of A2A and A2B receptors leads to the stimulation of AC, and therefore increases the formation of cAMP.

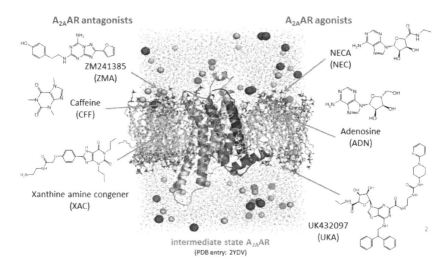

Fig. 2. The crystal structure of human A_{2A} adenosine structure bound with NECA (PDB ID: 2YDV) was adopted as the initial conformations of the receptor in all molecular dynamics simulations. The chemical structures of the ligands used in these simulations are depicted.

Adenosine (Figure 2) is the endogenous substrate of adenosine receptors. Except for the role in energy transfer, adenosine receptors also play important roles in the central nervous system (CNS), and thus are attractive targets for developing drugs for neurological diseases. The A_{2A} adenosine receptor ($A_{2A}AR$) has been considered a potential therapeutic target for the treatment of Parkinson's disease, Huntington's disease, schizophrenia, pain, depression, drug addiction, and so on.

Caffeine is the world's most widely consumed psychological stimulant which has profound antagonistic effects on adenosine receptors. Adenosine itself or its derivatives have been used clinically since the 1940s. The United States Food and Drug Administration (FDA) approved a selective $A_{2A}AR$ agonist Regadenoson for clinical use as a coronary vasodilator. A potent and selective $A_{2A}AR$ antagonist is currently in phase II/III clinical trials for Parkinson's disease. A dual-function compound that targets both $A_{2A}AR$ and the adenosine transporter has also been proved for its therapeutic potential in Huntington's disease (HD).

3   Markov Model Analysis

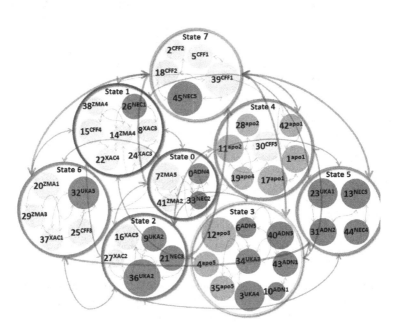

Fig. 3. The macrostates (hollow circles) and microstates (solid circles) emerged from the Markov model analysis of the 35 liganded and apo A_{2A} AR molecular dynamics simulations

4   Conformations in Different States

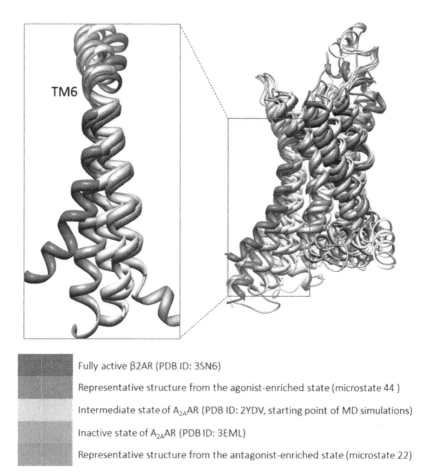

Fully active β2AR (PDB ID: 3SN6)

Representative structure from the agonist-enriched state (microstate 44)

Intermediate state of A$_{2A}$AR (PDB ID: 2YDV, starting point of MD simulations)

Inactive state of A$_{2A}$AR (PDB ID: 3EML)

Representative structure from the antagonist-enriched state (microstate 22)

Fig. 4. Superposition of the G-protein couple receptor (GPCR) structures in different states, namely, fully active, intermediate state, inactive state, and two "purer" states (agonist-enriched state and antagonist-enriched state) identified from Markov model analysis

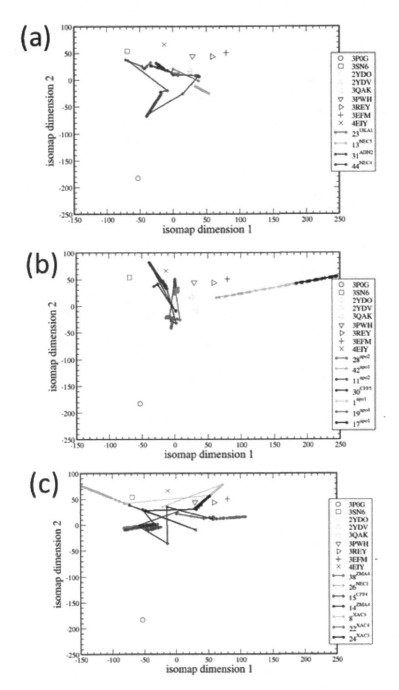

Fig. 5. Three "purer" states identified from Markov model analysis projected onto the first two dimensions with the isomap embedding method. (a) Agonist-enriched state (state 5) (b) apo-enriched state (state 4) and (c) antagonist-enriched state (state 1).

References

1. Kitao, A., Go, N.: Investigating protein dynamics in collective coordinate space. Current Opinion in Structural Biology 9, 164–169 (1999), doi:10.1016/s0959-440x(99)80023-2
2. Tenenbaum, J.B., de Silva, V., Langford, J.C.: A global geometric framework for nonlinear dimensionality reduction. Science 290, 2319–2323 (2000), doi:10.1126/science.290.5500.2319
3. Coifman, R.R., et al.: Geometric diffusions as a tool for harmonic analysis and structure definition of data: Diffusion maps. Proceedings of the National Academy of Sciences of the United States of America 102, 7426–7431 (2005), doi:10.1073/pnas.0500334102
4. Little, A.V., Lee, J., Jung, Y.-M., Maggioni, M.: Multiscale Estimation of Intrinsic Dimensionality of Data Sets. In: Proceedings of Association for the Advancement of Artificial Intelligence, 26–33 (2009)
5. Little, A.V., Lee, J., Jung, Y.-M., Maggioni, M.: Estimation of intrinsic dimensionality of samples from noisy low-dimensional manifolds in high dimensions with multiscale. In: SVD Proceedings of Statistical Signal Processing, pp. 85–88 (2009)
6. Laio, A., Parrinello, M.: Escaping free-energy minima. Proceedings of the National Academy of Sciences of the United States of America 99, 12562–12566 (2002), doi:10.1073/pnas.202427399
7. Darve, E., Rodriguez-Gomez, D., Pohorille, A.: Adaptive biasing force method for scalar and vector free energy calculations. Journal of Chemical Physics 128 (2008), doi:10.1063/1.2829861
8. Hamelberg, D., Mongan, J., McCammon, J.A.: Accelerated molecular dynamics: A promising and efficient simulation method for biomolecules. Journal of Chemical Physics 120, 11919–11929 (2004), doi:10.1063/1.1755656

Improving Activity Prediction of Adenosine A$_{2B}$ Receptor Antagonists by Nonlinear Models

Fahimeh Ghasemi[1,*], Alireza Mehri[2], Jorge Peña-García[3], Helena den-Haan[3], Alfonso Pérez-Garrido[3], Afshin Fassihi[3], and Horacio Péréz-Sánchez[3,*]

[1]Advanced Medical Technology Department, Medical University of Isfahan, Isfahan, Iran
[2]Computer Science Department, Universidad Católica San Antonio de Murcia (UCAM), E30107 Murcia, Spain
f_ghasemi_82@yahoo.com, mehri@med.mui.ac.ir,
{jpena,aperez,afassihi,hperez}@ucam.edu,
helenadenhaan@gmail.com

Abstract. This study deals on estimation of ligand activity with its descriptors. So, to achieve this goal, two different approaches were implemented. In the first one, the intervals between samples were determined. But in the second method, the intervals were clustered with k-means method. Afterwards, best descriptors of each ligands were extracted with genetic algorithm. Then, observations were classified with One-Against-All method. Finally, the activity of each ligands were estimated by forty percent of samples. In the first method, AUC values were between fifty four to ninety seven percent. For second approaches, there were about ninety seven percent.

Keywords: clustering, k-means, Genetic algorithm, classification, One-Against-All.

1 Introduction

Xanthine is a common structural framework, as part of both natural and synthetic compounds, with broad therapeutic interest. Accordingly, biological effects of xanthine derivatives include a wide range of pharmacological targets, the most important of which are: xanthine oxidase, cholinesterases, phosphodiesterases; dipeptidyl peptidase, and it should be emphasized that the target in which we can maybe find more examples of active ligands are adenosine receptors (ARs).

Adenosine is a widespread and endogenous nucleoside that acts as powerful neuromodulator in the nervous system. It is also an intermediate of the metabolic pathways responsible for adenine nucleotide salvage and recycling that are critical for the maintenance of ATP (itself an adenine nucleotide) levels in all types of cells, including neurones. The majority of the physiological effects of adenosine are believed to be mediated by interaction with specific extracellular ARs, which have been classified into four subtypes: A$_1$, A$_{2A}$, A$_{2B}$, and A$_3$.

---

[*] Corresponding author.

F. Ortuño and I. Rojas (Eds.): IWBBIO 2015, Part II, LNCS 9044, pp. 635–644, 2015.

The structural similarity of xanthine derivatives with neurotransmitter adenosine has led that derivatives of this heterocyclic ring are one of the most abundant chemical classes of ligands antagonists of adenosine receptor subtypes. Small changes in the xanthine scaffold have resulted in a wide array of ligands adenosine receptor antagonists, including some compounds with selectivity over a particular receptor subtype.

Among all adenosine receptor subtypes, specific ligands shortage A_{2B} adenosine receptor subtype caused to be less well characterized. Besides, this subtype is defined as "low affinity" because it requires high micromolar concentrations of adenosine to be activated. Nevertheless, A_{2B} AR regulates a number of pathological and physiological process involving vital organs lungs, kidneys, brain, mast cells, eyes, bladder, liver and adipose or other tissues.

Therefore, the search for ligands with affinity for A_{2B} has been intense as of late, being particularly noteworthy xanthine derivatives family.

Among natural compounds adenosine A_{2B} antagonists, caffeine and theofylline only showed moderately binding affinities to this receptor. Nonetheless, modifications on 1, 3 and 8 positions of xanthine with certain alkyl or aryl groups, has given many compound with affinity to this receptor subtype and recently some of these compounds have been patented for treating inflammatory diseases.

Despite having been synthesized and studied a large number of xanthine derivatives ligands, still no available guidelines for the rational design of new potent and selective A_{2B} AR antagonists. In order to prioritize synthesis and testing of a compound, a prediction of the binding affinity ligand-receptor is very valuable information. With this aim, non-linear models are a useful tool. The aim of this study is to obtain a non-linear model capable to predict the activity of A_{2B} AR antagonists.

Prediction of binding ligand to target is the main goal of drug discovery. For achieving this aim, various computational approaches are available, and among these, virtual screening is important tool for finding the best groups. There are two categories of computational techniques for virtual screening [1-4].

1. Ligand based: It lies on knowledge of molecules that bind to the target. There are different categories for this method. One of them is machine learning. The machine learning techniques are divided to two main groups [5,6].
   - Supervised: In this method, it is assumed that a set of training data are available and classifier is designed by using the priori known information about observation.
   - Unsupervised: In this technique, class labels are not available and clustering algorithms are used for dimension reduction to separate the molecules into different groups.
2. Structure based: It is possible when 3D structural information for the target is available. This method is used for predicting the interaction of ligand-protein. It calculates the binding of a small molecule drug candidate to a target protein [7-11].

In this study, both of categories of ligand based approaches were used for extracting best ligands. First of all, it was proposed that any information about activity are available. So, K-means method was used for clustering samples by their log activities to find interval between ligands. Second then, we used determined log-activity values for intervals. After that, OAA[1] method was used for classification. So, training and test

---

[1] One Against All.

groups were chosen. The size of them are sixty and forty percent for training and test group respectively.

GA_LDA[2] was used for classification of training data. After finding the best descriptors, test group was used to estimate activity values.

So, this paper is organized as follows:

- In section II, two main parts is presented in detail. At the first, molecular properties and target were used in this study. Secondly, our methods were used to find statistical model for best ligand extraction. It includes ligand based methods using unsupervised and supervised techniques.
- In section III, the algorithm and obtained results was developed and illustrated. Our proposed methods was shown good results.
- In last section, the obtained results with some conclusions was discussed.

2   Material and Methods

2.1   Material

Data set
The data set was retrieved from the literature. Among all of the binding data for specific targets, only measurements using human A_{2B} adenosine receptor subtype cloned in HEK-293 cells and [3H]DPCPX, as the radiolabeled ligands, were considered. A rigorous curation of structural data and elimination of questionable data points was performed. They include the removal of duplicates, detection of valence violations, ring aromatization and standardization of tautomeric forms. Finally, in order to obtain a reliable datasets, incomplete or unclear data was deleted. The output data set contains 413 xanthines and deazaxanthine.

Molecular Descriptors
A 2D molecular descriptors available in the DRAGON (version 6) and MOE (version 2008.10) software has been used in the present work. They include, for instance, pure topological descriptors, walk and path counts, connectivity indices, information indices, or 2D-autocorrelations. Taking into account the structural diversity of the compounds an initial subset of descriptors was computed for each molecule from the SMILES (Simplified Molecular Input Line Entry Specification) inputting of chemical structures. By disregarding descriptors with constant or near constant values inside each class, two final subset of 403 and 146 molecular descriptors were generated by using DRAGON and MOE, respectively.

2.2   Methods

As we mentioned in the introduction, the project is based on two main parts:

1. Sample interval selection
2. Classification and estimation the range of each ligand activity

---

[2]   Genetic Algorithm _ Linear Discriminant Analysis.

Sample Interval Selection

To achieve best results, two different interval were chosen.

Determined interval.
In this way, determined interval is intended. It is explained in the next session.

Undetermined interval.
Clustering method was used for sample clustering. A principal application of this method is the classification of compound databases into groups of similar compounds. This method depends on the calculation of log activity for all ligands. K-means method was used for clustering. It aims to partition n observations into k clusters in which each observation belongs to the cluster with the shortest path [12, 13].

K-means clustering aims to partition the n observations into k ($\leq n$) clusters and it wants to minimize the within-cluster sum of squares. The objective function is:

$$argmin_s \left(\sum_{i=1}^{k} \sum_{j=1}^{m} \|x_m - \mu_i\|^2 \right) \tag{1}$$

The k-means algorithm is below.

- Determine k (k is the number of cluster)
- Initialize means
- Assign each point to nearest mean.

$$S_i^{(t)} = \left\{ x_p \colon \|x_m - \mu_i\|^2 \leq \|x_m - \mu_j\|^2 \right\} \tag{2}$$

- Update all means

$$\mu_i = \frac{1}{\left|S_i^{(t)}\right|} \times \sum_{j=1}^{m} x_j \tag{3}$$

- Go back to third step

Classification and Estimation the Range of Each Ligand Activity

After choosing interval between ligands, we want to classify data. So, OAA method was used for this part.

OAA
The OAA modeling uses a system of M classifiers, where M is the number of classes. Each model is trained to define a discriminative boundary between a particular class of samples and the remaining ones. Every model is trained with the same dataset but different class labels. All models are trained independently. For classification in each step, GA_LDA was used [14, 15].

Ggenetic algorithm is adaptive heuristic search algorithm based on the evolutionary ideas of natural selection and genetics. It mimics some of the processes observed in natural evolution. The idea with GA is to use this power of evolution to solve optimization problems.

LDA is method used in machine learning to find a linear combination of features which characterizes some classes of objects. LDA is closely related to Bayesian classifier. The distance to each class is usually calculated using Euclidean distance as follows:

$$d_{ig}^2 = \left\| (x_i - \overline{x_g})(x_i - \overline{x_g})' \right\| \tag{4}$$

3 Results

We had 413 ligands and 416 different descriptors, activity and -logactivity were calculated for each of them.

As mentioned in prior section, two different parts were carried out.

3.1 Interval Selection

Two ways were used for interval suggestion. First method is choosing interval manually. But sizes of them didn't have uniform distribution and the results were not good. So, for choosing the best interval estimation, the kmeans clustering method was used. See Table 1 and Fig 1, 2.

Table 1. Range of sample intervals and size of each group. In below table, two different interval suggestion approaches are shown. (Manually selection and clustering selection). For the first approach, determined intervals were used but for the second one, clustering method was used. As you can see, in the first method, group numbers are very different but in second method, these are close together. (Log activity values are used for sample interval).

Interval Suggestion Method		First group	Second group	Third group	Forth group	Fifth group
Manually Selection	Size	5	20	82	210	96
	Sample Interval	-4:-3.4	-3.4:-2.9	-2.9:-1.9	-1.9:-0.9	-0.9: 0.1
Clustering Selection	Size	91	74	77	89	82
	Sample Interval	-4:-2.1	-2.1:-1.7	-1.7:-1.3	-1.3:-0.8	-0.8: 0.1

3.2 Classification and Estimation the Range of Each Ligand Activity

For classification and estimation ligand activities, GA_LAD was used. We have 416 different descriptors for each ligands but all of them could not be used because of time and speed program execution. So, for feature reduction and extraction step, one of the best optimization methods (Genetic Algorithm) was used. Then, six of the best

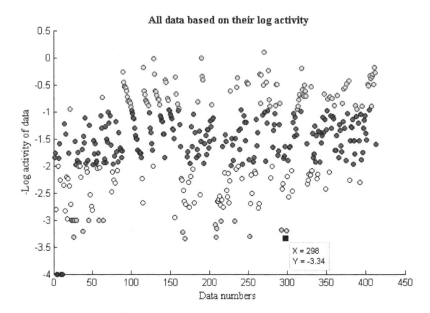

Fig. 1. Categorized data based on manually interval

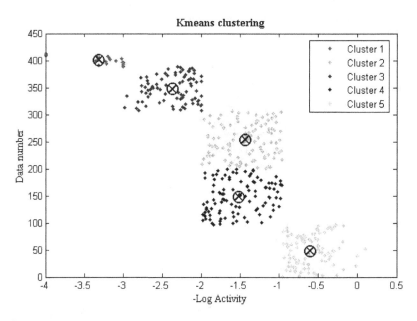

Fig. 2. Clustering data with kmeans method. It was shown that all of classes have the same members. For example first class has 91 members (red color) and forth class has 89 members (black color).

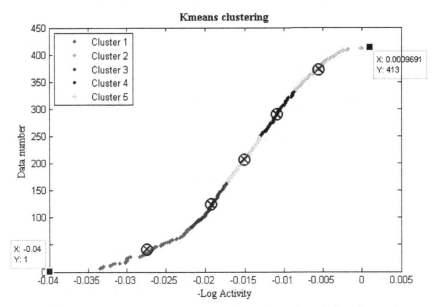

Fig. 3. Sort results after clustering. Data numbers are different from before figure. As you can see, it is interesting because this distribution is the same as Gaussian distribution and it show that other classes can also be used such as Bayesian classification.

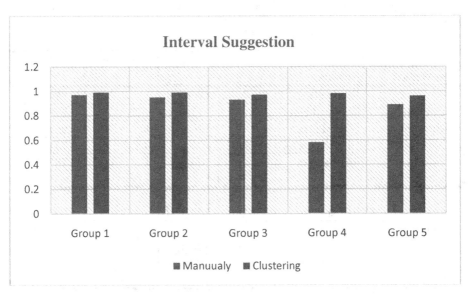

Fig. 4. AUC values for ROC curve obtained by GA_LDA for two ways interval suggestion. As expected, when members of classes are near together, the results would be better. So, we can better evaluate classification results when the number of class members are close together.

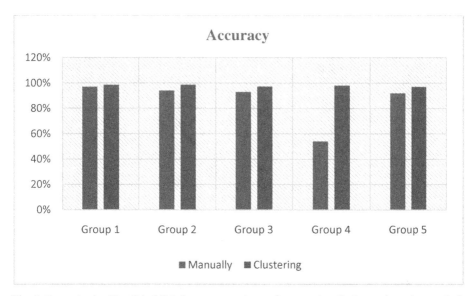

Fig. 5. Error obtained by GA_LDA for two ways interval suggestion. It shown that when optimization algorithm was used for feature selection, the results of classification would be excellent

Table 2. As it is shown in this table, some parameters were calculated for both methods. Output labels for test group is *P* (positive) or *N* (negative). If the classification prediction is *P* and the actual value is also *P*, then it is called *TP* (True Positive) but if both the prediction and the actual value are *N, TN* (Talse Negative) had been occured.

Interval Suggestion Method	Class number	TP/P	TN/N	Sensitivity	Specificity
Mannually suggestion	First class	2/2	145/163	1	0.889
	Second class	6/8	137/157	0.75	0.873
	Third class	33/33	126/132	1	0.954
	Forth class	72/84	60/81	0.857	0.741
	Fifth class	38/38	100/127	1	0.787
Clustering Selection	First class	36/36	124/130	1	0.954
	Second class	30/30	128/136	1	0.951
	Third class	31/31	126/135	1	0.933
	Forth class	34/36	128/130	0.944	0.985
	Fifth class	33/33	123/133	1	0.925

descriptors was estimated. For the objective function, error function was used. After that, for feature classification, LDA method was used. As we said in the previous section, sixty percent (60%) of data were used for training and forty percent (40%) were used for test randomly.

For comparison of two approaches, Error of output classifier, AUC value, Sensitivity and Specificity were used. These results are shown in Fig 4, 5 and Table 2.

We had 413 ligands that they were shown by their numbers in x axis. For example, the logactivity of 298th ligand is 3.34. In this figure, first interval suggestion was used. So, first class has five members only (dark blue color) and values of all of them are -4 but the forth class has 210 members (red color).

4   Discussion

When sample intervals are chosen manually, data distribution is not uniform because it is depend on target and ligands. So, size of training and test sets for all group will be different and results are not reliable. For example, group members for second and forth classes are 20 and 210 respectively. So, test group numbers for both mentioned classes are 8 and 84. Therefore, claims concerning the activity ligand prediction is not correct. Thus, depending on ligands that we have, clustering can be used for choosing interval.

Another important problem is ligand activity prediction with lowest descriptor numbers. There are lots of method can be used for optimization and classification but simplicity of model and runtime are important parameters to select method. GA is the famous method for optimization and LDA is linear classification, so all features can be found in this method.

5   Conclusions

In this study we have shown that how the GA algorithm can extract best descriptors and LDA can separate all classes. Also, we had shown clustering can give us good view to select best range interval to find active ligand. This model can be used for predicting range of activity for each ligand with no information about activity and only by its descriptors.

Acknowledgements. This work was partially supported by the Fundación Séneca del Centro de Coordinación de la Investigación de la Región de Murcia under Project 18946/JLI/13. This work has been funded by the Nils Coordinated Mobility under grant 012-ABEL-CM-2014A, in part financed by the European Regional Development Fund (ERDF).

References

1. Bleicher, K.H., Böhm, H.-J., Müller, K., Alanine, A.I.: Hit and lead generation: beyond high-throughput screening. Nature Reviews Drug Discovery 2, 369–378 (2003)
2. Bajorath, J.: Integration of virtual and high-throughput screening. Nature Reviews Drug Discovery 1, 882–894 (2002)
3. Schneider, G., Fechner, U.: Computer-based de novo design of drug-like molecules. Nature Reviews Drug Discovery 4, 649–663 (2005)

4. Hamza, A., Wei, N.-N., Zhan, C.-G.: Ligand-based virtual screening approach using a new scoring function. Journal of Chemical Information and Modeling 52, 963–974 (2012)

5. Huggins, D.J., Venkitaraman, A.R., Spring, D.R.: Rational methods for the selection of diverse screening compounds. ACS Chemical Biology 6, 208–217 (2011)

6. Butkiewicz, M., Lowe, E.W., Mueller, R., Mendenhall, J.L., Teixeira, P.L., Weaver, C.D., et al.: Benchmarking Ligand-Based Virtual High-Throughput Screening with the PubChem Database. Molecules 18, 735–756 (2013)

7. Ellingson, S.R., Baudry, J.: High-throughput virtual molecular docking with AutoDockCloud. Concurrency and Computation: Practice and Experience (2012)

8. Macarron, R., Banks, M.N., Bojanic, D., Burns, D.J., Cirovic, D.A., Garyantes, T., et al.: Impact of high-throughput screening in biomedical research. Nature Reviews Drug Discovery 10, 188–195 (2011)

9. Ellingson, S.R., Baudry, J.: High-throughput virtual molecular docking: Hadoop implementation of AutoDock4 on a private cloud. In: Proceedings of the Second International Workshop on Emerging Computational Methods for the Life Sciences, pp. 33–38 (2011)

10. Collignon, B., Schulz, R., Smith, J.C., Baudry, J.: Task-parallel message passing interface implementation of Autodock4 for docking of very large databases of compounds using high-performance super-computers. Journal of computational chemistry 32, 1202–1209 (2011)

11. Abdo, A., Salim, N.: Similarity-based virtual screening using bayesian inference network. Chemistry Central Journal 3, P44 (2009)

12. Jalali-Heravi, M., Mani-Varnosfaderani, A., Jahromi, P.E., Mahmoodi, M.M., Taherinia, D.: Classification of anti-HIV compounds using counterpropagation artificial neural networks and decision trees. SAR and QSAR in Environmental Research 22, 639–660 (2011)

13. Theodoridis, S., Pikrakis, A., Koutroumbas, K., Cavouras, D.: Introduction to Pattern Recognition: A Matlab Approach: A Matlab Approach. Access Online via Elsevier (2010)

14. Jalali-Heravi, M., Mani-Varnosfaderani, A., Valadkhani, A.: Integrated One-Against-One Classifiers as Tools for Virtual Screening of Compound Databases: A Case Study with CNS Inhibitors. Molecular Informatics (2013)

15. Plewczynski, D., Spieser, S.A., Koch, U.: Assessing different classification methods for virtual screening. Journal of Chemical Information and Modeling 46, 1098–1106 (2006)

Support Vector Machine Prediction
of Drug Solubility on GPUs

Gaspar Cano[1], José García-Rodríguez[1], Sergio Orts-Escolano[1], Jorge Peña-García[2], Dharmendra Kumar-Yadav[3], Alfonso Pérez-Garrido[2], and Horacio Péréz-Sánchez[2,*]

[1]Dept. of Computing Technology, University of Alicante, PO.Box. 99.
E03080. Alicante, Spain
{jgarcia,gcano}@dtic.ua.es
[2] Bioinformatics and High Performance Computing Research Group (BIO-HPC)
Computer Science Department, Catholic University of Murcia (UCAM)
E30107. Murcia, Spain
{jpena,aperez,hperez}@ucam.edu
[3]Department of Chemistry
University of Delhi, I110007. Delhi, India
dharmendra30oct@gmail.com

Abstract. The landscape in the high performance computing arena opens up great opportunities in the simulation of relevant biological systems and for applications in Bioinformatics, Computational Biology and Computational Chemistry. Larger databases increase the chances of generating hits or leads, but the computational time needed increases with the size of the database and with the accuracy of the Virtual Screening (VS) method and the model.

In this work we discuss the benefits of using massively parallel architectures for the optimization of prediction of compound solubility using computational intelligence methods such as Support Vector Machines (SVM) methods. SVMs are trained with a database of known soluble and insoluble compounds, and this information is being exploited afterwards to improve VS prediction.

We empirically demonstrate that GPUs are well-suited architecture for the acceleration of Computational Intelligence methods as SVM, obtaining up to a 15 times sustained speedup compared to its sequential counterpart version.

Keywords: SVM, GPU, CUDA, Bioinformatics, Computational Biology.

1 Introduction

The discovery of new drugs is a complicated process that can enormously profit, in the first stages, from the use of Virtual Screening (VS) methods. The limitations of VS predictions are directly related to a lack of computational resources, a major bottleneck that prevents the application from detailed, high-accuracy models to VS. However, the emergent massively parallel architectures, Graphics Processing Units

---

* Corresponding author.

F. Ortuño and I. Rojas (Eds.): IWBBIO 2015, Part II, LNCS 9044, pp. 645–654, 2015.

(GPU), are continuously demonstrating great performances in a wide variety of applications and, particularly, in such simulation methods [1].

The newest generations of GPUs are massively parallel processors that can support several thousand concurrent threads. Current NVIDIA [2] GPUs contain scalar processing elements per chip and are programmed using C language extensions called CUDA (Compute Unified Device Architecture) [3] . On GPUs, speedup increase reaches 100 times [4], while achieve a 200 times acceleration [5]. In NVIDIA Kepler architecture some GPUs models reached a peak performance above 3.52 TFLOPS [6].

In this paper, we focus on the optimization of the calculation of the solubility prediction using computational intelligence methods as support vector machines (SVM). SVMs are trained with a database of known soluble and insoluble compounds, and this information is being exploited afterwards to improve VS prediction.

The rest of the paper is organized as follows. Section 2 introduces the GPU architecture and CUDA programming model from NVIDIA. Section 3 shows intelligence computational methods as SVM, explains dataset and the molecular descriptors are the input for the prediction. Section 4 presents the experimentation SVM CPU vs. GPU. The performance evaluation is discussed in the Section 4. Finally, Section 5 ends with some conclusions and ideas for future work.

2   Methodology

This section describes the methods we used to improve the prediction of solubility. A computational intelligence technique such as SVM is used for the prediction of solubility and this method is trained with different datasets, and shows the GPU architecture used to accelerate the process, the dataset and molecular descriptor used.

2.1   GPU Architecture and CUDA Overview

As we mentioned previously, the process of apply computional intelligence methods to predict the solubility in large databases of elements is time consuming. To accelerate this process we propose the use of high performance hardware. GPUs are devices composed of a large number of processing units that were initially designed to assist the CPU in graphics-related calculations but have evolved in the course of time into programmable devices capable of tremendous performance in terms of standard integer and floating-point arithmetic.

When programmed correctly, GPUs can be used as general-purpose processors. Previous code for serial processors have to be readapted to these architectures, and this is not always a straightforward process. During the last four years, an increased programming feasibility and the introduction of new programming tools and languages has enabled a broader research community to exploit GPU programming.

GPUs are highly parallel devices with hundreds or thousands of cores. Special mathematical functions widely used in scientific calculations e.g. trigonometric and square root operations, are directly implemented in the hardware, allowing a fast performing of such calculations. The theoretical peak performance varies between the different GPU models, the NVIDIA GPU architecture is based on scalable processor array which streaming processors (SPs) cores organized as streaming multiprocessors (SMs) and off-chip memory called device memory. Each SM contains SPs with on-chip shared memory, which has very low access latency (see figure 1).

The CUDA programming model allows writting parallel programs for GPUs using some extensions of the C language. A CUDA program is divided into two main parts: the program which run on the CPU (host part) and the program executed on the GPU (device part), which is called kernel. In a kernel there are two main levels of parallelism: CUDA threads, and CUDA thread blocks[7] .

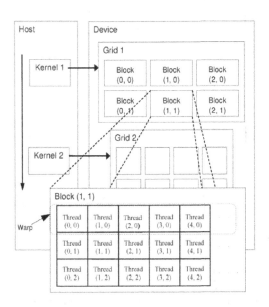

Fig. 1. Flowchart of NVIDIA GPU system

2.2   Support Vector Machines

Support vector machines (SVM) [8] are a group of supervised learning methods that can be applied to classification or regression. They represent the decision boundary in terms of a typically small subset of all training examples, called the support vectors. In a short period of time, SVM have found numerous applications in chemistry, such as drug design (discriminating between ligands and no ligands, inhibitors and no inhibitors, etc.) [9], drug discovery [10], quantitative structure-activity relationships (QSAR), where SVM regression is used to predict various physical, chemical, or biological properties) [11] chemometrics (optimization of chromatographic separation or compound concentration prediction from spectral data as examples), and sensors (for qualitative and quantitative prediction from sensor data), chemical engineering (fault detection and modelling of industrial processes) [12]. An excellent review of SVM applications in chemistry was published by Ivancicuc [13].

In our case, we exploit the idea that SVM produce a particular hyperplane in feature space that separates soluble or insoluble compounds, called the maximum margin hyperplane (see Figure 3). Most used kernels within SVM include: linear, Polynomial, Neural (sigmoid, Tanh), Anova, Fourier, Spline, B Spline, Additive, Tensor and Gaussian Radial Basis or Exponential Radial Basis.

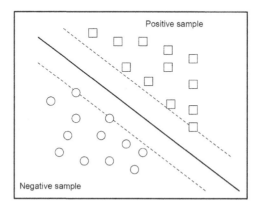

Fig. 2. Support Vector Machine margin hyperplanes

2.3   Ligand Databases and Molecular Properties

We employed standard VS benchmark tests, such as NCI (Release 4 – May 2012) is large database of molecules [14]. Next, using the ChemoPy package [15] we calculated for all ligands of the sets a diverse of molecular properties derived from the set of constitutional, CPSA (charged partial surface area) and fragment/fingerprint-based descriptors, as described in Table 1. Constitutional properties depend on very simple descriptors of the molecule that can be easily calculated just counting the number of molecular elements such as atoms, types of atoms, bonds, rings, etc. These descriptors should be able to differentiate very dissimilar molecules, but might have problems for separating closely related isomers. CPSA descriptors consider finer details of molecular structure, so they might be able to separate similar molecules, but might also have difficulties for separating isomers. Lastly, fragment and fingerprint-based descriptors take into account the presence of an exact structure (not a substructure) with limited specified attachment points. In the generation of fingerprints, the program assigns an initial code to each atom. The initial atom code is derived from the number of connections to the atom, the element type, atomic charge, and atomic mass. This corresponds to an ECFP with a neighbourhood size of zero. These atom codes are then updated in an iterative manner to reflect the codes of each atoms neighbour. In the next iteration, a hashing scheme is employed to incorporate information from each atom immediate neighbour. Each atom new code now describes a molecular structure with a neighbourhood size of one. This process is carried out for all atoms in the molecule. When the desired neighbourhood size is reached, the process is complete and the set of all features is returned as the fingerprint. For the ECFPs employed in this paper, neighbourhood sizes of two, four and six (ECFP 2, ECFP 4, ECFP 6) were used to generate the fingerprints. The resulting ECFPs can represent a much larger set of features than other fingerprints and contain a significant number of different structural units crucial for the molecular comparison, among the compounds.

Table 1. Molecular descriptors used in this study

CONSTITUTIONAL DESCRIPTORS	
Natom	Number of atoms
MolWe	Molecular Weight
NRing	Number of rings
NArRg	Numer of aromatic rings
NRotB	Number of rotatable bonds
NHDon	Number of H-bond donors
NHAcc	Number of H-bond acceptors
CPSA DESCRIPTORS	
Msurf	Molecular surface area
Mpola	Molecular polar surface area
Msolu	Molecular solubility
AlogP	Partition coefficient
FRAGMENT/FINGERPRINT-BASED DESCRIPTORS	
ECP2, ECP4, ECP6	Extended-connectivity fingerprints (ECFP)
EstCt	Estate counts
AlCnt	AlogP2 Estate counts
EstKy	Estate keys
MDLPK	MDL public keys

In order to discriminate which is the best option for discriminating between soluble and insoluble compounds in these datasets we use fingerprint-based descriptors and avoid the use of constitutional and CPSA descriptors. This is reasonable since fingerprint descriptors consider more details about the structure of molecules, being able to efficiently discriminate with more accuracy between active compounds and their decoys.

Next, we studied which combination of properties could lead to improvements on the predictive capability of these soft computing methods; it is clear that predictive capability increases (see Table 2).

Table 2. Combinations of molecular descriptors used in this study

COMBINATIONS OF CONSTITUTIONAL DESCRIPTORS	
MNBH	Molecular polar surface area (MPola)+ Number of rotatable bonds (NRotB) + Number of H-Bond acceptors (NHAcc)
MNB	Molecular polar surface area (MPola) + Number of rotatable bonds (NRotB)
NBH	Number of rotatable bonds (NRotB) + Number of H-Bond acceptors (NHAcc)
MoN	Molecular polar surface area (MPola) + Number of H-Bond acceptors (NHAcc)
MoN	Molecular polar surface area (MPola) + Number of H-Bond acceptors (NHAcc)
COMBINATIONS OF FRAGMENT/FINGERPRINT-BASED DESCRIPTORS	
EAE246	Estate counts (EstCt) + AlogP2 Estate counts (AlCnt) + Extended-connectivity fingerprints (ECFP)
EA	Estate counts (EstCt) + AlogP2 Estate counts (AlCnt)
AE246	AlogP2 Estate counts (AlCnt) + Extended-connectivity fingerprints (ECFP)
EE246	Estate counts (EstCt) + Extended-connectivity fingerprints (ECFP)

3   Experimentation SVM

In this section we outline the experimentation of SVM over R[16] packages for the computational intelligence methods both for CPU and GPU versions. We will compare the experimentation of the sequential version of SVM vs. parallel version of SVM in the prediction of solubility, with one simple molecular descriptor (ALOGP) and one complex molecular descriptor (EAE246).

3.1   Sequential version of SVM (CPU)

The package e1071 is based over the library LIBSVM [17] is currently working on some methods of efficient automatic parameter selection. It is a sequentially package that makes use of the CPU.

In R there are different implementations of SVM, in order to validate the results, we have chosen this package since it has a good performance on standardized dataset. In the Table 3, we compare the four SVM [18] implementations in terms of training time. In this comparison we only focus on the actual training time of the SVM excluding the time needed for estimating the training error or the cross-validation error. In these implementations we used the SVM package in R. The dataset used is the standard dataset from UCI Repository of Machine Learning Databases." University of California [19].

Table 3. Training times for the SVM implementations on different datasets in seconds. An AMD Athlon 1400 MHz computer running Linux was used.

Dataset	Kernlab	E1071	klaR	svmpath
spam	18.50	17.90	34.80	34.00
musk	1.40	1.30	4.65	13.80
Vowel	1.30	0.30	21.46	NA
DNA	22.40	23.30	116.30	NA
BreastCancer	0.47	0.36	1.32	11.55
HouseBoston	0.72	0.41	92.30	NA

3.2   Parallel Version of SVM (GPU)

The parallel version of SVM used is part of the package RPUDPRO [20], which is building round of a parallel version of SVMLIB. To process the parallel version of SVM, we consider that the problem of classification with SVM can be separated into two different tasks; the calculation of the kernel matrix (KM) and the core SVM task of decomposing and solving the classification model. The increasing size of the input data leads to a huge KM that cannot be calculated and stored in memory (see Algorithm 1). Therefore, the solver needs to calculate on-the-fly portions of the KM, which is a processing and memory-bandwidth intensive procedure. LIBSVM is using double precision for calculations but only the latest GPUs do support double precision. The pre-calculation is performed combining CPU and GPU to achieve maximum performance.

1: Pre-calculate on CPU the elements for each training vector.

2: Convert the training vectors array into columns wise format.

3: Allocate device memory on the GPU the training vectors array.

4: Load the training vectors array to the GPU memory.

5: FOR (each training vector) DO
- Load the training vector to the GPU.
- Perform operations with CUBLAS.
- Retrieve the dot products vector from the GPU.

END DO

6: De-allocate memory from GPU.

Algorithm 1. Parallel SVM pseudo code

As can be seen into pseudocode of the Algorithm 1, the proposed algorithm is based on recalculating on the CPU for each training vector calculation using the CUBLAS [15] library provided from NVIDIA.

4   Experimentation

The performance of sequential and GPU implementations are evaluated in a dual-core Intel E6400 (Conroe with 2 MB L2 cache), which acts as a host machine for our NVIDIA GeForce GT430 GPU. In previous papers, we show that selection of variables is very import step for the best prediction [21].

The benchmarks are executed by varying the number of molecules from the dataset and choosing different molecular descriptors as predictor for solver our classification problem (solubility or insolubility). From the simple molecular descriptors consisting of a number of small items such as CPSA descriptors (Msurf, Mpola, Msolu, AlogP) or constitutional descriptors (Natom, MolWe, NRing, NArRg, NRotB, NHDon, NHAcc) that have only a molecular features. The calculation of fragmement/fingerprint-based descriptors, vectors with some molecular characteristics, is computationally heavy. The combinations of constitutional descriptors (MNBH, MNB, NBH, MoN) or combinations of fragment/fingerprint-based descriptors (EAE246, EA, AE246) are the mix of the previous simples descriptors, and need more process time.

Results were obtained for different number of molecules from dataset, calculating the execution time with SVM with CPU and GPU versions. As predictor (input data), we choose a computational easy molecular descriptor (ALogp, Partition coefficient) and a computational heavy molecular descriptor (EAE246, Estate counts (EstCt) + AlogP2 Estate counts (AlCnt) + Extended-connectivity fingerprints (ECFP)).

We note that, the speedup factor between GPU and CPU increases faster e when the number of molecules in dataset is higher. This is because the number of molecules will be increasing the number of thread blocks running in parallel, and then the GPU

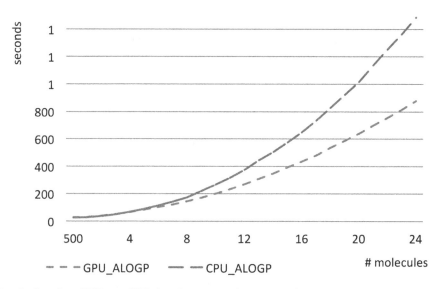

Fig. 3. Speedup GPU vs. CPU for simple descriptor (ALOGP). Increasing the runtime in seconds when increase the number of molecules (speeup x2).

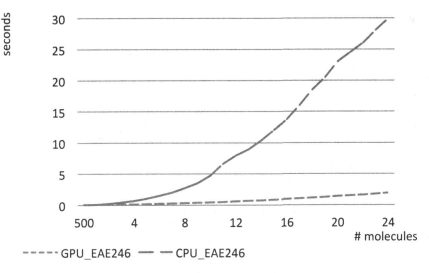

Fig. 4. Speedup GPU vs. CPU for complex descriptors (EAE246). Increasing the runtime in seconds when increase the number of molecules (speedup x15).

resources are fully used. However, it performs similarly compared to the GPU for the smallest benchmarks in which the GPU is not fully used, for the small molecular descriptor (ALOGP) and for a large number of molecules we have an speedup of 2x (GPU vs. CPU, see Figure 3). For complex molecular descriptors since a low number of molecules are use, the speedup to CPU vs. GPU is very significant. As shown in Figure 4 for large numbers of molecules and complex descriptor (EAE246), we have speedups of 15x (GPU vs. CPU).

5   Conclusions and Future Work

In this paper we have introduced the benefits of using massively parallel architectures for the optimization of prediction of solubility using computational intelligence methods as support vector machines (SVM). SVMs are trained with a database of known soluble and insoluble compounds, and this information is being exploited afterwards to improve VS prediction for different emergent parallel architectures. The results obtained for GPU are indeed promising, given the obtained speedup values up to 15, compared with the sequential version, and the progression is up.

The good results exposed open the way to use large databases for prediction of solubility, using computational intelligence methods as support vector machines (SVM). These methods fit well in the GPU architecture for a bigger numbers of molecules, or complex descriptors that are necessary to obtain good values in the prediction.

As future work we will use other computational intelligence methods, for classification, regression or selection of variables.

Acknowledgements. This work was partially funded by the projects: NILS Mobility Project 012-ABEL-CM-2014A and Fundacion Seneca 18946/JLI/13. Experiments were made possible with a generous donation of hardware from NVDIA.

References

1. Borkar, S.: Thousand core chips: A technology perspective. In: Proceedings of the 44th Annual Design Automation Conference, pp. 746–749 (2007)
2. Nvidia, W., Generation, N., Compute, C.: Whitepaper NVIDIA's Next Generation CUDA Compute Architecture, pp. 1–22
3. Nvidia, C.: Compute unified device architecture programming guide (2007)
4. Fan, X., Weber, W.-D., Barroso, L.A.: Power provisioning for a warehouse-sized computer. In: ACM SIGARCH Computer Architecture News, vol. 35(2), pp. 13–23 (2007)
5. Anderson, D.P.: Boinc: A system for public-resource computing and storage. In: Proceedings. Fifth IEEE/ACM International Workshop on Grid Computing, 2004, pp. 4–10 (2004)
6. Ruiz, A., Ujaldón, M.: Acelerando los momentos de Zernike sobre Kepler (2014)
7. Berl, A., Gelenbe, E., Di Girolamo, M., Giuliani, G., De Meer, H., Dang, M.Q., Pentikousis, K.: Energy-efficient cloud computing. Comput. J. 53(7), 1045–1051 (2010)
8. Cortes, C., Vapnik, V.: Support-Vector Networks. Mach. Learn. 20(3), 273–297 (1995)
9. Jorissen, R.N., Gilson, M.K.: Virtual Screening of Molecular Databases Using a Support Vector Machine. J. Chem. Inf. Model. 45(3), 549–561 (2005)
10. Warmuth, M.K., Liao, J., Rätsch, G., Mathieson, M., Putta, S., Lemmen, C.: Active learning with support vector machines in the drug discovery process. J. Chem. Inf. Comput. Sci. 43(2), 667–673 (2003)
11. Kriegl, J.M., Arnhold, T., Beck, B., Fox, T.: Prediction of Human Cytochrome P450 Inhibition Using Support Vector Machines. QSAR Comb. Sci. 24(4), 491–502 (2005)
12. Lee, D.E., Song, J.-H., Song, S.-O., Yoon, E.S.: Weighted Support Vector Machine for Quality Estimation in the Polymerization Process. Ind. Eng. Chem. Res. 44(7), 2101–2105 (2005)
13. Ivanciuc, O.: Applications of Support Vector Machines in Chemistry. In: Reviews in Computational Chemistry, pp. 291–400. John Wiley & Sons, Inc. (2007)
14. Voigt, J.H., Bienfait, B., Wang, S., Nicklaus, M.C.: Comparison of the NCI open database with seven large chemical structural databases. J. Chem. Inf. Comput. Sci 41(3), 702–712 (2001)
15. Cao, D.-S., Xu, Q.-S., Hu, Q.-N., Liang, Y.-Z.: ChemoPy: freely available python package for computational biology and chemoinformatics. Bioinforma 29(8), 1092–1094 (2013)
16. Team, R.C., et al.: R: A language and environment for statistical computing (2012)
17. Chang, C.-C., Lin, C.-J.: LIBSVM: A library for support vector machines. ACM Trans. Intell. Syst. Technol. 2(3), 27 (2011)
18. Hornik, K., Meyer, D., Karatzoglou, A.: Support vector machines in R. J. Stat. Softw. 15(9), 1–28 (2006)
19. Blake, C.L., Merz, C.J.: UCI repository of machine learning databases. University of California, Department of Information and Computer Science, Irvine (1998)
20. Yau: GPU Computing with R.' R Tutorial: An R Introduction to Statis, r - tutor.com/ (2014)
21. Pérez-Sánchez, H., Cano, G., García-Rodríguez, J.: Improving drug discovery using hybrid softcomputing methods. Appl. Soft Comput. 20, 119–126 (2014)

DIA-DB: A Web-Accessible Database for the Prediction of Diabetes Drugs

Antonia Sánchez-Pérez[1], Andrés Muñoz[1], Jorge Peña-García[1],
Helena den-Haan[1], Nick Bekas[2], Antigoni Katsikoudi[2], Andreas G. Tzakos[2,*],
and Horacio Péréz-Sánchez[2,**]

[1] Bioinformatics and High Performance Computing Research Group (BIO-HPC)
Computer Engineering Department, Universidad Católica San Antonio de Murcia
(UCAM), 30107 Guadalupe, Spain
hperez@ucam.edu
[2] Department of Chemistry, Section of Organic Chemistry and Biochemistry
University of Ioannina, Greece
agtzakos@gmail.com

Abstract. Diabetes mellitus is the 8th leading cause of death worldwide, with 1.5 million deaths in 2012, and medical costs that reached 245$ billion in the US. Moreover, it is estimated that roughly 387 million people worldwide suffer from diabetes mellitus, with numbers growing rapidly. These facts demonstrate the importance of creating a completely new and innovative method for the fast development of novel anti-diabetic compounds. In silico prediction methods represent an efficient approach for the prediction of diabetes drugs, aiming to explaining preclinical drug development and therefore enabling the reduction of associated time, costs and experiments. We present here DIA-DB, a web server for the prediction of diabetes drugs that uses two different approaches; a) comparison by similarity with a curated database of anti-diabetic drugs and experimental compounds, and b) inverse virtual screening of the input molecules chosen by the users against a set of protein targets identified as key elements in diabetes. The server is open to all users and is accessible at http://bio-hpc.eu/dia-db, where registration is not necessary, and a detailed report with the prediction results are sent to the user by email once calculations are finished. This is the first public domain database for diabetes drugs.

Keywords: Diabetes, Virtual Screening, Database, High Performance Computing.

1 Introduction

Diabetes mellitus is a group of chronic metabolic disorders defined by high blood sugar levels (a condition known as hyperglycemia) over a prolonged period. The two main types of Diabetes mellitus are:

---

* Corresponding author.
** Corresponding author.

F. Ortuño and I. Rojas (Eds.): IWBBIO 2015, Part II, LNCS 9044, pp. 655–663, 2015.
© Springer International Publishing Switzerland 2015

- Type I: Also known as insulin-dependent diabetes. Hyperglycemia is caused by the inability of the pancreas, due to the autoimmune destruction of beta cells, to produce sufficient amounts of insulin, a hormone that promotes the absorption of glucose by the cells and thus regulates sugar metabolism[1].
- Type II: Also known as non insulin-dependent diabetes, caused by an improper balance of glucose homeostasis and is characterized by peripheral insulin resistance (cellular insulin receptors have lost the ability of activation by insulin) combined with impaired insulin production [1].

Recent statistics show that type I diabetes accounts for 5-10% of the total cases of diabetes worldwide, whereas type II diabetes makes up to 90 % of cases. As of November 2014, an estimated 347 million people were suffering from diabetes mellitus, with rising tensions. Moreover, diabetes mellitus was the direct cause of death for roughly 1.5 million people in 2012 and it is estimated that it will be the 7th leading cause of death in 2030[2].

Because of these facts, discovering an efficient and innovative anti-diabetic medication is in the forefront of anti-diabetic research. Moreover, there is a great need for the assemble of all the important information not only about each authorized anti-diabetic drug, but also for every promising compound that is undergoing preclinical or clinical trials. This is the reason we developed DIA-DB, a pioneering online server that, on one hand, presents all the information about the activity of known anti-diabetic compounds and, on the other hand, can be an effective tool for drug design.

More specifically, DIA-DB offers 3 main options:

Compound Data: DIA-DB has a database of every compound that is either used for the treatment of diabetes or is under investigation and has shown promising results. In this database a number of crucial information about each compound is presented such as:

- Protein target
- Mechanism of action
- Compound family
- Side effects
- Binding affinity
- Crystal structure of protein + ligand
- Clinical trials among others

The database can also be used as a research tool. If a researcher has developed a new anti-diabetic drug, he will be able to compare its activity with that of the known compounds (for example, the binding affinity or the results from clinical trials) and derive useful information about its potential, as a result. Apart from that, the information mentioned above, and especially the crystallographic structures can act as a guide for drug design since the binding site-ligand interactions provides useful information for the development of specific substructures.

---

[1] http://autoimmune.pathology.jhmi.edu/diseases.cfm?
systemID=3&DiseaseID=23

[2] http://www.who.int/mediacentre/factsheets/fs312/en/

Compound Comparison: This tool has the ability of substructure comparison. As a result, a researcher can upload the structure of a potential anti-diabetic compound and DIA-DB is going to compare that structure with every compound of the database. That tool has limitless capabilities and many advantages since it is easy, quick and accurate and offers a great variety of results. As such , the researcher can learn about:

– The potential anti-diabetic activity of the desired compound
– The compounds and the drug family with which it is similar
– The protein target and the metabolic pathway
– Possible side-effects

In Silico Docking: The researcher can upload a pdb file of the desired compound and then run in silico prediction, during which docking experiments will be conducted between the compound and every known anti-diabetic protein target. This tool offers highly accurate results about the:

– Potential interaction between the compound and the protein binding site
– Exact location of the compound inside the binding site
– Specific groups of the compound and the amino-acids of the binding site that form interactions, and the exact nature of these intramolecular interactions (hydrogen bond, pi interactions etc.)
– Interactions between the compounds and other hetero-atoms inside the binding site (water if hydrophilic, metals etc.)

It is obvious that by running both of the two prediction tools mentioned above, the researcher will be able to have a full prediction of the potential anti-diabetic activity of the desired compound. In a nutshell, DIA-DB is a reliable, easily accessible free server that can be used both as an informative tool or as a guide for drug development. In the next sections we provide details about the database implementation, virtual screening protocols and examples of some use cases. The rest of the paper is as follows; section 2 explains the architecture of the database, while section 3 gives details about the model used in the creation of the database. Next, Virtual Screening techniques are explained in section 4, use cases are depicted in section 5, and principal conclusions outlined in section 6.

2   DIA-DB Architecture

In order to accomplish the different services offered by DIA-DB we have deployed a modular architecture. Figure 1 shows the main components of the DIA-DB architecture. Firstly, a main server is the responsible for distributing and coordinating user's requests received from a Web server. These requests are forwarded to the Database server either to register new experimental compounds to be evaluated or to compare the compound in the request with similar anti-diabetic drugs already registered in the database. Moreover, in the case that the request

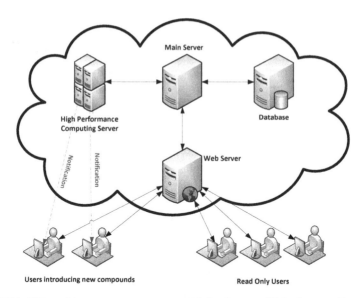

Fig. 1. DIA-DB architecture components: Main Server, Web Server, Database and High Performance Computing Server

demands an in-silico docking, the main server sends a job to the High Performance Computing (HPC) server. All these tasks are automatized by means of scripts.

The Web server acts as the interface between users and the DIA-DB system. It contains the different web pages to request comparing and/or docking services and to visualize results, as well as restricted access for administrators in order to update scripts, access the database, etc. Observe that in Figure 1 the Web server is separated from the main server for clarification reasons, and it could be perfectly integrated into the latter.

On the other hand, the Database server contains the database software and the data related to diabetes study. There could be different configuration at this point. The simplest configuration consists in a single relational database containing all the data. This is a typical solution when a small amount of data is needed and it is not distributed among different locations. A more usual intermediate configuration resides in a distributed database composed of several nodes. Note that in this case diabetes data could be stored in an heterogeneous form (for example, when using a federated database with different owners) and a common language should be defined to enable compound comparisons. A more advanced configuration combines relational databases (monolithic or distributed) with NoSQL databases [2] when very large volumes of data are available (hundreds of gigabytes) and further statistical relationships among compounds properties are to be studied. In the first version of our proposal, the database server contains a single node managed by the popular MySQL database software[3], while we are evaluating the creation of

---
[3] http://www.mysql.com

a distributed database that includes our own database along with other diabetes databases publicly available such as ZINC database[4]. In any case, the Database server is kept with a private IP for security reasons. Section 3 shows our database model proposed for representing diabetes data.

Finally, an in-house HPC server is the responsible for running the required in-silico docking calculations. Once the experiment is finished, the HPC server notifies the corresponding users directly (at this moment the notifications are sent by e-mail). Section 4 explains the fundamental features of the process executed in this HPC server.

3   Diabetes Database Model

One of the most important milestones in the DIA-DB project is the representation of the available information on anti-diabetes compounds and their trials. Based on previous studies and experiences on this field, we have designed an Entity-Relationship model following the traditional methodology in databases [3]. This section synthesizes the most relevant decisions and elements of our model.

Figure 2 depicts a simplified version of the database model used in our anti-diabetes study. One of the main entities in this model is *Compounds*, which contains the core information about them, as for example their name, structure, smiles (simplified molecular-input line-entry system), etc. A compound entity is related to several additional entities representing extended information on it, such as side-effects, the compound family, its ligands, binding protein affinity and protein targets along with its PDB associated, among others.

Another fundamental entity is *Clinical Trials*, which stores information about clinical experiments associated to each compound (note the relationship between the two main entities in the model, *Compounds* and *Clinical Trials*). This entity is also related to specific entities associated to clinical trials, such as the phases of the trial, possible interventions and conditions, the parameters involved in the study, sponsors defraying the trials and the outcomes obtained at the end of the experiment.

Although not depicted in Figure 2, there are other two main entities in the DIA-DB model aimed at representing user's experiments requested through the DIA-DB web page. These entities are *User*, containing at least the user's e-mail to contact her when the experiment is finished; and *In-Silico Experiment*, which contains data about the compound proposed by the user (reusing the *Compound* entity previously described), the state of the experiment, the HPC server where it is being executed and the prediction value obtained as a result of the experiment.

It is worth mentioning that the design of the DIA-DB knowledge model has been developed having in mind its use not only in anti-diabetes studies, but also for other diseases such as Parkinson, Cancer, etc. As a result, the core entities and relationships have been kept generic enough in order to be imported into new studies without additional effort.

---

[4] http://zinc.docking.org

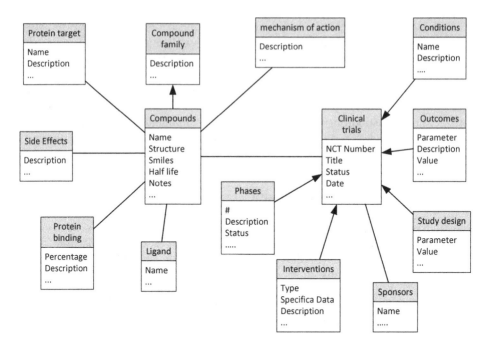

Fig. 2. Simplified Entity-Relationship model for DIA-DB

4   Virtual Screening

We describe in this section the main aspects of Virtual Screening techniques, and why and how are they used in our database.

Virtual Screening (VS) techniques [4] allow to provide predictions about which chemical compounds might interact with a given protein target in some specified way and thus achieving the desired biological function. They are mainly divided into Ligand Based Virtual Screening [5] (LBVS) and Structure Based Virtual Screening [6] (SBVS) and described below.

4.1   Ligand Similarity Based Virtual Screening

LBVS methods exploit all existing available information (structure, physico-chemical parameters, binding affinities, etc) about known active and inactive compounds. We will exploit the possibility of exploiting shape information for checking whether in our diabetes compound database are compound similar to the ones used in the input query. For that purpose our server uses internally the shape complementarity tool WEGA [7].

4.2   Structure Based Virtual Screening: Docking

The activity of many proteins strongly changes when small ligands dock into well defined cavities of protein receptors. These ligands can act as molecular switches of the protein and control its activity. For proteins involved in a metabolic pathway related to a disease, artificial ligands can act as drugs. As more and more metabolic pathways and associated key proteins are identified, the search for artificial ligands for such proteins has intensified to improve treatment of various diseases. The number of known protein structures continues to grow exponentially [8]. In addition, whole genome interaction screens and various large scale efforts to elucidate signal transduction, generate an ever increasing set of interesting target proteins.

Biomolecular screening has played an important role in drug discovery and experimental techniques are increasingly complemented by numerical simulation. One important application, so called Virtual Screening (VS), identifies lead compounds which can bind to a target protein with high affinity. This is achieved by determining the optimal binding position for each molecule in a large database of potential targets (docking) and then ranking the database according to the estimated affinity (scoring). Afterwards these candidates are subjected to refinement and selection processes which can evolve to in-vitro studies, animal investigations, and finally human trials. In addition to high-affinity the drug discovery process must address many additional issues, such as toxicity, bioactivity and specificity.

Most modern VS methods use an atomistic representation of the protein and the ligand. They permit the exploration of thousands of possible binding poses and ligand conformations in the docking process. As a result, binding modes are predicted reliably [9] for many complexes.

The VS protocol implemented in our database exploits the capabilities of the Autodock Vina [10] docking program. Vina finds well-binding ligands for a protein receptor of known structure in a database that contains the three-dimensional structures of many ligands. Each ligand of the database is docked into the whole surface of the protein using an all-atom representation of the protein and ligand.

5   Use Cases

This section describes the three most frequent use cases in DIA-DB: compound queries, compound comparisons and in-silico or virtual screening experiment. The first two involve the use of the database server, while the last one also requires the intervention of the HPC Server.

A compound query starts when a user introduces a protein sequence or other data in the DIA-DB web page. This query is sent to the main server, which forwards it to the Database server by means of a SQL query. The obtained results are sent back to the user and displayed in the web page. Likewise, compound comparison follows a similar process, but in this case the SQL query is more complex as it must search for similarity compounds according to the options selected by the user. Note that the database could be composed of a single node or several federated databases in a distributed manner, as explained in section 3.

The last use case regarding virtual screening is shown in Figure 3. In this case, the user submits a PDB file including the compound to be tested through the

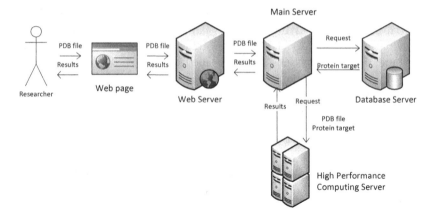

Fig. 3. Use case for in-silico docking (virtual screening) process in DIA-DB

web page. The data in the PDB file are registered in the database along with the applicant user data, and at the same time the PDB information is used for the main server to query the database in order to obtain the protein target to be evaluated in the in-silico experiment. Once this item has been retrieved, the main server sends it together with the PDB file to the HPC server as the input for the virtual screening process. After the experiment is completed, the results are registered in the database and the user is notified by e-mail so as she could view the results in detail in the web page.

6   Conclusion and Outlook

In this work we have stated the importance of creating a completely new and innovative method for the fast development of novel anti-diabetic compounds. With this purpose in mind we have created DIA-DB, a web server for the prediction of diabetes drugs that uses two different approaches grounded on similarity and structure based virtual screening. The server is open to all users and is accessible at `http://bio-hpc.eu/dia-db`, where registration is not necessary, and a detailed report with the prediction results are sent to the user by email once calculations are finished. This is the first public domain database for diabetes drugs. As future steps we are considering the extension of this database to other contexts of biological relevance such as Parkinson disease and cancer, and the implementation of additional and improved virtual screening tools for improved compound activity predictions.

Acknowledgements. This work was partially supported by the Fundación Séneca la Región de Murcia under Project 18946/JLI/13. This work has been funded by the Nils Coordinated Mobility under grant 012-ABEL-CM-2014A, in part financed by the European Regional Development Fund (ERDF).

References

1. Green, A., Christian Hirsch, N., Krøger Pramming, S.: The changing world demography of type 2 diabetes. Diabetes/Metabolism Research and Reviews 19(1), 3–7 (2003)
2. Manyam, G., Payton, M.A., Roth, J.A., Abruzzo, L.V., Coombes, K.R.: Relax with couchdb into the non-relational dbms era of bioinformatics. Genomics 100(1), 1–7 (2012)
3. Thalheim, B.: Entity-relationship modeling: foundations of database technology. Springer (2000)
4. Kitchen, D.B., Decornez, H., Furr, J.R., Bajorath, J.: Docking and scoring in virtual screening for drug discovery: methods and applications. Nature Reviews Drug Discovery 3(11), 935–949 (2004)
5. Geppert, H., Vogt, M., Bajorath, J.: Current trends in ligand-based virtual screening: Molecular representations, data mining methods, new application areas, and performance evaluation. Journal of Chemical Information and Modeling 50(2), 205–216 (2010)
6. Ghosh, S., Nie, A., An, J., Huang, Z.: Structure-based virtual screening of chemical libraries for drug discovery. Current Opinion in Chemical Biology 10(3), 194–202 (2006)
7. Ge, H., Wang, Y., Zhao, W., Lin, W., Yan, X., Xu, J.: Scaffold hopping of potential anti-tumor agents by wega: a shape-based approach. MedChemComm 5(6), 737–741 (2014)
8. Sussman, J.L., Lin, D., Jiang, J., Manning, N.O., Prilusky, J., Ritter, O., Abola, E.: Protein data bank (pdb): database of three-dimensional structural information of biological macromolecules. Acta Crystallographica Section D: Biological Crystallography 54(6), 1078–1084 (1998)
9. Huang, N., Shoichet, B.K., Irwin, J.J.: Benchmarking sets for molecular docking. Journal of Medicinal Chemistry 49(23), 6789–6801 (2006)
10. Trott, O., Olson, A.J.: Autodock vina: Improving the speed and accuracy of docking with a new scoring function, efficient optimization, and multithreading. Journal of Computational Chemistry 31(2), 455–461 (2010)

Molecular Docking and Biological Evaluation of Functionalized benzo[h]quinolines as Colon Cancer Agents

Ramendra Pratap[1,*], Dharmendra Kumar Yadav[1,*], Surjeet Singh[1], Reeta Rai[2], Naresh Kumar[3], Han-Sup Uhm[3], Harpreet Singh[4], and Horacio Péréz-Sánchez[5]

[1]Department of Chemistry, University of Delhi, North Campus, Delhi-110007, India
[2]Department of Biochemistry, All India Institute of Medical sciences, New Delhi-110029, India
[3]Plasma Bioscience Research Center, Kwangwoon University,
Nowon-Gu, Seoul 139-701, Korea
[4]Department of Bioinformatics, Indian Council of Medical Research, New Delhi-110029, India
[5]Computer Science Department, Catholic University of Murcia (UCAM) E30107,
Murcia, Spain
ramendrapratap@gmail.com, dharmendra30oct@gmail.com

Abstract. As a part of our drug discovery program, we have synthesized various 2-amino-benzo[h]quinoline-6-carbonitrile derivatives and analyzed them on human colon cancer cells in the form of percentage inhibition at different concentration gradient and time of incubation. Anticancer activity of these derivatives against the human HCT116 cancer cells using *in vitro* employing standard MTT assay. Compounds **3a-3e** showed significant anti-cancer activity especially compound **3b** and **3d** exhibit the good inhibitory activity on HCT116 cells. Additionally, docking study was performed on colon cancer target cyclindependent kinase-2 to understand the cytotoxic mechanism of action of active compounds.

Keywords: ADME/T, Benzo[h]quinoline, Colon cancer, Docking.

1   Introduction

Colorectal cancer is the fourth most commonly diagnosed malignant disease and is prevalent where the people have adopted western diets and also among the elderly. Its symptom is worsening constipation or bloody stool. It is a [1, 2] worldwide problem and is second most common cancer in women and the third in men. It is notable that a variety of research groups are paying attention on chemoprevention currently and working on new lead molecule development. It is well known that early detection of cancer and focus on improvement of current used drug can be very supportive for complete cure the disease. A careful literature search revels that the properly functionalized quinoline ring system and their fused derivatives are important structural unit

---

* Corresponding author.

F. Ortuño and I. Rojas (Eds.): IWBBIO 2015, Part II, LNCS 9044, pp. 664–673, 2015.
© Springer International Publishing Switzerland 2015

and present as substructure in various therapeutics and synthetic analogues, alkaloids, which demonstrate good biological activities [3-5]. Molecules with quinoline skelon are reported as anti-inflammatory, antimalarial, antiasthmatic, antihypertensive, antibacterial and platelet derived growth factor receptor tyrosine kinase (PDGF-RTK) inhibiting agents [6]. Various quinoline derivatives are reported to display considerable anticancer activities [7-8]. Quinoline skeleton act as anticancer agents through several mechanisms for example; inhibition of topoisomerase [9], cell cycle arrest in the G2 phase [10], and tubulin polymerization inhibition[11]. In this work, we report the five new compounds were synthesized and screened against human colon cancer cell lines. Most of compounds were found to exhibit potency compared to standard drug doxorubicin. The synthesis compounds **3a-3e** showed anti-proliferative activity, especially compound **3b** and **3d** exhibit best inhibition activity on HCT116 cell line. Molecular docking studies have been performed to evaluate possible mode of action of molecules in active site.

2 Material Methods

2.1 Chemical Synthesis

2.1.1 General Experimental Procedures
We have used commercially available reagent and solvent purchased by sigma Aldrich and Alfa aesar directly and the synthesis of all the compounds were carried out according to the procedure described in our recent publication (Figure 1). Synthetic procedure and characterization data of all the reported compound are available in the literature[12].

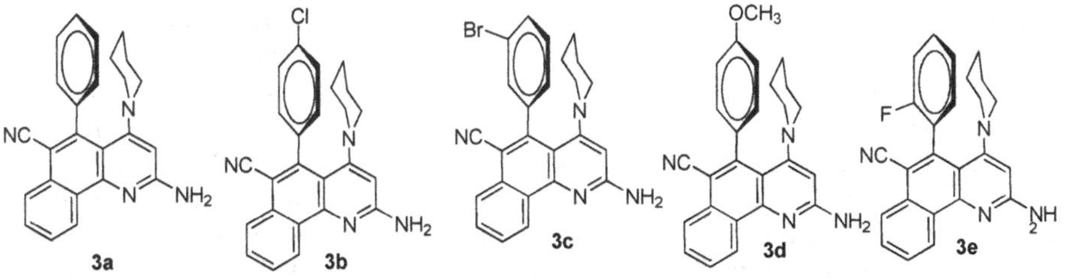

Fig. 1. Structures of benzo[h]quinoline-6-carbonitrile derivatives

2.2 Biology

2.2.1 Reagents and Consumables
Colon cancer HCT116 cell lines were procured from KCLB (Korean Cell Line Bank, South Korea), all cells were maintained in DMEM supplemented with 10% fetal bovine serum, 1% nonessential amino acids, 1% glutamine, 1% penicillin (100 IU/ml) and streptomycin (100 mg/ml) (all from Hyclone, USA). Human colon cells culture were maintained at 37°C, 95% relative humidity and 5% CO_2. The cells allowed to grow in 75 cm^2 tissue culture flasks until confluence were then sub-cultured for experimentation. The cell inhibiting activity of the compounds was determined using MTT assay

2.2.2 MTT Assay

The MTT assay was carried out as per the protocols described by us earlier [13, 14]. In brief cells were seeded in 96-well plates at a concentration of 2 x 10^4 cells/well in 200 µL of complete medium and incubated for 4 h at 37°C in a 5% CO_2 atmosphere to allow for cell adhesion. Stock solutions (2 mM/100µL) of the compounds made in DMSO were filter sterilized, then further diluted to 0.2 mg/ml incomplete medium for treatment against HCT116 colon cancer cells lines. A 100-µL solution of compound was added to a 100-µL solution of fresh medium in wells to give final concentrations of 20µM-0.156 µM /mL after serial dilution. MTT assays were performed in two independent sets of quadruplicate tests. A negative (no drug treatment) and positive control group used as vehicle that was used for dilution of compound were run in each assay. After incubation upto 48h, expose the cells to the drug and each well was carefully rinsed with 1ml phosphate buffered saline (PBS).Viability was assessed using MTT (3-[4,5-dimethylthiazol-2-yl]-2,5-diphenyltetrazolium bromide). 20 µL MTT solutions (5 mg/mL in PBS) along with 200 µL of fresh medium were added to each well in the plate and the plates were incubated for 3h. Following incubation, the medium was removed and the purple formazan precipitates in each well were released in the presence of 1ml DMSO (dimethyl sulphoxide). Absorbance was measured using a micro plate reader (Biotek, Techan, USA) at 540nm. The results are expressed as IC_{50} and percentage of viabilities of cell were assessed as the ratio of absorbance between compound treated and control cells, which was directly proportional to the metabolically active cell number.

2.3   Molecular Docking Study

Molecular modeling studies of of various benzo[h]quinolines were performed by using molecular modeling software Sybyl-X 2.0, (Tripos International, St. Louis, Missouri, 63144, USA). Drawing of structure and geometry cleaning of compounds were done through Chem Bio-Office suite Ultra v12.0 (2012) software (Cambridge Soft Corp., UK). All the five compounds were studied for their binding affinity into colon cancer target cyclin-dependent kinase 2. We have constructed a 3D model of the structures using the Surflexdoc module of Sybyl. Energy minimization was made through Tripos force field Optimization using a distance-dependent dielectric and Powell gradient algorithm with a convergence criterion of 0.001 kcal mol^{-1} for determining low energy conformations with the most favorable (lowest energy) geometry. Among various deposited crystallographic structures of human colon cancer cells in protein data bank (PDB), we have selected Cyclin-dependent kinase 2 (CDK2) [15]. The crystal structure of CDK2 in complex with 3-bromo-5-phenyl-N-(pyridin-3-ylmethyl)pyrazolo[1,5-*a*]pyrimidin-7-amine was obtained from the PDB http://www.rcsb.org/ pdb/explore/explore.do (ID: 2R3J). It was originated from Homo sapiens and its resolution was 1.65 Å. Hydrogen atoms and partial charges were added to the protein with the protonation 3D application in Sybyl. Gasteiger-Hückel method was used to assess the partial atomic charges. 2D structures were converted to 3D structures using the program Concord v4.0 and maximum number of iterations performed in the minimization was set to 2000. Geometry was further optimized through MOPAC-6 package using the semiempirical PM3 Hamiltonian method [16,17].

3   Result and Discussion

3.1   Biological Evaluation Study

3.1.1 Cytotoxicity

Benzo[*h*]quinolines (**3a- 3e**) were evaluated against HCT116 human colon cancer cell lines and only four compounds exhibited noteworthy cytotoxicity one compound shows low activity (Table 1). Rom structure activity relationship study, we have observed the presence of 4-substituted phenyl group enhances the activity. Compound **3b** and **3d** were significant amongst them exhibiting IC$_{50}$ 7.6 (±1.1) and 7.2 (±1.6) µg against HCT116. However, IC$_{50}$ of **3a, 3c** and **3e** was also good but less active than **3b** and **3d** probably due to functionalization at other position of phenyl ring, when incubation time was enhanced to 48h. The inhibitory activities for derivatives **3a–3e** were more than 50% at 0.612 µM (Figure 2).

Table 1. In-vitro cytotoxicity of benzo[*h*]quinolines by MTT assay

S. No.	Compound Name	HCT116 IC$_{50}$ (µM)
1.	3a	9.6 (±0.9)
2.	3b	7.6 (±1.1)
3.	3c	10.7 (±1)
4.	3d	7.2 (±1.6)
5.	3e	9.2 (±1.2)
6.	Doxorubicin	2.1 (±0.5)

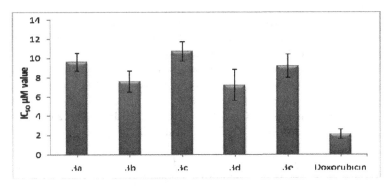

Fig. 2. Effect of active compounds (**3b, 3d** and doxorubicin) on HCT116 *in vitro* using standard endpoint MTT assay. The cells were treated at different concentration of 20 µM - 0.156 µM for different time intervals (48 h).

3.1.2 Molecular Modeling and Analysis of the Docked Results

The aim of the molecular docking study was to elucidate whether 2-amino-benzo[h]quinoline-6-carbonitrile derivatives and doxorubicin modulate the anticancer target or not. However, this study is used to identify the binding site pocket against the well-known anticancer molecular targets CDK2 apo-protein receptor. To find the possible bioactive conformations of **3b** and **3d**, the Sybyl X 2.0 interfaced with Surf-lex-Dock program was operated to dock the compounds into the active site of the known human colon cancer target protein CDK2 (PDB ID: 2R3J) [15]. To evaluate the binding affinity of the inhibitors within the protein molecules, docking study was performed. The docking energy in the form of total energy and hydrogen bonds between the ligand and amino acids in protein receptor was used to rank the binding affinity of all the synthesized compounds to colon cancer target CDK2. Evaluation of the hydrogen bonds was done by measuring length with in 3Å. This is most commonly used approach to see the molecular interaction and interlinked the experimental results very well. Through molecular docking simulations, we successfully explored the orientations and binding affinities (in terms of docking score refer here as 'total score') of 2-amino-benzo[h]quinoline-6-carbonitrile derivatives against the anticancer targets CDK2 apo-protein receptor.

Table 2. Comparison of binding affinity of standard drug doxorubicin and active benzo[h]quinoline derivatives against colon cancer target protein (PDB ID:2R3J)

Compound Name	Total Score	Amino acid involved in active pocket in 3Å	Involved group of Amino Acid	Length of H-bond Å	No. of Hydrogen Bond
Doxorubicin	5.1647	ILE-10, VAL-18, ALA-31, VAL-64, PHE-80, PHE-82, LEU-83, HIS-84, GLN-85, ASP-86, LYS-88, LYS-89, GLN-131, LEU-134, ALA-144	ASP-86 LYS-89 LEU-83	1.9 2.0 2.1	3
3b	5.5841	ILE-10, GLY-11, GLU-12, GLY-13, VAL-18, ALA-31, LYS-33, LEU-83, His-84, GLN-85, LYS-89, GLN-131, ASN-132, LEU-134, ALA-144, ASP-145	LEU-83 LYS-33	1.9 2.0	2
3d	5.2407	ILE-10, GLY-11, GLU-12, GLY-13, THR-14, VAL-18, ALA-31, LYS-33, GLU-81, PHE-82, LEU-83, HIS-84, GLN-85, ASP-86, LYS-89, LYS-129, GLN-131, ASN-132, LEU-134, ALA-144, ASP-145	LEU-83	2.0	1

Surflex-Dock scores (total scores) were expressed in $-\log 10(K_d)^2$ units to represent binding affinities

Table 3. Compliance of active benzo[h]quinoline derivatives of computational parameters of pharmacokinetics (ADME)

Compound Name	log S for aqueous solubility	log Khsa for Serum Protein Binding	log BB for brain/ blood	No. of metabolic reactions	Predicted CNS Activity	log HERG for K+ Channel Blockage	Apparent Caco-2 Permeability (nm/sec)	Apparent MDCK Permeability (nm/sec)	vdW Polar SA (PSA)	log Kp for skin permeability	% Human Oral Absorption in GI (+-20%)	Qual. Model for Human Oral Absorption
3a	-6.358	0.843	-0.69	0	0	-5.082	664.661	318.128	61.328	-2.37	100	Low
3b	-7.096	0.946	-0.55	0	0	-5.095	664.16	776.185	61.329	-2.532	100	Low
3c	-7.246	0.989	-0.56	0	0	-5.2	663.685	788.149	61.328	-2.495	100	Low
3d	-6.739	0.861	-0.81	1	-1	-5.241	664.264	317.923	69.614	-2.448	100	Low
3e	-6.632	0.873	-0.63	0	0	-5.1	663.921	453.657	61.332	-2.431	100	Low
Doxorubicin	-2.775	-0.543	-3.1	9	0	-6.268	2.15	0.716	215.962	-7.957	20	Low
Stand. Range*	(-6.5 / 0.5)	(-1.5 / 1.5)	(-3.0 / 1.2)	(1.0 / 8.0)	-2 (inactive) +2 active)	(concern below -5)	(<25 poor, >500 great)	(<25 poor, >500 great)	(7.0 / 200.0)	(-8.0 to -1.0, Kp in cm/hr)	(<25% is poor)	(>80% is high)

Note: * For 95% of known drugs, based on –Qikprop v.3.2 (Schrödinger, USA, 2011) software results # indicates a QSAR-based predicted active derivative

Table 4. Predicted ADME parameters (DS v3.5, Accelrys, USA)

Compound Name	Aqueous solubility	CYP2D6 binding	Hepatotoxicity	Intestinal absorption	Plasma Protein binding
3a	1	False	True (toxic)	0 (Good)	True (Highly bounded)
3b	1	True	True (toxic)	1 (moderate)	True (Highly bounded)
3c	0	False	True (toxic)	1 (moderate)	True (Highly bounded)
3d	1	False	True (toxic)	0 (Good)	True (Highly bounded)
3e	1	True	True (toxic)	0 (Good)	True (Highly bounded)
Doxorubicin	2	False	True (toxic)	3 (very poor)	False (Poorly bounded)

Doxorubicin (control compound) docked in to CDK2 apo-protein receptor and its docking score in the form of total score was 5.1647. Total fifteen residue in the binding site of the CDK2 had hydrophobic interaction with doxorubicin i.e. Ile-10, Val-18, Ala-31, Val-64, Phe-80, Phe-82, Leu-83, His-84, Gln-85, Asp-86, Lys-88, Lys-89, Gln-131, Leu-134 and Ala-144. Three residues i.e. Asp-86, Lys-89 and LEU-83 interacted with control compound doxorubicin via H-band. On the other hand docking result for colon cancer target CDK2 of benzo[h]quinoline derivatives **3b** and **3d** show significant binding affinity indicated by a total score 5.5841 and 5.2407 in figure 4A and 4B. There were formation of two H-bonds of length 1.9 and 2.0 Å between the protein and active derivative **3b** and one H-bond of length 2.0 Å with **3d** interact with nonpolar hydrophobic residue i.e. Leu-83 and polar basic uncharged residues Lys-83 participate in active pocket. In the docking poses of 3b, and 3d complex, the chemical nature of common residue in the binding site within radius 3Å was non polar hydrophobic residue, for example Ile-10 (Isoleucine), Val-18 (Valine), Ala-31 (Alanine),

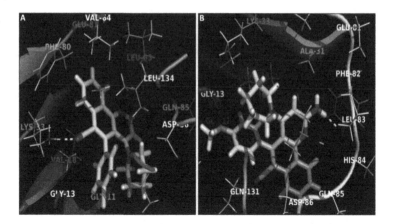

Fig. 3. Active compounds **3b** and **3d** docked on anticancer target CDK2 (PDB ID:2R3J) revealing respective binding site residues.

Leu-83, Leu-134 (Leucine); polar basic for example His-84 (Histidine), Lys-89 (Lysicine); and polar uncharged residue, for example Gln-85, Gln-131 (Glutamine) as a result the bound compounds show a strong interaction with CDK2 compare to stranded drug doxorubicin, thus leading to more stability and activity in this compounds (Table 2).

3.1.3 Pharmacokinetic Parameters Compliance

We considered several physiochemical properties related to pharmacokinetics, when screening for active derivatives. The results revealed that all of the derivatives followed Lipinski's rule of five. The hydrophilicity of each compound was measured through its logP value. Low hydrophilicity and therefore a high logP value may lead to poor absorption or permeation. The pharmacokinetics parameters such as ADME/T are important descriptors for human therapeutic use of any compounds.

These ADME descriptors were calculated and checked for compliance with their standard ranges. Screening for active and ADME compliant benzo[h]quinoline derivatives of namely, compounds **3a-3e** were evaluated through its logP (octanol-water partition coefficient), which has been implicated in logBB (blood-brain barrier penetration) penetration and permeability studies. The logP descriptor used to correlate passive molecular transport through membranes. All compounds showed less than five calculated logP and therefore may have high hydrophilicity and membrane permeation. The aqueous solubility (logS) of a compound significantly affects its absorption and distribution characteristics. The calculated logS values of the studied compounds were within the acceptable interval. The compound distribution in human was evaluated by following factors e.g., logBB, permeability (apparent Caco-2 and MDCK permeability, logKp for skin permeability), the volume of distribution and plasma protein binding refer by logKhsa (Schrödinger, USA, 2012). Process of excretion by which eliminates the

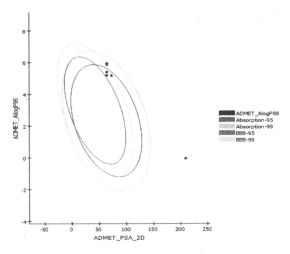

Fig. 4. Plot of polar surface area (PSA) versus ALogP for benzo[h]quinoline derivatives showing the 95% and 99% confidence limit ellipses corresponding to the blood brain barrier (BBB) and intestinal absorption

compound from the human body was evaluated by molecular weight and logP [16-17]. All derivatives have polarities that enabled better permeation and absorption, as revealed by the number of H-bond donors and H-bond acceptors. The calculated values of these ADME parameters showed close similarity with that of doxorubicin and lies within the standard range of values exhibited by 95% of all known drugs (Table 3). We have also evaluated through through Discovery Studio v3.5 molecular modeling & drug discovery software (Accelrys, USA). Compounds **3a-3e** were evaluated with standard descriptors and all the chemical descriptors and parameters of ADME were calculated (Table 4). The ADME results showed that there was no predictive hepatotoxicity was comparable to standard range.

The intestinal absorption and blood brain barrier penetration were predicted by developing an ADME model using descriptors 2D PSA and AlogP98 that include 95% and 99% confidence ellipses. These ellipses define regions where well absorbed compounds are expected to be found. The results of DS-ADME model screening showed that active benzo[*h*]quinoline derivatives **3b** and **3d** possess 99% confidence levels for human intestinal absorption and blood brain barrier (BBB) penetration. Similarly, another predicted active derivative also showed 99% confidence level for intestinal absorption and 95% confidence level for BBB penetration. Doxorubicin falls outside the ADME model ellipses filter, which indicate its poor intestinal absorption and BBB penetration ability. The plot of polar surface area and ALogP of benzo[*h*]quinoline derivatives are represented in Figure 5.

4   Conclusion

Developing promising colon cancer agents by synthesis of 2-amino-benzo[h]quinoline-6-carbonitrile derivatives functionalized at C-5 positions has been synthesized by reaction of

6-aryl-4-*sec*.amino-2-oxo-2*H*-pyran-3-carbonitriles and 2-cyanomethylbenzonitrile under basic conditions. Compounds **3a-3e** are not reported as anticancer agents and compounds 2-amino-5-(4-chlorophenyl)-4-(piperidin-1-yl)benzo[*h*]quinoline-6-carbonitrile (**3b**) and 2-amino-5-(4-methoxyphenyl)-4-(piperidin-1-yl)benzo[*h*]quinoline-6-carbonitrile (**3d**) showed promising cytotoxic activities, in *in vitro* mode, compared to the standard drug doxorubicin. This may act as source of lead anticancer drug. Binding affinity of studied derivatives revealed on known anticancer targets CDK2 through docking. All compounds showed compliance with standard range of known drug's ADME parameters

Conflict of interest: The author(s) confirm that this article content has no conflicts of interest.

Acknowledgement. Authors thank Council of Scientific and Industrial Research (CSIR, New Delhi) and UGC, New Delhi for financial support. DKY and SS thank Council of Scientific and Industrial Research (CSIR, New Delhi). RR thanks DST for supporting as woman scientist. Authors thank University of Delhi for providing research funding under R&D project and instrumentation facility. This work was partially supported by the Fundación Séneca del Centro de Coordinación de la Investigación de la Región de Murcia under Project 18946/JLI/13 and by the Nils Coordinated Mobility under grant 012-ABEL-CM-2014A, in part financed by the European Regional Development Fund (ERDF).

References

[1] Deming, D.A., Maher, M.E., Leystra, A.A., Grudzinski, J.P., Clipson, L., Albrecht, D.M., Washington, M.K., Matkowskyj, K.A., Hall, L.T., Lubner, S.J., Weichert, J.P., Halberg, R.B.: Phospholipid ether analogs for the detection of colorectal tumors. PloS One 9, e109668 (2014)

[2] Huang, W., Liu, G., Zhang, X., Fu, W., Zheng, S., Wu, Q., Liu, C., Liu, Y., Cai, S., Huang, Y.: Cost-effectiveness of colorectal cancer screening protocols in urban chinese populations. PloS One 9, e109150 (2014)

[3] Larsen, R.D., Corley, E.G., King, A.O., Carroll, J.D., Paul, D., Verhoeven, T.R., Reider, P.J., Labelle; Gauthier, J.Y., Xiang, Y.B., Zamboni, R.J.: Practical Route to a New Class of LTD4 Receptor Antagonists. The Journal of Organic Chemistry 61, 3398–3405 (1996)

[4] Roma, G., Di Braccio, M., Grossi, G., Mattioli, F., Ghia, M.: 1,8- naphthyridines IV. 9-substituted N,N-dialkyl-5-(alkylamino or cycloalkylamino) [1,2,4]triazolo[4,3-a][1,8]naphthyridine-6-carboxamides, new compounds with anti-aggressive and potent anti-inflammatory activities. Eur. J. Med. Chem. 35, 1021–1035 (2000)

[5] Chen, Y.L., Fang, K.C., Sheu, J.Y., Hsu, S.L., Tzeng, C.C.: Synthesis and antibacterial evaluation of certain quinolone derivatives. J. Med. Chem. 44, 2374–2377 (2001)

[6] Dubé, D., Blouin, M., Brideau, C., Chan, C.C., Desmarais, S., Ethier, D., Falgueyret, J.P., Friesen, R.W., Girard, M., Girard, Y., Guay, J., Riendeau, D., Tagari, P., Young, R.N.: Quinolines as potent 5-lipoxygenase inhibitors: synthesis and biological profile of L-746,530. Bioorganic & Medicinal Chemistry Letters 8, 1255–1260 (1998)

[7] Alqasoumi, S.I., Al-Taweel, A.M., Alafeefy, A.M., Noaman, E., Ghorab, M.M.: Novel quinolines and pyrimido[4,5-b]quinolines bearing biologically active sulfonamide moiety as a new class of antitumor agents. Eur. J. Med. Chem. 45, 738–744 (2010)

[8] Behforouz, M., Cai, W., Mohammadi, F., Stocksdale, M.G., Gu, Z., Ahmadian, M., Baty, D.E., Etling, M.R., Al-Anzi, C.H., Swiftney, T.M., Tanzer, L.R., Merriman, R.L., Behforouz, N.C.: Synthesis and evaluation of antitumor activity of novel N-acyllavendamycin analogues and quinoline- 5,8-diones. Bioorganic & Medicinal Chemistry 15, 495–510 (2007)

[9] Cheng, Y., An, L.K., Wu, N., Wang, X.D., Bu, X.Z., Huang, Z.S., Gu, L.Q.: Synthesis, cytotoxic activities and structure-activity relationships of topoisomerase I inhibitors: Indolizinoquinoline-5,12-dione derivatives. Bioorganic & Medicinal Chemistry 16, 4617–4625 (2008)

[10] Kim, Y.H., Shin, K.J., Lee, T.G., Kim, E., Lee, M.S., Ryu, S.H., Suh, P.G.: G2 arrest and apoptosis by 2-amino-N-quinoline-8-yl-benzenesulfonamide (QBS), a novel cytotoxic compound. Biochem. Pharmacol. 69, 1333–1341 (2005)

[11] Alqasoumi, S.I., Al-Taweel, A.M., Alafeefy, A.M., Hamed, M.M., Noaman, E., Ghorab, M.M.: Synthesis and biological evaluation of 2-amino-7,7- dimethyl 4-substituted-5-oxo-1-(3,4,5-trimethoxy)-1,4,5,6,7,8-hexahydro- quinoline-3-carbonitrile derivatives as potential cytotoxic agents. Bioorganic & Medicinal Chemistry Letters 19, 6939–6942 (2009)

[12] Singh, S., Yadav, P., Sahu, S.N., Sharon, A., Kumar, B., Ram, V.J., Pratap, R.: One-Pot Chemoselective Synthesis of Arylated Benzo[h]quinolines. Synlett 25, 2599–2604 (2014)

[13] Kumar, N., Kaushik, N.K., Park, G., Choi, E.H., Uhm, H.S.: Enhancement of glucose uptake in skeletal muscle L6 cells and insulin secretion in pancreatic hamster-insulinoma-transfected cells by application of non- thermal plasma jet. Appl. Phys. Lett. 103(20) (2013)

[14] Reth, M.: Hydrogen peroxide as second messenger in lymphocyte activation. Nat. Immunol. 3, 1129–1134 (2002)

[15] Yoon, H., Kim, T.W., Shin, S.Y., Park, M.J., Yong, Y., Kim, D.W., Islam, T., Lee, Y.H., Jung, K.Y., Lim, Y.: Design, synthesis and inhibitory activities of naringenin derivatives on human colon cancer cells. Bioorganic & Medicinal Chemistry Letters 23, 232–238 (2013)

[16] Yadav, D.K., Kalani, K., Singh, A.K., Khan, F., Srivastava, S.K., Pant, A.B.: Design, synthesis and in vitro evaluation of 18beta-glycyrrhetinic acid derivatives for anticancer activity against human breast cancer cell line MCF-7. Current Medicinal Chemistry 21, 1160–1170 (2014)

[17] Yadav, D.K., Dhawan, S., Chauhan, A., Qidwai, T., Sharma, P., Bhakuni, R.S., Dhawan, O.P., Khan, F.: QSAR and Docking Based Semi-Synthesis and In Vivo Evaluation of Artemisinin Derivatives for Antimalarial Activity. Curr. Drug Targets 15, 753–761 (2014)

Predicting Cross-Reactivity from Computational Studies for Pre-evaluation of Specific Hepatic Glycogen Phosphorylase Inhibitors

-*insilico* Screening of Drugs for Evaluating Specificity

V. Badireenath Konkimalla

School of Biological Sciences,
National Institute of Science Education and Research (NISER),
Bhubaneswar – 751 005, India
badireenath@niser.ac.in

Abstract. Over the past decade, only very few drugs made it to the market this bottleneck is mainly attributed due to their failures at the pre-clinical and clinical stages of drug development pipeline. Several bioinformatics tools and molecular docking approaches have enormously contributed to the rational design of cost-effective, efficacious and novel lead molecules. Research over past few decades resulted in the development of virtual screening methods, where several million compounds from various databases can be screened in limited time very effectively. Majority of the investigations performed so far focuses only on a single target ruling out the possibility that the same ligand can bind to an off-target with a high-sequence similarity (especially isozymes) leading to unwanted effects. Glycogen phosphorylases (GP) are phosphorylase enzymes that are implicated in several metabolic disorders such as type-2 diabetes and also other disease like cancer and coronary disease. GP are present both in muscle and hepatic tissue as isozymes and many inhibitors of GP (GPi) due to lack of specificity cross-react between both these isozymes and lead to delirious side-effects. Taking together the role and features of GP, the objective of the current work is to develop a strategy in molecular docking approach to pre-evaluate specificity of GPi using *in silico* derived pentacyclic triterpenes.

Keywords: *in silico* pharmacodynamics, glycogen phosphorylase, glycogen phosphorylase inhibitors, pentacyclic triterpenes, preclinical evaluation, cross-reactivity.

1   Introduction

Glycogen phosphorylase (GP) is a phosphorylase enzyme (EC 2.4.1.1) implicated in disease conditions such as type-2 diabetes, McArdle's disease [1], Hers' disease [2] and has been further proposed as a biomarker in gastric cancer [3]. In both liver and muscle, GP catalyzes the rate-limiting breakdown of glycogen to glucose-1-phosphate upon phosphorolytic cleavage of the R-1,4-linked glycosyl units. Type-2 diabetes is

F. Ortuño and I. Rojas (Eds.): IWBBIO 2015, Part II, LNCS 9044, pp. 674–682, 2015.
© Springer International Publishing Switzerland 2015

associated with elevated blood glucose levels as a result of disorder in the glucose metabolism by liver and periphery [4]. To counteract this condition, an ideal anti-diabetic agent is expected to be capable of lowering blood glucose in both fed and fasted states via insulin controlling the hepatic and peripheral glycogen metabolism [5]. Here, enzymes like GP being involved in gluconeogenesis and glycogenolysis, (the pathways by which glucose is formed in the liver), are potential targets for its treatment. GP exists in two interconvertible forms (*a* and *b*); the proportion that exists in each form is regulated by phosphorylation. In a study, Aiston et al. demonstrated that by inactivating glycogen phosphorylase GP*a* alone and not inhibiting glycogen kinase synthase-3 (GSK-3) could mimic insulin stimulation of hepatic glycogen synthesis. Further, signaling pathway involving dephosphorylation of GP*a* leading to both activation and translocation of glycogen synthase was an important mechanism by which insulin stimulate hepatic glycogen synthesis [6]. In this regard, inactivation of GP would not only reduce glycogenolysis but would also stimulate glycogen synthesis. GP inhibition has been regarded as a therapeutic strategy for blood glucose control in diabetes [7]. Several studies report the blood glucose lowering efficacy of GP inhibitors in animal models of diabetes [8, 9] and clinical trials [8]. A number of synthetic and natural GPi have shown good therapeutic potential for the treatment of type 2 diabetes.

Pharmacological inhibitors of GP have been developed so far are used as a potential drug for attenuating hyperglycemia associated with type 2 diabetes [10]. These compounds decrease blood glucose levels in vivo and thereby validate GP as a target for the treatment of diabetes [7]. Several structural classes of potent GP inhibitors (GPi) have been developed that bind either to the active site, the allosteric effector AMP (AMP site), the purine site or the indole site [11]. Pentacyclic triterpenes represented a new class of GPi that are widely distributed throughout the plant kingdom, and a variety of biological properties have been ascribed to this class of compounds [12]. Oleanolic acid and ursolic acid are two well-known members of the family of pentacyclic triterpenes [13, 14]. In a recent human test, it was reported for the first time that corosolic acid exhibited a glucose-lowering effect on post-challenge plasma glucose levels in human [15].

A major limitation for the current GPis is that they their lack specificity and have low degree of selectivity against muscle GP. This is due to significant sequence and structural homology between hepatic (hGP) and skeletal muscle GP (mGP) [16]. Now, it is question of great concern where any GPi, intended to inhibit hepatic glucose output, will also limit muscle glycogen degradation during contraction resulting in impaired muscle function. Therefore, novel GPi are required to be designed that are potent and highly selective hGPi. In the present study, structure based drug design strategies and molecular docking approaches are employed using pentacyclic triterpenes on the structure models of mGP and hGP. Apart from the binding energy other parameters like cluster orders, diversity in binding sites, sequence conservation were compared to obtain better lead compounds for specific targeting of hGP.

2   Materials and Methods

2.1   Data Set

The protein primary sequences for mGP (accession id: P11217, PYGM_HUMAN) and hGP (accession id : P06737, PYGL_HUMAN) were obtained from uniprot (www.uniprot.org). The X-ray crystal structure of hGP in complex with inhibitor GSK254 was obtained from Protein Data Bank (PDB) [17].

2.2   Sequence Analysis

Pairwise alignment of mGP and hGP were performed using ClustalW tool available at www.ebi.ac.uk/clustalw.

2.3   Homology Modelling

SWISS-MODEL tool was used to building and validation of homology models [18]. Pairwise sequence alignment of mGP and hGP primary sequences along with crystal structure of hGP (PDB id : 3DD1) was submitted to the SWISS-MODEL web-server. An optimized model of mGP was finally obtained upon satisfying the cut-off parameters to model the target based on the BLAST target-template alignment and the cut-off parameters to model the target based on a Search target-template alignment.

2.4   Computation Details

The chemical structures of 86 ursolic and corosolic acid derivatives were drawn using ACD/ Chemsketch (Suppl table 1). The 2D structures were converted to 3D structure using Balloon 1.1.0.800 [19]. The energy optimization was performed using MMFF94 force field for 300 generations. Such energy minimized ligand structures were taken for docking.

2.5   Molecular Docking

Autodock Vina was employed to dock the ligands and gain a better insight for the interactions between various ligands to mGP and hGP. The docking operation involves generation of grids, calculation of dock score and evaluation of conformers. A receptor and ligand file containing the coordinates in PDB or Mol2 format is a pre-requisite for performing docking using Autodock Vina. Prior to docking, the protein files were prepared by keeping the polar hydrogen atoms and adding Gasteiger charges as partial charges to the corresponding carbon atoms. The receptor PDB file was transformed to a PDBQT file containing the receptor atom coordinates, partial charges and solvation parameters. The ligand file was transformed into PDBQT file upon merging the nonpolar hydrogen atoms and defining the torsions. A blind docking was performed inorder to identify novel and/or specific ligand binding sites between mGP and hGP. The grid maps were the calculated for the entire protein. Inorder to perform a blind docking, the protein was placed in a grid box of 80 x 80 x 80 cells (with a spacing of 0.375 A°) to cover the entire protein surface. Structural visualizations and conformational analysis of rigid protein-flexible ligand docking of 86 different ligands was performed using PyMOL ver0.99.

3   Results and Discussion

Molecular targeting or specific targeting of isozymes with inhibitors is a major challenge in drug discovery. In this case, it is very important to develop specific inhibitors as GP isozymes are involved in the fundamental metabolic process (gluconeogenesis and glycogenolysis) localised in vital tissues or organs (muscle and liver). In a study on rabbit liver and muscle GP, Sprang et al crystallographic and molecular simulation data reported structural conservation irrespective of the extent of sequence similarity [20].

3.1   Sequence Analysis and Homology Modeling

In the present study, pairwise alignment of mGP (842 aa) and hGP (847 aa) primary protein sequences showed that the isozymes share approx 80% sequence identity (673 identical residues). Among the rest, 101 residues (~12%) are semi-conserved, 33 residues (~4%) are partially conserved and 40 residues (~5%) are unconserved. Designing ligands for molecular targeting in such cases are challenging as there are few residues that are specific among both the proteins. In order to understand the distribution of dissimilar residues among these two isoforms a homology model of mGP was prepared using the hGP structural template. The dissimilar residues (semi-conserved, partially conserved and unconserved) between both the protein sequences were mapped on the respective structures to identify potential specific pockets.

3.2   Molecular Docking Studies

The native inhibitor (GSK254) bound in the crystal structure of hGP was used in order to validate the mGP and hGP structure template prior to docking with 86 energy minimised ligand library using Autodock Vina. GSK254 docked to hGP and mGP structure with a best docking energy of -10.3 kcal/mol and -9.5 kcal/mol respectively. Docking analysis of 86 ligands to mGP and hGP template yielded an average docking energy -8.39 ± 0.45 kcal/mol in both cases (hGP: min docking energy = -9.7 kcal/mol, max docking energy = -7.4 kcal/mol and mGP: min docking energy = -9.4 kcal/mol, max docking energy = -7.3 kcal/mol).

As individual docking studies, the docking energies thus obtained for docking 86 library compounds for mGP and hGP are significant and closely match the docking energy of GSK254 for mGP and hGP. However, in order to address the issue about specificity for hGP the following parameters were considered to screen out potential lead ligands.

* Relatively high difference in binding energy.
* Bind to very few sites in the protein.
* Binding to unique binding pockets/sites in the protein.

Relatively High Difference in Binding Energy. A reasonable specificity of a ligand can be obtained when a ligand binds to hGP effectively and that the difference between the binding energy to mGP is relatively high. A binding energy of -8.39 kcal/mol was considered a cut-off, as this being the average binding energy obtained for docking the 86 ligands to mGP and hGP. All ligands that docked to mGP with binding energy more than -8.4 kcal/mol and the ligands that docked to hGP with binding energy less than -8.4 kcal/mol were considered as shown in table 1.

Table 1. Relative binding energy differences of ligand docked to mGP and hGP

Ligand	Binding to hGP (kcal/mol)	Binding to mGP (kcal/mol)
GSK254	-10.3	-9.5
18	-9.0	-8.0
67	-8.6	-7.8
8	-9.0	-8.4
46	-9.0	-8.4
59	-8.6	-8.1
62	-8.6	-8.1
74	-8.4	-7.9

Bind to Very Few Sites in the Protein. In molecular docking studies, specificity for a ligand is not restricted to ligand binding to its specific target protein alone but also the number of sites/pockets a ligand binds in the in the target protein. Visualisation of the number of binding sites preferred by a ligand to protein among the ligands was performed using PyMOL. Table 2 shows that about 7 ligands were selective for atleast 4 to 7 sites/pockets. Out of which, ligand 18 were observed to bind to 4 pockets in hGP and 5 pockets in mGP (fig 1).

Binding to Unique Binding Pockets/Sites in the Protein. The 20% unconserved residues (101 residues) obtained from the pairwise sequence alignment of mGP and hGP were mapped on to their 3D structure. It was observed that most of residues were distributed throughout the protein that might increase the chances of unspecific binding of the ligand. Structural investigations on the docking results showed that ligand 18 in fact bound only to those sites comprising of semi/partially conserved residues in other GP (fig 2). This is a very important parameter in dictating the specificity of a ligand for a given molecular target. Table 3 is a summary of those ligands that were observed to be docked either to the same site as in the crystal structure or to a new site. In some cases, the ligands 18, 46, 52 were found docked to two completely different binding sites/pockets in both mGP and hGP.

Table 2. Number of pockets docked by the ligands in mGP and hGP structure with the best docking energy less than -8.4 kcal/mol

Ligand	hGP	mGP
GSK254	6	5
18	4	5
67	6	4
8	5	5
46	6	4
59	7	6
62	4	5
74	6	5

Ligand-18 binding to 4 pockets in hGP   Ligand-18 binding to 5 pockets in mGP

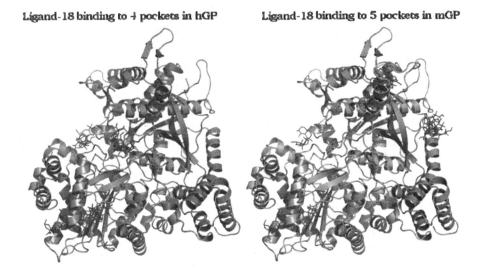

Fig. 1. Binding pockets for ligand **18**. *Right panel* : Four binding pockets for ligand **18** in hGP *Left panel* : Five binding pockets for ligand **18** in mGP

Table 3. List of ligands observed to be docked either in the same site or a new site in the structure

Ligand	hGP	mGP
GSK254	AMP site	AMP site
18	new	new
67	AMP site	new
8	AMP site	new
46	new	new
59	new	AMP site
62	new	new
74	AMP site	new

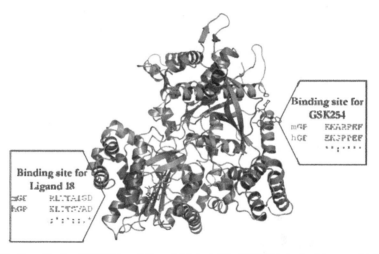

Fig. 2. Binding site for GSK254 and ligand-**18** in hGP. GSK254 docked best to highly conserved AMP binding pocket in all GP (highlighted in red on right) whereas ligand 18 docked best to novel site that is semi/partially conserved in other GP (highlighted in pink on left).

4   Conclusion

In this study, pentacyclic triterpenes were investigated using AutoDock VINA tool to delineate the parameters that are required in the ligand for designing specific inhibitors for hGP. From the molecular docking studies, it was observed that the docking energy for inhibitor GSK254 to hGP was low. However, other observations showed that GSK254 lacked specificity since the difference in the binding energy of the inhibitor to mGP and hGP was not very significant. Secondly, the number of pockets/sites docked by the inhibitor was more in hGP in comparison to mGP leading to a possibility that the inhibitor might alter several downstream pathways due to allosterism.

Thirdly, GSK254 was docked at the same site in both mGP and hGP where the residues are highly conserved.

On the other hand, in this study the results from the molecular docking of 86 ligands on the mGP and hGP showed that ligands 18, 67, 8, 46, 59, 62 and 74 fulfilled atleast two out of three parameters set. Among these ligand 18 although it did not have the best docking energy in comparison to GSK254, nevertheless, ranked highest in all the three parameters. From this study, it is clear that the specific inhibitor of hGP can be obtained if the difference in the docking energy for mGP and hGP can be increased with reduced binding sites and ligand binding to unconserved residues to that of its isozyme. Ligand 18 identified from this approach can indeed serve as a lead for comparative and specific targeting of hGP to mGP. Therefore, by drawing rules for addressing selectivity and specificity a computation strategy can be developed to address cross reactivity for better drug development.

Acknowledgements. Author (VBK) acknowledges NISER for its support

References

1. Bartram, C., Edwards, R.H., Beynon, R.J.: McArdle's disease-muscle glycogen phosphorylase deficiency. Biochim. Biophys. Acta 1272, 1–13 (1995)
2. Burwinkel, B., Bakker, H.D., Herschkovitz, E., Moses, S.W., Shin, Y.S., Kilimann, M.W.: Mutations in the liver glycogen phosphorylase gene (PYGL) underlying glycogenosis type VI. Am. J. Hum. Genet. 62, 785–791 (1998)
3. Shimada, S., Matsuzaki, H., Marutsuka, T., Shiomori, K., Ogawa, M.: Gastric and intestinal phenotypes of gastric carcinoma with reference to expression of brain (fetal)-type glycogen phosphorylase. J Gastroenterol. 36, 457–464 (2001)
4. Ramos, F.J., Langlais, P.R., Hu, D., Dong, L.Q., Liu, F.: Grb10 mediates insulin-stimulated degradation of the insulin receptor: a mechanism of negative regulation. Am. J. Physiol. Endocrinol. Metab. 290, E1262–E1266 (2006)
5. Bollen, M., Keppens, S., Stalmans, W.: Specific features of glycogen metabolism in the liver. Biochem. J. 336(pt. 1), 19–31 (1998)
6. Aiston, S., Coghlan, M.P., Agius, L.: Inactivation of phosphorylase is a major component of the mechanism by which insulin stimulates hepatic glycogen synthesis. Eur. J. Biochem. 270, 2773–2781 (2003)
7. Baker, D.J., Timmons, J.A., Greenhaff, P.L.: Glycogen phosphorylase inhibition in type 2 diabetes therapy: a systematic evaluation of metabolic and functional effects in rat skeletal muscle. Diabetes 54, 2453–2459 (2005)
8. Hoover, D.J., Lefkowitz-Snow, S., Burgess-Henry, J.L., Martin, W.H., Armento, S.J., Stock, I.A., McPherson, R.K., Genereux, P.E., Gibbs, E.M., Treadway, J.L.: Indole-2-carboxamide inhibitors of human liver glycogen phosphorylase. J. Med. Chem. 41, 2934–2938 (1998)
9. Martin, W.H., Hoover, D.J., Armento, S.J., Stock, I.A., McPherson, R.K., Danley, D.E., Stevenson, R.W., Barrett, E.J., Treadway, J.L.: Discovery of a human liver glycogen phosphorylase inhibitor that lowers blood glucose in vivo. Proc. Natl. Acad. Sci. USA 95, 1776–1781 (1998)
10. Treadway, J.L., Mendys, P., Hoover, D.J.: Glycogen phosphorylase inhibitors for treatment of type 2 diabetes mellitus. Expert Opin. Investig. Drugs 10, 439–454 (2001)

11. Henke, B.R., Sparks, S.M.: Glycogen phosphorylase inhibitors. Mini Rev. Med. Chem. 6, 845–857 (2006)

12. Dzubak, P., Hajduch, M., Vydra, D., Hustova, A., Kvasnica, M., Biedermann, D., Markova, L., Urban, M., Sarek, J.: Pharmacological activities of natural triterpenoids and their therapeutic implications. Nat. Prod. Rep. 23, 394–411 (2006)

13. Zhang, P., Hao, J., Liu, J., Lu, Q., Sheng, H., Zhang, L., Sun, H.: Synthesis of 3-deoxypentacyclic triterpene derivatives as inhibitors of glycogen phosphorylase. J. Nat. Prod. 72, 1414–1418 (2009)

14. Wen, X., Sun, H., Liu, J., Cheng, K., Zhang, P., Zhang, L., Hao, J., Ni, P., Zographos, S.E., Leonidas, D.D., Alexacou, K.M., Gimisis, T., Hayes, J.M., Oikonomakos, N.G.: Naturally occurring pentacyclic triterpenes as inhibitors of glycogen phosphorylase: synthesis, structure-activity relationships, and X-ray crystallographic studies. J. Med. Chem. 51, 3540–3554 (2008)

15. Fukushima, M., Matsuyama, F., Ueda, N., Egawa, K., Takemoto, J., Kajimoto, Y., Yonaha, N., Miura, T., Kaneko, T., Nishi, Y., Mitsui, R., Fujita, Y., Yamada, Y., Seino, Y.: Effect of corosolic acid on postchallenge plasma glucose levels. Diabetes Res. Clin. Pract. 73, 174–177 (2006)

16. Lerin, C., Montell, E., Nolasco, T., Garcia-Rocha, M., Guinovart, J.J., Gomez-Foix, A.M.: Regulation of glycogen metabolism in cultured human muscles by the glycogen phosphorylase inhibitor CP-91149. Biochem. J. 378, 1073–1077 (2004)

17. Thomson, S.A., Banker, P., Bickett, D.M., Boucheron, J.A., Carter, H.L., Clancy, D.C., Cooper, J.P., Dickerson, S.H., Garrido, D.M., Nolte, R.T., Peat, A.J., Sheckler, L.R., Sparks, S.M., Tavares, F.X., Wang, L., Wang, T.Y., Weiel, J.E.: Anthranilimide based glycogen phosphorylase inhibitors for the treatment of type 2 diabetes. Part 3: X-ray crystallographic characterization, core and urea optimization and in vivo efficacy. Bioorg. Med. Chem. Lett. 19, 1177–1182 (2009)

18. Arnold, K., Bordoli, L., Kopp, J., Schwede, T.: The SWISS-MODEL workspace: a web-based environment for protein structure homology modelling. Bioinformatics 22, 195–201 (2006)

19. Vainio, M.J., Johnson, M.S.: Generating conformer ensembles using a multiobjective genetic algorithm. J. Chem. Inf. Model 47, 2462–2474 (2007)

20. Withers, S.G., Madsen, N.B., Sprang, S.R., Fletterick, R.J.: Catalytic site of glycogen phosphorylase: structural changes during activation and mechanistic implications. Biochemistry 21, 5372–5382 (1982)

Externalising Moods and Psychological States to Smooth Pet-robot/Child Interaction through Bluetooth Communication

Ferran Larriba[1], Cristóbal Raya[1], Cecilio Angulo[1],
Jordi Albo-Canals[2], Marta Díaz[1], and Roger Boldú[2]

[1] GREC Research Group, Universitat Politècnica de Catalunya,
Pau Gargallo 5, 08028 Barcelona, Spain
flarribagil@gmail.com, {cristobal.raya,cecilio.angulo,marta.diaz}@upc.edu
[2] La Salle BCN, Universitat Ramon Llull,
Quatre Camins, 08022 Barcelona, Spain
albo@salleurl.edu, rboldu@media.mit.edu

Abstract. Nowadays have already passed more than two years since PATRICIA research project about using pet robots to reduce pain and anxiety in hospitalized children was started and the advances made are more than significant. Patients, parents, nurses, psychologists, engineers... all of them have adopted with illusion Pleo robot, a kind of robotic pet, working hard looking for new procedures and new solutions to the current pediatrics diseases. From this work, a technological contribution is provided going one step beyond to what seems a long path. Concretely, it is wanted to develop a system to wirelessly communicate with Pleo in order to help the coordinator who leads the therapy with the kid, to understand and control Pleo's behaviour at any moment. This article explains how this technological part is being developed and obtained technical results.

Keywords: ambient intelligence, bioemotional computing, communication, robotics.

1 Introduction

How far a robot can help overcome our traumas? Recent advances in the fields of robotics and medicine go further than physical healing. Now to diagnostic robots, to disability robots, to robots for rehabilitation or surgical robots, therapeutic robots are added, designed to motivate and assist the patient with psychological problems. Hence *robotherapy* consists of the interaction between human patients and robotic creatures (humanoids, pet toys, dummies...) in order to help patients build a positive attitude facing their disease [1].

Since pet robots are mainly considered in this study, we will focus on robots that can be used with a similar therapeutic effect to zootherapy (animal-assisted therapy) [2]. Ideally, an animal is the best solution. Unfortunately, these companion animals are not readily available. Concerns of dog bites, allergies, or disease

F. Ortuño and I. Rojas (Eds.): IWBBIO 2015, Part II, LNCS 9044, pp. 683–693, 2015.

have led to many nursing homes and hospitals to ban this therapy. When this therapy is offered it is a very regulated experience. The animal must be in the company of a trained professional at all times. Additionally, these sessions are scheduled activity and only occur for a few hours or less each day, once or twice a week. As a result of these restrictions this new form of therapy, robotherapy, has emerged.

The Program Child Life was started in 2004 at Hospital Sant Joan de Déu (HSJD), Barcelona, Spain, with the overall aim to design pioneering techniques to improve the children's experience when hospitalized by reducing pain and anxiety [3]. In 2010, more than 200 children and teenagers and their families have participated in this program in HSJD.

The PATRICIA project [4] is based on the use of social robots with the same aim. HSJD cooperates with Universitat Politècnica de Catalunya (UPC), together with La Salle Universitat Ramon Llull (La Salle URL) like technological partners. This kind of projects encourages easily to anyone to get interested in it: it is an *innovative* project with many unresolved horizons to explore, with a high diverge team, where so different concepts as robotics and human feelings are mixed, therefore *multidisciplinary*. Engineers, doctors, psychologists... working together looking for improve the life quality of those kids that were unlucky for living together with these disease.

1.1   Therapeutic Robots

It is not easy to build a machine able to satisfy the human need for companionship in hard times in the form of pet robots, however companies as Fujitsu, Innvo Labs or PARO Robots are working hard on it. In the last 5 years many studies worldwide have been presented with positive outcomes.

PARO is a baby seal shaped robot designed by Takanori Shibata in Japan in 1993, but did not begin to be commercialized until 2004 [5]. It is equipped with five kinds of sensors: temperature, touch, light, audio and position sensors. Additionally Paro is able to learn behaviors. This pet offers similar benefits as zootherapy, and is used in treatments to people with symptoms of Alzheimer and other disabilities.

Huggable Teddy Bear, is being developed by Fujitsu as a therapeutic companion for hospitals or nursing homes, for health care, education, and social communication applications [6]. Hides a dozen sensors to recognize facial expressions and movements of the patient by the camera on its nose. It is intended to record the patient's emotional state and react accordingly using a range of 300 shares scheduled actions to interact with the people around it.

ROMIBO is an open coded therapeutic robot [7] specially designed to the research and treatment of autism disease in kids. It brings all the wireless connections in order to be remotely controlled.

Finally, Pleo, our chosen platform, is a robot, imitation of a Camarasaurus dinosaur, that exhibits an appealing baby-likeness, expressiveness, and an array of different behavior and mood modes. Pleo has been tested in several research works [4,8,9,10] focused on the effect of Pleo in long-term interaction, especially

Fig. 1. Pleo the Camarasaurus

with children. Another kind of interesting researches with this robot platform is robot ethics [11], which plays a very important role in robotherapy.

Pleo is a commercial entertainment platform developed by Innvo Labs. It is equipped with 2 ARM7 CPUs, 14 motors, 8 touch sensors, IR transceivers, microphones, and 1 camera. It features also a pet like personality which develops in time, internals drives like hunger or sleep, and several mood modes: happy, extremely scared, curious...

The software running in Pleo is divided into three layers: the low-level software deals directly with hardware. Sensor information is provided to the mid-level layer through a blackboard system. The mid-level layer provides the application functional support to the high-level scripting layer. The script layer implements the highest-level functionality of Pleo. This is essentially Pleo's personality, determining how and when he responds to sensor input and internal goals.

1.2   Robot Autonomy

Four degrees of Pleo's autonomy can be deployed when interacting with children in the real scenario of the hospital:

1. Full autonomous behavior according to implicit –opaque to users– internal states: Pleo always acts according to its own criteria. The problem is that in this modality, Pleo's behavior is not totally predictable by the user at any time but may be inferred, anticipated or understood by the user according to previous experience in interaction, expectations and social comprehension of Pleos drives and situation awareness.
2. Full autonomous behavior according to observable internal states: The conductor of the intervention can see Pleo's internal states through a graphical interface that externalize or make transparent Pleos's internal states that facilitates the understanding and management of the interaction.
3. External control of Pleo's states: The coordinator is enabled to modify or control Pleo changing the internal states and letting Pleo perform the correlative activity.
4. External control of Pleo's behaviour: Fully tele-operated control of the movements and actions of Pleo. Always requires the presence of the coordinator to handle the Pleo.

Nowadays the best choice for testing Pleo in the hospital is located between the second and the third levels, hence this will be the approach of this research.

1.3   Technical Framework

Latest advances in the research project shows the capability of Pleo's platform helping the kids of HSJD and their families to improve their treatment. Then, the research group presents a prospective view of a new generation of health-care robots – combining cloud robotics and artificial intelligence – that provide children patients with an effective and individualized assistance [12].

The goal is to supply each young patient with a personal Pleo. Besides, a cloud multi-agent system able to perceive, collect and share hospitalized children status is build, and using artificial intelligence the behaviour of every patient's robot is modified. Finally, as all this information is in the cloud, the system can explore the most effective actions that the Pleo can carry to improve its own patient experience.

Until now, it is not achievable to modify the software system of Pleo due to it is not a full open source. However, it allows to modify some values or to commit some precise actions in a certain moment. For example, it can change or just show how hunger is it, how happy is it, or also to ask it for walking or "to give paw".

Hardware communication with the robot is not so easy as in previous Pleo commercial versions, as long as the new one has the USB port, the serial interface and the Bluetooth connectors behind the battery spot. In order to obtain data in "real time" from Pleo whereas it is powered, it was connected to an external power while stabling a gateway between the USB connector and remote computer.

It is possible to get in real time the distinct values for the different sensors of Pleo adopting the method described. After that, some Bluetooth connections appear connecting a Bluetooth module to the UART port [12] as a bridge between Pleo and a Raspberry Pi. Other processing platforms like Intel Galileo or Edison and wireless communications like ZigBee or WiFi could be also considered, depending on the power, processing or privacy needs. Then, the Raspberry Pi should be set into a little bag specially made for Pleo, and can access to the Internet thanks to a wifi dongle, uploading the data to the cloud. However, there were two problems in that communication: first a lack of space, because original battery of Pleo was employed compacting the wires with the battery and leaving the Bluetooth module in sight[1] (see Figure 2). It is a trouble taking in consideration that Pleo will be used by kids, then it will not accomplish the user specifications. Secondly, new Pleo robots are not equipped with this UART connector, so it could be exclusively tested in old versions.

Therefore, unfortunately each workshop that the psychologist/pediatric group has performed with the children of HSJD (the real scenario) to test the interaction between Pleo and the kids have been performed without any communication provided. And as this platform is new not only for the kids but also for the pediatric group, when Pleo has a bad mood or acts in a non-normal way there is no manner to know exactly what is happening to it. So in that cases the pediatric

---

[1] You can check a video at https://www.youtube.com/watch?v=2qNdZFt8by8.

Fig. 2. Pleo's Bluetooth communication [12]

group interact with Pleo based in their personal experience in order to correct the situation.

According to both, the desired robot autonomy, and the technical drawbacks when using the platform, this set of objectives will be pursued along this work:

- Bluetooth (Bth) communication for Pleo: to build a removable part that will be able to be mounted in any Pleo providing Bth communication.
- Modification of Pleo's states: the coordinator must have the ability to control Pleo to make a specific action because the situation requires it. Android application has been chosen as interface to do that, due to it is an open source easy to spread in the society.
- Bluetooth-battery package: the removable part would fit inside the battery hole, in order to assemble the module in one package.

2   The Proposed Solution

This section includes a description of the solution to be developed and other features that have been considered.

2.1   Hardware

The proposed solution for the hardware challenge, shown in Figure 3, is to switch Pleo's battery for a battery-Bluetooth package. Distinguishable components are listed below.

1. *PCB*: Its main function is to become the conductive element between the batteries to the springs that feed the robot. Based on the base of the battery, it must fix the 4 pins that establish contact to the Bluetooth output of Pleo.
2. *Connector pogo pins*: They catch the signal that Pleo sends and will be processed in the Bluetooth module.

Fig. 3. (Left) Layout of the proposed assembly; (Middle) Modules JY-MCU and RN-41; (Right) 3D print model

3. *Bluetooth module*: Receives the data signal from the robot and sends it to the device connected. The JY-MCU industrial serial port is one of the cheapest Bluetooth serial port modules in the market, but provided voltage is not enough. Then, a more sophisticated module is needed, for instance RN-41 microchip (see Figure 3(Middle)).
4. *Battery pack*: Specifications of Pleo's battery are for voltage (7.4V), charge (2800mAh), power (20.72Wh) and max temperature (60^oC). So, it should be replaced by a couple batteries of 3,7V of size AA without problems However, we will check in the Subsection 3.3 how this sentence is not really true.

Once all the components are assembled and tested, a case to pack it with a 3D printer is made[2] (see Figure 3(Right)).

2.2   Interface

The interface must be as simple and comfortable for the user as possible, using buttons and images to understand Pleo's behaviour. Moreover connecting to the Pleo robot must be user-friendly, Bluetooth must turn 'on' when the app is launched and a button should be present to search Pleo's signal and establish the communication. A first sketch of the main menu is shown in Figure 4. It should have a list of the different Pleo states, be able to modify the emotional status of Pleo, allowing to the user to ask for their value. If needed, it is possible to add more buttons for Pleo Actions, in order to make a specific action when the situation requires it.

3   Development of the Solution

In this section, due to lack of space, more that a detailed 'what has been done' description, different troubles and how to manage them will be presented.

---

[2] In fact, there are already some free models of Pleo's battery case in the Internet to download (http://www.thingiverse.com/thing:31721/#files

Fig. 4. Sketch of the proposed interface

3.1 Connection via Terminal

Pleo does not support Win 64 bits, so a virtual image for XP 32 bits was implemented. The easiest way to connect to the Pleo is via PuTTY and USB[3].

3.2 Bluetooth Assembly

One of hardest challenges in this work is the difficulty to take measures to build the Bluetooth assembly. On the rear face of the battery inside the hole it is impossible to insert a vernier caliper to measure. Moreover the rear is a little wider than the external perimeter of the hole.

To solve the problem of obtaining the correct pin positions, a thin sheet of transparent plastic have been cut, introduced inside the hole and marked with a fine-tipped pen. Then the plastic is extracted, measured with the vernier caliper and drawn with a CAD tool.

Once drawn the layout, the next step is to build the PCB.

Finally, a 10x1 female header is welded to connect the RN-41 without welding directly to the module, that could damage it.

3.3 Power Troubles

When the PCB is set inside the Pleo, batteries power the robot, but after pressing its button to turn it on, the robot does not move. It was thought that probably both batteries are not enough to power the robot and the Bluetooth module, so a button cell of 3,3V was used to power the RN-41, without taking up too much space. Then, after pressing the button, Pleo's makes the sound as wakening up, and sends some data to the screen of the terminal with it is communicating with, but afterwards the robot snores and turns off stopping the communication.

It seemed like a power failure, hence the minimum tension and intensity to make the robot work was measured. There it was discovered that the minimum power

---

[3] In order to do this, it is necessary to download the Pleo Development Kit, http://ipr10.wikidot.com/pleo

Fig. 5. In the first picture the battery works perfectly. Once Pleo has no power to make it run its battery have exactly 7,39V.

to supply to Pleo RB is 7,4V (equivalent to 3A), turning the robot to the sleep position in case of reducing its values a little bit. That explains why the same battery that Pleo carries provides more than 8,4V when it is completely full.

Another discover was that Pleo RB, unlike USB wired communication, does not allow Bluetooth communication if the robot is not running.

3.4   Android App

Figure 6 shows how the interface looks like. First scene is a presentation to introduce the user. Here the application itself turns 'on' the Bluetooth of the smartphone or device where the app is installed.

Fig. 6. Different scenes of the application "PleoSays": (Left) Presentation; (Middle) Pairing with Pleo; (Right) Communicate

At the next scene there is a button that user must push once the module RN-41 is powered in order to pair it with the device. Then the app searches the corresponding MAC direction and in case that it does not find it, displays an error message. On the other hand, if the communication is established the app redirects the user to the third scene or command window, where the user sends and receives data from Pleo.

When Pleo is switched on, unlike USB communication the robot starts sending data, information of the initialization of the source. Hence, the application developed cannot receive Pleo's states values or to send any motion command until all this initial info has been received and processed[4].

4 Results and Discussion

Wireless communication between Pleo RB and an Android device is finally achieved. With the Android app developed, the user is able to obtain any state of the robot without stopping its interaction with the patient. It also contains an editable textbox to allow the user to send orders to the robot, authorizing the user to a long permission list of actions, as modifying Pleo's states, ordering to commit any action or sound, camera's options, etc.

It has been a success to achieve the communication with the robot connecting the Bluetooth module to the output set behind the battery spot. As newest Pleo's versions have no UART port connection, the incoming works with Pleo should adapt to a solution similar to what has been explained in this article. From now it can be ensured that the bottom connection works properly and can be used in further research.

During the trials, it has been made evident that using a couple of 3,7V cells is not enough to make Pleo run due to this values are too close to the operating limit. Using a third cell is needed to make the Pleo robot move.

Moreover, the developed Android app can be installed in many devices and its interface is easily understandable for users.

After many hours working with Pleo, several observations can be expressed about it, which three have been highlighted and described below:

- Pleo is not a robust platform. It is a common fact for people that work with Pleo that one day without doing anything different of what they have already done before it does not turn on. Then, why to spend so many effort in working with this tool, if sometimes it breaks without knowing why? The answer is economy. If Pleo "survives" for more than one month working with the kids in therapy, investment can be considered as recovered, since it is very affordable.
- Pleo is selfish. It is not like a dog because Pleo does not react if the kid cries, laughs, plays,... This is a very important issue because it leads to an interest lost by the kid, and is known that the progress of the therapy is directly proportional to the motivation of the patient. If Pleo does not care of the kid, the kid will not care of Pleo and the therapy will be a failure.

---

[4] It is possible to hide all this data to the user.

– Related to the previous observation, and considering that the future of this research is the cloud net as explained, the cloud should collect the "states" of the kids.

5 Conclusions

The first one of the objectives of this project, the Bluetooth communication with Pleo, has been achieved completely. Using the PCB with the extensible pins, we can supply Bluetooth communication to any Pleo, even those without UART connection.

The second objective, to modify Pleo's states from an Android device, or any of their possible actions, has been successfully achieved too, so helping the coordinator to understand and to control the robot using an easy app.

The third objective, to assemble a Bluetooth + Batteries package has not been fully completed. Pleo will work using 3 (not 2) AA cells and it will fit to the designed package in this work, but it was not be completely developed, since 3D cage was build for two AA cells. However, it should be possible to build this assembly.

As far as Pleo is a commercial closed platform, alternative platforms should be considered in the near future. In this sense, two technical research lines are opened. The first one is either to consider commercial platforms as Pleo's alternative or to develop a pilot robotic platform to be certified for commercial use. The second line is working in the possibility to certify the employed device as a medical device for therapeutic purposes.

Acknowledgments. This research was supported in part by the PATRICIA Research Project (TIN2012-38416-C03-01,02,03), funded by the Spanish Ministry of Economy and Competitiveness. Thanks to Hospital Sant Joan de Déu, the team of nurses and volunteer corps by shared experience.

References

1. Libin, E., Libin, A.: New diagnostic tool for robotic psychology and robotherapy studies. Cyberpsy., Behavior, and Soc. Networking 6(4), 369–374 (2003)
2. López-Cepero, J., Rodríguez-Franco, L., Perea-Mediavilla, M.A., Blanco Piñero, N., Tejada Roldán, A., Blanco-Picabia, A.: Animal-assisted Interventions: Review of Current Status and Future Challenges. Int. J. Psych. and Psycholog. Therapy 14(1), 85–101 (2014)
3. Serrallonga-Tintoré, N., Cabré-Segarra, V.: El cuidado emocional en la prevención del dolor posquirúrgico en niños y adolescentes. Rev. Psicopat. Salud Mental Niño y Adol. 16, 49–56 (2010)
4. Angulo, C., Garriga, C., Luaces, C., Pérez, J., Albo-Canals, J., Díaz, M.: Pain and Anxiety Treatment based on Social Robot Interaction with Children to improve Patient Experience. Ongoing Research. In: Proceedings of the XIV ARCA days. Qualitative Systems and its Applications in Diagnose, Robotics and Ambient intelligence, pp. 25–32. Salou, Spain (2012)

5. Taggart, W., Turkle, S., Kidd, C.D.: An Interactive Robot in a Nursing Home: Preliminary Remarks. In: Toward Social Mechanisms of Android Science. Cognitive Science Society, Stresa, Italy (2005)
6. Stiehl, W.D., Lieberman, J., Breazeal, C., Basel. L., Lalla, L., Wolf, M.: The Design of the Huggable: A Therapeutic Robotic Companion for Relational, Affective Touch. In: Proceedings of AAAI Fall Symposium on Caring Machines, Washington, D.C. (2005)
7. Shick, A.: Romibo robot project: an open-source effort to develop a low-cost sensory adaptable robot for special needs therapy and education. In: Special Interest Group on Computer Graphics and Interactive Techniques, ACM SIGGRAPH 2013 Studio Talks, Anaheim, California (2013)
8. Díaz, M., Andrés, A.: Angulo. C.: Robots sociales en la escuela. Explorando la conducta interactiva con niñ@s en edad escolar. In: ROBOT 2011: Robótica Experimental, pp. 622–625, Sevilla, Spain (2011)
9. Chung-Chang, C.Y., Díaz, M., Angulo, C.: The Impact of Introducing Therapeutic Robots in Hospital's Organization. In: Bravo, J., Hervás, R., Rodríguez, M. (eds.) IWAAL 2012. LNCS, vol. 7657, pp. 312–315. Springer, Heidelberg (2012)
10. Heerink, M., Díaz, M., Albo-Canals, J., Angulo, C., Barco, A., Casacuberta, J., Garriga, C.: A field study with primary school children on perception of social presence and interactive behavior with a pet robot. In: 21st IEEE International Symposium on Robot and Human Interactive Communication (Ro-Man 2012), pp. 1045–1050. IEEE Press, New York (2011)
11. Darling, K.: Extending Legal Rights to Social Robots. In: We Robot Conference. University of Miami (2012)
12. Navarro, J., Sancho, A., Angulo, C., Garriga, C., Ortiz, J., Raya, C., Miralles, D., Albo-Canals, J.: A Cloud Robotics Architecture to Foster Individual Child Partnership in Medical Facilities. Workshop on Cloud Robotics, RSJ IEEE/RSJ International Conference on Intelligent Robots and Systems (IROS 2013). Tokyo Big Sight, Japan (2013)

Patient Lifecycle Management:
An Approach for Clinical Processes

Alberto Salido López [1,2], Carmelo Del Valle[2], María José Escalona[2], Vivian Lee[3],
and Masatomo Goto[3]

[1]Fidetia, Seville, Spain
alberto.salido@fidetia-plmlab.es
[2]University of Seville, Seville, Spain
{carmelo,mjescalona}@us.es
[3]Fujitsu Laboratories of Europe, Hayes, United Kingdom
{vivian.lee,masatomo.goto}@uk.fujitsu.com

Abstract. Clinical processes can be described, inside the Biomedical scope, like
a systematic guideline to assist practitioner and patient decisions about appropriate health care for specific clinical circumstances. In industry, Product Lifecycle Management (PLM) is the process of managing the entire lifecycle of a
product from inception, through engineering design and manufacture, to service
and disposal of manufactured products. Applying the concepts of PLM to Biomedical processes we create a synergy between the product's concept in the industrial case and the patient into the health care environment. This point of
view improves the actual clinical processes with a most specific treatment for
each patient, by modifying the statements to assist the patient according to
the needs of the patient and his illness. This research proposal tries to shift the
focus of the eHealth systems onto the patient, adapts the existing and defined
clinical processes or clinical paths to the patient's needs, applies Big Data
principles to bring even more attentions for the patient, and provides an easy to
use system for the medical staff.

Keywords: Biomedical, Biomedicine, Product Lifecycle Management, PLM,
Clinical Processes, Patient, eHealth, Big Data, Software Engineering.

1 Introduction

In the last decade, the health care sector has used clinical guidelines and protocols as
helpful instruments for decision-making. As defined by the Institute of Medicine,
clinical guidelines are systematically developed statements to assist practitioner and
patient decisions about appropriate health care for specific clinical circumstances [1].
They describe all the decision points and corresponding actions to be carried out depending on a specific patient's state or situation. Furthermore, clinical guidelines
identify the clinical tests to be performed in order to confirm or determine the
patient's state. Based on the test results, the guideline determines the treatment alternatives. Among the most important potential advantages of documenting and using

F. Ortuño and I. Rojas (Eds.): IWBBIO 2015, Part II, LNCS 9044, pp. 694–700, 2015.
© Springer International Publishing Switzerland 2015

clinical guidelines are assessing and improving the quality of care, providing support for medical decision-making, controlling health care costs and reducing both practice variability and the inappropriate use of resources [1,2].

In industry, product lifecycle management (PLM) is the process of managing the entire lifecycle of a product from inception, through engineering design and manufacture, to service and disposal of manufactured products [3]. PLM integrates people, data, processes and business systems and provides a product information backbone for companies and their extended enterprise.

The inspiration for the burgeoning business process now known as PLM came from American Motors Corporation (AMC). The automaker was looking for a way to speed up its product development process to compete better against its larger competitors in 1985 [4]. The first part in its quest for faster product development was computer-aided design (CAD) software system that makes engineers more productive [4]. The second part in this effort was the new communication system that allowed conflicts to be resolved faster, as well as reducing costly engineering changes because all drawings and documents were in a central database.

The main motivation for this research comes from several projects, regarding health care and PLM, in which the research group is involved. After analyzing and identifying the advantages and disadvantages of each research area, we focus on trying to improve health care systems using the main advantages that PLM paradigm brings. Our approach focuses on improving biomedical systems, by merging the main principle of clinical processes, or clinical pathways, with the Product Lifecycle Management paradigm in the industrial case. In this research proposal, we try to improve the attention of a patient with a chronicle illness, adapting the base clinical process, defined by specialist, for the patient. Biomedical informatics [5-12] incorporates a core set of methodologies that are applicable to data, information, and knowledge management across the translational medicine continuum, from bench biology to clinical care and research to public health [13].

The paper is structured as follows: Section 2 describes the PLM methodology as it is in the industrial environment. Section 3 explains our proposal to transform the idea of PLM to Patient Treatment. Finally, chapter 4 describes our conclusions and future work.

2   About PLM

Product Lifecycle Management (PLM) is the business activity of managing, in the most effective way, a company's product all the way across their lifecycle; from the very first idea for the product all the way through until it is retired and disposed of [14]. As it is shown in Fig 1, the lifecycle of a product, in most cases, is cyclical. From the extraction of the raw materials, passing through the manufacturing production and delivery, to the final customers; and once the product is useless, PLM covers the disposal of the product and its possible reutilization to recycling.

One of the most important advantages of using PLM is the interconnection of every phase in the product lifecycle across the World; being easily for the company to handle its products having different factories placed in different countries. Furthermore, all the information is shared, so the knowledge about the product manufacturing does not belong to a concrete sector.

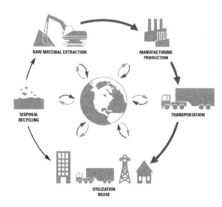

Fig. 1. PLM Overview

PLM manages both individual products and the Product Portfolio, the collection of all of a company's products.

PLM manages products from the beginning of their life, including development, through growth and maturity, to the end of life.

The objective of PLM is to increase product revenues, reduce product-related costs, maximize the value of the product portfolio, and maximize the value of current and future products for both customers and shareholders [14].

However, the benefits of operational PLM go far beyond incremental savings, yielding greater bottom line savings and top-line revenue growth not only by implementing tools and technologies, but also by making necessary, and often tough, changes in processes, practices and methods and gaining control over product lifecycle and lifecycle processes. The return on investment for PLM is based on a broader corporate business value, specifically the greater market share and increased profitability achieved by streamlining the business processes that help deliver innovative, winning products with high brand image quickly to market, while being able to make informed lifecycle decisions over the complete product portfolio during the lifecycle of each individual product.

The scope of product information being stored, refined, searched, and shared with PLM has expanded. PLM is a holistic business concept developed to manage a product and its lifecycle including not only items, documents, and BOM's (Bill Of Materials), but also analysis results, test specifications, environmental component information, quality standards, engineering requirements, change orders, manufacturing procedures, product performance information, component suppliers, and so forth.

On the other hand, modern PLM system capabilities include workflow, program management, and project control features that standardize, automate, and speed up product management operations. Web-based systems enable companies easily to connect their globally dispersed facilities with each other and with outside organizations such as suppliers, partners, and even customers. A PLM system is a collaborative backbone allowing people throughout extended enterprises to work together more effectively.

Operational efficiencies are improved with PLM because groups all across the value chain can work faster through advanced information retrieval, electronic information sharing, data reuse, and numerous automated capabilities, with greater information traceability and data security. This allows companies to process engineering change orders and respond to product support calls more quickly and with less labor. They can also work more effectively with suppliers in handling bids and quotes, exchange critical product information more smoothly with manufacturing facilities, and allow service technicians and spare part sales reps to quickly access required engineering data in the field [15]. Nowadays, PLM is used in most of industrial sectors, like automotive, naval and aeronautical industry, and architecture among many of them.

3   Patient Oriented Clinical Processes

Nowadays, all the biomedical systems are based on new technologies and the interaction with the doctors and medical staff, defining processes for different kinds of illness [16]. This kind of system improves the health systems by making them more efficient and easy to use for doctors. However, these systems overlook the most important factor in an eHealth system, the Patient. The next image shows the three vertex of a triangle of any eHealth system.

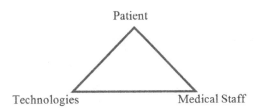

Fig. 2. eHealth System Organization

As it is shown on Fig. 2, an eHealth system can be represented by three vertex of a triangle. The base of the triangle is focused on Medical Staff and the Technologies used in the system. The third vertex, called Patient, provides the information for the system. Nowadays, all the studied systems are focused on the base of the triangle, and are trying to improve the communication between the system and the medical crew or the system itself, by doing it in a more efficient way.

As mentioned before, there are a large number of published guidelines since each guideline is focused on a desired health care outcome. Furthermore, guidelines may vary from hospital to hospital since they reflect variations in resources, staff and the design of the protocol, as well as in the working philosophy of the hospital in question [17]. Because of the vast amount of clinical guidelines, several organizations have undertaken efforts to publish them (using text formats such as HTML or PDF) in the literature and on the Internet to make them more accessible and to enable evidence-based knowledge to be reused [18, 19].

Our approach tries to explore the possibility of including the patient into the process, and shift the clinical pathway focus onto the patient. Using this proposal, called Patient Lifecycle Management, the first input of the clinical process is the patient - based on the assumption that the treatment is the proper one for this patient and this illness. Besides, this 'input' provided by the patient will be measured by using indicators, that show some characteristics of the disease (for example headache, nausea, etc.), providing to the system the state of the patient compared with all the patients with the same issue, using Big Data [20] techniques.

Big Data processing technologies allow sharing and using the patient information between different centers, providing to all the biomedical processes the possibility of evolves using the patients' needs. In this way, we get the most important factor of our research; the processes are adapted to the patient, instead of being the patient who has to adapt to the process for his illness.

The baseline for this research starts with the Product Lifecycle Management paradigm, which defines the whole lifecycle for a product into the industrial environment. The PLM paradigm has years of experience into the industrial sector, improving the management of the products, since the definition to the disposal phase, reducing time-to-market costs and production times. In this proposal, we will use some base concepts of the classical PLM paradigm to enrich the biomedical system to develop. Figure 3 summarizes the main overview of our initial research proposal, mapping it with the current PLM paradigm.

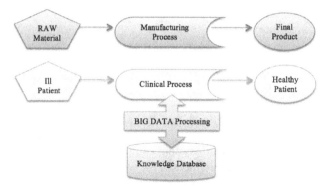

Fig. 3. Mapping between PLM and Patient Lifecycle Management

4   Conclusions and Future Work

As conclusion, our research team is working on this proposal, which applies the Product Lifecycle Management behavior to Clinical Processes, and tries to improve the flexibility of the biomedical systems for clinical pathways, by adapting them to the patient. This flexible pathway eventually improves communication between the patient and the process for his own illness, and hence the efficiency of the patient treatment.

One of the critical points of this proposal could be the adaption of the process to the patient, which is also the focus of our research; it is needed to define common indicators for specific cases on each clinical process. These indicators could be misinterpreted because the proper patient may indicate a wrong value for its pain. For example, we could use a one to five scale to measure patient's pain, if the patient is frightened about his illness, could indicate that its pain is five. Another factor is the patient thinks that the most painful is his state will be attended faster, so he will always say that his pain is the maximum value. To avoid this false value of the indicator, we will use Big Data principles to analyze the indicators of one patient according all patients with the same issue. Also it is very important to define the correct indicators and algorithms to avoid mistakes during the patient analysis, providing the best process for each patient.

Due to the high amount of data regarding clinical pathways, for our further research we will focus on three main illness and its treatments; Endometriosis, AIDS and Malformations of spinal cord. Thereby we could reduce the scope of this research project focusing into three concrete cases.

As for the future work for this research project, we will start analyzing the State-Of-the-Art technologies in this domain by conducting a Systematic Literature Review [21, 22], and based on the review result, we will try to improve the Biomedical System by applying the proposed method as mentioned previously.

Acknowledgements. This research project has been partially funded by MeGUS Project (TIN2013-46928-C3-3-R) and DExOFlex-BP Project (TIN2014-52382-R) of The Ministry of Economy and Competitiveness, Spain. We would like to thank Fujitsu Laboratory of Europe by the support to this research.

References

1. Field, M.J., Lohr, K.N. (eds.): Guidelines for Clinical Practice: From Development to Use. National Academies Press (1992)
2. Papadopoulos, C.: The development of Canadian clinical practice guidelines: a literature review and synthesis of findings: Discussion paper prepared for the CCA/CFCRB Task Force on Chiropractic Clinical Practice Guidelines June 15 2002. The Journal of the Canadian Chiropractic Association 47(1), 39 (2003)
3. Kurkin, O., Januska, M.: Product Life Cycl in Digital factory. In: 15th International-Business- Information-Management-Association Knowledge Management and Innovation: A Business Competitive Edge Perspective, vol. 1–3, Cairo, Egypt, November 06, 2007, ISBN: 978-0-9821489-4-5, pp. 1881–1886 (2010)
4. http://web.archive.org/web/20090213042744/, http://www.coe.org/coldfusion/newsnet/may03/technology.cfm. (last accessed: January 2015)
5. Cimino, J.J., Shortliffe, E.H.: Biomedical Informatics: Computer Applications in Health Care and Biomedicine (Health Informatics). Springer-Verlag New York, Inc. (2006)
6. Greenes, R.A., Shortliffe, E.H.: Commentary: Informatics in biomedicine and health care. Academic Medicine 84(7), 818–820 (2009)
7. Bernstam, E.V., Hersh, W.R., Johnson, S.B., Chute, C.G., Nguyen, H., Sim, I., Becich, M.J.: Synergies and distinctions between computational disciplines in biomedical research: perspective from the Clinical and Translational Science Award programs. Academic Medicine: Journal of the Association of American Medical Colleges 84(7), 964

8. Collen, M.F.: The origins of informatics. Journal of the American Medical Informatics Association 1(2), 91–107 (1994)
9. Collen, M.F.: Fifty years in medical informatics. Yearb. Med. Inform., 174–179 (2006)
10. Haux, R.: Individualization, globalization and health–about sustainable information technologies and the aim of medical informatics. International Journal of Medical Informatics 75(12), 795–808 (2006) .
11. Altman, R.B., Balling, R., Brinkley, J.F., Coiera, E., Consorti, F., Dhansay, M.A., Wiederhold, G.: Commentaries on "Informatics and medicine: from molecules to populations". Methods of Information in Medicine 47(4), 296 (2008)
12. Embi, P.J., Kaufman, S.E., Payne, P.R.: Biomedical Informatics and Outcomes Research Enabling Knowledge-Driven Health Care. Circulation 120(23), 2393–2399 (2009)
13. Sarkar, I.N.: Biomedical informatics and translational medicine. J. Transl. Med. 8(1), 22 (2010)
14. Stark, J.: Product lifecycle management, pp. 1–16. Springer, London (2011)
15. Saaksvuori, A., Immonen, A.: Product lifecycle management, p. 22. Springer, Berlin (2005)
16. Thorwarth, M., Arisha, A.: A simulation-based decision support system to model complex demand driven healthcare facilities. In: Proceedings of the Winter Simulation Conference (WSC), pp. 1–12. IEEE (December 2012)
17. Pérez, B., Porres, I.: Authoring and verification of clinical guidelines: A model driven approach. Journal of Biomedical Informatics 43(4), 520–536 (2010)
18. Agency for Healthcare Research and Quality, National guideline clearinghouse, guidelines for the prevention of intravascular catheter-related infections, http://www.guideline.gov (last accessed: January 2015)
19. SEIMC Sociedad Española de Enfermedades Infecciosas y Microbiología Clínica [On-line], Documentos científicos, http://www.seimc.org/documentos/ (last accessed: January 2015)
20. Berman, J.J.: Principles of big data: preparing, sharing, and analyzing complex information. Newnes (2013)
21. Petticrew, M., Roberts, H.: Systematic reviews in the social sciences: A practical guide. John Wiley & Sons (2008)
22. Domínguez-Mayo, F.J., Escalona, M.J., Mejías, M., Ross, M., Staples, G.: Towards a Homogeneous Characterization of the Model-Driven Web Development Methodologies. Journal of Web Engineering 13(1-2), 129–159 (2014)

Advertising Liking Recognition Technique Applied to Neuromarketing by Using Low-Cost EEG Headset

Luis Miguel Soria Morillo[1], Juan Antonio Alvarez García[1],
Luis Gonzalez-Abril[2], and J.A. Ortega Ramirez[1]

[1] Computer Languages and Systems Dept.,
University of Seville, 41012 Seville, Spain
[2] Applied Economics I Dept.,
University of Seville, 41018, Seville, Spain
{lsoria,jaalvarez,luisgon,jortega}@us.es
http://www.us.es

Abstract. In this paper a new neuroscience technique is applied into Marketing, which is becoming commonly known as the field of Neuromarketing. The aim of this paper is to recognize how brain responds during the visualization of short advertising movies. Using low cost electroencephalography (EEG), brain regions used during the presentation have been studied. We may wonder about how useful it is to use neuroscience knowledge in marketing, what can neuroscience add to marketing, or why use this specific technique. By using discrete techniques over EEG frequency bands of a generated labeled dataset, C4.5 and ANN learning methods have been applied to obtain the score assigned to each ads by the user. This techniques allows to reach more than 82% of accuracy, which is an excellent result taking into account the kind of low-cost EEG sensors used.

Keywords: neuromarketing, electroencephalography, advertising, brain, eeg, brain-computer interaction.

1 Introduction

The use of electroencephalography has been used over the last decade for the study of brain activity. In most of the studies is to determine the effects of certain thoughts on the cerebral cortex. Thus a cause-effect relationship that explores the different cerebral cortex is established. The combination of medicine and learning systems have been able to achieve milestones that were unthinkable years. Since the isolation of brain activity based on the thoughts and your use of electronic devices through a brain interface. This has led to a new era in the field of interaction human-computer, characterized by the absence of direct orders or actions by the user, to make way for management through direct thoughts. In an effort to isolate the actions, researchers have found associations for determining whether the music is heard or not the user's taste, or if a TV ad

F. Ortuño and I. Rojas (Eds.): IWBBIO 2015, Part II, LNCS 9044, pp. 701–709, 2015.

captures or not the individual's attention. This application is being exploited by the advertising industry. These actions, in this special paper attention will be paid to the relationship between the display of advertisements short video and the effects they cause on the mental state of the user. Specifically, this study aims to provide a solution to a recurring problem in the advertising field, as is determining cortical level marketing preferences without user interaction How does the cerebral cortex a given ad is user friendly? Are there effects electric level when the user is viewing advertising content relevant to your tastes? Determining these partnerships will allow high-level go a step further in neuromarketing [Lee *et al.(2007)Lee, Broderick, and Chamberlain, Ariely and Berns(2010), Morin(2011)*]. So far, most studies in the field of neuromarketing have been based on the use of electroencephalography sensors or sensors installed galvanic skin resistance of the user to determine the impact of ads brainwave level isolation [Senior and Lee(2008), Murphy *et al.(2008)Murphy, Illes, and Reiner*]. However, there have been few studies that have used this technology to classify information binomial how classes [I like] or [I hate]. This first step described in this article allow the development of a learning system based on neural networks [Hagan *et al.(1996)Hagan, Demuth, Beale, et al., Dreiseitl and Ohno-Machado(2002)*] and decision trees (C4.5) [Quinlan(1993)] to determine patterns of classification of content. These patterns, in a second level, given the possibility to determine independently whether an ad has not liked or user. The study of cortical signals at a high level can be transcendent to establish direct and automatic relationship between tastes and mental reflexes, which will make the user environment to suit user preferences [Morin(2011)].

2   Related Works

The first psychological studies done using EEG date as far as 1979. His studies and others later validated that electrical patterns were lateralized in the frontal region of the brain. Generally, the measure of alpha-band waves (813 Hz) in the left frontal lobe indicates positive emotions. A strong involvement of parietal areas during the observation of the TV commercials with an affective and cognitive content was also noted in a previous study, performed by using sophisticated MEG recordings. The magneto field tomography (MFT) results showed an increasing activity during the observation of cognitive stimuli rather than affective commercials in parietal and superior prefrontal areas. These regions are known to be associated with executive control of working memory and maintenance of highly processed representation of complex stimuli. Although the affect-related activations are more variable across subjects, these findings are consistent with previous PET and fMRI studies showing that stimuli with affective content modulates activity in the orbitofrontal and retrosplenial cortex, amygdale, and brainstem. However, in this study is to use a low-resolution sensor and an absequible price for this task [Vecchiato *et al.(2011)Vecchiato, Astolfi, De Vico Fallani, Toppi, Aloise, Bez, Wei, Kong, Dai, Cincotti, et al.*]. There are several startups that have products under development similar to Neurosky Mindwave (1) features, but with a much smaller size. Many of these products under development have a similar appearance to a wireless headphones, which makes the user can wear them continuously. This step in electroencephalograms wearable will require the use of simple devices and efficient algorithms to determine various aspects and relationships such as neuronal including a recognition of advertising tastes. There are also various applications of brain computer interfaces oriented to domestic use. Home automation and applications may be fostered by the massive use of this technology. Today, thanks to systems such as OSGi [Alliance(2003)] providers can connect context of this typology [Martín *et al.(2009)Martín, Seepold, Madrid, Álvarez, Fernández-Montes, and Ortega, Martínez Fernández* et al.(2010)Martínez Fernández, Seepold, Augusto, and Madrid, Soria-Morillo *et al.(2011)Soria-Morillo, Ortega-Ramírez, González-Abril, and Álvarez-García*].

Fig. 1. Neurosky mindwave device

3   Dataset

In order to carry out the taste ads recognition system, a dataset was produced. This dataset has been carried out by displaying a set of 14 ads for a total of 10 people of different age and sex. The distribution of subjects in the experiment is shown in Table 1.

The concentration level refers to the attitude presented by the person before conducting the test. This attribute can have, average high values (high interest in performing the test and little impact from external distractions) (high interest in performing the test and noticeable impact from external distractions) and low (average interest in conducting proof and high impact from external distractions). Users with different values in this attribute have been selected to test the robustness of face recognition patterns of middle- and high distraction, which are very common when users view the ads. The dataset was generated automatically by an Android app developed for this purpose. This application stores the videos which will be played. Once the videos have been played, the application prompts the user to issue a rating on them and, previously, indicating which video reminds (the latter can be understood as an indicator of the impact of the announcement on the user). The result is an XML file that contains the user's profile,

Table 1. Subject profile on the data recovery process

Subject	Age	Gender	Concentration level
Subject 1	28 years	man	high
Subject 2	31 years	woman	medium
Subject 3	27 years	woman	medium
Subject 4	26 years	man	high
Subject 5	54 years	man	low
Subject 6	59 years	woman	medium
Subject 7	60 years	man	high
Subject 8	63 years	woman	low
Subject 9	42 years	man	high
Subject 10	46 years	woman	medium

Listing 1.1. content of dataset description XML

```
1  <session>
2      <start_time>21:10:00</start_time>
3      <end_time>21:26:37</end_time>
4      <subject>
5          <name>Rosa Maria Rodriguez Calvo</name>
6          <age>32</age>
7          <country>spain</country>
8      </subject>
9      <times_viewed>0</times_viewed>
10     <conditions>In the living room. TV powered off. No sound headset connected.
           A child slept in the same room.</conditions>
11     <ads_time>
12         <ad_time>
13             <start>00:07</start>
14             <end>00:38</end>
15         </ad_time>
16
17         <ad_time>
18             <start>00:49</start>
19             <end>01:33</end>
20         </ad_time>
21     </ads_time>
22     <remembers>
23         <ad_remembered>15</ad_remembered>
24         <ad_remembered>1</ad_remembered>
25     </remembers>
26     <ads_score>
27         <ad_score>
28             <id>1</id>
29             <score>3</score>
30         </ad_score>
31
32         <ad_score>
33             <id>2</id>
34             <score>1</score>
35         </ad_score>
36     </ads_score>
37     <ads_viewed>
38         <ad>
39             <id>1</id>
40             <name>Car</name>
41             <link>https://www.youtube.com/watch?v=PBE98UMYsH0&amp;index=1&amp;list
                 =PLrHbIqWxx2UdIurxjIWmcfnr6ROsMP4tK</link>
42             <duration>00:32</duration>
43         </ad>
44
45         <ad>
46             <id>2</id>
47             <name>Coke</name>
48             <link>https://www.youtube.com/watch?v=pDSU6q6eD34&amp;list=
                 PLrHbIqWxx2UdIurxjIWmcfnr6ROsMP4tK&amp;index=2</link>
49             <duration>00:47</duration>
50         </ad>
51     </ads_viewed>
52  </session>
```

temporal information on the beginning and end of display ads, ads remembered by the user after completion of the test, score given for each ad and, finally, information for each ad. The score for each ad can be in the range [0, 5], with 0 being the minimum score (has not liked) and 5 high (liked it a lot). With this XML file, the information visualization session is stored. This information is important when learning process is conducting. On the other hand, while the user proceeds to displaying content, by Neurosky MindWave device information about brain activity associated is collected. All values delivered by the device are stored in order to obtain new statistics from them. The base signals are meditation, attention, delta amplitude, theta amplitude, low alpha amplitude, high alpha amplitude, low beta amplitude, high beta amplitude, low gamma amplitude and high gamma amplitude. This information can be seen in Figure 2.

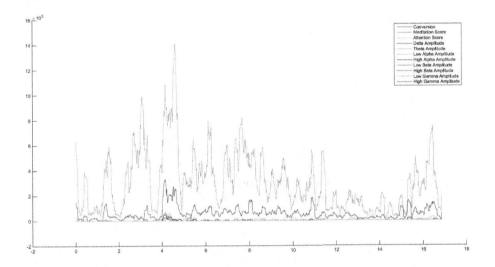

Fig. 2. EEG data from Neurosky Mindwave

From this database is possible to obtain a total of 480 statistical sampling parameters. These parameters will be considered hereafter as input variables for the classification system. Thanks to statistical signal derivatives from base values, discrete relationships can be established between instances with no temporary data processing needed. By one side, this allows to reduce the complexity of the learning and classification process and by other side, results will be more readable.

4   Advertising Recognition Method

In order to implement the recognition system a learning process on the data described above has been previously applied. With the purpose of allow comparison between different classification techniques, two methods have been studied:

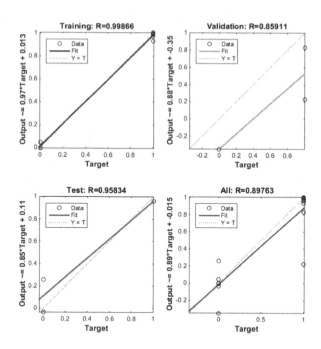

Fig. 3. Linear regression of targets relative to outputs

neural networks and decision tree. In the first case it has been decided to use a neural network where different statistical inputs are obtained from biometric parameters of the EEG, and targets are the score associated with each instance (advertising). For validation has established a set of 10-folds cross validation with 70% of training data, 15% of validation and 15% of testing. The network configuration, as shown in Figure 4, is composed of two layers (one hidden and one output). The hidden layer consistes on five neurons, while the output layer has only one neuron. This configuration has been obtained from experiments and their results are shown in Figure 5. Sigmoid type of neurons have been employed in both layers, the hidden and the output layers. The algorithm used for training process was the Scaled Conjugate Gradient and performance is optimized by the function Cross-Entropy. In Figure 3 regression line during the learning process can be seen.

On the other hand, a decision tree for classifying instances has been used. In this case, the classification algorithm is C4.5 [Quinlan(1993)] was applied from the dataset of statistical inputs. The structure of the resulting tree is shown in Figure 6. Used this method, the accuracy is around 71 %. The testing and validation process employed in this case has been the same to the previous setting using ANN (Artificial Neural Networks) [Hagan et al.(1996)Hagan, Demuth, Beale, et al.], so the results can be compared.

The results of the comparison made from EEG dataset using the above two methods is shown in Table 2. Rows show the different classes based on the

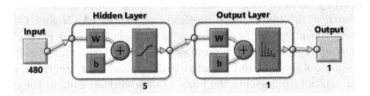

Fig. 4. Neural network configuration with 5 neurons in an unique hidden layer

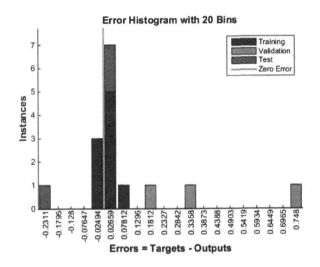

Fig. 5. Neural network training error histogram

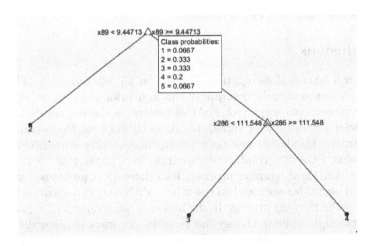

Fig. 6. Binary tree structure after the training process

score assigned by users to each advertisement (1-5). Columns present the details associated with each classification methods employed (ANN and C4.5). First column shows the number of advertisement well-classified for each class using ANN. the second one presents the number of advertisements misclassified. Same items for results obtained from C4.5 algorithm are shown in the third and fourth column. Next column represents the number of advertisements of the dataset on each class (score). Finally, the accuracy obtained for ANN and C4.5 is presented in the last two columns.

Table 2. Classification results for short advertising movies from EEG inputs

Class	ANN success	ANN errors	C4.5 success	C4.5 errors	Total instances	ANN accuracy (%)	C4.5 accuracy (%)
Class 1	15	3	13	5	18	83	72
Class 2	18	4	16	6	22	81	72
Class 3	12	4	11	5	16	75	68
Class 4	27	6	22	11	33	81	66
Class 5	19	5	16	8	24	79	66
Summary	91	22	78	35	113	80	69

5 Future Works

Future efforts can be oriented to improve the multimodal classification performance of this works. In one hand, data-driven approach, e.g., principal component analysis (Lin et al., 2009) and independent component analysis (Lin et al., 2010a), might be feasible to further elaborate the EEG spatio-spectral dynamics associated with implicit emotional responses. On the other hand, additional information could be included in the short advertising movies in order to determine low-level details influences on users liking.

6 Conclusions

In this paper a system of recognition of tastes on TV ads has made. The system proposed can automatically determine, through prior learning, whether users like the advertisements displayed. For this purpose a dataset consisting of over 400 statisticians from 11 basic signals obtained through electroencephalography device Neurosky Mindwave has been used. Subsequently a reduction process variables using PCA has conducted, resulting in a dataset of 15 variables to perform the statistical learning process. Two different algorithms, bye the one side artificial neural network and on the other side binary classification tree, have been used for the learning process. In the first case, an accuracy of approximately 80% was obtained, while in the second case the accuracy is lowered to 69% on average. Although the results are not excessively high, it is an important step towards inclusion of such systems on a daily life using low cost wearable sensors such as Neurosky Mindwave.

Acknowledgments. This research is partially supported by the projects of the Spanish Ministry of Economy and Competitiveness HERMES (TIN2013-46801-C4-1-r) and Simon(TIC-8052) of the Andalusian Regional Ministry of Economy, Innovation and Science.

References

[Lee et al.(2007)Lee, Broderick, and Chamberlain] Lee, N., Broderick, A.J., Chamberlain, L.: What is neuromarketing? A discussion and agenda for future research. International Journal of Psychophysiology 63, 199–204 (2007)

[Ariely and Berns(2010)] Ariely, D., Berns, G.S.: Neuromarketing: the hope and hype of neuroimaging in business. Nature Reviews Neuroscience 11, 284–292 (2010)

[Morin(2011)] Morin, C.: Neuromarketing: the new science of consumer behavior. Society 48, 131–135 (2011)

[Senior and Lee(2008)] Senior, C., Lee, N.: Editorial: A manifesto for neuromarketing science (2008)

[Murphy et al.(2008)Murphy, Illes, and Reiner] Murphy, E.R., Illes, J., Reiner, P.B.: Neuroethics of neuromarketing. Journal of Consumer Behaviour 7, 293–302 (2008)

[Hagan et al.(1996)Hagan, Demuth, Beale, et al.] Hagan, M.T., Demuth, H.B., Beale, M.H., et al.: Neural network design, vol. 1, Pws Boston (1996)

[Dreiseitl and Ohno-Machado(2002)] Dreiseitl, S., Ohno-Machado, L.: Logistic regression and artificial neural network classification models: a methodology review. Journal of Biomedical Informatics 35, 352–359 (2002)

[Quinlan(1993)] Quinlan, J.R.: C4. 5: programs for machine learning, vol. 1. Morgan kaufmann (1993)

[Vecchiato et al.(2011) Vecchiato, Astolfi, De Vico Fallani, Toppi, Aloise, Bez, Wei, Kong, Dai, Cincotti, et al.] Vecchiato, G., Astolfi, L., De Vico Fallani, F., Toppi, J., Aloise, F., Bez, F., Wei, D., Kong, W., Dai, J., Cincotti, F., et al.: On the use of EEG or MEG brain imaging tools in neuromarketing research. Computational Intelligence And Neuroscience 2011, 3 (2011)

[Alliance(2003)] Alliance, O.: Osgi service platform, release 3. IOS Press, Inc. (2003)

[Martín et al.(2009)Martín, Seepold, Madrid, Álvarez, Fernández-Montes, and Ortega] Martín, J., Seepold, R., Madrid, N.M., Álvarez, J.A., Fernández-Montes, A., Ortega, J.A.: A home e-Health System for Dependent people based on OSGi. In: Madrid, N.M., Seepold, R.E.D. (eds.) Intelligent Technical Systems. LNEE, vol. 38, pp. 117–130. Springer, Heidelberg (2009)

[Martínez Fernández et al.(2010)Martínez Fernández, Seepold, Augusto, and Madrid] Fernández, J.M., Seepold, R., Augusto, J.C., Madrid, N.M.: Sensors in trading process: A Stress Aware Trader. In: IEEE 8th Workshop on Intelligent Solutions in Embedded Systems (WISES), pp. 17–22 (2010)

[Soria-Morillo et al.(2011)Soria-Morillo, Ortega-Ramírez, González-Abril, and Álvarez-García] Soria-Morillo, L.M., Ortega-Ramírez, J.A., González-Abril, L., Álvarez-García, J.A.: Mobile architecture for communication and development of applications based on context. In: Mehrotra, K.G., Mohan, C.K., Oh, J.C., Varshney, P.K., Ali, M. (eds.) IEA/AIE 2011, Part II. LNCS, vol. 6704, pp. 48–57. Springer, Heidelberg (2011)

Heart Rate Variability Indicating Stress Visualized by Correlations Plots

Wilhelm Daniel Scherz[1], Juan Antonio Ortega[2],
Natividad Martínez Madrid[3], and Ralf Seepold[1]

[1]HTWG Konstanz, Brauneggerstr. 55, 78462 Konstanz, Germany
{wscherz,ralf.seepold}@htwg-konstanz.de
http://uc-lab.in.htwg-konstanz.de
[2] Universidad de Sevilla, Avda. Reina Mercedes s/n, 41012 Sevilla, Spain
jortega@us.es
http://www.us.es
[3]University Reutlingen, Alteburgstraße 150, 72762 Reutlingen, Germany
natividad.martinez@reutlingen-university.de
http://iotlab.reutlingen-university.de

Abstract. Stress is recognized as a factor of predominant disease and in the future the costs for treatment will increase. The presented approach tries to detect stress in a very basic and easy to implement way, so that the cost for the device and effort to wear it remain low. The user should benefit from the fact that the system offers an easy interface reporting the status of his body in real time. In parallel, the system provides interfaces to pass the obtained data forward for further processing and (professional) analyses, in case the user agrees. The system is designed to be used in every day's activities and it is not restricted to laboratory use or environments. The implementation of the enhanced prototype shows that the detection of stress and the reporting can be managed using correlation plots and automatic pattern recognition even on a very light-weighted microcontroller platform.

1 Introduction

For humans stress is regarded as a negative sensation and organisations like the World Health Organisation (WHO) recognise stress as a predominant disease [1] because of its continuous presence in modern life. Among well-known consequences of high and perdurable stress is failure to respond adequate to fiscal, mental and emotional demands [2, 3, 4]

As a result of high level of stress in modern society the amount of people who face long term limitations like burnouts or cardiac infarcts is increasing. This leads to the growth of the treating and healing costs for people suffering from stress illnesses. Assuming that many societies have an aging population, stress and the negative consequences of stress will influence the age and healthiness of people.

Stress appeared as a natural response that allowed people to react fast and effectively in dangerous situations. Stress triggers biological mechanisms that reorganise the priorities and works of the organism in order to reach the maximum performance

F. Ortuño and I. Rojas (Eds.): IWBBIO 2015, Part II, LNCS 9044, pp. 710–719, 2015.
© Springer International Publishing Switzerland 2015

when facing threat. This is also called the 'fight or flight response' [5].

Nowadays stress is the result of the exposure to high demands and pressure in daily life that can be both mental and physical [6], e. g. constant desertions demand or constant time pressure. Stress became permanent and due to this it causes variety of disorders. Among the symptoms of overabundance of stress are fatigue, sleep problems, etc. [7].

Stress can be self-induced or induced in a laboratory by using special exercises like the Trier test [8] or the Strop test. **Fig. 1** shows some of the symptoms of stress. The reaction to stress has not changed over the time although lifestyle, technologies and everyday habits changed a lot.

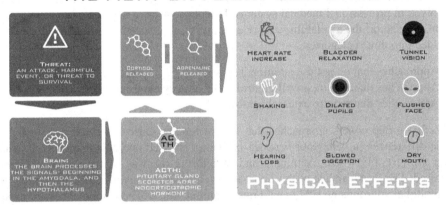

Fig. 1. Symptoms and physical response of stress

As shown in **Fig. 1** in case of a threat, the body activates several processes that prepare it for a fight against the threat or for fleeing from it. The brain is supporting this process, when releasing cortisol and adrenaline hormones, reducing the functionality of systems that are not necessity for imminent surviving like genitourinary system, digestion, hearing, peripheral view, etc. Furthermore, it is increasing the functionally of mechanisms supporting successful flee or fight strategies like an increased heart rate or dilated pupils.

One of the main reasons why stress is underestimated and at the same time one of the factors that complicate its detection is the subjective perception of stress: Some people demonstrate immediately symptoms of stress while others do not notice when passing the threshold of just 'being busy' to an objective high stress level [9]. This is why it is important to find an objective way to determine stress, or if possible, to determine the moment before a person passes the threshold.

The approach used in this work is based on the electrical characteristics of the heart (ECG). The system consists of hardware and software platform capable of hosting various algorithm and sensors for biological parameter measurement. The platform provides basic connectivity to a body area network and telemetric support for professional online analysis. The user is continuously informed about the current status.

2   State of the Art

The monitoring of indicators detecting stress has a long history but often it is used only to capture physical parameters like the heart rate without correlating the parameters directly. In only few cases the person will receive direct feedback but it is wearing a black-box. The results are analysed offline and diagnosis are reported later. However, due to miniaturization we notice a shift from purely professional and certified systems into a grey zone of recommendation and reporting devices not directly involved in professional medical systems. Independent from the area of expertise and from cost factor, the availability and relevant type of devices for our approach can be divided into three categories:

— Approaches that do not use additional sensors
— Approaches that require a well-controlled laboratory environment
— Approaches that require external sensors

The first group covers approaches that do not require sensors use and they analyse little differences in behaviour between not being stressed and being stressed. Examples are [10, 11] where examining and monitoring the way of typing while being stressed. The disadvantage of this approach is complexity and difficulty of adapting it to multiple environments and to calibrate to an individual. These approaches are most often context based and not human centred.

The second group are approaches requiring laboratory environments. Most often, they provide accurate and precise stress detection. The hormones cortisol, adrenaline and other stress hormones that are released in saliva and blood are measured for determining stress [12]. The limitation of these methods is that they lack mobility together with real time detection. Besides that these approaches are invasive and expensive due to the necessary equipment and precision.

The third group are characterized by approaches that use external psychological sensors like in [13]. For example, stress measurement while driving. The driver is monitored with an electrocardiogram (ECG) and an electromyogram that records the electrical activity of muscles (EMG), skin conductivity (SC), breath sensor and video camera that observes the driver. The main disadvantage is the limitation in the degree of movement and in this case the missing online analysis of all the data after it was collected. So, the driver is not receiving immediate feedback.

The approach that we develop uses an own developed low cost ECG which is compact, wearable, non-invasive and real time capable what allows different and usage in different contexts. The analyses results can be reported directly to the user via different very simple (or more detailed) user-interfaces, while the raw data is buffered or passed online for further processing by professionals. The buffer capability is relevant since mobile solutions do not offer connectivity in every place (e.g.

metro ride). This paper is also based on our previous studies and models for stress measurement [2, 4].

3   System Architecture

According to the previous studies mentioned above, we took the decision to develop a light weighted and low-cost system independent from smartphone sensor capabilities. In this approach, we concentrate on the ECG signals first because we would like to evaluate how accurate is the stress detection when capturing only one biological parameter. Of course, the system is open to capture more parameters and to cooperate with other external sensors (like smartphone sensors). For this enhanced prototype, we used the smartphone as the communication platform providing connectivity. The principle architecture is shown in **Fig. 2**. The system is composed of three parts: ECG sensors with the capability of continuous recording/extracting biological data, microprocessor for processing the signal and visualization device to give feedback to the user. A smartphone (or similar) to provide connectivity.

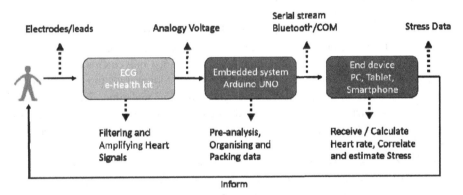

Fig. 2. System architecture for collecting, pre-processing and visualizing of biological data

The ECG board computes the signal obtained from the electrodes. Like in traditional ECG devices, three electrodes are used. The ECG unit composes the signal and send it as an analogue output to the microcontroller. In the moment, the microcontroller digitalizes the analogue signal and performs some pre-processing and filtering. Using the digitalized data we calculate the heart rate (HR) and the RR interval that is defined as the interval between two R peaks as shown in **Fig. 3**. The RR interval and the heart rate are later used for determining stress. With the help of the prototyping

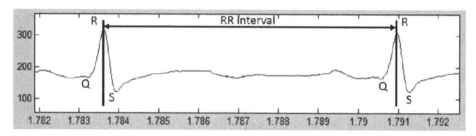

Fig. 3. Definition of an RR interval in a QRS segment

board or the connected smartphone, the user can be informed about his current status. As used in previous prototypes, a traffic light interface is suited to display easy to catch information. The interface display may be part of the board or the smartphone device implementing a small app. The more precise data is passed to the backend as stress data.

4   Stress Detection Method

Our method to detect stress is based on the ECG signal. As mentioned there are several possibilities to capture biological parameters, however, the ECG signal is easy to capture and the way to capture it is quite resistant against movements and thus guaranteeing an always-available signal. Based on the system architecture, the prototype supports direct access and processing of sensor data in real time. The ECG signal describes the electrical characteristic of the heart and is usually used for diagnostic proposes [14]. At the same time ECG data is unique enough to be used for identification of persons[15].

As shown in **Fig. 3**, we use the RR interval (the RR interval represents the duration between two consecutive heartbeats (1)) or the heart rate to calculate the heart rate variability (HRV). Later, the HRV is used for determining stress.

$$RR_{interval} = R_i - R_{i-1},\qquad(1)$$

HRV is calculated by examining the relations between two heart beats. Usually HRV correlates strongly with the respiration sinus. In [16] the influence of the breathing sinus on the heart rate is described. We can assume that HRV stays almost constant when a person is not stressed and the variation stops behaving regularly when the person is stressed.

Fig. 3 is normalised in the y-axis over a range between 0 to 350 mV (2) and the x-axis shows the time in ms (3).

$$0 \leq Y \geq 350,\qquad(2)$$

$$x \geq 0,\qquad(3)$$

To optimise and reduce the resources needed, for the detection of the R peaks and for the calculation of the RR interval (1) we define a threshold value of 250 mV

$$R_j > 250_{mv},\tag{4}$$

And so that the current R value is a peak, it has to be a maxima (5), R_i is the list of RR maxima

$$\left(R_{j-1} < R_j\right) \wedge \left(R_j > R_{j+1}\right) \xrightarrow{Then} R_i = R_j,\tag{5}$$

Only if both conditions (4, 5) are fulfilled, we have successfully detected R-peak (5). When we have detected two R peaks we can calculate the RR interval (1) by measuring the time difference between two peaks.

Next step is to correlate the RR and the HRV computed. If we visualise the obtained data in a two dimensional space, we obtain a correlation plot, e.g. **Fig. 4.**

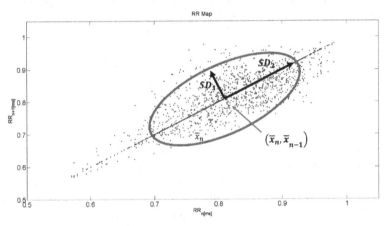

Fig. 4. Correlation plot of RR intervals that visualizes the HRV and their self-similarity

The x-axis and y-axis of the plot are defined as (6, 7).

$$X = \{RR_i, RR_{i+1}, \dots RR_n\},\tag{6}$$

$$Y = \{RR_{i+1}, RR_{i+2}, \dots RR_{n+1}\},\tag{7}$$

In the plot **Fig. 4** high concentration of values is notable in the centre of the plot. This concentration of the points indicates lower stress level. If the values would spared ao lot, it would indicate that the person is under stress.

The total variability can be expressed as the product of SD_1 and SD_2.

$$SD_1 = \sqrt{var(x_1)} \rightarrow x_1 = \frac{x_n - x_{n+1}}{\sqrt{2}},\tag{8}$$

$$SD_2 = \sqrt{var(x_2)} \rightarrow x_2 = \frac{x_n + x_{n+1}}{\sqrt{2}},\tag{9}$$

The current variance of the RR interval is calculated using the standard derivation like in (11) and (10).

$$\sigma = \left(\frac{1}{n-1} \sum_{i=0}^{n} (x_i - \overline{x})^2 \right)^{\frac{1}{2}}, \tag{10}$$

$$\sigma = \left(\frac{1}{n} \sum_{i=0}^{n} (x_i - \overline{x})^2 \right)^{\frac{1}{2}}, \tag{11}$$

5   Application of System and Results Obtained

The system has been implement on top of an Arduino Uno R3 prototyping board with the self-made ECG component stacked on top of the board. The ECG electrodes have been connected to the board and besides that a third board has been stacked in top of this providing a slot for a small SD memory card. The memory card may be integrated in the future directly to the ECG board as part of a future redesign. All candidates have been volunteer students aged between 23 and 28, none of them were smokers or alcoholics. The methods used in this work assume that none of the candidates suffered from cardiac problems or mental anomalies nor used a pacemaker.

As a method for inducing stress on the volunteers, we used the Trier Social Stress Test [8] because this method can be easily established in our laboratory environment and a driving simulator that has been used for the tests as well.

The Trier Social Stress Test was realised in three phases (anticipation period, presentation period and cool down period), the duration of each was 5 minutes. During the test, the volunteer has to prepare and make a small presentation on a random topic.

The driving simulator that we used in the experiment requires two mechanism to induce stress. A points system is introduced to reward the volunteer for fast and complex driving manoeuvres. The level of difficulty increases over time. As consequence, the fastest driver usually makes more complicated manoeuvres to increase the reward points but at the same time such driving manner increases the probability to lose all the points in case of mistake (accident).

The data that was obtained during the experiment has been processed and visualised using our approach. In the following, results from two volunteers are shown. The first person has a lower heart rate but is being under constant stress. The second person has a higher heart rate but is not under stress. Both datasets have the same length.

Fig. 5 and **Fig. 6** show the correlations plots of the RR intervals and **Fig. 7** and **Fig. 8** show the HRV.

Fig. 5 clearly shows that the values are more wide spread than in **Fig. 6**. This spreading of the values is caused by the stress and the conditions that were mentioned above (**Fig. 1**). Stress influences the heart rate and as the result the variation between two heart beats becomes bigger. Most of the values in **Fig. 5** belong to the interval between 0.7 and 0.9 sec for the RR interval (heart rate interval is between 66.7 bpm and 85.7 bpm).

Fig. 6 shows that the values of the second volunteer (relaxed and breathing calm and regularly) without stress does not spread as much as the values in **Fig. 5** although the heart rate is higher (the RR interval is shorter). Most of the values are between 0.45 sec and 0.6 sec (130 bpm and 85 bpm)

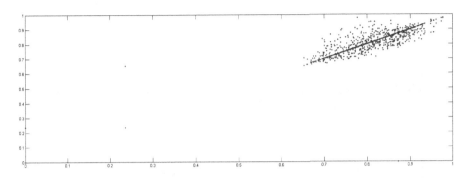

Fig. 5. Candidate under stress and with low HR

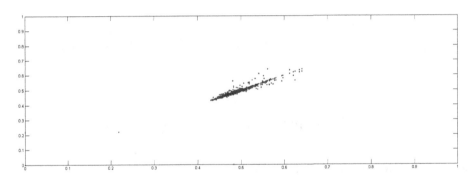

Fig. 6. Candidate no stress and with high HR

Comparing of the **Fig. 7** and **Fig. 8** shows that the values in **Fig. 8** vary less and more regularly than the values of **Fig. 7**.

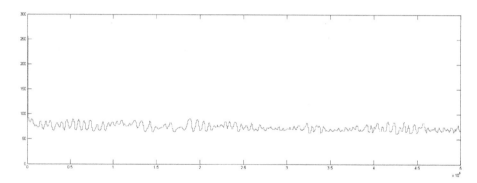

Fig. 7. Continuous HRV visualization of not uniform signal of a stressed candidate

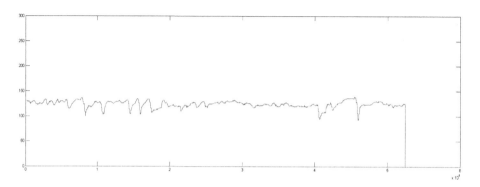

Fig. 8. Continuous HRV visualization of not stressed candidates

6   Conclusions and Future Work

This work presents a system architecture and an algorithm for calculating stress based on the HRV. The system is designed to be light-weighted, to integrate in mobile, cheap and easy to use devices like smartphone or even without the smartphone. The prototype can be currently connected to a different system like PC and smartphones. The algorithm for stress detection has been implemented with a small footprint and can be deployed in small prototyping board. For the experiment we used an Arduino Uno R3 development board.

The results of experiments has shown that stress can be detected even in cases when people have naturally low or high heart rate.

A future challenge of the system is the sensibility to abrupt and very strong movements, for example in some sporting activities. The movement of the electrodes during these measurements may cause strong interferences and artefacts in the signal. A solution to this problem could be a software reconstruction and filtering of the signal or using the same approach with a sensor that does not generate interferences. In the next step, it is planned to use a more powerful microcontroller boards in order to shift the processing functionality to the place where the data is captured, apply more sophisticated filters and overcome lacks in sensor signals during some sport activities. A third step will be the redesign of the ECG boards to tune it for low power and being more compact to be mobile.

References

[1] WHO, Cross-national comparisons of the prevalences and correlates of mental disorders. Bulletin of the World Health Organization, pp. 413–426 (2000)

[2] Fernández, J.M., Augusto, J.C., Seepold, R., Madrid, N.M.: A Sensor Technology Survey for a Stress Aware Trading Process. IEEE Trans. on Systems, Man and Cybernetics Part C: Applications and Reviews 42(6), 809–824 (2012)

[3] Fernández, J.M., Augusto, J.C., Seepold, R., Madrid, N.M.: Why Traders Need Ambient Intelligence. In: Augusto, J.C., Corchado, J.M., Novais, P., Analide, C. (eds.) ISAmI 2010. AISC, vol. 72, pp. 229–236. Springer, Heidelberg (2010)

[4] Fernández, J.M., Augusto, J.C., Trombino, G., Seepold, R., Madrid, N.M.: Self-Aware Trader: A New Approach to Safer Trading. Journal of Universal Computer Science (2013)

[5] Jansen, A.S.P., Van Nguyen, X., Karpitskiy, V., Thomas, C.: Central command neurons of the sympathetic nervous system: basis of the fight-or-flight response. Science 270, 644–646 (1995)

[6] Kidd, T., Carvalho, L.A., Steptoe, A.: The relationship between cortisol responses to laboratory stress andcortisol profiles in daily life. Biological Psychology, 34–40 (2014)

[7] Torbjörn, Å., John, A., Mats, L., Nicola, O., öran, K.G.: Do sleep, stress, and illness explain daily variations in fatigue?. Journal of Psychosomatic Research, 280–285 (2014)

[8] Kirschbaum, C., Pirke, K.-M., Hellhammer, D.H.: The 'Trier Social Stress Test'- A Tool for Investigating Pyschobiological Stress Responses in a Laboratory Settings. Neuropychobiologie, 78–81 (1993)

[9] Madrid, N.M., Fernandes, J.M., Seepold, R., Augusto, J.C.: Ambient assisted living (AAL) and smart homes. Springer Series on Chemical Sensors and Biosensors, vol. 13, pp. 39–71 (2013)

[10] Gunawardhane, S.D., De Silva, P.M., Kulathunga, D.S., Arunatileka, S.M.: Non Invasive Human Stress Detection Using Key Stroke Dynamics and Pattern Variations. In: International Conference on Advances in ICT for Emerging Regions (ICTer), Colombo (2013)

[11] Vizer, L., Zhou, L., Sears, A.: Automated stress detection using keystroke and linguistic features: an exploratory study. Int. J. of Human-Computer Studies 67(10), 870–886 (2009)

[12] Juliane, H., Melanie, S.: The physiological response to Trier Social Stress Test relates to subjective measures of stress during but not before or after the test. Psychoneuroendocrinology 37(1) (2012)

[13] Healey, J.A., Picard, R.W.: Detecting Stress During Real-World Driving Tasks Using Physiological Sensors. IEEE Transactions on Intelligent Transportation Systems 6(2) (2005)

[14] Dubin, D.: Rapid Interpretation of EKG's, Tampa. COVER Publishing Co., Florida (2000)

[15] Israel, S.A., Irvineb, J.M., Chengb, A., Wiederholdc, M.D.D., Wiederholdd, B.K.: ECG to identify individuals. Pattern Recognition, 133–142 (2004)

[16] Hirsch, J.A., Bishop, B.: Respiratory sinus arrhythmia in humans: how breathing pattern modulates heart rate. American Journal of Physiology - Heart and Circulatory Physiology (1981)

Emotions and Diabetes

Charrise M. Ramkissoon and Josep Vehí[*]

Institut d'Informatica I Aplicacions, University of Girona, Spain
(Charrise.Ramkissoon,Josep.Vehi)@udg.edu

Abstract. Diabetes Mellitus (DM) is a chronic metabolic disorder characterized by a deficiency in the function of β-cells, which are pancreatic cells responsible for the production of the hormone insulin. Psychological changes induce physical responses, which can be detected using sensors. It is believed that soon sensor arrays will be available which will be able to detect specific behavioral changes within a patient. The psychological changes that occur to those with DM can have significant effects on metabolic control. One of the major hurdles in this area is the need for a better understanding of psychological changes and their influence on blood glucose changes. However, these changes have yet to be quantified in an appropriate manner. Using additional sensors can lead to new ways of gathering data that will assist us in overcoming many of the obstacles of adequate diabetes care.

Keywords: Diabetes, Stress, Insulin Sensitivity.

1   Introduction

Diabetes Mellitus (DM) is a chronic metabolic disorder characterized by a deficiency in the function of β-cells, which are pancreatic cells responsible for the production of the hormone insulin. The diminished level of this hormone produces chronic hyperglycemia [1, 2]. DM is associated with an increased risk of premature birth, particularly because it is linked with a greater risk of cardiovascular diseases. In addition, DM patients are at greater risk of becoming blind, of suffering from renal insufficiency and losing their lower limbs to amputation [2, 3]. DM is broadly classified into two categories, type 1 diabetes mellitus (T1DM) and type 2 diabetes mellitus (T2DM). Both arise from complex interactions between genes and the environment and are thought to result in the reduction of insulin-producing β-cells and deficits in β-cell function—however their pathogenesis is distinct [1], [4].

Recently, T1DM research has increased as the potential for artificial pancreas (AP) has also increased. An AP is an electrochemical closed-loop device consisting of a continuous glucose monitor (CGM), insulin pump and controller that will alleviate many burdens encountered by DM patients. This research has led many researchers to focus on the physiological changes that occur in humans and their effect on DM patients as these changes will have to be accounted for by the controller in this device.

---

[*] Corresponding author.

F. Ortuño and I. Rojas (Eds.): IWBBIO 2015, Part II, LNCS 9044, pp. 720–727, 2015.
© Springer International Publishing Switzerland 2015

Many of these physiological changes can be induced by different emotions experienced by individuals that can trigger a hormonal response, which in turn, is reflected in blood glucose (BG) levels.

The influences of psychosocial and family factors on metabolic control have long been the subject of research, especially among children and youth with T1DM [5-7]. At the individual level, external locus of control, delayed intellectual and emotional development, impulsive and avoidant coping styles, and number of life events have been associated with poorer metabolic control [8-10].

Patients with DM must cope with a wide range of challenges specific not only with the disease but also to other areas of their lives, which influences their disease management and metabolic control. According to the American association of Diabetes educators: Psychological stress directly affects health and indirectly influences a person's motivation to keep their diabetes under control which can cause a deterioration in their ability to self-manage their disease [11].

Research on the effect of stress and how stress management could contribute to diabetes care began in the 1980s [12], including the role of depression and its relationships with metabolic control and psychological interventions to treat it [13-17]. This also included how psychological, behavioral, emotional, metabolic, genetic, and other biological factors interact in the expression of disease, course, complications, longevity, and quality of life. The American Association of Diabetes Educators has advanced recognition of the many connections among coping, behavior, emotions, and metabolism in diabetes management by identifying healthy coping as 1 of 7 key diabetes management behaviors.

Psychological changes can sometimes take a longer period of time to detect using conventional methods. In addition to healthy coping for the management of behavior changes, other methods can be employed to detect changes in behavior and counteract them.

2   Stress, Illness and Diabetes

Hyperglycemia resulting from medically stressful conditions is associated with a high degree of morbidity and mortality and is known to impair the healing of surgical and nonsurgical wounds. During periods of medical illness, emotional stress or treatment with medications such as corticosteroids, a fall in tissue sensitivity to insulin occurs. Such a change increases the requirement for insulin, and without compensation, predictability leads to hyperglycemia in those with DM [18-22]. Another cause of poorly controlled glycemia during periods of stress is the change in patient behavior that occurs. Changes in self-management with respect to diet, exercise and other behaviors may adversely affect glucose control [23].

Two studies on stress of note were done in 2010 and 2011 respectively both inducing stress with the use of corticoid steroids, which induce insulin resistance in peripheral tissues (muscle and fat) and in the liver [19], [24].

In 2010, a study by Finan et al. [24] uses a statistically based multivariate monitoring technique to attempt to distinguish between normal days (regular insulin sensitivity) and abnormal days (decrease in insulin sensitivity) using principal component analysis (PCA). This was based on statistical relationships among glucose measurements, insulin infusion rates, and meal information. Two metrics were used to test the model: sensitivity, which is the ability to detect normal days and abnormal day and specificity, the ability to classify days correctly when identified. This study had an 89% success rate with sensitivity and an 89% success rate with specificity. This study also investigated insulin pump occlusions and glucose sensor malfunction but did not report findings.

In 2011 a study done by El Youssef et al. [19] tested a novel adaptive algorithm's ability to detect and respond to reduced insulin sensitivity resulting from corticosteroid administration. The algorithm estimates insulin sensitivity at regular intervals and responds to those changes. This algorithm was named the adaptive proportional-derivative (APD) system and, based on insulin sensitivity, adjusts the gain factors used in a fading memory proportional-derivative (FMPD) algorithm that was previously developed. The FMPD algorithm incorporates the glycemic history, feeding exponentially-weighted errors of proportional and derivative components. This study compared glucose control using the APD algorithm vs. the FMPD algorithm.

This study found that in response to steroids, the APD algorithm consistently raised the total daily insulin requirement (TDIR), which was used to determine meal boluses, and raised the control gain factors, leading to higher insulin infusion rates, which led to lower glucose levels than those observed during the comparable period of the FMPD algorithm and the APD algorithm did not increase the incidence of hypoglycemia as compared to the FMPD algorithm. The APD algorithm however, has a very slow response rate which would not be appropriate when there are rapid changes in glucose levels such as in the scenario of intense exercise, which can quickly lead to large changes in insulin sensitivity and noninsulin-mediated glucose uptake [19].

One item of note in the use of steroids to study stress and the decrease in insulin sensitivity is the effect of steroids on the patient. Some patients may be more sensitive to the effect of steroids than others causing a larger decrease in insulin sensitivity. Also, the effects of corticosteroids are widespread, including profound alterations in carbohydrate, protein, and lipid metabolism, and the modulation of electrolyte and water balance. Corticosteroids affect all of the major systems of the body, including the cardiovascular, musculoskeletal, nervous, and immune systems, and play critical roles in fetal development including the maturation of the fetal lung. The direct effects of corticosteroids are sometimes difficult to separate from their complex relationship with other hormones, in part due to the permissive action of low levels of corticosteroid on the effectiveness of other hormones, including catecholamines and glucagon. Quantifying the effect of steroids on a system may be difficult because of its complex interactions with multiple systems in the body [25]. This leads to one of the major hurdles in exploring different emotions and diabetes: defining methodologies to induce different emotions that are accurate and ethically sound is a major challenge faced by investigators.

Several other stress related tests have been performed to measure the effect of stress on glycemic control in those with diabetes, although these tests found no significant effect of acute stress on glucose control. In 2005, a study done by Wiesli et al. [26], stress was assessed using the Trier Social Stress Test (TSST) both following food intake and in the fasting state. The TSST consists of a 5-min preparation task, a 5-min speech task where subjects have to introduce themselves and apply for a job, and a 5-min mental arithmetic task in front of an audience consisting of at least two members in white coats. To enhance stress, the session is videotaped, and the audience is trained to appear emotionally neutral. At the beginning of the stress test, subjects are informed that during their performance, nonverbal communication is particularly looked at and analyzed post hoc by means of the tape. In both studies, blood pressure increased in response to TSST from 122/77 ± 14/9 mmHg at baseline to a maximum of 152/93 ± 21/13 mmHg (P < 0.001), and heart rate increased from 80 ± 11 to 99 ± 19 bpm (P < 0.001). This study found that blood pressure and heart rate both increased in response to the TSST in both the fasting state and after food intake. However, it found the glucose concentrations were only affected following food intake, where patients experienced a significant delayed decrease of glucose concentrations, which became apparent 45 min after the onset of stress. A two-factor repeated-measures ANOVA revealed a significant difference of glucose concentrations over time (F = 646.65/P < 0.001).

In 2007 [27], a study on the effect of stress on diabetes was done where patients were exposed to acute mental stress by riding on two different rollercoasters within 15 min. It was found that following a meal the increase of glucose concentrations on stress testing day was slightly attenuated, and glucose concentrations tended to remain lower throughout the experiment, this is in agreement with the previous study. During the rides, heart rate rose from 82 ± 7 bpm at baseline up to a maximum of 158 ± 16 bpm (P < 0.001). Blood pressure increased from 124/79 ± 12/9 to 160/96 ± 17/14 mmHg between the two rides (P < 0.001). Salivary cortisol increased from 6.3 nmol/L (range 2.8 –11.4) to a maximum of 19.3 nmol/L (5.6–49.3) 60 min following the ride (P = 0.008).Glucose concentrations of 10 patients investigated in the fasting state remained fairly stable, both during the control and stress testing day. At the time of stress application, glucose concentrations were 6.2 ± 1.6 mmol/L on control and 6.7 ± 2.3 mmol/L on stress testing day (P = NS). This study found no significant effect on glucose control both in patients in the fasting state and with those following the intake of a meal.

In 2013 [28], glucose control was tested during driving training in both type 1 and type 2 diabetes patients. The driving training consisted of 3 consecutive exercises: first, a slalom track on dry and wet asphalt, secondly, a full braking exercise with water obstacles. Thirdly, the car was hurled around by a mechanical plate and the patients had to regain control over it. During the 2 h-training, the patients were driving continuously, with only brief interruptions to conduct the measurements. On the stress testing day, blood pressure rose from 142/86 ± 16/9 mmHg to 162/95 ± 22/11 mmHg (P < 0.001), heart rate from 72 ± 11 bpm to 86 ± 16 bpm (P < 0.001) and subjective stress perception from 1.4 ± 0.6 to 4.7 ± 2.5 points (P < 0.001). Salivary cortisol concentrations increased from a median of 5.1 nmol/L (Interquartile Range (IQR) 3.5–7.5 nmol/L) at baseline to 7.7 nmol/L (IQR 4.7–12.8 nmol/L, P < 0.001), all these measurements remained stable on the control day. Glucose showed no significant difference on the stress testing day compared to the control day (mean difference over time = 0.22 mmol/L, 95 %-CI − 1.5 to + 1.9 mmol/L, P = 0.794).

Systematic investigations are necessary to assess the effect size, reproducibility, and inter-individual variability on glucose concentrations and the remedial factors (e.g., increasing insulin delivery rates based on changes to insulin sensitivities) to inform and refine next-generation CLC algorithms [29].

3   Addition of Sensors

Psychological changes induce physical responses, which can be detected using sensors. It is believed that soon sensor arrays with be available which will be able to detect specific behavioral changes within a patient. Two studies of the AP have used additional sensors Turksoy et al. [30] which used a SenseWear® Pro3 armband which includes multiple sensors, and Breton et al. [31] which added a heart rate (HR) monitor, both studies used sensors as a means to detect exercise. These studies can be extended to detect different emotional responses using additional sensors.

Turksoy et al. [30] developed an adaptive closed-loop system for preventing hypoglycemia during and after exercise. The SenseWear® system has multiple sensors including: an accelerometer able to detect speed and type of motion, a galvanic skin response meter measures electrical conductivity of the skin which changes in response to sweat and emotional stimuli, a skin temperature sensor, and a heat flux sensor which measures the amount of heat dissipating from the body. The adaptive control system kept glucose concentration in the normal preprandial and postprandial range (70–180mg/dL) without any meal or activity announcements during the test period. After IOB estimation was added to the control system, mild hypoglycemic episodes were observed only in one of the four experiments. This was reflected in a plasma glucose value of 56mg/dL (YSI 2300 STAT; Yellow Springs Instrument, Yellow Springs, OH) and a CGM value of 63mg/dL).

Breton et al. [31] was a pilot test using a small sample size of 12 patients to test the efficacy of preventing hypoglycemia due to exercise using a HR monitor. The HR was used to inform a closed-loop system when exercise was occurring. The study found that BG levels were kept within a suitable range however, significant results were not obtained due to the small cohort of patients used. A larger number a patients is expected to achieve statistically significant results.

Ambient intelligence (AmI) is an emerging discipline that uses information from sensors, appliances, and other objects around us to bring intelligence to everyday environments. AmI research builds upon advances in sensors and sensor networks, pervasive computing and artificial intelligence [32]. The maturity of AmI combined with the expansion of wearable devices will result in technologies that revolutionize daily human life. In the specific case of DM, it will allow the development of new management tools and control algorithms that will take into account not only physical information but also emotion and the psychological state, leading to improved metabolic control.

A study done in 2009 [33] attempted a smart home-based health platform for behavioural monitoring and alteration for diabetic and obese individuals. This approach utilized personal connected devices such as commercially available wearable sensors and smart space test beds, which included sensors for motion, light, temperature, humidity and door usage. Activity recognition and a chewing classification algorithm were implemented into the design. Activity recognition used Markov models for the different activities. Using three-fold cross validation, the Markov model classification algorithm achieved 98% recognition accuracy. The chewing algorithm was able to automatically distinguish chewing with mouth closed versus chewing with mouth open. In one case, chewing was confused with talking.

These different studies show that additional sensors provide additional information that may be useful to the patient, healthcare practitioners and an AP system. This will allow for suitable changes to be made based on varying patient conditions.

4   Discussion

The psychological changes that occur to those with DM can have significant effects on metabolic control. One of the major hurdles in this area is the need for a better understanding psychological changes and their influence on BG changes. However, as mentioned briefly before, doing controlled studies on different psychological responses is also challenging because ethical concerns associated with the nature of these tests.

As seen in corticoid-steroid induced stress, which increases insulin sensitivity and increases the likelihood of hyperglycemia, other psychological responses could have similar effects on insulin sensitivity and BG levels. The next step is to determine changes induced by other psychological responses such as illness, depression, anxiety, etc. and the appropriate action that should be taken, especially when considering the controller in an AP.

Using additional sensors can lead to new ways to gathering data that will assist us in overcoming many of the obstacles of adequate diabetes care. Through advances in analysis and diagnosis, this information will be utilized to improve the effectiveness of diabetes self-management education as well as the information available to health care professionals providing diabetes care, and one day an AP which will use this information to make informed changes to patient care.

5   Conclusion

There is still much work that must be done to better understand the relationship between psychological changes and diabetes. Although, it is well understood that negative physiological changes also negatively affect the ability to adequately control glucose, these changes have yet to be quantified in an appropriate manner.

One way of addressing this issue is the application of additional technologies that will provide more information, which will lead the way to dramatic improvements in the lives of those living with diabetes and reduce the public and private health care costs associated with treating the disease.

References

1. Cobelli, C., Dalla Man, C., Sparacino, G., Magni, L., De Nicolao, G., Kovatchev, B.: Diabetes: models, signals, and control. IEE Rev. Biomed. Eng. 2, 54–96 (2009)
2. Pan American Health Organization (PAHO), http://www.paho.org
3. World Health Organization (WHO), http://www.who.int
4. Lanza, R., Langer, R., Vacanti, J.: Principles of Tissue Engineering, vol 3, pp. 619–643. Academic Press, USA (2007)
5. Anderson, B., Wolpert, H.: A developmental perspective on the challenges of diabetes education and care during the young adult period. Patient Educ. Couns. 53(3), 347–352 (2004)
6. Auslander, W.F., Bubb, J., Rogge, M., Santiago, J.: Family stress and resources: potential areas of intervention in recently diagnosed children with diabetes. Health Soc. Work. 18, 101–113 (1993)
7. Fisher, E.B., Thorpe, C.T., DeVellis, B.E., DeVellis, R.: Healthy coping, negative emotions, and diabetes management: a systematic review and appraisal. Diabetes Edu. 33(6), 1080–1103 (2007)
8. Hagglof, B., Fransson, P., Lernmark, B., Thernlund, G.: Psychosocial aspects of type 1 diabetes mellitus in children 0-14 years of age. Arctic. Med. Res. 53(suppl. 1), 20–29 (1994)
9. Hill-Briggs, F., Cooper, D., Loman, K., Brancati, F., Cooper, L.: A qualitative study of problem solving and diabetes control in type 2 diabetes self-management. Diabetes Educ. 29(6), 1018–1028 (2003)
10. Hill-Briggs, F., Gary, T., Yeh, H., et al.: Association of social problem solving with glycemic control in a sample of urban African Americans with type 2 diabetes. J. Behav. Med. 29(1), 69–78 (2006)
11. The Diabetes Educator, http://www.diabeteseducator.org/AADE7/index.shtml
12. Rosenbaum, L.: Biofeedback-assisted stress management for insulin-treated diabetes mellitus. Biofeedback Self Regul. 8(4), 519–532 (1983)
13. Lustman, P.J., Clouse, R.E., Carney, R.: Depression and the reporting of diabetes symptoms. Int. J. Psychiatry Med. 18(4), 295–303 (1988)
14. Lustman, P.J., Griffith, L.S., Clouse, R.E., Freedland, K.E., McGill, J.B., Carney, R.: Improvement in depression is associated with improvement in glycemic control. In: American Diabetes Association Annual Meeting, Atlanta, GA (1995)
15. Lustman, P.J., Griffith, L.S., Clouse, R.: Depression in adults with diabetes. Diabetes Care 11(8), 605–612 (1988)
16. Lustman, P.J., Griffith, L.S., Freedland, K.E., Kissel, S.S., Clouse, R.: Cognitive behavior therapy for depression in type 2 diabetes mellitus: a randomized, controlled trial. Ann. Intern. Med. 129(8), 613–621 (1988)
17. Lustman, P.J., Griffith, L.S., Gavard, J.A., Clouse, R.: Depression in adults with diabetes. Diabetes Care 15, 1631–1639 (1992)
18. Andrews, R.C., Walker, B.: Glucocorticoids and insulin resistance: old hormones, new targets. Clin. Sci (Lond.) 96(5), 513–523 (1999)
19. El Youssef, J., Castle, J.R., Branigan, D.L., et al.: A controlled study of the effectiveness of an adaptive closed-loop algorithm to minimize corticosteroid-induced stress hyperglycemia in type 1 diabetes. J. Diabetes Sci. Tech. 5, 1312–1326 (2011)
20. Henriksen, J.E., Alford, F., Ward, G.M., Beck-Nielsen, H.: Risk and mechanism of dexamethasone-induced deterioration of glucose tolerance in non-diabetic first-degree relatives of NIDDM patients. Diabetologia 40(12), 1439–1448 (1997)

21. Rizza, R.A., Mandarino, L.J., Gerich, J.: Cortisol-induced insulin resistance in man: impaired suppression of glucose production and stimulation of glucose utilization due to a postreceptor detect of insulin action. J. Clin. Endocrinol. Metab. 54(1), 131–138 (1982)

22. Venkatesan, N., Lim, J., Bouch, C., Marciano, D., Davidson, M.: Dexamethasone-induced impairment in skeletal muscle glucose transport is not reversed by inhibition of free fatty acid oxidation. Metabolism 45(1), 92–100 (1996)

23. Halford, W.K., Cuddihy, S.: Psychological stress and blood glucose regulation in type 1 diabetic patients. Health Psychology 9(5), 516–528 (1990)

24. Finan, D.A., Zisser, H., Jovanovic, L., Bevier, W.C., Seborg, D.: Automatic detection of stress states in type 1 diabetes subjects in ambulatory conditions. Ind. Eng. Chem. Res. 47(17), 7843–7848 (2010)

25. McKay, L.I., Cidlowski, J.: Physiologic and Pharmacologic Effects of Corticosteroids. In: Kufe, D.W., Pollock, R.E., Weichselbaum, R.R., et al. (eds.) Holland-Frei Cancer Medicine, vol. 6. BC Decker, Hamilton (2003)

26. Wiesli, P., Schmid, C., Kerwer, O., Nigg-Koch, C., Klaghofer, R., Seifert, B., Spinas, G.A., Schwegler, K.: Acute psychological stress affects glucose concentrations in patients with type 1 diabetes following food intake but not in the fasting state. Diabetes Care 28(8), 1910–1915 (2005)

27. Wiesli, P., Krayenbuhl, P., Kerwer, O., Seifert, B., Schmid, C.: Maintenance of glucose control in patients with type 1 diabetes during acute mental stress by riding high-speed rollercoasters. Diabetes Care 30(6), 1599–1601 (2007)

28. Truninger, R., Uthoff, H., Capraro, J., Frauchiger, B., Spinas, G.A., Wiesli, P.: Glucose control during a driving training in patients with type 1 and type 2 diabetes mellitus- a randomized, controlled trial. Exp. Clin. Endocrinol Diabetes 121, 420–424 (2013)

29. Kudva, Y.C., Carter, R.E., Cobelli, C., Basu, R., Basu, A.: Closed-Loop artificial pancreas systems: physiological input to enhance next-generation devices. Diabetes Care 37, 1184–1190 (2014)

30. Turksoy, K., Bayrak, E.S., Quinn, L., Littlejohn, E., Çinar, A.: Multivariable adaptive closed-loop control of an artificial pancreas without meal and activity announcement. Diabetes Technol. Ther. 15, 386–400 (2013)

31. Breton, M.D., Brown, S.A., Karvetski, C.H., Kollar, L., Topchyan, K.A., Anderson, S.M., Kovatchev, B.: Adding Heart Rate Signal to a Control-to-Range Artificial Pancreas System Improves the Protection Against Hypoglycemia During Exercise in Type 1 Diabetes. Diabetes Technol. Ther. 16(8), 506–511 (2014)

32. Cooka, D.J., Augusto, J.C., Jakkula, V.: Ambient intelligence: technologies, applications, and opportunities. Pervasive and Mobile Computing 5, 277–298 (2009)

33. Helal, A., Cook, D.J., Schmalz, M.: Smart home-based health platform for behavioral monitoring and alteration of diabetes patients. J. Diabetes Sci. Technol. 3(1), 141–148 (2009)

Author Index